Instructor's Solutic

Alison Hyslop
St. John's University

to accompany

CHEMISTRY
Matter and Its Changes

Fourth Edition

James E. Brady
St. John's University

Frederick Senese
Frostburg State University

Special Thanks to Michael J. Kenney, of Crabtree and Company, and Cynthia K. Anderson, of Robert E. Lee High School, for their contributions to previous editions of this manual.

Thank you to Nicholas Drapela, of Oregon State University, for his contributions to this edition of the manual.

JOHN WILEY & SONS, INC.

To order books or for customer service, please call 1-800-CALL-WILEY (225-5945).

ISBN 0-471-64903-1

10 9 8 7 6 5 4 3 2 1

Table of Contents

Chapter One

Practice Exercises

1.1 a) Fe_2O_3 contains iron (Fe), and oxygen (O)
b) Na_3PO_4 contains sodium (Na), phosphorus (P), and oxygen (O)
c) Al_2O_3 contains aluminum (Al), and oxygen (O)
d) $CaCO_3$ contains calcium (Ca), carbon (C), and oxygen (O)

1.2 The first sample has a ratio of

$$\frac{1.25 \text{ g Cd}}{0.357 \text{ g S}}$$

Therefore, the second sample must have the same ratio of Cd to S:

$$\frac{1.25 \text{ g Cd}}{0.357 \text{ g S}} = \frac{x}{3.50 \text{ g S}}$$

Cross-multiplication gives,

$$(1.25 \text{ g Cd})(3.50 \text{ g S}) = x(0.357 \text{ g S})$$

$$\frac{(1.25 \text{ g Cd})(3.50 \text{ g S})}{0.357 \text{ g S}} = x$$

$$x = 12.3 \text{ g Cd}$$

1.3 $2.24845 \times 12 \text{ u} = 26.9814 \text{ u}$

1.4 Copper is $63.546 \text{ u} \div 12 \text{ u} = 5.2955$ times as heavy as carbon

1.5 $(0.198 \times 10.0129 \text{ u}) + (0.802 \times 11.0093 \text{ u}) = 10.8 \text{ u}$

1.6 $^{240}_{94}Pu$

The bottom number is the atomic number, found on the periodic table (number of protons). The top number is the mass number (sum of the number of protons and the number of neutrons).

1.7 $^{35}_{17}Cl$ contains 17 protons, 17 electrons, and 18 neutrons.

The bottom number is the atomic number (17), found on the periodic table (number of protons). In a neutral atom, the number of electrons equals the number of protons, giving 17 electrons. The top number is the mass number (sum of the number of protons and the number of neutrons), so subtracting the protons gives 35 – 17 = 18 neutrons.

1.8 a) K, Ar, Al
b) Cl
c) Ba
d) Ne
e) Li
f) Ce

Thinking It Through

T1.1 First we must determine the amount of oxide that is formed. By the Law of Conservation of Mass this is simply the sum of the two masses that react: 15.0 g Al + 13.3 g O = 28.3 g aluminum oxide are formed. Since the same amount of oxygen is used in the second experiment, the amount of aluminum oxide formed should be the same as the first experiment, 28.3 g. There is excess aluminum in the second reaction. The Law of Conservation of Mass is the tool needed to solve this problem.

T1.2 We can simply divide one atomic mass by the other to solve this problem:
55.847 u ÷ 12 u = 4.6539. Thus, an average atom of iron is 4.6539 times heavier than an average carbon atom.

T1.3 Regardless of the starting definition of a mass scale, we would find the relative masses remain constant from one scale to another. The following conversion factor therefore gives us the ratio of the mass of a carbon to a phosphorus atom.

$$\left(\frac{12\text{ u C}}{30.97\text{ u P}}\right) = 0.3875\ \frac{\text{u C}}{\text{u P}}$$

If we start with 1/10 of the value assigned to phosphorus, namely 3.097 u, then the mass of carbon is obtained by using the above conversion factor.

$$(3.097\text{ u P})\times\ 0.3875\ \frac{\text{u C}}{\text{u P}} = 1.200\ \frac{\text{u C}}{\text{u P}}$$

T1.4 The number of protons is the same as the atomic number, 45. Thus we first deduce that we have element number 45, or rhodium, Rh. The number of electrons would also be equal to 45, since the particle is a neutral atom. The mass number is equal to the number of protons plus the number of neutrons, namely 45 + 58 = 103. The entire symbol is then:

$$^{103}_{45}\text{Rh}$$

T1.5 The number of protons is equal to the atomic number 49, or indium, In. This is a metal based on its position in the periodic table.

T1.6 The new element, atomic number 117, would fall below astatine, At, on the periodic table by counting over from element 112.

T1.7 From the formula we know that 1 mole of C combines with 4 moles of X. If we determine the number of moles of C present in 1.50 g, we can determine the number of moles of X present (4 × moles C). So, # moles C = (1.50 g C)(1mole C/12.01 g C) = 0.125 moles C. There are 4 × 0.125 moles = 0.500 moles X. The atomic mass of X is then (39.95 g X)/(0.500 moles X) = 79.9 g/mol X. X is probably bromine.

T1.8 The law of definite proportions assures us that the combining ratio of chlorine to phosphorus will be the same in the two separate preparations of PCl_3. Therefore, the ratio for one experiment:

$$\left(\frac{12.0\text{ g Cl}}{3.50\text{ g P}}\right) = \left(\frac{3.43\text{ g Cl}}{1.00\text{ g P}}\right)$$

will be obtained in the second experiment also. We have only to multiply the above conversion factor to obtain the answer:

$$8.50\text{ g P}\times\left(\frac{3.43\text{ g Cl}}{1.00\text{ g P}}\right) = 29.2\text{ g Cl}$$

T1.9 (a) For the number of grams of oxygen that would react in the second experiment, the law of definite proportions is needed. Since the ratio of the oxygen to phosphorous is constant, the ratio from the first experiment can be used to calculate the amount of oxygen needed to react in the second experiment:

$$\left(\frac{2.89 \text{ g O}}{2.40 \text{ g P}}\right) = \left(\frac{1.20 \text{ g O}}{1 \text{ g P}}\right)$$

Therefore the number of grams of oxygen needed for the second reaction would be the ratio times the number of grams of phosphorous used:

$$5.60 \text{ g P} \times \left(\frac{1.20 \text{ g O}}{1 \text{ g P}}\right) = 6.72 \text{ g O}$$

(b) The number of grams of compound that would be formed uses the law of conservation of mass. The mass of the compound is simply the mass of the two components added together:

$$6.72 \text{ g O} + 5.60 \text{ g P} = 12.32 \text{ g Compound}$$

T1.10 Compare the ratios of the number of grams arsenic to the number of grams of the compound in these two:

$$\left(\frac{3.62 \text{ g As}}{6.00 \text{ g compound}}\right) = 0.603 \qquad \left(\frac{2.89 \text{ g As}}{6.00 \text{ g compound}}\right) = 0.482$$

Because these two ratios are different, they are different compounds and demonstrate the Law of Multiple Proportions.

T1.11 Using law of conservation of mass, not all of the oxygen is consumed. Since all of the magnesium has reacted, then 1.0 g of the product is magnesium, this leaves 0.6 g of oxygen in the product

(1.6 g product – 1.0 g of Mg = 0.6 g of O).

For the reaction 2.0 g of oxygen was available, but only 1.4 g were used

(2.0 g O – 0.6 g O = 1.4 g O)

Review Questions

1.1 This answer will be student dependent.

1.2 Matter and its changes.

1.3 Matter is tangible, characteristically has mass and occupies space. Energy is the capacity or ability to do work. All items in the question are examples of matter.

1.4 This answer will be student dependent.

1.5 This answer will be student dependent

1.6 Observation, testing, and explanation.

1.7 A laboratory is one possible location where the scientific method is applied to solve problems.

1.8

(a)	Data:	Collection of facts
(b)	Hypothesis:	A tentative explanation of why nature behaves in certain ways.
(c)	Law:	A broad generalization based upon the results of a large number of experiments. Laws are often presented in mathematical form.
(d)	Theoretical Model:	A model of a tested explanation of the behavior of nature.

1.9 Frequently, experiments produce results that are unexpected yet lead to new and interesting products.

1.10 Physical properties are characteristic properties possessed by substances. A physical property can be measured and used to characterize a pure substance without reference to other substances. A chemical property governs the reactivity of a substance either with other substances or to give other substances. Intensive properties are independent of sample size. Extensive properties depend on sample size. Color and electrical conductivity are intensive properties. Mass and volume are extensive properties.

1.11 A physical change does not alter the chemical identity of a substance, whereas a chemical change does. A chemical change transforms the substance into another with a new set of characteristics. A change of state is a physical change.

1.12 Physical properties: electrical conductance, luster, hardness, melting point, boiling point, temperature, qualitative descriptions
Chemical Properties: reacts with water to give bubbles of hydrogen and calcium hydroxide.

1.13 A chemical reaction is an interaction between one or more different substances which results in a rearrangement of the atoms present creating new and different substances.

1.14 A reaction has occurred because sodium chloride, a different compound, is formed from the combination of lye (sodium hydroxide) and muriatic acid (hydrochloric acid).

1.15 A chemical reaction has occurred because we observe the formation of a new compound, carbon dioxide.

1.16 These are physical changes.

1.17 (a) Volume is an extensive property.
(b) Because volume is an extensive physical property, it cannot be used to distinguish the liquids in question.

1.18 Density, being equal to the ratio of the mass to the volume of the sample, is an intensive property. Additionally, the boiling points of the substances could be used to determine the identity of the substance. A chemical property which could be used to distinguish between water and gasoline is the reaction of gasoline with oxygen to produce heat and other products; no reaction occurs between oxygen and water.

1.19 a) The physical properties are given below beginning with color:

CdI_2 green-yellow
mp 387 °C
bp 796 °C

PbI_2 yellow
mp 402 °C
bp 954 °C

$BiBr_3$ yellow
mp 218 °C
bp 453 °C

(b) Only the melting points and boiling points could be used.

(c) This is PbI_2 based on a comparison of the melting points and boiling points.

1.20 The three physical states of matter are solid, liquid, and gas.

1.21 Solids are composed of atoms or molecules that are packed tightly together and are unable to move about so solids are rigid. Liquids have particles that are close together, but they are capable of moving past each other such that a liquid is able to flow. Gas particles are far enough apart that the empty space between them allows them to be compressed and transparent.

1.22
(a) An element is a pure substance that cannot be decomposed into something simpler.
(b) A compound is a pure substance that is composed of two or more elements in some fixed or characteristic proportion.
(c) Mixtures result from combinations of pure substances in varying proportions.
(d) A homogeneous mixture has one phase. It has the same properties throughout the sample.
(e) A heterogeneous mixture has more than one phase. The different phases have different properties.
(f) A phase is a region of a mixture that has properties that are different from other regions of the mixture.
(g) A solution is a homogeneous mixture.

1.23 A chemical change is needed to separate the compound into the elements that compose it.

1.24 A physical change is needed to separate a mixture into its components.

1.25

(a)	Cl	(b)	S	(c)	Fe	(d)	Ag	(e)	Na
(f)	P	(g)	I	(h)	Cu	(i)	Hg	(j)	Ca

1.26

(a)	potassium	(b)	zinc	(c)	silicon	(d)	tin
(e)	manganese	(f)	magnesium	(g)	nickel	(h)	aluminum
(i)	carbon	(j)	nitrogen				

1.27 Chemical symbols are used to identify elements and as an abbreviation for the chemical name.

1.28 The first law of chemical combination is the law of conservation of mass: no detectable gain or loss of mass occurs in chemical reactions. The other law is the law of definite proportions: in a given chemical compound, the elements are always combined in the same proportions by mass.

1.29 The Law of Conservation of Mass requires that atoms are indestructible. Elements combine in definite ratios, as atoms. This guarantees that elements combine in definite mass ratios, assuming the atoms are indestructible.

1.30 Isotopes of a particular element have nearly identical chemical properties and the average mass of an atom is independent, almost, of the source of the atom.

1.31 Conservation of mass derives from the postulate that atoms are not destroyed in normal chemical reactions. The Law of Definite Proportions derives from the notion that compound substances are always composed of the same types and numbers of atoms of the various elements in the compound.

1.32 This is the Law of Definite Proportions, which guarantees that a single pure substance is always composed of the same ratio of masses of the elements that compose it.

1.33 This is carbon-12 which has a mass of 12 u (exactly) by definition.

$$^{12}_{6}C$$

1.34 See text table 1.3.

1.35 Nearly all of the mass is located in the nucleus, because this is the portion of the atom where the proton and the neutron are located.

1.36 A nucleon is a subatomic particle found in the atomic nucleus. We have studies neutrons and protons.

1.37 The atomic number is equal to the number of protons in the nucleus of the atom, and the mass number is the sum of the neutrons and the protons.

1.38 The isotopes of an element have identical atomic numbers (number of protons) but differing number of neutrons, and thus differing masses.

1.39 Atoms of two different elements might have the same number of neutrons, or the same number of electrons but different charges. They might have the same mass number.

1.40 (a) mass number (b) atomic number

1.41 (a) ${}^{131}_{53}I$ (b) ${}^{90}_{38}Sr$ (c) ${}^{137}_{55}Ce$ (d) ${}^{18}_{9}F$

1.42 For all group IA elements (the alkali metals), the formula is MX, that is one Cl per atom of metal. For all group IIA elements (the alkaline earth metals), the formula is MX_2, that is two Cl atoms per atom of metal. The correspondence in formula and the similarities in chemical behavior allowed Mendeleev to locate theses two series into their separate groups on the periodic table.

1.43 Mendeleev constructed his periodic table by arranging the elements in order of increasing atomic weight, and grouping the elements by their recurring properties. The modern periodic table is arranged in order of increasing number of protons.

1.44 A period in the periodic table is a horizontal row of elements. A group is one of the vertical columns of the periodic table.

1.45 Not all of the elements had yet been discovered; therefore Mendeleev left spaces for the ones that he predicted would eventually be discovered because he grouped elements with similar chemical properties together.

1.46 These occur between cobalt and nickel, thorium and protactinium, uranium and neptunium.

1.47 The atomic number is better related to the chemistry of an element since the periodic table is based on atomic numbers, and the mass numbers vary with the number of neutrons in the atom which does not affect the chemistry of the elements as much as the number of protons.

1.48 Strontium and calcium are in the same Group of the periodic table, so they are expected to have similar chemical properties. Strontium should therefore form compounds that are similar to those of calcium, including the sorts of compounds found in bone.

1.49 Silver and gold are in the same periodic table group as copper, so they might well be expected to occur together in nature, because of their similar properties and tendencies to form similar compounds.

1.50 Cadmium is in the same periodic table group as zinc, but silver is not. Therefore cadmium would be expected to have properties similar to those of zinc, whereas silver would not.

1.51 See Figure in the margin of page 20.

1.52 (a) The representative elements are designated, using the traditional US system, as groups IA, IIA, and IIIA through VIIIA.
(b) The representative elements are designated, using IUPAC notation as groups 1, 2, and 13 through 18.

1.53 (a) Group IA = Group 1
(b) Group VIIA = Group 17

(c) Group IIIB = Group 3
(d) Group IB = Group 11
(e) Group IVA = Group 14

1.54 Germanium has an atomic mass of 72.6 u and an atomic number of 32. Next to it on the periodic table is arsenic which has an atomic number of 33. In order for there to be a new element with an atomic mass of 73, it would be expected to be next to germanium with one more proton, but arsenic has one more proton and has an atomic mass of 74.9 u.

1.55 (a) Li
(b) I
(c) W
(d) Xe
(e) Sm
(f) Pu
(g) Mg

1.56 Luster, electrical conductivity, thermal conductivity, ductility, malleability are the characteristic properties of metals.

1.57 Mercury is used in thermometers because it is a liquid and tungsten is used in light bulbs because is has such a high melting point.

1.58 Ductility

1.59 Malleability

1.60 Copper and gold

1.61 The noble gases: He, Ne, Ar, Kr, Xe, and Rn

1.62 Mercury and bromine

1.63 Metalloids are semiconductors

1.64 See text chapter 1, figure 1.17, or p 21.

1.65 Metals which are used to make jewelry are those that do not corrode, silver, gold, and platinum. Iron would be useless for jewelry because it is susceptible to rusting.

Review Problems

1.66 Compound (c). An authentic sample of laughing gas must have a mass ratio of nitrogen/oxygen of 1.75 to 1.00. The only possibility in this list is item (c), which has the ratio of mass of nitrogen to mass of oxygen of 8.84/5.05 = 1.75.

1.67 Compound (d). An authentic sample of calcium chloride must have a ratio of 1.00 to 1.77. The only possibility in the list is (d) which has the ratio of the mass of chlorine too calcium 2.39/1.35 = 1.77.

1.68 From the first ratio we see that there is a ratio of 4.66 to 1.00. Multiplying the mass of hydrogen by 4.66 we see that for every 6.28 g hydrogen there will be 29.3 g nitrogen.

1.69 From the ratio of phosphorous to chlorine, we see that for every 1.00 g phosphorus, there are 3.43 g chlorine. Dividing the mass of chlorine by 3.43, we find that for every 6.22 g chlorine there will be 1.81 g phosphorus.

1.70 5.54 g ammonia. From the ratio, Problem 1.68, above we see that for every 4.56 g nitrogen there needs to be 0.98 g hydrogen. According to the Law of Conservation of Mass 5.54 g ammonia will be produced.

1.71 55.0 g of the phosphorus chloride. From the ratio above, Problem 1.69, we see that for every 12.5 g phosphorus it is necessary to have 42.5 g chlorine. According to the Law of Conservation of Mass, 55.0 g of the phosphorus chloride compound was formed.

1.72 2.286 g of O. The first nitrogen-oxygen compound has an atom ratio of 1 atom of N for 1 atom of O, and a mass ratio of 1.000 g of N for 1.143 g of O. The second compound has an atom ratio of 1 atom of N for 2 atoms of O; therefore, the second compound has twice as many grams of O for each gram of N, or 2.286 g of O.

1.73 (a) This ratio should be 4/2 = 2/1, as required by the formulas of the two compounds.
(b) Twice 0.597 g Cl, or 1.19 g Cl.

1.74 $12 \times 1.6605402 \times 10^{-24}\ \text{g} = 1.9926482 \times 10^{-23}\ \text{g}$ for one C atom

1.75 $23 \times 1.660566520 \times 10^{-24}\ \text{g} = 3.819302996 \times 10^{-23}\ \text{g}$ for the average mass of one sodium atom

1.76 Since we know that the formula is CH_4, we know that one fourth of the total mass due to the hydrogen atom constitutes the mass that may be compared to the carbon. Hence we have 0.33597 g H ÷ 4 = 0.083993 g H and 1.00 g assigned to the amount of C-12 in the compound. Then it is necessary to realize that the ratio 1.00 g C ÷ 12 for carbon is equal to the ratio 0.083993 g H ÷ X, where X equals the relative atomic mass of hydrogen.

$$\left(\frac{1.000\ \text{g C}}{12\ \text{u C}}\right) = \left(\frac{0.083993\ \text{g H}}{\text{X}}\right) = 1.008\ \text{u}$$

1.77
$$\text{mol O} = (1.000\ \text{g O})\left(\frac{1\ \text{mol O}}{15.999\ \text{g O}}\right) = 0.06250\ \text{mol O}$$

$$\text{mol X} = (0.6250\ \text{mol O})\left(\frac{2\ \text{mol X}}{3\ \text{mol O}}\right) = 0.04167\ \text{mol X}$$

$$\#\ \text{g/mol} = \frac{1.125\ \text{g}}{0.04167\ \text{mol}} = 27.00\ \text{g/mol}$$

The element is aluminum.

1.78 Regardless of the definition, the ratio of the mass of hydrogen to that of carbon would be the same. If C–12 were assigned a mass of 24 (twice its accepted value), then hydrogen would also have a mass twice its current value, or 2.01588 u.

1.79 Taking the mass ratio of ^{109}Ag to ^{12}C and multiplying it by 12 we see that the mass of Ag-109 is 108.90 u.

1.80 $(0.6917 \times 62.9396\ \text{u}) + (0.3083 \times 64.9278\ \text{u}) = 63.55\ \text{u}$

1.81 $(0.7899 \times 23.9850\ \text{u}) + (0.1000 \times 24.9858\ \text{u}) + (0.1101 \times 25.9826\ \text{u}) = 24.31\ \text{u}$

1.82

	neutrons	protons	electrons
(a)	138	88	88
(b)	8	6	6
(c)	124	82	82
(d)	12	11	11

1.83

	electrons	protons	neutrons
(a)	55	55	82
(b)	53	53	78
(c)	92	92	146
(d)	79	79	118

Additional Questions

1.84 (a) Metal

(b) $^{55}_{25}Mn$

(c) 30 neutrons

(d) 25 electrons

(e) $\dfrac{54.9380}{12.000} = 4.578$ times heavier than a C-12 atom

1.85 First calculate the number of moles of oxygen that are combined with X:

$$(1.14 \text{ g oxygen})\left(\frac{1 \text{ mole O}}{16.00 \text{ g O}}\right) = 0.0713 \text{ moles oxygen}$$

Then calculate the moles of X in the oxygen compound:

$$(0.0713 \text{ moles O})\left(\frac{1 \text{ mole X}}{2 \text{ moles O}}\right) = 0.0356 \text{ moles X}$$

Finally, calculate the molar mass of X:

$$\frac{1.00 \text{ g X}}{0.0356 \text{ moles X}} = 28.1 \text{ g/mole X}$$

Note that X is Si, atomic number 14, and that its oxygen compound is SiO_2.

The same number of moles of X are also combined with Y:

$$(0.0356 \text{ moles X})\left(\frac{4 \text{ moles Y}}{1 \text{ moles X}}\right) = 0.142 \text{ moles Y}$$

The atomic mass of Y is thus:

$$\frac{5.07 \text{ g Y}}{0.142 \text{ moles Y}} = 35.6 \text{ g/mole Y}$$

Note that Y is Cl, atomic number 17, and that its compound with X is $SiCl_4$.

1.86 $(0.0580 \times 53.9396 \text{ u}) + (0.9172 \times 55.9349 \text{ u}) + (0.0220 \times 56.9354 \text{ u}) + (0.0028 \times 57.9333\text{u}) =$ 55.85 u

1.87 Let x equal the abundance of ^{79}Br and y equal the abundance of ^{81}Br. We know that $x + y = 1$ and $x(78.9183) + y(80.9163) = 79.904$. Substituting $y = 1 - x$ we get $x(78.9183) + (1 - x)80.9163 = 79.904$. Solving for x we get $x = 0.5067$ and $y = 0.4933$.

1.88 The mass ratio of Fe to O in the first compound is 2.325 g Fe to 1.000 g O and the atom ratio is 2 Fe atoms to 3 O atoms. Dividing the masses by the number of atoms gives the mass ratios of Fe to O

$$\frac{2.325 \text{ g Fe}}{3 \text{ atoms Fe}} = 1.162 \text{ g Fe/atom Fe}$$

$$\frac{1.000 \text{ g O}}{2 \text{ atoms O}} = 0.3333 \text{ g O/ atom O}$$

The second compound has a mass ratio of 2.616 g Fe to 1.000 g O. Dividing the mass of Fe by the mass/atom ratio determined for Fe and dividing the mass of O by the mass/atom ratio determined for O gives the ratio of number of atoms in the compound.

$$\frac{2.616 \text{ g Fe}}{1.162 \text{ g Fe/atom Fe}} = 2.251 \text{ atom Fe}$$

$$\frac{1.000 \text{ g O}}{0.3333 \text{ g O/atom O}} = 3.000 \text{ atom O}$$

Multiplying the atomic ratios by 4 gives whole numbers to the subscripts in the formula: Fe_9O_{12}. Both 9 and 12 are divisible by 3 to give: Fe_3O_4.

1.89 $(24.305)1.6605402 \times 10^{-24}\text{ g} = 4.0359 \times 10^{-23}\text{ g}$
$(55.847)1.6605402 \times 10^{-24}\text{ g} = 9.2736 \times 10^{-23}\text{ g}$
Comparing answers, we see that both numbers are on the order of 1×10^{-23}. We would expect 6×10^{23} atoms of Ca in 40.078 g of Ca.

Practice Exercises

2.1 a) 1 Ni, 2 Cl
b) 1 Fe, 1 S, 4 O
c) 3 Ca, 2 P, 8 O
d) 1 Co, 2 N, 12 O, 12 H

2.2 1 Mg, 2 O, 4 H, and 2 Cl (on each side)

2.3 $Mg(OH)_2(s) + 2\ HCl(aq) \rightarrow MgCl_2(aq) + 2\ H_2O(l)$

2.4 a) Colder temperatures mean lower molecular kinetic energies, so the kinetic energies of the molecules in the mixture are reduced.
b) Heat is absorbed into the mixture from the surroundings, as evidenced by a touch with the hand, so energy flows into the mixture and is stored as potential energy. The potential energy of the mixture increases.

2.5 $C_{10}H_{22}$
The formula for an alkane hydrocarbon is C_nH_{2n+2}. Here, n = 10, so 2n + 2 = 22.

2.6 a) C_3H_8O. Propane is C_3H_8, so removing one H and replacing it with OH gives C_3H_8O.

b) $C_4H_{10}O$. Butane is C_4H_{10}, so removing one H and replacing it with OH gives $C_4H_{10}O$.

2.7 a) phosphorus trichloride b) sulfur dioxide c) dichlorine heptaoxide

2.8 a) $AsCl_5$
b) SCl_6
c) S_2Cl_2

2.9 a) 8 protons, 8 electrons
b) 8 protons, 10 electrons
c) 13 protons, 10 electrons
d) 13 protons, 13 electrons

2.10 a) NaF b) Na_2O c) MgF_2 d) Al_4C_3

2.11 a) $CrCl_3$ and $CrCl_2$, Cr_2O_3 and CrO
b) CuCl, $CuCl_2$, Cu_2O and CuO

2.12 a) Na_2CO_3
b) $(NH_4)_2SO_4$
c) $KC_2H_3O_2$
d) $Sr(NO_3)_2$
e) $Fe(C_2H_3O_2)_3$

2.13 a) potassium sulfide
b) magnesium phosphide
c) nickel(II) chloride
d) iron(III) oxide

2.14 a) Al_2S_3 b) SrF_2 c) TiO_2 d) Au_2O_3

2.15 a) Lithium carbonate b) iron(III) hydroxide

2.16 a) $KClO_3$
b) $Ni_3(PO_4)_2$

2.17 diiodine pentaoxide

2.18 chromium(II) acetate

2.19 X is a nonmetal. Nonmetals combine with nonmetals to form molecular substances, which are characterized by low melting points such as this. Also, the inability of the substance to conduct electricity in the liquid state points to a molecular substance rather than an ionic substance, since molten ionic substances conduct electricity readily.

Thinking It Through

T2.1 We start by identifying elements based on their atomic numbers. Element 56 is barium, Ba and element 35 is chlorine, Cl. Both are representative elements based on their positions in the periodic table. Next we consult the table to determine the ion that each forms. These are Ba^{2+} and Cl^-. Finally we combine the two ions so as to obtain a neutral substance $BaCl_2$.

T2.2 The key to the problem is the list of physical properties. Here we have a compound that exists as a solid, liquid, or gas depending on temperature. When it does crystallize, the solid is soft. These are characteristics of a molecular compound. Ionic compounds tend to form hard brittle solids at ordinary temperatures.

T2.3 P_4O_{10} is molecular since it (1) is made up of two elements that are nonmetals and (2) for ionic compounds, the ratio of the atoms is reduced by the lowest common denominator, here the ratio is not divided by 2.

T2.4 Strontium selenide is an ionic compound because strontium is a metal and selenium is a nonmetal.

T2.5 Strontium selenate is not a binary compound because the anion, selenate, does not have the suffix –ide.

T2.6 The mass ratio of chlorine to phosphorus in phosphorous trichloride is:

$$\left(\frac{12.0 \text{ g Cl}}{3.50 \text{ g P}}\right) = \left(\frac{3.43 \text{ g Cl}}{1.00 \text{ g P}}\right)$$

The atomic ratio of chlorine to phosphorous is 3 Cl to 1 P. To determine the ratio of the masses of the atoms, divide each mass by the number of atoms in the compound:

$$\frac{\frac{3.43 \text{ g Cl}}{3 \text{ atoms Cl}}}{\frac{1.00 \text{ g P}}{1 \text{ atom P}}} = \frac{1.14 \text{ g Cl/atom Cl}}{1.00 \text{ g P/atom P}}$$

Therefore, one atom of Cl is 1.14 times heavier than one atom of P.

T2.7 AB_2 is an ionic compound, therefore A is a metal and B is a nonmetal. In the sample, the mass of B is 4.0 times heavier than the mass of A, but there is one atom of A for every two atoms of B. By dividing the mass of B by 2 (the number of atoms of B) and the mass of A by 1 (the number of atoms of A) would give atomic mass ratios of B to A as 2 to 1. By inspection of the atomic masses, the identity of the elements can be determined. Furthermore, the charge on A has to be opposite in sign and twice as large as the charge on B. The most likely candidates are A is Calcium (40.078g/mol) and B is F (18.998 g/mol).

Review Questions

2.1 This may stand for the name of an element or for the name of one atom of an element.

2.2 The smallest particle that is representative of a particular element is the atom of that element. A molecule is a representative unit that is made up of two or more atoms linked together.

2.3 Some substances which consist of molecules are elements, examples are H_2, hydrogen, and N_2, nitrogen.

2.4 H_2, hydrogen N_2, nitrogen, O_2, oxygen F_2, fluorine Cl_2, chlorine Br_2. bromine I_2, Iodine

2.5 When $MgSO_4 \cdot 7H_2O$ is completely dehydrated, $MgSO_4$ remains.

2.6 A chemical reaction is balanced when there is the same number of each kind of atom on both the reactant and product side of the equation. This condition must be met due to the law of conservation of matter.

2.7 Reactants are the substances to the left of the arrow in a reaction that are present before the reaction begins. Products are the substances to the right of the arrow in a reaction and they are formed during the reaction and are present when the reaction is over.

2.8 Coefficients

2.9 (a) Magnesium reacts with oxygen to give (yield) magnesium oxide.
(b) The reactants are Mg and O_2.
(c) The product is MgO.
(d) $2Mg(s) + O_2(g) \rightarrow 2MgO(s)$

2.10 $2C_8H_{18}(l) + 25\ O_2(g) \rightarrow 18\ CO_2(g) + 18H_2O(l)$

2.11 (a) Energy is the capacity or ability to do work.
(b) Work is the energy expended in moving an opposing force through some particular distance.

2.12 (a) Kinetic energy is the energy of motion (b) Stored energy. Its mass and velocity.

2.13 $KE = 1/2mv^2$, where m= mass and v= velocity of the object in motion.

2.14 The KE of the car increases by a factor of 4 since the car's speed has doubled and velocity is raised the power 2 in the KE equation.

2.15 Chemical energy is the potential energy in substances, which changes into other forms of energy when substances undergo chemical reactions.

2.16 The temperature is related to the average kinetic energy of a group of particles within an object. Heat is the total kinetic energy of all the atoms within an object. Also, temperature is an intensive quantity and heat is extensive.

2.17 The quart of boiling water has more heat, than one drop of boiling water because there is more water in a quart than in a drop and both waters are boiling.

2.18 Chemical energy is potential energy.

2.19 The law of conservation of energy is: energy cannot be created or destroyed; it can only be changed from one form to another. A ball rolling downhill is losing potential energy and gaining kinetic energy. These observations are consistent with the law of conservation of energy since the amount of potential energy the ball is losing should be equal to the amount of kinetic energy the ball is gaining.

2.20 The brakes become hot because the car is losing kinetic energy and the energy is transferred to the brakes which convert the kinetic energy into heat.

2.21 Heat is transferred from a hot object to a cold object through the transfer of the kinetic energy of the hot atoms to the cold atoms.

2.22 The temperature of the reacting chemicals must rise since there is a reduction in the chemical energy which is a potential energy. The energy must be converted into another form of energy, in this case, kinetic energy. The atoms move faster and the average kinetic energy of the atoms increases.

2.23 The average kinetic energy of substances in the cold packs decreases. The chemical energy of the water and the ammonium nitrate increases.

2.24 The potential energy of the gasoline and the oxygen is higher than the potential energy of the carbon dioxide and water vapor.

2.25 Thermal energy is the energy of motion of microscopic objects, or heat.

2.26 The noble gases, which are fairly unreactive, monatomic elements.

2.27 H_2, N_2, O_2, F_2, Cl_2, Br_2, I_2

2.28 Nonmetals

2.29 (a) CH_4 (b) NH_3 (c) H_2Te (d) HI

2.30 PH_3

2.31 HAt

2.32 SnH_4

2.33 (a) CH_4 (b) C_2H_6 (c) C_3H_8 (d) C_4H_{10}

2.34 (a) CH_3OH (b) C_2H_5OH

2.35

```
    H   H
    |   |
H — C — C — O — H
    |   |
    H   H
```

2.36 $C_{10}H_{22}$

2.37 $C_{23}H_{48}$

2.38 $C_6H_{12}O_6$ is expected to be a molecular compound because it (1) is composed of nonmetallic atoms and (2) the subscripts are not in the smallest whole number ratios.

2.39 (a) An ionic compound is formed by the transfer of electrons, and it is accompanied by the formation of ions of opposite charge.
(b) Molecular compounds arise from the sharing of electrons between atoms, rather than from the complete transfer of electrons as in (a).

2.40 Metals react with nonmetals.

2.41 Nonmetals react with metals, metalloids and nonmetals.

2.42 Nonmetals are more frequently found in compounds because of the large variety of ways they may combine. A particularly illustrative example is the combination of carbon, a nonmetal, with other elements. So many compounds are possible that there is one entire area of chemistry devoted to the study of carbon compounds, organic chemistry.

2.43 The atoms in nonmetal-nonmetal compounds are held to each other by a sharing of electrons between the two atoms. In compounds of nonmetal- metal compounds, ions are formed and the ions are held together in the solid by the electrostatic forces of attractions between oppositely charged ions.

2.44 An ionic compound contains ions which are held together by the electrostatic attraction between oppositely charged ions in an extended geometric array. Individual molecules cannot be found in ionic compounds because such substances are extended geometrical arrays of ions in the solid.

2.45 An ion is a charged particle. It can be monatomic or polyatomic, and it can have either a positive or a negative charge. It is derived from an atom or a molecule by gain or loss of electrons. Atoms and molecules are neutral.

2.46 In ionic substances, no molecules exist. Rather we have a continuous array of cations and anions, which are present in a constant ratio. The ratio is given by the formula unit.

2.47 Al_2Cl_6 is molecular because the formula is not written as an empirical formula, rather it is written with a specific number of atoms in the molecule as a molecular formula.

2.48
(a) Na and Na^+
(b) Yes, they have the same nuclei.
(c) Yes, they have the same number of protons.
(d) They could have the same number of neutrons, it depends on the isotopes of the atom and ion.
(e) No, they do not have the same number of electrons; Na^+ has one less electron (10 electrons) than Na (11 electrons)

2.49 A cation is an ion with a positive charge, whereas an anion is an ion with a negative charge.

2.50 These ions are Ca^{2+} and Cl^-.

2.51 This requires the loss of four electrons. The ion has 22 protons, as does the atom. The ion only has 18 electrons, however.

2.52 An atom that gains an electron acquires a negative charge.

2.53 This requires the gain of three electrons. There are seven protons and 10 electrons.

2.54
(1) The positive ion is given first in the formula.
(2) The subscripts in the formula must produce an electrically neutral formula unit.
(3) The subscripts should be the set of smallest whole numbers possible.
(4) The charges on the ions are not included in the finished formula for the substance.

2.55 Rubidium, Rb, is in group 1 and has a +1 charge. When it combines with Cl, possessing a –1 charge, the formula would be RbCl. In the second compound, the metal atom should be written first: Na_2S.

2.56 The simplest combining ratio, or formula unit is given for ionic compounds. This should be TiO_2.

2.57 The post transition metals appear in the periodic table row after a transition metal. Thus we have Ga and Ge, which follow Zn in the first transition metal row. In, Sn, and Sb which follow Cd in the second transition metal row, and Tl, Pb, Bi and Po, which follow Hg in the third transition metal row.

2.58 (a) Fe^{2+}, Fe^{3+}
(b) Co^{2+}, Co^{3+}
(c) Hg^{2+}, Hg_2^{2+}
(d) Cr^{2+}, Cr^{3+}
(e) Sn^{2+}, Sn^{4+}
(f) Pb^{2+}, Pb^{4+}
(g) Cu^{+}, Cu^{2+}

2.59 The incorrect ones are a, d, and e.

2.60 (a) CN^- (b) NH_4^+ (c) NO_3^-
(d) SO_3^{2-} (e) ClO_3^- (f) SO_4^{2-}

2.61 (a) OCl^- (b) HSO_4^- (c) PO_4^{3-}
(d) $H_2PO_4^-$ (e) MnO_4^- (f) $C_2O_4^{2-}$

2.62 (a) dichromate ion (b) hydroxide ion (c) acetate ion
(d) carbonate ion (e) cyanide ion (f) perchlorate ion

2.63 (a) $Ca(s) + Cl_2(g) \rightarrow CaCl_2(s)$
(b) $2Mg(s) + O_2(g) \rightarrow 2MgO(s)$
(c) $4Al(s) + 3O_2(g) \rightarrow 2Al_2O_3(s)$
(d) $S(s) + 2Na(s) \rightarrow Na_2S(s)$ or $S_8(s) + 16Na(s) \rightarrow 8Na_2S(s)$

2.64 The reason than ionic and molecular compounds have different physical properties is due to the bonding. Ionic compounds have ionic bonds which come from the electrostatic attractions of the cations and anions and form arrays. Molecular compounds for covalent bonds in which electrons are shared between two atoms and these form discreet units.

2.65 In the molten state, the ions of an ionic compound are free to move about, which is required for electrical conduction.

2.66 Ionic compounds are brittle because the slight movement of a layer of ions which occurs when the crystal is struck, suddenly places ions of the same charge next to each other and for that instant there are large repulsive forces that split the solid.

2.67 The ionic bond (the attraction of ions of opposite charge within a crystalline ionic compound) is very strong, and the melting of the solid requires large amounts of energy for the bonds to be disrupted, allowing individual ions to move freely. Naphthalene is a molecular compound; an ionic one would have a higher melting point.

2.68 Binary compounds, such as CCl_4 contain two elements only. A diatomic substance is composed of molecules having two atoms, such as HCl or N_2. In the latter, the two atoms may or may not be the same.

2.69 When naming the compound, molecular compounds need the prefixes to specify the number of atoms in the molecule. Ionic compounds made with transition metals or post-transition metals need to have the charge of the metal specified.

2.70 For faming ionic compounds of the transition elements it is essential to know the charge on the anion since that will help determine the charge on the transition element. Transition elements can have more than one charge.

Review Problems

2.71 (a) 2 K, 2 C, 4 O
(b) 2 H, 1 S, 3 O
(c) 12 C, 26 H
(d) 4 H, 2 C, 2 O
(e) 9 H, 2 N, 1 P, 4 O

2.72 (a) 3 Na, 1 P, 4 O
(b) 1 Ca, 4 H, 2 P, 8 O
(c) 4 C, 10 H
(d) 3 Fe, 2 As, 8 O
(e) 3 C, 8 H, 3 O

2.73 (a) 1 Ni, 2 Cl, 8 O
(b) 1 Cu, 1 C, 3 O
(c) 2 K, 2 Cr, 7 O
(d) 2 C, 4 H, 2 O
(e) 2 N, 9 H, 1 P, 4 O

2.74 (a) 6 C, 12 H, 2 O
(b) 14 H, 5 O, 1 Mg, 1 S
(c) 1 K, 1 Al, 2 S, 20 O, 24 H
(d) 1 Cu, 2 N, 6 O
(e) 4 C, 10 H, 1 O

2.75 1 Cr, 6 C, 9 H, 6 O

2.76 3 Ca, 5 Mg, 8 Si, 24 O, 2 H

2.77 (a) 6 N, 3 O
(b) 4 Na, 4 H, 4 C, 12 O
(c) 2 Cu, 2 S, 18 O, 20 H

2.78 (a) 14 C, 28 H, 14 O
(b) 4 N, 8 H, 2 C, 2 O
(c) 10 K, 10 Cr, 35 O

2.79 (a) 6 (b) 3 (c) 27

2.80 (a) 16 (b) 36 (c) 50

2.81 (a) K^+ (b) Br^- (c) Mg^{2+} (d) S^{2-} (e) Al^{3+}

2.82 (a) Ba^{2+} (b) O^{2-} (c) F^- (d) Sr^{2+} (e) Rb^+

2.83 (a) NaBr (b) KI (c) BaO (d) $MgBr_2$ (e) BaF_2

2.84 (a) $CrCl_2$ and $CrCl_3$
(b) $FeCl_2$ and $FeCl_3$
(c) $MnCl_2$ and $MnCl_3$
(d) CuCl and $CuCl_2$
(e) $ZnCl_2$

2.85 (a) KNO_3 (b) $Ca(C_2H_3O_2)_2$ (c) NH_4Cl
(d) $Fe_2(CO_3)_3$ (e) $Mg_3(PO_4)_2$

2.86 (a) $Zn(OH)_2$ (b) Ag_2CrO_4 (c) $BaSO_3$
(d) Rb_2SO_4 (e) $LiHCO_3$

2.87 (a) PbO and PbO_2 (b) SnO and SnO_2 (c) MnO and Mn_2O_3
(d) FeO and Fe_2O_3 (e) Cu_2O and CuO

2.88 (a) $CdCl_2$ (b) $AgCl$ (c) $ZnCl_2$ (d) $NiCl_2$

2.89 (a) silicon dioxide (b) xenon tetrafluoride
(c) tetraphosphorus decaoxide (d) dichlorine heptaoxide

2.90 (a) chlorine trifluoride (b) disulfur dichloride
(c) dinitrogen pentoxide (d) arsenic pentachloride

2.91 (a) calcium sulfide (b) aluminum bromide
(c) sodium phosphide (d) barium arsenide
(e) rubidium sulfide

2.92 (a) sodium fluoride (b) magnesium carbide
(c) lithium nitride (d) aluminum oxide
(e) potassium selenide

2.93 (a) iron(II) sulfide (b) copper(II) oxide
(c) tin(IV) oxide (d) cobalt(II) chloride hexahydrate

2.94 (a) manganese(III) oxide (b) mercury(I) chloride
(c) lead(II) sulfide (d) chromium(III) chloride tetrahydrate

2.95 (a) sodium nitrite
(b) potassium permanganate
(c) magnesium sulfate heptahydrate
(d) potassium thiocyanate

2.96 (a) potassium phosphate
(b) ammonium acetate
(c) iron(III) carbonate
(d) sodium thiosulfate pentahydrate

2.97 (a) ionic chromium(II) chloride
(b) molecular disulfur dichloride
(c) ionic ammonium acetate
(d) molecular sulfure trioxide
(e) ionic potassium iodate
(f) molecular tetraphosphorous hexaoxide
(g) ionic calcium sulfite
(h) ionic silver cyanide
(i) ionic zinc(II) bromide
(j) molecular hydrogen selenide

2.98 (a) ionic vanadium(III) nitrate
(b) ionic cobalt(II) acetate
(c) ionic gold(III) sulfide
(d) ionic gold(I) sulfide
(e) molecular germanium tetrabromide
(f) ionic potassium chromate

(g)	ionic	iron(II) hydroxide	
(h)	molecular	diiodine tetraoxide	
(i)	molecular	tetraiodine nonaoxide	
(j)	molecular	tetraphosphorus triselenide	

2.99	(a)	Na_2HPO_4	(b)	Li_2Se	(c)	$Cr(C_2H_3O_2)_3$
	(d)	S_2F_{10}	(e)	$Ni(CN)_2$	(f)	Fe_2O_3
	(g)	SbF_5				
2.100	(a)	Al_2Cl_6	(b)	As_4O_{10}	(c)	$Mg(OH)_2$
	(d)	$Cu(HSO_4)_2$	(e)	NH_4SCN	(f)	$K_2S_2O_3$
	(g)	I_2O_5				
2.101	(a)	$(NH_4)S$	(b)	$Cr_2(SO_4)_3{\cdot}6H_2O$	(c)	SiF_4
	(d)	MoS_2	(e)	$SnCl_4$	(f)	H_2Se
	(g)	P_4S_6				
2.102	(a)	$Hg(C_2H_3O_2)_2$	(b)	$Ba(HSO_3)_2$	(c)	BCl_3
	(d)	Ca_3P_2	(e)	$Mg(H_2PO_4)_2$	(f)	CaC_2O_4
	(g)	XeF_4				

2.103 diselenium hexasulfide and diselenium tetrasulfide

2.104 diphosphorous pentasulfide

Additional Excercises

2.105 $Hg_2(NO_3)_2{\cdot}2H_2O$ $Hg(NO_3)_2{\cdot}H_2O$

2.106	(a)	Cl_2	HBr	AsH_3	NO_2		
	(b)	CaO	$CuCl_2$	$NaNO_3$			
	(c)	Cl_2	CaO	HBr	$CuCl_2$	AsH_3	NO_2
	(d)	Cl_2	CaO	HBr			
	(e)	$CuCl_2$	NO_2				
	(f)	Cl_2	HBr	AsH_3	NO_2		
	(g)	CaO	$CuCl_2$				
	(h)	$NaNO_3$					

2.107	(a)	auric nitrate	(b)	cupric sulfate
	(c)	plumbic oxide	(d)	mercurous chloride
	(e)	mercuric chloride	(f)	cobaltous hydroxide
	(g)	stannous chloride	(h)	stannic sulfide
	(i)	auric sulfate		

2.108	(a)	chromium(II) chloride	$CrCl_2$
	(b)	manganese(II) sulfate	$MnSO_4$
	(c)	lead(II) acetate	$Pb(C_2H_3O_2)_2$
	(d)	copper(II) bromide	$CuBr_2$
	(e)	copper(I) iodide	CuI
	(f)	iron(II) sulfate	$FeSO_4$
	(g)	mercury(I) nitrate	$Hg_2(NO_3)_2$

2.109 ferric nitrate nonahydrate

Practice Exercises

3.1 a) $m \cdot v^2$ would have units of $kg \cdot (m/s)^2 = kg \cdot m^2/s^2$
b) mgh would have units of $kg \cdot (m/s^2) \cdot m = kg \cdot m^2/s^2$

3.2 a) mg
b) mm
c) ns

a) 1×10^{-9}
b) 1×10^{-2}
c) 1×10^{-12}

a) cg
b) Mm
c) ms

3.3 $$t_C = (t_F - 32\ ^\circ F)\left(\frac{5\ ^\circ C}{9\ ^\circ F}\right) = (50\ ^\circ F - 32\ ^\circ F)\left(\frac{5\ ^\circ C}{9\ ^\circ F}\right) = 10\ ^\circ C$$

To convert from °F to K we first convert to °C.

$$t_C = (t_F - 32\ ^\circ F)\left(\frac{5\ ^\circ C}{9\ ^\circ F}\right) = (68\ ^\circ F - 32\ ^\circ F)\left(\frac{5\ ^\circ C}{9\ ^\circ F}\right) = 20\ ^\circ C$$

$T_K = 273 + t_C = 273 + 20 = 293$ K

3.4 The data set from Worker C has the best precision.
The data set from Worker A has the best accuracy.

3.5 a) 42.0 g
b) 30.0 mL
c) 54.155 g
d) 11.3 g
e) 0.857 g/mL
f) 3.62 ft (1 and 12 are exact numbers)
g) 8.3 m^3

3.6 a) $$\#\ \text{in.} = (3.00\ \text{yd})\left(\frac{3\ \text{ft}}{1\ \text{yd}}\right)\left(\frac{12\ \text{in.}}{1\ \text{ft}}\right) = 108\ \text{in.}$$

b) $$\#\ \text{cm} = (1.25\ \text{km})\left(\frac{1000\ \text{m}}{1\ \text{km}}\right)\left(\frac{100\ \text{cm}}{1\ \text{m}}\right) = 1.25 \times 10^5\ \text{cm}$$

c) $$\#\ \text{ft} = (3.27\ \text{mm})\left(\frac{1\ \text{m}}{1000\ \text{mm}}\right)\left(\frac{100\ \text{cm}}{1\ \text{m}}\right)\left(\frac{1\ \text{in.}}{2.54\ \text{cm}}\right)\left(\frac{1\ \text{ft}}{12\ \text{in.}}\right) = 0.0107\ \text{ft}$$

d) $$\frac{\#\ \text{km}}{\text{L}} = \left(\frac{20.2\ \text{mile}}{1\ \text{gal}}\right)\left(\frac{1.609\ \text{km}}{1\ \text{mile}}\right)\left(\frac{1\ \text{gal}}{3.785\ \text{L}}\right) = 8.59\ \text{km/L}$$

3.7 density = mass/volume = $(1.24 \times 10^6\ g)/(1.38 \times 10^6\ cm^3) = 0.899\ g/cm^3$

3.8 $\text{mass (g)} = (250{,}000\ \text{cm}^3)\left(\dfrac{19.82\ \text{g}}{1\ \text{cm}^3}\right) = 5.0 \times 10^6\ \text{g (or 5.0 Mg)}$

3.9 $\text{sp. gr.} = \dfrac{d_{\text{aluminum}}}{d_{\text{water}}} = \dfrac{2.70\ \text{g/mL}}{1.00\ \text{g/mL}} = 2.70$

$$d_{\text{aluminum}} = \text{sp. gr.} \times d_{\text{water}} = 2.70 \times 62.4\ \text{lb/ft}^3 = 168\ \text{lb/ft}^3$$

3.10 $d_{\text{ethyl acetate}} = \text{sp. gr.} \times d_{\text{water}} = 0.902 \times 1.00\ \text{g/mL} = 0.902\ \text{g/mL}$

$$d_{\text{ethyl acetate}} = \text{sp. gr.} \times d_{\text{water}} = 0.902 \times 8.34\ \text{lb/gal} = 7.52\ \text{lb/gal}$$

Thinking it Through

T3.1 (a) First select a conversion factor that transforms cm to nm. Since 1 m is equal to both 100 cm and 10^9 nm (from Table 3.3), the following conversion factors may be determined:

$$\frac{1\ \text{m}}{100\ \text{cm}} \qquad \frac{10^9\ \text{nm}}{1\ \text{m}}$$

Setting up the calculation so that units cancel we have,

$$\#\ \text{nm} = 14.6\ \text{cm} \times \frac{1\ \text{m}}{100\ \text{cm}} \times \frac{10^9\ \text{nm}}{1\ \text{m}}$$

(b) First convert m to km. Since 1 km = 1000 m, a conversion factor is $\frac{1\ \text{km}}{1000\ \text{m}}$. We also need to convert s to hr. Since 3600 s = 1 hr, the conversion factor is: $\frac{3600\ \text{s}}{1\ \text{hr}}$. Setting up the calculation so that the units cancel we have:

$$\frac{\#\ \text{km}}{\text{hr}} = \frac{1350\ \text{m}}{\text{s}} \times \frac{1\ \text{km}}{1000\ \text{m}} \times \frac{3600\ \text{s}}{1\ \text{hr}}$$

(c) As shown in Table 3.3, there are 10^6 μg in 1 g. The first conversion factor is thus $\frac{10^6\ \mu\text{g}}{1\ \text{g}}$. Next, since 100 cm and 10^6 μm are both equal to 1 m (Table 3.3), we have 100 cm = 10^6 μm. Thus 1 cm = 10^4 μm and cubing each side of this equation gives us the desired units of volume: 1 $\text{cm}^3 = 10^{12}\ \mu\text{m}^3$. The conversion factor is thus $\frac{1\ \text{cm}^3}{10^{12}\ \mu\text{m}^3}$. The calculation is set up so that the units cancel and we convert first to μg and then to μm^3:

$$\frac{\#\ \mu\text{g}}{\mu\text{m}^3} = \frac{8.85\ \text{g}}{\text{cm}^3} \times \frac{10^6\ \mu\text{g}}{1\ \text{g}} \times \frac{1\ \text{cm}^3}{10^{12}\ \mu\text{m}^3}$$

(d) The length relationship that we must start with is 1 cm = 10mm. Since the question pertains to volume, these lengths must be cubed: 1 cm^3 = 1000 mm^3. The conversion factor is thus: $\frac{1000\ \text{mm}^3}{1\ \text{cm}^3}$. Setting up the calculation so that units cancel properly we have:

$$\#\ \text{mm}^3 = 4.20\ \text{cm}^3\left(\frac{1000\ \text{mm}^3}{1\ \text{cm}^3}\right)$$

(e) We first must convert to a height expressed in inches using the fact that 1 ft = 12 in:

$$\#\ \text{in} = 5\ \text{ft} \times \frac{12\ \text{in}}{1\ \text{ft}} + 9\ \text{in} = 69\ \text{in}$$

Next, since 1 in = 2.54 cm, we can construct a conversion factor to convert the height to cm:

$$\#\ \text{cm} = \left(5\ \text{ft} \times \frac{12\ \text{in}}{1\ \text{ft}} + 9\ \text{in}\right) \times \frac{2.54\ \text{cm}}{1\ \text{in}}$$

T3.2
- (a) One inch is exactly 2.54 cm.
- (b) The earth is approximately 5 billion years old.
- (c) Qualitative observations are just as useful as quantitative observations
- (d) Significant digits are those digits that are measured plus one that is estimated.
- (e) Counting does not always give an exact number.
- (f) The U.S. Census can estimate the population of the United States.
- (g) Errors are due to the inherent limitations in the measurement procedure.
- (h) An estimation is an educated guess.

T3.3 Two equations are needed, the first to convert °F to °C. This is equation 3.1 and we can substitute the values directly in to the equation:

$$98.6\ ^\circ\text{F} = \left(\frac{9\ ^\circ\text{F}}{5\ ^\circ\text{C}}\right)t_c + 32\ ^\circ\text{F}$$

rearranging gives us:

$$t_c = \left(\frac{5\ ^\circ\text{C}}{9\ ^\circ\text{F}}\right)\left(98.6\ ^\circ\text{F} - 32\ ^\circ\text{F}\right) = 37\ ^\circ\text{C}$$

We can convert to a temperature in Kelvin by using equation 3.3. Substituting the appropriate value gives:

$$T_K = \left(t_C + 273\ ^\circ\text{C}\right)\left(\frac{1\ \text{K}}{1\ ^\circ\text{C}}\right) = \left(37\ ^\circ\text{C} + 273\ ^\circ\text{C}\right)\left(\frac{1\ \text{K}}{1\ ^\circ\text{C}}\right) = 310\ \text{K}$$

T3.4 First convert m to km. Since 1 km = 1000 m, a conversion factor is $\frac{1\ \text{km}}{1000\ \text{m}}$. Then convert km to mi using the conversion factor $\frac{0.6215\ \text{mi}}{1\ \text{km}}$. We also need to convert s to hr. Since 3600 s = 1 hr, the conversion factor is: $\frac{3600\ \text{s}}{1\ \text{hr}}$. Setting up the calculation so that the units cancel we have:

$$\frac{\#\ \text{mi}}{\text{hr}} = \frac{2.9979\times10^8\ \text{m}}{\text{s}} \times \frac{1\ \text{km}}{1000\ \text{m}} \times \frac{0.6215\ \text{mi}}{1\ \text{km}} \times \frac{3600\ \text{s}}{1\ \text{hr}}$$

T3.5 The density in units of g/cm^3 may be calculated directly since we know the mass in grams and the dimensions of the cylinder. The volume of the cylinder may be calculated by multiplying the area of the base by the height of the cylinder: V = A × h. The area can be calculated as follows:

$$A = \pi\left(\frac{d}{2}\right)^2 = \pi\left(\frac{0.753\ \text{cm}}{2}\right)^2 = 0.445\ \text{cm}^2$$

$$V = A \times h = 0.445\ \text{cm}^2 \times 2.33\ \text{cm} = 1.04\ \text{cm}^3$$

Using equation 3.7, the density of the metal is:

$$d = \frac{m}{V} = \frac{8.423\ \text{g}}{1.04\ \text{cm}^3} = 8.12\ \text{g}/\text{cm}^3$$

To determine the specific gravity of the substance in lb/ft^3 we need to take the ratio of the density of the substance to the density of water. Since water has a density of 1.00 g/cm^3, the specific gravity of the metal is 8.12. The density of the metal is therefore 8.12 × the density of water (62.4 lb/ft^3) Hence the density of the metal is 507 lb/ft^3.

T3.6 We must rearrange equation 3.7 to:

$$V = \frac{m}{d} = \frac{3.5\text{ g}}{3.51\ \text{g}/_{\text{cm}^3}} = 1.0\text{ cm}^3$$

T3.7 The number of significant figures will come from the measurement of the mass of the diamond since it has the least number of significant figures, 2.

T3.8 To determine the mass in pounds of 3.56 gallons of jet fuel, use the equations 3.7 and 3.9:

$$d = \frac{m}{V} \qquad \text{and} \qquad d_{\text{substance}} = (\text{sp. gr.})_{\text{substnace}} \times d_{\text{water}}$$

and that the density of water is 8.34 lb/gal and the specific gravity of jet fuel is 0.75.
The specific gravity of jet fuel is 0.75

$$d_{\text{jet fuel}} = 0.75 \times 8.34\text{ lb/gal} = 6.3\text{ lb/gal}$$

$$6.3\text{ lb/gal} = \frac{m}{3.45\text{ gal}}$$

Rearrange to determine the mass of the jet fuel in pounds:

$$m = \left(6.3\ \text{lb}/_{\text{gal}}\right) \times 3.45\text{ gal}$$

T3.9 Using the equation 3.9 to determine the density of seawater:

$$d_{\text{substance}} = (\text{sp. gr.})_{\text{substnace}} \times d_{\text{water}}$$

$$d_{\text{sea water}} = 1.03 \times 62.4\text{ lb/ft}^3$$

$$d_{\text{sea water}} = 64.3\text{ lb/ft}^3$$

Convert the volume of the seawater from in^3 to ft^3, and then use equation 3.7 to determine the mass of the seawater:

$$(12\text{ in})^3 = (1\text{ ft})^3$$

$$1728\text{ in}^3 = 1\text{ ft}^3$$

$$146\text{ in}^3 \times \frac{1\text{ ft}^3}{1728\text{ in}^3} = 8.45 \times 10^{-4}\text{ ft}^3$$

$$d = \frac{m}{V} \qquad m = d \times V \qquad m = 64.3\text{ lb/ft}^3 \times \left(8.45 \times 10^{-4}\text{ ft}^3\right)$$

T3.10 To answer the question, convert the stated concentration to a concentration with units that are the same as the CDC standard. In this case, 2.5×10^{-4} g = 250 micrograms and 1 L = 10 deciliters. So, 2.5×10^{-4} g/L = 250 micrograms/10 deciliters = 25 micrograms/deciliter. This person has a high lead concentration and is in danger.

T3.11 Calculate the volume of gold needed to cover the sphere. To do this, determine the volume of the sphere which is gold coated and then subtract the volume of the original ball bearing. (Let gcb = gold coated bearing.)

$$V_{\text{gcb}} = \frac{4}{3}\pi r^3 = \frac{4}{3}\pi\left(\frac{0.700\text{ in}}{2} + 0.50\text{ mm}\right)^3$$

$$V_{\text{gcb}} = \frac{4}{3}\pi\left(\left(\frac{0.700\text{ in}}{2}\right)\left(\frac{2.54\text{ cm}}{1\text{ in}}\right) + (0.50\text{ mm})\left(\frac{1\text{ cm}}{10\text{ mm}}\right)\right)^3$$

$$V_{\text{gcb}} = \frac{4}{3}\pi(0.889\text{ cm} + 0.050\text{ cm})^3 = 3.468\text{ cm}^3$$

A similar calculation gives the volume of the uncoated ball bearing as V = 2.943 cm^3. Now;

$$V_{\text{gold}} = V_{\text{gcb}} - V = 3.468\text{ cm}^3 - 2.943\text{ cm}^3 = 0.525\text{ cm}^3$$

Finally, determine the mass of the gold using Equation 3.7, $m_{\text{gold}} = (19.31\text{ g/cm}^3)(0.525\text{ cm}^3) = 10.1\text{ g}$.

T3.12 First determine the density of kerosene in lb./ft^3, then convert the ft^3 to m^3:

$$d_{kerosene} = (\text{sp. gr.}) \times d_{water}$$
$$d_{kerosene} = 0.965 \times 52.4\ \text{lb/ft}^3$$
$$d_{kerosene} = 50.6\ \text{lb/ft}^3$$
$$50.6\ \text{lb/ft}^3 \times \left(\frac{3\ \text{ft}}{1\ \text{yd}}\right)^3 \times \left(\frac{1\ \text{yd}}{0.9144\ \text{m}}\right)^3 = 166\ \text{lb/m}$$

T3.13 Using the information provided, we can calculate the area of the hole that is the pore opening. We would need to know the thickness of the polycarbonate film to calculate the volume of gold needed to fill the pore. Once we know the volume, we multiply by the density of gold to determine the mass needed per pore opening.

Review Questions

3.1 Measurements involve a comparison. The unit gives the number meaning.

3.2 Le Système International d'Unités

3.3 Kilogram

3.4 Kilogram

3.5 force = mass × acceleration $\quad N = kg \times m/s^2$

3.6
(a) 0.01 10^{-2}
(b) 0.001 10^{-3}
(c) 1000 10^{3}
(d) 0.000001 10^{-6}
(e) 0.000000001 10^{-9}
(f) 0.000000000001 10^{-12}
(g) 1,000,000 10^{6}

3.7
(a) c (b) m
(c) k (d) μ
(e) n (f) p
(g) M

3.8
(a) centimeter or millimeter
(b) milliliter
(c) gram

3.9 Mass is measured by comparing the weight of an object to the weight of a set of standard masses. A balance compares the two masses directly, while a spring scale compares the force exerted by the object to a spring which possesses a known force.

3.10 The melting points and boiling points of water at 1 atmosphere pressure. On the Celsius scale these points correspond to 0 °C and 100 °C respectively.

3.11
(a) 1 Fahrenheit degree $<$ 1 Celsius degree
(b) 1 Celsius degree = 1 Kelvin
(c) 1 Fahrenheit degree $<$ 1 Kelvin

3.12 The digits that are significant figures in a quantity are those that are known (measured) with certainty plus the last digit, which contains some uncertainty.

3.13 The accuracy of a measured value is the closeness of that value to the true value of the quantity. The precision of a number of repeated measurements of the same quantity is the closeness of the measurements to one another.

3.14 The maximum concentration of the toxin would be 1.9 μg/ml. If the measurement had been written as 1.50 μg/ml then the maximum concentration would be 1.59 μg/ml.

3.15 The problem with using the fraction 3 yd/1 ft as a conversion factor is that there are 3 feet in one yard. The conversion factor should be 1 yd/3 ft. For the second part of the question, it is not possible to construct a valid conversion factor relating centimeters to meters from the equation 1 cm = 1000 m, since 10,000 cm = 1 m.

3.16 To convert 250 seconds to hours multiply 250 by:

$$\frac{1\ \text{h}}{3600\ \text{s}}$$

To convert 3.84 hours to seconds multiply 3.84 hours by:

$$\frac{3600\ \text{s}}{1\ \text{h}}$$

3.17 Four significant figures would be correct because the conversion factor contains exact values. The measured value determines the number of significant figures.

Review Problems

3.18	(a)	0.01 m	(b)	1000 m	(c)	10^{12} pm
	(d)	0.1 m	(e)	0.001 kg	(f)	0.01 g
3.19	(a)	10^{-9}	(b)	10^{-6}	(c)	10^{3}
	(d)	10^{6}	(e)	10^{-3}	(f)	0.1

3.20 (a) $t_F = (\frac{9\ °C}{5\ °F})(t_C) + 32\ °F = 9/5(50) + 32 = 120\ °F$ when rounded to the proper number of significant figures.

(b) $t_F = (\frac{9\ °C}{5\ °F})(t_C) + 32\ °F = 9/5(10) + 32 = 50\ °F$

(c) $t_{°C} = (\frac{5\ °F}{9\ °C})(t_F - 32\ °F) = 5/9(25.5 - 32) = 3.61\ °C$

(d) $t_C = (\frac{5\ °F}{9\ °C})(t_F - 32\ °F) = 5/9(49 - 32) = 9.4\ °C$

(e) $T_K = (t_C + 273\ °C)\frac{1\ K}{1\ °C} = 60 + 273 = 333K$

(f) $T_K = (t_C + 273\ °C)\frac{1\ K}{1\ °C} = -30 + 273 = 243K$

3.21 (a) $t_C = \left(\frac{5\ °C}{9\ °F}\right)(t_F - 32\ °F) = \left(\frac{5\ °C}{9\ °F}\right)(96\ °F - 32\ °F) = 36\ °C$

(b) $t_C = \left(\frac{5\ °C}{9\ °F}\right)(t_F - 32\ °F) = \left(\frac{5\ °C}{9\ °F}\right)(-6\ °F - 32\ °F) = -21\ °C$

(c) $t_F = \left(\dfrac{9\ °F}{5\ °C}\right)(t_C) + 32\ °F = \left(\dfrac{9\ °F}{5\ °C}\right)(-55\ °C) + 32\ °F = -67\ °F$

(b) $t_C = (T_K - 273\ K)\left(\dfrac{1\ °C}{1\ K}\right) = (273\ K - 273\ K)\left(\dfrac{1\ °C}{1\ K}\right) = 0\ °C$

(c) $t_C = (T_K - 273\ K)\left(\dfrac{1\ °C}{1\ K}\right) = (299\ K - 273\ K)\left(\dfrac{1\ °C}{1\ K}\right) = 26\ °C$

(d) $T_K = (t_C + 273\ °C)\left(\dfrac{1\ K}{1\ °C}\right) = (40\ °C + 273\ °C)\left(\dfrac{1\ K}{1\ °C}\right) = 313\ K$

3.22 $t_C = (t_F - 32\ °F)\left(\dfrac{5\ °C}{9\ °F}\right) = (103.5\ °F - 32\ °F)\left(\dfrac{5\ °C}{9\ °F}\right) = 39.7\ °C$

This dog has a fever; the temperature is out of normal canine range.

3.23 Convert 325 °F to t_C:

$$t_C = \left(\frac{5\ °C}{9\ °F}\right)(t_F - 32\ °F) = \left(\frac{5\ °C}{9\ °F}\right)(325\ °F - 32\ °F) = 163\ °C$$

This oven is not in the 200 to 225 °C range.

3.24 $t_C = (t_F - 32\ °F)\left(\dfrac{5\ °C}{9\ °F}\right) = (1500\ °F - 32\ °F)\left(\dfrac{5\ °C}{9\ °F}\right) = 830\ °C$

3.25 Convert –96 °F to t_c:

$$t_C = \left(\frac{5\ °C}{9\ °F}\right)(t_F - 32\ °F) = \left(\frac{5\ °C}{9\ °F}\right)(-96\ °F - 32\ °F) = -71\ °C$$

3.26 Range in Kelvins:

$K = (10\ MK)\left(\dfrac{1 \times 10^6\ K}{1\ MK}\right) = 1.0 \times 10^7\ K$ $\qquad$ $K = (25\ MK)\left(\dfrac{1 \times 10^6\ K}{1\ MK}\right) = 2.5 \times 10^7\ K$

Range in degrees Celsius:

$$t_C = (T_K - 273\ K)\left(\frac{1\ °C}{1\ K}\right) = (1.0 \times 10^7\ K - 273\ K)\left(\frac{1\ °C}{1\ K}\right) \approx 1.0 \times 10^7\ °C$$

$$t_C = (T_K - 273\ K)\left(\frac{1\ °C}{1\ K}\right) = (2.5 \times 10^7\ K - 273\ K)\left(\frac{1\ °C}{1\ K}\right) \approx 2.5 \times 10^7\ °C$$

Range in degrees Fahrenheit:

$t_F = 9/5(°C) + 32 = (9/5)(1.0 \times 10^7) + 32 \approx 1.8 \times 10^7\ °F$

$t_F = 9/5(°C) + 32 = (9/5)(2.5 \times 10^7) + 32 \approx 4.5 \times 10^7\ °F$

3.27 Convert 111 K to t_C:

$$t_C = (T_K - 273\ K)\left(\frac{1\ °C}{1\ K}\right) = (111\ K - 273K)\left(\frac{1\ °C}{1\ K}\right) = -162\ °C$$

3.28 $t_C = T_K - 273\text{ K} = (4\text{ K} - 273\text{ K})\left(\frac{1\text{ °C}}{1\text{ K}}\right) = -269\text{ °C}$

3.29 Convert 6000 K to t_C:

$$t_C = (T_K - 273\text{ K})\left(\frac{1\text{ °C}}{1\text{ K}}\right) = (6000\text{ K} - 273\text{K})\left(\frac{1\text{ °C}}{1\text{ K}}\right) = 5700\text{ °C}$$

This is hot enough to melt concrete, since it is hotter than 2000 °C.

3.30 $t_C = (T_K - 273)\left(\frac{1\text{ °C}}{1\text{ K}}\right) = 77 - 273 = -196\text{ °C}$

$t_F = 9/5\text{ (°C)} + 32 = (9/5)(-196) + 32 = -321\text{ °F}$

This temperature is *lower* than the boiling point of oxygen (–297 °F), so oxygen is a liquid at this temperature.

3.31 Convert 55,000 °F to t_C then convert to T_K:

$$t_C = \left(\frac{5\text{ °C}}{9\text{ °F}}\right)(t_F - 32\text{ °F}) = \left(\frac{5\text{ °C}}{9\text{ °F}}\right)(55{,}000\text{ °F} - 32\text{ °F}) = 31{,}000\text{ °C}$$

$$T_K = (t_C + 273\text{ °C})\left(\frac{1\text{ K}}{1\text{ °C}}\right) = (31{,}000\text{ °C} + 273\text{ °C})\left(\frac{1\text{ K}}{1\text{ °C}}\right) = 31{,}000\text{ K}$$

The air within a lightning bolt is hotter than the surface of the sun.

3.32
(a) 4 significant figures
(b) 5 significant figures
(c) 4 significant figures
(d) 2 significant figures
(e) 4 significant figures
(f) 2 significant figures

3.33
(a) 3 significant figures
(b) 6 significant figures
(c) 1 significant figures
(d) 5 significant figures
(e) 1 significant figures
(f) 5 significant figures

3.34
(a) 5 significant figures
(b) 5 significant figures
(c) 2 significant figures
(d) 5 significant figures
(e) 4 significant figures
(f) 1 significant figure

3.35
(a) 4 significant figures
(b) 4 significant figures
(c) 4 significant figures
(d) 2 significant figures
(e) 3 significant figures
(f) 2 significant figures

3.36
(a) 0.72 m^2
(b) 84.24 kg
(c) –0.465 g/cm^3 (dividing a number with 4 sig. figs by one with 3 sig. figs)
(d) 19.42 g/mL
(e) 857.7 cm^2

3.37
(a) 2.06 g/mL
(b) 4.02 mL
(c) 12.4 g/mL
(d) 0.276 g/mL
(e) 0.0006 m/s

3.38
(a) 4.34×10^3
(b) 3.20×10^7
(c) 3.29×10^{-3}
(d) 4.20×10^4
(e) 8.00×10^{-6}
(f) 3.24×10^5

3.39
(a) 4.89×10^2
(b) 3.75×10^{-3}
(d) 1.225×10^{-2}
(e) 2.43×10^0

(c) 8.23×10^4 (f) 2.732×10^4

3.40 (a) 310,000 (d) 0.000 000 000 004 4
(b) 0.000 004 35 (e) 0.000 000 356
(c) 3,900 (f) 88,000,000

3.41 (a) 0.0000000527 (d) 0.0235
(b) 712,000 (e) 40,000,000
(c) 0.0000000435 (f) 0.372

3.42 (a) 4.0×10^7 (think of it as $4.0 \times 10^7 - 0.021 \times 10^7$)
(b) 3.0×10^{-2}
(c) 4.4×10^{12}
(d) 2.5
(e) 2.3×10^{18}

3.43 (a) 6.3×10^9 (d) 1.1×10^5
(b) 2.0×10^{18} (e) 2.55×10^{-2}
(c) 5.5×10^{-9}

3.44 (a) $$\text{\# km} = (32.0\ \text{dm})\left(\frac{1\ \text{m}}{10\ \text{dm}}\right)\left(\frac{1\ \text{km}}{1000\ \text{m}}\right) = 3.20 \times 10^{-3}\ \text{km}$$

(b) $$\text{\# }\mu\text{g} = (8.2\ \text{mg})\left(\frac{1\ \text{g}}{1000\ \text{mg}}\right)\left(\frac{1 \times 10^6\ \mu\text{g}}{1\ \text{g}}\right) = 8.2 \times 10^3\ \mu\text{g}$$

(c) $$\text{\# kg} = (75.3\ \text{mg})\left(\frac{1\ \text{g}}{1000\ \text{mg}}\right)\left(\frac{1\ \text{kg}}{1000\ \text{g}}\right) = 7.53 \times 10^{-5}\ \text{kg}$$

(d) $$\text{\# L} = (137.5\ \text{mL})\left(\frac{1\ \text{L}}{1000\ \text{mL}}\right) = 0.1375\ \text{L}$$

(e) $$\text{\# mL} = (0.025\ \text{L})\left(\frac{1000\ \text{mL}}{1\ \text{L}}\right) = 25\ \text{mL}$$

(f) $$\text{\# dm} = (342\ \text{pm})\left(\frac{1 \times 10^{-12}\ \text{m}}{1\ \text{pm}}\right)\left(\frac{10\ \text{dm}}{1\ \text{m}}\right) = 3.42 \times 10^{-9}\ \text{dm}$$

3.45 (a) $$\text{\# km} = (32.0\ \text{dm})\left(\frac{1\ \text{m}}{10\ \text{dm}}\right)\left(\frac{1\ \text{km}}{1000\ \text{m}}\right) = 3.20 \times 10^{-3}\ \text{km}$$

(b) $$\text{\# }\mu\text{g} = (8.2\ \text{mg})\left(\frac{1\ \text{g}}{1000\ \text{mg}}\right)\left(\frac{1 \times 10^6\ \mu\text{g}}{1\ \text{g}}\right) = 8.2 \times 10^3\ \mu\text{g}$$

(c) $$\text{\# kg} = (75.3\ \text{mg})\left(\frac{1\ \text{g}}{1000\ \text{mg}}\right)\left(\frac{1\ \text{kg}}{1000\ \text{g}}\right) = 7.53 \times 10^{-5}\ \text{kg}$$

(d) $$\text{\# L} = (137.5\ \text{mL})\left(\frac{1\ \text{L}}{1000\ \text{mL}}\right) = 0.1375\ \text{L}$$

(e) $$\text{\# mL} = (0.025\ \text{L})\left(\frac{1000\ \text{mL}}{1\ \text{L}}\right) = 25\ \text{mL}$$

(f) $$\text{\# dm} = (342\ \text{pm})\left(\frac{1 \times 10^{-12}\ \text{m}}{1\ \text{pm}}\right)\left(\frac{10\ \text{dm}}{1\ \text{m}}\right) = 3.42 \times 10^{-9}\ \text{dm}$$

3.46 (a) $\# \text{ cm} = (230 \text{ km})\left(\frac{1\times10^{3}\text{ m}}{1\text{ km}}\right)\left(\frac{1\text{ cm}}{1\times10^{-2}\text{ m}}\right) = 2.3 \times 10^{7}\text{ cm}$

(b) $\# \text{ mg} = (423 \text{ kg})\left(\frac{1\times10^{3}\text{ g}}{1\text{ kg}}\right)\left(\frac{1\text{ mg}}{1\times10^{-3}\text{ g}}\right) = 4.23 \times 10^{8}\text{ mg}$

(c) $\# \text{ Mg} = (423 \text{ kg})\left(\frac{1\times10^{3}\text{ g}}{1\text{ kg}}\right)\left(\frac{1\text{ Mg}}{1\times10^{6}\text{ g}}\right) = 4.23 \times 10^{-1}\text{ Mg}$

(d) $\# \text{ mL} = (430\ \mu\text{L})\left(\frac{1\times10^{-6}\text{ L}}{1\ \mu\text{L}}\right)\left(\frac{1\text{ mL}}{1\times10^{-3}\text{ L}}\right) = 4.3 \times 10^{-1}\text{ mL}$

(e) $\# \text{ kg} = (27 \text{ ng})\left(\frac{1\times10^{-9}\text{ g}}{1\text{ ng}}\right)\left(\frac{1\text{ kg}}{1\times10^{3}\text{ g}}\right) = 2.7 \times 10^{-11}\text{ kg}$

3.47 (a) $\# \text{ cm} = (183 \text{ nm})\left(\frac{1\text{ m}}{1 \times 10^{9}\text{ nm}}\right)\left(\frac{100\text{ cm}}{1\text{ m}}\right) = 1.83 \times 10^{-5}\text{ cm}$

(b) $\# \text{ dg} = (3.55 \text{ g})\left(\frac{1\text{ dg}}{0.1\text{ g}}\right) = 35.5\text{ dg}$

(c) $\# \text{ nm} = (6.22 \text{ km})\left(\frac{1000\text{ m}}{1\text{ km}}\right)\left(\frac{1 \times 10^{9}\text{ nm}}{1\text{ m}}\right) = 6.22 \times 10^{12}\text{ nm}$

(d) $\# \text{ mm} = (33 \text{ dm})\left(\frac{1\text{ m}}{10\text{ dm}}\right)\left(\frac{1000\text{ mm}}{1\text{ m}}\right) = 3300\text{ mm}$

(e) $\# \text{ km} = (0.55 \text{ dm})\left(\frac{1\text{ m}}{10\text{ dm}}\right)\left(\frac{1\text{ km}}{1000\text{ m}}\right) = 5.5 \times 10^{-5}\text{ km}$

(b) $\# \text{ pg} = (53.8 \text{ ng})\left(\frac{1\text{ g}}{1 \times 10^{9}\text{ ng}}\right)\left(\frac{1 \times 10^{12}\text{ pg}}{1\text{ g}}\right) = 5.38 \times 10^{4}\text{ pg}$

3.48 (a) $\# \text{ cm} = (36 \text{ in.})\left(\frac{2.54\text{ cm}}{1\text{ in.}}\right) = 91\text{ cm}$

(b) $\# \text{ kg} = (5.0 \text{ lb})\left(\frac{1\text{ kg}}{2.205\text{ lb}}\right) = 2.3\text{ kg}$

(c) $\# \text{ mL} = (3.0 \text{ qt})\left(\frac{946.4\text{ mL}}{1\text{ qt}}\right) = 2800\text{ mL}$

(d) $\# \text{ mL} = (8 \text{ oz})\left(\frac{29.6\text{ mL}}{1\text{ oz}}\right) = 200\text{ mL}$

(e) $\# \text{ km/hr} = (55 \text{ mi/hr})\left(\frac{1.609\text{ km}}{1\text{ mi}}\right) = 88\text{ km/hr}$

(f) $\# \text{ km} = (50.0 \text{ mi})\left(\frac{1.609\text{ km}}{1\text{ mi}}\right) = 80.4\text{ km}$

3.49 (a) $\# \text{ qt} = (250 \text{ mL})\left(\frac{1\text{ qt}}{946.4\text{ mL}}\right) = 0.26\text{ qt}$

(b) $\# \text{ m} = (3.0 \text{ ft})\left(\frac{12 \text{ in.}}{1 \text{ ft}}\right)\left(\frac{2.54 \text{ cm}}{1 \text{ in.}}\right)\left(\frac{1 \text{ m}}{100 \text{ cm}}\right) = 0.91 \text{ m}$

(c) $\# \text{ lb} = (1.62 \text{ kg})\left(\frac{2.205 \text{ lb}}{1 \text{ kg}}\right) = 3.57 \text{ lb}$

(d) $\# \text{ oz} = (1.75 \text{ L})\left(\frac{1000 \text{ mL}}{1 \text{ L}}\right)\left(\frac{1 \text{ oz}}{29.6 \text{ mL}}\right) = 59.1 \text{ oz}$

(e) $\# \text{ mi/hr} = (35 \text{ km/hr})\left(\frac{1 \text{ mi}}{1.609 \text{ km}}\right) = 22 \text{ mi/hr}$

(f) $\# \text{ mi} = (80.0 \text{ km})\left(\frac{1 \text{ mi}}{1.609 \text{ km}}\right) = 49.7 \text{ mi}$

3.50 $\# \text{ mL} = (12 \text{ oz})\left(\frac{29.6 \text{ mL}}{1 \text{ oz}}\right) = 360 \text{ mL}$

3.51 $\# \text{ mL} = (32 \text{ oz})\left(\frac{29.6 \text{ mL}}{1 \text{ oz}}\right) = 950 \text{ mL}$

3.52 $\# \text{ qts} = (4700 \text{ mL})\left(\frac{1 \text{ qt}}{946.35 \text{ mL}}\right) = 5.0 \text{ qts}$

3.53 $\# \text{ oz} = (2.00 \text{ L})\left(\frac{1000 \text{ mL}}{1 \text{ L}}\right)\left(\frac{1 \text{ oz}}{29.6 \text{ mL}}\right) = 67.6 \text{ oz}$

3.54 $\# \text{ lb} = (1000 \text{ kg})\left(\frac{2.205 \text{ lb}}{1 \text{ kg}}\right) = 2205 \text{ lb}$

3.55 $\# \text{ kg} = 2000 \text{ lb}\left(\frac{1 \text{ kg}}{2.205 \text{ lb}}\right) = 907 \text{ kg}$

3.56 $\# \text{ mL} = (4.2 \text{ qts})\left(\frac{946.35 \text{ mL}}{1 \text{ qt}}\right) = 4.0 \times 10^3 \text{ mL}$ (stomach volume)

$4.0 \times 10^3 \text{ mL} \div 0.9 \text{ mL} = 4{,}000$ pistachios (don't try this at home)

3.57 To determine if 50 eggs will fit into 4.2 quarts, calculate the volume of fifty eggs, then compare the answer to the volume of the stomach:

$$\text{Volume of 50 eggs} = (50 \text{ eggs})(53 \text{ mL/egg})\left(\frac{1 \text{ L}}{1000 \text{ mL}}\right)\left(\frac{1.057 \text{ qt}}{1 \text{ L}}\right) = 2.8 \text{ qt}$$

$$2.8 \text{ qt} < 4.2 \text{ qt}$$

Luke can eat 50 eggs.

3.58 $\# \frac{\text{m}}{\text{s}} = \left(\frac{200 \text{ mi}}{1 \text{ hr}}\right)\left(\frac{5280 \text{ ft}}{1 \text{ mi}}\right)\left(\frac{30.48 \text{ cm}}{1 \text{ ft}}\right)\left(\frac{1 \times 10^{-2} \text{ m}}{1 \text{ cm}}\right)\left(\frac{1 \text{ hr}}{60 \text{ min}}\right)\left(\frac{1 \text{ min}}{60 \text{ s}}\right) = 90 \frac{\text{m}}{\text{s}}$

3.59 $\# \text{ km/h} = 2435 \frac{\text{ft}}{\text{s}}\left(\frac{1 \text{ yd}}{3 \text{ ft}}\right)\left(\frac{0.9144 \text{ m}}{1 \text{ yd}}\right)\left(\frac{1 \text{ km}}{1000 \text{ m}}\right)\left(\frac{3600 \text{ s}}{1 \text{ h}}\right) = 2672 \text{ km/h}$

3.60 $\# \text{ metric tons} = (2240 \text{ lb})\left(\frac{1 \text{ kg}}{2.205 \text{ lb}}\right)\left(\frac{1 \text{ metric ton}}{1000 \text{ kg}}\right) = 1.02 \text{ metric tons}$

3.61 $\# \text{ kg} = 75 \text{ kg}\left(\frac{2.205 \text{ lb}}{1 \text{ kg}}\right) = 165 \text{ kg}$

3.62 First, 6'2" = 74" $\# \text{ cm} = (74 \text{ in.})\left(\frac{2.54 \text{ cm}}{1 \text{ in.}}\right) = 190 \text{ cm}$

3.63 $\# \text{ kg} = (131 \text{ lb} + 12 \text{ oz}\left(\frac{1 \text{ lb}}{16 \text{ oz}}\right))\left(\frac{1 \text{ kg}}{2.205 \text{ lb}}\right) = 59.8 \text{ kg}$

3.64 (a) $\# \text{ cm}^2 = (8.4 \text{ ft}^2)\left(\frac{30.48 \text{ cm}}{1 \text{ ft}}\right)^2 = 7{,}800 \text{ cm}^2$

(b) $\# \text{ km}^2 = (223 \text{ mi}^2)\left(\frac{5280 \text{ ft}}{1 \text{ mi}}\right)^2\left(\frac{30.48 \text{ cm}}{1 \text{ ft}}\right)^2\left(\frac{1 \text{m}}{100 \text{ cm}}\right)^2\left(\frac{1 \text{ km}}{1\times10^3 \text{ m}}\right)^2 = 580 \text{ km}^2$

(c) $\# \text{ cm}^3 = (231 \text{ ft}^3)\left(\frac{30.48 \text{ cm}}{1 \text{ ft}}\right)^3 = 6.54\times10^6 \text{ cm}^3$

3.65 (a) $2.4 \text{ yd}^2\left(\frac{0.9144 \text{ m}}{1 \text{ yd}}\right)^2 = 2.0 \text{ m}^2$

(b) $8.3 \text{ in}^2\left(\frac{2.54 \text{ cm}}{1 \text{ in}}\right)^2\left(\frac{10 \text{ mm}}{1 \text{ cm}}\right)^2 = 5400 \text{ cm}^2$

(c) $9.1 \text{ ft}^3\left(\frac{1 \text{ yd}}{3 \text{ft}}\right)^3\left(\frac{0.9144 \text{ m}}{1 \text{ yd}}\right)^3\left(\frac{100 \text{ cm}}{1 \text{ m}}\right)^3\left(\frac{1 \text{ mL}}{1 \text{ cm}^3}\right)\left(\frac{1 \text{ L}}{1000 \text{ mL}}\right) = 260 \text{ L}$

3.66 (a) $\# \text{ m}^2 = (1.0 \text{ in}^2)\left(\frac{2.54 \text{ cm}}{1 \text{ in}}\right)^2\left(\frac{1\times10^{-2} \text{m}}{1 \text{ cm}}\right)^2 = 0.00065 \text{ m}^2$

(b) $\# \text{ km}^2 = (3.7 \text{ mi}^2)\left(\frac{5280 \text{ ft}}{1 \text{ mi}}\right)^2\left(\frac{30.48 \text{ cm}}{1 \text{ ft}}\right)^2\left(\frac{1 \text{m}}{100 \text{ cm}}\right)^2\left(\frac{1 \text{ km}}{1\times10^3 \text{ m}}\right)^2 = 9.6 \text{ km}^2$

(c) $\# \text{ mL} = (144 \text{ in}^3)\left(\frac{2.54 \text{ cm}}{1 \text{ in}}\right)^3\left(\frac{1 \text{ mL}}{1 \text{ cm}^3}\right) = 2{,}360 \text{ mL}$

3.67 (a) $9.3 \times 10^2 \text{ mm}^2\left(\frac{1 \text{ cm}}{10 \text{ mm}}\right)^2\left(\frac{1 \text{ in}}{2.54 \text{ cm}}\right)^2 = 1.4 \text{ in}^2$

(b) $8.6 \text{ m}^3\left(\frac{1 \text{ yd}}{0.9144 \text{ m}}\right)^3\left(\frac{3 \text{ ft}}{1 \text{ yd}}\right)^3 = 300 \text{ ft}^3$

(c) $314 \text{ in}^2\left(\frac{1 \text{ yd}}{36 \text{ in}}\right)^2\left(\frac{0.9144 \text{ m}}{1 \text{ yd}}\right)^2 = 0.203 \text{ m}^2$

3.68 $\# \frac{\text{mi}}{\text{hr}} = \left(\frac{3.5\text{ m}}{1\text{ s}}\right)\left(\frac{1\text{ cm}}{1\times10^{-2}\text{m}}\right)\left(\frac{1\text{ ft}}{30.48\text{ cm}}\right)\left(\frac{1\text{ mi}}{5280\text{ ft}}\right)\left(\frac{60\text{ s}}{1\text{ min}}\right)\left(\frac{60\text{ min}}{1\text{ hr}}\right) = 7.8\text{ mi/hr}$

3.69 $65\text{ mi/h}\left(\frac{1.609\text{ km}}{1\text{ mi}}\right) = 105\text{ km/hr}$

3.70 density = mass/ volume = 36.4 g/45.6 mL = 0.798 g/mL

3.71 $d = \frac{m}{V}$ $\qquad d = \frac{14.3\text{ g}}{8.46\text{ cm}^3} = 1.69\text{ g/cm}^3$

3.72 $\text{radius} = \frac{\text{diameter}}{2} = \left(\frac{1\times10^{-15}\text{ m}}{2}\right)\left(\frac{100\text{ cm}}{1\text{ m}}\right) = 5\times10^{-14}\text{cm}$

$\text{volume of sphere} = \frac{4}{3}\pi r^3 = \frac{4}{3}(3.1415)\left(5\times10^{-14}\right)^3 = 5.2\times10^{-40}\text{cm}^3$

density = mass/volume = 1.66×10^{-24} g/ $5.2\times10^{-40}\text{cm}^3 = 3\times10^{15}$ g/m^3

3.73 $d = \frac{m}{V}$ $\qquad V_{sphere} = \frac{4}{3}\pi r^3$

$V_{earth} = \frac{4}{3}\pi(6.38\times10^6\text{ m})^3 = 1.09\times10^{21}\text{ m}^3\left(\frac{100\text{ cm}}{1\text{ m}}\right)^3 = 1.09\times10^{27}\text{ cm}^3$

$m = 5.98\times10^{24}\text{ kg}\left(\frac{1000\text{ g}}{1\text{ kg}}\right) = 5.98\times10^{27}\text{ g}$

$d = \left(\frac{5.98\times10^{27}\text{ g}}{1.09\times10^{27}\text{ cm}^3}\right) = 5.49\text{ g/cm}^3$

3.74 $\#\text{ mL} = 25.0\text{g}\left(\frac{1\text{ mL}}{0.791\text{ g}}\right) = 31.6\text{ mL}$

3.75 $\#\text{mL} = (26.223\text{g})\left(\frac{1\text{mL}}{0.99704\text{g}}\right) = 26.301\text{ mL}$

3.76 $\#\text{ g} = 185\text{ mL}\left(\frac{1.492\text{ g}}{1\text{ mL}}\right) = 276\text{ g}$

3.77 $\#\text{ kg} = (34\text{ L})\left(\frac{1000\text{ mL}}{1\text{ L}}\right)\left(\frac{0.65\text{ g}}{1\text{ mL}}\right)\left(\frac{1\text{ kg}}{1000\text{ g}}\right) = 22\text{ kg}$

$\#\text{ lbs} = (22\text{ kg})\left(\frac{2.2\text{ lbs}}{1\text{ kg}}\right) = 48\text{ lbs}$

3.78 mass of silver = 62.00 g – 27.35 g = 34.65 g
volume of silver = 18.3 mL –15 mL = 3.3 mL
density of silver = (mass of silver)/(volume of silver) = (34.65 g)/(3.3 mL)
= 11 g/mL

3.79 volume of titanium = (1.84 cm)(2.24 cm)(2.44 cm) = 10.1 cm^3
density of titanium = 45.7 g/10.1 cm^3 = 4.54 g/cm^3

3.80 $$\text{sp.gr.} = \left(\frac{d_{\text{substance}}}{d_{\text{water}}}\right) = \left(\frac{0.715\ ^{g}/_{mL}}{1.00\ ^{g}/_{mL}}\right) = 0.715$$

3.81 $$\text{sp.gr.} = \left(\frac{d_{\text{substance}}}{d_{\text{water}}}\right) = \left(\frac{8.65\ ^{lb}/_{gal}}{8.34\ ^{lb}/_{gal}}\right) = 1.04$$

3.82 $d_{substance} = (\text{sp. gr.})_{substance} \times d_{water} = 1.47 \times 0.998 g/mL = 1.47\ g/mL$

$$\text{mass} = 1000\ mL\left(\frac{1.47 g}{1\ mL}\right) = 1470\ g$$

3.83 $$\text{density} = \left(\frac{227{,}641\ lb}{385{,}265\ gal}\right) = 0.5909\ ^{lb}/_{gal}$$

$$\text{sp. gr. of liquid hydrogen} = \frac{0.5909\ ^{lb}/_{gal}}{8.34\ ^{lb}/_{gal}} = 0.07085$$

$$\text{density} = 0.07085(1.00\ ^{g}/_{mL}) = 0.0709\ ^{g}/_{mL}$$

3.84 The density of gold is $1.2\dot{0}\times10^3\ lb/ft^3$

$$\#\ lbs = (1 ft^3)\left(\frac{1.20\times10^3 lb}{1\ ft^3}\right) = 1.2\dot{0}\times10^3\ lb$$

3.85 $$d_{H_2} = \left(\frac{0.0708\ ^{g}/_{mL}}{1.00\ ^{g}/_{mL}}\right)\left(\frac{62.4\ lb}{ft^3}\right) = 4.42\ ^{lb}/_{ft^3}$$

$$\#\ ft^3 = (565\ lbs)\left(\frac{1\ ft^3}{4.42\ lb}\right) = 128\ ft^3$$

Additional Exercises

3.86 Several possible answers for this one. Qualitatively, paint chips found on the fenders or in the dents of the cars could be held against the paint on the cars from which they presumably came for comparison. This will be less persuasive in court because many cars could have the same color paint. Quantitatively, the surface areas of the paint chips can be compared to the surface areas of missing paint on the cars. Also, the height of the fender of the red car may be compared to the height of the dent of the white car. These are much more persuasive arguments because it is unlikely that the same data for a randomly-chosen car would match that from a car at the accident site.

3.87 Hausberg Tarn $4350\ m\left(\frac{1\ yd}{0.9144\ m}\right) = 4960\ yd$

Mount Kenya $4600\ m\left(\frac{1\ yd}{0.9144\ m}\right) = 5000\ yd$ $4700\ m\left(\frac{1\ yd}{0.9144\ m}\right) = 5100\ yd$

Temperature $t_F = \left(\frac{9\ °F}{5\ °C}\right)(t_C) + 32\ °F = \left(\frac{9\ °F}{5\ °C}\right)(40\ °C) + 32\ °F = 130\ °F$

3.88 If the density is in metric tons…

$$\#\ g\ =\ 1\ \text{teaspoon}\left(\frac{4.93\ \text{mL}}{1\ \text{tsp}}\right)\left(\frac{1\ \text{cm}^3}{1\ \text{mL}}\right)\left(\frac{100{,}000{,}000\ \text{tons}}{1\ \text{cm}^3}\right)\left(\frac{1000\ \text{kg}}{1\ \text{ton}}\right)\left(\frac{1\times10^3\ \text{g}}{1\ \text{kg}}\right)$$

$= 4.93 \times 10^{14}$ g

If the density is in English tons…

$$\#\ g\ =\ 1\ \text{teaspoon}\left(\frac{4.93\ \text{mL}}{1\ \text{tsp}}\right)\left(\frac{1\ \text{cm}^3}{1\ \text{mL}}\right)\left(\frac{100{,}000{,}000\ \text{tons}}{1\ \text{cm}^3}\right)\left(\frac{2000\ \text{lbs}}{1\ \text{ton}}\right)\left(\frac{453.59\text{g}}{1\ \text{lb}}\right)$$

$= 4.47 \times 10^{14}$ g

3.89 As stated in the problem there are 365 days in one year.

$$\#\ \text{miles}\ =\ 1\ \text{light y}\left(\frac{365\ \text{days}}{1\ \text{y}}\right)\left(\frac{24\ \text{hr}}{1\ \text{day}}\right)\left(\frac{3600\text{s}}{1\ \text{hr}}\right)\left(\frac{3.0\times10^8\ \text{m}}{1\ \text{s}}\right)\left(\frac{1\ \text{km}}{1000\ \text{m}}\right)\left(\frac{0.6215\ \text{miles}}{1\ \text{km}}\right)\ =\ 5.88\times10^{12}\ \text{miles}$$

3.90 (a) In order to determine the volume of the pycnometer, we need to determine the volume of the water that fills it. We will do this using the mass of the water and its density.

mass of water = mass of filled pycnometer – mass of empty pycnometer
= 41.428g – 27.314g = 9.528g

$$\text{volume} = (9.528\ \text{g})\left(\frac{1\ \text{mL}}{0.99704\ \text{g}}\right)\ =\ 9.556\ \text{mL}$$

(b) We know the volume of chloroform from part (a). The mass of chloroform is determined in the same way that we determined the mass of water.

mass of chloroform = mass of filled pycnometer – mass of empty pycnometer
= 41.428g – 27.314g = 14.114g

$$\text{Density of chloroform} = \left(\frac{14.114\text{g}}{9.556\ \text{mL}}\right)\ =\ 1.477\text{g/mL}$$

3.91 For the message to get to the moon:

$$\#\text{s} = (239{,}000\ \text{miles})\left(\frac{1.609\ \text{km}}{1\ \text{mile}}\right)\left(\frac{1000\ \text{m}}{1\ \text{km}}\right)\left(\frac{1\ \text{s}}{3.00\ \times\ 10^8\ \text{m}}\right)\ =\ 1.3\ \text{s}$$

The reply would take the same amount of time, so the total time would be:

1.28 s × 2 = 2.56 s

3.92 (a) $4.50 = 30 min

(b) $$\#\$ = \left(\left(1\ \text{hr x}\ \frac{60\ \text{min}}{\text{hr}}\right) + 45\ \text{min}\right)\left(\frac{\$0.15}{\text{min}}\right)\ =\ \$15.75$$

(c) $$\#\ \text{min} = (\$17.35)\left(\frac{1\ \text{min}}{\$0.15}\right)\ =\ 116\ \text{min}$$

3.93 $d_{sea\ water} = (1.025)\left(\frac{62.4\ lb}{ft^3}\right) = 64.0\ lb/ft^3$

$$\#\ ft^3 = (4255\ tons)\left(\frac{2000\ lbs}{1\ ton}\right)\left(\frac{1\ ft^3}{64.0\ lb}\right) = 1.330 \times 10^5\ ft^3$$

3.94 $\#\ g = 2510\ cm^3\left(\frac{1\ in}{2.54\ cm}\right)^3\left(\frac{0.00011\ lbs}{1\ in^3}\right)\left(\frac{453.59\ g}{1\ lb}\right) = 0.76\ g$

3.95 Water will float on carbon tetrachloride since it is less dense.

3.96 Given both a mass and a volume we can determine density. Then, by experimentation, we could differentiate between two liquids.

3.97 The experimental density most closely matches the known density of methanol. Freezing, melting, and boiling points could also distinguish these two alcohols, but not color.

3.98 Since the density closely matches the known value, we conclude that this is an authentic sample of ethylene glycol.

3.99 (a) $\left(\frac{\#\ g}{L}\right) = \left(\frac{1.040\ g}{mL}\right)\left(\frac{1000\ mL}{L}\right) = 1040\ g/L = 1.040 \times 10^3\ g/L$

$$\frac{\#\ kg}{m^3} = \left(\frac{1.040\ g}{mL}\right)\left(\frac{1\ kg}{1000\ g}\right)\left(\frac{1\ mL}{1\ cm^3}\right)\left(\frac{100\ cm}{1\ m}\right)^3 = 1040\ kg/m^3 = 1.040 \times 10^3\ kg/m^3$$

3.100 a) $\frac{\#\ lb}{gal} = \left(\frac{785\ kg}{1\ m^3}\right)\left(\frac{2.205\ lb}{1\ kg}\right)\left(\frac{1\ m}{100\ cm}\right)^3\left(\frac{1000\ cm^3}{1\ L}\right)\left(\frac{3.785\ L}{1\ gal}\right)$

= 6.55 lb/gal

b) $\frac{\#\ lb}{gal} = \left(\frac{785\ kg}{1\ m^3}\right)\left(\frac{1000\ g}{1\ kg}\right)\left(\frac{1\ m}{100\ cm}\right)^3\left(\frac{1000\ cm^3}{1\ L}\right)$

= 785 g/L

3.101 $t_c = (T_K - 273)\left(\frac{1\ °C}{1\ K}\right) = (5800\ K - 273\ K)\left(\frac{1\ °C}{1\ K}\right) = 5500\ °C$

3.102 We solve by combining two equations:

$$t_F = \left(\frac{9\ °C}{5\ °F}\right)(t_C) + 32\ °F$$

$$t_F = t_C$$

If $t_F = t_C$, we can use the same variable for both temperatures:

$$t_C = \left(\frac{9\ °C}{5\ °F}\right)(t_C) + 32\ °F$$

$$\frac{5}{5}t_c = \left(\frac{9\ °C}{5\ °F}\right)(t_C) + 32\ °F$$

$$\frac{-4}{5}t_c = 32$$

$t_c = 32\dfrac{-5}{4} = -40$, therefore the answer is –40 °C.

3.103 Both the Rankine and the Kelvin scales have the same temperature at absolute zero: 0 R = 0 K. For converting from t_F to T_R:

$$t_C = \left(\frac{5\ °C}{9\ °F}\right)(t_F - 32\ °F) \quad \text{and} \quad t_C = (T_K - 273\ K)\left(\frac{1\ °C}{1\ K}\right)$$

therefore $$(T_K - 273\ K)\left(\frac{1\ °C}{1\ K}\right) = \left(\frac{5\ °C}{9\ °F}\right)(t_F - 32\ °F)$$

at T_K = 0 K = 0 R $$(0\ K - 273\ K)\left(\frac{1\ °C}{1\ K}\right) = \left(\frac{5\ °C}{9\ °F}\right)(t_F - 32\ °F)$$

$$-273\ °C = \left(\frac{5\ °C}{9\ °F}\right)(t_F - 32\ °F)$$

$$-491\ °F = t_F - 32\ °F$$

$$t_F = -459\ °F \text{ at absolute zero}$$

Also, T_R at absolute zero is 0 R and $$T_R = (t_F + 459\ °F)\left(\frac{1\ R}{1\ °F}\right)$$

So, the boiling point of water is 212 °F and in T_R:

$$T_R = (212\ °F + 459\ °F)\left(\frac{1\ R}{1\ °F}\right) = 671\ R$$

3.104 $$\#\,lb = 1\ pt\left(\frac{1\ qt}{2\ pts}\right)\left(\frac{946.56\ mL}{1\ qt}\right)\left(\frac{1.0\ g}{1\ mL}\right)\left(\frac{1\ lb}{453.59\ g}\right) = 1.0\ \text{pound}$$

Yes, the adage is accurate for water.

3.105 Sand d = 2.84 g/mL
Gold d = 19.3 g/mL
Mixture d = 3.10 g/mL

$$1.00\ \text{kg mixture}\left(\frac{1000\ g}{1\ kg}\right) = 1.00 \times 10^3\ \text{g of mixture}$$

1.00×10^3 g of mixture = $m_{sand} + m_{gold}$

$m_{sand} = (d_{sand})(V_{sand})$

$m_{gold} = (d_{gold})(V_{gold})$

1.00×10^3 g of mixture = $(d_{sand})(V_{sand}) + (d_{gold})(V_{gold})$

1.00×10^3 g of mixture = $(2.84\ g/mL)(V_{sand}) + (19.3\ g/mL)(V_{gold})$

$V_{mixture} = V_{sand} + V_{gold}$

$$d = \left(\frac{m}{V}\right)$$

$$\left(\frac{1.00 \times 10^3\ g}{3.10\ g/mL}\right) = 323\ mL$$

$V_{sand} + V_{gold} = 323$ mL

$V_{sand} = 323\ mL - V_{gold}$

1.00×10^3 g of mixture = $(2.84\ g/mL)(323\ mL - V_{gold}) + (19.3\ g/mL)(V_{gold})$

1.00×10^3 g of mixture = 917 g sand – (2.84 g/mL)(V_{gold}) + (19.3 g/mL)(V_{gold})

1.00×10^3 g of mixture – 917 g sand = (16.5 g/mL)(V_{gold})   5.0 mL = V_{gold}

1.00×10^3 g of mixture – 917 g sand = 83 g gold

$$\% \text{ mass of gold} = \frac{83 \text{ g gold}}{1000 \text{ g total}} \times 100\% = 8.3\% \text{ gold}$$

Practice Exercises

4.1 $\# \text{ atoms Au} = (15.0 \text{ g Au})\left(\frac{1 \text{ mol Au}}{197.0 \text{ g Au}}\right)\left(\frac{6.022 \times 10^{23} \text{ atoms Au}}{1 \text{ mol Au}}\right)$

$= 4.59 \times 10^{22}$ atoms Au

4.2 $CH_3(CH_2)_6CH_3 = 8 \text{ C} + 18 \text{ H}$ Formula mass = 8(12.01) + 18(1.01)
= 114.26

4.3 $C_{21}H_{23}NO_5$ Formula mass = 21(12.01) + 23(1.01) + 5(16.00)
= 369.41

4.4 Formula mass of heroin = 369.41 g/mol (see above exercise)

$\# \text{ molecules heroin} = (0.010 \text{ g heroin})\left(\frac{1 \text{ mol heroin}}{369.41 \text{ g heroin}}\right)\left(\frac{6.022\times10^{23} \text{ molecules heroin}}{1 \text{ mol heroin}}\right) = 1.6 \times$

10^{19} molecules heroin

4.5 $\# \text{ mol S} = (35.6 \text{ g S})\left(\frac{1 \text{ mol S}}{32.07 \text{ g S}}\right) = 1.11 \text{ mol S}$

4.6 $\# \text{ g Ag} = (0.263 \text{ mol Ag})\left(\frac{107.9 \text{ g Ag}}{1 \text{ mol Ag}}\right) = 28.4 \text{ g Ag}$

4.7 0.125 mol × 106 g/mol = 13.3 g

4.8 $\# \text{ mol } H_2SO_4 = (45.8 \text{ g } H_2SO_4)\left(\frac{1 \text{ mol } H_2SO_4}{98.1 \text{ g } H_2SO_4}\right) = 0.467 \text{ mol } H_2SO_4$

4.9 $\# \text{ atoms Pb} = (1.00\times10^{-9} \text{ g Pb})\left(\frac{1 \text{ mol Pb}}{207.2 \text{ g Pb}}\right)\left(\frac{6.022 \times 10^{23} \text{ atoms Pb}}{1 \text{ mol Pb}}\right)$

$= 2.91 \times 10^{12}$ atoms Pb

4.10 $\# \text{ mol N} = (8.60 \text{ mol O})\left(\frac{2 \text{ mol N}}{5 \text{ mol O}}\right) = 3.44 \text{ mol N atoms}$

4.11 $\# \text{ g Fe} = (25.6 \text{ g O})\left(\frac{1 \text{ mol O}}{16.0 \text{ g O}}\right)\left(\frac{2 \text{ mol Fe}}{3 \text{ mol O}}\right)\left(\frac{55.8 \text{ g Fe}}{1 \text{ mol Fe}}\right) = 59.5 \text{ g Fe}$

4.12 $\# \text{ g Fe} = (15.0 \text{ g } Fe_2O_3)\left(\frac{111.7 \text{ g Fe}}{159.7 \text{ g } Fe_2O_3}\right) = 10.5 \text{ g Fe}$

4.13 % N = 0.1417/0.5462 × 100 = 25.94 % N
% O = 0.4045/0.5462 × 100 = 74.06 % O
Since these two values constitute 100 %, there are no other elements present.

4.14 We first determine the number of grams of each element that are present in one mol of sample:
2 mol N × 14.01 g/mol = 28.02 g N
4 mol O × 16.00 g/mol = 64.00 g O

The percentages by mass are then obtained using the formula mass of the compound (92.02 g):

% N = 28.02/92.02 = 30.45 % N

% O = 64.00/92.02 = 69.55 % O

4.15 We first determine the number of mol of each element as follows:

$$\text{\# mol N} = (0.712 \text{ g N})\left(\frac{1 \text{ mol N}}{14.01 \text{ g N}}\right) = 0.0508 \text{ mol N}$$

We need to know the number of grams of O. Since there is a total of 1.525 g of compound and the only other element present is N, the mass of O = 1.525 g – 0.712 g = 0.813 g.

$$\text{\# mol O} = (0.813 \text{ g O})\left(\frac{1 \text{ mol O}}{16.00 \text{ g O}}\right) = 0.0508 \text{ mol O}$$

Since these two mol amounts are the same, the empirical formula is NO.

4.16 We first determine the number of mol of each element as follows:

$$\text{\# mol N} = (0.522 \text{ g N})\left(\frac{1 \text{ mol N}}{14.01 \text{ g N}}\right) = 0.0373 \text{ mol N}$$

We need to know the number of grams of O. Since there is a total of 2.012 g of compound and the only other element present is N, the mass of O = 2.012 g – 0.522 g = 1.490 g.

$$\text{\# mol O} = (1.490 \text{ g O})\left(\frac{1 \text{ mol O}}{16.00 \text{ g O}}\right) = 0.0931 \text{ mol O}$$

Next, we divide each of these mol amounts by the smallest in order to deduce the simplest whole number ratio:

For N: 0.0373 mol/0.0373 mol = 1.00

For O: 0.0931 mol/0.0373 mol = 2.50

There are 2.5 mol of oxygen atoms for every 1 mol of nitrogen atoms. Therefore, there are 2.5 O atoms for every 1 N atom. Empirical formulas must be whole number ratios: This corresponds to a whole number ratio of...5 O atoms for every 2 N atoms.

The empirical formula is therefore N_2O_5.

4.17 It is convenient to assume that we have 100 g of the sample, so that the % by mass values may be taken directly to represent masses. Thus there is 32.37 g of Na, 22.57 g of S and (100.00 – 32.37 – 22.57) = 45.06 g of O. Now, convert these masses to a number of mol:

$$\text{\# mol Na} = (32.37 \text{ g Na})\left(\frac{1 \text{ mol Na}}{23.00 \text{ g Na}}\right) = 1.407 \text{ mol Na}$$

$$\text{\# mol S} = (22.57 \text{ g S})\left(\frac{1 \text{ mol S}}{32.06 \text{ g S}}\right) = 0.7040 \text{ mol S}$$

$$\text{\# mol O} = (45.06 \text{ g O})\left(\frac{1 \text{ mol O}}{16.00 \text{ g O}}\right) = 2.816 \text{ mol O}$$

Next, we divide each of these mol amounts by the smallest in order to deduce the simplest whole number ratio:

For Na: 1.407 mol/0.7040 mol = 1.999

For S: 0.7040 mol/0.7040 mol = 1.000

For O: 2.816 mol/0.7040 mol = 4.000

The empirical formula is Na_2SO_4.

4.18 Since the entire amount of carbon that was present in the original sample appears among the products only as CO_2, we calculate the amount of carbon in the sample as follows:

$$\# \text{ g C} = (7.406 \text{ g CO}_2)\left(\frac{1 \text{ mol CO}_2}{44.01 \text{ g CO}_2}\right)\left(\frac{1 \text{ mol C}}{1 \text{ mol CO}_2}\right)\left(\frac{12.01 \text{ g C}}{1 \text{ mol C}}\right) = 2.021 \text{ g C}$$

Similarly, the entire mass of hydrogen that was present in the original sample appears among the products only as H_2O. Thus the mass of hydrogen in the sample is:

$$\# \text{ g H} = (3.027 \text{ g H}_2\text{O})\left(\frac{1 \text{ mol H}_2\text{O}}{18.02 \text{ g H}_2\text{O}}\right)\left(\frac{2 \text{ mol H}}{1 \text{ mol H}_2\text{O}}\right)\left(\frac{1.008 \text{ g H}}{1 \text{ mol H}}\right) = 0.3386 \text{ g H}$$

The mass of oxygen in the original sample is determined by difference:

5.048 g – 2.021 g– 0.3386 g = 2.688 g O

Next, these mass amounts are converted to the corresponding mol amounts:

$$\# \text{ mol C} = (2.021 \text{ g C})\left(\frac{1 \text{ mol C}}{12.01 \text{ g C}}\right) = 0.1683 \text{ mol C}$$

$$\# \text{ mol H} = (0.3386 \text{ g H})\left(\frac{1 \text{ mol H}}{1.008 \text{ g H}}\right) = 0.3359 \text{ mol H}$$

$$\# \text{ mol O} = (2.688 \text{ g O})\left(\frac{1 \text{ mol O}}{16.00 \text{ g O}}\right) = 0.1680 \text{ mol O}$$

The simplest formula is obtained by dividing each of these mol amounts by the smallest:

For C: 0.1683 mol/0.1680 mol= 1.002

for H: 0.3359 mol/0.1680 mol= 1.999

For O: 0.1680 mol/0.1680 mol = 1.000

These values give us the simplest formula directly, namely CH_2O.

4.19 The formula mass of the empirical unit is 1 N + 2 H = 16.03. Since this is half of the molecular mass, the molecular formula is N_2H_4.

4.20 $$\# \text{ mol H}_2\text{SO}_4 = (0.366 \text{ mol NaOH})\left(\frac{1 \text{ mol H}_2\text{SO}_4}{2 \text{ mol NaOH}}\right) = 0.183 \text{ mol H}_2\text{SO}_4$$

4.21 $$\# \text{ mol O}_2 = (0.575 \text{ mol CO}_2)\left(\frac{5 \text{ mol O}_2}{3 \text{ mol CO}_2}\right) = 0.958 \text{ mol O}_2$$

4.22 $$\# \text{ g Al}_2\text{O}_3 = (86.0 \text{ g Fe})\left(\frac{1 \text{ mol Fe}}{55.85 \text{ g Fe}}\right)\left(\frac{1 \text{ mol Al}_2\text{O}_3}{2 \text{ mol Fe}}\right)\left(\frac{102.0 \text{ g Al}_2\text{O}_3}{1 \text{ mol Al}_2\text{O}_3 \text{ mol Al}_2\text{O}_3}\right)$$

$= 78.5 \text{ g Al}_2\text{O}_3$

4.23 $3CaCl_2(aq) + 2K_3PO_4(aq) \rightarrow Ca_3(PO_4)_2(s) + 6KCl(aq)$

4.24 First determine the number of grams of O_2 that would be required to react completely with the given amount of ammonia:

$$\# \text{ g O}_2 = (30.00 \text{ g NH}_3)\left(\frac{1 \text{ mol NH}_3}{17.03 \text{ g NH}_3}\right)\left(\frac{5 \text{ mol O}_2}{4 \text{ mol NH}_3}\right)\left(\frac{32.00 \text{ g O}_2}{1 \text{ mol O}_2}\right)$$

$= 70.46 \text{ g O}_2$

Since this is more than the amount that is available, we conclude that oxygen is the limiting reactant. The rest of the calculation is therefore based on the available amount of oxygen:

$$\#\ \text{g NO} = (40.00\ \text{g } O_2)\left(\frac{1\ \text{mol } O_2}{32.00\ \text{g } O_2}\right)\left(\frac{4\ \text{mol NO}}{5\ \text{mol } O_2}\right)\left(\frac{30.01\ \text{g NO}}{1\ \text{mol NO}}\right)$$

$$= 30.01\ \text{g NO}$$

4.25 First determine the number of grams of C_2H_5OH that would be required to react completely with the given amount of sodium dichromate:

$$\#\ \text{g } C_2H_5OH = (90.0\ \text{g } Na_2Cr_2O_7)\left(\frac{1\ \text{mol } Na_2Cr_2O_7}{262.0\ \text{g } Na_2Cr_2O_7}\right)\left(\frac{3\ \text{mol } C_2H_5OH}{2\ \text{mol } Na_2Cr_2O_7}\right)\left(\frac{46.08\ \text{g } C_2H_5OH}{1\ \text{mol } C_2H_5OH}\right)$$

$$= 23.7\ \text{g } C_2H_5OH$$

Once this amount of C_2H_5OH is reacted the reaction will stop, even though there are 24.0 g C_2H_5OH present, because the $Na_2Cr_2O_7$ will be used up. Therefore $Na_2Cr_2O_7$ is the limiting reactant. The theoretical yield of acetic acid ($HC_2H_3O_2$) is therefore based on the amount of $Na_2Cr_2O_7$ added. This is calculated below:

$$\#\ \text{g } HC_2H_3O_2 = (90.0\ \text{g } Na_2Cr_2O_7)\left(\frac{1\ \text{mol } Na_2Cr_2O_7}{262.0\ \text{g } Na_2Cr_2O_7}\right)\left(\frac{3\ \text{mol } HC_2H_3O_2}{2\ \text{mol } Na_2Cr_2O_7}\right)\left(\frac{60.06\ \text{g } HC_2H_3O_2}{1\ \text{mol } HC_2H_3O_2}\right)$$

$$= 30.9\ \text{g } HC_2H_3O_2$$

Now the percentage yield can be calculated from the amount of acetic acid actually produced, 26.6 g:

$$\text{percent yield} = \left(\frac{\text{actual yield}}{\text{theoretical yield}}\right)\times 100 = \left(\frac{26.6\ \text{g } HC_2H_3O_2}{30.9\ \text{g } HC_2H_3O_2}\right)\times 100 = 86.1\%$$

Thinking It Through

T4.1

$$\#\ \text{mol C} = (0.60\ \text{mol H})\left(\frac{1\ \text{mol } (CH_3)_2N_2H_2}{8\ \text{mol H}}\right)\left(\frac{2\ \text{mol C}}{1\ \text{mol } (CH_3)_2N_2H_2}\right)$$

$$\#\ \text{mol } H_2 = (0.56\ \text{mol } (CH_3)_2N_2H_2)\left(\frac{8\ \text{mol H}}{1\ \text{mol } (CH_3)_2N_2H_2}\right)\left(\frac{1\ \text{mol } H_2}{2\ \text{mol H}}\right)$$

T4.2

$$\#\ \text{atoms S} = (0.400\ \text{mol } O_2)\left(\frac{2\ \text{mol O}}{1\ \text{mol } O_2}\right)\left(\frac{1\ \text{mol S}}{3\ \text{mol O}}\right)\left(\frac{6.02\times 10^{23}\ \text{atoms S}}{1\ \text{mol S}}\right)$$

T4.3 12 mol C/4 mol compound = 3 C and 48 atoms H/6 molecules compound = 8 H The formula must therefore be C_3H_8.

T4.4 (a)

$$\#\ \text{molecules } CHCl_3 = (18\ \text{atoms Cl})\left(\frac{1\ \text{molecule } CHCl_3}{3\ \text{atoms Cl}}\right)$$

(b)

$$\#\ \text{mol } CHCl_3 = (0.36\ \text{mol } Cl_2)\left(\frac{2\ \text{mol Cl}}{1\ \text{mol } Cl_2}\right)\left(\frac{1\ \text{mol } CHCl_3}{3\ \text{mol Cl}}\right)$$

T4.5 First determine the molar mass of C_5H_6OS, then convert the micrograms of 2-furylmethanethiol to grams, then convert to moles, followed by converting to moles of S, then grams and finally micrograms of S.

MM of C_5H_6OS = [(5 mol C)(12.01 g C/mol C) + (6 mol H)(1.01 g H/mol H) + (1 mol O)(10.00 g O/mol O) + (1 mol S)(32.06 g S/mol S) = 114.8 g C_5H_6OS/mol C_5H_6OS]

$$1.00\ \mu\text{g}\ C_5H_6OS\left(\frac{1\text{ g }C_5H_6OS}{10^6\ \mu\text{g }C_5H_6OS}\right)\left(\frac{1\text{ mol }C_5H_6OS}{114.8\text{ g }C_5H_6OS}\right)\left(\frac{1\text{ mol S}}{1\text{ mol }C_5H_6OS}\right)\left(\frac{32.07\text{ g S}}{1\text{ mol S}}\right)\left(\frac{10^6\ \mu\text{g S}}{1\text{ g S}}\right)$$

T4.6 2.05 g compound – 1.11 g Cl = 0.94 g Sn

Next convert each mass value to moles

$$\#\text{ mol Sn} = (0.94\text{ g Sn})\left(\frac{1\text{ mol Sn}}{118.69\text{ g Sn}}\right) = 0.0079\text{ mol Sn}$$

$$\#\text{ mol Cl} = (1.11\text{ g Cl})\left(\frac{1\text{ mol Cl}}{35.4527\text{ g Cl}}\right) = 0.0313\text{ mol Cl}$$

Finally compare the mole ratio

$\left(\frac{0.0079}{0.0079}\right) = 1$ $\left(\frac{0.0313}{0.0079}\right) = 3.96 \approx 4$ Therefore the formula must be $SnCl_4$.

T4.7 If we multiply the second piece of data by two, we determine there are 4 mol S per 10 mol O. The first piece of data tells us there are 4 mol Na per 10 mol O. This gives us the formula $Na_4S_4O_{10}$ for the compound. The empirical formula should be the formula expressed in smallest whole number ratios. We can reduce the formula $Na_4S_4O_{10}$ by dividing each subscript by 2. This gives $Na_2S_2O_5$. By inspection we can see that the number of moles of S is equivalent to the number of moles of Na.

$$2.7\text{ mol Na}\left(\frac{1\text{ mol S}}{1\text{ mol Na}}\right) = x\text{ mol S}$$

T4.8 $$\#\text{ g }H_2 = 10.0\text{ g }H_2O\left(\frac{1\text{ mol }H_2O}{18.02\text{ g }H_2O}\right)\left(\frac{2\text{ mol H}}{1\text{ mol }H_2O}\right)\left(\frac{1\text{ mol }H_2}{2\text{ mol H}}\right)$$

T4.9 Being that MO is a 1:1 compound, set up a ratio to determine the amount of M that combines with 16.0 g of O, i.e., the amount combining with 1 mole of O.

$$\frac{36\text{ g M}}{24\text{ g O}} = \frac{x\text{ g M}}{16.0\text{ g O}},$$ where x is the atomic mass of M.

T4.10 $$\#\text{ g of CuO} = 0.3433\text{ g Cu}\left(\frac{1\text{ mol Cu}}{63.55\text{ g Cu}}\right)\left(\frac{1\text{ mol CuO}}{1\text{ mol Cu}}\right)\left(\frac{79.55\text{ g CuO}}{1\text{ mol CuO}}\right)$$

T4.11 First, the mass of calcium in the binary compound is determined from the mass of CaC_2O_4 precipitated:

$$\#\text{ g Ca} = (0.650\text{ g }CaC_2O_4)\left(\frac{1\text{ mol }CaC_2O_4}{128.0976\text{ g }CaC_2O_4}\right)\left(\frac{1\text{ mol Ca}}{1\text{ mol }CaC_2O_4}\right)\left(\frac{40.078\text{ g Ca}}{1\text{ mol Ca}}\right) = 0.2034\text{ g Ca}$$

The mass of bromine in the sample is obtained by subtraction: 1.015 g sample – 0.2034 g Ca = 0.8116g Br. From the mass of calcium and that of bromine we can next calculate the percent compositions.

$$\%\text{ Ca} = \left(\frac{0.2034\text{ g}}{1.015\text{ g}}\right)100 = 0.2004\%$$

$$\%\text{ Br} = \left(\frac{0.8116\text{ g}}{1.015\text{ g}}\right)100 = 79.96\ \%$$

The empirical formula can now be calculated from the mass percentage compositions or from the masses directly.

$$\#\,\text{mol Ca} = (0.2034\text{ g})\left(\frac{1\text{ mol Ca}}{40.078\text{ g Ca}}\right) = 5.075\times10^{-3}\text{ mol Ca}$$

$$\#\,\text{mol Br} = (0.8116\text{ g})\left(\frac{1\text{ mol Br}}{79.904\text{ g Br}}\right) = 1.016\times10^{-2}\text{ mol Br}$$

The relative mole amounts are determined by dividing the smaller of these two mole amounts into each:
For Ca: 5.075×10^{-3} mol/5.075×10^{-3} mol = 1.000
For Br: 1.016×10^{-2} mol/5.075×10^{-3} mol = 2.001
The empirical formula is $CaBr_2$

T4.12 First it is necessary to determine the number of moles that are present in 12.5 g. For this we need the formula mass: 63.55 + 32.07 + (9 × 16.00) + (10 × 1.01) = 249.72 g/mol. Next we divide by the formula mass to determine the number of moles:

$$\#\,\text{mol CuSO}_4\cdot5\text{H}_2\text{O} = (12.5\text{ g CuSO}_4\cdot5\text{H}_2\text{O}\left(\frac{1\text{ mol CuSO}_4\cdot5\text{H}_2\text{O}}{249.72\text{ g CuSO}_4\cdot5\text{H}_2\text{O}}\right) = 5.01\times10^{-2}\text{ mol CuSO}_4\cdot5\text{H}_2\text{O}$$

If completely dehydrated, 5.01×10^{-2} mol of $CuSO_4$ remain and $5(5.01 \times 10^{-2}) = 0.250$ mol of water are released, according to the balanced equation:

$$CuSO_4\cdot5H_2O \rightarrow CuSO_4 + 5H_2O$$

The mass of $CuSO_4$ that is formed is calculated by using the formula mass of $CuSO_4$, 159.6 g/mol:

$$5.01\times 10^{-2}\text{ mol} \times 159.6\text{ g/mol} = 8.00\text{ g}$$

T4.13 (a) The stoichiometry of the equation and the molar mass of HBr are used to determine the mass of HBr needed:

$$\#\text{ g HBr} = (0.125\text{ mol CaBr}_2)\left(\frac{2\text{ mol HBr}}{1\text{ mol CaBr}_2}\right)\left(\frac{80.91\text{ g HBr}}{1\text{ mol HBr}}\right) = 20.2\text{ g HBr}$$

(b) This is a limiting reagent problem. The number of moles of $Ca(OH)_2$ and HBr used have to be compared to the balanced equation

$$\#\text{ mol Ca(OH)}_2 = (1.24\text{ g Ca(OH)}_2)\left(\frac{1\text{ mol Ca(OH)}_2}{74.09\text{ g Ca(OH)}_2}\right) = 0.0167\text{ mol Ca(OH)}_2$$

$$\#\text{ mol HBr} = (2.65\text{ g HBr})\left(\frac{1\text{ mol HBr}}{80.91\text{ g HBr}}\right) = 0.0328\text{ mol HBr}$$

mol HBr from mole of $Ca(OH)_2$

$$0.0167\text{ mol Ca(OH)}_2\left(\frac{2\text{ mol HBr}}{1\text{ mol Ca(OH)}_2}\right) = 0.0334\text{ mol HBr}$$

Although these values are very close to stoichiometric, the moles of HBr are slightly limiting. Therefore, the theoretical yield of $CaBr_2$ is:

$$\#\text{ g CaBr}_2 = (0.328\text{ mol HBr})\left(\frac{1\text{ mol CaBr}_2}{2\text{ mol HBr}}\right)\left(\frac{199.89\text{ g CaBr}_2}{1\text{ mol CaBr}_2}\right) = 3.28\text{ g CaBr}_2$$

(c) Since $Ca(OH)_2$ is in slight excess in part (b), the mass of $Ca(OH)_2$ used is:

$$\#\text{ g Ca(OH)}_2 = (0.0164\text{ mol Ca(OH)}_2)\left(\frac{74.09\text{ g Ca(OH)}_2}{1\text{ mol Ca(OH)}_2}\right) = 1.22\text{ g Ca(OH)}_2$$

The amount that remains is 1.24 g – 1.22 g = 0.02 g $Ca(OH)_2$.

T4.14 The percentage of mass of iron in pure Fe_3O_4 is needed first. For this we must divide the total mass of iron by the formula mass of Fe_3O_4

$$\% \text{ Fe} = \left(\frac{3 \text{ mol Fe} \times \dfrac{55.85 \text{ g Fe}}{1 \text{ mol Fe}}}{231.5 \text{ g Fe}_3\text{O}_4} \right) \times 100 = 72.38\%$$

Pure Fe_3O_4 would then have 72.38% by mass iron. The question thus becomes, what percent of 72.38 is 12.5?
12.5 = X% of 72.38
X = 17.3% of Fe_3O_4 in the ore

T4.15 The experimental % by mass of P should be compared with the theoretical % by mass of P in these three compounds. This will identify the compound. Then, the % by mass Na may be determined in the normal manner.

T4.16 Since a 1:1 mole ratio reacts, the mass amounts directly indicate the relative formula masses. We deduce that Y is heavier, by a ratio of 29.0/25.0.

T4.17 To solve this problem, treat it as two problems: first calculate the amount of aluminum oxide needed to make 127 g aluminum iodide, then calculate the amount of hydrogen iodide needed to make 127 g aluminum iodide.

$$\# \text{ g Al}_2\text{O}_3 = (127 \text{ g AlI}_3)\left(\frac{1 \text{ mol AlI}_3}{407.82 \text{ g AlI}_3}\right)\left(\frac{1 \text{ mol Al}_2\text{O}_3}{2 \text{ mol AlI}_3}\right)\left(\frac{101.96 \text{ g Al}_2\text{O}_3}{1 \text{ mol Al}_2\text{O}_3}\right) = 15.9 \text{ g Al}_2\text{O}_3$$

$$\# \text{ g HI} = (127 \text{ g AlI}_3)\left(\frac{1 \text{ mol AlI}_3}{407.82 \text{ g AlI}_3}\right)\left(\frac{6 \text{ mol HI}}{2 \text{ mol AlI}_3}\right)\left(\frac{127.95 \text{ g HI}}{1 \text{ mol HI}}\right) = 119.5 \text{ g HI}$$

All of the aluminum oxide reacts to form aluminum iodide, therefore the hydrogen iodide is in excess. By subtracting the amount of HI needed to make the aluminum iodide from the initial amount, we determine there is an excess of 20 g of HI.

T4.18 To solve the problem we need two equations. First, let x = grams of NaCl and y = g of $CaCl_2$. We know that x + y = 1.000 g. Using the mass of AgCl precipitated we can determine the mass of Cl.

$$\# \text{ g Cl} = (2.53 \text{ g AgCl})\left(\frac{1 \text{ mol AgCl}}{143.321 \text{ g AgCl}}\right)\left(\frac{1 \text{ mol Cl}}{1 \text{ mol AgCl}}\right)\left(\frac{35.453 \text{ g Cl}}{1 \text{ mol Cl}}\right) = 0.626 \text{ g Cl}$$

We know that the Cl originated in the NaCl and the $CaCl_2$. So we can set up the following equation:

$$\# \text{ g Cl} = (x \text{ g NaCl})\left(\frac{1 \text{ mol NaCl}}{58.44 \text{ g NaCl}}\right)\left(\frac{1 \text{ mol Cl}}{1 \text{ mol NaCl}}\right)\left(\frac{35.453 \text{ g Cl}}{1 \text{ mol Cl}}\right)$$

$$+ \ (y \text{ g CaCl}_2)\left(\frac{1 \text{ mol CaCl}_2}{110.984 \text{ g CaCl}_2}\right)\left(\frac{2 \text{ mol Cl}}{1 \text{ mol CaCl}_2}\right)\left(\frac{35.453 \text{ g Cl}}{1 \text{ mol Cl}}\right)$$

Substitute x = 1.000g – y and solve for y.

T4.19 There are three distinct empirical formulas represented CH_2, CH_3, and C_3H_8. There are two molecules with the empirical formula CH_3; CH_3 and C_2H_6. There is one C_3H_8, and there are two with the formula CH_2; C_5H_{10} and C_3H_6.

T4.20 Reaction 1

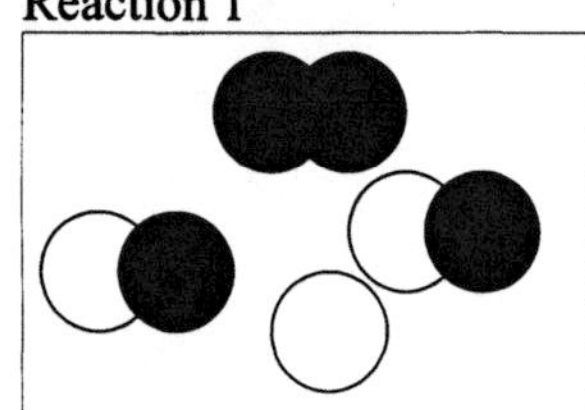

Reaction 2

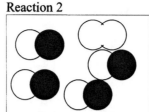

Review Questions

4.1 To estimate the number of atoms in a gram of iron, using atomic mass units,u, convert g to kg, then use the relationship, 1.661×10^{-27} kg = 1 u, finally using the atomic mass of Fe (55.85 u) to find the number of atoms:

$$1 \text{ g Fe}\left(\frac{1 \text{ kg}}{1000 \text{ g}}\right)\left(\frac{1 \text{ u}}{1.661\times10^{-27} \text{ kg}}\right)\left(\frac{1 \text{ molecule}}{55.85 \text{ u}}\right) = 1.08 \times 10^{22} \text{ atoms Fe}$$

4.2 The mole is the SI unit for the amount of a substance. A mole is equal in quantity to Avogadro's number (6.022×10^{23}) of particles, or the formula mass in grams of a substance.

4.3 Moles are used for calculations instead of atomic mass units because they have the right units for converting from grams to moles and vice versa.

4.4 There are the same number of molecules is 2.5 moles of H_2O and 2.5 moles of H_2.

4.5 There are 2 moles of iron atoms in 1 mole of Fe_2O_3. The stoichiometric equivalent between Fe and Fe_2O is 2 mol Fe ≡ 1 mol Fe_2O_3.
For the number of iron atoms in 1 mole of Fe_2O_3:

$$1 \text{ mol } Fe_2O_3\left(\frac{2 \text{ mol Fe}}{1 \text{ mole } Fe_2O_3}\right)\left(\frac{6.022 \times 10^{23} \text{ Fe atoms}}{1 \text{ mol Fe}}\right) = 1.204 \times 10^{24} \text{ atoms Fe}$$

4.6 (a) $\left(\frac{1 \text{ mol S}}{2 \text{ mol O}}\right)$ and $\left(\frac{2 \text{ mol O}}{1 \text{ mol S}}\right)$

(b) $\left(\frac{2 \text{ mol As}}{3 \text{ mol O}}\right)$ and $\left(\frac{3 \text{ mol O}}{2 \text{ mol As}}\right)$

(c) $\left(\frac{1 \text{ mol Pb}}{6 \text{ mol N}}\right)$ and $\left(\frac{6 \text{ mol N}}{1 \text{ mol Pb}}\right)$

(d) $\left(\frac{2 \text{ mol Al}}{6 \text{ mol Cl}}\right)$ and $\left(\frac{6 \text{ mol Cl}}{2 \text{ mol Al}}\right)$

4.7 (a) $\left(\frac{3 \text{ mol Mn}}{4 \text{ mol O}}\right)$ and $\left(\frac{4 \text{ mol O}}{3 \text{ mol Mn}}\right)$

(b) $\left(\frac{2 \text{ mol Sb}}{5 \text{ mol S}}\right)$ and $\left(\frac{5 \text{ mol S}}{2 \text{ mol Sb}}\right)$

(c) $\left(\frac{4 \text{ mol P}}{6 \text{ mol O}}\right)$ and $\left(\frac{6 \text{ mol O}}{4 \text{ mol P}}\right)$

(d) $\left(\frac{2 \text{ mol Hg}}{2 \text{ mol Cl}}\right)$ and $\left(\frac{2 \text{ mol Cl}}{2 \text{ mol Hg}}\right)$

4.8 The molecular mass is required to convert grams of a substance to moles of that same substance.

4.9 The statement, 1 mol O, does not indicate whether this is atomic oxygen O, or molecular oxygen O_2. The statement 64 g of oxygen is not ambiguous because the source of oxygen is not important.

4.10 $\left(\frac{26.98 \text{ g Al}}{1 \text{ mol Al}}\right)$ and $\left(\frac{1 \text{ mol Al}}{26.98 \text{ g Al}}\right)$

4.11 At a minimum, the identity and mass of each atomic element present must be known. If the total mass of the compound is known, then it is necessary to know all but one mass of the elements that compose the compound.

4.12 When balancing a chemical equation, changing the subscripts changes the identity of the substance.

4.13 To convert grams of a substance to molecules of the same substance, the molecular mass of the substance, and Avagadro's number are needed.

4.14 (a) Student B is correct.
(b) Student A wrote a properly balanced equation. However, by changing the subscript for the product of the reaction from an implied one, NaCl, to a two, $NaCl_2$, this student has changed the identity of the product. When balancing chemical equations, never change the value of the subscript given in the unbalanced equation.

4.15 Convert moles of B to moles of compound, A_5B_2; then using the stoichiometric ratio of moles of A to moles of A_5B_2, determine the moles of A; and finally convert the moles of A to grams of A using the molecular mass of A.

$$10 \text{ mol B}\left(\frac{1 \text{ mol } A_5B_2}{2 \text{ mol B}}\right)\left(\frac{5 \text{ mol A}}{1 \text{ mol } A_5B_2}\right)\left(\frac{100.0 \text{ g A}}{1 \text{ mol A}}\right)$$

4.16 Their formula weights must be identical.

4.17 To determine the number of grams of sulfur that would react with a gram of arsenic, the stoichiometric ratio of the arsenic to the sulfur in the compound is needed, as well as the molecular masses of sulfur and arsenic.

4.18 (a) The balanced equation describes the stoichiometry
(b) The scale of the reaction is determined by the mass of the reactants.

4.19 $2\ H_2O_2 \rightarrow 2\ H_2O + O_2$

4.20 Convert kg of NH_4NO_3 to grams then moles of NH_4NO_3, then use the stoichiometric ratio of moles of N_2 to moles of NH_4NO_3, and finally, calculate the number of molecules using Avagadro's number.

$$1.00 \text{ kg}\left(\frac{1000 \text{ g } NH_4NO_3}{1 \text{ kg } NH_4NO_3}\right)\left(\frac{1 \text{ mol } NH_4NO_3}{80.06 \text{ g } NH_4NO_3}\right)\left(\frac{1 \text{ mol } N_2}{1 \text{ mol } NH_4NO_3}\right)\left(\frac{6.022 \times 10^{23} \text{ molecules } N_2}{1 \text{ mol } N_2}\right)$$

$$= 7.52 \text{ molecules of } N_2$$

4.21 Avagadro's number would become 5×10^{23}.

$$\text{Avagadro's number} = [2 \times 10^{-27} \text{ kg}\left(\frac{1000 \text{ g}}{1 \text{ kg}}\right)]^{-1} = 5 \times 10^{23}$$

Review Problems

4.22 1:2

4.23 1:4, 1:4

4.24 1.56×10^{21} atoms Na $\left(\frac{1 \text{ mol Na}}{6.022 \times 10^{23} \text{ atoms Na}}\right) = 2.59 \times 10^{-3}$ mole Na

4.25 1.80×10^{24} molecules of N_2 $\left(\frac{1 \text{ mol molecules } N_2}{6.022 \times 10^{23} \text{ molecules } N_2}\right) = 2.99$ mole N_2

4.26 (a) 6 atom C:11 atom H
(b) 12 mole C:11 mole O
(c) 2 atom H:1 atom O
(d) 2 mole H:1 mole O

4.27 (a) 3 atom C: 1 atom O
(b) 3 mole C: 1 mole O
(c) 1 atom C:2 atom H
(d) 1 mole H:2 mole O

4.28 $\# \text{ mol Al} = (1.58 \text{ mol O})\left(\frac{2 \text{ mol Al}}{3 \text{ mol O}}\right) = 1.05 \text{ mol Al}$

4.29 $\# \text{ mol V} = (0.565 \text{ mol O})\left(\frac{2 \text{ mol V}}{5 \text{ mol O}}\right) = 0.226 \text{ mol V}$

4.30 $\# \text{ mol Al} = (2.16 \text{ mol } Al_2O_3)\left(\frac{2 \text{ mol Al}}{1 \text{ mol } Al_2O_3}\right) = 4.32 \text{ mol Al}$

4.31 $\# \text{ mol O} = (4.25 \text{ mol } CaCO_3)\left(\frac{3 \text{ mol O}}{1 \text{ mol } CaCO_3}\right) = 12.7 \text{ mol O}$

4.32 (a) $\left(\frac{2 \text{ mol Al}}{3 \text{ mol S}}\right)$ or $\left(\frac{3 \text{ mol S}}{2 \text{ mol Al}}\right)$

(b) $\left(\frac{3 \text{ mol S}}{1 \text{ mol } Al_2(SO_4)_3}\right)$ or $\left(\frac{1 \text{ mol } Al_2(SO_4)_3}{3 \text{ mol S}}\right)$

(c) $\# \text{ mol Al} = (0.900 \text{ mol S})\left(\frac{2 \text{ mol Al}}{3 \text{ mol S}}\right) = 0.600 \text{ mol Al}$

(d) $\# \text{ mol S} = (1.16 \text{ mol } Al_2(SO_4)_3)\left(\frac{3 \text{ mol S}}{1 \text{ mol } Al_2(SO_4)_3}\right) = 3.48 \text{ mol S}$

4.33 (a) $\left(\frac{3 \text{ mol Fe}}{1 \text{ mol } Fe_3O_4}\right)$ or $\left(\frac{1 \text{ mol } Fe_3O_4}{3 \text{ mol Fe}}\right)$

(b) $\left(\frac{3 \text{ mol Fe}}{4 \text{ molO}}\right)$ or $\left(\frac{4 \text{ mol O}}{3 \text{ mol Fe}}\right)$

(c) $\#\text{ mol Fe} = (2.75\text{ mol Fe}_3\text{O}_4)\left(\frac{3\text{ mol Fe}}{1\text{ mol Fe}_3\text{O}_4}\right) = 8.25\text{ mol Fe}$

(d) $\#\text{ mol Fe}_2\text{O}_3 = (4.50\text{ mol Fe}_3\text{O}_4)\left(\frac{3\text{ mol Fe}}{1\text{ mol Fe}_3\text{O}_4}\right)\left(\frac{1\text{ mol Fe}_2\text{O}_3}{2\text{ mol Fe}}\right) = 6.75\text{ mol Fe}_2\text{O}_3$

4.34 Based on the balanced equation:

$$2\text{ NH}_3(g) \rightarrow \text{N}_2(g) + 3\text{H}_2(g)$$

From this equation the conversion factors can be written:

$$\left(\frac{1\text{ mol N}_2}{2\text{ mol NH}_3}\right) \text{ and } \left(\frac{3\text{ mol H}_2}{2\text{ mol NH}_3}\right)$$

To determine the moles produced, simply convert from starting moles to end moles:

$$0.145\text{ mol NH}_3\left(\frac{1\text{ mol N}_2}{2\text{ mol NH}_3}\right) = 0.0725\text{ mol N}_2$$

The moles of hydrogen are calculated similarly:

$$0.145\text{ mol NH}_3\left(\frac{3\text{ mol H}_2}{2\text{ mol NH}_3}\right) = 0.218\text{ mol H}_2$$

4.35 Based on the balanced equation:

$$4\text{Al}(s) + 3\text{O}_2(g) \rightarrow 2\text{ Al}_2\text{O}_3(s)$$

From this equation the conversion factor can be written:

$$\left(\frac{3\text{ mol O}_2}{4\text{ mol Al}}\right)$$

To determine the moles of O_2 needed, simply convert from the moles of Al_2O_3 produced:

$$0.225\text{ mol Al}\left(\frac{3\text{ mol O}_2}{4\text{ mol Al}}\right) = 0.169\text{ mol O}_2$$

4.36 $\#\text{ moles UF}_6 = (1.25\text{ mol CF}_4)\left(\frac{4\text{ mol F}}{1\text{ mol CF}_4}\right)\left(\frac{1\text{ mol UF}_6}{6\text{ mol F}}\right) = 0.833\text{ moles UF}_6$

4.37 $\#\text{ moles Fe}_3\text{O}_4 = (0.260\text{ mol Fe}_2\text{O}_3)\left(\frac{2\text{ mol Fe}}{1\text{ mol Fe}_2\text{O}_3}\right)\left(\frac{1\text{ mol Fe}_3\text{O}_4}{3\text{ mol Fe}}\right) = 0.173\text{ moles Fe}_3\text{O}_4$

4.38 $\#\text{ atoms C} = (4.13\text{ mol H})\left(\frac{1\text{ mol C}_3\text{H}_8}{8\text{ mol H}}\right)\left(\frac{6.022 \times 10^{23}\text{ molecules C}_3\text{H}_8}{1\text{ mol C}_3\text{H}_8}\right)$

$$\times \left(\frac{3\text{ atoms C}}{1\text{ molecule C}_3\text{H}_8}\right) = 9.33 \times 10^{23}\text{ atoms C}$$

4.39 $\#\text{ atoms H} = (2.31\text{ mol C}_3\text{H}_8)\left(\frac{8\text{ mol H}}{1\text{ mol C}_3\text{H}_8}\right)\left(\frac{6.022\times10^{23}\text{ atoms H}}{1\text{ mol H}}\right) = 1.11\times10^{25}\text{ atoms H}$

4.40 # C, H and O atoms in glucose = 6 atoms C + 12 atoms H + 6 atoms O = 24 atoms

$$\#\ \text{atoms} = (0.260\ \text{mol glucose})\left(\frac{6.022\times10^{23}\ \text{molecules glucose}}{1\ \text{mol glucose}}\right)\times\left(\frac{24\ \text{atoms}}{1\ \text{molecule glucose}}\right) = 3.76\times10^{24}\ \text{atoms}$$

4.41 # N, H and O atoms in glucose = 2 atoms N + 4 atoms H + 3 atoms O = 9 atoms

$$\#\ \text{atoms} = (0.356\ \text{mol NH}_4\text{NO}_3)\left(\frac{6.022\times10^{23}\ \text{molecules NH}_4\text{NO}_3}{1\ \text{mol NH}_4\text{NO}_3}\right)\times\left(\frac{9\ \text{atoms}}{1\ \text{molecule glucose}}\right) = 1.93\times10^{24}\ \text{atoms}$$

4.42 $6\ \text{g}\times\left(\frac{1\ \text{mol C-12}}{12.00\ \text{g C-12}}\right) = 0.5\ \text{mol}$ $\quad 0.5\ \text{mol}\left(\frac{6.022\times10^{23}\ \text{atoms C-12}}{1\ \text{mol C-12}}\right) = 3.01\times10^{23}\ \text{atoms C-12}$

4.43 $1.5\ \text{mol C-12}\left(\frac{6.022\times10^{23}\ \text{atoms C-12}}{1\ \text{mol C-12}}\right) = 9.033\times10^{23}\ \text{atoms C-12}$

$1.5\ \text{mol C-12}\times\left(\frac{12.00\ \text{g C-12}}{1\ \text{mol C-12}}\right) = 18\ \text{g C-12}$

4.44 (a) 23.0 g Na
(b) 32.1 g S
(c) 35.5 g Cl

4.45 (a) 14.0 g N
(b) 55.8 g Fe
(c) 24.3 g Mg

4.46 (a) $\#\ \text{g Fe} = (1.35\ \text{mol Fe})\left(\frac{55.85\ \text{g Fe}}{1\ \text{mole Fe}}\right) = 75.4\ \text{g Fe}$

(b) $\#\ \text{g O} = (24.5\ \text{mol O})\left(\frac{16.0\ \text{g O}}{1\ \text{mole O}}\right) = 392\ \text{g O}$

(c) $\#\ \text{g Ca} = (0.876\ \text{mol Ca})\left(\frac{40.08\ \text{g Ca}}{1\ \text{mole Ca}}\right) = 35.1\ \text{g Ca}$

4.47 (a) $\#\ \text{g S} = (0.546\ \text{mol S})\left(\frac{32.07\ \text{g S}}{1\ \text{mole S}}\right) = 17.5\ \text{g S}$

(b) $\#\ \text{g N} = (4.29\ \text{mol N})\left(\frac{14.01\ \text{g N}}{1\ \text{mole N}}\right) = 46.1\ \text{g N}$

(c) $\#\ \text{g Al} = (8.11\ \text{mol Al})\left(\frac{26.98\ \text{g N}}{1\ \text{mole N}}\right) = 218\ \text{g Al}$

4.48 $\#\ \text{g K} = 2.00\times10^{12}\ \text{atoms K}\left(\frac{1\ \text{mol K}}{6.022\times10^{23}\ \text{atoms K}}\right)\left(\frac{39.10\ \text{g K}}{1\ \text{mol K}}\right) = 1.30\times10^{-10}\ \text{g K}$

4.49 $\# \text{ g Na} = 4.00 \times 10^{17} \text{ atoms Na}\left(\dfrac{1 \text{ mol Na}}{6.022\times10^{23} \text{ atoms Na}}\right)\left(\dfrac{22.99 \text{ g Na}}{1 \text{ mol Na}}\right) = 1.53 \times 10^{-5} \text{ g K}$

4.50 $\# \text{ mol Ni} = 17.7 \text{ g Ni}\left(\dfrac{1 \text{ mol Ni}}{58.69 \text{ g Ni}}\right) = 0.302 \text{ mol Ni}$

4.51 $\# \text{ mol Cr} = 85.7 \text{ g Cr}\left(\dfrac{1 \text{ mol Cr}}{52.00 \text{ g Cr}}\right) = 1.65 \text{ mol Cr}$

4.52 Note: all masses are in g/mole

(a)	$NaHCO_3$	=	1Na + 1H + 1C + 3O
		=	(22.99) + (1.01) + (12.01) + (3 × 16.00)
		=	84.0 g/mole
(b)	$K_2Cr_2O_7$	=	2K + 2Cr + 7O
		=	(2 × 39.10) + (2 × 52.00) + (7 × 16.00)
		=	294.2 g/mole
(c)	$(NH_4)_2CO_3$	=	2N + 8H + C + 3O
		=	(2 × 14.01) + (8 × 1.01) + (12.01) + (3 × 16.00)
		=	96.1 g/mole
(d)	$Al_2(SO_4)_3$	=	2Al + 3S + 12O
		=	(2 × 26.98) + (3 × 32.07) + (12 × 16.00)
		=	342.2 g/mole
(e)	$CuSO_4{\bullet}5H_2O$	=	1Cu + 1S +9O + 10H
		=	63.55 + 32.07 + (9 × 16.00) + (10 × 1.01)
		=	249.7 g/mole

4.53 Note: all masses are in g/mole

(a)	$Ca(NO_3)_2$	=	1Ca + 2N + 6O
		=	(40.08) + (2 × 14.01) + (6 × 16.00)
		=	164.1 g/mole
(b)	$Pb(C_2H_5)_4$	=	1Pb + 8C + 20H
		=	(207.2) + (8 × 12.01) + (20 × 1.01)
		=	323.5 g/mole (Since the mass of Pb is known exactly.)
(c)	$Mg_3(PO_4)_2$	=	3Mg + 2P + 8O
		=	(3 × 24.31) + (2 × 30.97) + (8 × 16.0)
		=	262.9 g/mole
(d)	$Fe_4[Fe(CN)_6]_3$	=	7Fe + 18C + 18N
		=	(7 × 55.85) + (18 × 12.01) + (18 × 14.01)
		=	859.3 g/mole
(e)	$Na_2SO_4{\bullet}10H_2O$	=	2Na + 1S +14O + 20H
		=	(2 × 22.99) + 32.07 + (14 × 16.00) + (20 × 1.01)
		=	322.3 g/mole

4.54
(a) # g =(1.25 mol $Ca_3(PO_4)_2$)(310.18 g $Ca_3(PO_4)_2$/1 mol $Ca_3(PO_4)_2$) = 388 g $Ca_3(PO_4)_2$
(b) # g =(0.625 mol $Fe(NO_3)_3$)(241.87 g $Fe(NO_3)_3$/1 mol $Fe(NO_3)_3$ = 151 g $Fe(NO_3)_3$
(c) # g = (0.600 mol C_4H_{10})(58.12 g C_4H_{10}/1 mole C_4H_{10}) = 34.9 g C_4H_{10}
(d) # g = (1.45 mol $(NH_4)_2CO_3$)(96.11 g/mole) = 139 g $(NH_4)_2CO_3$

4.55
(a) # g = (0.754 mol $ZnCl_2$)(136.28 g $ZnCl_2$/1 mol $ZnCl_2$) = 103 g $ZnCl_2$
(b) # g = (0.194 mol KIO_3)(214.00 g KIO_3/1 mol KIO_3) = 41.5 g KIO_3
(c) # g = (0.322 mol $POCl_3$)(153.32 g/mole) = 49.4 g $POCl_3$
(d) # g = (4.31×10^{-5} mol $(NH_4)_2HPO_4$)(132.1 g/mole) = 5.69×10^{-3} g $(NH_4)_2HPO_4$

4.56 (a) $\# \text{ moles } CaCO_3 = (21.5 \text{ g } CaCO_3)\left(\frac{1 \text{ mole } CaCO_3}{100.09 \text{ g } CaCO_3}\right) = 0.215 \text{ moles } CaCO_3$

(b) $\# \text{ moles } NH_3 = (1.56 \text{ g } NH_3)\left(\frac{1 \text{ mole } NH_3}{17.03 \text{ g } NH_3}\right) = 9.16 \times 10^{-2} \text{ moles } NH_3$

(c) $\# \text{ moles } Sr(NO_3)_2 = (16.8 \text{ g } Sr(NO_3)_2)\left(\frac{1 \text{ mole } Sr(NO_3)_2}{211.6 \text{ g } Sr(NO_3)_2}\right) = 7.94 \times 10^{-2} \text{ moles } Sr(NO_3)_2$

(d) $\# \text{ moles } Na_2CrO_4 = (6.98 \times 10^{-6} \text{ g } Na_2CrO_4)\left(\frac{1 \text{ mole } Na_2CrO_4}{162.0 \text{ g } Na_2CrO_4}\right)$

$= 4.31 \times 10^{-8} \text{ moles } Na_2CrO_4$

4.57 (a) $\# \text{ moles } Ca(OH)_2 = (9.36 \text{ g } Ca(OH)_2)\left(\frac{1 \text{ mole } Ca(OH)_2}{74.10 \text{ g } Ca(OH)_2}\right) = 0.126 \text{ moles } Ca(OH)_2$

(b) $\# \text{ moles } PbSO_4 = (38.2 \text{ g } PbSO_4)\left(\frac{1 \text{ mole } PbSO_4}{303.3 \text{ g } PbSO_4}\right) = 0.126 \text{ moles } PbSO_4$

(c) $\# \text{ moles } H_2O_2 = (4.29 \text{ g } H_2O_2)\left(\frac{1 \text{ mole } H_2O_2}{34.02 \text{ g } H_2O_2}\right) = 0.126 \text{ moles } H_2O_2$

(d) $\# \text{ mol } NaAuCl_4 = (0.465 \text{ mg } NaAuCl_4)\left(\frac{1 \text{ g}}{1000 \text{ mg}}\right)\left(\frac{1 \text{ mol } NaAuCl_4}{361.8 \text{ g } NaAuCl_4}\right)$

$= 1.29 \times 10^{-5} \text{ mol } NaAuCl_4$

4.58 The formula CaC_2 indicates that there is 1 mole of Ca for every 2 moles of C. Therefore, if there are 0.150 moles of C there must be 0.0750 moles of Ca.

$\# \text{ g Ca} = (0.075 \text{ mol Ca})\left(\frac{40.078 \text{ g Ca}}{1 \text{ mole Ca}}\right) = 3.01 \text{ g Ca}$

4.59 $\# \text{ mol I} = (0.500 \text{ mol } Ca(IO_3)_2)\left(\frac{2 \text{ moles I}}{1 \text{ mole } Ca(IO_3)_2}\right) = 1.00 \text{ moles I}$

$\# \text{ g } Ca(IO_3)_2 = (0.500 \text{ mol } Ca(IO_3)_2)\left(\frac{390 \text{ g } Ca(IO_3)_2}{1 \text{ mole } Ca(IO_3)_2}\right) = 195 \text{ g } Ca(IO_3)_2$

4.60 $\# \text{ mol N} = (0.650 \text{ mol } (NH_4)_2CO_3)\left(\frac{2 \text{ moles N}}{1 \text{ mole } (NH_4)_2CO_3}\right) = 1.30 \text{ moles N}$

$\# \text{ g } (NH_4)_2CO_3 = (0.650 \text{ mol } (NH_4)_2CO_3)\left(\frac{96.09 \text{ g } (NH_4)_2CO_3}{1 \text{ mole } (NH_4)_2CO_3}\right)$

$= 62.5 \text{ g } (NH_4)_2CO_3$

4.61 $\# \text{ mol N} = (0.556 \text{ mol } NH_4NO_3)\left(\frac{2 \text{ moles N}}{1 \text{ mole } NH_4NO_3}\right) = 1.11 \text{ moles N}$

$\# \text{ g } NH_4NO_3 = (0.556 \text{ mol } NH_4NO_3)\left(\frac{80.05 \text{ g } NH_4NO_3}{1 \text{ mole } NH_4NO_3}\right) = 44.5 \text{ g } NH_4NO_3$

4.62 $\# \text{ kg fertilizer} = (1 \text{ kg N})\left(\frac{1000 \text{ g N}}{1 \text{ kg N}}\right)\left(\frac{1 \text{ mol N}}{14.01 \text{ g N}}\right)\left(\frac{1 \text{ mol (NH}_4)_2\text{CO}_3}{2 \text{ mol N}}\right)$

$$\times\left(\frac{96.11 \text{ g (NH}_4)_2\text{CO}_3}{1 \text{ mol (NH}_4)_2\text{CO}_3}\right)\left(\frac{1 \text{ kg (NH}_4)_2\text{CO}_3}{1000 \text{ g (NH}_4)_2\text{CO}_3}\right) = 3.43 \text{ kg fertilizer}$$

4.63 $1 \text{ kg P}\left(\frac{1000 \text{ g P}}{1 \text{ kg P}}\right)\left(\frac{1 \text{ mol P}}{30.97 \text{ g P}}\right)\left(\frac{1 \text{ mol P}_2\text{O}_5}{2 \text{ mol P}}\right)\left(\frac{93.94 \text{ g P}_2\text{O}_5}{1 \text{ mol P}_2\text{O}_5}\right)\left(\frac{1 \text{ kg P}_2\text{O}_5}{1000 \text{ g P}_2\text{O}_5}\right) = 1.5 \text{ kg P}_2\text{O}_5$

4.64 $\% \text{ O in morphine} = \frac{48.00 \text{ g O}}{285.36 \text{ g C}_{17}\text{H}_{19}\text{NO}_3} \times 100\,\% = 16.82\,\% \text{ O}$

$\% \text{ O in heroin} = \frac{80.00 \text{ g O}}{369.44 \text{ g C}_{21}\text{H}_{23}\text{NO}_5} \times 100\,\% = 21.65\,\% \text{ O}$

Therefore heroin has a higher percentage oxygen.

4.65 Carbamazepine

4.66 $\% \text{ Cl in Freon-12} = \frac{70.90 \text{ g Cl}}{285.36 \text{ g CCl}_2\text{F}_2} \times 100\,\% = 24.85\,\% \text{ Cl}$

$\% \text{ Cl in Freon 141b} = \frac{106.35 \text{ g Cl}}{187.37 \text{ g C}_2\text{Cl}_3\text{F}_3} \times 100\,\% = 56.76\,\% \text{ Cl}$

Therefore Freon 141b has a higher percentage chlorine.

4.67 Freon-12

4.68 Assume one mole total for each of the following.

(a) The molar mass of NaH_2PO_4 is 119.98 g/mol.

$\% \text{ Na} = \frac{23.0 \text{ g Na}}{119.98 \text{ g NaH}_2\text{PO}_4} \times 100\,\% = 19.2\,\%$

$\% \text{ H} = \frac{2.02 \text{ g H}}{119.98 \text{ g NaH}_2\text{PO}_4} \times 100\,\% = 1.68\,\%$

$\% \text{ P} = \frac{31.0 \text{ g P}}{119.98 \text{ g NaH}_2\text{PO}_4} \times 100\,\% = 25.8\,\%$

$\% \text{ O} = \frac{64.0 \text{ g O}}{119.98 \text{ g NaH}_2\text{PO}_4} \times 100\,\% = 53.3\,\%$

(b) The molar mass of $NH_4H_2PO_4$ is 115.05 g/mol.

$\% \text{ N} = \frac{14.0 \text{ g N}}{115.05 \text{ g NH}_4\text{H}_2\text{PO}_4} \times 100\,\% = 12.2\,\%$

$\% \text{ H} = \frac{6.05 \text{ g H}}{115.05 \text{ g NH}_4\text{H}_2\text{PO}_4} \times 100\,\% = 5.26\,\%$

$\% \text{ P} = \frac{31.0 \text{ g P}}{115.05 \text{ g NH}_4\text{H}_2\text{PO}_4} \times 100\,\% = 26.9\,\%$

$\% \text{ O} = \frac{64.0 \text{ g O}}{115.05 \text{ g NH}_4\text{H}_2\text{PO}_4} \times 100\,\% = 55.6\,\%$

(c) The molar mass of $(CH_3)_2CO$ is 58.05 g/mol

$$\% C = \frac{36.0 \text{ g C}}{58.05 \text{ g } (CH_3)_2CO} \times 100\,\% = 62.0\,\%$$

$$\% H = \frac{6.05 \text{ g H}}{58.05 \text{ g } (CH_3)_2CO} \times 100\,\% = 10.4\,\%$$

$$\% O = \frac{16.0 \text{ g O}}{58.05 \text{ g } (CH_3)_2CO} \times 100\,\% = 27.6\,\%$$

(d) The molar mass of $CaSO_4$ is 136.2 g/mol.

$$\% Ca = \frac{40.1 \text{ g Ca}}{136.2 \text{ g } CaSO_4} \times 100\,\% = 29.4\,\%$$

$$\% S = \frac{32.1 \text{ g S}}{136.2 \text{ g } CaSO_4} \times 100\,\% = 23.6\,\%$$

$$\% O = \frac{64.0 \text{ g O}}{136.2 \text{ g } CaSO_4} \times 100\,\% = 47.0\,\%$$

(e) The molar mass of $CaSO_4{\bullet}2H_2O$ is 172.2 g/mol.

$$\% Ca = \frac{40.1 \text{ g Ca}}{172.2 \text{ g } CaSO_4{\bullet}2H_2O} \times 100\,\% = 23.3\,\%$$

$$\% S = \frac{32.1 \text{ g S}}{172.2 \text{ g } CaSO_4{\bullet}2H_2O} \times 100\,\% = 18.6\,\%$$

$$\% O = \frac{96.0 \text{ g O}}{172.2 \text{ g } CaSO_4{\bullet}2H_2O} \times 100\,\% = 55.7\,\%$$

$$\% H = \frac{4.03 \text{ g H}}{172.2 \text{ g } CaSO_4{\bullet}2H_2O} \times 100\,\% = 2.34\,\%$$

4.69 (a) The molar mass of $(CH_3)_2N_2H_2$ is 60.12 g/mol.

$$\% C = \frac{24.02 \text{ g C}}{60.12 \text{ g } (CH_3)_2N_2H_2} \times 100\,\% = 40.0\,\%$$

$$\% H = \frac{8.06 \text{ g H}}{60.12 \text{ g } (CH_3)_2N_2H_2} \times 100\,\% = 13.4\,\%$$

$$\% N = \frac{28.0 \text{ g N}}{60.12 \text{ g } (CH_3)_2N_2H_2} \times 100\,\% = 46.6\,\%$$

(b) The molar mass of $CaCO_3$ is 100.09 g/mol.

$$\% Ca = \frac{40.1 \text{ g Ca}}{100.1 \text{ g } CaCO_3} \times 100\,\% = 40.0\,\%$$

$$\% C = \frac{12.01 \text{ g C}}{100.1 \text{ g } CaCO_3} \times 100\,\% = 12.0\,\%$$

$$\% O = \frac{48.0 \text{ g O}}{100.1 \text{ g } CaCO_3} \times 100\,\% = 48.0\,\%$$

(c) The molar mass of $Fe(NO_3)_3$ is 241.9 g/mol.

$$\% \text{ Fe} = \frac{55.85 \text{ g Fe}}{241.9 \text{ g Fe(NO}_3)_3} \times 100\% = 23.1\%$$

$$\% \text{ N} = \frac{42.0 \text{ g N}}{241.9 \text{ g Fe(NO}_3)_3} \times 100\% = 17.4\%$$

$$\% \text{ O} = \frac{144 \text{ g O}}{241.9 \text{ g Fe(NO}_3)_3} \times 100\% = 59.5\%$$

(d) The molar mass of C_3H_8 is 44.1 g/mol.

$$\% \text{ C} = \frac{36.03 \text{ g C}}{44.1 \text{ g C}_3\text{H}_8} \times 100\% = 81.7\%$$

$$\% \text{ H} = \frac{8.08 \text{ g H}}{44.1 \text{ g C}_3\text{H}_8} \times 100\% = 18.3\%$$

(e) The molar mass of $Al_2(SO_4)_3$ is 342.2 g/mol.

$$\% \text{ Al} = \frac{54.0 \text{ g Al}}{342.2 \text{ g Al}_2(\text{SO}_4)_3} \times 100\% = 15.8\%$$

$$\% \text{ S} = \frac{96.2 \text{ g S}}{342.2 \text{ g Al}_2(\text{SO}_4)_3} \times 100\% = 28.1\%$$

$$\% \text{ O} = \frac{192 \text{ g O}}{342.2 \text{ g Al}_2(\text{SO}_4)_3} \times 100\% = 56.1\%$$

4.70 $$\% \text{ P} = \frac{0.539 \text{ g P}}{2.35 \text{ g compound}} \times 100\% = 22.9\%$$

% Cl = 100% Ğ 22.9% = 77.1%

4.71 $$\% \text{ N} = \frac{1.47 \text{ g N}}{5.67 \text{ g compound}} \times 100\% = 25.9\%$$

$\%\text{O} = 100\% - 25.9\% = 74.1\%$

4.72 For $C_{17}H_{25}N$, the molar mass (17C + 25H + 1N) equals 243.43 g/mole, and the three theoretical values for % by weight are calculated as follows:

$$\% \text{ C} = \frac{204.2 \text{ g C}}{243.4 \text{ g C}_{17}\text{H}_{25}\text{N}} \times 100\,\% = 83.89\,\%$$

$$\% \text{ H} = \frac{25.20 \text{ g H}}{243.4 \text{ g C}_{17}\text{H}_{25}\text{N}} \times 100\,\% = 10.35\,\%$$

$$\% \text{ N} = \frac{14.01 \text{ g N}}{243.4 \text{ g C}_{17}\text{H}_{25}\text{N}} \times 100\,\% = 5.76\,\%$$

These data are consistent with the experimental values cited in the problem.

4.73 For $C_{20}H_{25}N_3O$, the molar mass (20C + 25H + 3N + O) equals 323.44 g/mole, and the theoretical values for % by weight are calculated as follows:

$$\% C = \frac{240.22 \text{ g C}}{323.44 \text{ g } C_{20}H_{25}N_3O} \times 100\,\% = 74.27\,\%$$

$$\% H = \frac{25.20 \text{ g H}}{323.44 \text{ g } C_{20}H_{25}N_3O} \times 100\,\% = 7.791\,\%$$

$$\% N = \frac{42.02 \text{ g N}}{323.44 \text{ g } C_{20}H_{25}N_3O} \times 100\,\% = 12.99\,\%$$

$$\% O = \frac{16.00 \text{ g O}}{323.44 \text{ g } C_{20}H_{25}N_3O} \times 100\,\% = 4.947\,\%$$

(a) The % by mass oxygen in the suspected sample may be determined by difference: 100% – (74.07 + 7.95 + 9.99)% = 7.99 %.

(b) These data are not consistent with the theoretical formula for LSD.

4.74 $$\#\text{ g O} = (7.14 \times 10^{21} \text{ atoms N})\left(\frac{1 \text{ mol N}}{6.02 \times 10^{23} \text{ atoms N}}\right)\left(\frac{5 \text{ mol O}}{2 \text{ mol N}}\right)\left(\frac{16.0 \text{ g O}}{1 \text{ mol O}}\right) = 0.474 \text{ g O}$$

4.75 $$\#\text{ g C} = (4.25 \times 10^{23} \text{ atoms H})\left(\frac{1 \text{ mol H}}{6.02 \times 10^{23} \text{ atoms H}}\right)\left(\frac{5 \text{ mol C}}{12 \text{ mol H}}\right)\left(\frac{12.01 \text{ g C}}{1 \text{ mol C}}\right) = 3.53 \text{ g C}$$

4.76 The molecular formula is some integer multiple of the empirical formula. This means that we can divide the molecular formula by the largest possible whole number that gives an integer ratio among the atoms in the empirical formula.

(a) SCl (b) CH_2O (c) NH_3 (d) AsO_3 (e) HO

4.77 (a) CH_3O (b) HSO_4 (c) C_2H_5 (d) BH_3 (e) C_2H_6O

4.78 We begin by realizing that the mass of oxygen in the compound may be determined by difference: 0.896 g total – (0.111 g Na + 0.477 g Tc) = 0.308 g O. Next we can convert each mass of an element into the corresponding number of moles of that element as follows:

$$\#\text{ moles Na} = (0.111 \text{ g Na})\left(\frac{1 \text{ mole Na}}{23.00 \text{ g Na}}\right) = 4.83 \times 10^{-3} \text{ moles Na}$$

$$\#\text{ moles Tc} = (0.477 \text{ g Tc})\left(\frac{1 \text{ mole Tc}}{98.9 \text{ g Tc}}\right) = 4.82 \times 10^{-3} \text{ moles Tc}$$

$$\#\text{ moles O} = (0.308 \text{ g O})\left(\frac{1 \text{ mole O}}{16.0 \text{ g O}}\right) = 1.93 \times 10^{-2} \text{ moles O}$$

Now we divide each of these numbers of moles by the smallest of the three numbers, in order to obtain the simplest mole ratio among the three elements in the compound:

for Na, 4.83×10^{-3} moles / 4.82×10^{-3} moles = 1.00
for Tc, 4.82×10^{-3} moles / 4.82×10^{-3} moles = 1.00
for O, 1.93×10^{-2} moles / 4.82×10^{-3} moles = 4.00

These relative mole amounts give us the empirical formula: $NaTcO_4$.

4.79 $\# \text{ mol C} = (0.423 \text{ g C})\left(\frac{1 \text{ mol C}}{12.01 \text{ g C}}\right) = 0.0352 \text{ mol C}$

$\# \text{ mol Cl} = (2.50 \text{ g Cl})\left(\frac{1 \text{ mol Cl}}{35.45 \text{ g Cl}}\right) = 0.0705 \text{ mol Cl}$

$\# \text{ mol F} = (1.34 \text{ g F})\left(\frac{1 \text{ mol F}}{19.00 \text{ g F}}\right) = 0.0705 \text{ mol F}$

Now we divide each of these numbers of moles by the smallest of the three numbers, in order to obtain the simplest mole ratio among the three elements in the compound:

for C, 0.0352 moles / 0.0352 moles = 1.00
for Cl, 0.0705 moles / 0.0352moles = 2.000
for F, 0.0705 moles / 0.0352 moles = 2.00

These relative mole amounts give us the empirical formula CCl_2F_2

4.80 Assume a 100g sample:

$\# \text{ mol C} = (14.5 \text{ g C})\left(\frac{1 \text{ mol C}}{12.01 \text{ g C}}\right) = 1.21 \text{ mol C}$

$\# \text{ mol Cl} = (85.5 \text{ g Cl})\left(\frac{1 \text{ mol Cl}}{35.45 \text{ g Cl}}\right) = 2.41 \text{ mol Cl}$

Now we divide each of these numbers of moles by the smallest of the three numbers, in order to obtain the simplest mole ratio among the three elements in the compound:

for C, 1.21 moles / 1.21 moles = 1.00
for Cl, 2.41 moles / 1.21moles = 2.000

These relative mole amounts give us the empirical formula CCl_2

4.81 To solve this problem we will assume that we have a 100 g sample. This implies that we have 77.26 g Hg, 9.25 g C, 1.17 g H and 12.32 g O. The amount of oxygen was determined by subtracting the total amounts of the other three elements from the total assumed mass of 100 g. Convert each of these masses into a number of moles:

$\# \text{ moles Hg} = (77.26 \text{ g Hg})\left(\frac{1 \text{ mole Hg}}{200.59 \text{ g Hg}}\right) = 0.3852 \text{ moles Hg}$

$\# \text{ moles C} = (9.25 \text{ g C})\left(\frac{1 \text{ mole C}}{12.011 \text{ g C}}\right) = 0.770 \text{ moles C}$

$\# \text{ moles H} = (1.17 \text{ g H})\left(\frac{1 \text{ mole H}}{1.008 \text{ g H}}\right) = 1.16 \text{ moles H}$

$\# \text{ moles O} = (12.32 \text{ g O})\left(\frac{1 \text{ mole O}}{15.999 \text{ g O}}\right) = 0.7700 \text{ moles O}$

The relative mole amounts are determined as follows:

for Hg, 0.3852 moles / 0.3852 moles = 1.000
for C, 0.770 moles / 0.3852 moles = 2.00
for H, 1.16 moles / 0.3852 moles = 3.01
for O, 0.7700 moles / 0.3852 moles = 1.999

and the empirical formula is $HgC_2H_3O_2$. The empirical formula weight is 260 g/mole, which must be multiplied by 2 in order to obtain the molecular weight. This means that the molecular formula is twice the empirical formula, or $Hg_2C_4H_6O_4$.

4.82 Assume a 100g sample:

$$\# \text{ mol Na} = (22.9 \text{ g Na})\left(\frac{1 \text{ mol Na}}{22.99 \text{ g Na}}\right) = 0.996 \text{ mol Na}$$

$$\# \text{ mol B} = (21.5 \text{ g B})\left(\frac{1 \text{ mol B}}{10.81 \text{ g B}}\right) = 1.99 \text{ mol B}$$

$$\# \text{ mol O} = (55.7 \text{ g O})\left(\frac{1 \text{ mol O}}{16.00 \text{ g O}}\right) = 3.48 \text{ mol O}$$

Now we divide each of these numbers of moles by the smallest of the three numbers, in order to obtain the simplest mole ratio among the three elements in the compound:

for Na, 0.996 moles / 0.996 moles = 1.00
for B, 1.99 moles / 0.996 moles = 2.00
for O, 3.48 moles / 0.996 moles = 3.49
These relative mole amounts give us the empirical formula $Na_2B_4O_7$

4.83 Assume a 100g sample:

$$\# \text{ mol C} = (64.2 \text{ g C})\left(\frac{1 \text{ mol C}}{12.01 \text{ g C}}\right) = 5.35 \text{ mol C}$$

$$\# \text{ mol H} = (5.26 \text{ g H})\left(\frac{1 \text{ mol H}}{1.008 \text{ g H}}\right) = 5.22 \text{ mol H}$$

$$\# \text{ mol O} = (31.6 \text{ g O})\left(\frac{1 \text{ mol O}}{16.00 \text{ g O}}\right) = 1.98 \text{ mol O}$$

Now we divide each of these numbers of moles by the smallest of the three numbers, in order to obtain the simplest mole ratio among the three elements in the compound:

for C, 5.26 moles / 1.98 moles = 2.66 = (8/3)
for H, 5.22 moles / 1.98 moles = 2.66 = (8/3)
for O, 1.98 moles / 1.98 moles = 1.00 = (3/3)

These relative mole amounts give us the empirical formula $C_8H_8O_3$

4.84 All of the carbon is converted to carbon dioxide so,

$$\# \text{ g C} = (1.312 \text{ g } CO_2)\left(\frac{1 \text{ mol } CO_2}{44.01 \text{ g } CO_2}\right)\left(\frac{1 \text{ mol C}}{1 \text{ mol } CO_2}\right)\left(\frac{12.01 \text{ g C}}{1 \text{ mol C}}\right) = 0.358 \text{ g C}$$

$$\# \text{ moles C} = (0.358 \text{ g C})\left(\frac{1 \text{ mol C}}{12.01 \text{ g C}}\right) = 2.98 \times 10^{-2} \text{ mol C}$$

All of the hydrogen is converted to H_2O, so

$$\# \text{ g H} = (0.805 \text{ g } H_2O)\left(\frac{1 \text{ mol } H_2O}{18.02 \text{ g } H_2O}\right)\left(\frac{2 \text{ mol H}}{1 \text{ mol } H_2O}\right)\left(\frac{1.008 \text{ g H}}{1 \text{ mol H}}\right) = 0.0901 \text{ g H}$$

$$\# \text{ mol H} = (0.0901 \text{ g H})\left(\frac{1 \text{ mol H}}{1.008 \text{ g H}}\right) = 8.93 \times 10^{-2} \text{ moles H}$$

The amount of O in the compound is determined by subtracting the mass of C and the mass of H from the sample.

g O = 0.684g – 0.358 g – 0.0901 g = 0.236 g

$$\# \text{ mol O} = (0.236 \text{ g O})\left(\frac{1 \text{ mol O}}{16 \text{ g O}}\right) = 1.48 \times 10^{-2} \text{ moles}$$

The relative mole ratios are:

for C, 0.0298 moles / 0.0148 moles = 2.01
for H, 0.0893 moles/ 0.0148 moles = 6.03
for O, 0.0148 moles / 0.0148 moles = 1.00

The relative mole amounts give the empirical formula C_2H_6O

4.85 All of the carbon is converted to carbon dioxide so,

$$\# \text{ g C} = (2.01 \text{ g CO}_2)\left(\frac{1 \text{ mol CO}_2}{44.01 \text{ g CO}_2}\right)\left(\frac{1 \text{ mol C}}{1 \text{ mol CO}_2}\right)\left(\frac{12.01 \text{ g C}}{1 \text{ mol C}}\right) = 0.549 \text{ g C}$$

$$\# \text{ moles C} = (0.549 \text{ g C})\left(\frac{1 \text{ mol C}}{12.01 \text{ g C}}\right) = 0.0457 \text{ mol C}$$

All of the hydrogen is converted to H_2O, so

$$\# \text{ g H} = (0.827 \text{ g H}_2\text{O})\left(\frac{1 \text{ mol H}_2\text{O}}{18.02 \text{ g H}_2\text{O}}\right)\left(\frac{2 \text{ mol H}}{1 \text{ mol H}_2\text{O}}\right)\left(\frac{1.008 \text{ g H}}{1 \text{ mol H}}\right) = 0.0925 \text{ g H}$$

$$\# \text{ mol H} = (0.0925 \text{ g H})\left(\frac{1 \text{ mol H}}{1.008 \text{ g H}}\right) = 0.0918 \text{ moles H}$$

The amount of O in the compound is determined by subtracting the mass of C and the mass of H from the sample.

g O = 0.822g – 0.549 g – 0.0925 g = 0.181 g

$$\# \text{ mol O} = (0.181 \text{ g O})\left(\frac{1 \text{ mol O}}{16.00 \text{ g O}}\right) = 0.0113 \text{ moles}$$

The relative mole ratios are:
for C, 0.0457 moles / 0.0113 moles = 4.04
for H, 0.0918 moles/ 0.0113 moles = 8.12
for O, 0.0113 moles / 0.0113 moles = 1.00

The relative mole amounts give the empirical formula C_4H_8O

4.86 This type of combustion analysis takes advantage of the fact that the entire amount of carbon in the original sample appears as CO_2 among the products. Hence the mass of carbon in the original sample must be equal to the mass of carbon that is found in the CO_2.

$$\# \text{ g C} = (19.73 \times 10^{-3} \text{ g CO}_2)\left(\frac{1 \text{ mole CO}_2}{44.01 \text{ g CO}_2}\right)\left(\frac{1 \text{ mole C}}{1 \text{ mole CO}_2}\right)\left(\frac{12.011 \text{ g C}}{1 \text{ mole C}}\right)$$

$$= 5.385 \times 10^{-3} \text{ g C}$$

Similarly, the entire mass of hydrogen that was present in the original sample ends up in the products as H_2O:

$$\# \text{ g H} = (6.391 \times 10^{-3} \text{ g H}_2\text{O})\left(\frac{1 \text{ mole H}_2\text{O}}{18.02 \text{ g H}_2\text{O}}\right)\left(\frac{2 \text{ mole H}}{1 \text{ mole H}_2\text{O}}\right)\left(\frac{1.008 \text{ g H}}{1 \text{ mole H}}\right)$$

$$= 7.150 \times 10^{-4} \text{ g H}$$

The mass of oxygen is determined by subtracting the mass due to C and H from the total mass: 6.853 mg total – (5.385 mg C + 0.7150 mg H) = 0.753 mg O. Now, convert these masses to a number of moles:

$$\# \text{ moles C} = (5.385 \times 10^{-3} \text{ g C})\left(\frac{1 \text{ mole C}}{12.011 \text{ g C}}\right) = 4.483 \times 10^{-4} \text{ moles C}$$

$$\# \text{ moles H} = (7.150 \times 10^{-4} \text{ g H})\left(\frac{1 \text{ mole H}}{1.0079 \text{ g H}}\right) = 7.094 \times 10^{-4} \text{ moles H}$$

$$\# \text{ moles O} = (7.53 \times 10^{-4} \text{ g O})\left(\frac{1 \text{ mole O}}{15.999 \text{ g O}}\right) = 4.71 \times 10^{-5} \text{ moles O}$$

The relative mole amounts are:

for C, 4.483×10^{-4} mol / 4.71×10^{-5} mol = 9.52
for H, 7.094×10^{-4} mol / 4.71×10^{-5} mol = 15.1
for O, 4.71×10^{-5} mol / 4.71×10^{-5} mol = 1.00

The relative mole amounts are not nice whole numbers as we would like. However, we see that if we double the relative number of moles of each compound, there are approximately 19 moles of C, 30 moles of H and 2 moles of O. If we assume these numbers are correct, the empirical formula is $C_{19}H_{30}O_2$, for which the formula weight is 290 g/mole. Since the molar mass is equal to the mass of this empirical formula, the molecular formula is the same as this empirical formula, $C_{19}H_{30}O_2$, and we have performed the analysis correctly.

In most problems where we attempt to determine an empirical formula, the relative mole amounts should work out to give a "nice" set of values for the formula. Rarely will a problem be designed that gives very odd coefficients. With experience and practice, you will recognize when a set of values is reasonable.

4.87 This type of combustion analysis takes advantage of the fact that the entire amount of carbon in the original sample appears as CO_2 among the products. Hence the mass of carbon in the original sample must be equal to the mass of carbon that is found in the CO_2.

$$\# \text{ g C} = (18.490 \times 10^{-3} \text{ g CO}_2)\left(\frac{1 \text{ mole CO}_2}{44.010 \text{ g CO}_2}\right)\left(\frac{1 \text{ mole C}}{1 \text{ mole CO}_2}\right)\left(\frac{12.011 \text{ g C}}{1 \text{ mole C}}\right) = 5.0462 \times 10^{-3} \text{ g C}$$

Similarly, the entire mass of hydrogen that was present in the original sample ends up in the products as H_2O:

$$\# \text{ g H} = (6.232 \times 10^{-3} \text{ g H}_2\text{O})\left(\frac{1 \text{ mole H}_2\text{O}}{18.02 \text{ g H}_2\text{O}}\right)\left(\frac{2 \text{ mole H}}{1 \text{ mole H}_2\text{O}}\right)\left(\frac{1.008 \text{ g H}}{1 \text{ mole H}}\right) = 6.972 \times 10^{-4} \text{ g H}$$

The mass of oxygen is determined by subtracting the mass due to C and H from the total mass: 5.983 mg total – (5.0462 mg C + 0.6972 mg H) = 0.240 mg O. Now, convert these masses to a number of moles:

$$\# \text{ moles C} = (5.0462 \times 10^{-3} \text{ g C})\left(\frac{1 \text{ mole C}}{12.011 \text{ g C}}\right) = 4.2013 \times 10^{-4} \text{ moles C}$$

$$\# \text{ moles H} = (6.972 \times 10^{-4} \text{ g H})\left(\frac{1 \text{ mole H}}{1.0079 \text{ g H}}\right) = 6.917 \times 10^{-4} \text{ moles H}$$

$$\# \text{ moles O} = (2.40 \times 10^{-4} \text{ g O})\left(\frac{1 \text{ mole O}}{15.999 \text{ g O}}\right) = 1.50 \times 10^{-5} \text{ moles O}$$

The relative mole amounts are:

for C, 4.2013×10^{-4} mol / 1.50×10^{-5} mol = 28.0
for H, 6.917×10^{-4} mol / 1.50×10^{-5} mol = 46.1
for O, 1.50×10^{-5} mol / 1.50×10^{-5} mol = 1.00

and the empirical formula is $C_{28}H_{46}O$.

4.88 (a) Formula mass = 135.1 g

$$\frac{270.4 \text{ g/mol}}{135.1 \text{ g/mol}} = 2.001$$

The molecular formula is $Na_2S_4O_6$

(b) Formula mass = 73.50 g

$$\frac{147.0 \text{ g/mol}}{73.50 \text{ g/mol}} = 2.000$$

The molecular formula is $C_6H_4Cl_2$

(c) Formula mass = 60.48 g

$$\frac{181.4 \text{ g/mol}}{60.48 \text{ g/mol}} = 2.999$$

The molecular formula is $C_6H_3Cl_3$

4.89 (a) Formula mass = 122.1 g

$$\frac{732.6 \text{ g/mol}}{122.1 \text{ g/mol}} = 6.000$$

The molecular formula is $Na_{12}Si_6O_{18}$

(b) Formula mass = 102.0 g

$$\frac{305.9 \text{ g/mol}}{102.0 \text{ g/mol}} = 2.999$$

The molecular formula is $Na_3P_3O_9$

(c) Formula mass = 31.03 g

$$\frac{62.1 \text{ g/mol}}{31.03 \text{ g/mol}} = 2.00$$

The molecular formula is $C_2H_6O_2$

4.90 The formula mass for the compound $C_{19}H_{30}O_2$ is 290 g/mol. Thus, the empirical and molecular formulas are equivalent.

4.91 The formula mass for the compound $C_{28}H_{46}O$ is 399 g/mol. Thus, the empirical and molecular formulas are equivalent.

4.92 From the information provided, we can determine the mass of mercury as the difference between the total mass and the mass of bromine:

$$\text{\# g Hg} = 0.389 \text{ g compound} - 0.111 \text{ g Br} = 0.278 \text{ g Hg}$$

To determine the empirical formula, first convert the two masses to a number of moles.

$$\text{\# moles Hg} = (0.278 \text{ g Hg})\left(\frac{1 \text{ mole Hg}}{200.59 \text{ g Hg}}\right) = 1.39 \times 10^{-3} \text{ moles Hg}$$

$$\text{\# moles Br} = (0.111 \text{ g Br})\left(\frac{1 \text{ mole Br}}{79.904 \text{ g Br}}\right) = 1.39 \times 10^{-3} \text{ moles Br}$$

Now, we would divide each of these values by the smaller quantity to determine the simplest mole ratio between the two elements. By inspection, though, we can see there are the same number of moles of Hg and Br. Consequently, the simplest mole ratio is 1:1 and the empirical formula is HgBr.

To determine the molecular formula, recall that the ratio of the molecular mass to the empirical mass is equivalent to the ratio of the molecular formula to the empirical formula. Thus, we need to calculate an empirical mass: (1 mole Hg)(200.59 g Hg/mole Hg) + (1 mole Br)(79.904 g Br/mole Br) = 280.49 g/mole HgBr. The molecular mass, as reported in the problem is 561 g/mole. The ratio of these is:

$$\frac{561 \text{ g/mole}}{280.49 \text{ g/mole}} = 2.00$$

So, the molecular formula is two times the empirical formula or Hg_2Br_2.

4.93 From the information provided, the mass of sulfur is the difference between the total mass and the mass of antimony:

$$\# \text{ g S} = 0.6662 \text{ g compound} - 0.4017 \text{ g Sb} = 0.2645 \text{ g S}$$

To determine the empirical formula, first convert the two masses to a number of moles.

$$\# \text{ moles S} = (0.2645 \text{ g S})\left(\frac{1 \text{ mole S}}{32.066 \text{ g S}}\right) = 8.249 \times 10^{-3} \text{ moles S}$$

$$\# \text{ moles Sb} = (0.4017 \text{ g Sb})\left(\frac{1 \text{ mole Sb}}{121.76 \text{ g Sb}}\right) = 3.299 \times 10^{-3} \text{ moles Sb}$$

Now, divide each of these values by the smaller quantity to determine the simplest mole ratio between the two elements:

For Sb: 3.299×10^{-3} moles/3.299×10^{-3} moles = 1.000 mol Sb
For S: 8.249×10^{-3} moles/3.299×10^{-3} moles = 2.500 mol S

Hence the empirical formula is Sb_2S_5, and the empirical mass is $(2 \times Sb) + (5 \times S) = 403.85$ g/mol. Since the molecular mass reported in the problem is the same as the calculated empirical mass, the empirical formula is the same as the molecular formula.

4.94 First, determine the amount of oxygen in the sample by subtracting the masses of the other elements from the total mass: 0.6216 g – (0.1735 g C + 0.01455 g H + 0.2024 g N) = 0.2312 g O. Now, convert these masses into a number of moles for each element:

$$\# \text{ moles C} = (0.1735 \text{ g C})\left(\frac{1 \text{ mole C}}{12.011 \text{ g C}}\right) = 1.445 \times 10^{-2} \text{ moles C}$$

$$\# \text{ moles H} = (0.01455 \text{ g H})\left(\frac{1 \text{ mole H}}{1.0079 \text{ g H}}\right) = 1.444 \times 10^{-2} \text{ moles H}$$

$$\# \text{ moles N} = (0.2024 \text{ g N})\left(\frac{1 \text{ mole N}}{14.007 \text{ g N}}\right) = 1.445 \times 10^{-2} \text{ moles N}$$

$$\# \text{ moles O} = (0.2312 \text{ g O})\left(\frac{1 \text{ mole O}}{15.999 \text{ g O}}\right) = 1.445 \times 10^{-2} \text{ moles O}$$

These are clearly all the same mole amounts, and we deduce that the empirical formula is CHNO, which has a formula weight of 43. It can be seen that the number 43 must be multiplied by the integer 3 in order to obtain the molar mass ($3 \times 43 = 129$), and this means that the empirical formula should similarly be multiplied by 3 in order to arrive at the molecular formula, $C_3H_3N_3O_3$.

4.95 To solve this problem we will assume that we have a 100 g sample. This implies that we have 75.42 g C, 6.63 g H, 8.38 g N and 9.57 g O. The amount of oxygen was determined by subtracting the total amounts of the other three elements from the total assumed mass of 100 g. Convert each of these masses into a number of moles:

$$\text{\# moles C} = (75.42 \text{ g C})\left(\frac{1 \text{ mole C}}{12.011 \text{ g C}}\right) = 6.279 \text{ moles C}$$

$$\text{\# moles H} = (6.63 \text{ g H})\left(\frac{1 \text{ mole H}}{1.008 \text{ g H}}\right) = 6.58 \text{ moles H}$$

$$\text{\# moles N} = (8.38 \text{ g N})\left(\frac{1 \text{ mole N}}{14.007 \text{ g N}}\right) = 0.598 \text{ moles N}$$

$$\text{\# moles O} = (9.57 \text{ g O})\left(\frac{1 \text{ mole O}}{15.999 \text{ g O}}\right) = 0.598 \text{ moles O}$$

The relative mole amounts are determined as follows:

for C, 6.279 mol / 0.598 mol = 10.5
for H, 6.58 mol / 0.598 mol = 11.0
for N, 0.598 mol / 0.598 mol = 1.00
for O, 0.598 mol / 0.598 mol = 1.00

In order to obtain nice whole numbers, each of these values is multiplied by 2 and we determine the empirical formula is $C_{21}H_{22}N_2O_2$. The empirical formula weight is 334 g/mole. This means that the molecular formula is the same as the empirical formula, or $C_{21}H_{22}N_2O_2$.

4.96 36 mol H

4.97 24 moles of O

4.98 $4Fe(s) + 3O_2(g) \rightarrow 2Fe_2O_3(s)$

4.99 $2NO(g) + O_2(g) \rightarrow 2NO_2(g)$

4.100
(a) $Ca(OH)_2 + 2HCl \rightarrow CaCl_2 + 2H_2O$
(b) $2AgNO_3 + CaCl_2 \rightarrow Ca(NO_3)_2 + 2AgCl$
(c) $2Fe_2O_3 + 3C \rightarrow 4Fe + 3CO_2$
(d) $2NaHCO_3 + H_2SO_4 \rightarrow Na_2SO_4 + 2H_2O + 2CO_2$
(e) $2C_4H_{10} + 13O_2 \rightarrow 8CO_2 + 10H_2O$

4.101
(a) $2SO_2 + O_2 \rightarrow 2SO_3$
(b) $P_4O_{10} + 6H_2O \rightarrow 4H_3PO_4$
(c) $Pb(NO_3)_2 + Na_2SO_4 \rightarrow PbSO_4 + 2NaNO_3$
(d) $Fe_2O_3 + 3H_2 \rightarrow 2Fe + 3H_2O$
(e) $2Al + 3H_2SO_4 \rightarrow Al_2(SO_4)_3 + 3H_2$

4.102
(a) $Mg(OH)_2 + 2HBr \rightarrow MgBr_2 + 2H_2O$
(b) $2HCl + Ca(OH)_2 \rightarrow CaCl_2 + 2H_2O$
(c) $Al_2O_3 + 3H_2SO_4 \rightarrow Al_2(SO_4)_3 + 3H_2O$
(d) $2KHCO_3 + H_3PO_4 \rightarrow K_2HPO_4 + 2H_2O + 2CO_2$
(e) $C_9H_{20} + 14O_2 \rightarrow 9CO_2 + 10H_2O$

4.103 (a) $CaO + 2HNO_3 \rightarrow Ca(NO_3)_2 + H_2O$

(b) $Na_2CO_3 + Mg(NO_3)_2 \rightarrow MgCO_3 + 2NaNO_3$

(c) $(NH_4)_3PO_4 + 3NaOH \rightarrow Na_3PO_4 + 3NH_3 + 3H_2O$

(d) $2LiHCO_3 + H_2SO_4 \rightarrow Li_2SO_4 + 2H_2O + 2CO_2$

(e) $C_4H_{10}O + 6O_2 \rightarrow 4CO_2 + 5H_2O$

4.104 (a) $\#\text{ moles } Na_2S_2O_3 = (0.12 \text{ moles } Cl_2)\left(\frac{1 \text{ mole } Na_2S_2O_3}{4 \text{ mole } Cl_2}\right) = 0.030 \text{ mol } Na_2S_2O_3$

(b) $\#\text{ moles HCl} = (0.12 \text{ moles } Cl_2)\left(\frac{8 \text{ mole HCl}}{4 \text{ mole } Cl_2}\right) = 0.24 \text{ moles HCl}$

(c) $\#\text{ moles } H_2O = (0.12 \text{ moles } Cl_2)\left(\frac{5 \text{ mole } H_2O}{4 \text{ mole } Cl_2}\right) = 0.15 \text{ moles } H_2O$

(d) $\#\text{ moles } H_2O = (0.24 \text{ moles HCl})\left(\frac{5 \text{ mole } H_2O}{8 \text{ mole HCl}}\right) = 0.15 \text{ moles } H_2O$

4.105 (a) $\#\text{ moles } O_2 = (6 \text{ moles } C_8H_{18})\left(\frac{25 \text{ mole } O_2}{2 \text{ mole } C_8H_{18}}\right) = 80 \text{ moles } O_2$

(Note: This calculation is limited due to sig figs.)

(b) $\#\text{ moles } CO_2 = (0.5 \text{ moles } C_8H_{18})\left(\frac{16 \text{ mole } CO_2}{2 \text{ mole } C_8H_{18}}\right) = 4 \text{ moles } CO_2$

(c) $\#\text{ moles } H_2O = (8 \text{ moles } C_8H_{18})\left(\frac{18 \text{ mole } H_2O}{2 \text{ mole } C_8H_{18}}\right) = 70 \text{ moles } H_2O$

(d) $\#\text{ moles } O_2 = (6.00 \text{ moles } CO_2)\left(\frac{25 \text{ mole } O_2}{16 \text{ mole } CO_2}\right) = 9.38 \text{ moles } O_2$

$\#\text{ moles } C_8H_{18} = (6.00 \text{ moles } CO_2)\left(\frac{2 \text{ mole } C_8H_{18}}{16 \text{ mole } CO_2}\right) = 0.750 \text{ moles } C_8H_{18}$

4.106 (a) $0.11 \text{ mol } Au(CN)_2^- \left(\frac{1 \text{ mol Zn}}{2 \text{ mol } Au(CN)_2^-}\right)\left(\frac{65.39 \text{ g Zn}}{1 \text{ mol Zn}}\right) = 3.6 \text{ g Zn}$

(b) $0.11 \text{ mol } Au(CN)_2^- \left(\frac{2 \text{ mol Au}}{2 \text{ mol } Au(CN)_2^-}\right)\left(\frac{197.0 \text{ g Au}}{1 \text{ mol Au}}\right) = 22 \text{ g Zn}$

(c) $0.11 \text{ mol Zn} \left(\frac{2 \text{ mol } Au(CN)_2^-}{1 \text{ mol Zn}}\right)\left(\frac{249.0 \text{ g } Au(CN)_2^-}{1 \text{ mol } Au(CN)_2^-}\right) = 55 \text{ g } Au(CN)_2^-$

4.107 (a) $3 \text{ mol } C_3H_8 \left(\frac{5 \text{ mol } O_2}{1 \text{ mol } C_3H_8}\right)\left(\frac{32.00 \text{ g } O_2}{1 \text{ mol } O_2}\right) = 500 \text{ g } O_2$

(b) $0.1 \text{ mol } C_3H_8 \left(\frac{3 \text{ mol } CO_2}{1 \text{ mol } C_3H_8}\right)\left(\frac{44.01 \text{ g } CO_2}{1 \text{ mol } CO_2}\right) = 13 \text{ g } CO_2$

(c) $4 \text{ mol } C_3H_8 \left(\frac{4 \text{ mol } H_2O}{1 \text{ mol } C_3H_8}\right)\left(\frac{18.01 \text{ g } H_2O}{1 \text{ mol } H_2O}\right) = 300 \text{ g } H_2O$

4.108 (a) $4P + 5O_2 \rightarrow P_4O_{10}$

(b) $$\#\ g\ O_2 = (6.85\ g\ P)\left(\frac{1\ mole\ P}{30.97\ g\ P}\right)\left(\frac{5\ mole\ O_2}{4\ mole\ P}\right)\left(\frac{32.0\ g\ O_2}{1\ mol\ O_2}\right) = 8.85\ g\ O_2$$

(c) $$\#\ g\ P_4O_{10} = (8.00\ g\ O_2)\left(\frac{1\ mole\ O_2}{32.00\ g\ O_2}\right)\left(\frac{1\ mole\ P_4O_{10}}{5\ mole\ O_2}\right)\left(\frac{283.9\ g\ P_4O_{10}}{1\ mole\ P_4O_{10}}\right) = 14.2\ g\ P_4O_{10}$$

(d) $$\#\ g\ P = (7.46\ g\ P_4O_{10})\left(\frac{1\ mole\ P_4O_{10}}{283.9\ g\ P_4O_{10}}\right)\left(\frac{4\ mole\ P}{1\ mole\ P_4O_{10}}\right)\left(\frac{30.97\ g\ P}{1\ mole\ P}\right) = 3.26\ g\ P$$

4.109 (a) $2C_4H_{10} + 13O_2 \rightarrow 8CO_2 + 10H_2O$

(b) $$\#\ g\ C_4H_{10} = (4.46\ g\ H_2O)\left(\frac{1\ mole\ H_2O}{18.02\ g\ H_2O}\right)\left(\frac{2\ moles\ C_4H_{10}}{10\ mole\ H_2O}\right)\left(\frac{58.12\ g\ C_4H_{10}}{1\ mole\ C_4H_{10}}\right) = 2.88\ g\ C_4H_{10}$$

(c) $$\#\ g\ O_2 = (2.88\ g\ C_4H_{10})\left(\frac{1\ mole\ C_4H_{10}}{58.12\ g\ C_4H_{10}}\right)\left(\frac{13\ moles\ O_2}{2\ moles\ C_4H_{10}}\right)\left(\frac{32.0\ g\ O_2}{1\ mole\ O_2}\right) = 10.3\ g\ O_2$$

(d) $$\#\ g\ CO_2 = (0.0496\ moles\ C_4H_{10})\left(\frac{8\ moles\ CO_2}{2\ moles\ C_4H_{10}}\right)\left(\frac{44.01\ g\ CO_2}{1\ mole\ CO_2}\right) = 8.72\ g\ CO_2$$

4.110 $$\#\ g\ of\ HNO_3 = (11.45\ g\ Cu)\left(\frac{1\ mole\ Cu}{63.546\ g\ Cu}\right)\left(\frac{8\ moles\ HNO_3}{3\ moles\ Cu}\right)\left(\frac{63.08\ g\ HNO_3}{1\ mole\ HNO_3}\right) = 30.31\ g\ HNO_3$$

4.111 $$\#\ g\ H_2O_2 = (852\ g\ N_2H_4)\left(\frac{1\ mole\ N_2H_4}{32.05\ g\ N_2H_4}\right)\left(\frac{7\ moles\ H_2O_2}{1\ mole\ N_2H_4}\right)\left(\frac{34.02\ g\ H_2O_2}{1\ mole\ H_2O_2}\right) = 6330\ g\ H_2O_2$$

4.112 $$\#\ kg\ O_2 = 1.0\ kg\ H_2O_2\left(\frac{1000\ g\ H_2O_2}{1\ kg\ H_2O_2}\right)\left(\frac{1\ mole\ H_2O_2}{34.01\ g\ H_2O_2}\right)\left(\frac{1\ moles\ O_2}{2\ mole\ H_2O_2}\right)\left(\frac{32.00\ g\ O_2}{1\ mole\ O_2}\right)$$
$$\times\left(\frac{1\ kg\ O_2}{1000\ g\ O_2}\right) = 0.47\ kg\ O_2$$

4.113 $$\#\ kg\ O_2 = 1.0\ kg\ KClO_2\left(\frac{1000\ g\ KClO_3}{1\ kg\ KClO_3}\right)\left(\frac{1\ mole\ KClO_3}{122.6\ g\ KClO_3}\right)\left(\frac{3\ moles\ O_2}{2\ mole\ KClO_3}\right)\left(\frac{32.00\ g\ O_2}{1\ mole\ O_2}\right)$$
$$\times\left(\frac{1\ kg\ O_2}{1000\ g\ O_2}\right) = 0.39\ kg\ O_2$$

4.114 (a) First determine the amount of Fe_2O_3 that would be required to react completely with the given amount of Al:

$$\#\ moles\ Fe_2O_3 = (4.20\ moles\ Al)\left(\frac{1\ mole\ Fe_2O_3}{2\ moles\ Al}\right) = 2.10\ moles\ Fe_2O_3$$

Since only 1.75 mol of Fe_2O_3 are supplied, it is the limiting reactant. This can be confirmed by calculating the amount of Al that would be required to react completely with all of the available Fe_2O_3:

$$\#\ moles\ Al = (1.75\ moles\ Fe_2O_3)\left(\frac{2\ moles\ Al}{1\ mole\ Fe_2O_3}\right) = 3.50\ moles\ Al$$

Since an excess (4.20 mol – 3.50 mol = 0.70 mol) of Al is present, Fe_2O_3 must be the limiting reactant, as determined above.

(b) $$\# \text{ g Fe} = (1.75 \text{ moles Fe}_2\text{O}_3)\left(\frac{2 \text{ moles Fe}}{1 \text{ mole Fe}_2\text{O}_3}\right)\left(\frac{55.847 \text{ g Fe}}{1 \text{ mole Fe}}\right) = 195 \text{ g Fe}$$

4.115 (a) First determine the amount of C_2H_4 that would be required to react completely with the given amount of H_2O:

$$\# \text{ g H}_2\text{O} = 1.0 \text{ kg C}_2\text{H}_4\left(\frac{1000 \text{ g C}_2\text{H}_4}{1 \text{ kg C}_2\text{H}_4}\right)\left(\frac{1 \text{ mole C}_2\text{H}_4}{56.02 \text{ g C}_2\text{H}_4}\right)\left(\frac{1 \text{ moles H}_2\text{O}}{1 \text{ mole C}_2\text{H}_4}\right)$$
$$\left(\frac{18.02 \text{ g H}_2\text{O}}{1 \text{ moles H}_2\text{O}}\right)\left(\frac{1 \text{ kg H}_2\text{O}}{1000 \text{ g H}_2\text{O}}\right) = 0.32 \text{ kg H}_2\text{O}$$

Since only 0.010 kg of H_2O are supplied, it is the limiting reactant. This can be confirmed by calculating the amount of C_2H_4 that would be required to react completely with all of the available H_2O:

$$\# \text{ g C}_2\text{H}_4 = 0.01 \text{ kg H}_2\text{O}\left(\frac{1000 \text{ g H}_2\text{O}}{1 \text{ kg H}_2\text{O}}\right)\left(\frac{1 \text{ mol H}_2\text{O}}{18.02 \text{ g H}_2\text{O}}\right)\left(\frac{1 \text{ mol C}_2\text{H}_4}{1 \text{ mol H}_2\text{O}}\right)$$
$$\left(\frac{56.02 \text{ g C}_2\text{H}_4}{1 \text{ mol C}_2\text{H}_4}\right)\left(\frac{1 \text{ kg C}_2\text{H}_4}{1000 \text{ g C}_2\text{H}_4}\right) = 0.031 \text{ kg C}_2\text{H}_4$$

(b) $$\# \text{ g C}_2\text{H}_5\text{OH} = (0.010 \text{ kg H}_2\text{O})\left(\frac{1000 \text{ g H}_2\text{O}}{1 \text{ kg H}_2\text{O}}\right)\left(\frac{1 \text{ mol H}_2\text{O}}{18.02 \text{ g H}_2\text{O}}\right)$$
$$\times\left(\frac{1 \text{ mol C}_2\text{H}_5\text{OH}}{1 \text{ mol H}_2\text{O}}\right)\left(\frac{46.08 \text{ g C}_2\text{H}_5\text{OH}}{1 \text{ mole C}_2\text{H}_5\text{OH}}\right) = 26 \text{ g C}_2\text{H}_5\text{OH}$$

4.116 $3AgNO_3 + FeCl_3 \rightarrow 3AgCl + Fe(NO_3)_3$
Calculate the amount of $FeCl_3$ that are required to react completely with all of the available silver nitrate:

$$\# \text{ g Fe Cl}_3 = (18.0 \text{ g AgNO}_3\left(\frac{1 \text{ mole AgNO}_3}{169.87 \text{ g AgNO}_3}\right)\left(\frac{1 \text{ mole FeCl}_3}{3 \text{ mole AgNO}_3}\right)\left(\frac{162.21 \text{ g FeCl}_3}{1 \text{ mole FeCl}_3}\right)$$
$$= 5.73 \text{ g FeCl}_3$$

Since more than this minimum amount is available, $FeCl_3$ is present in excess, and $AgNO_3$ must be the limiting reactant.

We know that only 5.73 g $FeCl_3$ will be used. Therefore, the amount left unused is: 32.4 g total – 5.73 g used = 26.7 g $FeCl_3$

4.117 First, calculate the amount of H_2O needed to completely react with the available ClO_2;

$$\# \text{ g H}_2\text{O} = 142.0 \text{ g ClO}_2\left(\frac{1 \text{ mole ClO}_2}{67.45 \text{ g ClO}_2}\right)\left(\frac{3 \text{ moles H}_2\text{O}}{6 \text{ moles ClO}_2}\right)\left(\frac{18.02 \text{ g H}_2\text{O}}{1 \text{ mole H}_2\text{O}}\right) = 18.97 \text{ g H}_2\text{O}$$

So, there is excess H_2O present. The amount that remains is 38.0 g – 18.97 g = 19.0 g H_2O.

4.118 First calculate the number of moles of water that are needed to react completely with the given amount of NO_2:

$$\# \text{ g H}_2\text{O} = 0.0010 \text{ g NO}_2\left(\frac{1 \text{ mol NO}_2}{46.01 \text{ g NO}_2}\right)\left(\frac{1 \text{ mol H}_2\text{O}}{2 \text{ mol NO}_2}\right)\left(\frac{18.02 \text{ g H}_2\text{O}}{1 \text{ mol H}_2\text{O}}\right) = 2.0 \times 10^4 \text{ g H}_2\text{O}$$

Since this is less than the amount of water that is supplied, the limiting reactant must be NO_2. Therefore, to calculate the amount of HNO_3:

$$\text{\# g } HNO_3 = 0.0010 \text{ g } NO_2\left(\frac{1 \text{ mol } NO_2}{46.01 \text{ g } NO_2}\right)\left(\frac{3 \text{ mol } HNO_3}{2 \text{ mol } NO_2}\right)\left(\frac{63.02 \text{ g } HNO_3}{1 \text{ mol } HNO_3}\right) = 2.1 \times 10^3 \text{ g } HNO_3$$

4.119 (a) First calculate the number of moles of water that are needed to react completely with the given amount of PCl_5:

$$\text{\# moles } H_2O = (0.360 \text{ moles } PCl_5)\left(\frac{4 \text{ mole } H_2O}{1 \text{ moles } PCl_5}\right) = 1.44 \text{ moles } H_2O$$

Since this is less than the amount of water that is supplied, the limiting reactant must be PCl_5. This can be confirmed by the following calculation:

$$\text{\# moles } PCl_5 = (2.88 \text{ moles } H_2O)\left(\frac{1 \text{ moles } PCl_5}{4 \text{ mole } H_2O}\right) = 0.720 \text{ moles } PCl_5$$

which also demonstrates that the limiting reactant is PCl_5.

(b) $$\text{\# g HCl} = (0.360 \text{ moles } PCl_5)\left(\frac{5 \text{ mole HCl}}{1 \text{ moles } PCl_5}\right)\left(\frac{36.5 \text{ g HCl}}{1 \text{ mole HCl}}\right) = 65.7 \text{ g HCl}$$

4.120 First determine the theoretical yield:

$$\text{\# g } BaSO_4 = (75.00 \text{ g } Ba(NO_3)_2\left(\frac{1 \text{ mole } Ba(NO_3)_2}{261.34 \text{ g } Ba(NO_3)_2}\right)\left(\frac{1 \text{ mole } BaSO_4}{1 \text{ mole } Ba(NO_3)_2}\right)$$
$$\times\left(\frac{233.39 \text{ g } BaSO_4}{1 \text{ mole } BaSO_4}\right) = 66.98 \text{ g } BaSO_4$$

Then calculate a % yield:

$$\% \text{ yield} = \frac{\text{actual yield}}{\text{theoretical yield}} \times 100 = \frac{64.45 \text{ g}}{66.98 \text{ g}} \times 100 = 96.22\ \%$$

4.121 The theoretical yield is

$$\text{\# g } Na_2CO_3 = 120 \text{ g NaCl}\left(\frac{1 \text{ mole NaCl}}{58.44 \text{ g NaCl}}\right)\left(\frac{1 \text{ mole } NaHCO_3}{1 \text{ mole NaCl}}\right)\left(\frac{1 \text{ mole } Na_2CO_3}{2 \text{ mole } NaHCO_3}\right)$$
$$\times\left(\frac{105.99 \text{ g } Na_2CO_3}{1 \text{ mole } Na_2CO_3}\right) = 108.8 \text{ g } Na_2CO_3$$

$$\% \text{ yield} = \frac{\text{actual yield}}{\text{theoretical yield}} \times 100 = \frac{85.4 \text{ g}}{108.8 \text{ g}} \times 100 = 78.5\ \%$$

4.122 First, determine how much H_2SO_4 is needed to completely react with the $AlCl_3$

$$\text{\# g } H_2SO_4 = 25 \text{ g } AlCl_3\left(\frac{1 \text{ mole } AlCl_3}{133.34 \text{ g } AlCl_3}\right)\left(\frac{3 \text{ mole } H_2SO_4}{2 \text{ mole } AlCl_3}\right)$$
$$\times\left(\frac{98.08 \text{ g } H_2SO_4}{1 \text{ mole } H_2SO_4}\right) = 27.58 \text{ g } H_2SO_4$$

There is an excess of H_2SO_4 present.

Determine the theoretical yield:

$$\# \text{ g } Al_2(SO_4)_3 = 25.00 \text{ g } AlCl_3\left(\frac{1 \text{ mole } AlCl_3}{133.34 \text{ g } AlCl_3}\right)\left(\frac{1 \text{ mole } Al_2(SO_4)_3}{2 \text{ mole } AlCl_3}\right)$$

$$\times\left(\frac{342.15 \text{ g } Al_2(SO_4)_3}{1 \text{ mole } Al_2(SO_4)_3}\right) = 32.07 \text{ g } Al_2(SO_4)_3$$

$$\% \text{ yield } = \frac{\text{actual yield}}{\text{theoretical yield}} \times 100 = \frac{28.46 \text{ g}}{32.07 \text{ g}} \times 100 = 88.74\ \%$$

4.123 Assume there is excess oxygen present and determine the theoretical yield of carbon dioxide.

$$\# \text{ g } CO_2 = (6.40 \text{ g } CH_3OH)\left(\frac{1 \text{ mol } CH_3OH}{32.04 \text{ g } CH_3OH}\right)\left(\frac{2 \text{ mol } CO_2}{2 \text{ mol } CH_3OH}\right)\left(\frac{44.01 \text{ g } CO_2}{1 \text{ mol } CO_2}\right) = 8.79 \text{ g } CO_2$$

$$\% \text{ yield } = \frac{6.12 \text{ g}}{8.79\text{g}} \times 100\% = 69.6\%$$

4.124 If the yield for this reaction is only 71 % and we need to have 11.5 g of product, we will attempt to make 16 g of product. This is determined by dividing the actual yield by the percent yield. Recall that;

$\% \text{ yield } = \frac{\text{actual yield}}{\text{theoretical yield}} \times 100$. If we rearrange this equation we can see that

$\text{theoretical yield } = \frac{\text{actual yield}}{\% \text{ yield}} \times 100$. Substituting the values from this problem gives the 16 g of product mentioned above.

$$\# \text{ g } C_7H_8 = 16 \text{ g } KC_7H_5O_2\left(\frac{1 \text{ mole } KC_7H_5O_2}{160.21 \text{ g } KC_7H_5O_2}\right)\left(\frac{1 \text{ mole } C_7H_8}{1 \text{ mole } KC_7H_5O_2}\right)\left(\frac{92.14 \text{ g } C_7H_8}{1 \text{ mole } C_7H_8}\right) = 9.2 \text{ g } C_7H_8$$

4.125 Note that we have a certain number of moles of MnI_2 present. According to the problem statement, we will only prepare 75% of this number of moles of MnF_3. Calculate the mass of F_2 accounting for the yield,

$$\# \text{ g } F_2 = (12.0 \text{ g } MnI_2)\left(\frac{1 \text{ mole } MnI_2}{308.83 \text{ g } MnI_2}\right)\left(\frac{13 \text{ mole } F_2}{2 \text{ mole } MnI_2}\right)\left(\frac{38.00 \text{ g } F_2}{1 \text{ mole } F_2}\right)(0.75)$$

$$= 7.20 \text{ g } F_2$$

Additional Exercises

4.126 1.0 % of 263 tons = 2.6 tons Hg

Since there are 200.59 g Hg in one mol of $(CH_3)_2Hg$ (230.67 g), it follows that:

$$\# \text{ lbs } (CH_3)_2Hg = (2.6 \text{ tons Hg})\left(\frac{230.67 \text{ tons } (CH_3)_2Hg}{200.59 \text{ tons Hg}}\right)$$

$$\times\left(\frac{2{,}000 \text{ pounds } (CH_3)_2Hg}{1 \text{ ton } (CH_3)_2Hg}\right) = 6{,}000 \text{ pounds (or } 6.0\times10^3 \text{ pounds)}$$

4.127 (a) $\text{FM } (CH_3)_2Hg = (2 \text{ mol C})(12.01 \text{ g C/mol C}) + (6 \text{ mol H})(1.0079 \text{ g H/mol H})$
$+ (1 \text{ mol Hg})(200.59 \text{ g Hg/mol Hg}) = 230.66 \text{ g } (CH_3)_2Hg/\text{mol}$

(b) $$\# \text{ g } (CH_3)_2Hg = 25.0 \text{ g Hg}\left(\frac{1 \text{ mol Hg}}{200.59 \text{ g Hg}}\right)$$

$$\times\left(\frac{1 \text{ mol } (CH_3)_2Hg}{1 \text{ mol Hg}}\right)\left(\frac{230.66 \text{ g } (CH_3)_2Hg}{1 \text{ mol } (CH_3)_2Hg}\right) = 28.7 \text{ g } (CH_3)_2Hg$$

4.128 Assume 100 g of magnesium boron compound, therefore there are 52.9 g of Mg and 47.1 g of B.

$$\# \text{ mol of Mg} = 52.9 \text{ g Mg}\left(\frac{1 \text{ mol Mg}}{24.305 \text{ g Mg}}\right) = 2.18 \text{ mol Mg}$$

$$\# \text{ mol of B} = 47.1\text{g B}\left(\frac{1 \text{ mol B}}{10.811 \text{ g B}}\right) = 4.36 \text{ mol Mg}$$

Divide the number of moles of each element by the least number of moles:

$$\frac{2.18 \text{ mol Mg}}{2.18 \text{ mol Mg}} = 1$$

$$\frac{4.26 \text{ mol B}}{2.18 \text{ mol Mg}} = 2$$

Formula is MgB_2.

4.129 Volume of 1 mole of M&Ms:

$$0.5 \text{ cm}^3\text{/M\&Ms}\left(\frac{1 \text{ m}}{100 \text{ cm}}\right)^3 6.022 \times 10^{23} \text{ M\&Ms} = 3.01 \times 10^{23} \text{ m}^3$$

No, 18 tractor trailers would not be enough.

4.130 $$\# \text{ yrs} = (6.022 \times 10^{23} \text{ pennies})\left(\frac{1 \text{ dollar}}{100 \text{ pennies}}\right)\left(\frac{1 \text{ second}}{5.00 \times 10^8 \text{ dollars}}\right)$$

$$\times\left(\frac{1 \text{ min}}{60 \text{ sec}}\right)\left(\frac{1 \text{ hr}}{60 \text{ min}}\right)\left(\frac{1 \text{ day}}{24 \text{ hrs}}\right)\left(\frac{1 \text{ yr}}{365 \text{ days}}\right) = 382{,}000 \text{ yrs}$$

4.131 First, we determine the number of grams of chlorine in the original sample:

$$\# \text{ g Cl} = (0.3383 \text{ g AgCl}(\left(\frac{1 \text{ mole AgCl}}{143.32 \text{ g AgCl}}\right)\left(\frac{1 \text{ mole Cl}}{1 \text{ mole AgCl}}\right)\left(\frac{35.453 \text{ g Cl}}{1 \text{ mole Cl}}\right) = 0.08369 \text{ g Cl}$$

The mass of Cr in the original sample is thus 0.1246 – 0.08369 g = 0.0409 g Cr. Converting to moles, we have:

$$\text{for Cl: } 0.08369 \text{ g}\left(\frac{1 \text{ mol Cl}}{35.453 \text{ g Cl}}\right) = 2.361 \times 10^{-3} \text{ mol Cl}$$

$$\text{for Cr: } 0.0409 \text{ g Cr}\left(\frac{1 \text{ mol Cr}}{51.996 \text{ g Cr}}\right) = 7.866 \times 10^{-4} \text{ mol Cr}$$

The relative mole amounts are:

for Cl: 2.361×10^{-3} mol / 7.87×10^{-4} mol = 3.00

for Cr: 7.87×10^{-4} mol / 7.87×10^{-4} mol = 1.00

The empirical formula is thus $CrCl_3$.

4.132 First determine the percentage by weight of each element in the respective original samples. This is done by determining the mass of the element in question present in each of the original samples. The percentage by weight of each element in the unknown will be the same as the values we calculate.

$$\text{\# g Ca} = (0.160 \text{ g CaCO}_3)\left(\frac{1 \text{ mole CaCO}_3}{100.09 \text{ g CaCO}_3}\right)\left(\frac{1 \text{ mole Ca}}{1 \text{ mole CaCO}_3}\right)\left(\frac{40.1 \text{ g Ca}}{1 \text{ mole Ca}}\right) = 0.0641 \text{ g Ca}$$

% Ca = (0.0641/0.250) × 100% = 25.6 % Ca

$$\text{\# g S} = (0.344 \text{ g BaSO}_4)\left(\frac{1 \text{ mole BaSO}_4}{233.8 \text{ g BaSO}_4}\right)\left(\frac{1 \text{ mole S}}{1 \text{ mole BaSO}_4}\right)\left(\frac{32.07 \text{ g S}}{1 \text{ mole S}}\right) = 0.0472 \text{ g S}$$

% S = (0.0472/0.115) × 100% = 41.0 % S

$$\text{\# g N} = (0.155 \text{ g NH}_3\left(\frac{1 \text{ mole NH}_3}{17.03 \text{ g NH}_3}\right)\left(\frac{1 \text{ mole N}}{1 \text{ mole NH}_3}\right)\left(\frac{14.01 \text{ g N}}{1 \text{ mole N}}\right) = 0.128 \text{ g N}$$

% N = (0.128/0.712) × 100% = 18.0 % N

% C = 100.0 – (25.6 + 41.0 + 18.0) = 15.4 % C. Next, we assume 100 g of the compound, and convert these weight percentages into mole amounts:

$$\text{\# moles Ca} = (25.6 \text{ g Ca})\left(\frac{1 \text{ mole Ca}}{40.08 \text{ g Ca}}\right) = 0.639 \text{ moles Ca}$$

$$\text{\# moles S} = (41.0 \text{ g S})\left(\frac{1 \text{ mole S}}{32.07 \text{ g S}}\right) = 1.28 \text{ moles S}$$

$$\text{\# moles N} = (18.0 \text{ g N})\left(\frac{1 \text{ mole N}}{14.07 \text{ g N}}\right) = 1.28 \text{ moles N}$$

$$\text{\# moles C} = (15.4 \text{ g C})\left(\frac{1 \text{ mole C}}{12.01 \text{ g C}}\right) = 1.28 \text{ moles C}$$

Dividing each of these mole amounts by the smallest, we have:

For Ca: 0.639 mol / 0.639 mol = 1.00
For S: 1.28 mol / 0.639 mol = 2.00
For N: 1.28 mol / 0.639 mol = 2.00
For C: 1.28 mol / 0.639 mol = 2.00

The empirical formula is therefore $CaC_2S_2N_2$, and the mass of the empirical unit is Ca + 2S + 2N + 2C = 156 g/mol. Since the molecular mass is the same as the empirical mass, the molecular formula is $CaC_2S_2N_2$.

4.133 (a) One mole of N_2, 2 moles of H_2O and 1/2 mole of O_2 for a total of 3 1/2 moles of gases.

(b)

$$\text{\# mol of gases} = (1.00 \text{ ton NH}_4\text{NO}_3)\left(\frac{2000 \text{ lb}}{1 \text{ ton}}\right)\left(\frac{453.59 \text{ g}}{1 \text{ lb}}\right)$$
$$\times\left(\frac{1 \text{ mole NH}_4\text{NO}_3}{80.04 \text{ g NH}_4\text{NO}_3}\right)\left(\frac{3.5 \text{ moles gas}}{1 \text{ mole NH}_4\text{NO}_3}\right) = 3.97\times10^4 \text{ mol gas}$$

4.134 $$\text{\# g (NH}_2)_2\text{CO} = 6.00 \text{ g N}\left(\frac{1 \text{ mole N}}{14.007 \text{ g N}}\right)\left(\frac{1 \text{ mole (NH}_2)_2\text{CO}}{2 \text{ moles N}}\right)\left(\frac{60.06 \text{ g (NH}_2)_2\text{CO}}{1 \text{ mole (NH}_2)_2\text{CO}}\right) = 12.9 \text{ g (NH}_2)_2\text{CO}$$

4.135 $\# \text{ g Na}_2\text{C}_2\text{O}_4 = (125 \text{ g C}_6\text{H}_6)\left(\frac{1 \text{ mole C}_6\text{H}_6}{78.11 \text{ g C}_6\text{H}_6}\right)\left(\frac{6 \text{ moles C}}{1 \text{ moles C}_6\text{H}_6}\right)$

$$\times\left(\frac{1 \text{ mole Na}_2\text{C}_2\text{O}_4}{2 \text{ moles C}}\right)\left(\frac{134.00 \text{ g Na}_2\text{C}_2\text{O}_4}{1 \text{ mole Na}_2\text{C}_2\text{O}_4}\right) = 643 \text{ g NaC}_2\text{O}_4$$

4.136 Assume the hydrogen is the limiting reactant.

$$\text{lb O}_2 = 227{,}641 \text{ lb H}_2\left(\frac{453.59237 \text{ g}}{1 \text{ lb}}\right)\left(\frac{1 \text{ mol H}_2}{2.01588 \text{ g H}_2}\right)\left(\frac{1 \text{ mol O}_2}{2 \text{ mol H}_2}\right)$$

$$\times\left(\frac{31.9988 \text{ g O}_2}{1 \text{ mol O}_2}\right)\left(\frac{1 \text{ lb O}_2}{453.59237 \text{ g O}_2}\right) = 1{,}806{,}714 \text{ lb O}_2$$

Since this is more than the amount of O_2 that is supplied, the limiting reactant must be O_2.Next calculate the amount of H_2 needed to react completely with all of the available O_2.

$$\text{lb H}_2 = 1{,}361{,}936 \text{ lb O}_2\left(\frac{453.59237 \text{ g}}{1 \text{ lb}}\right)\left(\frac{1 \text{ mol O}_2}{31.9988 \text{ g O}_2}\right)\left(\frac{2 \text{ mol H}_2}{1 \text{ mol O}_2}\right)$$

$$\times\left(\frac{2.01588 \text{ g H}_2}{1 \text{ mol H}_2}\right)\left(\frac{1 \text{ lb H}_2}{453.59237 \text{ g H}_2}\right) = 171{,}600 \text{ lb H}_2$$

Since only 171,600 lb. of H_2 reacted, there are 227,641 lb. – 171,600 lb. = 56,041 lb. of unreacted H_2.

4.137 Since 5.00 g represents 86.0 % of the required amount, we can solve for the amount that should be made: 5.00 g = 86.0 % of X; X = 5.81 g $Pb(NO_3)_2$.

$$\# \text{ g PbO} = (5.81 \text{ g Pb(NO}_3)_2)\left(\frac{1 \text{ mole Pb(NO}_3)_2}{331.21 \text{ g Pb(NO}_3)_2}\right)\left(\frac{1 \text{ mole PbO}}{1 \text{ mole Pb(NO}_3)_2}\right)\left(\frac{223.2 \text{ g PbO}}{1 \text{ mole PbO}}\right) = 3.92 \text{ g PbO}$$

4.138 $\# \text{ g F} = 1.0 \times 10^{-9} \text{ g Cl}\left(\frac{1 \text{ mol Cl}}{35.453 \text{ g Cl}}\right)\left(\frac{1 \text{ mol F}}{1 \text{ mol Cl}}\right)\left(\frac{18.998 \text{ g F}}{1 \text{ mol F}}\right) = 5.4 \times 10^{-10} \text{ g F}$

Practice Exercises

5.1 a) $MgCl_2(s) \rightarrow Mg^{2+}(aq) + 2Cl^-(aq)$
b) $Al(NO_3)_3(s) \rightarrow Al^{3+}(aq) + 3NO_3^-(aq)$
c) $Na_2CO_3(s) \rightarrow 2Na^+(aq) + CO_3^{2-}(aq)$

5.2 molecular: $CdCl_2(aq) + Na_2S(aq) \rightarrow CdS(s) + 2NaCl(aq)$
ionic: $Cd^{2+}(aq) + 2Cl^-(aq) + 2Na^+(aq) + S^{2-}(aq) \rightarrow CdS(s) + 2Na^+(aq) + 2Cl^-(aq)$
net ionic: $Cd^{2+}(aq) + S^{2-}(aq) \rightarrow CdS(s)$

5.3 a) molecular: $AgNO_3(aq) + NH_4Cl(aq) \rightarrow AgCl(s) + NH_4NO_3(aq)$
ionic: $Ag^+(aq) + NO_3^-(aq) + NH_4^+(aq) + Cl^-(aq) \rightarrow AgCl(s) + NH_4^+(aq) + NO_3^-(aq)$
net ionic: $Ag^+(aq) + Cl^-(aq) \rightarrow AgCl(s)$

b) molecular: $Na_2S(aq) + Pb(C_2H_3O_2)_2(aq) \rightarrow 2NaC_2H_3O_2(aq) + PbS(s)$
ionic: $2Na^+(aq) + S^{2-}(aq) + Pb^{2+}(aq) + 2C_2H_3O_2^-(aq) \rightarrow 2Na^+(aq) + 2C_2H_3O_2^-(aq) + PbS(s)$
net ionic: $S^{2-}(aq) + Pb^{2+}(aq) \rightarrow PbS(s)$

5.4 $HCHO_2(aq) + H_2O \rightarrow H_3O^+(aq) + CHO_2^-(aq)$

5.5 $H_3C_6H_5O_7(s) + H_2O \rightarrow H_3O^+(aq) + H_2C_6H_5O_7^-(aq)$
$H_2C_6H_5O_7^-(aq) + H_2O \rightarrow H_3O^+(aq) + HC_6H_5O_7^{2-}(aq)$
$HC_6H_5O_7^{2-}(aq) + H_2O \rightarrow H_3O^+(aq) + C_6H_5O_7^{3-}(aq)$

5.6 $HONH_2(aq) + H_2O \rightarrow HONH_3^+(aq) + OH^-(aq)$

5.7 HF: Hydrofluoric acid, sodium salt = sodium fluoride (NaF)
HBr: Hydrobromic acid, sodium salt = sodium bromide (NaBr)

5.8 Sodium arsenate

5.9 Calcium formate

5.10 $H_3PO_4(aq) + NaOH(aq) \rightarrow NaH_2PO_4(aq) + H_2O$
$NaH_2PO_4(aq) + NaOH(aq) \rightarrow Na_2HPO_4(aq) + H_2O$
$Na_2HPO4(aq) + NaOH(aq) \rightarrow Na_3PO_4(aq) + H_2O$

5.11 $NaHSO_3$ (sodium hydrogen sulfite)

5.12 $HNO_2(aq) + H_2O \rightleftharpoons H_3O^+(aq) + NO_2^-(aq)$

5.13 $CH_3NH_2(aq) + H_2O \rightleftharpoons CH_3NH_3^+(aq) + OH^-(aq)$

5.14 molecular: $2HCl(aq) + Ca(OH)_2(aq) \rightarrow CaCl_2(aq) + 2H_2O$
ionic: $2H^+(aq) + 2Cl^-(aq) + Ca^{2+}(aq) + 2OH^-(aq) \rightarrow Ca^{2+}(aq) + 2Cl^-(aq) + 2H_2O$
net ionic: $2H^+(aq) + 2OH^-(aq) \rightarrow 2H_2O$

5.15 a) molecular: $HCl(aq) + KOH(aq) \rightarrow H_2O + KCl(aq)$
ionic: $H^+(aq) + Cl^-(aq) + K^+(aq) + OH^-(aq) \rightarrow H_2O + K^+(aq) + Cl^-(aq)$
net ionic: $H^+(aq) + OH^-(aq) \rightarrow H_2O$

b) molecular: $HCHO_2(aq) + LiOH(aq) \rightarrow H_2O + LiCHO_2(aq)$
ionic: $HCHO_2(aq) + Li^+(aq) + OH^-(aq) \rightarrow H_2O + Li^+(aq) + CHO_2^-(aq)$
net ionic: $HCHO_2(aq) + OH^-(aq) \rightarrow H_2O + CHO_2^-(aq)$

c) molecular: $N_2H_4(aq) + HCl(aq) \rightarrow N_2H_5Cl(aq)$
ionic: $N_2H_4(aq) + H^+(aq) + Cl^-(aq) \rightarrow N_2H_5^+(aq) + Cl^-(aq)$
net ionic: $N_2H_4(aq) + H^+(aq) \rightarrow N_2H_5^+(aq)$

5.16 molecular: $CH_3NH_2(aq) + HCHO_2(aq) \rightarrow CH_3NH_3CHO_2(aq)$
ionic: $CH_3NH_2(aq) + HCHO_2(aq) \rightarrow CH_3NH_3^+(aq) + CHO_2^-(aq)$
net ionic: $CH_3NH_2(aq) + HCHO_2(aq) \rightarrow CH_3NH_3^+(aq) + CHO_2^-(aq)$

5.17 molecular: $Al(OH)_3(s) + 3HCl(aq) \rightarrow AlCl_3(aq) + 3H_2O$
ionic: $Al(OH)_3(s) + 3H^+(aq) + 3Cl^-(aq) \rightarrow Al^{3+}(aq) + 3Cl^-(aq) + 3H_2O$
net ionic: $Al(OH)_3(s) + 3H^+(aq) \rightarrow Al^{3+}(aq) + 3H_2O$

5.18 a) Formic acid, a weak acid will form.
Net ionic equation: $H^+(aq) + CHO_2^-(aq) \rightleftharpoons HCHO_2(aq)$
b) Carbonic acid will form and it will further dissociate to water and carbon dioxide:
$CuCO_3(s) + 2H^+(aq) \rightarrow 2CO_2(g) + 2H_2O + Cu^{2+}(aq)$
c) NR
d) Insoluble nickel hydroxide will precipitate.
$Ni^{2+}(aq) + 2OH^-(aq) \rightarrow Ni(OH)_2(s)$

5.19 $$\#\text{ mol Na}_2\text{SO}_4 = (3.550\text{ g Na}_2\text{SO}_4)\left(\frac{1\text{ mol Na}_2\text{SO}_4}{142.1\text{ g Na}_2\text{SO}_4}\right) = 0.02498\text{ mol Na}_2\text{SO}_4$$

$$\#\text{ L solution} = (100.0\text{ mL})\left(\frac{1\text{ L}}{1000\text{ mL}}\right) = 0.1000\text{ L solution}$$

$$M = \left(\frac{\text{moles solute}}{\text{L solution}}\right) = \left(\frac{0.02498\text{ mol Na}_2\text{SO}_4}{0.1000\text{ L solution}}\right) = 0.2498\text{ M}$$

5.20 $$\#\text{ mL solution} = (0.0500\text{ mol KCl})\left(\frac{1\text{ L solution}}{0.150\text{ mol KCl}}\right)\left(\frac{1000\text{ mL solution}}{1\text{ L solution}}\right) = 333\text{ mL}$$

5.21 $$\#\text{ g AgNO}_3 = (250\text{ mL solution})\left(\frac{1\text{ L solution}}{1000\text{ mL solution}}\right)\left(\frac{0.0125\text{ mol AgNO}_3}{1\text{ L solution}}\right) \times \left(\frac{169.90\text{ g AgNO}_3}{1\text{ mol AgNO}_3}\right) = 0.53\text{ g AgNO}_3$$

5.22 $(V_{dil})(M_{dil}) = (V_{conc})(M_{conc})$

$(100\text{ mL})(0.125M) = (V_{conc})(0.500M)$

$V_{conc} = (100\text{ mL})(0.125M)/(0.500M) = 25.0\text{ mL}$

Therefore, mix 25.0 mL of 0.500 M H_2SO_4 with enough water to make 100 mL of total solution.

5.23 $$\#\text{ mL NaOH} = (15.4\text{ mL H}_2\text{SO}_4)\left(\frac{1\text{ L H}_2\text{SO}_4}{1000\text{ mL H}_2\text{SO}_4}\right)\left(\frac{0.108\text{ mol H}_2\text{SO}_4}{1\text{ L H}_2\text{SO}_4}\right) \times \left(\frac{2\text{ mol NaOH}}{1\text{ mol H}_2\text{SO}_4}\right)\left(\frac{1\text{ L NaOH}}{0.124\text{ mol NaOH}}\right)\left(\frac{1000\text{ mL NaOH}}{1\text{ L NaOH}}\right) = 26.8\text{ mL NaOH}$$

5.24 $FeCl_3 \rightarrow Fe^{3+} + 3Cl^-$

$$\text{M } Fe^{3+} = \left(\frac{0.40 \text{ mol } FeCl_3}{1 \text{ L } FeCl_3 \text{ soln}}\right)\left(\frac{1 \text{ mol } Fe^{3+}}{1 \text{ mol } FeCl_3}\right) = 0.40 \text{ M } Fe^{3+}$$

$$\text{M } Cl^- = \left(\frac{0.40 \text{ mol } FeCl_3}{1 \text{ L } FeCl_3 \text{ soln}}\right)\left(\frac{3 \text{ mol } Cl^-}{1 \text{ mol } FeCl_3}\right) = 1.2 \text{ M } Cl^-$$

5.25 $$\text{M } Na^+ = \left(\frac{0.250 \text{ mol } PO_4^{3-}}{1 \text{ L } Na_3PO_4 \text{ soln}}\right)\left(\frac{3 \text{ mol } Na^+}{1 \text{ mol } PO_4^{3-}}\right) = 0.750 \text{ M } Na^+$$

5.26 The balanced net ionic equation is: $Fe^{2+}(aq) + 2OH^-(aq) \rightarrow Fe(OH)_2(s)$.
First determine the number of moles of Fe^{2+} present.

$$\text{\# moles } Fe^{2+} = (60.0 \text{ mL } FeCl_2 \text{ solution})\left(\frac{0.250 \text{ mol } FeCl_2}{1000 \text{ mL solution}}\right)\left(\frac{1 \text{ mol } Fe^{2+}}{1 \text{ mol } FeCl_2}\right)$$

$$= 1.50 \times 10^{-2} \text{ mol } Fe^{2+}$$

Now, determine the amount of KOH needed to react with the Fe^{2+}.

$$\text{\# mL KOH} = (1.50 \times 10^{-2} \text{ mol } Fe^{2+})\left(\frac{2 \text{ mol } OH^-}{1 \text{ mol } Fe^{2+}}\right)\left(\frac{1 \text{ mol KOH}}{1 \text{ mol } OH^-}\right)\left(\frac{1000 \text{ mL solution}}{0.500 \text{ mol KOH}}\right)$$

$$= 60.0 \text{ mL KOH}$$

5.27 The net ionic equation is $Ba^{2+}(aq) + SO_4^{2-}(aq) \rightarrow BaSO_4(s)$
First, determine the initial number of moles of Ba^{2+} ion that are present:

$$\text{\# mol } Ba^{2+} = (20.0 \text{ mL } BaCl_2 \text{ soln})\left(\frac{0.600 \text{ mol } BaCl_2}{1000 \text{ mL } BaCl_2 \text{ soln}}\right)\left(\frac{1 \text{ mol } Ba^{2+}}{1 \text{ mol } BaCl_2}\right)$$

$$= 1.20 \times 10^{-2} \text{ mol } Ba^{2+}$$

Next, determine the initial number of moles of sulfate ion that are present:

$$\text{\# mol } SO_4^{2-} = (30.0 \text{ mL } MgSO_4 \text{ soln})\left(\frac{0.500 \text{ mol } MgSO_4}{1000 \text{ mL } MgSO_4 \text{ soln}}\right)\left(\frac{1 \text{ mol } SO_4^{2-}}{1 \text{ mol } MgSO_4}\right)$$

$$= 1.50 \times 10^{-2} \text{ mol } SO_4^{2-}$$

Now determine the number of moles of barium ion that are required to react with this much sulfate ion, and compare the result to the amount of barium ion that is available:

$$\text{\# mol } Ba^{2+} = (1.50 \times 10^{-2} \text{ mol } SO_4^{2-})\left(\frac{1 \text{ mol } Ba^{2+}}{1 \text{ mol } SO_4^{2-}}\right)$$

$$= 1.50 \times 10^{-2} \text{ mol } Ba^{2+}$$

Since there is not this much Ba^{2+} available according to the above calculation, then we can conclude that Ba^{2+} must be the limiting reactant, and that subsequent calculations should be based on the number of moles of it that are present:

Since this reaction is 1:1, we know that 1.20×10^{-2} mole of $BaSO_4$ will be formed.

If we assume that the $BaSO_4$ is completely insoluble, then the concentration of barium ion will be essentially zero. The concentrations of the other ions are determined as follows:

$$\# \text{ M Cl}^- = \frac{(20.0 \text{ mL BaCl}_2 \text{ soln})\left(\frac{0.600 \text{ mol BaCl}_2}{1000 \text{ mL BaCl}_2 \text{ soln}}\right)\left(\frac{2 \text{ mol Cl}^-}{1 \text{ mol BaCl}_2}\right)}{((20.0 + 30.0) \text{ mL soln})\left(\frac{1 \text{ L soln}}{1000 \text{ mL soln}}\right)}$$

$$= 0.480 \text{ M Cl}^-$$

$$\# \text{ M Mg}^{2+} = \frac{(30.0 \text{ mL MgSO}_4 \text{ soln})\left(\frac{0.500 \text{ mol MgSO}_4}{1000 \text{ mL MgSO}_4 \text{ soln}}\right)\left(\frac{1 \text{ mol Mg}^{2+}}{1 \text{ mol MgSO}_4}\right)}{((30.0 + 20.0) \text{ mL soln})\left(\frac{1 \text{ L soln}}{1000 \text{ mL soln}}\right)}$$

$$= 0.300 \text{ M Mg}^{2+}$$

For sulfate, we subtract the amount that reacted with the Ba^{2+}:
$\# \text{ mol } SO_4^{2-} = 1.50 \times 10^{-2} \text{ mol} - 1.20 \times 10^{-2} \text{ mol} = 3.0 \times 10^{-3} \text{ mol}$

This allows a calculation of the final sulfate concentration:

$$\# \text{ M SO}_4^{2-} = \frac{3.0 \times 10^{-3} \text{ mol SO}_4^{2-}}{((30.0 + 20.0) \text{ mL soln})\left(\frac{1 \text{ L soln}}{1000 \text{ mL soln}}\right)}$$

$$= 6.0 \times 10^{-2} \text{ M SO}_4^{2-}$$

5.28 a) $$\# \text{ mol Ca}^{2+} = (0.736 \text{ g CaSO}_4)\left(\frac{1 \text{ mol CaSO}_4}{136.14 \text{ g CaSO}_4}\right)\left(\frac{1 \text{ mol Ca}^{2+}}{1 \text{ mol CaSO4}}\right)$$

$$= 5.41 \times 10^{-3} \text{ mol Ca}^{2+}$$

b) Since all of the Ca^{2+} is precipitated as $CaSO_4$, there were originally 5.41×10^{-3} moles of Ca^{2+} in the sample.

c) All of the Ca^{2+} comes from $CaCl_2$, so there were 5.41×10^{-3} moles of $CaCl_2$ in the sample.

d) $$\# \text{ g CaCl2} = (5.41 \times 10^{-3} \text{ mol CaCl}_2)\left(\frac{110.98 \text{ g CaCl}_2}{1 \text{ mol CaCl}_2}\right) = 0.600 \text{ g CaCl}_2$$

e) $$\% \text{ CaCl}_2 = \frac{0.600 \text{ g CaCl}_2}{2.000 \text{ g sample}} \times 100 = 30.0 \% \text{ CaCl}_2$$

5.29

$$\text{M H}_s\text{S}_4 = \frac{(36.42 \text{ mL NaOH soln})\left(\frac{1 \text{ L NaOH soln}}{1000 \text{ mL NaOH soln}}\right)\left(\frac{0.147 \text{ mol NaOH}}{1 \text{ L NaOH soln}}\right)\left(\frac{1 \text{ mol H}_2\text{SO}_4}{2 \text{ mol NaOH}}\right)}{(15.00 \text{ ml H}_2\text{SO}_4 \text{ soln})\left(\frac{1 \text{ L H}_2\text{SO}_4 \text{ soln}}{1000 \text{ mL H}_2\text{SO}_4 \text{ soln}}\right)}$$

$$= 0.178 \text{ M H}_s\text{SO}_4$$

5.30

$$\text{M HCl} = \left[\frac{(11.00\text{ mL})\left(\dfrac{1\text{ L KOH soln}}{1000\text{ mL NaOH soln}}\right)\left(\dfrac{0.100\text{ mol KOH soln}}{1\text{ L KOH soln}}\right)\left(\dfrac{1\text{ mol HCl}}{1\text{ mol KOH}}\right)}{(5.00\text{ mL HCl son})\left(\dfrac{1\text{ L HCl soln}}{1000\text{ mL HCl soln}}\right)}\right] = 0.0220\text{ M HCl}$$

$$\#\text{ g HCl} = (5.00\text{ mL HCl})\left(\frac{0.0220\text{ mol HCl}}{1000\text{ mL HCl}}\right)\left(\frac{36.5\text{ g HCl}}{1\text{ mol HCl}}\right) = 4.02 \times 10^{-3}\text{ g HCl}$$

$$\text{weight \%} = \frac{4.02 \times 10^{-3}\text{ g}}{5.00\text{ g}} \times 100\% = 0.0803\%$$

Thinking It Through

T5.1 We need to choose metal hydroxides that are both strong bases and highly soluble in water. Here the only two possibilities are NaOH and KOH. $Mg(OH)_2$ is a strong base, but is not very soluble in water.

T5.2 To determine the number of grams of $CuSO_4{\cdot}5H_2O$ required to prepare 250 mL of a 0.20 *M* $CuSO_4$ solution, determine the number of moles of $CuSO_4$ in the solution. Then determine the number of moles of $CuSO_4{\cdot}5H_2O$ that will give the desired number of moles of $CuSO_4$ and finally the number of grams of $CuSO_4{\cdot}5H_2O$.

$$\#\text{ g } CuSO_4{\cdot}5H_2O = 250\text{ mL}\left(\frac{1\text{ L } CuSO_4}{1000\text{ mL } CuSO_4}\right)\left(\frac{0.20\text{ mol } CuSO_4}{1\text{ L } CuSO_4}\right)\left(\frac{1\text{ mol } CuSO_4\cdot 5H_2O}{1\text{ mol } CuSO_4}\right) \times\left(\frac{249.7\text{ g } CuSO_4\cdot 5H_2O}{1\text{ mol } CuSO_4\cdot 5H_2O}\right)$$

T5.3 To determine the number of grams of Na_3PO_4 required to prepare 20.0 mL of 0.16 *M* Na^+ solution, determine the number of moles of Na^+ in the solution. Then determine the number of moles of Na_3PO_4 that will give the desired number of moles of Na^+ and finally the number of grams of Na_3PO_4 required.

$$\#\text{ g } Na_3PO_4 = 20.0\text{ mL}\left(\frac{1\text{ L } Na^+}{1000\text{ mL } Na^+}\right)\left(\frac{0.160\text{ mol } Na^+}{1\text{ L } Na^+}\right)\left(\frac{1\text{ mol } Na_3PO_4}{3\text{ mol } Na^+}\right)\left(\frac{163.9\text{ g } Na_3PO_4}{1\text{ mol } Na_3PO_4}\right)$$

T5.4 The final molariaty of the solute can be determined using the equation $M_1V_1 = M_2V_2$, where M_1 = 150 mL, V_1 = 0.20 M and V_2 = (25 mL + 150 mL)

(150 mL)(0.20 M) = M_2(175 mL)

Solve for M_2.

T5.5 HBr is a strong acid, and is 100 % ionized. Magnesium carbonate, although insoluble in water, does react with HBr to form soluble $MgBr_2$ and CO_2 gas. We should therefore see the solid dissolve with the evolution of a gas, as long as HBr remains un-neutralized. If an excess of $MgCO_3$ remains unreacted, it will not dissolve. The molecular equation is:

$$MgCO_3(s) + 2HBr(aq) \rightarrow MgBr_2(aq) + CO_2(g) + H_2O(\ell)$$

The net ionic equation is:

$$MgCO_3(s) + 2H^+(aq) \rightarrow Mg^{2+}(aq) + CO_2(g) + H_2O(\ell)$$

To answer the question, first determine which compound is the limiting reactant. If the $MgCO_3$ is in excess, determine how much was consumed in the reaction and subtract it from the initial amount. The amount consumed is determined by the amount of HBr added. The concentration of the ions present at the end of the reaction is also dependent on the limiting reactant.

T5.6 First, the balanced equation is needed:

$$Ag^+(aq) + X^-(aq) \rightarrow AgX(s)$$

which shows that the stoichiometry is 1:1. Next, we determine the number of moles of Ag^+ ion used in the titration, by multiplying molarity by volume in L. The molecular mass of the unknown is then determined by dividing mass (0.546 g) by number of moles (determined by titration). The formula mass should match one of the possibilities.

T5.7 Both solutions are soluble.

$Fe(NO_3)_2$: $[NO_3^-] = 0.240$ M
$[Fe^{2+}] = [NO_3^-]/2 = 0.120$ M

$Fe_2(SO_4)_3$: $[SO_4^{2-}] = 0.210$ M
$[Fe^{3+}] = 2/3[SO_4^{2-}] = 0.140$ M

T5.8 Both solutions are soluble and the net ionic equation is
$Ag^+(aq) + Cl^-(aq) \rightarrow AgCl(s)$
The solution is found by performing a limiting reactant analysis using volume and molarity to determine the number of moles of each reactant.

T5.9 The precipitate that forms is $Cr(OH)_3(s)$. From the starting mass of the chromium sulfate we can determine the mass of chromium hydroxide we expect to form. Since chromium hydroxide is tribasic, it will require 3 moles of acid for each mole of it for neutralization. Since we know the concentration of the acid and the amount of base, we can calculate the volume of acid needed to neutralize the base.

T5.10 First determine the number of moles of hydroxide used in the neutralization:

$$\text{\# moles OH}^- = (7.58\text{ mL})\left(\frac{0.150\text{ mol}}{1000\text{ mL}}\right) = 1.14 \times 10^{-3}\text{ mol OH}^-$$

Then determine the number of moles of acid

$$\text{\# moles acid} = (0.100\text{g acid})\left(\frac{1\text{ mol acid}}{176.1\text{ g acid}}\right) = 5.68 \times 10^{-4}\text{ moles acid}$$

The number of moles of acidic protons may be determined by the ratio of the moles hydroxide to moles acid.

$$\left(\frac{1.14 \times 10^{-3}\text{ moles}}{5.68 \times 10^{-4}\text{ moles}}\right) = 2.01\text{ therefore, the acid is diprotic}$$

T5.11 Because there are two unknown masses, we need a pair of equations to solve this problem. The first equation is relatively obvious;

$$m_{HC_3H_5O_3} + m_{HC_6H_{11}O_2} = 0.1000\text{ g}$$

Let $x = m_{HC_3H_5O_3}$ and $y = m_{HC_6H_{11}O_2}$

$$x + y = 0.1000\text{g}$$

We know the number of moles of OH^- used in the neutralization and this is equal to the moles H^+ present.

$$\text{\# mol H}^+ = 20.4\text{ mL NaOH}\left(\frac{0.0500\text{ mol NaOH}}{1000\text{ mL}}\right)\left(\frac{1\text{ mol OH}^-}{1\text{ mol NaOH}}\right)\left(\frac{1\text{ mol H}^+}{1\text{ mol OH}^-}\right) = 1.02 \times 10^{-3}\text{ mol H}^+$$

We can set up the following equation:

$$x\left(\frac{1\text{ mol HC}_3\text{H}_5\text{O}_3}{90.08\text{ g HC}_3\text{H}_5\text{O}_3}\right)\left(\frac{1\text{ mol H}^+}{1\text{ mol HC}_3\text{H}_5\text{O}_3}\right) + y\left(\frac{1\text{ mol HC}_6\text{H}_{11}\text{O}_2}{116.12\text{ g HC}_6\text{H}_{11}\text{O}_2}\right)\left(\frac{1\text{ mol H}^+}{1\text{ mol HC}_6\text{H}_{11}\text{O}_2}\right)$$

$= 1.02 \times 10^{-3}$ mol H^+

Substituting: x = 0.1000 – y we can solve for y and, finally determine x.

(the answers are 0.0639g = g $HC_3H_5O_3$ 0.0361 g = g $HC_6H_{11}O_2$)

T5.12 Diagram (d) is the best. The net ionic equation for the precipitation of lead chloride is:

$Pb^{2+} + 2Cl^- \rightarrow PbCl_2(s)$

Diagram (d) shows the presence of lead and chloride ions. It also shows them in the proper mole ratio of 2 moles chloride ion for every mole lead ion.

The other diagrams are wrong because:
(a) The diagram shows lead atoms and chlorine molecules
(b) The anion is illustrated having a 2– charge
(c) The cation is illustrated having a 1+ charge
(e) There is not enough Cl^- present.

Review Questions

5.1 (a) Solvent – the medium into which something (a solute) is dissolved to make a solution
(b) Solute – Something dissolved in a solvent to make a solution
(c) Concentration – the ratio of the quantity of solute to the quantity of solution or quantity of solvent

5.2 (a) Concentrated – a solution that has a large ratio of the amounts of solute to solvent
(b) Dilute – a solution in which the ratio of the quantities of solute to solvent is small
(c) Saturated – a solution that holds as much solute as it can at a given temperature
(d) Unsaturated – Any solution with a concentration less than that of a saturated solution of the same solute and solvent
(e) Supersaturated – a solution whose concentration of solute exceeds the equilibrium concentration

5.3 Solubility – the ratio of the quantity of solute to the quantity of solvent in a saturated solution

5.4 Chemical reactions are often carried out using solutions because this allows the reactants to move about and come in contact with each other. Furthermore, solutions can be made with a high enough concentration to allow the reaction to proceed at a reasonable rate.

5.5 Percentage concentration – a ratio of the amount of solute to the amount of solution expressed as a percent
Molar concentration – the number of moles of solute per liter of solution

5.6 When water is added to a solution of sugar, the average distance between the sugar molecules increases because there are more water molecules to fill in the spaces between the sugars.

5.7 When a sugar crystal is added to
(a) a saturated sugar solution, the sugar crystal will not dissolve.
(b) a supersaturated sugar solution, the sugar crystal will cause the extra sugar in solution to precipitate forming more sugar crystals.
(c) an unsaturated sugar solution, the added sugar crystal will dissolve.

5.8 Precipitate – a solid that separates from a solution usually as the result of a chemical reaction

For a precipitate to form spontaneously in a solution, the equilibrium must be disrupted. A supersaturated solution may form a precipitate spontaneously, or if the temperature changes in the direction that will cause a precipitate to form.

5.9 Electrolyte – a substance that dissolves in water to give an electrically conducting solution.

Nonelectrolyte – a substance that dissolves in water to give a solution that does not conduct electricity.

5.10 An electrolyte is able to conduct electricity since it has a charge and can allow an electron to flow, a nonelectrolyte does not have a charge, so it cannot carry a flow of electrons. An ion is hydrated when it is surrounded by water molecules.

5.11 Dissociation – the dissolving of an ionic compound in water such that the individual ions that compose the ionic compound become separated from one another (via hydration), and move about freely in solution, acting more or less independently of one another.

5.12 A strong electrolyte becomes completely dissociated or ionized in water solution and is a strong conductor of electricity. A weak electrolyte becomes only partially ionized in water solution and is a weak conductor of electricity.

5.13 For a weak electrolyte, there is a strong tendency for the ions to recombine to form their parent molecular compound. For a strong electrolyte, the converse is true; a strong electrolyte has a great tendency to be fully ionized in water solution.

5.14 A molecular equation is a chemical equation in which one has written complete molecular formulas for all reactants and products of a reaction. An ionic equation is a chemical equation in which all of the soluble strong electrolytes are listed in their dissociated form. A net ionic equation is written by listing only those ions and molecules that are involved in the reaction at hand. A net ionic equation differs from an ionic equation in that all of the spectator ions are omitted from the former. Spectator ions do not take part in a reaction. They are the ions that result from strong electrolytes.

5.15 The spectator ions are Na^+ and Cl^-. The net ionic equation is:

$$Co^{2+} + 2OH^- \rightarrow Co(OH)_2(s)$$

5.16 In a balanced ionic equation, both the mass and the electrical charge must be balanced.

5.17 The charge is not balanced.

5.18 A metathesis reaction is also called a double replacement reaction. A precipitate is an insoluble ionic solid that forms when two ionic solutions are mixed.

5.19 Lead iodide is not soluble in water.

5.20 (a) $CaCl_2(aq) \rightarrow Ca^{2+}(aq) + 2Cl^-(aq)$

(b) $(NH_4)_2SO_4(aq) \rightarrow 2NH_4^+(aq) + SO_4^{2-}(aq)$

(c) $NaC_2H_3O_2(aq) \rightarrow Na^+(aq) + C_2H_3O_2^-(aq)$

5.21 $3Ca^{2+}(aq) + 2PO_4^{3-}(aq) \rightarrow Ca_3(PO_4)_2(s)$

$3Mg^{2+}(aq) + 2PO_4^{3-}(aq) \rightarrow Mg_3(PO_4)_2(s)$

5.22 The substance is an electrolyte that dissolves readily in water:

$$Na_2CO_3 \bullet \bullet 10H_2O(s) \rightarrow 2Na^+(aq) + CO_3^{2-}(aq) + 10H_2O(\ell)$$

The carbonate anion then serves to cause the precipitation of calcium cations:

$$Ca^{2+}(aq) + CO_3^{2-}(aq) \rightarrow CaCO_3(s)$$

5.23 $Na_2C_2O_4 + CaCl_2 \rightarrow CaC_2O_4(s) + 2NaCl$

5.24 Acid – sour taste, turns litmus red, corrode some metals, etc...
Base – bitter taste, turns litmus blue, soapy "feel", etc...

5.25 If a solution is believed to be basic, red litmus paper should be used so that it would turn blue. The blue litmus paper may not change color if the solution is neutral.

5.26 According to the definition of Arrhenius, an acid gives H^+ ions in water, and a base gives OH^- ions in water.

5.27 Ionic substances already contain the ions that dissociate when the substance dissolves in water. On the other hand, molecular compounds such as HCl, form ions by reaction with water (ionization), although such substances do not contain the resulting ions to start with. Both dissociation of an ionic compound and ionization of a molecular compound in water solutions lead to the formation of solutions that are electrically conducting.

5.28 (a) NaOH dissociation
(b) HNO_3 ionization
(c) NH_3 ionization
(d) H_2SO_4 ionization

5.29 $HClO_4$, $HClO_3$, HCl, HBr, HI, HNO_3, H_2SO_4

5.30 (a) P_4O_{10} acidic solutions
(b) K_2O basic solutions
(c) SeO_3 acidic solutions
(d) Cl_2O acidic solutions

5.31 Dynamic equilibrium is a condition in which two opposing processes are occurring at equal rates.

5.32 Double arrows are not used for the reaction of a strong acid with water because the reaction is not in equilibrium. These are not reversible reactions, i.e., the reverse reaction has practically no tendency to occur.

5.33 The electrical conductivity would decrease regularly, until one solution had neutralized the other, forming a nonelectrolyte:

$$Ba^{2+}(aq) + 2OH^-(aq) + H^+(aq) + HSO_4^-(aq) \rightarrow BaSO_4(s) + 2H_2O(\ell)$$

Once the point of neutralization had been reached, the addition of excess sulfuric acid would cause the conductivity to increase, because sulfuric acid is a strong electrolyte itself.

5.34

H
H_3C—N
H

H
H_3C—N—H +
H

5.35 The products of the reaction of a strong acid with a strong base are a salt and water.

5.36 $HCHO_2$ will react with the following:
(a) KOH $HCHO_2 + KOH \rightarrow KCHO_2 + H_2O$
(b) MgO $2HCHO_2 + MgO \rightarrow Mg(CHO_2)_2 + 2H_2O$
(c) $CuCO_3$ $2HCHO_2 + CuCO_3 \rightarrow Cu(CHO_2)_2 + 2H_2CO_3$
$H_2CO_3 \rightarrow H_2O + CO_2$
(d) NH_3 $HCHO_2 + NH_3 \rightarrow NH_4^+ + CHO_2^-$

5.37 (a) hydrogen selenide
(b) hydroselenic acid

5.38 (a) periodic acid
(b) iodic acid
(c) iodous acid
(d) hypoiodous acid
(e) hydroiodic acid

5.39 (a) IO_4^- IO_3^- IO_2^- IO^- I^-

(b) periodate iodate iodite hypoiodite iodide

5.40 (a) H_2CrO4
(b) H_2CO_3
(c) $H_2C_2O_4$

5.41 (a) sodium bicarbonate or sodium hydrogen carbonate
(b) potassium dihydrogen phosphate
(c) ammonium hydrogen phosphate

5.42 NaH_2PO_4 Na_2HPO_4 Na_3HPO_4

5.43

(a)	hypochlorous acid	sodium hypochlorite	$NaOCl$
(b)	iodous acid	sodium iodite	$NaIO_2$
(c)	bromic acid	sodium bromate	$NaBrO_3$
(d)	perchloric acid	sodium perchlorate	$NaClO_4$

5.44 H_3AsO_3

5.45 sodium butyrate

5.46 propionic acid

5.47 Any solution containing ammonium ion will react with a strong base to yield ammonia. The presence of ammonia is easily detected by its odor.

5.48 (a) $HCl(aq) + NaHCO_3(aq) \rightarrow NaCl(aq) + H_2O(\ell) + CO_2(g)$
(b) $2HCl(aq) + Na_2S(aq) \rightarrow 2NaCl(aq) + H_2S(g)$
(c) $2HCl(aq) + K_2SO_3(aq) \rightarrow 2KCl(aq) + H_2O(\ell) + SO_2(g)$

5.49 Formation of a weak electrolyte, water, a gas, or an insoluble solid.

5.50 (a) $CuCl_2(aq) + Na_2CO_3(aq) \rightarrow 2NaCl(aq) + CuCO_3(s)$
The formation of $CuCO_3(s)$ drives the reaction.
(b) $K_2SO_3(aq) + H_2SO_4(aq) \rightarrow K_2SO_4(aq) + H_2O(\ell) + SO_2(g)$
The formation of $SO_2(g)$ drives the reaction.
(c) $2NaClO(aq) + H_2SO_4(aq) \rightarrow Na_2SO_4(aq) + 2HClO(aq)$
The formation of a weak electrolyte, $HClO(aq)$, drives the reaction.
(d) $Ba(OH)_2(aq) + 2HNO_3(aq) \rightarrow Ba(NO_3)_2(aq) + 2H_2O(\ell)$
The formation of water drives this reaction.

5.51 Since AgBr is insoluble, the concentrations of $Ag^+(aq)$ and $Br^-(aq)$ in a saturated solution of AgBr are very small, practically zero. When solutions of $AgNO_3$ and NaBr are mixed, the concentrations of Ag^+ and Br^- are momentarily larger than those in a saturated AgBr solution. Since this solution is immediately supersaturated in the moment of mixing, a precipitate of AgBr forms spontaneously.

5.52 The two conversion factors are:

$$\left(\frac{0.150 \text{ mol KOH}}{1 \text{ L KOH}}\right) \qquad \left(\frac{1 \text{ L KOH}}{0.150 \text{ mol KOH}}\right)$$

5.53 Molarity is the number of moles of solute per liter of solution, also known as molar concentration.

$$\left(\frac{\text{mmol}}{\text{mL}}\right)\left(\frac{1 \text{ mol}}{1000 \text{ mmol}}\right)\left(\frac{1000 \text{ mL}}{1 \text{ L}}\right) = \left(\frac{1000 \text{ mol}}{1000 \text{ L}}\right) = \left(\frac{\text{mol}}{\text{L}}\right)$$

5.54 The number of moles of HNO_3 in the solution has not changed because none of the original sample was removed. Instead, the concentration has decreased since more water was added.

5.55 The number of moles of $CaCl_2$ is the same in both solutions, but A is 0.1 M $CaCl_2$ and B is 0.2 M $CaCl_2$. The volume of solution A is 50 mL, therefore the volume of B is 25 mL:

$$\text{\# mol CaCl}_2 \text{ present} = 50 \text{ mL}\left(\frac{1 \text{ L}}{1000 \text{ mL}}\right)\left(\frac{0.1 \text{ mol CaCl}_2}{1 \text{ L CaCl}_2}\right) = 5 \times 10^{-3} \text{ mol CaCl}_2$$

$$\text{volume of solution B} = 5 \times 10^{-3} \text{ mol CaCl}_2 \left(\frac{1 \text{ L CaCl}_2}{0.2 \text{ M CaCl}_2}\right)\left(\frac{1000 \text{ mL}}{1 \text{ L}}\right) = 25 \text{ mL solution B}$$

5.56 $\text{M} \times \text{L} = \left(\frac{\text{mol}}{\text{L}}\right)\text{L} = \text{mol}$

5.57 Qualitative analysis is the use of experimental procedures to determine what elements are present in a substance.
Quantitative analysis determines the percentage composition of a compound or the percentage of a component in a mixture.
Qualitative analysis answers the question, "what is in the sample?" Quantitative analysis answers the question, "how much is in the sample?"

5.58 (a) Buret – a long glass tube fitted with a stopcock, graduated in mL, and used for the controlled, measured addition of a volume of a solution to a receiving flask.
(b) Titration – a procedure for obtaining quantitative information about a reactant by a controlled addition of one substance to another until a signal (usually a color change of an indicator) shows that equivalent quantities have reacted.
(c) Titrant – the solution delivered from a buret during a titration.
(d) End point – that point during a titration when the indicator changes color, the titration is stopped, and the total added volume of the titrant is recorded.

5.59 The indicator provides a visible signal that the solution has changed from an acid to a base.
(a) Phenolphthalein is colorless in acid solution.
(b) Phenolphthalein is pink in base solution.

Review Problems

5.60 (a) $LiCl(s) \rightarrow Li^{+}(aq) + Cl^{-}(aq)$
(b) $BaCl_2(s) \rightarrow Ba^{2+}(aq) + 2Cl^{-}(aq)$
(c) $Al(C_2H_3O_2)_3(s) \rightarrow Al^{3+}(aq) + 3C_2H_3O_2^{-}(aq)$
(d) $(NH_4)_2CO_3(s) \rightarrow 2NH_4^{+}(aq) + CO_3^{2-}(aq)$
(e) $FeCl_3(s) \rightarrow Fe^{3+}(aq) + 3Cl^{-}(aq)$

5.61 (a) $CuSO_4(s) \rightarrow Cu^{2+}(aq) + SO_4^{2-}(aq)$
(b) $Al_2(SO_4)_3(s) \rightarrow 2Al^{3+}(aq) + 3SO_4^{2-}(aq)$

(c) $CrCl_3(s) \rightarrow Cr^{3+}(aq) + 3Cl^-(aq)$
(d) $(NH_4)_2HPO_4(s) \rightarrow 2NH_4^+(aq) + HPO_4^{2-}(aq)$
(e) $KMnO_4(s) \rightarrow K^+(aq) + MnO_4^-(aq)$

5.62 (a) ionic: $2NH_4^+(aq) + CO_3^{2-}(aq) + Mg^{2+}(aq) + 2Cl^-(aq) \rightarrow 2NH_4^+(aq) + 2Cl^-(aq) + MgCO_3(s)$
net: $Mg^{2+}(aq) + CO_3^{2-}(aq) \rightarrow MgCO_3(s)$

(b) ionic: $Cu^{2+}(aq) + 2Cl^-(aq) + 2Na^+(aq) + 2OH^-(aq) \rightarrow (OH)_2(s) + 2Na^+(aq) + 2Cl^-(aq)$
net: $Cu^{2+}(aq) + 2OH^-(aq) \rightarrow Cu(OH)_2(s)$

(c) ionic: $3Fe^{2+}(aq) + 3SO_4^{2-}(aq) + 6Na^+(aq) + 2PO_4^{3-}(aq) \rightarrow Fe_3(PO_4)_2(s) + 6Na^+(aq) + 3SO_4^{2-}(aq)$
net: $3Fe^{2+}(aq) + 2PO_4^{3-}(aq) \rightarrow Fe_3(PO_4)_2(s)$

(d) ionic: $2Ag^+(aq) + 2C_2H_3O_2^-(aq) + Ni^{2+}(aq) + 2Cl^-(aq) \rightarrow 2AgCl(s) + Ni^{2+}(aq) + 2C_2H_3O_2^-(aq)$
net: $2Ag^+(aq) + 2Cl^-(aq) \rightarrow 2AgCl(s)$

5.63 (a) ionic: $Cu^{2+}(aq) + SO_4^{2-}(aq) + Ba^{2+}(aq) + 2Cl^-(aq) \rightarrow BaSO_4(s) + Cu^{2+}(aq) + 2Cl^-(aq)$
net: $Ba^{2+}(aq) + SO_4^{2-}(aq) \rightarrow BaSO_4(s)$

(b) $Fe^{3+}(aq) + 3NO_3^-(aq) + 3Li^+(aq) + 3OH^-(aq) \rightarrow (OH)_3(s) + 3Li^+(aq) + 3NO_3^-(aq)$
net: $Fe^{3+}(aq) + 3OH^-(aq) \rightarrow Fe(OH)_3(s)$

(c) $6Na^+(aq) + 2PO_4^{3-}(aq) + 3Ca^{2+}(aq) + 6Cl^-(aq) \rightarrow Ca_3(PO_4)_2(s) + 6Na^+(aq) + 6Cl^-(aq)$
net: $3Ca^{2+}(aq) + 2PO_4^{3-}(aq) \rightarrow Ca_3(PO_4)_2(s)$

(d) $2Na^+(aq) + S^{2-}(aq) + 2Ag^+(aq) + 2C_2H_3O_2^-(aq) \rightarrow 2Na^+(aq) + 2C_2H_3O_2^-(aq) + Ag_2S(s)$
net: $2Ag^+(aq) + S^{2-}(aq) \rightarrow Ag_2S(s)$

5.64 molecular: $Na_2S(aq) + Cu(NO_3)_2(aq) \rightarrow CuS(s) + 2NaNO_3(aq)$
ionic: $2Na^+(aq) + S^{2-}(aq) + Cu^{2+}(aq) + 2NO_3(aq) \rightarrow CuS(s) + 2Na^+(aq) + 2NO_3(aq)$
net: $Cu^{2+}(aq) + S^{2-}(aq) \rightarrow CuS(s)$

5.65 molecular: $Fe_2(SO_4)_3(aq) + 3BaCl_2(aq) \rightarrow 3BaSO_4(s) + 2FeCl_3(aq)$
ionic: $2Fe^{3+}(aq) + SO_4^{2-}(aq) + 3Ba^{2+}(aq) + 6Cl^-(aq) \rightarrow 3BaSO_4(s) + 2Fe^{3+}(aq) + 6Cl^-(aq)$
net: $3Ba^{2+}(aq) + SO_4^{2-}(aq) \rightarrow 3BaSO_4(s)$

5.66 molecular: $AgNO_3(aq) + NaBr(aq) \rightarrow AgBr(s) + NaNO_3(aq)$
ionic: $Ag^+(aq) + NO_3^-(aq) + Na^+(aq) + Br^-(aq) \rightarrow AgBr(s) + Na^+(aq) + NO_3^-(aq)$
net: $Ag^+(aq) + Br^-(aq) \rightarrow AgBr(s)$

5.67 molecular: $2Na_3PO_4(aq) + 3CaCl_2(aq) \rightarrow 6NaCl(aq) + Ca_3(PO_4)_2(s)$
ionic: $6Na^+(aq) + 2PO_4^{3-}(aq) + 3Ca^{2+}(aq) + 6Cl^-(aq) \rightarrow Ca_3(PO_4)_2\ (s) + 6Na^+(aq) + 6Cl^-(aq)$
net: $3Ca^{2+}(aq) + 2PO_4^{3-}(aq) \rightarrow Ca_3(PO_4)_2\ (s)$

5.68 This is an ionization reaction: $HClO_4(\ell) + H_2O(\ell) \rightarrow H_3O^+(aq) + ClO_4^-(aq)$

5.69 $HBr(\ell) + H_2O(\ell) \rightarrow H_3O^+(aq) + Br^-(aq)$

5.70 $N_2H_4(aq) + H_2O(\ell) \leftrightarrows N_2H_5^+(aq) + OH^-(aq)$

5.71 $CH_3NH_2(aq) + H_2O(\ell) \leftrightarrows CH_3NH_3^+(aq) + OH^-(aq)$

5.72 $HNO_2(aq) + H_2O(\ell) \leftrightarrows H_3O^+(aq) + NO_2^-(aq)$

5.73 $HCHO_2(aq) + H_2O(\ell) \leftrightarrows H_3O^+(aq) + CHO_2^-(aq)$

5.74 $H_2CO_3(aq) + H_2O(\ell) \leftrightarrows H_3O^+(aq) + HCO_3^-(aq)$

$HCO_3^-(aq) + H_2O(\ell) \leftrightarrows H_3O^+(aq) + CO_3^{2-}(aq)$

5.75 $H_3PO_4(aq) + H_2O(\ell) \leftrightarrows H_3O^+(aq) + H_2PO_4^-(aq)$

$H_2PO_4^-(aq) + H_2O(\ell) \leftrightarrows H_3O^+(aq) + HPO_4^{2-}(aq)$

$HPO_4^{2-}(aq) + H_2O(\ell) \leftrightarrows H_3O^+(aq) + PO_4^{3-}(aq)$

5.76 (a) molecular: $Ca(OH)_2(aq) + 2HNO_3(aq) \rightarrow Ca(NO_3)_2(aq) + 2H_2O(\ell)$

ionic: $Ca^{2+}(aq) + 2OH^-(aq) + 2H^+(aq) + 2NO_3^-(aq) \rightarrow {}^+(aq) + 2NO_3^-(aq) + 2H_2O(\ell)$

net: $H^+(aq) + OH^-(aq) \rightarrow H_2O(\ell)$

(b) molecular: $Al_2O_3(s) + 6HCl(aq) \rightarrow 2AlCl_3(aq) + 3H_2O(\ell)$

ionic: $Al_2O_3(s) + 6H^+(aq) + 6Cl^-(aq) \rightarrow 2Al^{3+}(aq) + 6Cl^-(aq) + 3H_2O(\ell)$

net: $Al_2O_3(s) + 6H^+(aq) \rightarrow 2Al^{3+}(aq) + 3H_2O(\ell)$

(c) molecular: $Zn(OH)_2(s) + H_2SO_4(aq) \rightarrow ZnSO_4(aq) + 2H_2O(\ell)$

ionic: $Zn(OH)_2(s) + 2H^+(aq) + SO_4^{2-}(aq) \rightarrow Zn^{2+}(aq) + SO_4^{2-}(aq) + 2H_2O(\ell)$

net: $Zn(OH)_2(s) + 2H^+(aq) \rightarrow Zn^{2+}(aq) + 2H_2O(\ell)$

5.77 (a) molecular: $2HC_2H_3O_2(aq) + Mg(OH)_2(s) \rightarrow Mg(C_2H_3O_2)_2(aq) + 2H_2O(\ell)$

ionic: $2HC_2H_3O_2(aq) + Mg(OH)_2(s) \rightarrow Mg^{2+}(aq) + 2C_2H_3O_2^-(aq) + 2H_2O(\ell)$

net: $2HC_2H_3O_2(aq) + Mg(OH)_2(s) \rightarrow Mg^{2+}(aq) + 2C_2H_3O_2^-(aq) + 2H_2O(\ell)$

(b) molecular: $HClO_4(aq) + NH_3(aq) \rightarrow NH_4ClO_4(aq)$
ionic: $H^+(aq) + ClO_4^-(aq) + NH_3(aq) \rightarrow NH_4^+(aq) + ClO_4^-(aq)$
net: $H^+(aq) + NH_3(aq) \rightarrow NH_4^+(aq)$

(c) molecular: $H_2CO_3(aq) + 2NH_3(aq) \rightarrow (NH_4)_2CO_3(aq)$
ionic: $H_2CO_3(aq) + 2NH_3(aq) \rightarrow 2NH_4^+(aq) + CO_3^{2-}(aq)$
net: $H_2CO_3(aq) + 2NH_3(aq) \rightarrow 2NH_4^+(aq) + CO_3^{2-}(aq)$

5.78 (a) $2H^+(aq) + CO_3^{2-}(aq) \rightarrow H_2O(\ell) + CO_2(g)$

(b) $NH_4^+(aq) + OH^-(aq) \rightarrow NH_3(aq) + H_2O(\ell)$

5.79 (a) $H^+(aq) + HSO_4^-(aq) + Na^+(aq) + HSO_3^-(aq) \rightarrow H_2O(\ell) + SO_2(g) + Na^+(aq) + HSO_4^-(aq)$

net: $H^+(aq) + HSO_3^-(aq) \rightarrow H_2O(\ell) + SO_2(g)$

(b) net: $2H^+(aq) + CO_3^{2-}(aq) \rightarrow H_2O(\ell) + CO_2(g)$

5.80 These reactions have the following "driving forces":
(a) formation of insoluble $Cr(OH)_3$
(b) formation of water, a weak electrolyte

5.81 These reactions have the following "driving forces":
(c) formation of a gas, CO_2
(d) formation of a weak electrolyte, $H_2C_2O_4$

5.82 The soluble ones are (a), (b), and (d).

5.83 The soluble ones are (a), (b), (d), and (f).

5.84 The insoluble ones are (a), (d), and (f).

5.85 The insoluble ones are (b), (c), (e), (f), and (i)

5.86 a) molecular: $3HNO_3(aq) + Cr(OH)_3(s) \rightarrow Cr(NO_3)_3(aq) + 3H_2O(\ell)$

ionic: $3H^+(aq) + 3NO_3^-(aq) + Cr(OH)_3(s) \rightarrow Cr^{3+}(aq) + 3NO_3^-(aq) + 3H_2O(\ell)$

net: $3H^+(aq) + Cr(OH)_3(s) \rightarrow Cr^{3+}(aq) + 3H_2O(\ell)$

(b) molecular: $HClO_4(aq) + NaOH(aq) \rightarrow NaClO_4(aq) + H_2O(\ell)$

ionic: $H^+(aq) + ClO_4^-(aq) + Na^+(aq) + OH^-(aq) \rightarrow Na^+(aq) + ClO_4^-(aq) + H_2O(\ell)$

net: $H^+(aq) + OH^-(aq) \rightarrow H_2O(\ell)$

(c) molecular: $Cu(OH)_2(s) + 2HC_2H_3O_2(aq) \rightarrow Cu(C_2H_3O_2)_2(aq) + 2H_2O(\ell)$

ionic: $Cu(OH)_2(s) + 2H^+ + 2C_2H_3O_2^-(aq) \rightarrow Cu^{2+}(aq) + 2C_2H_3O_2^-(aq) + 2H_2O(\ell)$

net: $Cu(OH)_2(s) + 2H^+(aq) \rightarrow Cu^{2+}(aq) + 2H_2O(\ell)$

(d) molecular: $ZnO(s) + 2HBr(aq) \rightarrow ZnBr_2(aq) + H_2O(\ell)$

ionic: $ZnO(s) + 2H^+(aq) + 2Br^-(aq) \rightarrow Zn^{2+}(aq) + 2Br^-(aq) + H_2O(\ell)$

net: $ZnO(s) + 2H^+(aq) \rightarrow Zn^{2+}(aq) + H_2O(\ell)$

5.87 (a) molecular: $NaHSO_3(aq) + HBr(aq) \rightarrow SO_2(g) + NaBr(aq) + H_2O(\ell)$

ionic: $Na^+(aq) + HSO_3^-(aq) + H^+(aq) + Br^-(aq) \rightarrow (g) + Na^+(aq) + Br^-(aq) + H_2O(\ell)$

net: $HSO_3^-(aq) + H^+(aq) \rightarrow SO_2(g) + H_2O(\ell)$

(b) molecular: $(NH_4)_2CO_3(aq) + 2NaOH(aq) \rightarrow 2NH_3(g) + Na_2CO_3(aq) + 2H_2O(\ell)$

ionic: $2NH_4^+(aq) + CO_3^{2-}(aq) + 2Na^+(aq) + 2OH^-(aq) \rightarrow$
$2NH_3(g) + 2Na^+(aq) + CO_3^{2-}(aq) + 2H_2O(\ell)$

net: $NH_4^+(aq) + OH^-(aq) \rightarrow NH_3(g) + H_2O(\ell)$

(c) molecular: $(NH_4)_2CO_3(aq) + Ba(OH)_2(aq) \rightarrow BaCO_3(s) + 2NH_3(g) + 2H_2O(\ell)$

ionic: $2NH_4^+(aq) + CO_3^{2-}(aq) + Ba^{2+}(aq) + 2OH^-(aq) \rightarrow BaCO_3(s) + 2NH_3(g) + 2H_2O(\ell)$

net: $2NH_4^+(aq) + CO_3^{2-}(aq) + Ba^{2+}(aq) + 2OH^-(aq) \rightarrow BaCO_3(s) + 2NH_3(g) + 2H_2O(\ell)$

(d) molecular: $FeS(s) + 2HCl(aq) \rightarrow FeCl_2(aq) + H_2S(g)$
ionic: $FeS(s) + 2H^+(aq) + 2Cl^-(aq) \rightarrow Fe^{2+}(aq) + 2Cl^-(aq) + H_2S(g)$
net: $FeS(s) + 2H^+(aq) \rightarrow Fe^{2+}(aq) + H_2S(g)$

5.88 (a) molecular: $Na_2SO_3(aq) + Ba(NO_3)_2(aq) \rightarrow BaSO_3(s) + 2NaNO_3(aq)$
ionic: $2Na^+(aq) + SO_3^{2-}(aq) + Ba^{2+}(aq) + 2NO_3^-(aq) \rightarrow BaSO_3(s) + 2Na^+(aq) + 2NO_3^-(aq)$
net: $Ba^{2+}(aq) + SO_3^{2-}(aq) \rightarrow BaSO_3(s)$

(b) molecular: $2HCHO_2(aq) + K_2CO_3(aq) \rightarrow CO_2(g) + H_2O(\ell) + 2KCHO_2(aq)$

ionic: $2H^+(aq) + 2CHO_2^-(aq) + 2K^+(aq) + CO_3^{2-}(aq) \rightarrow (g) + H_2O(\ell) + 2K^+(aq) + 2CHO_2^-(aq)$

net: $2H^+(aq) + CO_3^{2-}(aq) \rightarrow CO_2(g) + H_2O(\ell)$

(c) molecular: $2NH_4Br(aq) + Pb(C_2H_3O_2)_2(aq) \rightarrow 2NH_4C_2H_3O_2(aq) + PbBr_2(s)$
ionic: $2NH_4^+(aq) + 2Br^-(aq) + Pb^{2+}(aq) + 2C_2H_3O_2^-(aq) \rightarrow 2NH_4^+(aq) + 2C_2H_3O_2^-(aq) + PbBr_2(s)$
net: $Pb^{2+}(aq) + 2Br^-(aq) \rightarrow PbBr_2(s)$

(d) molecular: $2NH_4ClO_4(aq) + Cu(NO_3)_2(aq) \rightarrow Cu(ClO_4)_2(aq) + 2NH_4NO_3(aq)$
ionic: $2NH_4^+(aq) + 2ClO_4^-(aq) + Cu^{2+}(aq) + 2NO_3^-(aq) \rightarrow$
$Cu^{2+}(aq) + 2ClO_4^-(aq) + 2NO_3^-(aq) + 2NH_4^+(aq)$
net: N.R.

5.89 (a) molecular: $(NH_4)_2S(aq) + 2NaOH(aq) \rightarrow 2NH_3(g) + 2H_2O(\ell) + Na_2S(aq)$

ionic: $2NH_4^+(aq) + S^{2-}(aq) + 2Na^+(aq) + 2OH^-(aq) \rightarrow (g) + 2H_2O(\ell) + 2Na^+(aq) + S^{2-}(aq)$

net: $NH_4^+(aq) + OH^-(aq) \rightarrow NH_3(g) + H_2O(\ell)$

(b) molecular: $Cr_2(SO_4)_3(aq) + 3K_2CO_3(aq) \rightarrow Cr_2(CO_3)_3(s) + 3K_2SO_4(aq)$
ionic: $2Cr^{3+}(aq) + 3SO_4^{2-}(aq) + 6K^+(aq) + 3CO_3^{2-}(aq) \rightarrow Cr_2(CO_3)_3(s) + 6K^+(aq) + 3SO_4^{2-}(aq)$
net: $2Cr^{3+}(aq) + 3CO_3^{2-}(aq) \rightarrow Cr_2(CO_3)_3(s)$

(c) molecular: $3AgNO_3(aq) + Cr(C_2H_3O_2)_3(aq) \rightarrow 3AgC_2H_3O_2(aq) + Cr(NO_3)_3(aq)$
ionic: $3Ag^+(aq) + 3NO_3(aq) + Cr^{3+}(aq) + 3C_2H_3O_2(aq)$
$\rightarrow 3Ag^+(aq) + 3C_2H_3O_2(aq) + Cr^{3+}(aq) + 3NO_3(aq)$
net: NR

(d) molecular: $Sr(OH)_2(aq) + MgCl_2(aq) \rightarrow SrCl_2(aq) + Mg(OH)_2(s)$
ionic: $Sr^{2+}(aq) + 2OH^-(aq) + Mg^{2+}(aq) + 2Cl^-(aq) \rightarrow Mg(OH)_2(s) + Sr^{2+}(aq) + 2Cl^-(aq)$
net: $Mg^{2+}(aq) + 2OH^-(aq) \rightarrow Mg(OH)_2(s)$

5.90 There are numerous possible answers. One of many possible sets of answers would be:
(a) $NaHCO_3(aq) + HCl(aq) \rightarrow NaCl(aq) + CO_2(g) + H_2O(\ell)$
(b) $FeCl_2(aq) + 2NaOH(aq) \rightarrow Fe(OH)_2(s) + 2NaCl(aq)$
(c) $Ba(NO_3)_2(aq) + K_2SO_3(aq) \rightarrow BaSO_3(s) + 2KNO_3(aq)$
(d) $2AgNO_3(aq) + Na_2S(aq) \rightarrow Ag_2S(s) + 2NaNO_3(aq)$
(e) $ZnO(s) + 2HCl(aq) \rightarrow ZnCl_2(aq) + H_2O(\ell)$

5.91 We need to choose a set of reactants that are both soluble and that react to yield only one solid product. Choose (a), (b), (c) and (d).

5.92 The two conversion factors are:

$$\left(\frac{0.25 \text{ mol HCl}}{1 \text{ L HCl}}\right) \qquad \left(\frac{1 \text{ L HCl}}{0.25 \text{ mol HCl}}\right)$$

5.93 The two conversion factors are:

$$\left(\frac{0.75 \text{ mol } CuSO_4}{1 \text{ L } CuSO_4}\right) \qquad \left(\frac{1 \text{ L } CuSO_4}{0.75 \text{ mol } CuSO_4}\right)$$

5.94 (a) $NaOH \rightarrow Na^+ + OH^-$

$$\# \text{ mol NaOH} = 4.00 \text{ g NaOH}\left(\frac{1 \text{ mol NaOH}}{40.00 \text{ g NaOH}}\right) = 0.100 \text{ mol NaOH}$$

$$\text{M NaOH solution} = \left(\frac{0.100 \text{ mol NaOH}}{100.0 \text{ mL NaOH}}\right)\left(\frac{1000 \text{ mL NaOH}}{1 \text{ L NaOH}}\right) = 1.00 \text{ M NaOH}$$

(b) $CaCl_2 \rightarrow Ca^{2+} + 2Cl^-$

$$\# \text{ mol } CaCl_2 = 16.0 \text{ g } CaCl_2\left(\frac{1 \text{ mol } CaCl_2}{110.98 \text{ g } CaCl_2}\right) = 0.144 \text{ mol } CaCl_2$$

$$\text{M } CaCl_2 \text{ solution} = \left(\frac{0.144 \text{ mol } CaCl_2}{250.0 \text{ mL } CaCl_2}\right)\left(\frac{1000 \text{ mL } CaCl_2}{1 \text{ L } CaCl_2}\right) = 0.577 \text{ M } CaCl_2$$

(c) $KOH \rightarrow K^+ + OH^-$

$$\# \text{ mol KOH} = 14.0 \text{ g KOH}\left(\frac{1 \text{ mol KOH}}{56.106 \text{ g KOH}}\right) = 0.250 \text{ mol KOH}$$

$$\text{M KOH solution} = \left(\frac{0.250 \text{ mol KOH}}{75.0 \text{ mL KOH}}\right)\left(\frac{1000 \text{ mL KOH}}{1 \text{ L KOH}}\right) = 3.33 \text{ M KOH}$$

(d) $H_2C_2O_4 \rightarrow 2H^+ + C_2O_4^{2-}$

$$\# \text{ mol } H_2C_2O_4 = 6.75 \text{ g } H_2C_2O_4\left(\frac{1 \text{ mol } H_2C_2O_4}{90.04 \text{ g } H_2C_2O_4}\right) = 0.0750 \text{ mol } H_2C_2O_4$$

$$\text{M } H_2C_2O_4 \text{ solution} = \left(\frac{0.0750 \text{ mol } H_2C_2O_4}{500 \text{ mL } H_2C_2O_4}\right)\left(\frac{1000 \text{ mL } H_2C_2O_4}{1 \text{ L } H_2C_2O_4}\right) = 0.150 \text{ M } H_2C_2O_4$$

5.95 (a) $H_2SO_4 \rightarrow 2H^+ + SO_4^{2-}$

$$\# \text{ mol } H_2SO_4 = 3.60 \text{ g } H_2SO_4\left(\frac{1 \text{ mol } H_2SO_4}{98.08 \text{ g } H_2SO_4}\right) = 0.0367 \text{ mol } H_2SO_4$$

$$\text{M } H_2SO_4 \text{ solution} = \left(\frac{0.0367 \text{ mol } H_2SO_4}{450 \text{ mL } H_2SO_4}\right)\left(\frac{1000 \text{ mL } H_2SO_4}{1 \text{ L } H_2SO_4}\right) = 0.0816 \text{ M } H_2SO_4$$

(b) $Fe(NO_3)_2 \rightarrow Fe^{2+} + 2NO_3^-$

$$\text{M } Fe(NO_3)_2 \text{ solution} = \left(\frac{2.0\times10^{-3} \text{ mol } Fe(NO_3)_2}{12.0 \text{ mL } Fe(NO_3)_2}\right)\left(\frac{1000 \text{ mL } Fe(NO_3)_2}{1 \text{ L } Fe(NO_3)_2}\right) = 0.17 \text{ M } Fe(NO_3)_2$$

(c) $HCl \rightarrow H^+ + Cl^-$

$$\text{M HCl solution} = \left(\frac{1.65 \text{ mol HCl}}{2.16 \text{ L HCl}}\right) = 0.764 \text{ M HCl}$$

(d) $HCl \rightarrow H^+ + Cl^-$

$$\# \text{ mol HCl} = 18.0 \text{ g HCl}\left(\frac{1 \text{ mol HCl}}{36.46 \text{ g HCl}}\right) = 0.494 \text{ mol HCl}$$

$$\text{M HCl solution} = \left(\frac{0.494 \text{ mol HCl}}{0.375 \text{ L HCl}}\right) = 1.32 \text{ M HCl}$$

5.96 (a) $$\# \text{ g NaCl} = 125 \text{ mL soln}\left(\frac{1 \text{ L}}{1000 \text{ mL}}\right)\left(\frac{0.200 \text{ mol NaCl}}{1 \text{ L soln}}\right)\left(\frac{58.44 \text{ g NaCl}}{1 \text{ mol NaCl}}\right) = 1.46 \text{ g NaCl}$$

(b) $$\# \text{ g } C_2H_{12}O_6 = 250 \text{ mL soln}\left(\frac{1 \text{ L}}{1000 \text{ mL}}\right)\left(\frac{0.360 \text{ mol } C_6H_{12}O_6}{1 \text{ L soln}}\right)\left(\frac{180.2 \text{ g } C_6H_{12}O_6}{1 \text{ mol } C_6H_{12}O_6}\right)$$
$$= 16.2 \text{ g } C_2H_{12}O_6$$

(c) $$\# \text{ g } H_2SO_4 = 250 \text{ mL soln}\left(\frac{1 \text{ L}}{1000 \text{ mL}}\right)\left(\frac{0.250 \text{ mol } H_2SO_4}{1 \text{ L soln}}\right)\left(\frac{98.08 \text{ g } H_2SO_4}{1 \text{ mol } H_2SO_4}\right) = 6.13 \text{ g } H_2SO_4$$

5.97 (a) $$\# \text{ g } K_2SO_4 = 250 \text{ mL soln}\left(\frac{1 \text{ L}}{1000 \text{ mL}}\right)\left(\frac{0.100 \text{ mol } K_2SO_4}{1 \text{ L soln}}\right)\left(\frac{174.3 \text{ g } K_2SO_4}{1 \text{ mol } K_2SO_4}\right) = 4.36 \text{ g } K_2SO_4$$

(b) $$\# \text{ g } K_2CO_3 = 100 \text{ mL soln}\left(\frac{1 \text{ L}}{1000 \text{ mL}}\right)\left(\frac{0.250 \text{ mol } K_2CO_3}{1 \text{ L soln}}\right)\left(\frac{138.2 \text{ g } K_2CO_3}{1 \text{ mol } K_2CO_3}\right) = 3.46 \text{ g } K_2CO_3$$

(c) $$\# \text{ g KOH} = 500 \text{ mL soln}\left(\frac{1 \text{ L}}{1000 \text{ mL}}\right)\left(\frac{0.400 \text{ mol KOH}}{1 \text{ L soln}}\right)\left(\frac{56.11 \text{ g KOH}}{1 \text{ mol KOH}}\right) = 11.22 \text{ g } K_2CO_3$$

5.98 $$\text{mol of } H_2SO_4 = 25.0 \text{ mL } H_2SO_4\left(\frac{1 \text{ L soln}}{1000 \text{ mL soln}}\right)\left(\frac{0.56 \text{ mol } H_2SO_4}{1 \text{ L soln}}\right) = 0.014 \text{ mol } H_2SO_4$$

$$\text{M of final solution} = \left(\frac{0.014 \text{ mol } H_2SO_4}{125 \text{ mL } H_2SO_4}\right)\left(\frac{1000 \text{ mL } H_2SO_4}{1 \text{ L } H_2SO_4}\right) = 0.11 \text{ M } H_2SO_4$$

5.99 $$\text{mol of } HNO_3 = 150 \text{ mL } HNO_3\left(\frac{1 \text{ L soln}}{1000 \text{ mL soln}}\right)\left(\frac{0.45 \text{ mol } HNO_3}{1 \text{ L soln}}\right) = 0.067 \text{ mol } HNO_3$$

$$\text{M of final solution} = \left(\frac{0.067 \text{ mol } HNO_3}{450 \text{ mL } HNO_3}\right)\left(\frac{1000 \text{ mL } HNO_3}{1 \text{ L } HNO_3}\right) = 0.15 \text{ M } HNO_3$$

5.100 $M_1V_1 = M_2V_2$ $\qquad V_2 = \left(\frac{M_1V_1}{M_2}\right)$

$$V_2 = \frac{(18.0 \text{ M } H_2SO_4)(25.0 \text{ mL})}{1.50 \text{ M } H_2SO_4} = 300 \text{ mL } H_2SO_4$$

The 25.0 mL of H_2SO_4 must be diluted to 300 mL.

5.101 $M_1V_1 = M_2V_2$ $\qquad V_2 = \left(\frac{M_1V_1}{M_2}\right)$

$$V_2 = \frac{(1.50 \text{ M HCl})(50.0 \text{ mL})}{0.200 \text{ M HCl}} = 375 \text{ mL HCl}$$

The 50.0 mL of HCl must be diluted to 375 mL.

5.102 $M_1V_1 = M_2V_2$ $\qquad V_2 = \left(\frac{M_1V_1}{M_2}\right)$

$$V_2 = \frac{(2.5\text{ M KOH})(150.0\text{ mL})}{1.0\text{ M KOH}} = 375\text{ mL KOH}$$

The 150.0 mL of KOH must be diluted to 375 mL. The volume of water to be added is:
375 mL of V_2 – 150 mL of V_1 = 225 mL water

5.103 $M_1V_1 = M_2V_2$ $\qquad V_2 = \left(\frac{M_1V_1}{M_2}\right)$

$$V_2 = \frac{(1.5\text{ M HCl})(120\text{ mL})}{1.00\text{ M HCl}} = 180\text{ mL HCl}$$

The 120.0 mL of HCl must be diluted to 180 mL. The volume of water to be added is:
180 mL of V_2 – 120 mL of V_1 = 60. mL water

5.104 (a) $KOH \rightarrow K^+ + OH^-$

\# mol KOH = 1.25 mol/L × 0.0350 L = 0.0437 mol KOH

$$0.0437\text{ mol KOH}\left(\frac{1\text{ mol }K^+}{1\text{ mol KOH}}\right) = 0.0437\text{ mol }K^+$$

$$0.0437\text{ mol KOH}\left(\frac{1\text{ mol }OH^-}{1\text{ mol KOH}}\right) = 0.0437\text{ mol }OH^-$$

(b) $CaCl_2 \rightarrow Ca^{2+} + 2Cl^-$

\# mol $CaCl_2$ = 0.45 mol/L × 0.0323 L = 0.0145 mol $CaCl_2$

$$0.0145\text{ mol }CaCl_2\left(\frac{1\text{ mol }Ca^{2+}}{1\text{ mol }CaCl_2}\right) = 0.0145\text{ mol }Ca^{2+}$$

$$0.0145\text{ mol }CaCl_2\left(\frac{2\text{ mol }Cl^-}{1\text{ mol }CaCl_2}\right) = 0.0290\text{ mol }Cl^-$$

(c) $AlCl_3 \rightarrow Al^{3+} + 3Cl^-$

\# mol $AlCl_3$ = 0.40 mol/L × 0.0500 L = 0.020 mol $AlCl_3$

$$0.020\text{ mol }AlCl_3\left(\frac{1\text{ mol }Al^{3+}}{1\text{ mol }AlCl_3}\right) = 0.020\text{ mol }Al^{3+}$$

$$0.020\text{ mol }AlCl_3\left(\frac{3\text{ mol }Cl^-}{1\text{ mol }AlCl_3}\right) = 0.060\text{ mol }Cl^-$$

5.105 (a) $(NH_4)_2CO_3 \rightarrow 2NH_4^+ + CO_3^{2-}$

\# mol $(NH_4)_2CO_3$ = 0.40 mol/L × 0.0185 L = 0.0074 mol $(NH_4)_2CO_3$
0.0074 mol $(NH_4)_2CO_3$ × 2 mol NH_4^+/mol $(NH_4)_2CO_3$ = 0.015 mol NH_4^+
0.0074 mol $(NH_4)_2CO_3$ × 1 mol CO_3^{2-}/mol $(NH_4)_2CO_3$ = 0.0074 mol CO_3^{2-}

(b) $Al_2(SO_4)_3 \rightarrow 2Al^{3+} + 3SO_4^{2-}$

\# mol $Al_2(SO_4)_3$ = 0.35 mol/L × 0.0300 L = 0.011 mol $Al_2(SO_4)_3$
0.0105 mol $Al_2(SO_4)_3$ × 2 mol Al^{3+}/mol $Al_2(SO_4)_3$ = 0.021 mol Al^{3+}
0.0105 mol $Al_2(SO_4)_3$ × 3 mol SO_4^{2-}/mol $Al_2(SO_4)_3$ = 0.033 mol SO_4^{2-}

(c) $Zn(C_2H_3O_2) \rightarrow Zn^{2+} + 2C_2H_3O_2^-$

$$\text{\# moles } Zn(C_2H_3O_2)_2 = (60.0 \text{ mL } Zn(C_2H_3O_2)_2)\left(\frac{0.12 \text{ mol } Zn(C_2H_3O_2)_2}{1000 \text{ mL } Zn(C_2H_3O_2)_2}\right)$$

$$= 7.2 \times 10^{-3} \text{ moles } Zn(C_2H_3O_2)_2$$

$$\text{\# moles } Zn^{2+} = (7.2 \times 10^{-3} \text{ mol } Zn(C_2H_3O_2)_2)\left(\frac{1 \text{ mol } Zn^{2+}}{1 \text{ mol } Zn(C_2H_3O_2)_2}\right)$$

$$= 7.2 \times 10^{-3} \text{ moles } Zn^{2+}$$

$$\text{\# moles } C_2H_3O_2^- = (7.2 \times 10^{-3} \text{ moles } Zn(C_2H_3O_2)_2)\left(\frac{2 \text{ moles } C_2H_3O_2^-}{1 \text{ mol } Zn(C_2H_3O_2)_2}\right)$$

$$= 1.4 \times 10^{-2} \text{ moles } C_2H_3O_2^-$$

5.106 (a) $Cr(NO_3)_2 \rightarrow Cr^{2+} + 2NO_3^-$

$$\text{M } Cr^{2+} = \left(\frac{0.25 \text{ mol } Cr(NO_3)_2}{1 \text{ L } Cr(NO_3)_2 \text{ soln}}\right)\left(\frac{1 \text{ mol } Cr^{2+}}{1 \text{ mol } Cr(NO_3)_2}\right) = 0.25 \text{ M } Cr^{2+}$$

$$\text{M } NO_3^- = \left(\frac{0.25 \text{ mol } Cr(NO_3)_2}{1 \text{ L } Cr(NO_3)_2 \text{ soln}}\right)\left(\frac{2 \text{ mol } NO_3^-}{1 \text{ mol } Cr(NO_3)_2}\right) = 0.50 \text{ M } NO_3^-$$

(b) $CuSO_4 \rightarrow Cu^{2+} + SO_4^{2-}$

$$\text{M } Cu^{2+} = \left(\frac{0.10 \text{ mol } CuSO_4}{1 \text{ L } CuSO_4 \text{ soln}}\right)\left(\frac{1 \text{ mol } Cu^{2+}}{1 \text{ mol } CuSO_4}\right) = 0.10 \text{ M } Cu^{2+}$$

$$\text{M } SO_4^{2-} = \left(\frac{0.10 \text{ mol } CuSO_4}{1 \text{ L } CuSO_4 \text{ soln}}\right)\left(\frac{1 \text{ mol } SO_4^{2-}}{1 \text{ mol } CuSO_4}\right) = 0.10 \text{ M } SO_4^{2-}$$

(c) $Na_3PO_4 \rightarrow 3Na^+ + PO_4^{3-}$

$$\text{M } Na^+ = \left(\frac{0.16 \text{ mol } Na_3PO_4}{1 \text{ L } Na_3PO_4 \text{ soln}}\right)\left(\frac{3 \text{ mol } Na^+}{1 \text{ mol } Na_3PO_4}\right) = 0.48 \text{ M } Na^+$$

$$\text{M } PO_4^{3-} = \left(\frac{0.16 \text{ mol } Na_3PO_4}{1 \text{ L } Na_3PO_4 \text{ soln}}\right)\left(\frac{1 \text{ mol } PO_4^{3-}}{1 \text{ mol } Na_3PO_4}\right) = 0.16 \text{ M } PO_4^{3-}$$

(d) $Al_2(SO_4)_3 \rightarrow 2Al^{3+} + 3SO_4^{2-}$

$$\text{M } Al^{3+} = \left(\frac{0.075 \text{ mol } Al_2(SO_4)_3}{1 \text{ L } Al_2(SO_4)_3 \text{ soln}}\right)\left(\frac{2 \text{ mol } Al^{3+}}{1 \text{ mol } Al_2(SO_4)_3}\right) = 0.15 \text{ M } Al^{3+}$$

$$\text{M } SO_4^{2-} = \left(\frac{0.075 \text{ mol } Al_2(SO_4)_3}{1 \text{ L } Al_2(SO_4)_3 \text{ soln}}\right)\left(\frac{3 \text{ mol } SO_4^{2-}}{1 \text{ mol } Al_2(SO_4)_3}\right) = 0.22 \text{ M } SO_4^{2-}$$

$$\text{M } Ca^{2+} = \left(\frac{0.060 \text{ mol } Ca(OH)_2}{1 \text{ L } Ca(OH)_2 \text{ soln}}\right)\left(\frac{1 \text{ mol } Ca^{2+}}{1 \text{ mol } Ca(OH)_2}\right) = 0.060 \text{ M } Ca^{2+}$$

$$\text{M } OH^- = \left(\frac{0.060 \text{ mol } Ca(OH)_2}{1 \text{ L } Ca(OH)_2 \text{ soln}}\right)\left(\frac{2 \text{ mol } OH^-}{1 \text{ mol } Ca(OH)_2}\right) = 0.12 \text{ M } OH^-$$

5.107 (a) $Ca(OH)_2 \rightarrow Ca^{2+} + 2OH^-$

$$\text{M } Ca^{2+} = \left(\frac{0.060 \text{ M } Ca(OH)_2}{1 \text{ L } Ca(OH)_2 \text{ solution}}\right)\left(\frac{1 \text{ mol } Ca^{2+}}{1 \text{ mol } Ca(OH)_2}\right) = 0.06 \text{ M } Ca^{2+}$$

$$\text{M } OH^- = \left(\frac{0.060 \text{ M } Ca(OH)_2}{1 \text{ L } Ca(OH)_2 \text{ solution}}\right)\left(\frac{1 \text{ mol } OH^-}{1 \text{ mol } Ca(OH)_2}\right) \ 0.12 \text{ M } OH^-$$

(b) $FeCl_3 \rightarrow Fe^{3+} + 3Cl^-$

$$\text{M } Fe^{3+} = \left(\frac{0.15 \text{ mol } FeCl_3}{1 \text{ L } FeCl_3 \text{ soln}}\right)\left(\frac{1 \text{ mol } Fe^{3+}}{1 \text{ mol } FeCl_3}\right) = 0.15 \text{ M } Fe^{3+}$$

$$\text{M } Cl^- = \left(\frac{0.15 \text{ mol } FeCl_3}{1 \text{ L } FeCl_3 \text{ soln}}\right)\left(\frac{3 \text{ mol } Cl^-}{1 \text{ mol } FeCl_3}\right) = 0.45 \text{ M } Cl^-$$

(c) $Cr_2(SO_4)_3 \rightarrow 2Cr^{3+} + 3SO_4^{2-}$

$$\text{M } Cr^{3+} = \left(\frac{0.22 \text{ mol } Cr_2(SO_4)_3}{1 \text{ L } Cr_2(SO_4)_3 \text{ soln}}\right)\left(\frac{2 \text{ mol } Cr^{3+}}{1 \text{ mol } Cr_2(SO_4)_3}\right) = 0.44 \text{ M } Cr^{3+}$$

$$\text{M } SO_4^{2-} = \left(\frac{0.22 \text{ mol } Cr_2(SO_4)_3}{1 \text{ L } Cr_2(SO_4)_3 \text{ soln}}\right)\left(\frac{3 \text{ mol } SO_4^{2-}}{1 \text{ mol } Cr_2(SO_4)_3}\right) = 0.66 \text{ M } SO_4^{2-}$$

(d) $(NH_4)_2SO_4 \rightarrow 2NH_4^+ + SO_4^{2-}$

$$\text{M } NH_4^+ = \left(\frac{0.60 \text{ mol } (NH_4)_2SO_4}{1 \text{ L } (NH_4)_2SO_4 \text{ soln}}\right)\left(\frac{2 \text{ mol } NH_4^+}{1 \text{ mol } (NH_4)_2SO_4}\right) = 1.2 \text{ M } NH_4^+$$

$$\text{M } SO_4^{2-} = \left(\frac{0.60 \text{ mol } (NH_4)_2SO_4}{1 \text{ L } (NH_4)_2SO_4 \text{ soln}}\right)\left(\frac{1 \text{ mol } SO_4^{2-}}{1 \text{ mol } (NH_4)_2SO_4}\right) = 0.60 \text{ M } SO_4^{2-}$$

5.108 $$\text{M } Na_3PO_4 = \left(\frac{0.21 \text{ mol } Na^+}{1 \text{ L } Na_3PO_4 \text{ soln}}\right)\left(\frac{1 \text{ mol } Na_3PO_4}{3 \text{ mol } Na^+}\right) = 0.070 \text{ M } Na_3PO_4$$

5.109 $$\text{M } SO_4^{2-} = \left(\frac{0.48 \text{ mol } Fe^{3+}}{1 \text{ L } Fe_2(SO_4)_3 \text{ soln}}\right)\left(\frac{3 \text{ mol } SO_4^{2-}}{2 \text{ mol } Fe^{3+}}\right) = 0.72 \text{ M } SO_4^{2-}$$

5.110 $$\#\text{ g } Al_2(SO_4)_3 = (50.0 \text{ mL solution})\left(\frac{0.12 \text{ mol } Al^{3+}}{1000 \text{ mL solution}}\right)\left(\frac{1 \text{ mol } Al_2(SO_4)_3}{2 \text{ mol } Al^{3+}}\right)\left(\frac{342.14 \text{ g } Al_2(SO_4)_3}{1 \text{ mol } Al_2(SO_4)_3}\right)$$

$= 1.0 \text{ g } Al_2(SO_4)_3$

5.111 $$\#\text{ g } NiCl_2 = (250 \text{ mL soln})\left(\frac{0.055 \text{ mol } Cl^-}{1000 \text{ mL solution}}\right)\left(\frac{1 \text{ mol } NiCl_2}{2 \text{ mol } Cl^-}\right)\left(\frac{129.6 \text{ g } NiCl_2}{1 \text{ mole } NiCl_2}\right) = 0.89 \text{ g } NiCl_2$$

5.112 $$\text{M KOH} = \frac{(20.78 \text{ mL HCl soln})\left(\frac{1 \text{ L HCl soln}}{1000 \text{ mL HCl soln}}\right)\left(\frac{0.116 \text{ mol HCl}}{1 \text{ L HCl}}\right)\left(\frac{1 \text{ mol HCl}}{1 \text{ mol KOH}}\right)}{21.34 \text{ mL KOH}\left(\frac{1 \text{ L KOH}}{1000 \text{ mL KOH}}\right)}$$

$= 0.113$ M KOH

5.113 $$\text{M } H_2SO_4 = \frac{(26.04 \text{ mL NaOH soln})\left(\frac{1 \text{ L NaOH soln}}{1000 \text{ mL NaOH soln}}\right)\left(\frac{0.1024 \text{ mol NaOH}}{1 \text{ L NaOH}}\right)\left(\frac{1 \text{ mol } H_2SO_4}{2 \text{ mol NaOH}}\right)}{12.88 \text{ mL } H_2SO_4\left(\frac{1 \text{ L } H_2SO_4}{1000 \text{ mL } H_2SO_4}\right)}$$

$= 0\ 1035$ M H_2SO_4

5.114 $$\# \text{ mL } NiCl_2\text{soln} = 20.0 \text{ mL soln}\left(\frac{0.15 \text{ mol } Na_2CO_3}{1000 \text{ mL soln}}\right)\left(\frac{1 \text{ mol } NiCl_2}{1 \text{ mol } Na_2CO_3}\right)\left(\frac{1000 \text{ mL soln}}{0.25 \text{ mol } NiCl_2}\right)$$

$= 12.0$ mL $NiCl_2$ soln

$$\# \text{ g } NiCO_3 = 12.0 \text{ mL } NiCl_2 \text{ soln}\left(\frac{0.25 \text{ mol } NiCl_2}{1000 \text{ mL soln}}\right)\left(\frac{1 \text{ mol } NiCO_3}{1 \text{ mol } NiCl_2}\right)\left(\frac{118.7 \text{ g } NiCO_3}{1 \text{ mol } NiCO_3}\right) = 0.36 \text{ g } NiCO_3$$

5.115 $$\# \text{ g } Al_2(SO_4)_3 = 85.0 \text{ mL soln}\left(\frac{0.0500 \text{ mol } Ba(OH)_2}{1000 \text{ mL soln}}\right)\left(\frac{1 \text{ mol } Al_2(SO_4)_3}{3 \text{ mol } Ba(OH)_2}\right)\left(\frac{171.3 \text{ g } Al_2(SO_4)_3}{1 \text{ mol } Al_2(SO_4)_3}\right)$$

$= 0.243$ g $Al_2(SO_4)_3$

5.116 $$\# \text{ mL NaOH} = 25.0 \text{ mL } H_3PO_4\left(\frac{0.25 \text{ mol } H_3PO_4}{1000 \text{ mL soln}}\right)\left(\frac{3 \text{ mol NaOH}}{1 \text{ mol } H_3PO_4}\right)\left(\frac{1000 \text{ mL soln}}{0.100 \text{ mol NaOH}}\right)$$

$= 188$ mL NaOH soln

5.117 $NaHCO_3 + HCl \rightarrow NaCl + H_2O + CO_2$

$$\# \text{ g } NaHCO_3 = (162 \text{ ml HCl soln})\left(\frac{0.052 \text{ mol HCl}}{1000 \text{ mL HCl soln}}\right)$$

$$\times \left(\frac{1 \text{ mol } NaHCO_3}{1 \text{ mol HCl}}\right)\left(\frac{84.01 \text{ g } NaHCO_3}{1 \text{ mol } NaHCO_3}\right)$$

$= 0.71$ g $NaHCO_3$

5.118

$$\text{M } Ba(OH)_2 = \frac{(20.78 \text{ mL HCl soln})\left(\frac{1 \text{ L HCl soln}}{1000 \text{ mL HCl soln}}\right)\left(\frac{0.116 \text{ mol HCl}}{1 \text{ L HCl soln}}\right)\left(\frac{1 \text{ mol } Ba(OH)_2}{2 \text{ mol HCl}}\right)}{(21.34 \text{ mL } Ba(OH)_2)\left(\frac{1 \text{ L } Ba(OH)_2 \text{ soln}}{1000 \text{ mL } Ba(OH)_2 \text{ soln}}\right)}$$

$= 0.0565$ M $Ba(OH)_2$

5.119

$$\text{M } H_3PO_4 = \frac{(26.04 \text{ mL NaOH soln})\left(\frac{1 \text{ L NaOH soln}}{1000 \text{ mL NaOH soln}}\right)\left(\frac{0.1024 \text{ mol NaOH}}{1 \text{ L NaOH soln}}\right)\left(\frac{1 \text{ mol } H_3PO_4}{3 \text{ mol NaOH}}\right)}{(12.88 \text{ mL } H_3PO_4)\left(\frac{1 \text{ L } H_3PO_4 \text{ soln}}{1000 \text{ mL } H_3PO_4 \text{ soln}}\right)}$$

$= 0.06901$ M H_3PO_4

5.120 $\# \text{ mL FeCl}_3 \text{ soln} = 20.0 \text{ mL AgNO}_3\left(\frac{0.0450 \text{ mol AgNO}_3}{1000 \text{ mL soln}}\right)\left(\frac{1 \text{ mol Ag}^+}{1 \text{ mol AgNO}_3}\right)\left(\frac{1 \text{ mol Cl}^-}{1 \text{ mol Ag}^+}\right)$

$$\times\left(\frac{1 \text{ mol FeCl}_3}{3 \text{ mol Cl}^{\bar{G}}}\right)\left(\frac{1000 \text{ mL soln}}{0.150 \text{ mol FeCl}_3}\right) = 2.00 \text{ mL FeCl}_3 \text{ soln}$$

$$\# \text{ g AgCl} = (20.0 \text{ mL AgNO}_3)\left(\frac{1 \text{ L AgNO}_3}{1000 \text{ mL AgNO}_3}\right)\left(\frac{0.0450 \text{ mol AgNO}_3}{1 \text{ L AgNO}_3}\right)$$

$$\times\left(\frac{3 \text{ mol AgCl}}{3 \text{ mol AgNO}_3}\right)\left(\frac{143.32 \text{ AgCl}}{1 \text{ mol AgCl}}\right) = 0.129 \text{ g AgCl}$$

5.121 $\# \text{ g CoCl}_2 = 60.0 \text{ mL KOH soln}\left(\frac{0.200 \text{ mol KOH}}{1000 \text{ mL soln}}\right)\left(\frac{1 \text{ mol OH}^-}{1 \text{ mol KOH}}\right)\left(\frac{1 \text{ mol Co}^{2+}}{2 \text{ mol OH}^-}\right)$

$$\times\left(\frac{1 \text{ mol CoCl}_2}{1 \text{ mol Co}^{2+}}\right)\left(\frac{129.8 \text{ g CoCl}_2}{1 \text{ mol CoCl}_2}\right) = 0.779 \text{ g CoCl}_2$$

5.122 $\# \text{ mL Ba(OH)}_2 = 35.0 \text{ mL H}_2\text{PO}_4 \text{ soln}\left(\frac{0.125 \text{ mol H}_3\text{PO}_4}{1000 \text{ mL soln}}\right)\left(\frac{3 \text{ mol Ba(OH)}_2}{2 \text{ mol H}_3\text{PO}_4}\right)\left(\frac{1000 \text{ mL Ba(OH)}_2}{0.0500 \text{ mol Ba(OH)}_2}\right)$

$$= 131 \text{ mL Ba(OH)}_2$$

5.123 $\# \text{ g NaHCO}_3 = 145 \text{ mL H}_2\text{SO}_4 \text{ soln}\left(\frac{0.0364 \text{ mol H}_2\text{SO}_4}{1000 \text{ mL soln}}\right)\left(\frac{2 \text{ mol NaHCO}_3}{1 \text{ mol H}_2\text{SO}_4}\right)\left(\frac{84.01 \text{ g NaHCO}_3}{1 \text{ mol NaHCO}_3}\right)$

$$= 0.887 \text{ g NaHCO}_3$$

5.124 $\text{Ag}^+ + \text{Cl}^- \rightarrow \text{AgCl}(s)$

$$\# \text{ mL AlCl}_3 = (20.0\text{AgC}_2\text{H}_3\text{O}_2)\left(\frac{0.500 \text{ mol AgC}_2\text{H}_3\text{O}_2}{1000 \text{ mL AgC}_2\text{H}_3\text{O}_2}\right)\left(\frac{1 \text{ mol Ag}^+}{1 \text{ mol AgC}_2\text{H}_3\text{O}_2}\right)$$

$$\left(\frac{1 \text{ mol Cl}^-}{1 \text{ mol Ag}^+}\right)\left(\frac{1 \text{ mol AlCl}_3}{3 \text{ mol Cl}^-}\right)\left(\frac{1000 \text{ mL AlCl}_3}{0.250 \text{ moles AlCl}_3}\right) = 13.3 \text{ mL AlCl}_3$$

5.125

$$\# \text{ mL (NH}_4)_2\text{SO}_4 \text{ soln} =$$

$$(50.0 \text{ ml NaOH soln})\left(\frac{1.00 \text{ mol NaOH}}{1000 \text{ mL NaOH soln}}\right)\left(\frac{1 \text{ mol OH}^-}{1 \text{ mol NaOH}}\right)$$

$$\times\left(\frac{1 \text{ mol NH}_4^+}{1 \text{ mol OH}^-}\right)\left(\frac{1 \text{ mol (NH}_4)_2\text{SO}_4}{2 \text{ mol NH}_4^+}\right)\left(\frac{1000 \text{ mL (NH}_4)_2\text{SO}_4 \text{ soln}}{0.250 \text{ mol (NH}_4)_2\text{SO}_4}\right)$$

$$= 1.00 \times 10^2 \text{ mL (NH}_4)_2\text{SO}_4 \text{ soln}$$

5.126 $Fe_2O_3 + 6HCl \rightarrow 2FeCl_3 + 3H_2O$
$0.0250 \text{ L HCl} \times 0.500 \text{ mol/L} = 1.25 \times 10^{-2} \text{ mol HCl}$

$$\# \text{ mol Fe}^{3+} = (1.25 \times 10^{-2} \text{ mol HCl})\left(\frac{1 \text{ mol Fe}_2\text{O}_3}{6 \text{ mol HCl}}\right)\left(\frac{2 \text{ mol Fe}^{3+}}{1 \text{ mol Fe}_2\text{O}_3}\right)$$

$$= 4.17 \times 10^{-3} \text{ mol Fe}^{3+}$$

$$\text{M Fe}^{3+} = \frac{4.17 \times 10^{-3} \text{ mol Fe}^{3+}}{0.0250 \text{ L soln}} = 0.167 \text{ M Fe}^{3+}$$

$$\# \text{ g Fe}_2\text{O}_3 = (4.17 \times 10^{-3} \text{ mol Fe}^{3+})\left(\frac{1 \text{ mol Fe}_2\text{O}_3}{2 \text{ mol Fe}^{3+}}\right)\left(\frac{159.69 \text{ g Fe}_2\text{O}_3}{1 \text{ mol Fe}_2\text{O}_3}\right)$$

$$= 0.333 \text{ g Fe}_2\text{O}_3$$

Therefore, the mass of Fe_2O_3 that remains unreacted is:
(4.00 g – 0.333 g) = 3.67 g

5.127 First determine the moles of H_2SO_4 present

$$\# \text{ moles H}_2\text{SO}_4 = (30.0 \text{ mL H}_2\text{SO}_4)\left(\frac{0.500 \text{ mol H}_2\text{SO}_4}{1000 \text{ mL H}_2\text{SO}_4}\right) = 0.015 \text{ moles H}_2\text{SO}_4$$

Next determine the number of moles of $Mg(OH)_2$ present:

$$\# \text{ mol Mg(OH)}_2 = (3.50 \text{ g Mg(OH)}_2)\left(\frac{1 \text{ mol Mg(OH)}_2}{58.32 \text{ g Mg(OH)}_2}\right) = 0.0600 \text{ mol Mg(OH)}_2$$

Thus, from the reaction, $H_2SO_4(aq) + Mg(OH)_2(s) \rightarrow 2H_2O(\ell) + MgSO_4(aq)$, we see that 0.015 mol $Mg(OH)_2$ will react with 0.015 mol H_2SO_4. This produces 0.015 mol of $MgSO_4(aq)$ in 30.0 mL of solution and leaves 0.0600 – 0.015 = 0.045 mol $Mg(OH)_2$ unreacted. The concentration of Mg^{2+} is:

$$[\text{Mg}^{2+}] = \left(\frac{0.015 \text{ mol MgSO}_4}{0.0300 \text{ L soln}}\right)\left(\frac{1 \text{ mol Mg}^{2+}}{1 \text{ mol MgSO}_4}\right) = 0.50 \text{ M}$$

The number of grams of $Mg(OH)_2$ not dissolved is:

$$\# \text{ g Mg(OH)}_2 = (0.045 \text{ mol Mg(OH)}_2)\left(\frac{58.32 \text{ g Mg(OH)}_2}{1 \text{ mol Mg(OH)}_2}\right) = 2.6 \text{ g Mg(OH)}_2$$

5.128 The equation for the reaction indicates that the two materials react in equimolar amounts, i.e. the stoichiometry is 1 to 1:

$$AgNO_3(aq) + NaCl(aq) \rightarrow AgCl(s) + NaNO_3(aq)$$

(a) Because this reaction is 1:1, we can see by inspection that the $AgNO_3$ is the limiting reagent. We know this because the concentration of the $AgNO_3$ is lower than the NaCl. Since we start with equal volumes, there are fewer moles of the $AgNO_3$.

$$\# \text{ mol AgCl} = (25.0 \text{ mL AgNO}_3 \text{ soln})\left(\frac{0.320 \text{ mol AgNO}_3}{1000 \text{ mL AgNO}_3 \text{ soln}}\right) \times \left(\frac{1 \text{ mol AgCl}}{1 \text{ mol AgNO}_3}\right)$$

$$= 8.00 \times 10^{-3} \text{ mol AgCl}$$

(b) Assuming that AgCl is essentially insoluble, the concentration of silver ion can be said to be zero since all of the $AgNO_3$ reacted. The number of moles of chloride ion would be reduced by the precipitation of 8.00×10^{-3} mol AgCl, such that the final number of moles of chloride ion would be:

$$0.0250 \text{ L} \times 0.440 \text{ mol/L} - 8.00 \times 10^{-3} \text{ mol} = 3.0 \times 10^{-3} \text{ mol Cl}^-$$

The final concentration of Cl^- is, therefore:

$$3.0 \times 10^{-3} \text{ mol} \div 0.0500 \text{ L} = 0.060 \text{ M Cl}^-$$

All of the original number of moles of NO_3^- and of Na^+ would still be present in solution, and their concentrations would be:

For NO_3^-:

$$\# \text{ M NO}_3^- = \frac{(25.0 \text{ mL AgNO}_3 \text{ soln})\left(\frac{0.320 \text{ mol AgNO}_3}{1000 \text{ mL AgNO}_3 \text{ soln}}\right)\left(\frac{1 \text{ mol NO}_3^-}{1 \text{ mol AgNO}_3}\right)}{(50.0 \text{ mL soln})\left(\frac{1 \text{ L soln}}{1000 \text{ mL soln}}\right)}$$

$$= 0.160 \text{ M NO}_3^-$$

For Na^+:

$$\# \text{ M Na}^+ = \frac{(25.0 \text{ mL NaCl soln})\left(\frac{0.440 \text{ mol NaCl}}{1000 \text{ mL NaCl soln}}\right)\left(\frac{1 \text{ mol Na}^+}{1 \text{ mol NaCl}}\right)}{(50.0 \text{ mL soln})\left(\frac{1 \text{ L soln}}{1000 \text{ mL soln}}\right)}$$

$$= 0.220 \text{ M Na}^+$$

5.129 (a) $3Ca(NO_3)_2(aq) + 2Na_3PO_4(aq) \rightarrow Ca_3(PO_4)_2(s) + 6NaNO_3(aq)$
First, determine the initial number of moles of Ca^{2+} ion that are present:

$$\# \text{ mol Ca}^{2+} = (34.0 \text{ mL Ca(NO}_3)_2 \text{ soln})\left(\frac{0.140 \text{ mol Ca(NO}_3)_2}{1000 \text{ mL Ca(NO}_3)_2 \text{ soln}}\right)$$

$$\times \left(\frac{1 \text{ mol Ca}^{2+}}{1 \text{ mol Ca(NO}_3)_2}\right) = 4.76 \times 10^{-3} \text{ mol Ca}^{2+}$$

Next, determine the initial number of moles of phosphate ion that are present:

$$\# \text{ mol PO}_4^{3-} = (25.0 \text{ mL Na}_3\text{PO}_4 \text{ soln})\left(\frac{0.185 \text{ mol Na}_3\text{PO}_4}{1000 \text{ mL Na}_3\text{PO}_4 \text{ soln}}\right)$$

$$\times \left(\frac{1 \text{ mol PO}_4^{3-}}{1 \text{ mol Na}_3\text{PO}_4}\right) = 4.63 \times 10^{-3} \text{ mol PO}_4^{3-}$$

Now determine the number of moles of calcium ion that are required to react with this much phosphate ion, and compare the result to the amount of calcium ion that is available:

$$\# \text{ mol Ca}^{2+} = (4.63 \times 10^{-3} \text{ mol PO}_4^{\ 3-})\left(\frac{3 \text{ mol Ca}^{2+}}{2 \text{ mol PO}_4^{\ 3-}}\right)$$

$$= 6.94 \times 10^{-3} \text{ mol Ca}^{2+}$$

Since there is not this much Ca^{2+} available according to the above calculation, then we can conclude that Ca^{2+} must be the limiting reagent, and that subsequent calculations should be based on the number of moles of it that are present:

$$\# \text{ g Ca}_3(\text{PO}_4)_2 = (4.76 \times 10^{-3} \text{ mol Ca}^{2+})\left(\frac{1 \text{ mol Ca}_3(\text{PO}_4)_2}{3 \text{ mol Ca}^{2+}}\right)$$

$$\times \left(\frac{310.2 \text{ g Ca}_3(\text{PO}_4)_2}{1 \text{ mol Ca}_3(\text{PO}_4)_2}\right) = 0.492 \text{ g Ca}_3(\text{PO}_4)_2$$

(b) If we assume that the $Ca_3(PO_4)_2$ is completely insoluble, then its concentration may be said to be essentially zero. The concentrations of the other ions are determined as follows:

For nitrate:

$$\# \text{ M NO}_3^{\ -} = \frac{(34.0 \text{ mL Ca(NO}_3)_2 \text{ soln})\left(\frac{0.140 \text{ mol Ca(NO}_3)_2}{1000 \text{ mL Ca(NO}_3)_2 \text{ soln}}\right)\left(\frac{2 \text{ mol NO}_3^{\ -}}{1 \text{ mol Ca(NO}_3)_2}\right)}{((34.0 + 25.0) \text{ mL soln})\left(\frac{1 \text{ L soln}}{1000 \text{ mL soln}}\right)}$$

$$= 0.161 \text{ M NO}_3^{\ -}$$

For Na^+:

$$\# \text{ M Na}^+ = \frac{(25.0 \text{ mL Na}_3\text{PO}_4 \text{ soln})\left(\frac{0.185 \text{ mol Na}_3\text{PO}_4}{1000 \text{ mL Na}_3\text{PO}_4 \text{ soln}}\right)\left(\frac{3 \text{ mol Na}^+}{1 \text{ mol Na}_3\text{PO}_4}\right)}{((34.0 + 25.0) \text{ mL soln})\left(\frac{1 \text{ L soln}}{1000 \text{ mL soln}}\right)} = 0.235 \text{ M Na}^+$$

For phosphate, we determine the number of moles that react with calcium:

$$\# \text{ mol PO}_4^{\ 3-} = (4.76 \times 10^{-3} \text{ mol Ca}^{2+})\left(\frac{2 \text{ mol PO}_4^{\ 3-}}{3 \text{ mol Ca}^{2+}}\right)$$

$$= 3.17 \times 10^{-3} \text{ mol PO}_4^{\ 3-}$$

and subtract from the original number of moles that were present:

$$\# \text{ mol PO}_4^{\ 3-} = 4.63 \times 10^{-3} \text{ mol PO}_4^{\ 3-} - 3.17 \times 10^{-3} \text{ mol PO}_4^{\ 3-}$$

$$= 1.46 \times 10^{-3} \text{ mol PO}_4^{\ 3-}$$

This allows a calculation of the final phosphate concentration:

$$\# \text{ M } PO_4^{3-} = \frac{1.45 \times 10^{-3} \text{ mol } PO_4^{3-}}{((34.0 + 25.0) \text{ mL soln})\left(\frac{1 \text{ L soln}}{1000 \text{ mL soln}}\right)}$$

$$= 0.0247 \text{ M } PO_4^{3-}$$

5.130 $\# \text{ g Pb} = (1.081 \text{ g } PbSO_4)\left(\frac{1 \text{ mol } PbSO_4}{303.27 \text{ g } PbSO_4}\right)\left(\frac{1 \text{ mol Pb}}{1 \text{ mol } PbSO_4}\right)\left(\frac{207.2 \text{ g Pb}}{1 \text{ mol Pb}}\right) = 0.7386 \text{ g Pb}$

The percentage of Pb in the sample can be calculated as

$$\% \text{ Pb} = \left(\frac{\text{mass of Pb}}{\text{mass of sample}}\right)\times 100\% = \frac{0.7386 \text{ g Pb}}{1.526 \text{ g } PbSO_4}\times 100\% = 48.40\% \text{ Pb}$$

5.131 $\# \text{ g Ba} = (1.159 \text{ g } BaSO_4)\left(\frac{1 \text{ mol } BaSO_4}{233.39 \text{ g } BaSO_4}\right)\left(\frac{1 \text{ mol Ba}}{1 \text{ mol } BaSO_4}\right)\left(\frac{137.33 \text{ g Ba}}{1 \text{ mol Ba}}\right) = 0.6820 \text{ g Ba}$

%Ba in the original ore = (0.6820 g Ba/1.542 g ore) × 100% = 44.23%

5.132 First, calculate the number of moles HCl based on the titration according to the following equation:

$NaOH(aq) + HCl(aq) \rightarrow NaCl(aq) + H_2O(\ell)$

$$\# \text{ mol HCl} = (23.25 \text{ mL NaOH})\left(\frac{0.105 \text{ mol NaOH}}{1000 \text{ mL NaOH}}\right)\left(\frac{1 \text{ mol HCl}}{1 \text{ mol NaOH}}\right)$$

$$= 2.44 \times 10^{-3} \text{ mol HCl}$$

Next, determine the concentration of the HCl solution:
2.44×10^{-3}mol ÷ 0.02145 L = 0.114 M HCl

5.133 (a) The balanced equation for the titration is:

$NaOH(aq) + HC_2H_3O_2(aq) \rightarrow NaC_2H_3O_2(aq) + H_2O(\ell)$

$$\# \text{ mol } HC_2H_3O_2 = (20.65 \text{ mL NaOH})\left(\frac{0.504 \text{ mol NaOH}}{1000 \text{ mL NaOH}}\right)\left(\frac{1 \text{ mol } HC_2H_3O_2}{1 \text{ mol NaOH}}\right) = 1.04 \times 10^{-2} \text{ mol } HC_2H_3O_2$$

1.04×10^{-2}mol/0.0125 L = 0.833 M $HC_2H_2O_2$

(b) First convert the density of vinegar to a value appropriate for one liter of solution:
1.0 g/mL × 1000 mL/L = 1000 g/L
We know that one liter of this vinegar contains 0.833 mol of acetic acid so we can determine the mass of acetic acid that is present in one liter of this vinegar:

$$\# \text{ g } HC_2H_3O_2 = (0.833 \text{ mol } HC_2H_3O_2)\left(\frac{60.05 \text{ g } HC_2H_3O_2}{1 \text{ mol } HC_2H_3O_2}\right)$$

$$= 50.0 \text{ g } HC_2H_3O_2$$

The % by weight of acetic acid in vinegar solution is then given by the following:

(50.0 g $HC_2H_3O_2$/L ÷ 1000 g/L) × 100 = 5.00 % acetic acid

This is the mass of acetic acid in one L of solution divided by the total mass of one L of solution, multiplied by 100%.

5.134 Since lactic acid is monoprotic, it reacts with sodium hydroxide on a one to one mole basis:

$$\# \text{ mol } HC_3H_5O_3 = (17.25 \text{ mL NaOH})\left(\frac{0.155 \text{ mol NaOH}}{1000 \text{ mL NaOH}}\right)\left(\frac{1 \text{ mol } HC_3H_5O_3}{1 \text{ mol NaOH}}\right)$$

$$= 2.67 \times 10^{-3} \text{ mol } HC_3H_5O_3$$

5.135 Note that ascorbic acid is diprotic.

$$\# \text{ mol } H_2C_6H_6O_6 = (15.20 \text{ mL NaOH})\left(\frac{0.0200 \text{ mol NaOH}}{1000 \text{ mL NaOH}}\right)\left(\frac{1 \text{ mol } H_2C_6H_6O_6}{2 \text{ mol NaOH}}\right)\left(\frac{176.13 \text{ g } H_2C_6H_6O_6}{1 \text{ mol } H_2C_6H_6O_6}\right)$$

$$= 2.68 \times 10^{-2} \text{ mol } H_2C_6H_6O_6$$

$$\% \ H_2C_6H_6O_6 = \frac{2.68 \times 10^{-2} \text{ g}}{0.1000 \text{ g}} \times 100\% = 26.8\%$$

5.136 $MgSO_4 + BaCl_2 \rightarrow BaSO_4(s) + Mg^{2+} + 2Cl^-$

$$\# \text{ mol } BaSO_4 = (1.174 \text{ g } BaSO_4)\left(\frac{1 \text{ mol } BaSO_4}{233.39 \text{ g } BaSO_4}\right) = 5.030 \times 10^{-3} \text{ mol } BaSO_4$$

There are as many moles of $MgSO_4$ as of $BaSO_4$, namely 5.030×10^{-3}mol $MgSO_4$. We know this because we can see that there is one mole of SO_4^{2-} in both the Ba and Mg compounds. Since both of these elements are in the same family, the reaction to produce the barium salt from the magnesium salt must be 1:1.

First determine the mass of $MgSO_4$ that was present:

5.030×10^{-3} mol $MgSO_4 \times 120.37$ g/mol = 0.6055 g $MgSO_4$

and subtract this from the total mass of the sample to find the mass of water in the original sample:

1.24 g – 0.6055 g = 0.63 g H_2O

We need to know the number of moles of water that are involved:

0.63 g ÷ 18.0 g/mol = 3.5×10^{-2}mol H_2O

and the relative mole amounts of water and $MgSO_4$:
For $MgSO_4$, 5.030×10^{-3}moles/5.030×10^{-3} moles = 1.000
For water, $3.5 \times 10^{-2}/5.030 \times 10^{-3} = 7.0$
Hence the formula is $MgSO_4 \cdot 7H_2O$

5.137 From the Law of Conservation of Mass we know that the amount of Cl we ended with must equal the amount initially:

$$\# \text{ g Cl} = (0.678 \text{ g AgCl})\left(\frac{1 \text{ mol AgCl}}{143.3 \text{ g AgCl}}\right)\left(\frac{1 \text{ mol Cl}}{1 \text{ mol AgCl}}\right)\left(\frac{35.45 \text{ g Cl}}{1 \text{ mol Cl}}\right)$$

$$= 0.168 \text{ g Cl}$$

Determine moles and then relative moles of Fe and Cl:

0.168 g Cl × 35.45 g/mol = 4.73×10^{-3} mol Cl

(0.300 – 0.168) g Fe ÷ 55.85 g/mol Fe = 2.36×10^{-3} mol Fe

For Cl, the relative number of moles is:

4.74×10^{-3} mol/2.36×10^{-3} mol = 2.00

For Fe, the relative number of moles is:

2.36×10^{-3} mol/2.36×10^{-3} mol = 1.00

and the formula is seen to be $FeCl_2$.

5.138 The HCl that is added will react with the Na_2CO_3. Therefore, the amount of Na_2CO_3 in the mixture may be determined from the amount of HCl needed to react with it. We determine this amount by measuring the amount of HCl added in excess. First determine the number of moles HCl added:

$$\# \text{ moles HCl} = (50.0 \text{ mL HCl})\left(\frac{0.240 \text{ mol HCl}}{1000 \text{ mL HCl}}\right) = 1.20 \times 10^{-2} \text{ moles HCl}$$

Next determine how many moles of HCl remain after the reaction:

$$\# \text{ moles HCl} = (22.90 \text{ mL NaOH})\left(\frac{0.100 \text{ mol NaOH}}{1000 \text{ mL NaOH}}\right)\left(\frac{1 \text{ mol HCl}}{1 \text{ mol NaOH}}\right)$$

$$= 2.29 \times 10^{-3} \text{ moles HCl}$$

Therefore, we know that;
1.20×10^{-2} moles HCl – 2.29×10^{-3} moles HCl = 9.71×10^{-3} moles HCl reacted with the Na_2CO_3 present in the mixture. The balanced equation for the reaction is:

$2 HCl + Na_2CO_3 \rightarrow 2NaCl + H_2O + CO_2$

$$\# \text{ g } Na_2CO_3 = (9.71 \times 10^{-}3 \text{ mol HCl})\left(\frac{1 \text{ mol } Na_2CO_3}{2 \text{ mol HCl}}\right)\left(\frac{105.99 \text{ g } Na_2CO_3}{1 \text{ mol } Na_2CO_3}\right) = 0.515 \text{ g } Na_2CO_3$$

$$\# \text{ g NaCl} = 1.243 \text{ g} - 0.515\text{g} = 0.728 \text{ g}$$

$$\% \text{ NaCl} = \frac{0.728 \text{ g}}{1.243 \text{ g}} \times 100\% = 58.6\,\%$$

5.139 First determine the amount of unreacted HCl:

$$\# \text{ moles HCl} = (13.11 \text{ mL KOH})\left(\frac{0.100 \text{ mol KOH}}{1000 \text{ mL KOH}}\right)\left(\frac{1 \text{ mol HCl}}{1 \text{ mol KOH}}\right)$$

$$= 1.31 \times 10^{-3} \text{ moles HCl}$$

This is the total amount of unreacted HCl. Remember that a 25.00 mL sample was taken from a solution of 100.0 mL. So, the total amount of unreacted HCl is 5.24×10^{-3} mol HCl. Next, determine the amount of HCl added:

$$\# \text{ moles HCl} = (50.0 \text{ mL HCl})\left(\frac{0.150 \text{ mol HCl}}{1000 \text{ mL HCl}}\right) = 7.50 \times 10^{-3} \text{ moles HCl}$$

So, the amount of HCl that reacted with the K_2SO_3 is:

7.50×10^{-3} mol – 5.24×10^{-3} mol = 2.26×10^{-3} mol HCl

The reaction is:

$2HCl + K_2SO_3 \rightarrow 2KCl + H_2O + SO_2$

$$\# \text{ g } K_2SO_3 = (2.26 \times 10^{-3} \text{ mol HCl})\left(\frac{1 \text{ mol } K_2SO_3}{2 \text{ mol HCl}}\right)\left(\frac{158.261 \text{ g } K_2SO_3}{1 \text{ mol } K_2SO_3}\right) = 0.179 \text{ g } K_2SO_3$$

$$\% \text{ } K_2SO_3 = \frac{0.179 \text{ g}}{0.486 \text{ g}} \times 100\% = 36.8\%$$

Additional Exercises

5.140 (a) strong electrolyte
(b) nonelectrolyte
(c) strong electrolyte
(d) non electrolyte
(e) weak electrolyte
(f) nonelectrolyte
(g) strong electrolyte
(h) weak electrolyte

5.141 (a) not strongly conducting – it is not soluble in water
(b) not strongly conducting – it is a weak electrolyte
(c) not strongly conducting – it is a nonelectrolyte
(d) strongly conducting – it is a strong electrolyte
(e) not strongly conducting – it is a weak electrolyte

5.142 (a) $CaCO_3(s) + 2H^+(aq) \rightarrow Ca^{2+}(aq) + CO_2(g) + H_2O(\ell)$

(b) $CaCO_3(s) + 2H^+(aq) + SO_4^{2-}(aq) \rightarrow CaSO_4(s) + CO_2(g) + H_2O(\ell)$

(c) $FeS(s) + 2H^+(aq) \rightarrow Fe^{2+}(aq) + H_2S(g)$

(d) $Sn^{2+}(aq) + 2OH^-(aq) \rightarrow Sn(OH)_2(s)$

5.143 (a) molecular: $Na_2S(aq) + H_2SO_4(aq) \rightarrow H_2S(g) + Na_2SO_4(aq)$
net ionic: $S^{2-}(aq) + 2H^+(aq) \rightarrow H_2S(g)$

(b) molecular: $LiHCO_3(aq) + HNO_3(aq) \rightarrow LiNO_3(aq) + H_2O(\ell) + CO_2(g)$
net ionic: $H^+(aq) + HCO_3^-(aq) \rightarrow H_2O(\ell) + CO_2(g)$

(c) molecular: $(NH_4)_3PO_4(aq) + 3KOH(aq) \rightarrow K_3PO_4(aq) + 3NH_3(aq) + 3H_2O(\ell)$
net ionic: $NH_4^+(aq) + OH^-(aq) \rightarrow NH_3(aq) + H_2O(\ell)$

(d) molecular: $K_2SO_3(aq) + HCl(aq) \rightarrow KHSO_3(aq) + KCl(aq)$
net ionic: $SO_3^{2-}(aq) + H^+(aq) \rightarrow HSO_3^-(aq)$

(e) molecular: $BaCO_3(s) + 2HBr(aq) \rightarrow BaBr_2(aq) + CO_2(g) + H_2O(\ell)$
net ionic: $BaCO_3(s) + 2H^+(aq) \rightarrow Ba^{2+}(aq) + CO_2(g) + H_2O(\ell)$

(f) no reaction

5.144

$$\#\text{ g NaCl} = (0.277\text{ g AgCl})\left(\frac{1\text{ mol AgCl}}{143.32\text{ g AgCl}}\right)\left(\frac{1\text{ mol NaCl}}{1\text{ mol AgCl}}\right)\left(\frac{58.44\text{ g NaCl}}{1\text{ mol NaCl}}\right)$$
$$= 0.113\text{ g NaCl}$$

The entire sample was NaCl

5.145 $0.02940\text{ L KOH} \times 0.0300\text{ mol/L} = 8.82 \times 10^{-4}\text{ mol KOH}$
$8.82 \times 10^{-4}\text{ mol KOH} \times (1\text{ mol aspirin}/1\text{ mol KOH}) = 8.82 \times 10^{-4}\text{ mol aspirin}$
$8.82 \times 10^{-4}\text{ mol aspirin} \times 180.2\text{ g/mol} = 0.159\text{ g aspirin}$
$0.159\text{ g aspirin}/0.250\text{ g total} \times 100\% = 63.6\ \%\text{ aspirin}$

5.146 (a) $3Ba^{2+}(aq) + 2Al^{3+}(aq) + 6OH^-(aq) + 3SO_4^{2-}(aq) \rightarrow 3BaSO_4(s) + 2Al(OH)_3(s)$

(b) Because we know the amounts of both starting materials this is a limiting reactant problem. So start by assuming that the barium hydroxide is the limiting reactant.

$$\#\ \text{g BaSO}_4 = (40.0\ \text{mL Ba(OH)}_2)\left(\frac{0.270\ \text{mol Ba(OH)}_2}{1000\ \text{mL Ba(OH)}_2}\right)\left(\frac{1\ \text{mol Ba}^{2+}}{1\ \text{mol Ba(OH)}_2}\right)$$
$$\left(\frac{1\ \text{mol BaSO}_4}{1\ \text{mol Ba}^{2+}}\right)\left(\frac{233.39\ \text{g BaSO}_4}{1\ \text{mol BaSO}_4}\right) = 2.52\ \text{g BaSO}_4$$

Now assume $Al_2(SO_4)_3$ is the limiting reactant.

$$\#\ \text{g BaSO}_4 = (25.0\ \text{mL Al}_2(\text{SO}_4)_3)\left(\frac{0.330\ \text{mol Al}_2(\text{SO}_4)_3}{1000\ \text{mL Al}_2(\text{SO}_4)_3}\right)\left(\frac{3\ \text{mol SO}_4^{2-}}{1\ \text{mol Al}_2(\text{SO}_4)_3}\right)$$
$$\left(\frac{1\ \text{mol BaSO}_4}{1\ \text{mol SO}_4^{2-}}\right)\left(\frac{233.39\ \text{g BaSO}_4}{1\ \text{mol BaSO}_4}\right) = 5.78\ \text{g BaSO}_4$$

Therefore the barium hydroxide is the limiting reactant. Now we can calculate the mass of aluminum hydroxide that is produced.

$$\#\ \text{g Al(OH)}_3 = (40.0\ \text{mL Ba(OH)}_2)\left(\frac{0.270\ \text{mol Ba(OH)}_2}{1000\ \text{mL Ba(OH)}_2}\right)\left(\frac{2\ \text{mol OH}^-}{1\ \text{mol Ba(OH)}_2}\right)$$
$$\left(\frac{1\ \text{mol Al(OH)}_3}{3\ \text{mol OH}^-}\right)\left(\frac{78.00\ \text{g Al(OH)}_3}{1\ \text{mol Al(OH)}_3}\right) = 0.562\ \text{g Al(OH)}_3$$

The total mass of the precipitate is 2.52 g + 0.562 g = 3.08 g

(c) All of the barium ion and hydroxide ion are reacted so the concentration of each is 0. We started with the following:

$$\#\ \text{mol Al}^{3+} = (25.0\text{mL Al}_2(\text{SO}_4)_3)\left(\frac{0.330\ \text{mol Al}_2(\text{SO}_4)_3}{1000\ \text{mL Al}_2(\text{SO}_4)_3}\right)\left(\frac{2\ \text{mol Al}^{3+}}{1\ \text{mol Al}_2(\text{SO}_4)_3}\right)$$
$$= 1.65 \times 10^{-2}\ \text{moles Al}^{3+}$$

$$\#\ \text{mol SO}_4^{2-} = (25.0\ \text{mL Al}_2(\text{SO}_4)_3)\left(\frac{0.330\ \text{mol Al}_2(\text{SO}_4)_3}{1000\ \text{mL Al}_2(\text{SO}_4)_3}\right)\left(\frac{3\ \text{mol SO}_4^{2-}}{1\ \text{mol Al}_2(\text{SO}_4)_3}\right)$$
$$= 2.48 \times 10^{-2}\ \text{moles SO}_4^{2-}$$

In precipitating the $Al(OH)_3$ above, we used 7.2×10^{-3} mol Al leaving

$(1.65 \times 10^{-2} - 7.2 \times 10^{-3}) = 9.3 \times 10^{-3}$ mol Al^{3+} in solution, i.e., the resulting concentration of Al^{3+} is

$$9.3 \times 10^{-3}\ \text{mol} / (0.0400 + 0.0250)\ \text{L} = 0.143\ \text{M Al}^{3+}.$$

Similarly for SO_4^{2-}, the concentration of SO_4^{2-} remaining is

$$\frac{2.48 \times 10^{-2}\ \text{mol} - 1.08 \times 10^{-2}\ \text{mol}}{(0.0400 + 0.250)\ \text{L}} = 0.215\ \text{M SO}_4^{2-},$$

where 1.08×10^{-2} mol SO_4^{2-} represents the amount precipitated as $BaSO_4$.

5.147 (a) The mass of carbon in the original sample must be equal to the mass of carbon that is found in the CO_2.

$$\# \text{ g C} = (22.17 \times 10^{-3} \text{ g } CO_2)\left(\frac{1 \text{ mole } CO_2}{44.01 \text{ g } CO_2}\right)\left(\frac{1 \text{ mole C}}{1 \text{ mole } CO_2}\right)\left(\frac{12.011 \text{ g C}}{1 \text{ mole C}}\right)$$

$$= 6.051 \times 10^{-3} \text{ g C}$$

Similarly, the entire mass of hydrogen that was present in the original sample ends up in the products as H_2O:

$$\# \text{ g H} = (3.40 \times 10^{-3} \text{ g } H_2O)\left(\frac{1 \text{ mole } H_2O}{18.02 \text{ g } H_2O}\right)\left(\frac{2 \text{ mole H}}{1 \text{ mole } H_2O}\right)\left(\frac{1.008 \text{ g H}}{1 \text{ mole H}}\right)$$

$$= 3.80 \times 10^{-4} \text{ g H}$$

The mass of oxygen is determined by subtracting the mass due to C and H from the total mass:

10.46 mg total – (6.051 mg C + 0.380 mg H) = 4.03 mg O.

Convert these masses to mass percents:

$$\% \text{C} = \frac{6.051 \text{ mg}}{10.46 \text{ mg}} \times 100 = 57.85 \%$$

$$\% \text{H} = \frac{0.380 \text{ mg}}{10.46 \text{ mg}} \times 100 = 3.63 \%$$

$$\% \text{O} = \frac{4.03 \text{ mg}}{10.46 \text{ mg}} \times 100 = 38.5 \%$$

(b) Now, convert these masses to a number of moles:

$$\# \text{ moles C} = \left(6.051 \times 10^{-3} \text{ g C}\right)\left(\frac{1 \text{ mole C}}{12.011 \text{ g C}}\right) = 5.038 \times 10^{-4} \text{ moles C}$$

$$\# \text{ moles H} = \left(3.80 \times 10^{-4} \text{ g H}\right)\left(\frac{1 \text{ mole H}}{1.0079 \text{ g H}}\right) = 3.77 \times 10^{-4} \text{ moles H}$$

$$\# \text{ moles O} = \left(4.03 \times 10^{-3} \text{ g O}\right)\left(\frac{1 \text{ mole O}}{15.999 \text{ g O}}\right) = 2.52 \times 10^{-4} \text{ moles O}$$

The relative mole amounts are:

for C, $5.038 \times 10^{-4} \div 2.52 \times 10^{-4} = 2.00$

for H, $3.77 \times 10^{-4} \div 2.52 \times 10^{-4} = 1.50$

for O, $2.52 \times 10^{-4} \div 2.52 \times 10^{-4} = 1.00$

The empirical formula is $C_4H_3O_2$.

(c) Since the empirical mass is 83 and the molecular mass is twice this amount (166), we conclude that the molecular formula must be $C_8H_6O_4$.

(d) The number of moles of unknown acid used in the titration are:

$$\# \text{ mol acid} = (0.1680 \text{ g acid})\left(\frac{1 \text{ mol acid}}{166 \text{ g acid}}\right)$$

$$= 1.01 \times 10^{-3} \text{ mol acid}$$

The number of moles of base used in the titration are:

$$\# \text{ mol base} = (16.18 \text{ mL base})\left(\frac{0.1250 \text{ mol base}}{1000 \text{ mL base}}\right)$$

$$= 2.023 \times 10^{-3} \text{ mol base}$$

We therefore conclude that the unknown acid is diprotic since we used twice as many moles of base as moles of acid.

5.148 Since the number of moles in the final solution must be equal to the number of moles contributed by both solutions, the equation $M_fV_f = M_iV_i$ may be used, and the volumes of the final solution must equal the volumes of the two solution combined.

$V_f = V_1 + V_2$

$V_f = 50.0 \text{ mL} + V_2$

$$\left(\frac{0.25 \text{ mol}}{1000 \text{ mL}}\right)(V_f) = \left(\frac{0.10 \text{ mol}}{1000 \text{ mL}}\right)(V_2) + \left(\frac{0.40 \text{ mol}}{1000 \text{ mL}}\right)(50 \text{ mL})$$

$$\left(\frac{0.25 \text{ mol}}{1000 \text{ mL}}\right)(50 \text{ mL} + V_2) = \left(\frac{0.10 \text{ mol}}{1000 \text{ mL}}\right)(V_2) + \left(\frac{0.40 \text{ mol}}{1000 \text{ mL}}\right)(50 \text{ mL})$$

$$\left(\frac{0.25 \text{ mol}}{1000 \text{ mL}}\right)(50 \text{ mL}) + \left(\frac{0.25 \text{ mol}}{1000 \text{ mL}}\right)(V_2) = \left(\frac{0.10 \text{ mol}}{1000 \text{ mL}}\right)(V_2) + \left(\frac{0.40 \text{ mol}}{1000 \text{ mL}}\right)(50 \text{ mL})$$

$$\left(\frac{0.40 \text{ mol}}{1000 \text{ mL}}\right)(50 \text{ mL}) - \left(\frac{0.25 \text{ mol}}{1000 \text{ mL}}\right)(50 \text{ mL}) = \left(\frac{0.25 \text{ mol}}{1000 \text{ mL}}\right)(V_2) - \left(\frac{0.10 \text{ mol}}{1000 \text{ mL}}\right)(V_2)$$

multiply through by 1000 mL

$(0.40 \text{ mol})(50 \text{ mL}) - (0.25 \text{ mol})(50 \text{ mL}) = (0.25 \text{ mol})(V_2) - (0.10 \text{ mol})(V_2)$

$7.5 \text{ mol mL} = (0.15 \text{ mol})V_2$

$(7.5 \text{ mol mL})/(0.15 \text{ mol}) = V_2$

$V_2 = 50 \text{ mL}$

Practice Exercises

6.1 $2Al(s) + 3Cl_2(g) \rightarrow 2AlCl_3(aq)$
Aluminum is oxidized and is, therefore, the reducing agent.
Chlorine is reduced and is, therefore, the oxidizing agent.

6.2 If H_2O_2 acts as an oxidizing agent, it gets reduced itself in the process. Examining the oxidation numbers:

H_2O_2	H = +1, O = -1
H_2O	H = +1, O = -2
O_2	O = 0

If H_2O_2 is reduced it must form water, since the oxidation number of oxygen drops from –1 to –2 in the formation of water (a reduction).

The product is therefore water.

6.3 a) Ni +2; Cl –1
b) Mg +2; Ti +4; O –2
c) K +1; Cr +6; O –2
d) H +1; P +5, O –2
e) V +3; C 0; H +1; O –2

6.4 There is a total charge of +8, divided over three atoms, so the average charge is +8/3.

6.5 First the oxidation numbers of all atoms must be found.

$$Cl_2 + 2NaClO_2 \rightarrow 2ClO_2 + 2NaCl$$

Reactants:	Products:
Cl = 0	Cl = +4
	O = –2
Na = +1	Na = +1
Cl = +3	Cl = –1
O = –2	

The oxidation numbers for O and Na do not change. However, the oxidation numbers for all chlorine atoms change. There is no simple way to tell which chlorines are reduced and which are oxidized in this reaction.

One analysis would have the Cl in Cl_2 end up as the Cl in NaCl, while the Cl in $NaClO_2$ ends up as the Cl in ClO_2. In this case Cl_2 is reduced and is the oxidizing agent, while $NaClO_2$ is oxidized and is the reducing agent.

6.6 First the oxidation numbers of all atoms must be found.

$$KClO_3 + 6HNO_2 \rightarrow KCl + HNO_3$$

Reactants:	Products:
K = +1	K = +1
Cl = +5	Cl = -1
O = -2	
H = +1	H = +1
N = +3	N = +5
O = -2	O = -2

The oxidation numbers for K and Na do not change. However, the oxidation numbers for all chlorines atoms drop. The oxidation numbers for nitrogen increase.

Therefore, $KClO_3$ is reduced and HNO_2 is oxidized.
This means $KClO_3$ is the oxidizing agent and HNO_2 is the reducing agent.

6.7 $Al(s) + Cu^{2+}(aq) \rightarrow Al^{3+}(aq) + Cu(s)$

First, we break the reaction above into half-reactions:

$$Al(s) \rightarrow Al^{3+}(aq)$$
$$Cu^{2+}(aq) \rightarrow Cu(s)$$

Each half-reaction is already balanced with respect to atoms, so next we add electrons to balance the charges on both sides of the equations:

$$Al(s) \rightarrow Al^{3+}(aq) + 3e^-$$
$$2e^- + Cu^{2+}(aq) \rightarrow Cu(s)$$

Next, we multiply both equations so that the electrons gained equals the electrons lost,

$$2(Al(s) \rightarrow Al^{3+}(aq) + 3e^-)$$
$$3(2e^- + Cu^{2+}(aq) \rightarrow Cu(s))$$

which gives us:

$$2Al(s) \rightarrow 2Al^{3+}(aq) + 6e^-$$
$$6e^- + 3Cu^{2+}(aq) \rightarrow 3Cu(s)$$

Now, by adding the half-reactions back together, we have our balanced equation:

$$2Al(s) + 3Cu^{2+}(aq) \rightarrow 2Al^{3+}(aq) + 3Cu(s)$$

6.8 $TcO_4^- + Sn^{2+} \rightarrow Tc^{4+} + Sn^{4+}$

First, we break the reaction above into half-reactions:

$$TcO_4^- \rightarrow Tc^{4+}$$
$$Sn^{2+} \rightarrow Sn^{4+}$$

Each half-reaction is already balanced with respect to atoms other than O and H, so next we balance the O atoms by using water:

$$TcO_4^- \rightarrow Tc^{4+} + 4H_2O$$
$$Sn^{2+} \rightarrow Sn^{4+}$$

Now we balance H by using H^+:

$$8H^+ + TcO_4^- \rightarrow Tc^{4+} + 4H_2O$$
$$Sn^{2+} \rightarrow Sn^{4+}$$

Next, we add electrons to balance the charges on both sides of the equations:

$$3e^- + 8H^+ + TcO_4^- \rightarrow Tc^{4+} + 4H_2O$$
$$Sn^{2+} \rightarrow Sn^{4+} + 2e^-$$

We multiply the equations so that the electrons gained equals the electrons lost,

$$2(3e^- + 8H^+ + TcO_4^- \rightarrow Tc^{4+} + 4H_2O)$$
$$3(Sn^{2+} \rightarrow Sn^{4+} + 2e^-)$$

which gives us:

$$6e^- + 16H^+ + 2TcO_4^- \rightarrow 2Tc^{4+} + 8H_2O$$
$$3Sn^{2+} \rightarrow 3Sn^{4+} + 6e^-$$

Now, by adding the half-reactions back together, we have our balanced equation:

$$3Sn^{2+} + 16H^+ + 2TcO_4^- \rightarrow 2Tc^{4+} + 8H_2O + 3Sn^{4+}$$

6.9 $(Cu \rightarrow Cu^{2+} + 2e^-) \times 4$
$2NO_3^- + 10H^+ + 8e^- \rightarrow N_2O + 5H_2O$
$4Cu + 2NO_3^- + 10H^+ \rightarrow 4Cu^{2+} + N_2O + 5H_2O$

6.10 $(MnO_4^- + 4H^+ + 3e^- \rightarrow MnO_2 + 2H_2O) \times 2$
$(C_2O_4^{2-} + 2H_2O \rightarrow 2CO_3^{2-} + 4H^+ + 2e^-) \times 3$

$2MnO_4^- + 3C_2O_4^{2-} + 2H_2O \rightarrow 2MnO_2 + 6CO_3^{2-} + 4H^+$

Adding $4OH^-$ to both sides of the above equation we get:

$$2MnO_4^- + 3C_2O_4^{2-} + 2H_2O + 4OH^- \rightarrow 2MnO_2 + 6CO_3^{2-} + 4H_2O$$

which simplifies to give:

$$2MnO_4^- + 3C_2O_4^{2-} + 4OH^- \rightarrow 2MnO_2 + 6CO_3^{2-} + 2H_2O$$

6.11 a) molecular: $Mg(s) + 2HCl(aq) \rightarrow MgCl_2(aq) + H_2(g)$
ionic: $Mg(s) + 2H^+(aq) + 2Cl^-(aq) \rightarrow Mg^{2+}(aq) + 2Cl^-(aq) + H_2(g)$
net ionic: $Mg(s) + 2H^+(aq) \rightarrow Mg^{2+}(aq) + H_2(g)$

b) molecular: $2Al(s) + 6HCl(aq) \rightarrow 2AlCl_3(aq) + 3H_2(g)$
ionic: $2Al(s) + 6H^+(aq) + 6Cl^-(aq) \rightarrow 2Al^{3+}(aq) + 6Cl^-(aq) + 3H_2(g)$
net ionic: $2Al(s) + 6H^+(aq) \rightarrow 2Al^{3+}(aq) + 3H_2(g)$

6.12 a) $2Al(s) + 3Cu^{2+}(aq) \rightarrow 2Al^{3+}(aq) + 3Cu(s)$
b) N.R.

6.13 $2C_4H_{10}(l) + 13O_2(g) \rightarrow 8CO_2(g) + 10H_2O(g)$

6.14 $C_2H_5OH(l) + 3O_2(g) \rightarrow 2CO_2(g) + 3H_2O(g)$

6.15 $4Fe(s) + 3O_2(g) \rightarrow 2Fe_2O_3(s)$

6.16 First we need a balanced equation:
$Cl_2 + 2e^- \rightarrow 2Cl^-$
$S_2O_3^{2-} + 5H_2O \rightarrow 2SO_4^{2-} + 10H^+ + 8e^-$
$4Cl_2 + S_2O_3^{2-} + 5H_2O \rightarrow 8Cl^- + 2SO_4^{2-} + 10H^+$

$$\#\text{ g } Na_2S_2O_3 = (4.25\text{ g } Cl_2)\left(\frac{1\text{ mol } Cl_2}{70.906\text{ g } Cl_2}\right)\left(\frac{1\text{ mol } Na_2S_2O_3}{4\text{ mol } Cl_2}\right)\left(\frac{158.132\text{ g } Na_2S_2O_3}{1\text{ mol } Na_2S_2O_3}\right)$$

$$= 2.37\text{ g } Na_2S_2O_3$$

6.17 a) $(Sn^{2+} \rightarrow Sn^{4+} + 2e^-) \times 5$
$(MnO_4^- + 8H^+ + 5e^- \rightarrow Mn^{2+} + 4H_2O) \times 2$
$5Sn^{2+} + 2MnO_4^- + 16H^+ \rightarrow 5Sn^{4+} + 2Mn^{2+} + 8H_2O$

b)

$$\#\text{ g Sn} = (8.08\text{ mL } KMnO_4\text{ soln})\left(\frac{0.0500\text{ mol } KMnO_4}{1000\text{ mL } KMnO_4}\right)\left(\frac{1\text{ mol } MnO_4^-}{1\text{ mol } KMnO_4}\right)$$

$$\times\left(\frac{5\text{ mol } Sn^{2+}}{2\text{ mol } MnO_4^-}\right)\left(\frac{1\text{ mol Sn}}{1\text{ mol } Sn^{2+}}\right)\left(\frac{118.71\text{ g Sn}}{1\text{ mol Sn}}\right)$$

$$= 0.120\text{ g Sn}$$

c) $$\%\text{ Sn} = \frac{0.120\text{ g Sn}}{0.300\text{ g sample}} \times 100 = 40.0\ \%\text{ Sn}$$

d)

$$\#\text{ g Sn} = (8.08\text{ mL } KMnO_4\text{ soln})\left(\frac{0.0500\text{ mol } KMnO_4}{1000\text{ mL } KMnO_4}\right)\left(\frac{1\text{ mol } MnO_4^-}{1\text{ mol } KMnO_4}\right)$$

$$\times\left(\frac{5\text{ mol } Sn^{2+}}{2\text{ mol } MnO_4^-}\right)\left(\frac{1\text{ mol } SnO_2}{1\text{ mol } Sn^{2+}}\right)\left(\frac{150.71\text{ g } SnO_2}{1\text{ mol } SnO_2}\right) = 0.152\text{ g } SnO_2$$

$$\% \text{ SnO}_2 = \frac{0.152 \text{ g SnO}_2}{0.300 \text{ g sample}} \times 100 = 50.7 \% \text{ SnO}_2$$

Thinking It Through

T6.1 The oxidation states of the elements in the reactants need to be compared to the oxidation states of the elements in the products. If there is a change, then the reaction is an oxidation reduction reaction. For $HIO_4 + 2H_2O \rightarrow H_5IO_6$:

Element	Reactant	Product
H	+1	+1
I	–2	–2
O	+7	+7

T6.2 After finding both nickel and cadmium in the activity series, we decide which will be more readily oxidized, that is, which metal is higher in the activity series. If Cd is more easily oxidized than Ni, then Cd metal will react with Ni^{2+}, giving Ni metal.

T6.3 Find Au and Fe in the activity series and determine which one is more easily oxidized. The more active metal will react with the ions of the other metal and the metallic strip will corrode.

T6.4 The combustion of hydrocarbon fuels forms water as one of its products. It is formed at a high temperature and then condenses in the cold exhaust pipe. After the exhaust system becomes hot enough, the water is expelled as vapor (gas).

T6.5 First, the overall equation should be balanced using the ion-electron method.

$$IO_3^-(aq) + 6NO_2(g) + 3H_2O(l) \rightarrow I^-(aq) + 6NO_3^-(aq) + 6H^+(aq)$$

The molecular mass of NO_2 must be used to convert grams of NO_2 to moles of NO_2. The coefficients in the balanced equation will be needed to convert from moles of NO_2 to moles of IO_3^-. Finally, moles of IO_3^- are converted to volume of IO_3^- solution by dividing moles by molarity.

$$\# \text{mL IO}_3^- = (0.230 \text{ g NO}_2)\left(\frac{1 \text{ mol NO}_2}{46.01 \text{ g NO}_2}\right)\left(\frac{1 \text{ mol IO}_3^-}{6 \text{ mol NO}_2}\right)\left(\frac{1000 \text{ mL IO}_3^-}{0.0200 \text{ mol IO}_3^-}\right)$$

T6.6 It is first necessary to balance the equation for the reaction.

$$4Cl_2(g) + S_2O_3^{2-}(aq) + 5H_2O(l) \rightarrow 8Cl^-(aq) + 2SO_4^{2-}(aq) + 10H^+(aq)$$

The number of moles of $Na_2S_2O_3$ can be calculated using the coefficients in the balanced equation. Next, these numbers of moles can be converted to a volume by dividing by molarity.

$$\# \text{ mL Na}_2\text{S}_2\text{O}_3 = (0.020 \text{ mol Cl}_2)\left(\frac{1 \text{ mol S}_2\text{O}_3^-}{4 \text{ mol Cl}_2}\right)$$

$$\times \left(\frac{1 \text{ mol Na}_2\text{S}_2\text{O}_3}{1 \text{ mol S}_2\text{O}_3^-}\right)\left(\frac{1000 \text{ mL Na}_2\text{S}_2\text{O}_3}{0.500 \text{ mol Na}_2\text{S}_2\text{O}_3}\right)$$

T6.7 It is first necessary to balance the equation for the reaction.

$$I_3^-(aq) + 2S_2O_3^{2-}(aq) \rightarrow 3I^-(aq) + S_4O_6^{2-}(aq)$$

The number of moles of $Na_2S_2O_3$ can be calculated using the coefficients in the balanced equation. Next, these numbers of moles can be converted to a volume by dividing by molarity.

$$\# \text{ mL } Na_2S_2O_3 = \left(0.020 \text{ mol } I_3^-\right)\left(\frac{2 \text{ mol } S_2O_3^-}{1 \text{ mol } I_3^-}\right) \times \left(\frac{1 \text{ mol } Na_2S_2O_3}{1 \text{ mol } S_2O_3^-}\right)\left(\frac{1000 \text{ mL } Na_2S_2O_3}{0.200 \text{ mol } Na_2S_2O_3}\right)$$

T6.8 The balanced equation is the place to start.

$$6Fe^{2+}(aq) + Cr_2O_7^{2-}(aq) + 14H^+(aq) \rightarrow 6Fe^{3+}(aq) + 2Cr^{3+}(aq) + 7H_2O(l)$$

Because amounts of both reagents are specified, we must work a limiting reactant problem to find out which of the two reactants is completely consumed. From the number of moles of this reactant that disappear, we can calculate the number of moles of H^+ that react. This amount is subtracted from the initial number of moles of hydrogen ion, and the amount of titrant is calculated by dividing moles by molarity of NaOH solution.

If Fe^{2+} is the limiting reactant:

$$\# \text{ mL NaOH} = \left(0.280 \text{ mol } H^+ - \left(400 \text{ mL } Fe^{2+}\right)\left(\frac{0.060 \text{ mol } Fe^{2+}}{1000 \text{ mL } Fe^{2+}}\right)\left(\frac{14 \text{ mol } H^+}{6 \text{ mol } Fe^{2+}}\right)\right)\left(\frac{1 \text{ mol NaOH}}{1 \text{ mol } H^+}\right) \times \left(\frac{1000 \text{ mL NaOH}}{0.0100 \text{ mol NaOH}}\right)$$

If $Cr_2O_7^{2-}$ is the limiting reactant:

$$\# \text{ mL NaOH} = \left(0.280 \text{ mol } H^+ - \left(300 \text{ mL } Cr_2O_7^{2-}\right)\left(\frac{0.0200 \text{ mol } Cr_2O_7^{2-}}{1000 \text{ mL } Cr_2O_7^{2-}}\right)\left(\frac{14 \text{ mol } H^+}{1 \text{ mol } Cr_2O_7^{2-}}\right)\right)\left(\frac{1 \text{ mol NaOH}}{1 \text{ mol } H^+}\right) \times \left(\frac{1000 \text{ mL NaOH}}{0.0100 \text{ mol NaOH}}\right)$$

T6.9 To begin, determine balanced equations for the reaction of SO_3^{2-} with CrO_4^{2-} and for $S_2O_3^{2-}$ with CrO_4^{2-}.

$$3SO_3^{2-} + 2CrO_4^{2-} + H_2O \rightarrow 3SO_4^{2-} + 2CrO_2^- + 2OH^-$$

$$3S_2O_3^{2-} + 8CrO_4^{2-} + H_2O \rightarrow 6SO_4^{2-} + 8CrO_2^- + 2OH^-$$

We know that the amount of CrO_4^{2-} that reacted is

$$\# \text{ moles } CrO_4^{2-} = (80.0 \text{ mL } CrO_4^{2-})\left(\frac{0.0500 \text{ moles } CrO_4^{2-}}{1000 \text{ mL } CrO_4^{2-}}\right)$$

$$= 4.00 \times 10^{-3} \text{ moles } CrO_4^{2-}$$

We also know the amount of SO_4^{2-} produced

$$\#\text{ moles } SO_4^{2-} = (0.9336\text{ g }BaSO_4)\left(\frac{1\text{ mol }BaSO_4}{233.39\text{ g }BaSO_4}\right)\left(\frac{1\text{ mol }SO_4^{2-}}{1\text{ mol }BaSO_4}\right)$$

$$= 4.00\times10^{-3}\text{ moles }SO_4^{2-}$$

Let x = # moles SO_3^{2-} and y = # moles $S_2O_3^{2-}$, from the balanced equations and the known quantities we can write:

$$4.00\times10^{-3}\text{ moles }SO_4^{2-} = x\left(\frac{1\text{ mol }SO_4^{2-}}{1\text{ mol }SO_3^{2-}}\right) + y\left(\frac{2\text{ mol }SO_4^{2-}}{1\text{ mol }S_2O_3^{2-}}\right)$$

$$4.00\times10^{-3}\text{ moles }CrO_4^{2-} = x\left(\frac{2\text{ mol }CrO_4^{2-}}{3\text{ mol }SO_3^{2-}}\right) + y\left(\frac{8\text{ mol }CrO_4^{2-}}{3\text{ mol }S_2O_3^{2-}}\right)$$

We can solve for x and y from these and calculate the initial concentration of SO_3^{2-} and $S_2O_3^{2-}$.

T6.10 It is first necessary to write a balanced equation for the reaction of MnO_4^- with Sn^{2+}. Then, we calculate the original moles of Sn^{2+} in the 50.0 mL of 0.0300 M $SnCl_2$-solution. The moles of Sn^{2+} that were titrated are calculated by multiplying (remembering to include stoichiometry) molarity (0.0100 M) by volume of titrant (0.02728 L). This number of moles of tin ion remaining is subtracted from the total that was available in the 50.0 mL portion that was titrated, and the answer is converted to the number of moles of MnO_4^- that had reacted with this number of moles of Sn^{2+}. Multiply this number by 10 to get the moles of MnO_4^- that had not reacted with the SO_2 in the original 500 mL of 0.0200 M $KMnO_4$. By difference, calculate the moles of SO_2 that had reacted, which is equal to the number of moles of S in the original sample. The mass of S is calculated by dividing moles by atomic mass, and the percentage of S in the original sample is the mass of S divided by the total sample mass, times 100.

T6.11 The reaction that occurs is $2Ag^+(aq) + Cu(s) \rightarrow 2Ag(s) + Cu^{2+}(aq)$. If we assume that there is excess copper available, we need to determine the number of moles of Ag that will be produced.

The number of moles of Ag^+ available for the reaction is

$$0.250\text{ M} \times 0.0500\text{ L} = 0.0125\text{ mol }Ag^+$$

We can determine the amount of copper consumed from the balanced equation. Since the stoichiometry is 2/1, the number of moles of Cu^{2+} ion that are consumed is $0.0125 \div 2 = 0.00625$ mol. Convert this number of moles to a number of grams: $0.00625\text{ mol} \times 63.546\text{ g/mol} = 0.397$ g. The amount of unreacted copper is thus: $32.00\text{ g} - 0.397\text{ g} = 31.60$ g Cu.

The mass of Ag that is formed is: $0.0125\text{ mol} \times 107.9\text{ g/mol} = 1.35$ g Ag.

The final mass of the bar will include the unreacted copper and the silver that is formed: $31.60\text{ g} + 1.35\text{ g} = 32.95$ g.

T6.12 Place a clean metal bar of each metal in each of the other metal nitrate solutions. The more active metals will react (dissolve) in the nitrate solution. The most active metal will react with each of the other three solutions, the least active metal will react with none of the solutions, and the other two metals will react with one or two of the solutions. Arrange the metals in the order that corresponds to the number of solutions in which it reacted.

Review Questions

6.1 (a) Oxidation is the loss of one or more electrons.
Reduction is the gain of one or more electrons.

(b) The oxidation number decreases in a reduction and increases in an oxidation.

6.2 The oxidation number of the Mg changes from 0 to +2. The oxidation number of the oxygen changes from 0 in O_2 to –2 in MgO. The magnesium is oxidized and the oxygen is reduced. Consequently, Mg is the reducing agent and the O_2 is the oxidizing agent.

6.3 The number of electrons involved in both the reduction and the oxidation must be the same; only those electrons that come from the reductant and go to the oxidant are involved. No electrons from external, or uninvolved sources are allowed to enter the process, and there cannot be any electrons left unreacted at the end. An oxidizing agent is the species that is reduced or gains electrons in an oxidation reduction reaction. A reducing agent is the species that is oxidized or loses electrons in an oxidation reduction reaction.

6.4 Oxidation state and oxidation number are synonyms. Thus, the As is in the +3 oxidation state.

6.5 This is not a redox reaction since there is no change in oxidation number.

6.6 The oxidation state of Cr is +6 in both reactants and products. No other element changes oxidation state. Therefore this is not a redox reaction.

6.7 This change in oxidation number represents a reduction of nitrogen by 5 units, and it requires that nitrogen gain 5 electrons.

6.8 The equation is not balanced since the charge is different on either side of the arrow. It is easily balanced by inspection to give:

$$2Ag^+ + Fe \rightarrow 2Ag + Fe^{2+}$$

6.9 By inspection,

$$2Cr^{3+} + 3Zn \rightarrow 2Cr + 3Zn^{2+}$$

6.10 (a) +9 charge on the left, +1 charge on the right; add 8 electrons to the left side.
(b) 0 charge on the left, +6 charge on the right; add 6 electrons to the right side.

6.11 (a) is a reduction
(b) is an oxidation

6.12 It is a reaction in which one element replaces another in a compound.

6.13 A "nonoxidizing acid" is one in which the H^+ ion is the strongest oxidizing agent. That is, the anion of the acid is not itself a better oxidizing agent than H^+. Examples are HCl and H_2SO_4. The oxidizing agent in a nonoxidizing acid is the H^+.

6.14 $NO_3^-(aq)$

6.15 The metal must be below hydrogen in the activity series in order for it to react with HCl.

6.16 The best reducing agents are those that are most easily oxidized and are found at the bottom of the activity series. The best oxidizing agents are those that are most easily reduced and are found at the top of the activity series. (See Table 6.2 for clarification.)

6.17 This would be any metal higher (less reactive) than hydrogen, i.e. gold, mercury, silver, and copper.

6.18 The most active types of metals will react with water for example:
$2Cs(s) + 2H_2O(l) \rightarrow 2CsOH(aq) + H_2(g)$
$2Rb(s) + 2H_2O(l) \rightarrow 2RbOH(aq) + H_2(g)$
$Ba(s) + 2H_2O(l) \rightarrow Ba(OH)_2(aq) + H_2(g)$
Up to
$Pb(s) + 2H_2O(l) \rightarrow Pb(OH)_2(aq) + H_2(g)$

6.19 Based on the activity series, the manganese is oxidized and therefore is the reducing agent.

6.20 Combustion is the rapid reaction of a substance with oxygen, which is accompanied by the evolution of light and heat.

6.21 Historically, the reaction of a substance with oxygen was termed oxidation. Now we realize that reaction with oxygen most typically means that oxygen acquires electrons from the substance with which it reacts. The oxidation of a substance is, therefore, taken to represent the loss of electrons by a substance, whether the substance has reacted with oxygen or with another oxidizing agent.

6.22 (a) $CO_2(g)$ and $H_2O(g)$
(b) $CO(g)$, $CO_2(g)$ and $H_2O(g)$
(c) $C(s)$ and $H_2O(g)$

6.23 The products will be CO_2, H_2O, and SO_2:
$2C_2H_6S(l) + 9O_2(g) \rightarrow 4CO_2(g) + 2SO_2(g) + 6H_2O(g)$

6.24 The other product is water:
$4NH_3(g) + 3O_2(g) \rightarrow 2N_2(g) + 6H_2O(g)$

Review Problems

6.25 (a) substance reduced (and oxidizing agent): HNO_3
substance oxidized (and reducing agent): H_3AsO_3
(b) substance reduced (and oxidizing agent): HOCl
substance oxidized (and reducing agent): NaI
(c) substance reduced (and oxidizing agent): $KMnO_4$
substance oxidized (and reducing agent): $H_2C_2O_4$
(d) substance reduced (and oxidizing agent): H_2SO_4
substance oxidized (and reducing agent): Al

6.26 (a) substance reduced (and oxidizing agent): H_2SO_4
substance oxidized (and reducing agent): Cu
(b) substance reduced (and oxidizing agent): HNO_3
substance oxidized (and reducing agent): SO_2
(c) substance reduced (and oxidizing agent): H_2SO_4
substance oxidized (and reducing agent): Zn
(d) substance reduced (and oxidizing agent): HNO_3
substance oxidized (and reducing agent): I_2

6.27 Recall that the sum of the oxidation numbers must equal the charge on the molecule or ion.
(a) –2
(b) +4; we know that O is usually in a –2 oxidation state.
(c) The oxidation state of an element is always zero, by definition.
(d) –3; hydrogen is usually in a +1 oxidation state.

6.28 Recall that the sum of the oxidation numbers must equal the charge on the molecule or ion.
(a) +7; oxygen is usually in a –2 oxidation state and the charge is 1–.
(b) +3; chlorine is usually in a –1 oxidation state and the charge is 0.
(c) +4; sulfur is usually in a –2 oxidation state and the charge is 0.
(d) +3; the nitrate ion has a charge of 1– and the total charge is 0.

6.29 The sum of the oxidation numbers should be zero:

(a)	Na:	+1	(c)	Na:	+1
	H:	+1		S:	+2.5

	P:	+5		O:	–2
	O:	–2			
(b)	Ba:	+2	(d)	Cl:	+3
	Mn:	+6		F:	–1
	O:	–2			

6.30 The sum of the oxidation numbers should be equal to the total, or overall charge on the ion:

(a)	N	+5	(c)	N	+3
	O	–2		O	–2
(b)	S	+4	(d)	Cr	+6
	O	–2		O	–2

6.31 The sum of the oxidation numbers should be zero:

(a)	+2	(d)	+4
(b)	+5	(e)	–2
(c)	–1		

6.32 The sum of the oxidation numbers should be zero:

(a)	–3	(d)	–1/3
(b)	+3	(e)	+3
(c)	0		

6.33 The sum of the oxidation numbers should be zero:

(a)	O: –2	(c)	O: –2
	Na: +1		Na: +1
	Cl: +1		Cl: +5
(b)	O: –2	(d)	O: –2
	Na: +1		Na: +1
	Cl: +3		Cl: +7

6.34 The sum of the oxidation numbers should be equal to the total charge.

(a)	O: –2	(c)	O: –2
	Ca: +2		Mn: +6
	V: +5		
(b)	Cl: –1	(d)	O: –2
	Sn: +4		Mn: +4

6.35 The sum of the oxidation numbers should be zero:

(a)	S: –2	(c)	O: –2
	Pb: +2		Sr: +2
			I: +5
(b)	Cl: –1	(d)	S: –2
	Ti: +4		Cr: +3

6.36

(a)	O +2	(c)	Cs +1
	F –1		O –1/2 (The Cs can only have an oxidation number of +1 or 0.)
(b)	H +1	(d)	F –1
	F –1		O +1
	O 0		

6.37 $Cl_2(aq) + H_2O \leftrightarrows H^+(aq) + Cl^-(aq) + HOCl(aq)$

In the forward direction: The oxidation number of the chlorine atoms decreases from 0 to -1. Therefore **Cl_2 is reduced**. However, in HOCl, chlorine has an oxidation number of +1, so **Cl_2 also oxidized!** (One atom is reduced, the other is oxidized.)

In the reverse direction: The Cl^- ion begins with an oxidation number of -1 and ends with an oxidation number of 0. Therefore the Cl^- ion is oxidized: This means **Cl^- is the reducing agent.** Since the oxidation number of H^+ does not change, **HOCl must be the oxidizing agent.**

6.38 N is reduced from +4 to +2 and N is oxidized from +4 to +5.

6.39 (a) $BiO_3^- + 6H^+ + 2e^- \rightarrow Bi^{3+} + 3H_2O$ This is reduction of BiO_3^{2-}.

(b) $Pb^{2+} + 2H_2O \rightarrow PbO_2 + 4H^+ + 2e^-$ This is oxidation of Pb^{2+}.

6.40 (a) $NO_3^- + 10H^+ + 8e^- \rightarrow NH_4^+ + 3H_2O$ This constitutes reduction of NO_3^-.

(b) $6H_2O + Cl_2 \rightarrow 2ClO_3^- + 12H^+ + 10e^-$ This constitutes oxidation of Cl_2.

6.41 (a) $Fe + 2OH^- \rightarrow Fe(OH)_2 + 2e^-$ This is oxidation of Fe.

(b) $2e^- + 2OH^- + SO_2Cl_2 \rightarrow SO_3^{2-} + 2Cl^- + H_2O$ This is reduction of SO_2Cl_2.

6.42 (a) $6OH^- + Mn(OH)_2 \rightarrow MnO_4^{2-} + 4H_2O + 4e^-$ Oxidation of $Mn(OH)_2$.

(b) $14e^- + 4H_2O + 2H_4IO_6^- \rightarrow I_2 + 16OH^-$ Reduction of $H_4IO_6^-$.

6.43 (a)
$2S_2O_3^{2-} \rightarrow S_4O_6^{2-} + 2e^-$
$OCl^- + 2H^+ + 2e^- \rightarrow Cl^- + H_2O$
$OCl^- + 2S_2O_3^{2-} \rightarrow 2H^+ + S_4O_6^{2-} + Cl^- + H_2O$

(b)
$(NO_3^- + 2H^+ + e^- \rightarrow NO_2 + H_2O) \times 2$
$Cu \rightarrow Cu^{2+} + 2e^-$
$2NO_3^- + Cu + 4H^+ \rightarrow 2NO_2 + Cu^{2+} + 2H_2O$

(c)
$IO_3^- + 6H^+ + 6e^- \rightarrow I^- + 3H_2O$
$(H_2O + AsO_3^{3-} \rightarrow AsO_4^{3-} + 2H^+ + 2e^-) \times 3$
$IO_3^- + 3AsO_3^{3-} + 6H^+ + 3H_2O \rightarrow I^- + 3AsO_4^{3-} + 3H_2O + 6H^+$
which simplifies to give:
$3AsO_3^{3-} + IO_3^- \rightarrow I^- + 3AsO_4^{3-}$

(d)
$SO_4^{2-} + 4H^+ + 2e^- \rightarrow SO_2 + 2H_2O$
$Zn \rightarrow Zn^{2+} + 2e^-$
$Zn + SO_4^{2-} + 4H^+$ Æ $Zn^{2+} + SO_2 + 2H_2O$

(e)
$NO_3^- + 10H^+ + 8e^- \rightarrow NH_4^+ + 3H_2O$
$(Zn \rightarrow Zn^{2+} + 2e^-) \times 4$
$NO_3^- + 4Zn + 10H^+ \rightarrow 4Zn^{2+} + NH_4^+ + 3H_2O$

(f)
$2Cr^{3+} + 7H_2O \rightarrow Cr_2O_7^{2-} + 14H^+ + 6e^-$
$(BiO_3^- + 6H^+ + 2e^- \rightarrow Bi^{3+} + 3H_2O) \times 3$
$2Cr^{3+} + 3BiO_3^- + 18H^+ + 7H_2O \rightarrow Cr_2O_7^{2-} + 14H^+ + 3Bi^{3+} + 9H_2O$
which simplifies to give:
$2Cr^{3+} + 3BiO_3^- + 4H^+ \rightarrow Cr_2O_7^{2-} + 3Bi^{3+} + 2H_2O$

(g)
$I_2 + 6H_2O \rightarrow 2IO_3^- + 12H^+ + 10e^-$
$(OCl^- + 2H^+ + 2e^- \rightarrow Cl^- + H_2O) \times 5$
$I_2 + 5OCl^- + H_2O \rightarrow 2IO_3^- + 5Cl^- + 2H^+$

(h)
$(Mn^{2+} + 4H_2O \rightarrow MnO_4^- + 8H^+ + 5e^-) \times 2$
$(BiO_3^- + 6H^+ + 2e^- \rightarrow Bi^{3+} + 3H_2O) \times 5$

$2Mn^{2+} + 5BiO_3^- + 30H^+ + 8H_2O \rightarrow 2MnO_4^- + 5Bi^{3+} + 16H^+ + 15H_2O$
which simplifies to:
$2Mn^{2+} + 5BiO_3^- + 14H^+ \rightarrow 2MnO_4^- + 5Bi^{3+} + 7H_2O$

(i) $(H_3AsO_3 + H_2O \rightarrow H_3AsO_4 + 2H^+ + 2e^-) \times 3$
$Cr_2O_7^{2-} + 14H^+ + 6e^- \rightarrow 2Cr^{3+} + 7H_2O$
$3H_3AsO_3 + Cr_2O_7^{2-} + 3H_2O + 14H^+ \rightarrow 3H_3AsO_4 + 2Cr^{3+} + 6H^+ + 7H_2O$
which simplifies to give:
$3H_3AsO_3 + Cr_2O_7^{2-} + 8H^+ \rightarrow 3H_3AsO_4 + 2Cr^{3+} + 4H_2O$

(j) $2I^- \rightarrow I_2 + 2e^-$
$HSO_4^- + 3H^+ + 2e^- \rightarrow SO_2 + 2H_2O$
$2I^- + HSO_4^- + 3H^+ \rightarrow I_2 + SO_2 + 2H_2O$

6.44 (a) $(Sn + 2H_2O \rightarrow SnO_2 + 4H^+ + 4e^-) \times 3$
$(NO_3^- + 4H^+ + 3e^- \rightarrow NO + 2H_2O) \times 4$
$3Sn + 4NO_3^- + 16H^+ + 6H_2O \rightarrow 3SnO_2 + 12H^+ + 4NO + 8H_2O$
which simplifies to:
$3Sn + 4NO_3^- + 4H^+ \rightarrow 3SnO_2 + 4NO + 2H_2O$

(b) $PbO_2 + 2Cl^- + 4H^+ + 2e^- \rightarrow PbCl_2 + 2H_2O$
$2Cl^- \rightarrow Cl_2 + 2e^-$
$PbO_2 + 4Cl^- + 4H^+ \rightarrow PbCl_2 + Cl_2 + 2H_2O$

(c) $Ag \rightarrow Ag^+ + e^-$
$NO_3^- + 2H^+ + e^- \rightarrow NO_2 + H_2O$
$Ag + 2H^+ + NO_3^- \rightarrow Ag^+ + NO_2 + H_2O$

(d) $(Fe^{3+} + e^- \rightarrow Fe^{2+}) \times 4$
$2NH_3OH^+ \rightarrow N_2O + H_2O + 6H^+ + 4e^-$
$4Fe^{3+} + 2NH_3OH^+ \rightarrow 4Fe^{2+} + N_2O + 6H^+ + H_2O$

(e) $2I^- \rightarrow I_2 + 2e^-$
$(HNO_2 + H^+ + e^- \rightarrow NO + H_2O) \times 2$
$2I^- + 2HNO_2 + 2H^+ \rightarrow I_2 + 2NO + 2H_2O$

(f) $C_2O_4^{2-} \rightarrow 2CO_2 + 2e^-$
$(HNO_2 + H^+ + e^- \rightarrow NO + H_2O) \times 2$
$C_2O_4^{2-} + 2HNO_2 + 2H^+ \rightarrow 2CO_2 + 2NO + 2H_2O$

(g) $(HNO_2 + H_2O \rightarrow NO_3^- + 3H^+ + 2e^-) \times 5$
$(MnO_4^- + 8H^+ + 5e^- \rightarrow Mn^{2+} + 4H_2O) \times 2$
$5HNO_2 + 2MnO_4^- + 16H^+ + 5H_2O \rightarrow 5NO_3^- + 2Mn^{2+} + 15H^+ + 8H_2O$
which simplifies to give:
$5HNO_2 + 2MnO_4^- + H^+ \rightarrow 5NO_3^- + 2Mn^{2+} + 3H_2O$

(h) $(H_3PO_2 + 2H_2O \rightarrow H_3PO_4 + 4H^+ + 4e^-) \times 3$
$(Cr_2O_7^{2-} + 14H^+ + 6e^- \rightarrow 2Cr^{3+} + 7H_2O) \times 2$
$3H_3PO_2 + 2Cr_2O_7^{2-} + 28H^+ + 6H_2O \rightarrow 3H_3PO_4 + 4Cr^{3+} + 14H_2O + 12H^+$
which simplifies to:
$3H_3PO_2 + 2Cr_2O_7^{2-} + 16H^+ \rightarrow 3H_3PO_4 + 4Cr^{3+} + 8H_2O$

(i) $(VO_2^+ + 2H^+ + e^- \rightarrow VO^{2+} + H_2O) \times 2$
$Sn^{2+} \rightarrow Sn^{4+} + 2e^-$
$2VO_2^+ + Sn^{2+} + 4H^+ \rightarrow 2VO^{2+} + Sn^{4+} + 2H_2O$

(j) $XeF_2 + 2e^- \rightarrow Xe + 2F^-$
$2Cl^- \rightarrow Cl_2 + 2e^-$
$XeF_2 + 2Cl^- \rightarrow Xe + Cl_2 + 2F^-$

6.45 For redox reactions in basic solution, we proceed to balance the half reactions as if they were in acid solution, and then add enough OH^- to each side of the resulting equation in order to neutralize (titrate) all of the H^+. This gives a corresponding amount of water ($H^+ + OH^- \rightarrow H_2O$) on one side of the equation, and an excess of OH^- on the other side of the equation, as befits a reaction in basic solution.

(a) $(CrO_4^{2-} + 4H^+ + 3e^- \rightarrow CrO_2^- + 2H_2O) \times 2$
$(S^{2-} \rightarrow S + 2e^-) \times 3$
$2CrO_4^{2-} + 3S^{2-} + 8H^+ \rightarrow 2CrO_2^- + 2S + 4H_2O$
Adding $8OH^-$ to both sides of the above equation we obtain:
$2CrO_4^{2-} + 3S^{2-} + 8H_2O \rightarrow 2CrO_2^- + 8OH^- + 3S + 4H_2O$
which simplifies to:
$2CrO_4^{2-} + 3S^{2-} + 4H_2O \rightarrow 2CrO_2^- + 3S + 8OH^-$

(b) $(C_2O_4^{2-} \rightarrow 2CO_2 + 2e^-) \times 3$
$(MnO_4^- + 4H^+ + 3e^- \rightarrow MnO_2 + 2H_2O) \times 2$
$3C_2O_4^{2-} + 2MnO_4^- + 8H^+ \rightarrow 6CO_2 + 2MnO_2 + 4H_2O$
Adding $8OH^-$ to both sides of the above equation we get:
$3C_2O_4^{2-} + 2MnO_4^- + 8H_2O \rightarrow 6CO_2 + 2MnO_2 + 4H_2O + 8OH^-$
which simplifies to give:
$3C_2O_4^{2-} + 2MnO_4^- + 4H_2O \rightarrow 6CO_2 + 2MnO_2 + 8OH^-$

(c) $(ClO_3^- + 6H^+ + 6e^- \rightarrow Cl^- + 3H_2O) \times 4$
$(N_2H_4 + 2H_2O \rightarrow 2NO + 8H^+ + 8e^-) \times 3$
$4ClO_3^- + 3N_2H_4 + 24H^+ + 6H_2O \rightarrow 4Cl^- + 6NO + 12H_2O + 24H^+$
which needs no OH^-, because it simplifies directly to:
$4ClO_3^- + 3N_2H_4 \rightarrow 4Cl^- + 6NO + 6H_2O$

(d) $NiO_2 + 2H^+ + 2e^- \rightarrow Ni(OH)_2$
$2Mn(OH)_2 \rightarrow Mn_2O_3 + H_2O + 2H^+ + 2e^-$
$NiO_2 + 2Mn(OH)_2 \rightarrow Ni(OH)_2 + Mn_2O_3 + H_2O$

(e) $(SO_3^{2-} + H_2O \rightarrow SO_4^{2-} + 2H^+ + 2e^-) \times 3$
$(MnO_4^- + 4H^+ + 3e^- \rightarrow MnO_2 + 2H_2O) \times 2$
$3SO_3^{2-} + 3H_2O + 8H^+ + 2MnO_4^- \rightarrow 3SO_4^{2-} + 6H^+ + 2MnO_2 + 4H_2O$
Adding $8OH^-$ to both sides of the equation we obtain:
$3SO_3^{2-} + 11H_2O + 2MnO_4^- \rightarrow 3SO_4^{2-} + 10H_2O + 2MnO_2 + 2OH^-$
which simplifies to:
$3SO_3^{2-} + 2MnO_4^- + H_2O \rightarrow 3SO_4^{2-} + 2MnO_2 + 2OH^-$

6.46 (a) $(CrO_2^- + 2H_2O \rightarrow CrO_4^{2-} + 4H^+ + 3e^-) \times 2$
$(S_2O_8^{2-} + 2e^- \rightarrow 2SO_4^{2-}) \times 3$
$3S_2O_8^{2-} + 2CrO_2^- + 4H_2O \rightarrow 2CrO_4^{2-} + 6SO_4^{2-} + 8H^+$
Adding $8OH^-$ to both sides of this equation:
$3S_2O_8^{2-} + 2CrO_2^- + 4H_2O + 8OH^- \rightarrow 2CrO_4^{2-} + 6SO_4^{2-} + 8H_2O$
which simplifies to give:
$3S_2O_8^{2-} + 2CrO_2^- + 8OH^- \rightarrow 2CrO_4^{2-} + 6SO_4^{2-} + 4H_2O$

(b) $(SO_3^{2-} + H_2O \rightarrow SO_4^{2-} + 2H^+ + 2e^-) \times 3$
$(CrO_4^{2-} + 4H^+ + 3e^- \rightarrow CrO_2^- + 2H_2O) \times 2$
$3SO_3^{2-} + 2CrO_4^{2-} + 8H^+ + 3H_2O \rightarrow 3SO_4^{2-} + 2CrO_2^- + 6H^+ + 4H_2O$
Adding $8OH^-$ to both sides of the equation we get:
$3SO_3^{2-} + 2CrO_4^{2-} + 11H_2O \rightarrow 3SO_4^{2-} + 2CrO_2^- + 2OH^- + 10H_2O$
which simplifies to:
$3SO_3^{2-} + 2CrO_4^{2-} + H_2O \rightarrow 3SO_4^{2-} + 2CrO_2^- + 2OH^-$

(c) $(O_2 + 2H^+ + 2e^- \rightarrow H_2O_2) \times 2$
$N_2H_4 \rightarrow N_2 + 4H^+ + 4e^-$
$2O_2 + 4H^+ + N_2H_4 \rightarrow 2H_2O_2 + N_2 + 4H^+$
which simplifies to:
$2O_2 + N_2H_4 \rightarrow 2H_2O_2 + N_2$

(d) $(Fe(OH)_2 + OH^- \rightarrow Fe(OH)_3 + e^-) \times 4$
$O_2 + 2H_2O + 4e^- \rightarrow 4OH^-$
$4Fe(OH)_2 + O_2 + 4OH^- + 2H_2O \rightarrow 4Fe(OH)_3 + 4OH^-$
which simplifies to:
$4Fe(OH)_2 + O_2 + 2H_2O \rightarrow 4Fe(OH)_3$

(e) $(Au + 4CN^- \rightarrow Au(CN)_4^- + 3e^-) \times 4$
$(O_2 + 2H_2O + 4e^- \rightarrow 4OH^-) \times 3$
$4Au + 16CN^- + 3O_2 + 6H_2O \rightarrow 4Au(CN)_4^- + 12OH^-$

6.47 $(OCl^- + 2H^+ + 2e^- \rightarrow Cl^- + H_2O) \times 4$
$S_2O_3^{2-} + 5H_2O \rightarrow 2SO_4^{2-} + 10H^+ + 8e^-$
$4OCl^- + S_2O_3^{2-} + 5H_2O + 8H^+ \rightarrow 4Cl^- + 2SO_4^{2-} + 10H^+ + 4H_2O$
which simplifies to:
$4OCl^- + S_2O_3^{2-} + H_2O \rightarrow 4Cl^- + 2SO_4^{2-} + 2H^+$

6.48 $(H_2C_2O_4 \rightarrow 2CO_2 + 2H^+ + 2e^-) \times 3$
$K_2Cr_2O_7 + 14H^+ + 6e– \rightarrow 2K^+ + 2Cr^{3+} + 7H_2O$
$3H_2C_2O_4 + K_2Cr_2O_7 + 14H^+ \rightarrow 6CO_2 + 2K^+ + 2Cr^{3+} + 6H^+ + 7H_2O$
which simplifies to:
$3H_2C_2O_4 + K_2Cr_2O_7 + 8H^+ \rightarrow 6CO_2 + 2K^+ + 2Cr^{3+} + 7H_2O$

6.49 $O_3 + 6H^+ + 6e^- \rightarrow 3H_2O$
$Br^- + 3H_2O \rightarrow BrO_3^- + 6H^+ + 6e^-$
$O_3 + Br^- + 3H_2O + 6H^+ \rightarrow BrO_3^- + 3H_2O + 6H^+$
which simplifies to:
$O_3 + Br^- \rightarrow BrO_3^-$

6.50 $(Cl_2 + 2e^- \rightarrow 2Cl^-) \times 4$
$S_2O_3^{2-} + 5H_2O \rightarrow 2SO_4^{2-} + 10H^+ + 8e^-$
$4Cl_2 + S_2O_3^{2-} + 5H_2O \rightarrow 8Cl^- + 2SO_4^{2-} + 10H^+$

6.51 (a) m: $Mn(s) + 2HCl(aq) \rightarrow MnCl_2(aq) + H_2(g)$
I: $Mn(s) + 2H^+(aq) + 2Cl^-(aq) \rightarrow Mn^{2+}(aq) + 2Cl^-(aq) + H_2(g)$
NI: $Mn(s) + 2H^+(aq) \rightarrow Mn^{2+}(aq) + H_2(g)$

(b) m: $Cd(s) + 2HCl(aq) \rightarrow CdCl_2(aq) + H_2(g)$
I: $Cd(s) + 2H^+(aq) + 2Cl^-(aq) \rightarrow Cd^{2+}(aq) + Cl^-(aq) + H_2(g)$
NI: $Cd(s) + 2H^+(aq) \rightarrow Cd^{2+}(aq) + H_2(g)$

(c) m: $Sn(s) + 2HCl(aq) \rightarrow SnCl_2(aq) + H_2(g)$
I: $Sn(s) + 2H^+(aq) + 2Cl^-(aq) \rightarrow Sn^{2+}(aq) + 2Cl^-(aq) + H_2(g)$
NI: $Sn(s) + 2H^+(aq) \rightarrow Sn^{2+}(aq) + H_2(g)$

(d) m: $Ni(s) + 2HCl(aq) \rightarrow NiCl_2(aq) + H_2(g)$
I: $Ni(s) + 2H^+(aq) + 2Cl^-(aq) \rightarrow Ni^{2+}(aq) + 2Cl^-(aq) + H_2(g)$
NI: $Ni(s) + 2H^+(aq) \rightarrow Ni^{2+}(aq) + H_2(g)$

(e) m: $2Cr(s) + 6HCl(aq) \rightarrow 2CrCl_3(aq) + 3H_2(g)$
I: $2Cr(s) + 6H^+(aq) + 6Cl^-(aq) \rightarrow 2Cr^{3+}(aq) + 6Cl^-(aq) + 3H_2(g)$
NI: $2Cr(s) + 6H^+(aq) \rightarrow 2Cr^{3+}(aq) + 3H_2(g)$

6.52 (a) m: $Mn(s) + H_2SO_4(aq) \rightarrow MnSO_4(aq) + H_2(g)$
I: $Mn(s) + 2H^+(aq) + SO_4^{2-}(aq) \rightarrow Mn^{2+}(aq) + SO_4^{2-}(aq) + H_2(g)$
NI: $Mn(s) + 2H^+(aq) \rightarrow Mn^{2+}(aq) + H_2(g)$

(b) m: $Cd(s) + H_2SO_4(aq) \rightarrow CdSO_4(aq) + H_2(g)$
I: $Cd(s) + 2H^+(aq) + SO_4^{2-}(aq) \rightarrow Cd^{2+}(aq) + SO_4^{2-}(aq) + H_2(g)$
NI: $Cd(s) + 2H^+(aq) \rightarrow Cd^{2+}(aq) + H_2(g)$

(c) m: $Sn(s) + H_2SO_4(aq) \rightarrow SnSO_4(aq) + H_2(g)$
I: $Sn(s) + 2H^+(aq) + SO_4^{2-}(aq) \rightarrow Sn^{2+}(aq) + SO_4^{2-}(aq) + H_2(g)$
NI: $Sn(s) + 2H^+(aq) \rightarrow Sn^{2+}(aq) + H_2(g)$

(d) m: $Ni(s) + H_2SO_4(aq) \rightarrow NiSO_4(aq) + H_2(g)$
I: $Ni(s) + 2H^+(aq) + SO_4^{2-}(aq) \rightarrow Ni^{2+}(aq) + SO_4^{2-}(aq) + H_2(g)$
NI: $Ni(s) + 2H^+(aq) \rightarrow Ni^{2+}(aq) + H_2(g)$

(e) m: $2Cr(s) + 3H_2SO_4(aq) \rightarrow Cr_2(SO_4)_3(aq) + 3H_2(g)$
I: $2Cr(s) + 6H^+(aq) + 3SO_4^{2-}(aq) \rightarrow 2Cr^{3+}(aq) + 3SO_4^{2-}(aq) + 3H_2(g)$
NI: $2Cr(s) + 6H^+(aq) \rightarrow 2Cr^{3+}(aq) + 3H_2(g)$

6.53 (a) $3Ag(s) + 4HNO_3(aq) \rightarrow 3AgNO_3(aq) + 2H_2O(l) + NO(g)$
(b) $Ag(s) + 2HNO_3(aq) \rightarrow AgNO_3(aq) + H_2O(l) + NO_2(aq)$

6.54 $Cu(s) \rightarrow Cu^{2+}(aq) + 2e^-$
$H_2SO_4(aq) + 2H^+(aq) + 2e^- \rightarrow SO_2(g) + 2H_2O(l)$
$Cu(s) + H_2SO_4(aq) + 2H^+(aq) \rightarrow Cu^{2+}(aq) + SO_2(g) + 2H_2O(l)$

6.55 First, we write the half-reactions for the equation:

$$Sn^{2+} \rightarrow Sn^{4+}$$
$$BrO_3^- \rightarrow Br^-$$

Now we may begin balancing each half-reaction, assume an acidic solution. All atoms other than O and H are already balanced, so we begin by balancing oxygen atoms using water:

$$Sn^{2+} \rightarrow Sn^{4+}$$
$$BrO_3^- \rightarrow Br^- + 3H_2O$$

Next, we balance hydrogen atoms using H^+ ions:

$$Sn^{2+} \rightarrow Sn^{4+}$$
$$6H^+ + BrO_3^- \rightarrow Br^- + 3H_2O$$

At this point, we add electrons to balance the charges on both sides of the equations:

$$Sn^{2+} \rightarrow Sn^{4+} + 2e^-$$
$$6e^- + 6H^+ + BrO_3^- \rightarrow Br^- + 3H_2O$$

We multiply the first equation by 3 so that the electrons gained equals the electrons lost,

$$3(Sn^{2+} \rightarrow Sn^{4+} + 2e^-)$$
$$6e^- + 6H^+ + BrO_3^- \rightarrow Br^- + 3H_2O$$

which gives us:

$$3Sn^{2+} \rightarrow 3Sn^{4+} + 6e^-$$
$$6e^- + 6H^+ + BrO_3^- \rightarrow Br^- + 3H_2O$$

Adding the two equations together produces the following:

$$3Sn^{2+} + 6e^- + 6H^+ + BrO_3^- \rightarrow Br^- + 3H_2O + 3Sn^{4+} + 6e^-$$

Canceling like species on each side gives us the final, balanced equation:

$$3Sn^{2+} + 6H^+ + BrO_3^- \rightarrow Br^- + 3H_2O + 3Sn^{4+}$$

6.56 $(H_2C_2O_4 \rightarrow 2CO_2 + 2H^+ + 2e^-) \times 3$
$Cr_2O_7^{2-} + 14H^+ + 6e^- \rightarrow 2Cr^{3+} + 7H_2O$
$3H_2C_2O_4 + Cr_2O_7^{2-} + 8H^+ \rightarrow 6CO_2 + 2Cr^{3+} + 7H_2O$

6.57 In each case, the reaction should proceed to give the less reactive of the two metals, together with the ion of the more reactive of the two metals. The reactivity is taken from the reactivity series table 6.2.
(a) N.R.
(b) $2Cr(s) + 3Pb^{2+}(aq) \rightarrow 2Cr^{3+}(aq) + 3Pb(s)$
(c) $2Ag^+(aq) + Fe(s) \rightarrow 2Ag(s) + Fe^{2+}(aq)$
(d) $3Ag(s) + Au^{3+}(aq) \rightarrow Au(s) + 3Ag^+(aq)$

6.58 In each case, the reaction should proceed to give the less reactive of the two metals, together with the ion of the more reactive of the two metals. The reactivity is taken from the reactivity series table 6.2.
(a) $Mn(s) + Fe^{2+} \rightarrow Mn^{2+} + Fe(s)$
(b) N.R.
(c) $Mg(s) + Co^{2+} \rightarrow Mg^{2+} + Co(s)$
(d) $2Cr(s) + 3Sn^{2+} \rightarrow 2Cr^{3+} + 3Sn(s)$

6.59 Increasing ease of oxidation: Pt, Ru, Tl, Pu

6.60 Increasing ease of oxidation: Ga, Be, Pu

6.61 The equation given shows that Cd is more active than Ru. Coupled with the information in Review Problem 6.59, we also see that Cd is more active than Pt. This means that in a mixture of Cd and Pt, Cd will be oxidized and Pt will be reduced:

$$Cd(s) + PtCl_2(aq) \rightarrow CdCl_2(aq) + Pt(s)$$

(The Pt(s) and the $Cd(NO_3)_2$(aq) will not react.)

6.62 The observation shows that Mg is more active than Be. With the information from Review Problem 6.60, we see that Ga is less active than Be. This means that Ga^{3+} will oxidize Mg, and reaction (a) will occur spontaneously:

$$2Ga^{3+} + 3Mg \rightarrow 3Mg^{2+} + 2\ Ga$$

6.63 (a) $2C_6H_6(l) + 15O_2(g) \rightarrow 12CO_2(g) + 6H_2O(g)$
(b) $C_3H_8(g) + 5O_2(g) \rightarrow 3CO_2(g) + 4H_2O(g)$
(c) $C_{21}H_{44}(s) + 32O_2(g) \rightarrow 21CO_2(g) + 22H_2O(g)$

6.64 (a) $2C_{12}H_{26}(l) + 37O_2(g) \rightarrow 24CO_2(g) + 26H_2O(g)$
(b) $C_{18}H_{36}(l) + 27O_2(g) \rightarrow 18CO_2(g) + 18H_2O(g)$
(c) $C_7H_8(l) + 9O_2(g) \rightarrow 7CO_2(g) + 4H_2O(g)$

6.65 (a) $2C_6H_6(l) + 9O_2(g) \rightarrow 12CO(g) + 6H_2O(g)$
$2C_3H_8(g) + 7O_2(g) \rightarrow 6CO(g) + 8H_2O(g)$
$2C_{21}H_{44}(s) + 43O_2(g) \rightarrow 42CO(g) + 44H_2O(g)$

(b) $2C_6H_6(l) + 3O_2(g) \rightarrow 12C(s) + 6H_2O(g)$
$C_3H_8(g) + 2O_2(g) \rightarrow 3(s) + 4H_2O(g)$
$C_{21}H_{44}(s) + 11O_2(g) \rightarrow 21C(s) + 22H_2O(g)$

6.66 (a) $C_{12}H_{26}(l) + 25/2O_2(g) \rightarrow 12CO(g) + 13H_2O(g)$
$C_{18}H_{36}(l) + 18O_2(g) \rightarrow 18CO(g) + 18H_2O(g)$
$C_7H_8(l) + 11/2\ O_2(g) \rightarrow 7\ CO_2(g) + 4H_2O(g)$

(b) $C_{12}H_{26}(l) + 13/2O_2(g) \rightarrow 12C(s) + 13H_2O(g)$
$C_{18}H_{36}(l) + 9O_2(g) \rightarrow 18C(s) + 18H_2O(g)$
$C_7H_8(l) + 2O_2(g) \rightarrow 7C(s) + 4H_2O(g)$

6.67 $2CH_3OH(l) + 3O_2(g) \rightarrow 2CO_2(g) + 4H_2O(g)$

6.68 $C_6H_{12}O_6(s) + 6O_2(g) \rightarrow 6CO_2(g) + 6H_2O(g)$

6.69 (a) $IO_3^- + 6H^+ + 6e^- \rightarrow I^- + 3H_2O$
$[SO_3^{2-} + H_2O \rightarrow SO_4^{2-} + 2H^+ + 2e^-] \times 3$
$IO_3^- + 3SO_3^{2-} + 6H^+ + 3H_2O \rightarrow I^- + 3SO_4^{2-} + 3H_2O + 6H^+$
Which simplifies to:
$IO_3^- + 3SO_3^{2-} \rightarrow I^- + 3SO_4^{2-}$

(b)

$$\#\ \text{g Na}_2\text{SO}_3 = (5.00\text{ g NaIO}_3)\left(\frac{1\text{ mol NaIO}_3}{197.9\text{ g NaIO}_3}\right)\left(\frac{3\text{ mol Na}_2\text{SO}_3}{1\text{ mol NaIO}_3}\right)\left(\frac{126.0\text{ g Na}_2\text{SO}_3}{1\text{ mol Na}_2\text{SO}_3}\right)$$

$$= 9.55\text{ g Na}_2\text{SO}_3$$

6.70 (a) $[Mn^{2+}(aq) + 4H_2O(l) \rightarrow MnO_4^-(aq) + 8H^+(aq) + 5e^-] \times 2$
$[BiO_3^-(aq) + 6H^+(aq) + 2e^- \rightarrow Bi^{3+}(aq) + 3H_2O(l)] \times 5$
$2Mn^{2+}(aq) + 5BiO_3^-(aq) + 8H_2O(l) + 30H^+(aq) \rightarrow$
$2MnO_4^-(aq) + 5Bi^{3+}(aq) + 15H_2O(l) + 16H^+(aq)$
which simplifies to give:
$2Mn^{2+}(aq) + 5BiO_3^-(aq) + 14H^+(aq) \rightarrow 2MnO_4^-(aq) + 5Bi^{3+}(aq) + 7H_2O(l)$

(b)

$$\#\ \text{g NaBiO}_3 = (18.5\text{ mg MnSO}_4)\left(\frac{1\text{ g}}{1000\text{ mg}}\right)\left(\frac{1\text{ mol MnSO}_4}{151.0\text{ g MnSO}_4}\right)\left(\frac{1\text{ mol Mn}^{2+}}{1\text{ mol MnSO}_4}\right)$$

$$\times \left(\frac{5\text{ mol BiO}_3^-}{2\text{ mol Mn}^{2+}}\right)\left(\frac{1\text{ mol NaBiO}_3}{1\text{ mol BiO}_3^-}\right)\left(\frac{280.0\text{ g NaBiO}_3}{1\text{ mol NaBiO}_3}\right)$$

$$= 0.08585\text{ g NaBiO}_3$$

6.71 $Cu + 2Ag^+ \rightarrow Cu^{2+} + Ag$

$$\#\ \text{g Cu} = (12.0\text{ g Ag})\left(\frac{1\text{ mol Ag}}{107.868\text{ g Ag}}\right)\left(\frac{1\text{ mol Cu}}{2\text{ mol Ag}}\right)\left(\frac{63.54\text{ g Cu}}{1\text{ mol Cu}}\right) = 3.53\text{ g Cu}$$

6.72 $Al(s) + 3AgNO_3(aq) \rightarrow 3Ag(s) + Al(NO_3)_3(aq)$

$$\#\ \text{g Al} = (25.0\text{ g AgNO}_3)\left(\frac{1\text{ mol AgNO}_3}{169.9\text{ g AgNO}_3}\right)\left(\frac{1\text{ mol Al}}{3\text{ mol AgNO}_3}\right)\left(\frac{26.98\text{ g Al}}{1\text{ mol Al}}\right) = 1.32\text{ g Al}$$

6.73 (a) $[MnO_4^- + 8H^+ + 5e^- \rightarrow Mn^{2+} + 4H_2O] \times 2$
$[Sn^{2+} \rightarrow Sn^{4+} + 2e^-]$ x 5
$2MnO_4^- + 5Sn^{2+} + 16H^+ \rightarrow 2Mn^{2+} + 5Sn^{4+} + 8H_2O$

(b) $$\text{\# mL KMnO}_4 = (40.0\text{ mL SnCl}_2)\left(\frac{0.250\text{ mol SnCl}_2}{1000\text{ mL SnCl}_2}\right)\left(\frac{1\text{ mol Sn}^{2+}}{1\text{ mol SnCl}_2}\right)$$
$$\times\left(\frac{2\text{ mol MnO}_4^-}{5\text{ mol Sn}^{2+}}\right)\left(\frac{1\text{ mol KMnO}_4}{1\text{ mol MnO}_4^-}\right)\left(\frac{1000\text{ mL KMnO}_4}{0.230\text{ mol KMnO}_4}\right) = 17.4\text{ mL}$$

6.74 (a) $[HSO_3^- + H_2O \rightarrow SO_4^{2-} + 3H^+ + 2e^-] \times 3$

$ClO_3^- + 6H^+ + 6e^- \rightarrow Cl^- + 3H_2O$

$3HSO_3^- + ClO_3^- + 3H_2O + 6H^+ \rightarrow 3SO_4^{2-} + 9H^+ + Cl^- + 3H_2O$

Which simplifies to:

$3HSO_3^- + ClO_3^- \rightarrow 3SO_4^{2-} + 3H^+ + Cl^-$

(b) $$\text{\# mL NaClO}_3 = (30.0\text{ mL NaHSO}_3)\left(\frac{0.450\text{ mol NaHSO}_3}{1000\text{ mL NaHSO}_3}\right)\left(\frac{1\text{ mol HSO}_3^-}{1\text{ mol NaHSO}_3}\right)$$
$$\times\left(\frac{1\text{ mol ClO}_3^-}{3\text{ mol HSO}_3^-}\right)\left(\frac{1\text{ mol NaClO}_3}{1\text{ mol ClO}_3^-}\right)\left(\frac{1000\text{ mL NaClO}_3}{0.150\text{ mol NaClO}_3}\right) = 30.0\text{ mL NaClO}_3$$

6.75 (a) $$\text{Molarity of I}_3^-\text{ solution} = 0.462\text{ g KIO}_3\left(\frac{1\text{ mol KIO}_3}{214.00\text{ g KIO}_3}\right)\left(\frac{1\text{ mol IO}_3^-}{1\text{ mol KIO}_3}\right)\left(\frac{3\text{ mol I}_3^-}{1\text{ mol IO}_3^-}\right)$$
$$\times\left(\frac{1}{0.250\text{ L}}\right) = 0.0259\text{ M I}_3^-$$

(b) $$\text{\# g (NH}_4)_2\text{S}_2\text{O}_3 = 27.99\text{ mL I}_3^-\text{ solution}\left(\frac{1\text{ L}}{1000\text{ mL}}\right)\left(\frac{0.0259\text{ mol I}_3^-}{1\text{ L}}\right)\left(\frac{2\text{ mol S}_2\text{O}_3^{2-}}{1\text{ mol I}_3^-}\right)$$
$$\times\left(\frac{1\text{ mol (NH}_4)_2\text{S}_2\text{O}_3}{1\text{ mol S}_2\text{O}_3^{2-}}\right)\left(\frac{148.24\text{ g (NH}_4)_2\text{S}_2\text{O}_3}{1\text{ mol (NH}_4)_2\text{S}_2\text{O}_3}\right) = 0.2150\text{ g (NH}_4)_2\text{S}_2\text{O}_3$$

(c) $$\%\text{ by mass of (NH}_4)_2\text{S}_2\text{O}_3 = \left(\frac{0.2150\text{ g (NH}_4)_2\text{S}_2\text{O}_3}{0.218\text{ g sample}}\right)\times 100\% = 98.6\%\text{ (NH}_4)_2\text{S}_2\text{O}_3\text{ in sample}$$

6.76 (a) $$\text{Mol of I}_3^- = 0.0421\text{ g NaIO}_3\left(\frac{1\text{ mol NaIO}_3}{197.89\text{ g NaIO}_3}\right)\left(\frac{1\text{ mol IO}_3^-}{1\text{ mol NaIO}_3}\right)\left(\frac{3\text{ mol I}_3^-}{1\text{ mol IO}_3^-}\right) = 6.38\times 10^{-4}\text{ mol I}_3^-$$

$$\text{Molarity of I}_3^- = \left(\frac{6.38\times 10^{-4}\text{ mol I}_3^-}{100\text{ mL}}\right)\left(\frac{1000\text{ mL}}{1\text{ L}}\right) = 6.38\times 10^{-3}\text{ M I}_3^-$$

(b) $$\text{\# g SO}_2 = 2.47\text{ mL I}_3^-\left(\frac{1\text{ L I}_3^-}{1000\text{ mL I}_3^-}\right)\left(\frac{6.38\times 10^{-3}\text{ mol I}_3^-}{1\text{ L I}_3^-}\right)$$
$$\times\left(\frac{1\text{ mol SO}_2}{1\text{ mol I}_3^-}\right)\left(\frac{48.07\text{ g SO}_2}{1\text{ mol SO}_2}\right) = 7.58\times 10^{-4}\text{ g SO}_2$$

(c) The density of the wine was 0.96 g/mL and the SO_2 concentration was 1.52×10^{-4} g SO_2/mL

$$\text{concentration SO}_2 = \frac{7.58\times 10^{-4}\text{ g SO}_2}{50\text{ mL}} = 1.52\times 10^{-5}\text{ g SO}_2\text{/mL}$$

In 1 mL of solution there are 96 g of wine and 1.52×10^{-5} g SO_2

Therefore the percentage of SO_2 in the wine is

$$\frac{1.52\times10^{-5}\text{ g }SO_2}{96\text{ g wine}}\times 100\% = 1.58\times10^{-5}\,\%$$

(d) $$\text{ppm } SO_2 = \frac{1.52\times10^{-5}\text{ g }SO_2}{96\text{ g wine}}\times 10^6\text{ ppm} = 0.158\text{ ppm}$$

6.77 $$\#\text{ mol }Cu^{2+} = (29.96\text{ mL }S_2O_3{}^{2-})\left(\frac{0.02100\text{ mol }S_2O_3{}^{2-}}{1000\text{ mL }S_2O_3{}^{2-}}\right)$$

$$\times\left(\frac{1\text{ mol }I_3{}^-}{2\text{ mol }S_2O_3{}^{2-}}\right)\left(\frac{2\text{ mol }Cu^{2+}}{1\text{ mol }I_3{}^-}\right) = 6.292\times10^{-4}\text{ mol }Cu^{2+}$$

$$\#\text{ g Cu} = (6.292\times10^{-4}\text{ mol Cu})\times(63.546\text{ g Cu/mol Cu})$$
$$= 3.998\times10^{-2}\text{ g Cu}$$

$$\%\text{ Cu} = (3.998\times10^{-2}\text{ g Cu}/0.4225\text{ g sample})\times 100 = 9.463\ \%$$

(b) $$\#\text{ g }CuCO_3 = (6.292\times10^{-4}\text{ mol Cu})\left(\frac{1\text{ mol }CuCO_3}{1\text{ mol Cu}}\right)\left(\frac{123.56\text{ g }CuCO_3}{1\text{ mol }CuCO_3}\right) = 0.07774\text{ g }CuCO_3$$

$$\%\ CuCO_3 = \left(\frac{0.07774\text{ g }CuCO_3}{0.4225\text{ g sample}}\right)\times 100 = 18.40\ \%$$

6.78 (a) $$\#\text{ g }Fe^{2+} = 39.42\text{ mL }MnO_4{}^{2-}\left(\frac{0.0281\text{ mol }MnO_4{}^{2-}}{1000\text{ mL }MnO_4{}^{2-}}\right)\left(\frac{5\text{ mol }Fe^{2+}}{1\text{ mol }MnO_4{}^{2-}}\right)$$

$$\times\left(\frac{55.845\text{ g }Fe^{2+}}{1\text{ mol }Fe^{2+}}\right) = 0.309\text{ g }Fe^{2+}$$

$$\%\text{ Fe} = \frac{0.309\text{ g}}{1.362\text{ g}}\times 100\% = 22.7\%\text{ Fe}$$

(b) $$\#\text{ g }Fe_3O_4 = (0.309\text{ g Fe})\left(\frac{1\text{ mol Fe}}{55.845\text{ g Fe}}\right)\left(\frac{1\text{ mol }Fe_3O_4}{3\text{ mol Fe}}\right)\left(\frac{231.55\text{ g }Fe_3O_4}{1\text{ mol }Fe_3O_4}\right)$$

$$= 0.427\text{ g }Fe_3O_4$$

$$\%\text{ Fe} = \frac{0.427\text{ g}}{1.362\text{ g}}\times 100\% = 31.4\%$$

6.79 (a) $$\#\text{ g }H_2O_2 = (17.60\text{ mL }KMnO_4)\left(\frac{0.02000\text{ mol }KMnO_4}{1000\text{ mL }KMnO_4}\right)\left(\frac{1\text{ mol }MnO_4{}^-}{1\text{ mol }KMnO_4}\right)$$

$$\times\left(\frac{5\text{ mol }H_2O_2}{2\text{ mol }MnO_4{}^-}\right)\left(\frac{34.02\text{ g }H_2O_2}{1\text{ mol }H_2O_2}\right)$$
$$= 0.02994\text{ g }H_2O_2$$

(b) $$(0.02994\text{ g}/1.000\text{ g})\times 100\% = 2.994\%\ H_2O_2$$

6.80 $$\#\text{ g }NaNO_2 = (12.15\text{ mL }KMnO_4)\left(\frac{0.01000\text{ mol }KMnO_4}{1000\text{ mL }KMnO_4}\right)\left(\frac{1\text{ mol }MnO_4{}^{2-}}{1\text{ mol }KMnO_4}\right)$$

$$\times\left(\frac{5\text{ mol }HNO_2}{2\text{ mol }MnO_4{}^{2-}}\right)\left(\frac{1\text{ mol }NaNO_2}{1\text{ mol }HNO_2}\right)\left(\frac{68.995\text{ g }NaNO_2}{1\text{ mol }NaNO_2}\right)$$
$$= 2.096\times10^{-2}\text{ g }NaNO_2$$

% $NaNO_2$ = (2.096×10^{-2} g $NaNO_2$ / 1.000 g sample) × 100% = 2.096 %

6.81 (a) $2CrO_4^{2-} + 3SO_3^{2-} + H_2O \rightarrow 2CrO_2^{-} + 3SO_4^{2-} + 2OH^-$

(b) $$\#\ \text{mol } CrO_4^{2-} = (3.18\text{ g } Na_2SO_3)\left(\frac{1\text{ mol } Na_2SO_3}{126.04\text{ g } Na_2SO_3}\right) \times \left(\frac{1\text{ mol } SO_3^{2-}}{1\text{ mol } Na_2SO_3}\right)\left(\frac{2\text{ mol } CrO_4^{2-}}{3\text{ mol } SO_3^{2-}}\right)$$
$$= 1.68 \times 10^{-2}\text{ mol } CrO_4^{2-}$$

Since there is one mole of Cr in each mole of CrO_4^{2-}, then the above number of moles of CrO_4^{2-} is also equal to the number of moles of Cr that were present:

0.0168 mol Cr × 52.00 g/mol = 0.875 g Cr in the original alloy.

(c) (0.875 g/3.450 g) × 100 = 25.4 % Cr

6.82 (a) $(Sn^{2+} \rightarrow Sn^{4+} + 2e^-) \times 3$

$Cr_2O_7^{2-} + 14H^+ + 6e^- \rightarrow 2Cr^{3+} + 7H_2O$

$3Sn^{2+} + Cr_2O_7^{2-} + 14H^+ \rightarrow 3Sn^{4+} + 2Cr^{3+} + 7H_2O$

(b) $$\#\ \text{g Sn} = (0.368\text{ g } Na_2Cr_2O_7)\left(\frac{1\text{ mol } Na_2Cr_2O_7}{262.0\text{ g } Na_2Cr_2O_7}\right)\left(\frac{1\text{ mol } Cr_2O_7^{2-}}{1\text{ mol } Na_2Cr_2O_7}\right) \times \left(\frac{3\text{ mol } Sn^{2+}}{1\text{ mol } Cr_2O_7^{2-}}\right)\left(\frac{1\text{ mol Sn}}{1\text{ mol } Sn^{2+}}\right)\left(\frac{118.7\text{ g Sn}}{1\text{ mol Sn}}\right)$$

0.500 g Sn

(c) (0.500 g Sn / 1.50 g solder) × 100% = 33.3 %

6.83 (a) $$\#\ \text{mol } C_2O_4^{2-} = (21.62\text{ mL } KMnO_4)\left(\frac{0.1000\text{ mol } KMnO_4}{1000\text{ mL } KMnO_4}\right)\left(\frac{5\text{ mol } C_2O_4^{2-}}{2\text{ mol } KMnO_4}\right)$$
$$= 5.405 \times 10^{-3}\text{ mol } C_2O_4^{2-}$$

(b) The stoichiometry for calcium is as follows:

1 mol $C_2O_4^{2-}$ = 1 mol Ca^{2+} = 1 mol $CaCl_2$

Thus the number of grams of $CaCl_2$ is given simply by:

5.405×10^{-3} mol $CaCl_2$ × 110.98 g/mol = 0.5999 g $CaCl_2$

(c) (0.5999 g/2.463 g) × 100 = 24.35 % $CaCl_2$

6.84 (a) $$\#\ \text{mol } I_3^- = (29.25\text{ mL } Na_2S_2O_3)\left(\frac{0.3000\text{ mol } Na_2S_2O_3}{1000\text{ mL } Na_2S_2O_3}\right)\left(\frac{1\text{ mol } I_3^-}{2\text{ mol } Na_2S_2O_3}\right)$$
$$=4.388 \times 10^{-3}\text{ mol } I_3^-$$

(b) $$\#\ \text{mol } NO_2^- = (4.388 \times 10^{-3}\text{ mol } I_3^-)\left(\frac{2\text{ mol } NO_2^-}{1\text{ mol } I_3^-}\right) = 8.776 \times 10^{-3}\text{ mol } NO_2^-$$

(c) $$\#\ \text{g } NaNO_2 = (8.776 \times 10^{-3}\text{ mol } NO_2^-)\left(\frac{1\text{ mol } NaNO_2}{1\text{ mol } NO_2^-}\right)\left(\frac{68.995\text{ g } NaNO_2}{1\text{ mol } NaNO_2}\right) = 0.6055\text{ g } NaNO_2$$

% $NaNO_2$ = (0.6055 g $NaNO_2$ / 1.104 g sample) × 100% = 54.85 % $NaNO_2$

Additional Exercises

6.85 Total charge = –2 = (charge of sulfur atoms) + (charge of oxygen atoms)
Charge of oxygens = 6(–2) = –12
Charge of sulfur atoms = –2 –(–12) = +10
+10 spread out over 4 sulfur atoms gives a charge of: +10/4 per S atom, or:
Oxidation number of S = +2.5

6.86

Element oxidized:	Cl
Element reduced:	Cl
Oxidizing agent:	NaOCl
Reducing agent:	$NaClO_2$

6.87 (a) –2 (b) 0 (c) +4 (d) +4

6.88 The first reaction demonstrates that Al is more readily oxidized than Cu. The second reaction demonstrates that Al is more readily oxidized than Fe. Reaction 3 demonstrates that Fe is more readily oxidized than Pb. Reaction 4 demonstrates that Fe is more readily oxidized than Cu. The fifth reaction demonstrates that Al is more readily oxidized than Pb. The last reaction demonstrates that Pb is more readily oxidized than Cu.

Altogether, the above facts constitute the following trend of increasing ease of oxidation:
Cu < Pb < Fe < Al

6.89 No, the first, fourth and fifth reactions were not necessary.

6.90 Any metal that is lower than hydrogen in the activity series shown in Table 5.2 of the text will react with H^+: (c) zinc and (d) magnesium.

6.91 We choose the metal that is lower (more reactive) in the activity series shown in Table 5.2: (a) aluminum (b) zinc (c) magnesium

6.92
$$\#\text{ mg } KIO_3 = 22.61\text{ mL } S_2O_3^- \left(\frac{0.0500\text{ mol } S_2O_3^{2-}}{1000\text{ mL}}\right)\left(\frac{1\text{ mol } I_3^-}{2\text{ mol } S_2O_3^{2-}}\right)\left(\frac{1\text{ mol } IO_3^-}{3\text{ mol } I_3^-}\right)$$
$$\times\left(\frac{1\text{ mol } KIO_3}{1\text{ mol } IO_3^-}\right)\left(\frac{214.00\text{ g } KIO_3}{1\text{ mol } KIO_3}\right)\left(\frac{1000\text{ mg } KIO_3}{1\text{ g } KIO_3}\right) = 40.32\text{ mg } KIO_3$$

6.93 In each case, the reaction should proceed to give the less reactive of the two metals, together with the ion of the more reactive of the two metals. The reactivity is taken from the reactivity series.
(a) $Zn(s) + Sn^{2+}(aq) \rightarrow Zn^{2+}(aq) + Sn(s)$
(b) $2Cr(s) + 6H^+(aq) \rightarrow 2Cr^{3+}(aq) + 3H_2(g)$
(c) N.R.
(d) $Mn(s) + Pb^{2+}(aq) \rightarrow Mn^{2+}(aq) + Pb(s)$
(e) $Zn(s) + Co^{2+}(aq) \rightarrow Zn^{2+}(aq) + Co(s)$

6.94 $C_{12}H_{22}O_{11}(s) + 12O_2(g) \rightarrow 12CO_2(g) + 11H_2O(l)$

6.95
(a) $2Zn(s) + O_2(g) \rightarrow 2ZnO(s)$
(b) $4Al(s) + 3O_2(g) \rightarrow 2Al_2O_3(s)$
(c) $2Mg(s) + O_2(g) \rightarrow 2MgO(s)$
(d) $2Fe(s) + O_2(g) \rightarrow 2FeO(s)$
alternatively we have: $4Fe(s) + 3O_2(g) \rightarrow 2Fe_2O_3(s)$
(e) $2Ca(s)+ O_2(g) \rightarrow 2CaO(s)$

6.96 (a)

$2NBr_3 + 6e^- \rightarrow N_2 + 6Br^-$
$2NBr_3 + 6H_2O \rightarrow N_2 + 6HOBr + 6H^+ + 6e^-$
$4NBr_3 + 6H_2O \rightarrow 2N_2 + 6HOBr + 6Br^- + 6H^+$
Add $6OH^-$ to both sides of the above equation:
$4NBr_3 + 6H_2O + 6OH^- \rightarrow 2N_2 + 6HOBr + 6Br^- + 6H_2O$
which simplifies to give:
$4NBr_3 + 6OH^- \rightarrow 2N_2 + 6HOBr + 6Br^-$
or $2NBr_3 + 3OH^- \rightarrow N_2 + 3HOBr + 3Br^-$

(b)

$(Cl_2 + 2e^- \rightarrow 2Cl^-) \times 5$
$Cl_2 + 6H_2O \rightarrow 2ClO_3^- + 12H^+ + 10e^-$
$6Cl_2 + 6H_2O \rightarrow 2ClO_3^- + 10Cl^- + 12H^+$
Adding $12OH^-$ to both sides gives:
$6Cl_2 + 6H_2O + 12OH^- \rightarrow 2ClO_3^- + 10Cl^- + 12H_2O$
which simplifies to:
$6Cl_2 + 12OH^- \rightarrow 2ClO_3^- + 10Cl^- + 6H_2O$
or $3Cl_2 + 6OH^- \rightarrow ClO_3^- + 5Cl^- + 3H_2O$

(c)

$H_2SeO_3 + 4H^+ + 4e^- \rightarrow Se + 3H_2O$
$(H_2S \rightarrow S + 2H^+ + 2e^-) \times 2$
$2H_2S + H_2SeO_3 + 4H^+ \rightarrow 2S + 4H^+ + Se + 3H_2O$
which simplifies to give:
$2H_2S + H_2SeO_3 \rightarrow 2S + Se + 3H_2O$

(d)

$MnO_2 + 4H^+ + 2e^- \rightarrow Mn^{2+} + 2H_2O$
$2SO_3^{2-} \rightarrow S_2O_6^{2-} + 2e^-$
$2SO_3^{2-} + MnO_2 + 4H^+ \rightarrow Mn^{2+} + S_2O_6^{2-} + 2H_2O$

(e)

$XeO_3 + 6H^+ + 6e^- \rightarrow Xe + 3H_2O$
$(2I^- \rightarrow I_2 + 2e^-) \times 3$
$XeO_3 + 6I^- + 6H^+ \rightarrow 3I_2 + Xe + 3H_2O$

(f)

$(CN)_2 + 2e^- \rightarrow 2CN^-$
$(CN)_2 + 2H_2O \rightarrow 2OCN^- + 4H^+ + 2e^-$
$2(CN)_2 + 2H_2O \rightarrow 2CN^- + 2OCN^- + 4H^+$
Adding $4OH^-$ to both sides gives:
$2(CN)_2 + 4OH^- + 2H_2O \rightarrow 2CN^- + 2OCN^- + 4H_2O$
which simplifies to:
$(CN)_2 + 2OH^- \rightarrow CN^- + OCN^- + H_2O$

6.97 $$\#\text{ g } PbO_2 = 15.0 \text{ g } Cl_2\left(\frac{1 \text{ mol } Cl_2}{70.91 \text{ g } Cl_2}\right)\left(\frac{1 \text{ mol } PbO_2}{1 \text{ mol } Cl_2}\right)\left(\frac{239.2 \text{ g } PbO_2}{1 \text{ mol } PbO_2}\right) = 50.6 \text{ g } PbO_2$$

6.98 The oxidation state of cerium in the reactant ion is +4. The number of moles of this ion in the reactant solution is:
$0.0150 \text{ M} \times 0.02500 \text{ L} = 3.75 \times 10^{-4} \text{ mol } Ce^{4+}$
The number of moles of electrons that come from the Fe^{2+} reducing agent is:
$0.0320 \text{ M} \times 0.02344 \text{ L} = 7.50 \times 10^{-4} \text{ mol } e^-$
The ratio of moles of electrons to moles of Ce^{4+} reactant is therefore 2:1, and we conclude that the product is Ce^{2+}.

6.99 $Cu + 2Ag^+ \rightarrow 2Ag + Cu^{2+}$

The number of moles of Ag^+ available for the reaction is

$0.125\ M \times 0.255\ L = 0.0319\ mol\ Ag^+$

Since the stoichiometry is 2/1, the number of moles of Cu^{2+} ion that are consumed is $0.0319 \div 2 = 0.0159$ mol. The mass of copper consumed is $0.0159\ mol \times 63.546\ g/mol = 1.01\ g$. The amount of unreacted copper is thus: $12.340\ g - 1.01\ g = 11.33\ g\ Cu$. The mass of Ag that is formed is: $0.0319\ mol \times 108\ g/mol = 3.45\ g\ Ag$. The final mass of the bar is: $11.33\ g + 3.45\ g = 14.78\ g$.

6.100 The balanced equation for the reaction is:

$5H_2C_2O_4 + 2MnO_4^- + 6H^+ \rightarrow 10CO_2 + 2Mn^{2+} + 8H_2O$

$$M\ KMnO_4 = \frac{(0.1244\ g\ K_2C_2O_4)\left(\frac{1\ mol\ K_2C_2O_4}{166.2\ g\ K_2C_2O_4}\right)\left(\frac{2\ mol\ KMnO_4}{5\ mol\ K_2C_2O_4}\right)}{(13.93\ mL\ KMnO_4)\left(\frac{1\ L}{1000\ mL}\right)}$$

$$= 0.02149\ M\ KMnO_4$$

6.101 The balanced equation for the oxidation-reduction reaction is:

$3H_2C_2O_4 + Cr_2O_7^{2-} + 8H^+ \rightarrow 6CO_2 + 2Cr^{3+} + 7H_2O$

$$\#\ moles\ H_2C_2O_4 = (6.25\ mL\ K_2Cr_2O_7)\left(\frac{0.200\ moles\ K_2Cr_2O_7}{1000\ mL\ K_2Cr_2O_7}\right)\left(\frac{3\ moles\ H_2C_2O_4}{1\ mole\ K_2Cr_2O_7}\right)$$

$$= 3.75 \times 10^{-3}\ moles\ H_2C_2O_4$$

So, if we titrate the same oxalic acid solution using NaOH we will need:

$$\#\ ml\ NaOH = (3.75 \times 10^{-3}\ moles\ H_2C_2O_4)\left(\frac{2\ moles\ NaOH}{1\ mole\ H_2C_2O_4}\right)\left(\frac{1000\ mL\ NaOH}{0.450\ moles\ NaOH}\right)$$

$$= 16.7\ ml\ NaOH$$

Practice Exercises

7.1 $\# \text{ dietary calories} = (54\text{ kJ})\left(\frac{1\text{ Cal}}{4.184\text{ kJ}}\right) = 13\text{ Cal per pound of body weight}$

7.2 For 50.0 g water, the energy needed is 50 times that for 1.0 g water, or 50 cal:

$$\#\text{ cal} = (100\text{ °C})\left(\frac{50\text{ cal}}{1\text{ °C}}\right) = 5.00\times10^3\text{ cal}$$

In joules, this is:

$$\#\text{ J} = (5.00\times10^3\text{ cal})\left(\frac{4.184\text{ J}}{1\text{ cal}}\right) = 20{,}900\text{ J}$$

7.3 $\text{q gained by water} = (5.0\text{ °C})\left(\frac{250\text{ cal}}{1\text{ °C}}\right)\left(\frac{4.184\text{ J}}{1\text{ cal}}\right) = 5230\text{ J}$

q lost by ball bearing = –q gained by water = –5,230 J

$C = q/\Delta T$

$C = -5{,}230\text{ J}/(30.0\text{ °C}-220\text{ °C}) = 27.5\text{ J/°C}$

7.4 The amount of heat transferred into the water is:

$$\#\text{ J} = (250\text{ g H}_2\text{O})\left(4.184\ \text{J}/\text{g °C}\right)(30.0\text{ °C} - 25.0\text{ °C}) = 5200\text{ J}$$

$$\#\text{ kJ} = (5200\text{ J})\left(\frac{1\text{ kJ}}{1000\text{ J}}\right) = 5.2\text{ kJ}$$

$$\#\text{ cal} = (5200\text{ J})\left(\frac{1\text{ cal}}{4.184\text{ J}}\right) = 1200\text{ cal}$$

$$\#\text{ kcal} = (1200\text{ cal})\left(\frac{1\text{ kcal}}{1000\text{ cal}}\right) = 1.2\text{ kcal}$$

7.5 Supplied. It takes energy to separate particles which are attracted to one another.

7.6 $\text{Heat absorbed by calorimeter} = (25.51\text{°C} - 20.00\text{ °C})\left(\frac{8.930\text{ kJ}}{1\text{ °C}}\right) = 49.2\text{ kJ}$

$$\#\text{ mol C} = (1.50\text{ g C})\left(\frac{1\text{ mol C}}{12.01\text{ g C}}\right) = 0.125\text{ mol C}$$

ΔE = energy/mol = 49.2 kJ/0.125 mol C = 394 kJ/mol C

7.7 q = specific heat × mass × temperature change
= 4.184 J/g °C × (175 g + 4.90 g) × (14.9 °C – 10.0 °C)
= 3.7×10^3 J = 3.7 kJ of heat released by the process.

This should then be converted to a value representing kJ per mole of reactant, remembering that the sign of ΔH is to be negative, since the process releases heat energy to surroundings. The number of moles of sulfuric acid is:

$$\#\text{ moles } H_2SO_4 = (4.90\text{ g } H_2SO_4)\left(\frac{1\text{ mole } H_2SO_4}{98.08\text{ g } H_2SO_4}\right) = 5.00 \times 10^{-2}\text{ moles } H_2SO_4$$

and the enthalpy change in kJ/mole is given by:

3.7 kJ ÷ 0.0500 moles = 74 kJ/mole

7.8 We can proceed by multiplying both the equation and the thermochemical value of Example 7.9 by the fraction 2.5/2:

$2.5\ H_2(g) + 2.5/2\ O_2(g) \rightarrow 2.5\ H_2O(g)$, $\Delta H = (-517.8\text{ kJ}) \times 2.5/2$
$2.5\ H_2(g) + 1.25\ O_2(g) \rightarrow 2.5\ H_2O(g)$, $\Delta H = -647.3\text{ kJ}$

7.9

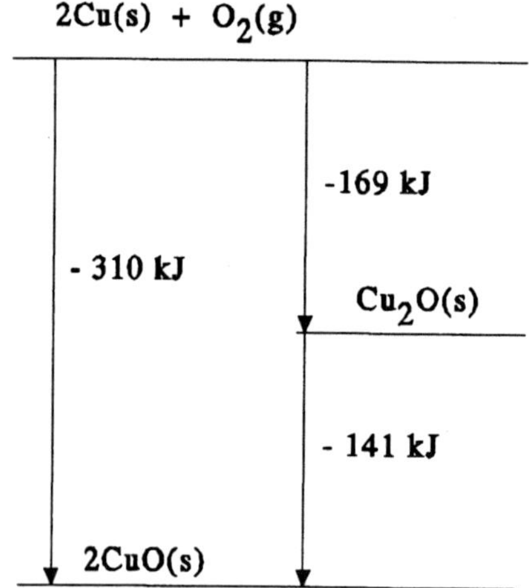

7.10 This problem requires that we add the reverse of the second equation (remembering to change the sign of the associated ΔH value) to the first equation:

$C_2H_4(g) + 3O_2(g) \rightarrow 2CO_2(g) + 2H_2O(l)$, $\Delta H° = -1411.1\text{ kJ}$
$2CO_2(g) + 3H_2O(l) \rightarrow C_2H_5OH(l) + 3O_2(g)$, $\Delta H° = +1367.1\text{ kJ}$

which gives the following net equation and value for ΔH°:

$C_2H_4(g) + H_2O(l) \rightarrow C_2H_5OH(l)$ $\Delta H° = -44.0\text{ kJ}$

7.11 $\#\text{ kJ} = (480\text{ moles } C_8H_{18})\left(\frac{5450.5\text{ kJ/mol}}{1\text{ moles } C_8H_{18}}\right) = 2.62 \times 10^6\text{ kJ}$

7.12 $Na(s) + 1/2H_2(g) + C(s) + 3/2O_2(g) \rightarrow NaHCO_3(s)$, $\Delta H_f^\circ = -947.7\text{ kJ/mol}$

7.13 a) $\Delta H° = \text{sum } \Delta H_f^\circ[\text{products}] - \text{sum } \Delta H_f^\circ[\text{reactants}]$
$= 2\,\Delta H_f^\circ[NO_2(g)] - \{2\,\Delta H_f^\circ[NO(g)] + \Delta H_f^\circ[O_2(g)]\}$
$= 2\text{ mol} \times 33.8\text{ kJ/mol} - [2\text{ mol} \times 90.37\text{ kJ/mol} + 1\text{ mol} \times 0\text{ kJ/mol}]$
$= -113.1\text{ kJ}$

b) $\Delta H° = \{ \Delta H_f^\circ[H_2O(l)] + \Delta H_f^\circ[NaCl(s)]\} - \{ \Delta H_f^\circ[NaOH(s)] + \Delta H_f^\circ[HCl(g)]\}$

$= [(-285.9 \text{ kJ/mol}) + (-411.0 \text{ kJ/mol})] - [(-426.8 \text{ kJ/mol}) + (-92.30 \text{ kJ/mol})]$

$= -177.8 \text{ kJ}$

Thinking it Through

T7.1 We need the equation for kinetic energy, K.E. = $1/2mv^2$. Since kinetic energy is directly proportional to mass but is directly proportional to the square of velocity, velocity has a larger influence on mass. This means that vehicle Y has the greater kinetic energy (even though it has half the mass of vehicle X) because of its doubled velocity.

T7.2 First, convert the energy in a candy bar from dietary calories to kilojoules, 1 Cal = 1 kcal = 4.184 kJ. Next determine how many 25 μJ are in the available kJ, 10^6 μJ = 1 J = 10^{-3} kJ.

$$\text{\# of pulses} = 250 \text{ Cal}\left(\frac{1\,\text{kcal}}{1\,\text{Cal}}\right)\left(\frac{4.184\,\text{kJ}}{1\,\text{kcal}}\right)\left(\frac{1\,\text{J}}{10^{-3}\,\text{kJ}}\right)\left(\frac{10^6\,\mu\text{J}}{1\,\text{J}}\right)\left(\frac{1\,\text{pulse}}{25\,\mu\text{J}}\right)$$

T7.3 First determine the heat lost by the iron by using equation 7.6

$$\text{heat lost} = \text{specific heat} \times \text{g} \times {}^\circ\text{C}$$

$$\text{heat lost} = (0.4498 \text{J/g}^\circ\text{C})(85.0 \text{ g})(55^\circ\text{C}) = 2100 \text{ J}$$

The heat gained by water is equal to the heat lost by iron. By rearranging equation 7.6 we find the mass of water.

$$\text{\# g } H_2O = \frac{2100 \text{ J}}{(10^\circ\text{C})(4.1796 \text{ J/g}^\circ\text{C})} = 50.3 \text{ g } H_2O$$

T7.4 It is necessary to know the mass of the substance.

T7.5 It is necessary to first convert from kg to g and from kJ to J

$$\text{\# g} = (4.50 \text{ kg})\left(\frac{1000 \text{ g}}{1 \text{ kg}}\right) = 4.5 \times 10^3 \text{ g}$$

$$\text{\# J} = (1.44 \text{ kJ})\left(\frac{1000 \text{ J}}{1 \text{ kJ}}\right) = 1.44 \times 10^3 \text{ J}$$

$$\text{specific heat} = \left(\frac{1.44 \times 10^3 \text{ J/}^\circ\text{C}}{4.5 \times 10^3 \text{g}}\right)\left(\frac{1 \text{ cal}}{4.184 \text{ J}}\right) = 7.65 \times 10^{-2} \text{ cal/g}^\circ\text{C}$$

T7.6
(a) Heat is the amount of energy that is transferred, not the amount of internal energy.
(b) Heat is the amount of energy that is transferred, not the amount of energy an object contains.
(c) Heat will flow from an object of higher temperature to an object with a lower temperature.
(d) When an object is heated, the internal energy of the object has increased, but some molecules may move faster, and some slower, the total internal energy is higher.
(e) Internal energy is a state function. Heat is not a state function.
(f) The heat flow between two objects will stop when the temperature of the two objects becomes the same.
(g) Heat is an intensive property and depends on the amount of matter present.

T7.7 We need the initial and final temperatures, the mass of the bar, and the specific heat of iron.

T7.8 $\Delta H = \Delta E + P\Delta V$

For this reaction, ΔV is approximately zero so $\Delta H = \Delta E$

T.7.9 For the first reaction,

$\Delta H° = \Delta H_f°[H_2O(\ell)] - \{\Delta H_f°[H_2(g)] + 1/2\Delta H_f°[O_2(g)]\} = \Delta H_f°[H_2O(\ell)]$

Since $\Delta H_f°[H_2(g)] = \Delta H_f°[O_2(g)] = 0$

Thus, for the second reaction,

$\Delta H° = \{\Delta H_f°[H_2(g)] + 3/2\Delta H_f°[O_2(g)]\} - 3\Delta H_f°[H_2O(\ell)]$

$= -3\Delta H_f°[H_2O(\ell)]$

T7.10 First we need to balance the thermochemical equation and obtain the $\Delta H°$ value for the reaction:

$$CH_3OH(\ell) + 3/2O_2(g) \rightarrow CO_2(g) + 2H_2O(g).$$

$\Delta H°_{rxn} = \Delta H_f°[CO_2(g)] + 2 \times \Delta H_f°[H_2O(g)] - \Delta H_f°[CH_3OH(\ell)]$

$= [1 \text{ mol } CO_2(g) \times (-394 \text{ kJ/mol})] + [2 \text{ mol } H_2O(g) \times (-241.8 \text{ kJ/mol})]$

$- [1 \text{ mol } CH_3OH(\ell) \times (-238 \text{ kJ/mol})]$

$\Delta H°_{rxn} = -640 \text{ kJ}$

This value applies to the combustion of one mole of methanol, as in the given equation. We must therefore finally convert grams to moles and multiply by the $\Delta H°$ value:

$$\# \text{ moles } CH_3OH = (35.5 \text{ g } CH_3OH)\left(\frac{1 \text{ mole } CH_3OH}{32.04 \text{ g } CH_3OH}\right) = 1.11 \text{ moles } CH_3OH$$

$$\Delta H = (1.11 \text{ moles } CH_3OH)\left(\frac{-640 \text{ kJ}}{1 \text{ mole } CH_3OH}\right) = -709 \text{ kJ}$$

T7.11 Let x = velocity of the 20 kg rock and y = the velocity of the 10 kg rock. Since $KE = 1/2\ mv^2$ and KE is conserved we can write:

$1/2(20 \text{ kg})(x)^2 = 1/2(10 \text{ kg})(y)^2$

Rearranging gives us:

$$\left(\frac{y}{x}\right)^2 = \frac{20}{10} \qquad \frac{y}{x} = \left(\frac{20}{10}\right)^{\frac{1}{2}} = 1.414$$

The 10 kg rock will have a velocity that is 1.4 times that of the 20 kg rock.

T7.12 Heat lost = (2000g)(0.803 J/g °C)(95 °C – x)

$$\text{Heat gained} = (2.00\text{L})\left(\frac{1000 \text{ mL}}{1\text{L}}\right)\left(\frac{1 \text{ g}}{1 \text{ mL}}\right)\left(\frac{4.1796 \text{ J}}{\text{g °C}}\right)(x - 22°\text{C})$$

Where x = final temperature we can solve for x since heat lost = heat gained

T7.13 We first calculate the amount of heat that was transferred to the water:

4.18 J/g °C × 10.0 g × (35.2 °C – 24.6 °C) = 443 J

Next, we realize that this is the amount of heat energy that was transferred out of the lead. We can therefore calculate the specific heat of lead using equation 7.6:

X J/g °C × 51.36 g × (100.0 °C – 35.2 °C) = 443 J

X = 0.133 J/g °C

T7.14 We must begin by determining the amount of heat energy that the truck has delivered, by first converting mass into kg and velocity into m/s:

mass = 28.0×10^3 lb × 454 g/lb = 1.27×10^7 g = 1.27×10^4 kg

velocity = 45.0 mi/hr × 5280 ft/mi × 0.3048 m/ft × (2.78×10^{-4} hr/s)
= 20.1 m/s

K.E. = $1/2mv^2 = (1.27 \times 10^4 \text{ kg})(20.1 \text{ m/s})^2 = 2.57 \times 10^6 \text{ kg m}^2/\text{s}^2$
= 2.57×10^6 J

The energy is transferred into 5 gal of water, which must be converted to a mass in g:

5.00 gal × 3.79 L/gal × 1000 mL/L × 1.00 g/mL = 1.90×10^4 g

Finally, the temperature change can be calculated using equation 7.6:

$2.57 \times 10^6 \text{ J} = 4.18 \text{ J/g °C} \times 1.90 \times 10^4 \text{ g} \times \Delta T$

$\Delta T = 32.4$ °C

T7.15 $\Delta H° = \Sigma\Delta H_f°(\text{products}) - \Sigma\Delta H_f°(\text{reactants})$
Since $\Delta H_f°(C_2H_2(g))$ and $\Delta H_f°(C_2H_4(g)) < 0$ (both are stable compounds), the $|\Delta H_f°(C_2H_4(g))| > |\Delta H_f°(C_2H_2(g))|$. If this were not true, the $\Delta H°$ of the reaction would be >0.

T7.16 (a) The atoms in both compartments would be moving within the compartment. The atoms in the left compartment, the lower temperature compartment, would be moving more slowly (on average) than those in the right compartment. There would be more atom-atom collisions as well as more collisions with the walls of the chamber in the right compartment.

(b) The high temperature atoms will slow down and the low temperature atoms will speed up. Eventually, all atoms in both compartments will be traveling at the same average speed. The atoms in the higher temperature compartment will collide with the conducting partition and energy would be transferred to atoms in the lower temperature compartment. The high temperature atoms will slow down and the low temperature atoms will speed up. Eventually, all atoms in both compartments will be traveling at the same average speed.

(c) Over time, the temperature of the two compartments would average to 55 °C (assuming there are equal quantities of gas in each compartment.) The thermometer in the left compartment would show a rise in temperature and the thermometer in the right compartment would show a decrease in temperature.

T7.17 The temperature rises because the two solutions react in an exothermic manner. The gradual cooling to room temperature occurs after the reaction is completed and the system equilibrates with the surroundings.

T7.18 The final temperature will be about 65 °C. The mixing of the cold water with the hot water will raise the temperature of the cold water and lower the temperature of the hot water. The block of copper, initially at 100 °C will also decrease to the final temperature. The final temperature will be closer to the hot temperature because there is more hot water than cold water. The block of copper will also affect the final temperature.

To solve this problem exactly we need the heat capacity of water and copper. Then set up the equation showing that heat lost = heat gained, i.e.,

Heat lost by hot water + Heat lost by copper = Heat gained by cold water

$$(m_{H_2O})(\text{specific heat of } H_2O)(100°C - T_{\text{final}}) + (m_{Cu})(\text{specific heat of Cu})(100°C - T_{\text{final}}) = (m_{H_2O})(\text{specific heat of } H_2O)(T_{\text{final}} - 10°C)$$

Review Questions

7.1 The sum of the total kinetic energy and the total potential energy is equal to the total energy of the system and this is a fixed quantity. If the total kinetic energy increases, then the total potential energy decreases.

7.2 Chemical energy is the potential energy in substances, which changes into other forms of energy when substances undergo chemical reactions.

7.3 (a) Kinetic energy must increase.
(b) The temperature of the system must increase because kinetic energy increased.

7.4 The energy that is released when a bond form is in the form of kinetic energy.

7.5 (a) increase
(b) increase
(c) increase
(d) decrease

7.6 (a) The potential energy decreases.
(b) The potential energy increases due to the separation of positive and negative charges. Also the decrease in temperature indicates a decrease in kinetic energy.

7.7 $1J = 1(kg\ m^2)/s^2$

7.8 According to SI units 1 cal = 4.184 J. Traditionally, 1 cal equals the amount of energy needed to raise the temperature of 1.0 g of water by 1 Celsius degree.

7.9 1 nutritional Calorie = 1000 thermochemical calories or kcal. So:

$$\#\ J = (1000\ cal)\left(\frac{4.184\ J}{1\ cal}\right) = 4.184 \times 10^3\ J$$

7.10 Thermal equilibrium is when two objects in contact with each other are at the same temperature.

7.11 The internal energy is the sum of the molecular kinetic energy and the potential energy.

7.12 Temperature is related to the average kinetic energy of the particles within a system, whereas internal energy is related to the total molecular kinetic energy.

7.13 The state of a system in chemistry is usually specified by its current conditions such as its chemical composition, its pressure, its temperature and its volume.

7.14 (a) point c
(b) point d
(c) It will increase
(d) It will decrease

7.15 Heat can be conducted from one object to another through molecular collisions.

7.16 a, c

7.17 (a) specific heat
(b) molar heat capacity
(c) heat capacity

7.18 The energy depends directly on the specific heat, so the material with the higher specific heat requires the higher energy input for a given rise in temperature.

7.19 Heat capacity is an extensive property and is proportional to the mass of the sample. Specific heat is an intensive property.

7.20 A substance with a high specific heat needs more energy. A substance with a high specific heat has a greater capacity to hold the energy and not undergo an increase in T. Therefore, a substance with a low specific heat capacity will experience the larger increase in temperature when it absorbs 100 J.

7.21 The numerical values would not change because both the mass unit (g to kg) and the energy unit (J to kJ) are increasing 1000 fold.

7.22 An isolated system cannot exchange matter or energy with its surroundings, and a closed system can absorb or release energy, but not mass across its boundaries.

7.23 A negative value for heat means that heat is released from the system, it is exothermic.

7.24 Since the amount of heat is related to the change in temperature by:

$$q = C\Delta T$$

If the temperature increase 10 fold, then the heat must be increased 10 times.

7.25 The information necessary to determine the heat capacity of an object is amount of heat exchanged and the change in temperature. To determine the specific heat of an object, its mass must be known.

7.26 If object A has twice the specific heat and twice the mass of object B, and the same amount of heat is applied to both objects, the temperature change of A will be one-fourth the temperature change in B.

$$q = ms\Delta T$$

q = amount of heat applied

s = specific heat of B 2s = specific heat of A

m = mass of B 2m = mass of B

$$\Delta t_B = \frac{q}{ms} \qquad \Delta t_A = \frac{q}{2m \times 2s} = \frac{1}{4}\left(\frac{q}{ms}\right)$$

7.27 A spring is often used as a model for chemical bonds.

7.28 Exothermic, chemical energy decreases.

7.29 Endothermic, chemical energy increases.

7.30 A state function is a function whose value is independent of the path (route) that is taken to achieve it. A state function has a value that depends only on the difference between the final and initial states, not on the path taken to change from the initial to the final state. Energy (E) and enthalpy (H) are state functions.

7.31 E = KE + PE, i.e. the total energy

7.32 $\Delta E = q + w$, the change in energy of a system equals the amount of heat added plus the amount of work done on it by the surroundings.

7.33 The heat of reaction at constant volume is ΔE. The heat of reaction at constant pressure is ΔH. $\Delta E = \Delta H$ if there is no change in volume.

7.34 $w = -p\Delta V$

7.35 $H = E + pV$

7.36 $\Delta H = H_{final} - H_{initial}$, for chemical reactions $\Delta H = H_{products} - H_{reactants}$. Since most reactions occur under conditions of constant pressure, ΔH is a more meaningful quantity than ΔE.

7.37 Since energy is conserved, the enthalpy of the surroundings must decrease by 100 kJ.

7.38 $\Delta H < 0$ for an exothermic change.

7.39 Since the heats of reaction will in general depend on temperature and pressure, we need some standard set of values for temperature and pressure so that comparisons of various heats of reaction are made under identical conditions. The standard temperature is 25 °C, approximate room temperature, and the standard pressure is 1 atmosphere.

7.40 A thermochemical equation contains the value for the associated ΔH.

7.41 The coefficient always signifies the number of moles.

7.42 Fractional coefficients are permitted because thermochemical properties are extensive (they depend on the amount of material present). If 1/2 the amount of materials reacts, 1/2 the amount of heat will be generated or required.

7.43 ΔH is a state function.

7.44 The reaction must produce one mole of a compound at 25°C and 1 atm from the elements in their standard states.

7.45 Hess's Law states that for any reaction that can be written in steps, the standard heat of reaction for the overall process is the sum of the standard heats of reaction of the individual steps. Hess's Law is dependent upon the fact that enthalpy is a state function; its value does not depend on the path or on the kinds of steps used to proceed from the initial state to the final state.

7.46 $$\Delta H^\circ_{\text{reaction}} = \left[\text{Sum of } \Delta H^\circ_c \text{ of all of the products}\right] - \left[\text{Sum of } \Delta H^\circ_c \text{ of all of the reactants}\right]$$

Review Problems

7.47 $$KE = \tfrac{1}{2}mv^2$$

$$= 1/2(2150\text{ kg})\left(80\ \frac{\text{km}}{\text{hr}}\right)^2\left(\frac{1\text{ hr}}{3600\text{s}}\right)^2\left(\frac{1000\text{m}}{1\text{ km}}\right)^2$$

$$= (5.31 \times 10^5\text{ kg m}^2/\text{s}^2)\left(\frac{1\text{ kJ}}{1000\text{ kg m}^2/\text{s}^2}\right)$$

$$= 531\text{ kJ}$$

7.48 $$KE = \tfrac{1}{2}mv^2$$

$$= 1/2(2.04 \times 10^4\text{ kg})\left(80\ \frac{\text{km}}{\text{hr}}\right)^2\left(\frac{1\text{ hr}}{3600\text{s}}\right)^2\left(\frac{1000\text{m}}{1\text{ km}}\right)^2$$

$$= (5.04 \times 10^6\text{ kgm}^2/\text{s}^2)\left(\frac{1\text{ kJ}}{1000\text{ kgm}^2/\text{s}^2}\right)$$

$$= 504 \times 10^3\text{ kJ}$$

7.49 (a) $$\#\text{ kcal} = (347\text{ kJ})\left(\frac{1\text{ kcal}}{4.184\text{ kJ}}\right) = 82.9\text{ kJ}$$

(b) $\# \text{ kJ} = (308 \text{ kcal})\left(\frac{4.184 \text{ kJ}}{1 \text{ kcal}}\right) = 1290 \text{ kJ}$

7.50 (a) $\# \text{ cal} = (26 \text{ J})\left(\frac{1 \text{ cal}}{4.184 \text{ J}}\right) = 6.2 \text{ cal}$

(b) $\# \text{ J} = (78 \text{ cal})\left(\frac{4.184 \text{ J}}{1 \text{ cal}}\right) = 3.3 \times 10^2 \text{ J}$

7.51 $\# \text{ of cal} = 1 \text{ h}\left(\frac{3600 \text{ s}}{1 \text{ h}}\right)\left(\frac{100 \text{ J}}{\text{s}}\right)\left(\frac{1 \text{ cal}}{4.184 \text{ J}}\right) = 8.60 \times 10^4 \text{ cal}$

$\# \text{ of dietary calories} = 8.60 \times 10^4 \text{ cal}\left(\frac{1 \text{ kcal}}{1000 \text{ cal}}\right)\left(\frac{1 \text{ dietary Calorie}}{1 \text{ kcal}}\right) = 86.0 \text{ dietary Cal}$

7.52 $\# \text{ of cal} = 1 \text{ day}\left(\frac{24 \text{ h}}{1 \text{ day}}\right)\left(\frac{3600 \text{ s}}{1 \text{ h}}\right)\left(\frac{70 \text{ J}}{\text{s}}\right)\left(\frac{1 \text{ cal}}{4.184 \text{ J}}\right) = 1.4 \times 10^6 \text{ cal}$

$\# \text{ of dietary calories} = 1.4 \times 10^6 \text{ cal}\left(\frac{1 \text{ kcal}}{1000 \text{ cal}}\right)\left(\frac{1 \text{ dietary Calorie}}{1 \text{ kcal}}\right) = 1.4 \times 10^3 \text{ dietary Cal}$

7.53 $\Delta E = q + w = 28\text{J} - 45\text{J} = -17 \text{ J}$

7.54 $\Delta E = q + w = 48\text{J} - 22\text{J} = 26 \text{ J}$

7.55 Here, ΔE must = 0 in order for there to be no change in energy for the cycle.

$$\Delta E = q + w$$
$$0 = q + (-100\text{J})$$
$$q = +100\text{J}$$

7.56 If the engine absorbs 250 J of heat, the maximum amount of work it can do is 250 J.

7.57 $\# \text{kJ} = (175 \text{ g } H_2O)\left(4.184 \frac{\text{J}}{\text{g °C}}\right)(25.0 \text{ °C} - 15.0 \text{ °C})\left(\frac{1 \text{ kJ}}{1000 \text{ J}}\right) = 7.32 \text{ kJ}$

7.58 $\# \text{kJ} = (1.0 \times 10^3 \text{ g } H_2O)\left(4.184 \frac{\text{J}}{\text{g °C}}\right)(99 \text{ °C} - 25 \text{ °C})\left(\frac{1 \text{ kJ}}{1000 \text{ J}}\right) = 310 \text{ kJ}$

7.59 $\# \text{ J} = (0.4498 \text{ J g}^{-1} \text{ °C}^{-1})(15.0 \text{ g})(20.0 \text{ °C}) = 135 \text{ J}$

7.60 $0.387 \text{ J g}^{-1} \text{ °C}^{-1}$

7.61 a) $\# \text{ J} = 4.184 \text{ J g}^{-1} \text{ °C}^{-1} \times 100 \text{ g} \times 4.0\text{°C} = 1.67 \times 10^3 \text{ J}$
b) $1.67 \times 10^3 \text{ J}$
c) $1.67 \times 10^3 \text{ J}/(100 - 28.0)\text{°C} = 23.2 \text{ J °C}^{-1}$
d) $23.2 \text{ J °C}^{-1} \div 5.00 \text{ g} = 4.64 \text{ J g}^{-1}\text{°C}^{-1}$

7.62 a) $\# \text{ J} = 4.184 \text{ J g}^{-1}\text{°C}^{-1} \times 200 \text{ g} \times 1.50\text{°C} = 1.26 \times 10^3 \text{ J}$
b) $1.26 \times 10^3 \text{ J}$
c) $1.26 \times 10^3 \text{ J} = 0.387 \text{ J g}^{-1} \text{ °C}^{-1} \times (120 - 26.50)\text{°C} \times X \text{ g}$
$X = 34.7 \text{ g}$

7.63 (a) Since fat tissue is 85% fat, there are only 0.85 lbs of fat lost for every pound of tissue lost in the weight reduction program. The water, which is a part of the fat tissue, is not lost. Therefore;

$$\#\text{ kcal} = (0.85\text{ lb fat})\left(\frac{456\text{ g}}{1\text{ lb}}\right)\left(\frac{9.0\text{ kcal}}{1\text{ g fat}}\right) = 3.5 \times 10^3\text{ kcal}$$

(b) In part (a) we determined that we would need to expend 3500 kcal in order to burn off 1 lb of fat tissue.

$$\#\text{ miles} = (1.0\text{ lb fat tissue})\left(\frac{3.5 \times 10^3\text{ kcal}}{1.0\text{ lb fat tissue}}\right)\left(\frac{1\text{ hr}}{5.0 \times 10^2\text{ kcal}}\right)\left(\frac{8.0\text{ miles}}{\text{hr}}\right)$$

$$= 56\text{ miles}$$

7.64 We must first determine the amount of fat consumed each day:

$$\#\text{ g fat} = (0.25\text{ lb fat})\left(\frac{1\text{ kg}}{2.2\text{ lbs}}\right)\left(\frac{1000\text{ g}}{1\text{ kg}}\right) = 110\text{ g fat}$$

Now using the energy equivalence stated in the problem, determine the amount of fat tissue which is added if we consume 110 g of fat in a given day:

$$\#\text{ g fat tissue} = (110\text{ g fat})\left(\frac{9.0\text{ kcal}}{1\text{ g fat}}\right)\left(\frac{0.50\text{ lb fat tissue}}{3.5 \times 10^2\text{ kcal}}\right) = 1.4\text{ lb fat tissue}$$

Finally, how many days will it take to lose one kg of fat tissue if we eliminate butter and oil from our diet:

$$\#\text{ days} = 1.0\text{ kg fat tissue}\left(\frac{2.2\text{ lb}}{1\text{ kg}}\right)\left(\frac{1\text{ day}}{1.4\text{ lb fat tissue}}\right) = 1.6\text{ days}$$

7.65 $$\frac{\#\text{ J}}{\text{mol °C}} = \left(\frac{0.4498\text{ J}}{\text{g °C}}\right)\left(\frac{55.847\text{ g Fe}}{1\text{ mol Fe}}\right) = 25.12\ \text{J}/\text{mol °C}$$

7.66 $$\frac{\#\text{ J}}{\text{mol °C}} = \left(\frac{0.586\text{ cal}}{\text{g °C}}\right)\left(\frac{4.184\text{ J}}{\text{cal}}\right)\left(\frac{46.1\text{ g }C_2H_5OH}{1\text{ mol }C_2H_5OH}\right) = 113\ \text{J}/\text{mol °C}$$

7.67 $4.18\text{ J g}^{-1}\text{ °C}^{-1} \times (4.54 \times 10^3\text{ g}) \times (58.65 - 60.25)\text{ °C} = -3.04 \times 10^4\text{ J} = -30.4\text{ kJ}$

7.68 $4.18\text{ J g}^{-1}\text{ °C}^{-1} \times (2.46 \times 10^3\text{ g}) \times (27.31 - 25.24)\text{ °C} = 2.1 \times 10^4\text{ J} = 21\text{ kJ}$

7.69 Keep in mind that the total mass must be considered in this calculation, and that both liquids, once mixed, undergo the same temperature increase:

heat = (4.18 J/g °C) × (55.0 g + 55.0 g) × (31.8 °C – 23.5 °C)
= 3.8×10^3 J of heat energy released

Next determine the number of moles of reactant involved in the reaction:
0.0550 L × 1.3 mol/L = 0.072 mol of acid and of base.

Thus the enthalpy change is: $$\frac{\#\text{ kJ}}{\text{mol}} = \frac{(3.8\times10^3\text{ J})\left(\frac{1\text{ kJ}}{1000\text{ J}}\right)}{(0.072\text{ mol})} = 53\ \text{kJ}/\text{mol}$$

7.70 The heat of neutralization is released to three "independent" components of the system, all of which undergo the same temperature increase: ΔT = 20.610 °C – 16.784 °C = 3.826 °C. Also, the total heat capacity of the system is the sum of the three heat capacities:

heat capacity$_{HCl}$ + heat capacity$_{NaOH}$ + heat capacity$_{calorimeter}$
= $(4.031\text{ J g}^{-1}\text{ °C}^{-1} \times 610.29\text{ g}) + (4.046\text{ J g}^{-1}\text{ °C}^{-1} \times 615.31\text{ g}) + 77.99\text{ J °C}^{-1}$
= 5028 J °C^{-1}

The heat flow to the system is thus:
heat = 5028 J °C^{-1} × 3.826 °C = 1.924 × 10^4 J = 19.24 kJ

and the heat of neutralization is the negative of this value, since the neutralization process is exothermic:
ΔH = –19.24 kJ ÷ 0.33183 mol = –57.98 kJ/mol

7.71 (a) $\# J = (97.1 \text{ kJ/°C})(27.282 \text{ °C} - 25.000 \text{ °C}) = 222 \text{ kJ} = 2.22 \times 10^5 \text{ J}$
(b) ΔH° = – 222 kJ/mol

7.72 (a) $\# J = (45.06 \text{ kJ/°C})(26.413 \text{ °C} - 25.000 \text{ °C}) = 63.67 \text{ kJ} = 6.367 \times 10^4 \text{ J}$

(b) $$\# J = 1 \text{ mol } C_7H_8 \left(\frac{6.367 \times 10^4 \text{ J}}{1.500 \text{ g}}\right)\left(\frac{92.14 \text{ g } C_7H_8}{1 \text{ mol } C_7H_8}\right) = 3.911 \times 10^6 \text{ J}$$

7.73 (a) Multiply the given equation by the fraction 2/3.
$2CO(g) + O_2(g) \rightarrow 2CO_2(g)$, ΔH° = –566 kJ
(b) To determine ΔH for 1 mol, simply multiply the original ΔH by 1/3; –283 kJ/mol.

7.74 (a) Divide the given equation by 4.
$NH_3(g) + 7/4\ O_2(g) \rightarrow NO_2(g) + 3/2H_2O(g)$ ΔH° = –1132 kJ/4 = –283 kJ

(b) Divide the given equation by 6
$2/3NH_3(g) + 7/6\ O_2(g) \rightarrow 2/3\ NO_2(g) + H_2O(g)$ ΔH° = –1132 kJ/6 = –189 kJ

7.75 $4Al(s) + 2Fe_2O_3(s) \rightarrow 2Al_2O_3(s) + 4Fe(s)$ ΔH° = –1708 kJ

7.76 $3C_6H_6(l) + 42/2O_2(g) \rightarrow 18CO_2(g) + 9H_2O(l)$ ΔH°= –9813 kJ

7.77 $$\# \text{ kJ} = (6.54 \text{ g Mg})\left(\frac{-1203 \text{ kJ}}{2 \text{ mol Mg}}\right)\left(\frac{1 \text{ mol Mg}}{24.305 \text{ g Mg}}\right) = -162 \text{ kJ}$$
162 kJ of heat are evolved

7.78 $$\# \text{ kJ} = (46.0 \text{ g } CH_3OH)\left(\frac{1 \text{ mol } CH_3OH}{32.04 \text{ g } CH_3OH}\right)\left(\frac{-1199 \text{ kJ}}{2 \text{ mol } CH_3OH}\right) = -861 \text{ kJ}$$ 7.79

7.79

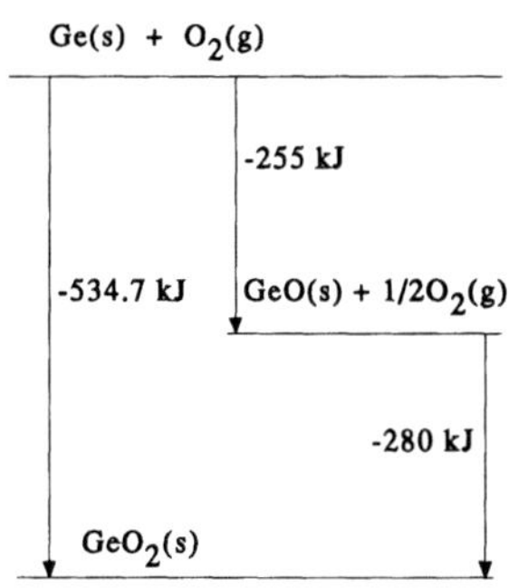

The enthalpy change for the reaction $GeO(s) + 1/2O_2(g) \rightarrow GeO_2(s)$ is –280 kJ as seen in the figure above.

7.80

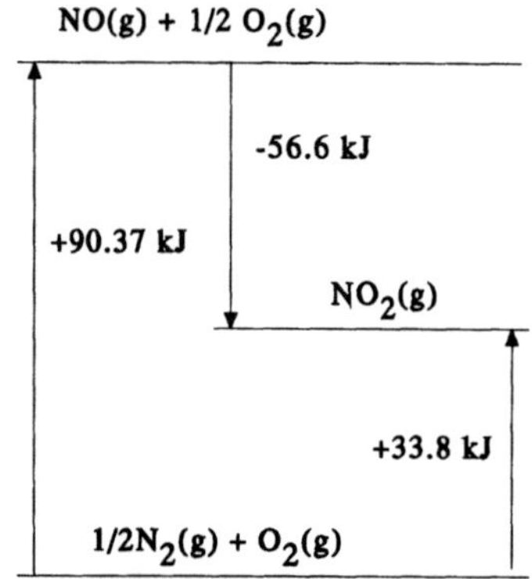

The enthalpy change for the reaction $NO(g) + 1/2O_2(g) \rightarrow NO_2(g)$ is –56.6 kJ as seen in the figure above.

7.81 Since NO_2 does not appear in the desired overall reaction, the two steps are to be manipulated in such a manner so as to remove it by cancellation. Add the second equation to the inverse of the first, remembering to change the sign of the first equation, since it is to be reversed:

$2NO_2(g) \rightarrow N_2O_4(g)$, $\Delta H° = -57.93$ kJ
$2NO(g) + O_2(g) \rightarrow 2NO_2(g)$, $\Delta H° = -113.14$ kJ

Adding, we have:

$2NO(g) + O_2(g) \rightarrow N_2O_4(g)$, $\Delta H° = -171.07$ kJ

7.82 Reverse the first equation, multiply the result by two, and add it to the second equation:

$2KCl(s) + 2H_2O(\ell) \rightarrow 2HCl(g) + 2KOH(s)$, $\Delta H° = 407.2$ kJ

$H_2SO_4(aq) + 2KOH(s) \rightarrow K_2SO_4(s) + 2H_2O(\ell)$, $\Delta H° = -342.4$ kJ

Adding gives us:

$2KCl(s) + H_2SO_4(\ell) \rightarrow 2HCl(g) + K_2SO_4(s)$, $\Delta H° = 64.8$ kJ

7.83 If we label the four known thermochemical equations consecutively, 1, 2, 3, and 4, then the sum is made in the following way: Divide equation #3 by two, and reverse all of the other equations (#1, #2, and #4), while also dividing each by two:

$\frac{1}{2}Na_2O(s) + HCl(g) \rightarrow \frac{1}{2}H_2O(\ell) + NaCl(s)$, $\Delta H° = -253.66$ kJ
$NaNO_2(s) \rightarrow \frac{1}{2}Na_2O(s) + \frac{1}{2}NO_2(g) + \frac{1}{2}NO(g)$, $\Delta H° = +213.57$ kJ
$\frac{1}{2}NO(g) + \frac{1}{2}NO_2(g) \rightarrow \frac{1}{2}N_2O(g) + \frac{1}{2}O_2(g)$, $\Delta H° = -21.34$ kJ
$\frac{1}{2}H_2O(\ell) + \frac{1}{2}O_2(g) + \frac{1}{2}N_2O(g) \rightarrow HNO_2(\ell)$, $\Delta H° = -17.18$ kJ

Adding gives:

$HCl(g) + NaNO_2(s) \rightarrow NaCl(s) + HNO_2(\ell)$, $\Delta H° = -78.61$ kJ

7.84 Add the reverse of the first equation to the second equation:

$H_2SO_4(\ell) \rightarrow SO_3(g) + H_2O(\ell)$, $\Delta H° = 78.2$ kJ

$BaO(s) + SO_3(g) \rightarrow BaSO_4(s)$, $\Delta H° = -213$ kJ

$BaO(s) + H_2SO_4(\ell) \rightarrow BaSO_4(s) + H_2O(\ell)$, $\Delta H° = -135$ kJ

7.85 Reverse the second equation, and then divide each by two before adding:

$CO(g) + \frac{1}{2}O_2(g) \rightarrow CO_2(g)$, $\Delta H° = -283.0$ kJ
$CuO(s) \rightarrow Cu(s) + \frac{1}{2}O_2(g)$, $\Delta H° = 155.2$ kJ

$CuO(s) + CO(g) \rightarrow Cu(s) + CO_2(g)$, $\Delta H° = -127.8$ kJ

7.86 Reverse the second and the third thermochemical equations and add them to the first:

$CaO(s) + 2HCl(aq) \rightarrow CaCl_2(aq) + H_2O(\ell)$, $\Delta H° = -186$ kJ
$Ca(OH)_2(s) \rightarrow CaO(s) + H_2O(\ell)$, $\Delta H° = 65.1$ kJ
$Ca(OH)_2(aq) \rightarrow Ca(OH)_2(s)$, $\Delta H° = 12.6$ kJ

$Ca(OH)_2(aq) + 2HCl(aq) \rightarrow CaCl_2(aq) + 2H_2O(\ell)$, $\Delta H° = -108$ kJ

7.87 Multiply all of the equations by $\frac{1}{2}$ and add them together.

$\frac{1}{2}CaO(s) + \frac{1}{2}Cl_2(g) \rightarrow \frac{1}{2}CaOCl_2(s)$ $\Delta H° = \frac{1}{2}(-110.9\text{kJ})$

$\frac{1}{2}H_2O(\ell) + \frac{1}{2}CaOCl_2(s) + NaBr(s) \rightarrow NaCl(s) + \frac{1}{2}Ca(OH)_2(s) + \frac{1}{2}Br_2(\ell)$
$\Delta H° = \frac{1}{2}(-60.2\text{ kJ})$

$\frac{1}{2}Ca(OH)_2(s) \rightarrow \frac{1}{2}CaO(s) + \frac{1}{2}H_2O(\ell)$ $\Delta H° = \frac{1}{2}(+65.1\text{ kJ})$

$\frac{1}{2}Cl_2(g) + NaBr(s) \rightarrow NaCl(s) + \frac{1}{2}Br_2(\ell)$ $\Delta H° = \frac{1}{2}(-106\text{ kJ}) = -53$ kJ

7.88 The equation we want is $Cu(s) + \frac{1}{2}O_2(g) \rightarrow CuO(s)$. If we multiply all reactions by $\frac{1}{2}$ and reverse the second reaction we get:

$Cu(s) + \frac{1}{2}S(s) \rightarrow \frac{1}{2}Cu_2S(s)$ $\Delta H° = \frac{1}{2}(-79.5\text{ kJ})$
$\frac{1}{2}SO_2(g) \rightarrow \frac{1}{2}S(s) + \frac{1}{2}O_2(g)$ $\Delta H° = \frac{1}{2}(+297\text{ kJ})$
$\frac{1}{2}Cu_2S(s) + O_2(g) \rightarrow CuO(s) + \frac{1}{2}SO_2(g)$ $\Delta H° = \frac{1}{2}(-527.5\text{ kJ})$

$Cu(s) + \frac{1}{2}O_2(g) \rightarrow CuO(s)$ $\Delta H° = -155$ kJ

7.89 We need to eliminate the NO_2 from the two equations. To do this, multiply the first reaction by 3 and the second reaction by two and add them together.

$12NH_3(g) + 21O_2(g) \rightarrow 12NO_2(g) + 18H_2O(g)$ $\Delta H° = 3(-1132\text{ kJ})$

$12NO_2(g) + 16NH_3(g) \rightarrow 14N_2(g) + 24H_2O(g)$ $\Delta H° = 2(-2740\text{ kJ})$

$28NH_3(g) + 21O_2(g) \rightarrow 14N_2(g) + 42O_2(g)$ $\Delta H° = -8876$ kJ

Now divide this equation by 7 to get
$4NH_3(g) + 3O_2(g) \rightarrow 2N_2(g) + 6H_2O(g)$ $\Delta H° = 1/7(-8876\text{ kJ}) = -1268$ kJ

7.90 Multiply the second equation by two and add them together:

$3Mg(s) + 2NH_3(g) \rightarrow Mg_3N_2(s) + 3H_2(g)$ $\Delta H° = -371$ kJ

$N_2(g) + 3H_2(g) \rightarrow 2NH_3(g)$ $\Delta H° = 2(-46$ kJ)

$3Mg(s) + N_2(g) \rightarrow Mg_3N_2(s)$ $\Delta H° = -463$ kJ

7.91 The equation we want is:

$Mg(s) + N_2(g) + 3O_2(g) \rightarrow Mg(NO_3)_2(s)$

Reverse all three reactions AND multiply the third equation by three

$Mg_3N_2(s) + 6MgO(s) \rightarrow 8Mg(s) + Mg(NO_3)_2(s)$ $\Delta H° = +3884$ kJ

$3Mg(s) + N_2(g) \rightarrow Mg_3N_2(s)$ $\Delta H° = -463$ kJ

$6Mg(s) + 3O_2(g) \rightarrow 6MgO(s)$ $\Delta H° = 3(-1203$ kJ)

$Mg(s) + N_2(g) + 3O_2(g) \rightarrow Mg(NO_3)_2(s)$ $\Delta H° = -188$ kJ

7.92 Reverse the first equation:
$2H_2O(\ell) \rightarrow 2H_2(g) + O_2(g)$ $\Delta H° = +571.5$ kJ

Reverse the second equation AND multiply by two:
$4HNO_3(\ell) \rightarrow 2N_2O_5(g) + 2H_2O(\ell)$ $\Delta H° = +153.2$ kJ

Multiply the last equation by 4:
$2N_2(g) + 6O_2(g) + 2H_2(g) \rightarrow 4HNO_3(\ell)$ $\Delta H° = -696$ kJ

Now add the three equations:
$2N_2(g) + 5O_2(g) \rightarrow 2N_2O_5(g)$ $\Delta H° = +28.7$ kJ

7.93 Only (b) should be labeled with ΔH_f°.

7.94 Only (c) should be labeled with ΔH_f°.

7.95 (a) $2C(graphite) + 2H_2(g) + O_2(g) \rightarrow HC_2H_3O_2(\ell)$, $\Delta H_f^\circ = -487.0$ kJ

(b) $Na(s) + 1/2H_2(g) + C(graphite) + 3/2O_2(g) \rightarrow NaHCO_3(s)$ $\Delta H_f^\circ = -947.7$ kJ

(c) $Ca(s) + 1/8S_8(s) + 3O_2(g) + 2H_2(g) \rightarrow CaSO_4 \cdot 2H_2O(s)$ $\Delta H_f^\circ = -2021.1$ kJ

7.96 (a) $C(graphite) + 1/2O_2(g) + N_2(g) + 2H_2(g) \rightarrow CO(NH_2)_2(s)$, $\Delta H_f^\circ = -333.19$ kJ

(b) $Ca(s) + 1/8S_8(s) + 18/4O_2(g) + 1/2H_2(g) \rightarrow CaSO_4 \cdot \frac{1}{2} H_2O(s)$ $\Delta H_f^\circ = -1575.2$ kJ

(c) $C(graphite) + 2H_2(g) + 1/2O_2(g) \rightarrow CH_3OH(\ell)$, $\Delta H_f^\circ = -238.6$ kJ

7.97 (a) $\Delta H_f^\circ = \Delta H_f^\circ[O_2(g)] + 2\Delta H_f^\circ[H_2O(\ell)] - 2\Delta H_f^\circ[H_2O_2(\ell)]$

$\Delta H_f^\circ = 0\ \text{kJ/mol} + 2\ \text{mol} \times (-285.9\ \text{kJ/mol}) - 2 \times (-187.6\ \text{kJ/mol})$

$= -196.6\ \text{kJ}$

(b) $\Delta H_f^\circ = \Delta H_f^\circ[H_2O(\ell)] + \Delta H_f^\circ[NaCl(s)] - \Delta H_f^\circ[HCl(g)] - \Delta H_f^\circ[NaOH(s)]$

$= 1\ \text{mol} \times (-285.9\ \text{kJ/mol}) + 1\ \text{mol} \times (-411.0\ \text{kJ/mol})$

$- 1\ \text{mol} \times (-92.30\ \text{kJ/mol}) - 1\ \text{mol} \times (-426.8\ \text{kJ/mol})$

$= -177.8\ \text{kJ}$

7.98 (a) $\Delta H_f^\circ = \Delta H_f^\circ[HCl(g)] + \Delta H_f^\circ[CH_3Cl(g)] - \Delta H_f^\circ[CH_4(g)] - \Delta H_f^\circ[Cl_2(g)]$

$= 1\ \text{mol} \times (-92.30\ \text{kJ/mol}) + 1\ \text{mol} \times (-82.0\ \text{kJ/mol})$

$- 1\ \text{mol} \times (-74.848\ \text{kJ/mol}) - 1\ \text{mol} \times (0.0\ \text{kJ/mol})$

$= -99.5\ \text{kJ}$

(b) $\Delta H_f^\circ = \Delta H_f^\circ[H_2O(\ell)] + \Delta H_f^\circ[CO(NH_2)_2(s)] - 2\Delta H_f^\circ[NH_3(g)] - \Delta H_f^\circ[CO_2(g)]$

$= 1\ \text{mol} \times (-285.9\ \text{kJ/mol}) + 1\ \text{mol} \times (-333.19\ \text{kJ/mol})$

$- 2\ \text{mol} \times (-46.19\ \text{kJ/mol}) - 1\ \text{mol} \times (-393.5\ \text{kJ/mol})$

$= -133.2\ \text{kJ}$

7.99 (a) $\frac{1}{2}H_2(g) + \frac{1}{2}Cl_2(g) \rightarrow HCl(g)$, $\Delta H_f^\circ = -92.30\ \text{kJ/mol}$

(b) $\frac{1}{2}N_2(g) + 2H_2(g) + \frac{1}{2}Cl_2(g) \rightarrow NH_4Cl(s)$, $\Delta H_f^\circ = -315.4\ \text{kJ/mol}$

7.100 (a) $2C(\textit{graphite}) + 2H_2(g) + O_2(g) \rightarrow HC_2H_3O_2(\ell)$, $\Delta H_f^\circ = -487.0\ \text{kJ/mol}$

(b) $2Na(s) + C(\textit{graphite}) + 3/2O_2(g) \rightarrow Na_2CO_3(s)$, $\Delta H_f^\circ = -1131\ \text{kJ/mol}$

7.101 $C_{12}H_{22}O_{11}(s) + 12O_2(g) \rightarrow 12CO_2(g) + 11H_2O(\ell)$ $\Delta H^\circ_{\text{combustion}} = -5.65 \times 10^3\ \text{kJ/mol}$

$\Delta H^\circ_{\text{combustion}} = \Sigma\Delta H_f^\circ(\text{products}) - \Sigma\Delta H_f^\circ(\text{reactants})$

$=[12\ \text{mol}\ CO_2 \times \Delta H_f^\circ(CO_2(g)) + 11\ \text{mol}H_2O \times \Delta H_f^\circ(H_2O(\ell))]$

$- [1\ \text{mol}\ C_{12}H_{22}O_{11}\ \Delta H_f^\circ(C_{12}H_{22}O_{11}(s)) + 12\ \text{mol}\ O_2\Delta H_f^\circ(O_2(g))]$

Rearranging and realizing the $\Delta H_f^\circ O_2(g) = 0$ we get

$\Delta H_f^\circ(C_{12}H_{22}O_{11}(s)) = 12\Delta H_f^\circ(CO_2(g)) + 11\Delta H_f^\circ(H_2O(\ell)) - \Delta H^\circ_{\text{combustion}}$

$= 12(-393\text{kJ}) + 11(-285.9\ \text{kJ}) - (-5.65 \times 10^3\ \text{kJ}) = -2.22 \times 10^3\ \text{kJ}$

7.102 $\Delta H_f^\circ(C_2H_2(g)) = \frac{1}{2}\{4\Delta H_f^\circ(CO_2(g)) + 2\Delta H_f^\circ(H_2O(\ell)) - \Delta H^\circ_{\text{combustion}}\}$

$= \frac{1}{2}\{4(-393.5\text{kJ}) + 2(-285.9\ \text{kJ}) - (-2599.3\ \text{kJ}) = +226.8\ \text{kJ}$

Additional Problems

7.103 (a) Either heat was subtracted from the system or work was done by the system.

(b) $\#\ \text{kJ} = (750\text{g})(4.184\text{J/g}^\circ\text{C})(19.50 - 25.50)^\circ\text{C}\left(\frac{1\ \text{kJ}}{1000\ \text{J}}\right) = -18.8\ \text{kJ}$

7.104 Heat lost = – Heat gained

$\text{Heat lost} = (1.000\text{kg})\left(\frac{1000\ \text{g}}{1\ \text{kg}}\right)(0.4498\ \text{J/g}\ ^\circ\text{C})(100.0\ ^\circ\text{C} - x)$

$$\text{Heat gained} = (2.000\text{ kg})\left(\frac{1000\text{ g}}{1\text{ kg}}\right)(4.184\text{ J/g °C})(25.00\text{ °C} - x)$$

$449.8\text{ J/°C}(100.0\text{ °C} - x) = -(8368.\text{ J/°C})(25.00\text{ °C} - x)$

$254180\text{ J} = (7918.2\text{ J/ °C})(x)$

$x = 254180\text{ J}/881.8\text{ J/g °C} = 28.8\text{ °C}$

7.105 $\Delta H° = \Delta H_f^\circ[H_2SO_4(\ell)] - \Delta H_f^\circ(SO_3(g)) + \Delta H_f^\circ(H_2O(\ell))]$
$= -811.32\text{ kJ} - [(-395.2\text{ kJ}) + (-285.9\text{ kJ})]$
$= -130.2\text{ kJ}$

7.106 $\Delta H_{reaction} = 3\Delta H_f^\circ[CO_2(g)] + 4\Delta H_f^\circ[Fe(s)] - 2\Delta H_f^\circ[Fe_2O_3(s)] - 3\Delta H_f^\circ[C(s)]$
$= 3\text{ mol} \times (-393.5\text{ kJ/mol}) + 4\text{ mol} \times (0.0\text{ kJ/mol})$
$- 2\text{ mol} \times (-822.2\text{ kJ/mol}) - 3\text{ mol} \times (0.0\text{ kJ/mol})$
$= 463.9\text{ kJ}$
This reaction is endothermic.

7.107 Multiply the first reaction by 1/2:
$1/2Fe_2O_3(s) + 3/2CO(g) \rightarrow Fe(s) + 3/2CO_2(g)$ $\Delta H° = 1/2(-28\text{ kJ})$

Reverse the second reaction AND multiply by 1/6:
$1/3Fe_3O_4(s) + 1/6CO_2(g) \rightarrow 1/2\ Fe_2O_3(s) + 1/6CO(g)$ $\Delta H° = 1/6(+59\text{ kJ})$

Reverse the third reaction AND multiply by 1/3:
$FeO(s) + 1/3CO_2(g) \rightarrow 1/3Fe_3O_4(s) + 1/3CO(g)$ $\Delta H° = 1/3(-38\text{ kJ})$

Now add the equations together:
$FeO(s) + CO(g) \rightarrow Fe(s) + CO_2(g)$ $\Delta H° = -16.8\text{ kJ}$

7.108 $\Delta H°_{reaction} = [\Delta H_f^\circ(CO_2(g)) + \Delta H_f^\circ(Fe(s))] - [\Delta H_f^\circ(FeO(s)) + \Delta H_f^\circ(CO(g))]$
Rearranging, and remembering that $\Delta H_f^\circ\ Fe(s) = 0$
$\Delta H_f^\circ(FeO(s)) = \Delta H_f^\circ(CO_2(g)) - \Delta H_f^\circ(CO(g)) - \Delta H°_{reaction}$
$\Delta H_f^\circ = -393.5\text{ kJ} - (-110.5\text{ kJ}) - (-16.8\text{ kJ}) = -266.2\text{ kJ}$

7.109 These two thermochemical equations are added along with six times that for the formation of liquid water:
$P_4O_{10}(s) + 6H_2O(\ell) \rightarrow 4H_3PO_4(\ell)$, $\Delta H° = -257.2\text{ kJ}$
$4P(s) + 5O_2(g) \rightarrow P_4O_{10}(s)$, $\Delta H° = -3062\text{ kJ}$
$6H_2(g) + 3O_2(g) \rightarrow 6H_2O(\ell)$, $\Delta H° = -1715.4\text{ kJ}$

Adding gives:

$4P(s) + 6H_2(g) + 8O_2(g) \rightarrow 4H_3PO_4(\ell)$, $\Delta H° = -5035\text{ kJ}$

The above result should then be divided by four:
$P(s) + 3/2H_2(g) + 2O_2(g) \rightarrow H_3PO_4(\ell)$, $\Delta H° = -1259\text{ kJ/mol}$

7.110 $\Delta H° = 8\text{ mol} \times \Delta H_f°[CO_2(g)] + 10\text{ mol} \times \Delta H_f°[H_2O(\ell)] + 2\text{ mol} \times \Delta H_f°[N_2(g)]$
$- 4\text{ mol} \times \Delta H_f°[C_2H_5NO_2(s)] - 9\text{ mol} \times \Delta H_f°[O_2(g)]$

The value of $\Delta H°$ for the given combustion reaction is:
$\Delta H° = 4\text{ mol} \times (-973.49\text{ kJ/mol}) = -3894.0\text{ kJ}$

The standard enthalpy of formation for each substance is taken from Table 7.2:

$$-3894.0 \text{ kJ} = 8 \text{ mol} \times (-393.5 \text{ kJ/mol}) + 10 \text{ mol} \times (-285.9 \text{ kJ/mol}) - 4 \text{ mol} \times \Delta H_f^\circ[C_2H_5NO_2]$$

Solving for the desired enthalpy of formation, we have:

$$\Delta H_f^\circ [C_2H_5NO_2] = -528.3 \text{ kJ/mol}$$

7.111 The desired net equation is obtained by adding together the reverse of the two thermochemical equations:

$Zn(NO_3)_2(aq) + Cu(s) \rightarrow Cu(NO_3)_2(aq) + Zn(s)$, $\Delta H^\circ = 258$ kJ
$Cu(NO_3)_2(aq) + 2Ag(s) \rightarrow 2AgNO_3(aq) + Cu(s)$, $\Delta H^\circ = 106$ kJ

$2Ag(s) + Zn(NO_3)_2(aq) \rightarrow Zn(s) + 2AgNO_3(aq)$, $\Delta H^\circ = 364$ kJ

Since this ΔH° has a positive value, it is the reverse reaction that occurs spontaneously, and the reaction as written in endothermic.

7.112 The first thermochemical equation is reversed and multiplied by two, before adding it to the second equation:

$2LiOH(s) \rightarrow 2Li(s) + O_2(g) + H_2(g)$, $\Delta H^\circ = 974.0$ kJ
$2Li(s) + Cl_2(g) \rightarrow 2LiCl(s)$, $\Delta H^\circ = -815.0$ kJ

Next we add twice the standard heat of formation of liquid water:

$2H_2(g) + O_2(g) \rightarrow 2H_2O(\ell)$, $\Delta H^\circ = -571.8$ kJ

and also we add the reverse of twice the enthalpy of formation of HCl(g):

$2HCl(g) \rightarrow H_2(g) + Cl_2(g)$, $\Delta H^\circ = 184.6$ kJ

The last three equations that we need are obtained from those listed in the problem:

$2LiOH(aq) \rightarrow 2LiOH(s)$, $\Delta H^\circ = 38.4$ kJ
$2HCl(aq) \rightarrow 2HCl(g)$, $\Delta H^\circ = 154$ kJ
$2LiCl(s) \rightarrow 2LiCl(aq)$, $\Delta H^\circ = -72.0$ kJ

where two of the last three have been reversed, and where all of the last three have been multiplied by two. The sum of all of the above equations is:

$2LiOH(aq) + 2HCl(aq) \rightarrow 2LiCl(aq) + 2H_2O(\ell)$, $\Delta H^\circ = -107.8$ kJ

The desired thermochemical equation is half that given above:

$LiOH(aq) + HCl(aq) \rightarrow LiCl(aq) + H_2O(\ell)$, $\Delta H^\circ = -53.9$ kJ

7.113 The equation may be written as: $1/2\ H_2(g) + 1/2\ Br_2(\ell) \rightarrow HBr(g)$; $\Delta H_f^\circ = -36$ kJ

To obtain ΔH, combine the equations in the following manner:

$Br_2(aq) + 2KCl(aq) \rightarrow Cl_2(g) + 2KBr(aq)$ $\Delta H^\circ = 96.2$ kJ
$H_2(g) + Cl_2(g) \rightarrow 2HCl(g)$ $\Delta H^\circ = -184$ kJ

$2HCl(aq) + 2KOH(aq)$	$\rightarrow 2KCl(aq) + 2H_2O(\ell)$	$\Delta H° = -115$ kJ
$2KBr(aq) + 2H_2O(\ell)$	$\rightarrow 2HBr(aq) + 2KOH(aq)$	$\Delta H° = 115$ kJ
$2HCl(g)$	$\rightarrow 2HCl(aq)$	$\Delta H° = -154$ kJ
$2HBr(aq)$	$\rightarrow 2HBr(g)$	$\Delta H° = 160$ kJ
$Br_2(\ell)$	$\rightarrow Br_2(aq)$	$\Delta H° = -4.2$ kJ

Add all of the above to get;

$H_2(g) + Br_2(\ell) \rightarrow 2HBr(g)$; $\Delta H° = -86$ kJ

Now divide this equation by two to give the thermochemical equation for the formation of 1 mol of HBr(g):

$1/2\ H_2(g) + 1/2\ Br_2(\ell) \rightarrow HBr(g)$; $\Delta H° = -43$ kJ

Comparing this value to the $\Delta H_f°$ value listed in Appendix E and at the outset of this problem, we see that this experimental data indicates a value that is close to the reported value.

7.114 The equation we want is:
$2C(s) + H_2(g) \rightarrow C_2H_2(g)$

Take this one step at a time. Start with the fourth equation:
$CaC_2(s) + 2H_2O(\ell) \rightarrow Ca(OH)_2(s) + C_2H_2(g)$ $\Delta H° = -126$ kJ

Add the reverse of the first equation and get rid of calcium hydroxide:
$Ca(OH)_2(s) \rightarrow CaO(s) + H_2O(\ell)$ $\Delta H° = +65.3$ kJ

Add the second equation to eliminate the $CaO(s)$:
$CaO(s) + 3C(s) \rightarrow CaC_2(s) + CO(g)$ $\Delta H° = +462.3$ kJ

This also eliminates the $CaC_2(s)$ we had from the first equation. To eliminate the H_2O, reverse the last equation AND multiply by 1/2:
$H_2(g) + 1/2O_2(g) \rightarrow H_2O(\ell)$ $\Delta H° = -286$ kJ

Now get rid of the CO by reversing the fifth equation AND multiplying by 1/2:
$CO(g) \rightarrow C(s) + 1/2O_2(g)$ $\Delta H° = +110$ kJ

Add the equations together to get:
$2C(s) + H_2(g) \rightarrow C_2H_2(g)$ $\Delta H° = +226$ kJ

7.115 $C_{12}H_{22}O_{11}(s) + 12O_2(g) \rightarrow 12CO_2(g) + 11H_2O(\ell)$ $\Delta H_{combustion} = -5.63 \times 10^3$ kJ/mol (from 7.101)

$\Delta H_{combustion} = \Sigma\Delta H_f°(\text{products}) - \Sigma\Delta H_f°(\text{reactants})$
$=[12\ \Delta H_f°(CO_2(g)) + 11\ \Delta H_f°(H_2O(\ell))] - [\Delta H_f°(C_{12}H_{22}O_{11}(s)) + 12\Delta H_f°(O_2(g))]$
$=[12(-393\text{kJ}) + 11(-285.9\text{ kJ})] - [(-2230\text{ kJ}) + 12(0\text{ kJ})] = -5.63 \times 10^3$ J/mol
This is the amount of heat liberated for 1 mol of sucrose. Thus, for 28.4 g we have,

$$\#\text{kJ} = (28.4\text{ g})\left(\frac{1\text{ mol}}{342.3\text{ g}}\right)\left(\frac{-5630\text{ kJ}}{1\text{ mol}}\right) = -467\text{ kJ}$$

7.116 The combustion reaction is $C_2H_5OH(\ell) + 3O_2(g) \rightarrow 2CO_2(g) + 3H_2O(\ell)$

$\Delta H° = [3\Delta H_f°(H_2O(\ell)) + 2\Delta H_f°(CO_2(g))] - [\Delta H_f°(C_2H_5OH(\ell) + 3\Delta H_f°(O_2(g))]$
$= 3(-285.9 \text{ kJ}) + 2(-393.5 \text{ kJ}) - (-277.63 \text{ kJ}) = -1367.1 \text{ kJ/mol}$

$$\# \text{ kJ} = (1 \text{ gal})\left(\frac{3.785 \text{ L}}{1 \text{ gal}}\right)\left(\frac{1000 \text{ mL}}{1 \text{ L}}\right)\left(\frac{0.787 \text{ g}}{1 \text{ mL}}\right)\left(\frac{1 \text{ mol}}{46.07 \text{ g}}\right)\left(\frac{-1367 \text{ kJ}}{1 \text{ mol}}\right) = -88400 \text{ kJ}$$

7.117 $1/2HCHO_2(\ell) + 1/2H_2O(\ell) \rightarrow 1/2CH_3OH(\ell) + 1/2O_2(g)$ $\Delta H° = +206\text{kJ}$

$1/2CO(g) + H_2(g) \rightarrow 1/2CH_3OH(\ell)$ $\Delta H° = -64 \text{ kJ}$

$1/2HCHO_2(\ell) \rightarrow 1/2CO(g) + 1/2H_2O(\ell)$ $\Delta H° = -17 \text{ kJ}$

Add these together:
$HCHO_2(\ell) + H_2(g) \rightarrow CH_3OH(\ell) + 1/2O_2(g)$ $\Delta H° = +125 \text{ kJ}$

7.118 Simply add the two together to get:
$O_3(g) + O(g) \rightarrow 2O_2(g)$ $\Delta H° = -394 \text{ kJ}$

7.119 Add together the fourth equation, the first equation, the second equation and the reverse of the third equation:
$4CuS(s) + 2CuO(s) \rightarrow 3Cu_2S(s) + SO_2(g)$ $\Delta H° = -13.1 \text{ kJ}$
$2Cu(s) + O_2(g) \rightarrow 2CuO(s)$ $\Delta H° = -155 \text{ kJ}$
$Cu(s) + S(s) \rightarrow CuS(s)$ $\Delta H° = -53.1 \text{ kJ}$
$SO_2(g) \rightarrow S(s) + O_2(g)$ $\Delta H° = +297 \text{ kJ}$

The net reaction is:
$3CuS(s) + 3Cu(s) \rightarrow 3Cu_2S(s)$ $\Delta H° = +76 \text{ kJ}$

We want 1/3 of this:
$CuS(s) + Cu(s) \rightarrow Cu_2S(s)$ $\Delta H° = +25 \text{ kJ}$

Practice Exercises

8.1 $\nu = 10.0\ \mu m\left(\frac{1\times10^{-6}\,m}{1\ \mu m}\right) = 1.00\times10^{-5}\,m$

$\nu = c/\lambda = (3.00\times10^{8}\ m/s)/(1.00\times10^{-5}\ m) = 3.00\times10^{13}\ s^{-1} = 3.00\times10^{13}\ Hz$

8.2 $\lambda = \frac{c}{\nu} = \frac{2.9979\times10^{8}\ m\ s^{-1}}{104.3\times10^{6}\ s^{-1}} = 2.874\ m$

8.3

$\frac{1}{\lambda} = 109{,}678\ cm^{-1}\times\left(\frac{1}{2^2}-\frac{1}{3^2}\right) = 109{,}678\ cm^{-1}\times(0.2500-0.1111)$

$\frac{1}{\lambda} = 1.523\times10^{4}\ cm^{-1}$

$1 = 6.566\times10^{-5}$ cm = 656.6 nm, which is red.

8.4 When $n = 3$, $\ell = 0, 1, 2$. Thus we have s, p and d subshells.

When $n = 4$, $\ell = 0, 1, 2, 3$. Thus we have s, p, d and f subshells.

8.5 For the g subshell, $\ell = 4$ and there are 9 possible values of m_ℓ; –4, –3, –2, –1, 0, +1, +2, +3, +4. There are therefore 9 orbitals.

8.6 a) Mg: $1s^22s^22p^63s^2$
b) Ge: $1s^22s^22p^63s^23p^63d^{10}4s^24p^2$
c) Cd: $1s^22s^22p^63s^23p^63d^{10}4s^24p^64d^{10}5s^2$
d) Gd: $1s^22s^22p^63s^23p^63d^{10}4s^24p^64d^{10}4f^75s^25p^65d^16s^2$

8.7

(a) Na: (↑↓) (↑↓) (↑↓)(↑↓)(↑↓) (↑) ( )( )( ) ( ) ( )( )( )( )( )
1s 2s 2p 3s 3p 4s 3d

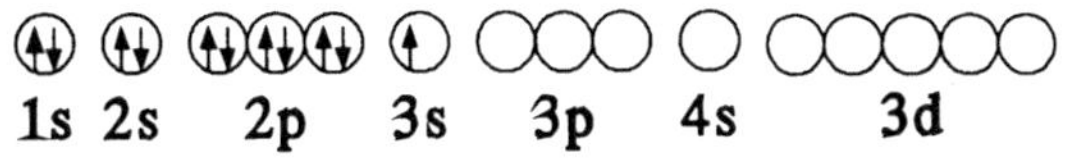

(b) S: (↑↓) (↑↓) (↑↓)(↑↓)(↑↓) (↑↓) (↑↓)(↑)(↑) ( ) ( )( )( )( )( )
1s 2s 2p 3s 3p 4s 3d

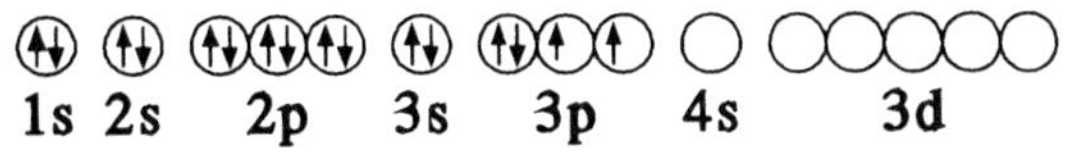

(c) Fe: (↑↓) (↑↓) (↑↓)(↑↓)(↑↓) (↑↓) (↑↓)(↑↓)(↑↓) (↑↓) (↑↓)(↑)(↑)(↑)(↑)
1s 2s 2p 3s 3p 4s 3d

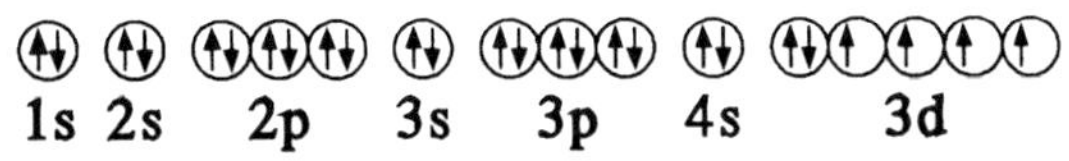

8.8 a) P: $[Ne]3s^23p^3$

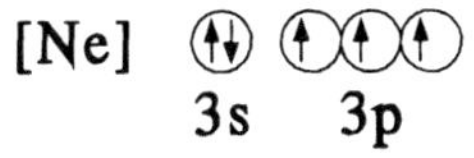

[Ne] (↑↓) (↑)(↑)(↑)
3s 3p
(3 unpaired electrons)

b) Sn: $[Kr]4d^{10}5s^25p^2$

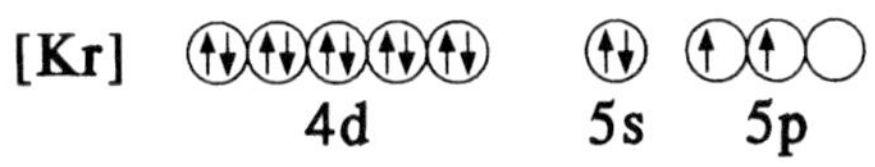

[Kr] (↑↓)(↑↓)(↑↓)(↑↓)(↑↓) (↑↓) (↑)(↑)( )
4d 5s 5p
(2 unpaired electrons)

8.9 a) Se: $4s^24p^4$ b) Sn: $5s^25p^2$ c) I: $5s^25p^5$

8.10 a) Sn b) Ga c) Cr d) S^{2-}

8.11 a) Be b) C

Thinking It Through

T8.1 We must convert to wavelength using the equation $\lambda = c/\nu$, and we must use the relationship 1 Hz = 1 s^{-1} to convert to frequency in the proper units. Remember also that 1 MHz = 10^6 Hz.

$$\lambda = \frac{c}{\nu} = \frac{2.9979 \times 10^8 \text{ m s}^{-1}}{101.9 \times 10^6 \text{ s}^{-1}}$$

The frequency (101.9×10^6 s^{-1}) tells us how many peaks pass a given point in space in a second.

T8.2 We must convert to wavelength using the equation $\lambda = v/\nu$, where v = speed of sound, and we must use the relationship 1 Hz = 1 s^{-1} to convert to frequency in the proper units. Remember also that 1 KHz = 10^3 Hz.

$$\lambda = \frac{v}{\nu} = \frac{330 \text{ m s}^{-1}}{20 \times 10^3 \text{ s}^{-1}}$$

The frequency (20×10^3 s^{-1}) tells us how many peaks pass a given point in space in a second.

T8.3 The equation for energy is $E = hc/\lambda$. Here we must use wavelength in meters, or 150×10^{-9} m:

$$E = \frac{hc}{\lambda} = \frac{(6.626\times10^{-34}\text{ J s})(3.00\times10^{8}\text{ m/s})}{(150\times10^{-9}\text{ m})}$$

We can calculate the energy of the 60 Hz alternating current using the equation $E = h\nu$:

$$E = h\nu = (6.626\times10^{-34}\text{ J s})(60\text{ s}^{-1})$$

T8.4 Here we must use the Rydberg equation:

$$\frac{1}{\lambda} = 109{,}678\text{ cm}^{-1} \times \left(\frac{1}{2^2} - \frac{1}{6^2}\right)$$

Taking the reciprocal gives us wavelength. To determine the color, the calculated wavelength should be compared to the spectrum in Figure 8.4.

T8.5 In order to answer this question, we must examine the ground state orbital diagrams for the elements and find any that contain unpaired electrons. The ones that are paramagnetic are As and Ag.

T8.6 This will be the element that lies farther to the right in a given row: S. The reasoning for this is that as we move to the right on the Periodic Table within the same period, valence electrons are essentially being fed into the same volume at the same time as that the nuclear charge is increasing as the number of protons increases. Thus, the effective nuclear charge on the valence electrons increases to the right.

T8.7 (a) Germanium atoms have a larger principal quantum number and are, therefore, larger.

(b) Sulfur probably has a larger first ionization energy since it has a larger effective nuclear charge and smaller radius.

(c) Sulfur is likely to have a more exothermic electron affinity since it is more electronegative. On the other hand, phosphorus has a stable half–filled *p*–subshell.

T8.8 When an atom is ionized, an electron is completely removed from the atom. This could be modeled by assuming the principal quantum number of an ionized electron is infinity. Substituting this value into the Rydberg equation would enable a calculation of the ionization energy of hydrogen from a variety of initial *n* levels.

T8.9 The removal of two electrons gives us Mg^{2+}, and requires the sum of the first and second ionization potentials: 737 + 1450 kJ/mol = 2187 kJ/mol. The thermochemical calculation is done as in Chapter 7, remembering to use the units joules:

$$\text{heat} = \text{specific heat} \times \text{mass} \times \text{temperature change}$$
$$2187 \times 10^3 \text{ J} = 4.184 \text{ J/g °C} \times (\text{X g}) \times (100 \text{ °C} - 25 \text{ °C})$$

Solving for X will give the desired result.

T8.10 Ionization of a fluoride ion is simply the reverse process of the creation of the fluoride ion from a neutral fluorine atom. Thus, the ionization energy will simply be the negative of the electron affinity of fluorine. Because the electron affinity of F is exothermic, the ionization of F^- is endothermic.

T8.11 As in the previous question, the electron affinity of an Al^{3+} ion is the reverse process of ionization. We determine the third ionization energy of an Al atom. The electron affinity for Al^{3+} is simply the negative of this value.

T8.12 In order to make Na(*g*) and Cl(*g*) we need to add energy to both Na(*s*) and $Cl_2(g)$. The energy added to the Na is used to overcome the internuclear forces of attraction. Energy is added to Cl_2 to break the Cl–Cl bond. In order to make $Na^+(g)$, we need to add additional energy to ionize the gaseous sodium atom. Formation of $Cl^-(g)$ from Cl(*g*) is an exothermic process. If the magnitude of the ionization energy of sodium is larger than the electron affinity of chlorine, $Na^+(g)$ and $Cl^-(g)$ will be less stable (higher energy) than Na(*g*) and Cl(*g*). If the opposite is true, i.e., the magnitude of the electron affinity is greater then the magnitude of the ionization energy, then Na(*g*) and Cl(*g*) will be the less stable situation. The minimum distance at which the ions would be more stable would be the bond length determined from experimentation.

T8.13 $E = \dfrac{hc}{\lambda}$ $\qquad \lambda = \dfrac{hc}{E} = \dfrac{hc}{\text{ionization energy}}$

<u>Review Exercises</u>

8.1 Light is a form of energy that results from small oscillations in the electrical and magnetic properties of particles.

8.2 In general, frequency describes the number of times an event occurs in a finite time period. The frequency of light is the number of times a wave crest passes a specific point in space in a given time interval. The symbol for frequency is the Greek letter "nu", ν, and has the units of inverse seconds, s^{-1}.

8.3 Wavelength is the distance between consecutive maxima of a wave. The symbol is the Greek letter "lambda", λ.

8.4 See Figure 8.3.

8.5 The amplitude affects the brightness of light. The color of light is affected by the wavelength or frequency. The energy of the light is affected by the frequency or wavelength.

8.6 gamma rays < X rays < ultraviolet < visible < infrared < microwaves < TV waves

8.7 By the visible spectrum, we mean that narrow portion of the electromagnetic spectrum to which our eyes are sensitive. These are the wavelengths from about 400 nm to 750 nm.

8.8 violet < blue < green < yellow < orange < red

8.9 $\lambda\nu = c$, where c is a constant equal to the speed of light, λ is the wavelength and ν is the frequency.

8.10 $E = h\nu$, where E is the energy, h is Planck's constant and ν is the frequency.

8.11 A photon is one unit of electromagnetic radiation whose energy is the product of h and ν.

8.12 Since $\nu = c/\lambda$, we can substitute into equation 8.2 to get $E = hc/\lambda$.

8.13
(a) infrared
(b) visible light
(c) X-rays
(d) ultraviolet light

8.14 The quantum is the lowest possible packet of energy, that of a single photon.

8.15 An atomic spectrum consists of a series of discrete (selected, definite and reproducible) frequencies (and therefore of discrete energies) that are emitted by atoms that have been excited. The particular values for the emission frequencies are characteristic of the element at hand. In contrast, a continuous spectrum, such as that emitted by the sun or another hot, glowing object, contains all frequencies and, therefore, photons of all energies. An electron in an atom can have only certain specific values for energy. Aside from these discrete energies, other energies are not allowed.

8.16 When an excited atom loses energy, not just any arbitrary amount can be lost, only specific amounts of energy can be lost. In other words, the energy of an electron is quantized.

8.17 Bohr proposed a model similar in design to a solar system. The nucleus is at the center and the electrons orbit the nucleus in specific orbits that are a constant fixed distance from the nucleus.

8.18 When an electron falls from an orbit of higher energy (larger radius) to an orbit of lower energy (smaller radius), the energy that is released appears as a photon with the appropriate frequency. The energy of the photon is the same as the difference in energy between the two orbits.

8.19 The lowest energy state of an atom is termed the ground state.

8.20 Bohr's model was a success because it accounted for the spectrum of the hydrogen atom, but it failed to account for spectra of more complex atoms.

8.21 Very small particles have properties that are reminiscent of both a particle and a wave. Massive particles also have this but the wavelike properties are too small to be observed.

8.22 Diffraction is a phenomenon caused by the constructive or the destructive interference of two or more waves. The fact that electrons and other subatomic particles exhibit diffraction supports the theory that matter is correctly considered to have wave nature.

8.23 To determine whether a beam was behaving as a wave or as a stream of particles, a diffraction experiment would have to be done.

8.24 Wave/particle duality is that light and matter have both wave-like properties and particle-like properties.

8.25 In a traveling wave, the positions of the peaks and nodes change with time. In a standing wave, the peaks and nodes remain in the same positions.

8.26 The collapsing atom paradox comes from classical mechanics and asks, "Why doesn't the electron fall into the nucleus?"

8.27 Quantum mechanics resolves the collapsing atom paradox by allowing only certain energy levels for the electron.

8.28 Wave mechanics and quantum mechanics.

8.29 This is the orbital of the electron.

8.30 First, we are interested in the energies of orbitals, because it is the energies of the various orbitals that determine which orbitals are occupied by the electrons of the atom. Secondly, we are interested in the shapes and orientations of the various orbitals, because this is important in determining how atoms form bonds in chemical compounds.

8.31 $n = 1, 2, 3, 4, 5, \ldots$

8.32 (a) $n = 1$ (b) $n = 3$

8.33 Every shell contains the possibility that $\ell = 0$.

8.34 (a) 1 (b) 3 (c) 5 (d) 7

8.35 Yes, if the value for ℓ for this electron is 3 or larger.

8.36 The only impossible set of legitimate quantum numbers would be an electron having the exact same values for the four quantum numbers. Recall the Pauli Exclusion Principle.

8.37 The electron behaves like a magnet, because the revolving charge (spin) of the electron creates a magnetic field. Recall that the electron does not actually revolve but behaves in a fashion reminiscent of an electro-magnet.

8.38 Atoms with unpaired electrons are termed paramagnetic.

8.39 No two electrons in the same atom can have exactly the same set of values for all of the four quantum numbers. This limits the allowed number of electrons per orbital to two, since with other quantum numbers being necessarily the same, two electrons in the same orbital must at least have different values of m_s.

8.40 $m_s = +1/2$ or $-1/2$

8.41 The distribution of electrons among the orbitals of an atom.

8.42 The orbitals within a given shell are arranged in the following order of increasing energy: $s < p < d < f$.

8.43 The energies of the subshells are quantized.

8.44 The orbitals of a given subshell have the same energy.

8.45

Li	$1s^22s^1$	N	$1s^22s^22p^3$
Be	$1s^22s^2$	O	$1s^22s^22p^4$

B	$1s^22s^22p^1$	F	$1s^22s^22p^5$
C	$1s^22s^22p^2$	Ne	$1s^22s^22p^6$

8.46 (a) Cr $[Ar]4s^13d^5$ or $[Ar]3d^54s^1$
(b) Cu $[Ar]4s^13d^{10}$ or $[Ar]3d^{10}4s^1$

8.47 $[Kr]5s^14d^{10}$

8.48 Elements in a given group generally have the same electron configuration except that the value for *n* is different, and corresponds to the row in which the element is found.
O: $[He]2s^22p^4$
S: $[Ne]3s^23p^4$
Se: $[Ar]4s^23d^{10}4p^4$
Te: $[Kr]5s^24d^{10}5p^4$
Po: $[Xe]6s^24f^{14}5d^{10}6p^4$

8.49 The *valence shell* is the occupied shell having the largest value of n. The *valence electrons* are those electrons in the valence shell.

8.50 As explained by the Heisenberg Uncertainly Principle, the position and momentum of an electron cannot be know with 100% certainty, therefore it is impossible to know where an electron is, but we can determine the probability of finding an electron in a given space. An electron is visualized as being within a cloud around the nucleus. This electron cloud defines a volume in space where the probability of finding an electron is high.

8.51 (a) See Figures 8.23 and 8.24 (b) See Figures 8.25 and 8.26.

8.52 As *n* becomes larger, the orbital becomes larger.

8.53 The three *p* orbitals of a given *p* subshell are oriented at right angles (90°) to one another.

8.54 A nodal plane is a plane in which there is zero probability of finding an electron.

8.55 The presence of a node increases the energy of an orbital

8.56 A *p* orbital has 1 nodal plane; a *d* orbital has 2 nodal planes.

8.57 See Figure 8.28.

8.58 The effective nuclear charge is the net nuclear charge that an electron actually experiences. It is different from the formal nuclear charge because of the varying imperfect ways in which one electron is shielded from the nuclear charge by the other electrons that are present. The effective nuclear charge remains nearly constant from top to bottom in any one group of the periodic table, and it increases from left to right in any one row of the periodic table.

8.59 The larger atoms are found in the lower left corner of the periodic table; the smaller atoms are found in the upper right corner of the periodic table.

8.60 The size changes within a transition series are more gradual because, whereas the "outer" electrons are in an *s* subshell, the electrons that are added from one element to another enter an inner (n - 1)*d* subshell.

8.61 Ionization energy is the energy that is needed in order to remove an electron from a gaseous atom or ion. These are positive values because the force of attraction between an electron and the nucleus of either an atom or a positive ion (cation) must be overcome in order to remove the electron.

8.62 (a) $O(g) \rightarrow O^+(g) + e^-$

(b) $O^{2+}(g) \rightarrow O^{3+}(g) + e^-$

8.63 Ionization energy increases from left to right in a row of the periodic table because the effective nuclear charge increases from left to right. The latter trend occurs because of the consequences of the increasingly imperfect shielding of electrons by other electrons within the same level. The ionization energy decreases down a periodic table group because the electrons reside farther from the nucleus with each successive quantum level that is occupied. The farther the electrons are from the nucleus, the less tightly they are held by the nucleus.

8.64 Removing a second electron involves pulling it away from a greater positive charge because of the positive charge created by the removal of the first electron. Hence, more energy must be spent to ionize the second electron than the first.

8.65 An electron must be removed from an ion having a 4+ rather than only a 3+ charge. Also, the fifth electron must be removed from a different, lower *n* shell.

8.66 The last valence electron of aluminum begins the occupation of the 3*p* set of orbitals. This electron is therefore well shielded from the nuclear charge by the 3*s* electrons.

8.67 The last valence electron of sulfur is placed in an orbital with another electron, and the two electrons are required to be spin-paired by the Pauli exclusion principle. This destabilizes the last electron, making it easier to ionize than is the case for the last electron of phosphorus.

8.68 Electron affinity is the enthalpy change associated with the addition of an electron to a gaseous atom:

$$X(g) + e^- \rightarrow X^-(g)$$

8.69 $S(g) + e^- \rightarrow S^-(g)$, first electron affinity
$S^-(g) + e^- \rightarrow S^{2-}(g)$, second electron affinity
The second electron affinity should be more endothermic than the first because work must be done to force the electron into the negative S^- ion.

8.70 The value for fluorine is low because of the especially small size of the atom, which causes addition of an electron to be relatively unfavorable. The comparison between chlorine and bromine follows the normal trend on descent of a group, the electron affinity decreasing with ionization energy.

8.71 The second electron affinity is always unfavorable (endothermic) because it requires that a second electron be forced onto an ion that is already negative.

8.72 The electron affinity becomes more negative (exothermic) as effective nuclear charge increases. Fluorine therefore has the more exothermic electron affinity because it has the larger effective nuclear charge.

Review Problems

8.73 $$\nu = \frac{c}{\lambda} = \frac{3.00 \times 10^8 \text{ m/s}}{430 \times 10^{-9} \text{ m}} = 6.98 \times 10^{14} \text{ s}^{-1} = 6.98 \times 10^{14} \text{ Hz}$$

8.74 $$\nu = \frac{c}{\lambda} = \frac{3.00 \times 10^8 \text{ m/s}}{280 \times 10^{-9} \text{ m}} = 1.07 \times 10^{15} \text{ s}^{-1} = 1.07 \times 10^{14} \text{ Hz}$$

8.75 $$\nu = \frac{c}{\lambda} = \frac{3.00 \times 10^8 \text{ m/s}}{6.85 \times 10^{-6} \text{ m}} = 4.38 \times 10^{13} \text{ s}^{-1} = 4.38 \times 10^{13} \text{ Hz}$$

8.76 $\nu = \dfrac{c}{\lambda} = \dfrac{3.00 \times 10^8 \text{ m/s}}{0.48 \times 10^{-6} \text{ m}} = 6.25 \times 10^{14} \text{ s}^{-1} = 6.25 \times 10^{14} \text{ Hz}$

8.77 $295 \text{ nm} = 295 \times 10^{-9} \text{ m}$

$\nu = \dfrac{c}{\lambda} = \dfrac{3.00 \times 10^8 \text{ m/s}}{295 \times 10^{-9} \text{ m}} = 1.02 \times 10^{15} \text{ s}^{-1} = 1.02 \times 10^{15} \text{ Hz}$

8.78 $\nu = \dfrac{c}{\lambda} = \dfrac{2.99792458 \times 10^8 \text{ m/s}}{632.99139822 \times 10^{-9} \text{ m}} = 4.73670145 \times 10^{14} \text{ s}^{-1} = 4.73670145 \times 10^{14} \text{ Hz}$

The speed of light is only given to 9 significant figures in the text.

8.79 $101.1 \text{ MHz} = 101.1 \times 10^6 \text{ Hz} = 101.1 \times 10^6 \text{ s}^{-1}$

$\lambda = \dfrac{c}{\nu} = \dfrac{3.00 \times 10^8 \text{ m/s}}{101.1 \times 10^6 \text{ s}^{-1}} = 2.98 \text{ m}$

8.80 $5.09 \times 10^{14} \text{ Hz} = 5.09 \times 10^{14} \text{ s}^{-1}$

$\lambda = \dfrac{c}{\nu} = \dfrac{3.00 \times 10^8 \text{ m/s}}{5.09 \times 10^{14} \text{ s}^{-1}} = 5.89 \text{ x } 10^{-7} \text{ m} \times \dfrac{1 \text{ nm}}{1 \text{ x } 10^{-9} \text{ m}} = 589 \text{ nm}$

8.81 $\lambda = \dfrac{c}{\nu} = \dfrac{3.00 \times 10^8 \text{ m/s}}{60 \text{ s}^{-1}} = 5.0 \times 10^6 \text{ m} = 5.0 \times 10^3 \text{ km}$

8.82 $1.50 \times 10^{18} \text{ Hz} = 1.50 \times 10^{18} \text{ s}^{-1}$

$\lambda = \dfrac{c}{\nu} = \dfrac{3.00 \times 10^8 \text{ m/s}}{1.50 \times 10^{18} \text{ s}^{-1}} = 2.00 \times 10^{-10} \text{ m} = 2.00 \times 10^{-1} \text{ nm} = 2.00 \times 10^2 \text{ pm}$

8.83 $E = h\nu = 6.63 \times 10^{-34} \text{ J s} \times 4.0 \times 10^{14} \text{ s}^{-1} = 2.7 \times 10^{-19} \text{ J}$

$\dfrac{\# \text{ J}}{\text{mol}} = \left(\dfrac{2.7 \times 10^{-19} \text{ J}}{1 \text{ photon}}\right)\left(\dfrac{6.022 \times 10^{23} \text{ photons}}{1 \text{ mol}}\right) = 1.6 \times 10^5 \text{ J mol}^{-1}$

8.84 $E = h\nu = hc/\lambda$, and $560 \text{ nm} = 560 \times 10^{-9} \text{ m}$

$$E = \frac{hc}{\lambda} = \frac{(6.626 \times 10^{-34} \text{ J s})(3.00 \times 10^8 \text{ m/s})}{(560 \times 10^{-9} \text{ m})} = 3.55 \times 10^{-19} \text{ J}$$

8.85 (a) violet (see Figure 8.8)

(b) $\nu = c/\lambda = 3.00 \times 10^8 \text{ m s}^{-1} / 410.3 \times 10^{-9} \text{ m} = 7.31 \times 10^{14} \text{ s}^{-1}$

(c) $E = h\nu = 6.63 \times 10^{-34} \text{ J s x } 7.31 \times 10^{14} \text{ s}^{-1} = 4.85 \times 10^{-19} \text{ J}$

8.86 (a) yellow (See Figure 8.8)

(b) $\nu = \dfrac{c}{\lambda} = \dfrac{3.00 \times 10^8 \text{ m/s}}{589 \times 10^{-9} \text{ m}} = 5.09 \times 10^{14} \text{ s}^{-1}$

(c) $E = h\nu = (6.626 \times 10^{-34} \text{ Js})(5.09 \times 10^{14} \text{ s}^{-1}) = 3.37 \times 10^{-19} \text{ J}$

8.87 $\frac{1}{\lambda} = 109{,}678 \text{ cm}^{-1} \times \left(\frac{1}{3^2} - \frac{1}{6^2}\right) = 109{,}678 \text{ cm}^{-1} \times (0.1111 - 0.02778)$

$\frac{1}{\lambda} = 9.140 \times 10^3 \text{ cm}^{-1}$

$\lambda = 1.094 \times 10^{-4}$ cm = 1090 nm, which is not in the visible region.

8.88 $\frac{1}{\lambda} = 109{,}678 \text{ cm}^{-1} \times \left(\frac{1}{2^2} - \frac{1}{5^2}\right) = 109{,}678 \text{ cm}^{-1} \times (0.250 - 0.0400)$

$\frac{1}{\lambda} = 2.303 \times 10^4 \text{ cm}^{-1}$

$\lambda = 434.2$ nm, this will be purple and will just be visible

8.89

$\frac{1}{\lambda} = 109{,}678 \text{ cm}^{-1} \times \left(\frac{1}{4^2} - \frac{1}{10^2}\right)$

$\frac{1}{\lambda} = 5.758 \times 10^3 \text{ cm}^{-1}$

$\lambda = 1.737 \times 10^{-6}$ m, this is in the infrared region

8.90

$\frac{1}{\lambda} = 109{,}678 \text{ cm}^{-1} \times \left(\frac{1}{1^2} - \frac{1}{4^2}\right) = 109{,}678 \text{ cm}^{-1} \times (1 - 0.0625)$

$\frac{1}{\lambda} = 102{,}823 \text{ cm}^{-1}$

$\nu = 9.73 \times 10^{-6}$ cm = 97.3 nm, which is in the ultraviolet region.

8.91 (a) *p* (b) *f*

8.92 (a) 3 (b) 2

8.93 (a) $n = 3, \ell = 0$ (b) $n = 5, \ell = 2$

8.94 (a) $n=4, \ell=1$ (b) $n = 6, \ell = 3$

8.95 0, 1, 2, 3, 4, 5

8.96 $n = 8$

8.97 (a) $m_\ell = 1, 0, \text{ or } -1$ (b) $m_\ell = 3, 2, 1, 0, -1, -2, \text{ or } -3$

8.98 –5, –4, –3, –2, –1, 0, 1, 2, 3, 4, 5

8.99 When $m_\ell = -4$ the minimum value of ℓ is 4 and the minimum value of n is 5.

8.100 There are eleven values for m_ℓ: –5, –4, –3, –2, –1, 0, 1, 2, 3, 4, and 5. Thus there are eleven orbitals.

8.101

n	ℓ	m_ℓ	m_s
2	1	−1	+1/2
2	1	−1	−1/2
2	1	0	+1/2
2	1	0	−1/2
2	1	+1	+1/2
2	1	+1	−1/2

8.102

n	ℓ	m_ℓ	m_s
3	2	−2	+1/2
3	2	−2	−1/2
3	2	−1	+1/2
3	2	−1	−1/2
3	2	0	+1/2
3	2	0	−1/2
3	2	1	+1/2
3	2	1	−1/2
3	2	2	+1/2
3	2	2	−1/2

8.103 21 electrons have $\ell = 1$, 20 electrons have $\ell = 2$

8.104 12 electrons have $\ell = 0$, 12 electrons have $m_\ell = 1$

8.105
(a) S $1s^22s^22p^63s^23p^4$
(b) K $1s^22s^22p^63s^23p^64s^1$
(c) Ti $1s^22s^22p^63s^23p^63d^24s^2$
(d) Sn $1s^22s^22p^63s^23p^64s^23d^{10}4p^64d^{10}5s^25p^2$

8.106
(a) As $1s^22s^22p^63s^23p^64s^23d^{10}4p^3$
(b) Cl $1s^22s^22p^63s^23p^5$
(c) Ni $1s^22s^22p^63s^23p^64s^23d^8$
(d) Si $1s^22s^22p^63s^23p^2$

8.107
(a) Mn is $[Ar]4s^23d^5$, ∴ five unpaired electrons, paramagnetic
(b) As is $[Ar]\ 3d^{10}4s^24p^3$, ∴ three unpaired electrons, paramagnetic
(c) S is $[Ne]3s^23p^4$, ∴ two unpaired electrons, paramagnetic
(d) Sr is $[Kr]5s^2$, ∴ zero unpaired electrons, not paramagnetic
(e) Ar is $1s^22s^22p^63s^23p^6$, ∴ zero unpaired electrons, not paramagnetic

8.108
(a) Ba is $[Xe]6s^2$, zero unpaired electrons: diamagnetic
(b) Se is $[Ar]4s^23d^{10}4p^4$, two unpaired electrons: paramagnetic
(c) Zn is $[Ar]4s^23d^{10}$, zero unpaired electrons: diamagnetic
(d) Si is $[Ne]3s^23p^2$, two unpaired electrons: paramagnetic

8.109
(a) Mg is $1s^22s^22p^63s^2$, zero unpaired electrons
(b) P is $1s^22s^22p^63s^23p^3$, three unpaired electrons
(c) V is $1s^22s^22p^63s^23p^63d^34s^2$, three unpaired electrons

8.110
(a) Cs is $[Xe]6s^1$, 1 unpaired electron
(b) S is $[Ne]3s^23p^4$, 2 unpaired electrons
(c) Ni is $[Ar]4s^23d^8$, 2 unpaired electrons

8.111 (a) Ni $[Ar]3d^8 4s^2$
(b) Cs $[Xe]6s^1$
(c) Ge $[Ar]\ 3d^{10}4s^2 4p^2$
(d) Br $[Ar]\ 3d^{10}4s^2 4p^5$
(e) Bi $[Xe]\ 4f^{14}5d^{10}6s^2 6p^3$

8.112 (a) Al $[Ne]3s^2 3p^1$
(b) Se $[Ar]4s^2 3d^{10}4p^4$ or $[Ar]3d^{10}4s^2 4p^4$
(c) Ba $[Xe]6s^2$
(d) Sb $[Kr]5s^2 4d^{10}5p^3$ or $[Kr]4d^{10}5s^2 5p^3$
(e) Gd $[Xe]6s^2 4f^7 5d^1$ or $[Xe]\ 5d^1 4f^7 6s^2$

8.113

(a) **Mg**: (↑↓) (↑↓) (↑↓)(↑↓)(↑↓) (↑↓) ( )( )( ) ( ) ( )( )( )( )( )
1s 2s 2p 3s 3p 4s 3d

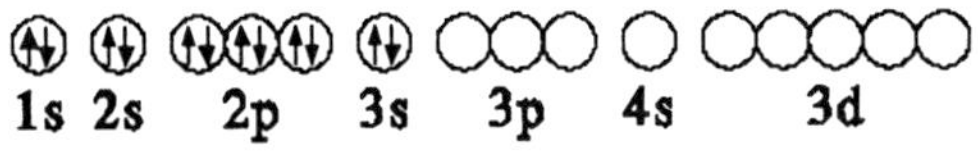

(b) **Ti**: (↑↓) (↑↓) (↑↓)(↑↓)(↑↓) (↑↓) (↑↓)(↑↓)(↑↓) (↑↓) (↑)(↑)( )( )( )
1s 2s 2p 3s 3p 4s 3d

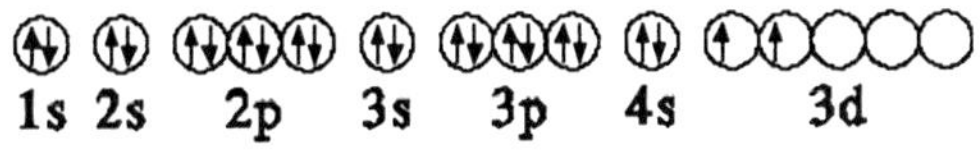

8.114

(a) **As**: (↑↓) (↑↓) (↑↓)(↑↓)(↑↓) (↑↓) (↑↓)(↑↓)(↑↓)
1s 2s 2p 3s 3p

(↑↓) (↑↓)(↑↓)(↑↓)(↑↓)(↑↓) (↑)(↑)(↑)
4s 3d 4p

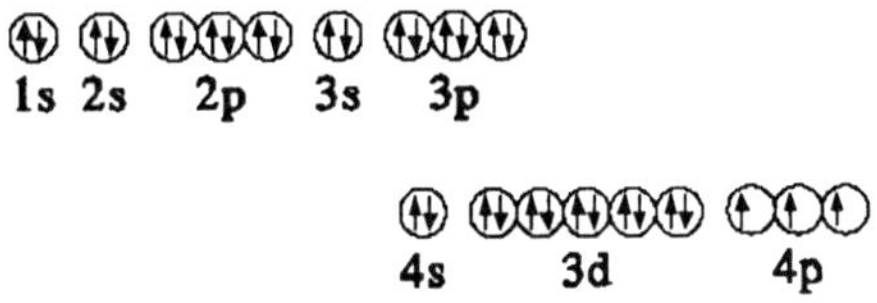

(b) **Ni**: (↑↓) (↑↓) (↑↓)(↑↓)(↑↓) (↑↓) (↑↓)(↑↓)(↑↓) (↑↓) (↑↓)(↑↓)(↑↓)(↑)(↑)
1s 2s 2p 3s 3p 4s 3d

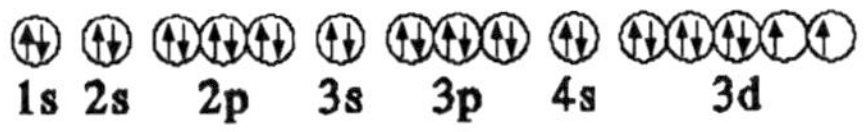

8.115

(a) Ni: [Ar] (↑↓) (↑↓)(↑↓)(↑↓)(↑)(↑)
4s 3d

(b) Cs: [Xe] (↑)
6s

(c) Ge: [Ar] (↑↓) (↑↓)(↑↓)(↑↓)(↑↓)(↑↓) (↑)(↑)( )
4s 3d 4p

(d) Br: [Ar] (↑↓) (↑↓)(↑↓)(↑↓)(↑↓)(↑↓) (↑↓)(↑↓)(↑)
4s 3d 4p

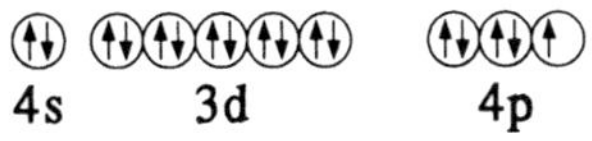

8.116

(a) Al : [Ne] (↑↓) (↑)( )( )
3s 3p

(b) Se : [Ar] (↑↓) (↑↓)(↑↓)(↑↓)(↑↓)(↑↓) (↑↓)(↑)(↑)
4s 3d 4p

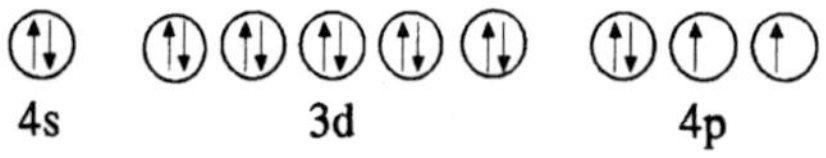

(c) Ba : [Xe] (↑↓)
6s

(d) Sb : [Kr] (↑↓) (↑↓)(↑↓)(↑↓)(↑↓)(↑↓) (↑)(↑)(↑)
5s 4d 5p

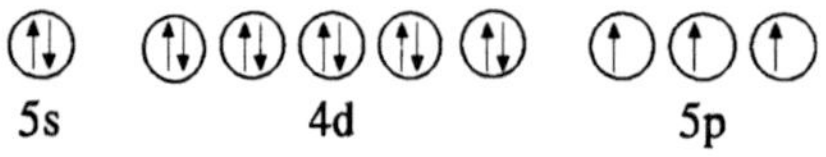

8.117 The value corresponds to the row in which the element resides:
(a) 5 (b) 4 (c) 4 (d) 6

8.118 The value corresponds to the row in which the element resides:
(a) 3 (b) 4 (c) 6 (d) 5

8.119 (a) Na $3s^1$ (b) Al $3s^23p^1$ (c) Ge $4s^24p^2$ (d) P $3s^23p^3$

8.120 (a) Mg $3s^2$ (b) Br $4s^24p^5$ (c) Ga $4s^24p^1$ (d) Pb $6s^26p^2$

8.121

(a) Na: (↑)
3s

(b) Al: (↑↓) (↑)( )( )
3s 3p

(c) Ge: (↑↓) (↑)(↑)( )
4s 4p

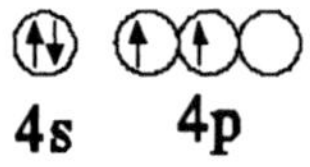

(d) P: (↑↓) (↑)(↑)(↑)
3s 3p

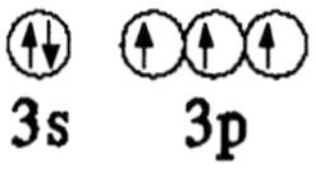

8.122

(a) Mg: (↑↓)
3s

(b) Br: (↑↓) (↑↓)(↑↓)(↑)
4s 4p

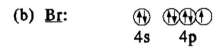

(c) Ga: (↑↓) (↑)( )( )
4s 4p

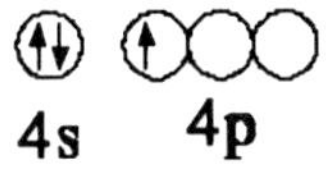

(d) Pb: (↑↓) (↑)(↑)( )
6s 6p

8.123 (a) 1 (b) 6 (c) 7

8.124 (a) 2 (b) 4 (c) 7

8.125 (a) Na (b) Sb

8.126 (a) Al (b) In

8.127 Sb is a bit larger than Sn, according to Figure 8.30. (Based upon trends, Sn would be predicted to be larger, but this is one area in which an exception to the trend exists. See Figure 8.30.)

8.128 Since these atoms and ions all have the same number of electrons, the size should be inversely related to the positive charge:

$$Mg^{2+} < Na^{+} < Ne < F^{-} < O^{2-} < N^{3-}$$

8.129 Cations are generally smaller than the corresponding atom, and anions are generally larger than the corresponding atom:
(a) Na (b) Co^{2+} (c) Cl^{-}

8.130 Cations are generally smaller than the corresponding atom, and anions are generally larger than the corresponding atom:
(a) S^{2-} (b) Al (c) Au^{+}

8.131 (a) C (b) O (c) Cl

8.132 (a) Li (b) Si (c) F

8.133 (a) Cl (b) Br

8.134 (a) As (b) Si

8.135 Mg

8.136 Si

Additional Exercises

8.137 (speed = wavelength × frequency $v = \lambda \times \nu$

$$\lambda = \frac{v}{\nu}$$

(a) longest wavelength $= \dfrac{330\ \text{m/s}}{20\ \text{s}^{-1}} = 16.\ \text{m}$

shortest wavelength $= \dfrac{330\ \text{m/s}}{20{,}000\ \text{s}^{-1}} = 0.016\ \text{m}$

(b) longest wavelength $= \dfrac{1500\ \text{m/s}}{20\ \text{s}^{-1}} = 75\ \text{m}$

shortest wavelength $= \dfrac{1500\ \text{m/s}}{20{,}000\ \text{s}^{-1}} = 0.075\ \text{m}$

8.138 We proceed by calculating the wavelength of a single photon:

$$E = \frac{hc}{\lambda} = \frac{(6.626\times10^{-34}\ \text{J s})(3.00\times10^{8}\ \text{m/s})}{(3.00\times10^{-3}\ \text{m})} = 6.63\times10^{-23}\ \text{J}$$

It requires 4.184 J to increase the temperature of 1.00 g of water by 1 Celsius degree. So, # photons = 4.184J ÷ 6.63×10^{-23} J/photon = 6.32×10^{22} photons

8.139

$$\frac{1}{\lambda} = 109{,}678\ \text{cm}^{-1} \times \left(\frac{1}{n_2^{\,2}} - \frac{1}{n_1^{\,2}}\right)$$

$$= \left(\frac{1}{410.3\ \text{nm}}\right)\left(\frac{1\times10^{-9}\ \text{m}}{1\ \text{nm}}\right)\left(\frac{100\ \text{cm}}{1\ \text{m}}\right)$$

$$\left(\frac{1}{4.103\times10^{-5}\ \text{cm}}\right) = 109{,}678\ \text{cm}^{-1} \times \left(\frac{1}{2^2} - \frac{1}{x^2}\right)$$

$$\frac{2.437\times10^{4}\ \text{cm}^{-1}}{109678\ \text{cm}^{-1}} = \frac{1}{4} - \frac{1}{x^2} = 0.222$$

$$\frac{1}{x^2} = \frac{1}{4} - 0.222$$

$$x = 6$$

8.140 This corresponds to the special case in the Rydberg equation for which $n_1 = 1$ and $n_2 = \infty$.

$$\frac{1}{\lambda} = 109{,}678\ \text{cm}^{-1} \times \left(\frac{1}{1^2} - \frac{1}{\infty^2}\right) = 109{,}678\ \text{cm}^{-1} \times (1 - 0)$$

$$\frac{1}{\lambda} = 109{,}678\ \text{cm}^{-1}$$

$\lambda = 9.12 \times 10^{-6}$ cm = 91.2 nm.

8.141 A transition from high energy to low energy may result in light emission. The transition from $4p \rightarrow 3d$, $5f \rightarrow 4d$, and $4d \rightarrow 2p$ are the only possibilities.

8.142 $E = 1/2\ mv^2$

$v = \sqrt{\dfrac{2E}{m}}$ first determine E for a single particle

$$E = (2080 \times 10^3 \text{ J/mol})\left(\frac{1 \text{ mole}}{6.022 \times 10^{23} \text{ particles}}\right) = 3.454 \times 10^{-18} \text{ J/particle}$$

$$v = \sqrt{\frac{2(3.454 \times 10^{-18} \text{ J})}{9.109 \times 10^{-31} \text{ kg}}} = 2.754 \times 10^6 \text{ m/sec}$$

8.143 (a) We must first calculate the energy in joules of a mole of photons.

$$E = \frac{hc}{\lambda} = \frac{(6.63 \times 10^{-34} \text{ J s})(3.00 \times 10^8 \text{ m/s})}{600 \times 10^{-9} \text{ m}} = 3.32 \times 10^{-19} \text{ J/photon}$$

$$(3.32 \times 10^{-19} \text{ J/photon})(6.02 \times 10^{23} \text{ photons/mol}) = 2.00 \times 10^5 \text{ J/mol}$$

Next, we calculate the heat transfer problem as in Chapter 7:

Heat = (specific heat)(mass)(change in temperature)
$2.00 \times 10^5 \text{ J} = 4.18 \text{ J g}^{-1} \text{ °C}^{-1} \times X \text{ g} \times 5.0 \text{ °C}$
$X = 9.57 \times 10^3 \text{ g}$

(b) $E = 6.63 \times 10^{-19}$ J/photon
$E = 3.99 \times 10^5$ J/mol
$X = 1.91 \times 10^4$ g

8.144 To solve this problem use $E = hc/\lambda$, where λ is our unknown quantity.

$$\lambda = \frac{hc}{E} = \frac{(6.626 \times 10^{-34} \text{ J s})(3.00 \times 10^8 \text{ m/s})}{(328 \times 10^3 \text{ J/mol})}(6.02 \times 10^{23} \text{ photons/mol})$$
$$= 3.65 \times 10^{-7} \text{ m} = 365 \text{ nm}$$

8.145 (a) Start by calculating λ

$$\frac{1}{\lambda} = 109{,}678 \text{ cm}^{-1} \times \left(\frac{1}{1^2} - \frac{1}{5^2}\right) = 109{,}678 \text{ cm}^{-1} \times (0.1000 - 0.04000)$$

$$\frac{1}{\lambda} = 1.053 \times 10^5 \text{ cm}^{-1}$$

$\lambda = 9.498 \times 10^{-6}$ cm = 94.98 nm, which is in the ultraviolet region of the visible spectrum.

(b)

$$\frac{1}{\lambda} = 109{,}678 \text{ cm}^{-1} \times \left(\frac{1}{2^2} - \frac{1}{4^2}\right) = 109{,}678 \text{ cm}^{-1} \times (0.2500 - 0.0625)$$

$$\frac{1}{\lambda} = 2.056 \times 10^4 \text{ cm}^{-1}$$

$\lambda = 4.863 \times 10^{-5}$ cm = 486.3 nm, which is in the visible region of the spectrum.

(c)

$$\frac{1}{\lambda} = 109{,}678\ \text{cm}^{-1} \times \left(\frac{1}{4^2} - \frac{1}{6^2}\right) = 109{,}678\ \text{cm}^{-1} \times (0.0625 - 0.0278)$$

$$\frac{1}{\lambda} = 3.808 \times 10^3\ \text{cm}^{-1}$$

$\lambda = 2.626 \times 10^{-4}$ cm = 2626 nm, which is in the infrared region of the spectrum.

8.146 (a) The *s* shell is not completely filled.
(b) Using the shorthand notation of [Kr], we imply that the 4*s*, 3*d* and 4*p* shells are filled but the rest of the configuration negates this assumption.
(c) There is nothing wrong.
(d) The *s* shell should be completely filled. If moving an electron from the *s* subshell would have filled the *d* subshell, then we could have this case but the *d* subshell is still two electrons short of filling.

8.147 (a) This diagram violates the aufbau principle. Specifically, the *s*-orbital should be filled before filling the higher energy *p*-orbitals.
(b) This diagram violates the aufbau principle. Specifically, the lower energy *s*-orbital should be filled completely before filling the *p*-orbitals.
(c) This diagram violates the aufbau principle. Specifically, the lower energy *s*-orbital should be filled completely before filling the *p*-orbitals.
(d) This diagram violates the Pauli Exclusion Principle since 2 electrons have the same set of quantum numbers. This electron distribution is impossible.

8.148 14

8.149 The 4*s* electrons are lost; $n = 4$, $\ell = 0$, $m_\ell = 0$, $m_s = \pm 1/2$

8.150 (a) This corresponds to the special case in the Rydberg equation for which $n_1 = 1$ and $n_2 = \infty$.
For a single atom, we have:

$$\frac{1}{\lambda} = 109{,}678\ \text{cm}^{-1} \times \left(\frac{1}{1^2} - \frac{1}{\infty^2}\right) = 109{,}678\ \text{cm}^{-1} \times (1 - 0)$$

$$\frac{1}{\lambda} = 109{,}678\ \text{cm}^{-1}$$

$\lambda = 9.12 \times 10^{-6}$ cm = 91.2 nm.

Converting to energy, we have:

E = hc/λ

$$E = \frac{hc}{\lambda} = \frac{(6.626 \times 10^{-34}\ \text{J s})(3.00 \times 10^8\ \text{m/s})}{(91.2 \times 10^{-9}\ \text{m})} = 2.18 \times 10^{-18}\ \text{J}$$

(b) Conversion to kJ/mol gives us:

$$\#\ \text{kJ/mole} = \left(\frac{2.18 \times 10^{-18}\ \text{J}}{\text{photon}}\right)\left(\frac{1\ \text{kJ}}{1000\ \text{J}}\right)\left(\frac{6.02 \times 10^{23}\ \text{photons}}{\text{mole}}\right)$$

$$= 1.31 \times 10^3\ \text{kJ/mol}$$

8.151

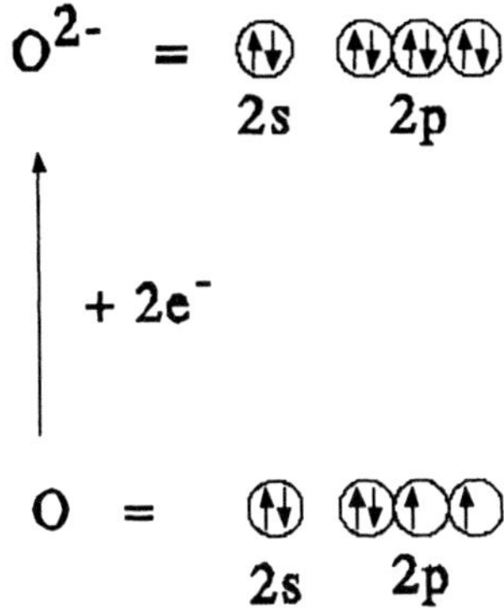

Thus the oxide ion, O^{2-}, has the electron configuration of [Ne].

Since the oxide ion has a closed shell electron configuration, the addition of a third electron is extremely difficult.

8.152 We simply reverse the electron affinities of the corresponding ions.

(a) $F^-(g) \rightarrow F(g) + e^-$, $\Delta H° = 328$ kJ/mol

(b) $O^-(g) \rightarrow O(g) + e^-$, $\Delta H° = 141$ kJ/mol

(c) $O^{2-}(g) \rightarrow O^-(g) + e^-$, $\Delta H° = -844$ kJ/mol

The last of these is exothermic, meaning that loss of an electron from the oxide ion is favorable from the standpoint of enthalpy.

8.153 In problem 8.152 parts b and c we can determine that

$$O(g) + 2e^- \rightarrow O^{2-}(g)\ \Delta H = +703\text{kJ}$$

From table 8.2 we can see that $O(g) \rightarrow O^+(g) + e^-$ $\Delta H = +1314$kJ/mol.
It takes more energy to ionize oxygen than to create $O^{2-}(g)$

Practice Exercises

9.1 Cr: [Ar] $3d^4 4s^2$
a) Cr^{2+}: [Ar]$3d^4$
b) Cr^{3+}: [Ar]$3d^3$
c) Cr^{6+}: [Ar]

9.2 S^{2-}: [Ne] $3s^2 3p^6$
Cl^-: [Ne] $3s^2 3p^6$
The electron configurations are identical.

9.3

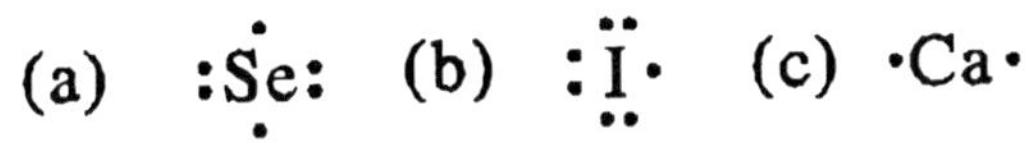

(a) :Se: (b) :I· (c) ·Ca·

9.4

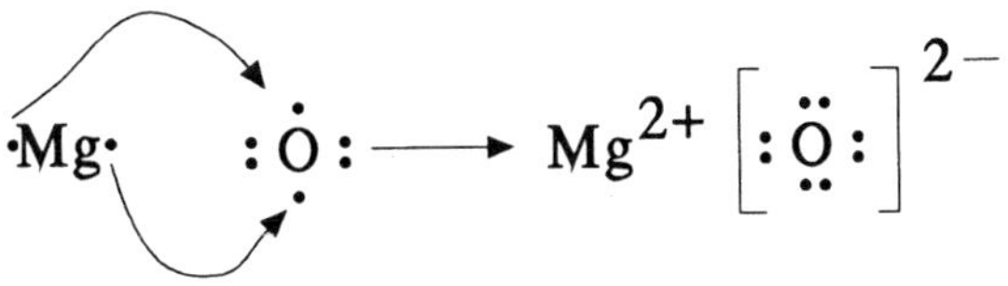

·Mg· :O: ⟶ Mg^{2+} [:O:]$^{2-}$

9.5

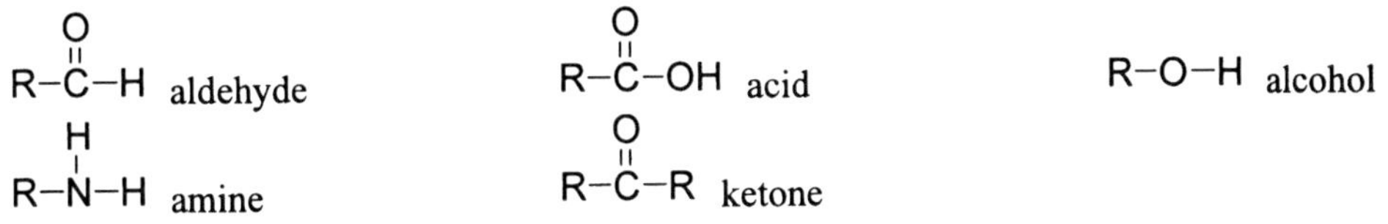

R–C(=O)–H aldehyde R–C(=O)–OH acid R–O–H alcohol
R–N(H)–H amine R–C(=O)–R ketone

9.6 a) Br b) Cl c) Cl

9.7

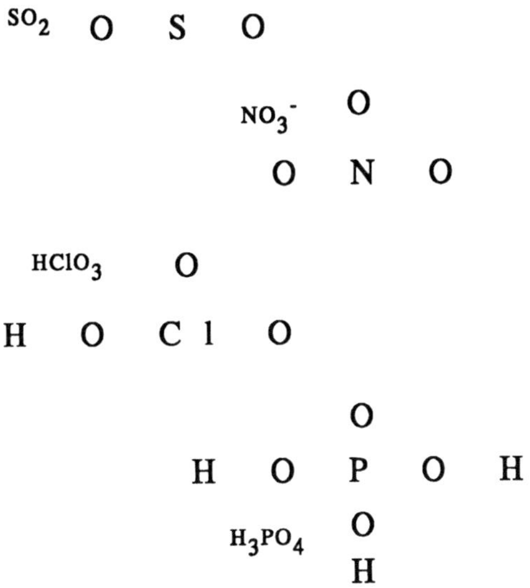

SO_2 O S O

NO_3^- O
O N O

$HClO_3$ O
H O Cl O

O
H O P O H
H_3PO_4 O
H

9.8 SO_2 has 18 valence electrons
PO_4^{3-} has 32 valence electrons
NO^+ has 10 valence electrons

9.9

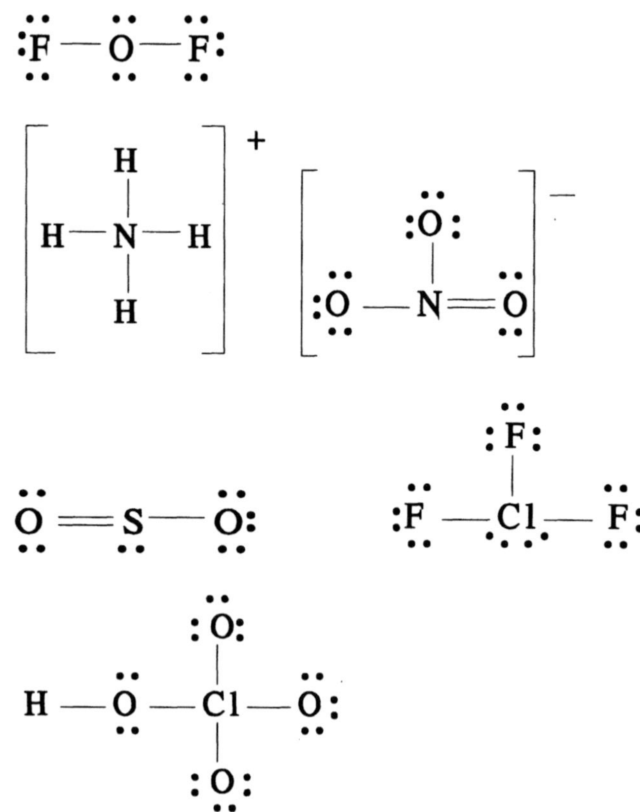

9.10 a)

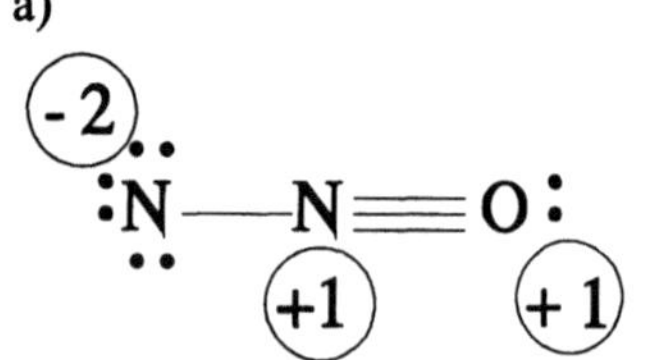

b)

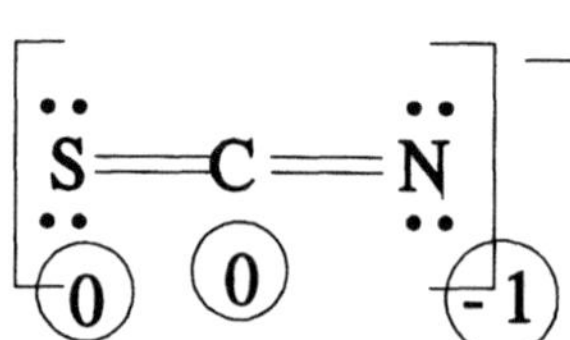

9.11 In each of these problems, we try to minimize the formal charges in order to determine the preferred Lewis structure. This frequently means violating the octet rule by expanding the octet. Of course, this can only be done for atoms beyond the second period as the atoms in the first and second periods will never expand the octet.

(a) O=S=O

(b) O=Cl(=O)–O–H

(c) (HO)₃P=O

9.12

$\left[\begin{array}{c} H \\ | \\ :\ddot{O}: \\ | \\ \ddot{O}=C-\ddot{O}: \end{array}\right]^- \longleftrightarrow \left[\begin{array}{c} H \\ | \\ :\ddot{O}: \\ | \\ :\ddot{O}-C=\ddot{O} \end{array}\right]^-$

9.13

$\left[\begin{array}{c} :\ddot{O}: \\ | \\ \ddot{O}=P-\ddot{O}: \\ | \\ :\ddot{O}: \end{array}\right]^{3-} \longleftrightarrow \left[\begin{array}{c} :\ddot{O}: \\ | \\ :\ddot{O}-P=\ddot{O} \\ | \\ :\ddot{O}: \end{array}\right]^{3-} \longleftrightarrow \left[\begin{array}{c} :\ddot{O}: \\ | \\ :\ddot{O}-P-\ddot{O}: \\ \| \\ :\ddot{O}: \end{array}\right]^{3-} \longleftrightarrow \left[\begin{array}{c} :\ddot{O}: \\ \| \\ :\ddot{O}-P-\ddot{O}: \\ | \\ :\ddot{O}: \end{array}\right]^{3-}$

9.14

$H-\ddot{O}:^- \quad H^+ \longrightarrow H-\ddot{O}-H$

coordinate covalent bond

Thinking It Through

T9.1 We need the enthalpy changes for (a) the first ionization energy of Na, (b) the first and second electron affinities for O, (c) the vaporization of Na(s), (d) the bond dissociation energy for–one half a mole of $O_2(g)$, and (e) the lattice energy, U, of Na_2O. The ΔH_f is the sum of these quantities, since enthalpy is a state function, as defined in Chapter 7.

T9.2 The transfer of one electron from each of two K atoms to one sulfur atom can be diagramed as follows:

K •   K • $\rightarrow$ $\cdot\ddot{S}:$

$2K^+ + [:\ddot{S}:]^{2-}$

T9.3 The electron configuration of Pb is:

Pb: $[Xe]6s^2 4f^{14} 5d^{10} 6p^2$.

The first two electrons are removed from the orbital having the highest value of the principal quantum number, which in this case is the 6s or 6p orbital. We need to decide which orbital will lose the two electrons. Since the 6p orbital is higher in energy, take the electrons from that orbital:

Pb^{2+}: $[Xe]6s^2 4f^{14} 5d^{10}$

The next two electrons are removed from the 6s orbitals:

Pb^{4+}: $[Xe]4f^{14} 5d^{10}$

T9.4 We must compare the ionization energies of K and H and determine which one gives rise to a favorable charge separation to form an ionic compound. Evidently for K_2S the lattice energy is a large negative number, whereas for H_2S the sharing of electrons to form a covalent bond is more favorable.

T9.5 This question requires us to compare the electronegativities of P (2.1) and S (2.4). Since P has the lower electronegativity, the difference in electronegativity of P and Cl in PCl_3 is greater than the difference in electronegativity between S and Cl in SCl_2. This causes the P–Cl bonds to be more polar. The chlorine atoms carry the partial negative charge, since they are most electronegative.

T9.6 No double bonds are needed in order to draw a proper Lewis diagram:

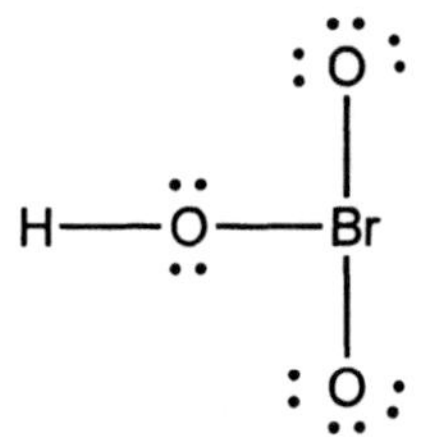

However, formal charge is reduced if we use two double bonds:

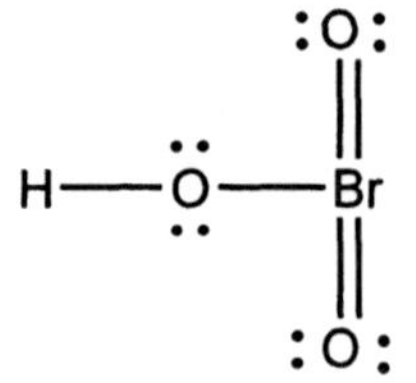

T9.7

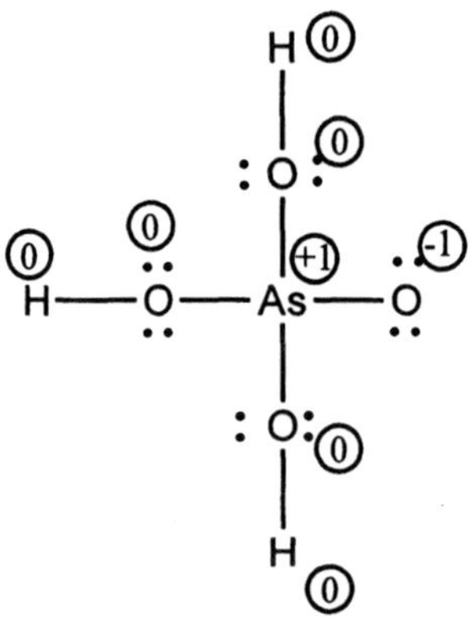

This is the Lewis structure that is obtained initially. By making a double bond between the oxygen atom and the arsenic atom we can eliminate all formal charges.

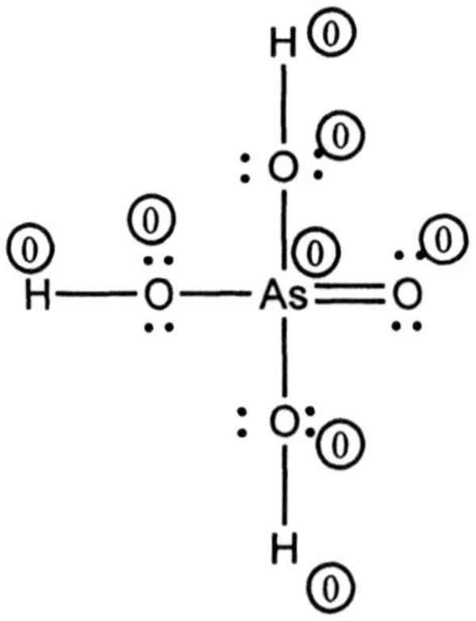

T9.8 These are not resonance structures since atoms have been moved. One resonance structure may differ from another only in the positions of electrons. The figure on the left with P as the central atom is the more likely arrangement. P should be the central atom since it is the least electronegative atom.

T9.9 First it is necessary to draw the best Lewis structures for these two ions:

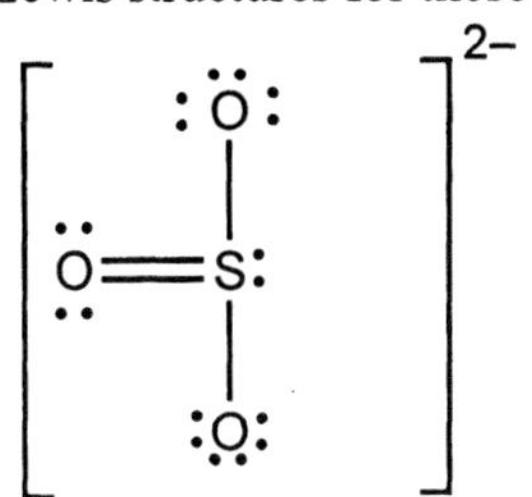

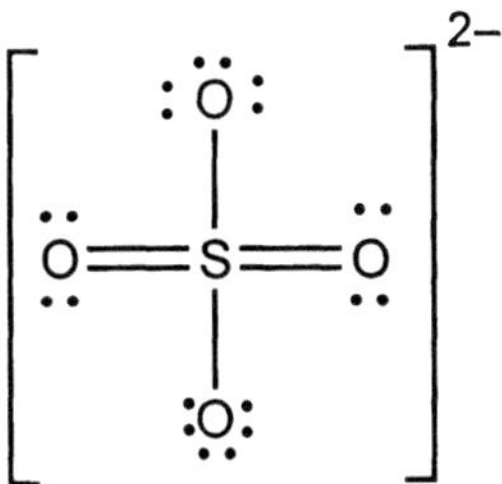

Since more double bonds in SO_4^{2-} lead to a greater bond order of 1.5 as opposed to 1.33 in SO_3^{2-}, the SO bonds in SO_4^{2-} should be shorter.

T9.10 The bond energy of the molecule is 435 kJ/mol (see figure 9.3). This is equivalent to 7.22×10^{-19} J/molecule. Using equation 8.1 and 8.2, $E = h\nu = hc/\lambda$, and rearranging

$$\lambda = \frac{hc}{E} = \frac{(6.626 \times 10^{-34}\ \text{Jsec})(3.00 \times 10^{8}\ \text{m/sec})}{7.22 \times 10^{-19}\ \text{J}}$$

$= 2.75 \times 10^{-7}$ m
$= 275$ nm

T9.11 The net reaction is $H_2(g) \rightarrow H^+(g) + H^-(g)$
The total energy change for this reaction has three parts

$H_2(g) \rightarrow 2H(g)$	$\Delta H = 435$ kJ/mol
$H(g) + e^- \rightarrow H^-(g)$	$\Delta H = -73$ kJ/mol
$H(g) \rightarrow H^+(g) + e^-$	$\Delta H = 1312$ kJ/mol

Adding these three together gives the net reaction and $\Delta H = 1674$ kJ/mol $= 2.780 \times 10^{-18}$ J/molecule. Using equations 9.1 and 9.2 and rearranging:

$$\lambda = \frac{hc}{E} = \frac{(6.626 \times 10^{-34}\ \text{Jsec})(3.00 \times 10^{8}\ \text{m/sec})}{2.780 \times 10^{-18}\ \text{J}}$$

$= 7.151 \times 10^{-8}$ m
$= 71.51$ nm

T9.12 The charge on the atoms (q) in a polar compound is calculated from the dipole moment (μ) and the bond length (r).

$$\mu = q * r$$

Calculate q:

$$\mu = 1.83\ \text{D}\left(\frac{3.34\times10^{-30}\ \text{C}\cdot\text{m}}{1\,\text{D}}\right) = q\times 91.7\ \text{pm}\left(\frac{1\,\text{m}}{10^{12}\ \text{pm}}\right)$$

Hydrogen has the charge of $+qe^-$ and fluorine has the charge of $-qe^-$.

T9.13 KBr < NaCl < CaO < MgO
The univalent (+1 or –1) ions will have a weaker attraction than the divalent ions. Hence KBr and NaCl have lower lattice energies. The lattice energy for KBr < NaCl because the ions are larger in KBr and the attractive forces are weaker. The same is true for CaO and MgO.

T9.14 Determine the electronegativity difference, using figure 9.5 and estimate the percent ionic character from figure 9.6

CO_2	1.0	15%
CS_2	0.1	0%
CSe_2	0	0%
CTe_2	0.5	2%
MgO	2.2	80%
MgS	1.1	15%
MgSe	1.2	20%
MgTe	0.7	5%

The data above and Figure 9.6 both demonstrate that most compounds possess some amount of covalent characteristics.

Review Questions

9.1 A stable compound forms from a collection of atoms when bonding results in a net lowering of the energy. The process of bonding the atoms together must release energy to make the bonded compound more stable than the original collection.

9.2 The ionic bond is the attraction between positive and negative ions in an ionic compound. It is largely an electrostatic attraction, and it gives rise to the lattice energy of the ionic compound. The lattice energy is the net gain in stability when the gaseous ions are brought together to form the crystalline ionic compound.

9.3 Ionic bonds tend to form upon combining an element having a high EA with an element having a low IE.

9.4 The lattice energy is the energy necessary to separate a mole of an ionic solid into its constituent ions in the gas phase. It is also the energy that is released on forming the ionic solid from the gaseous ions. As discussed in the answer to questions 9.1 and 9.2, it is the lattice energy that is primarily responsible for the stability of ionic compounds.

9.5 Magnesium ([Ne]$3s^2$) can achieve the electron configuration of the nearest noble gas (Ne) by losing only two electrons: Mg^{2+} $1s^22s^22p^6$. Magnesium will not form the Mg^{3+} ion because an extremely high amount of energy would be required to break into the $2s^22p^6$ core to remove an electron.

9.6 When chlorine gains one electron to form Cl^-, it has filled an orbital and achieved a noble gas configuration. To make the Cl^{2-} ion, an electron would have to be placed in the next higher shell. The amount of energy required for this to occur is extremely high and makes the creation of a Cl^{2-} ion energetically unfavorable.

9.7 Many of the transition metals have an ns^2 outer-shell electron configuration. Since these characteristically are the first electrons to be lost when a transition metal atom is ionized, it is common that a 2+ ion should be formed.

9.8 The largest difference in ionization energies would be between the second and the third successive ionization energies because of the extremely high amount of energy that would be required to break into the $2s^22p^6$ core to remove an electron.

9.9 (a) Al_2O_3
(b) BeO
(c) NaCl

9.10 The valence electrons are primarily responsible for chemical bonds.

9.11 The number of dots is the same as the group number in the representative group elements.

9.12 (a) correct
(b) incorrect
(c) correct
(d) incorrect

9.13 As two hydrogen atoms approach each other in forming the H_2 molecule, the electron density of the two atoms shifts to the region between the two nuclei.

9.14 Ionic bonding does not occur between two nonmetal elements because more energy must be provided in the form of IE and EA than can be recovered from the lattice energy.

9.15 Bond formation is always exothermic.

9.16 The energy drops to some optimum or minimum value when the nuclei have become separated by the distance called the bond distance. The electron spins become paired.

9.17 When a covalent bond is formed, the electrons that are shared become paired. Also, the normal, single covalent bond is created by the sharing of two (a pair of) electrons.

9.18 The force that holds the nuclei together in a covalently bonded pair of atoms is the attraction that the two nuclei have for the negatively charged electron density that is found between the nuclei.

9.19 The bond distance in a covalent bond is determined by a balance (compromise) between the separate attractions of the nuclei for the electron density that is found between them, and the repulsions between the like-charged nuclei and those between the like-charged electrons. These attractions and repulsions oppose one another, and a bond distance is achieved that maximizes the attraction while minimizing the repulsions.

9.20 The octet rule is the expectation that atoms tend to lose or gain electrons until they achieve a noble gas-type electron configuration, namely eight electrons in the valence, or outer-most, shell. It is the stability of the closed-shell electron configuration of a noble gas that accounts for this.

9.21 (a) one (b) four (c) two
(d) three (e) one

9.22 The valence shell of a period 2 element can hold only eight electrons since only two types of subshells, *s* and *p*, are available in this period. The valence shell of elements in row three can hold as many as eighteen electrons. This results because there are three types of subshells, *s*, *p* and *d*, that can hold electrons. The *s* subshells hold 2 electrons, *p* subshells hold 6 electrons and *d* subshells hold 10 electrons. The total is, therefore, 18 electrons.

9.23 (a) single bond: a covalent bond formed by the sharing of one pair of electrons.
(b) double bond: a covalent bond formed by the sharing of two pairs of electrons.
(c) triple bond: a covalent bond formed by the sharing of three pairs of electrons.

9.24 A structural formula shows which atoms are attached to one another in a molecule or polyatomic ion.

9.25

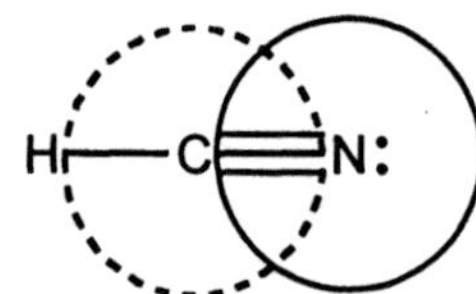

9.26 Since the outer shell (or valence shell) of hydrogen can hold only two electrons, hydrogen is not said to obey the octet rule. It does, however, readily satisfy its requirement for a closed shell electron configuration through the formation of one covalent bond.

9.27 Carbon is interesting because it almost always makes four covalent bonds. And it can form chains of atoms.

9.28

```
               H
               |
methane      H—C—H
               |
               H

               H  H
               |  |
ethane       H—C—C—H
               |  |
               H  H

               H  H  H
               |  |  |
propane      H—C—C—C—H
               |  |  |
               H  H  H
```

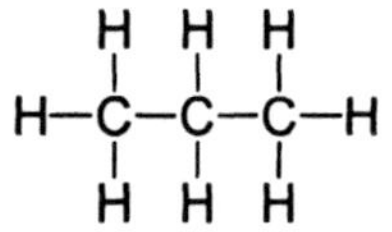

9.29 This molecule pentane has 12 H atoms and the formula is C_5H_{12}

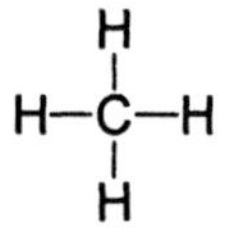

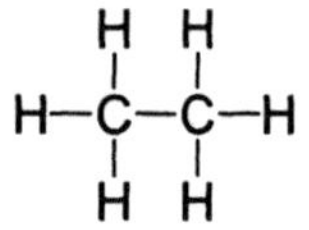

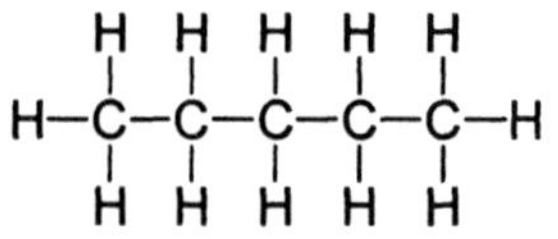

```
  H  H  H  H  H
  |  |  |  |  |
H—C—C—C—C—C—H
  |  |  |  |  |
  H  H  H  H  H
```

9.30

```
  H  H  H  H  H
  |  |  |  |  |
H—C—C—C—C—C—H
  |  |  |  |  |
  H  H  H  H  H
```

```
       H
       |
     H—C—H
       |
  H    |  H  H
  |    |  |  |
H—C————C——C——C—H
  |    |  |  |
  H    H  H  H
```

```
           H
           |
  H      H—C—H     H
  |        |       |
H—C————————C———————C—H
  |        |       |
  H      H—C—H     H
           |
           H
```

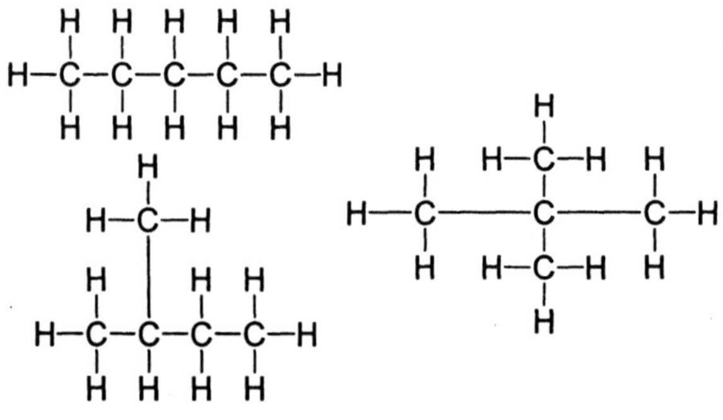

9.31 A carbonyl group is an oxygen atom double bonded to a carbon atom. Carbonyl groups are found in aldehydes, ketone, and acids.

9.32

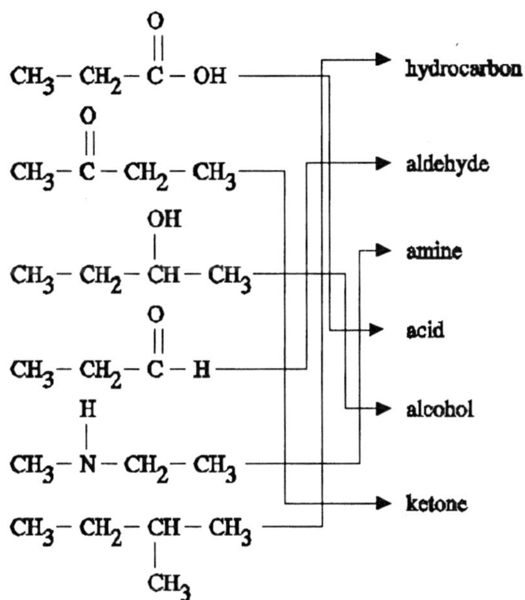

9.33

$$CH_3CH_2\overset{\displaystyle O}{\overset{\|}{C}}OH \rightleftharpoons CH_3CH_2\overset{\displaystyle O}{\overset{\|}{C}}O^- + H^+$$

This is a weak acid

9.34

$$CH_3\overset{\displaystyle H}{\overset{|}{N}}CH_2CH_3 + H_2O \longrightarrow CH_3\overset{\displaystyle H_2^+}{\overset{|}{N}}CH_2CH_3 + OH^-$$

9.35

$$CH_3CH_2\overset{\displaystyle O}{\overset{\|}{C}}OH + CH_3\overset{\displaystyle H}{\overset{|}{N}}CH_2CH_3 \longrightarrow$$

$$CH_3CH_2\overset{\displaystyle O}{\overset{\|}{C}}\underset{\underset{\displaystyle CH_3}{|}}{N}CH_2CH_3 + H_2O$$

9.36

(a) ethylene $H-\overset{H}{\overset{|}{C}}=\overset{H}{\overset{|}{C}}-H$

(b) acetylene $H-C\equiv C-H$

9.37 A polar covalent bond is one in which the electrons of the bond are not shared equally by the atoms of the bond, and this causes one end of the linkage to carry a partial negative charge while the other end carries a corresponding partial positive charge. In other words, there is a dipole in a polar bond.

9.38 A dipole moment is the product of the amount of charge on one end of a polar bond (which behaves as a dipole) and the distance between the two partial charges that compose the dipole. It is also normally taken to be the product of the charge in the dipole of a polar bond and the internuclear distance in the polar bond.

1 debye = 3.34×10^{-34} coulomb × meter

9.39 Electronegativity is the attraction that an atom has for the electrons in chemical bonds to that atom.

9.40 Pauling based his scale of electronegativity on the greater bond energy polar bonds have than would be expected if the opposite ends of the bonds were electrically neutral.

9.41 Fluorine has the largest electronegativity, whereas oxygen has the second largest electronegativity.

9.42 The noble gases are assigned electronegativity values of zero because they have a complete shell of electrons.

9.43 b and d are the only ones with an electronegativity difference greater than 1.7.

9.44 Elements having low electronegativities are metals. Elements with low ionization potentials and low electron affinities tend to have low electronegativities.

9.45 The dipole moment of NO is larger than the dipole moment of CO.

9.46 This has to do with the ease of oxidation.

9.47 The most reactive metals are in the left-most groups of the periodic table, namely the metals of Groups IA and IIA. The least reactive metals are found in the second and third rows of the transition elements.

9.48 The lower the electronegativity, the more reactive is the metal.

9.49 calcium > iron > silver > iridium

9.50
(a) NR
(b) $2NaI(aq) + Cl_2(g) \rightarrow 2NaCl(aq) + I_2(s)$
(c) $2KCl(aq) + F_2(aq) \rightarrow 2KF(aq) + Cl_2(g)$
(d) $CaBr_2(aq) + Cl_2(g) \rightarrow CaCl_2(aq) + Br_2(l)$
(e) $2AlBr_3(aq) + 3F_2(g) \rightarrow 2AlF_3(aq) + 3Br_2(l)$
(f) NR

9.51
(a) F_2 (b) P_4 (c) Br_2
(d) S_8 (e) Cl_2 (f) S_8

9.52
(a) 4 electrons or two bonding pairs
(b) six electrons, all in bonding pairs
(c) two electrons for each hydrogen

9.53 We expect a minimum of ten electrons, in bonding pairs between As and each of five Cl atoms.

9.54 N is in the second period, it can only have an octet of electrons, so it cannot form five bonds. Whereas, As is in the fourth row, it can have an expanded octet and form five bonds; therefore, it can form $AsCl_3$ and $AsCl_5$.

9.55 Bond length - the distance between the nuclei of two atoms that are linked by a covalent bond.
Bond energy - the energy that is needed to break a chemical bond; conversely, the energy that is released when a chemical bond is formed.

9.56 The number of electron pairs that comprise a covalent linkage (bond) between two atoms is called the bond order. As bond order increases, so does the strength of the bond. Therefore, as bond order increases, bond energy increases and bond length decreases, in keeping with increased bond strength.

9.57 The H — Cl bond energy is defined to be the energy required to break the bond to give atoms of H and Cl, not the ions H^+ and Cl^-. In other words, when the bond is broken between the H and the Cl atoms, one of the two electrons of the bond must go to each of the atoms. When ions are obtained, however, both electrons of the bond are given to the chlorine atom and the lattice energy would be the more appropriate term to use.

9.58 The formal charge on an atom is its apparent charge in a particular Lewis diagram. For any particular atom in a Lewis diagram, the formal charge is calculated by subtracting the number of bonds to the atom, and the number of unshared electrons on the atom, from the number of valence electrons that the unbonded atom normally has in the ground state.

9.59 It is generally held that the Lewis structure having the smallest formal charges is most favored. This is based on the notion that either an accumulation of too much charge, or too much charge separation, is destabilizing.

9.60 The formal charges on the atoms in HCl are 0. The actual charges are H: $+0.17e^-$ and Cl: $-0.17e^-$. The formal charges arise from the bookkeeping in Lewis structures, and are not the same as actual charges.

9.61 Lewis structures are often inadequate, without the concept of resonance, to describe the distribution of electrons in certain molecules and ions.

9.62 A resonance hybrid is the true structure of a molecule or polyatomic ion, whereas the various resonance structures that are used to depict the hybrid do not individually have any reality. The hybrid is a mix, or average, of the various resonance structures that compose it.

9.63

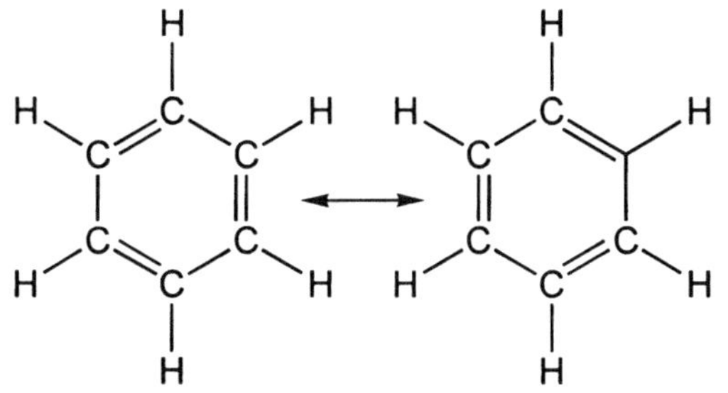

The resonance structures of benzene are more stable than the ring containing three carbon-carbon double bonds because six bonds with a bond order of 1.5 is more stable than three single bond and three double bonds together.

9.64

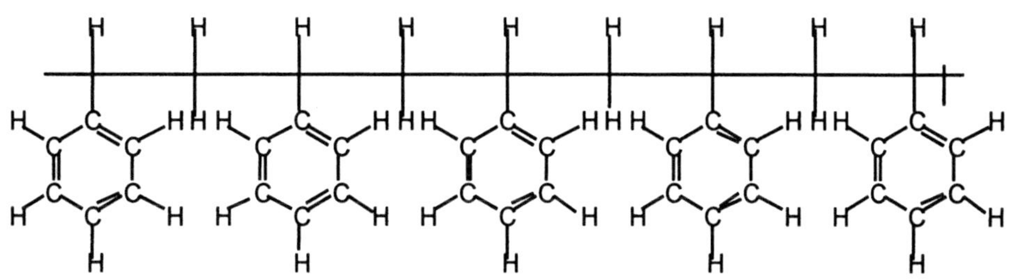

9.65 A coordinate covalent bond is one in which both electrons of the bond are contributed by (or donated by) only one of the atoms that are linked by the bond.

9.66 Once formed, a coordinate covalent bond is no different than any other covalent bond.

9.67

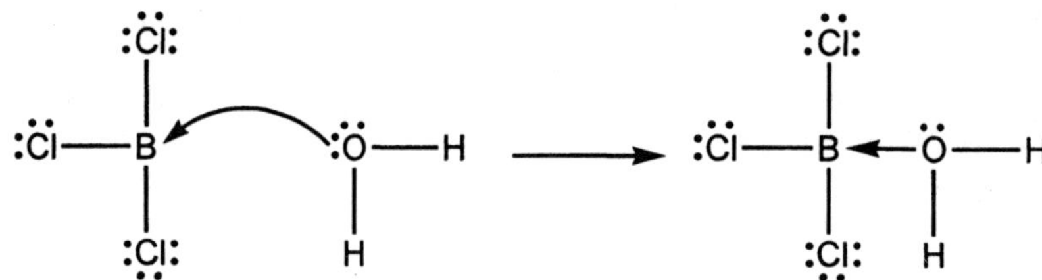

Review Problems

9.68 Magnesium loses two electrons:
$Mg \rightarrow Mg^{2+} + 2e^-$
$[Ne]3s^2 \rightarrow [Ne]$

Bromine gains an electron:
$Br + e^- \rightarrow Br^-$
$[Ar]3d^{10}4s^24p^5 \rightarrow [Kr]$

To keep the overall change of the molecule neutral, two Br^- ions combine with one Mg^{2+} ion to form $MgBr_2$:
$Mg^{2+} + 2Br^- \rightarrow MgBr_2$

9.69 Nitrogen ($1s^22s^22p^3$) gains three electrons to achieve the electron configuration of the next noble gas, neon: N^{3-} $1s^22s^22p^6$
Lithium ($1s^22s^1$) loses an electron to achieve the electron configuration of the closest noble gas helium: Li^{1+} $1s^2$

9.70 Pb^{2+}: $[Xe]\,4f^{14}5d^{10}6s^2$
Pb^{4+}: $[Xe]4f^{14}5d^{10}$

9.71 Bi^{3+}: $[Xe]6s^24f^{14}5d^{10}$
Bi^{5+}: $[Xe]4f^{14}5d^{10}$

9.72 Mn^{3+}: $[Ar]3d^4$ 4 unpaired electrons

9.73 Co^{3+}: $[Ar]3d^6$ 4 unpaired electrons

9.74

(a) ·Si· (b) :Sb· (c) ·Ba· (d) ·Al· (e) :S:

9.75

(a) ·K (b) ·Ge· (c) ·As· (d) :Br· (e) :Se:

9.76

(a) $[K]^+$ (b) $[Al]^{3+}$ (c) $[:S:]^{2-}$ (d) $[:Si:]^{4-}$ (e) $[Mg]^{2+}$

9.77

(a) $[:\ddot{\underset{..}{Br}}:]^-$ (b) $[:\ddot{\underset{..}{Se}}:]^{2-}$ (c) $[Li]^+$ (d) $[:\ddot{\underset{..}{C}}:]^{4-}$ (e) $[:\ddot{\underset{..}{As}}:]^{3-}$

9.78

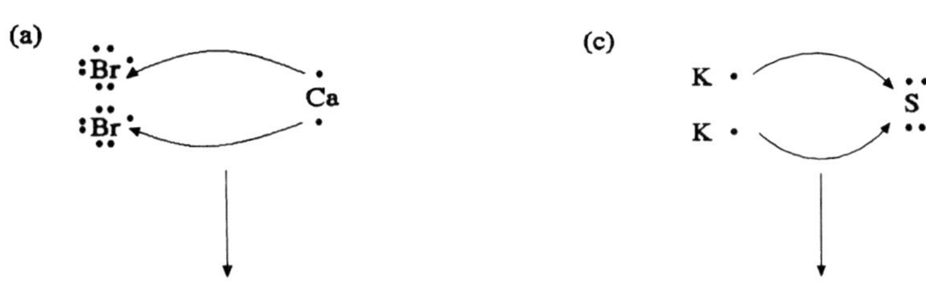

(a) $2[:\ddot{\underset{..}{Br}}:]^- \quad + \quad [Ca]^{2+}$

(c) $2\,[K]^+ \quad + \quad [:\ddot{\underset{..}{S}}:]^{2-}$

(b)

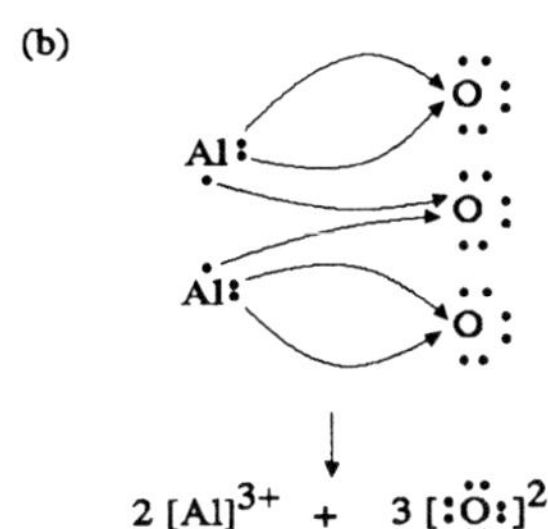

$2\,[Al]^{3+} \quad + \quad 3\,[:\ddot{\underset{..}{O}}:]^{2-}$

9.79

(a)

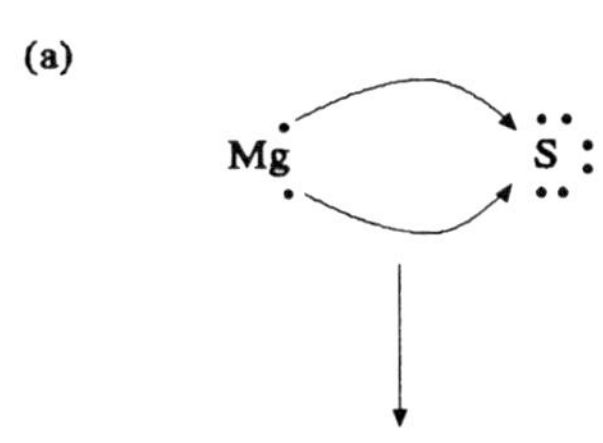

$[Mg]^{2+} \quad + \quad [:\ddot{\underset{..}{S}}:]^{2-}$

(b)

$[Mg]^{2+} \quad + \quad 2\,[:\ddot{\underset{..}{Cl}}:]^-$

(c)

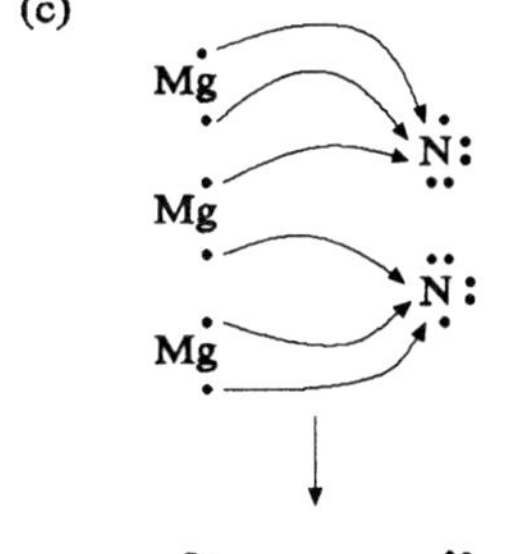

$3\,[Mg]^{2+} \quad + \quad 2\,[:\ddot{\underset{..}{N}}:]^{3-}$

9.80 $\mu = q \times r$

$$\mu = 0.16\text{ D}\left(\frac{3.34\times10^{-30}\text{ C}\cdot\text{m}}{1\text{ D}}\right) = q\,(115\text{ pm})\left(\frac{1\text{ m}}{10^{12}\text{ pm}}\right)$$

$5.34 \times 10^{-31}\text{ C}\cdot\text{m} = q\,(115 \times 10^{-12}\text{ m})$

$q = 4.64 \times 10^{-21}\text{ C}$

$$q = 4.64 \times 10^{-21}\ C\left(\frac{1e^-}{1.60\times10^{-19}\ C}\right) = 0.029\ e^-$$

The charge on the oxygen is $-0.029e^-$ and the charge on the nitrogen is $+0.029e^-$. The nitrogen atom is positive.

9.81 $\mu = q \times r$

$$\mu = 1.42\ D\left(\frac{3.34\times10^{-30}\ C\cdot m}{1\ D}\right) = q\ (176\ pm)\left(\frac{1\ m}{10^{12}\ pm}\right)$$

$4.74 \times 10^{-30}\ C{\cdot}m = q\ (176 \times 10^{-12}\ m)$
$q = 2.69 \times 10^{-20}\ C$

$$q = 2.69 \times 10^{-20}\ C\left(\frac{1e^-}{1.60\times10^{-19}\ C}\right) = 0.17\ e^-$$

The charge on the fluorine is $-0.17e^-$ and the charge on the bromine is $+0.17e^-$. The bromine atom is positive.

9.82 Let q be the amount of energy released in the formation of 1 mol of H_2 molecules from H atoms: 435 kJ/mol, the single bond energy for hydrogen.

$q = \text{specific heat} \times \text{mass} \times \Delta T$
$\therefore\ \text{mass} = q \div (\text{specific heat} \times \Delta T)$

$$\#\,g\ H_2O = \frac{(435\times10^3\ J)}{\left(4.184\ {}^{J}\!/\!_{g\,^\circ C}\right)\left(100\ ^\circ C - 25\ ^\circ C\right)} = 1.4\times10^3\ g$$

9.83 $$\#\ J/molecule = (242.6 \times 10^3\ J/mole)\left(\frac{1\ mole}{6.022 \times 10^{23}\ molecules}\right)$$
$$= 4.029 \times 10^{-19}\ J/molecule$$

9.84 $E = h\nu = \dfrac{hc}{\lambda} \qquad \lambda = \dfrac{hc}{E}$

$$E = (348 \times 10^3\ J/mol)\left(\frac{1\ mol}{6.022 \times 10^{23}\ molecules}\right) = 5.78 \times 10^{-19}\ J/molecule$$

$\lambda = (6.626 \times 10^{-34}\ Jsec)(3.00 \times 10^8\ m/s)/5.78 \times 10^{-19}\ J = 3.44 \times 10^{-7}\ m = 344\ nm$
Ultraviolet

9.85 $E = h\nu = \dfrac{hc}{\lambda} \qquad \lambda = \dfrac{hc}{E}$

$$E = (242.6 \times 10^3\ J/mol)\left(\frac{1\ mol}{6.022 \times 10^{23}\ molecules}\right) = 4.029 \times 10^{-19}\ J/molecule$$

$\lambda = (6.626 \times 10^{-34}\ Jsec)(3.00 \times 10^8\ m/s)/4.029 \times 10^{-19}\ J = 4.91 \times 10^{-7}\ m = 491\ nm$

9.86

(a) $:\ddot{\underset{..}{Br}}\cdot + \cdot\ddot{\underset{..}{Br}}: \rightarrow :\ddot{\underset{..}{Br}}-\ddot{\underset{..}{Br}}:$

(b) $2\,H\cdot + \cdot\ddot{\underset{..}{O}}\cdot \rightarrow H-\ddot{O}:-H$ (H bonded below O)

(c) $3\,H\cdot + \cdot\ddot{\underset{\cdot}{N}}\cdot \rightarrow H-\ddot{N}-H$ (H bonded below N)

9.87

(a) NCl_3: :N— bonded to three :Cl: atoms (each Cl with three lone pairs, N with one lone pair)

(b) CCl_4: C bonded to four :Cl: atoms (each Cl with three lone pairs)

(c) SCl_2: :S: bonded to two :Cl: atoms (each Cl with three lone pairs, S with two lone pairs)

(d) $:\ddot{\underset{..}{Br}}-\ddot{\underset{..}{Cl}}:$

9.88 (a) We predict the formula H_2Se because selenium, being in Group VIA, needs only two additional electrons (one each from two hydrogen atoms) in order to complete its octet.

(b) Arsenic, being in Group VA, needs three electrons from hydrogen atoms in order to complete its octet, and we predict the formula H_3As.

(c) Silicon is in Group IVB, and it needs four electrons (and hence four hydrogen atoms) to complete its octet: SiH_4.

9.89 (a) Each chlorine atom needs one further electron in order to achieve an octet, and the phosphorus atom requires three electrons from an appropriate number of chlorine atoms. We conclude that a phosphorus atom is bonded to three chlorine atoms, and that each chlorine atom is bonded only once to the phosphorus atom: PCl_3.

(b) Since carbon needs four additional electrons, the formula must be CF_4. In this arrangement, each fluorine acquires the one additional electron that is needed to reach its octet.

(c) Each halogen atom needs only one additional electron from the other: ICl.

9.90 Here we choose the atom with the smaller electronegativity:

(a) S (b) Si (c) Br (d) C

9.91 (a) I (b) I (c) F (d) N

9.92 Here we choose the linkage that has the greatest difference in electronegativities between the atoms of the bond: N—S.

9.93 The least polar bond of the four is P—I because it is the bond that has the smallest difference in electronegativities between the linked atoms.

9.94

(a)

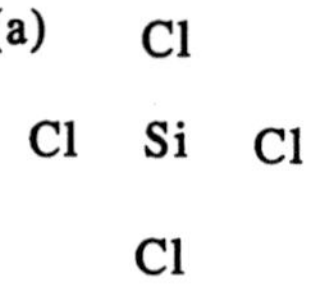

Cl

Cl Si Cl

Cl

(b)

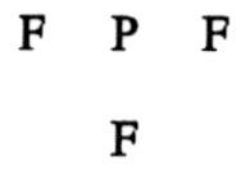

F P F

F

(c)

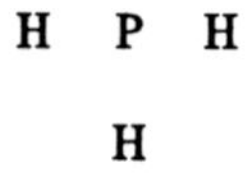

H P H

H

(d)

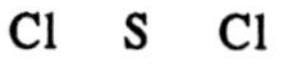

Cl S Cl

9.95

(a)

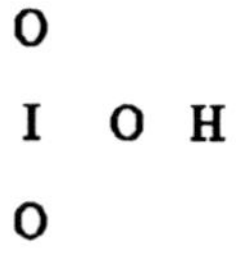

O

I O H

O

(b)

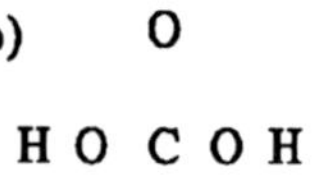

O

H O C O H

(c)

O

O C O H

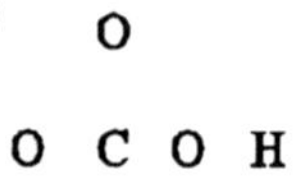

(d)

Cl

Cl P Cl

Cl

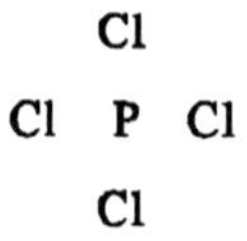

9.96 (a) 32 (b) 26 (c) 8 (d) 20

9.97 (a) 26 (b) 24 (c) 24 (d) 32

9.98 (a)

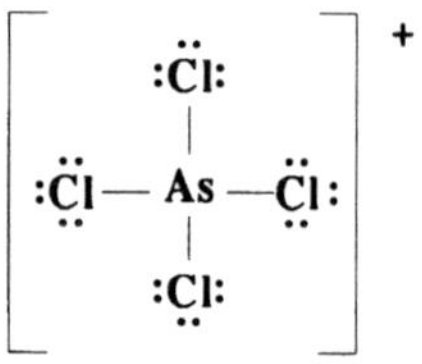

$[AsCl_4]^+$ (Lewis structure: :C̈l: bonded to As on four sides)

(b)

$[:\ddot{O}—\ddot{Cl}—\ddot{O}:]^-$

(c)

H–Ö–N̈=Ö

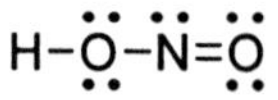

(d)

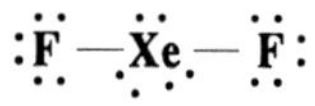

:F̈—Xe—F̈:

9.99 (a)

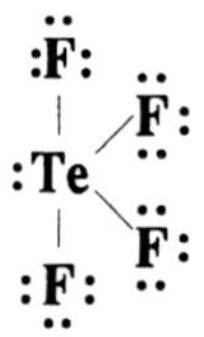

:Te with four F atoms (TeF$_4$ Lewis structure)

(b)

Cl with five F atoms (ClF$_5$ Lewis structure)

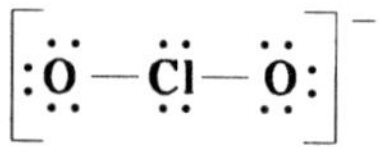

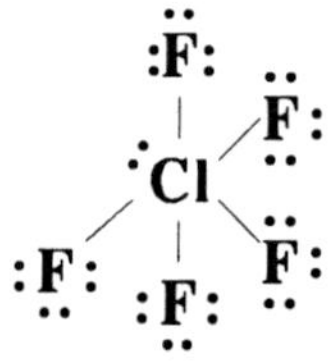

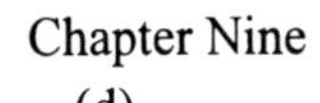

(c)

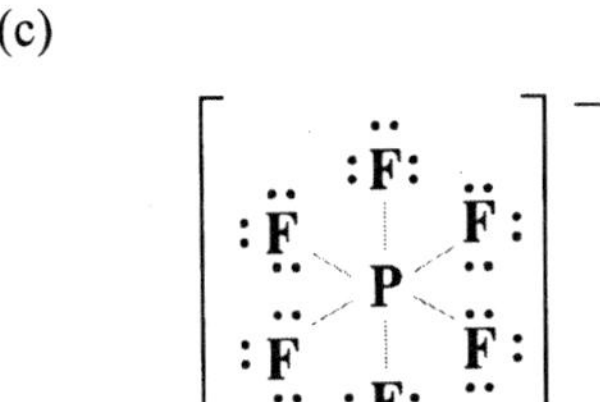

(d)

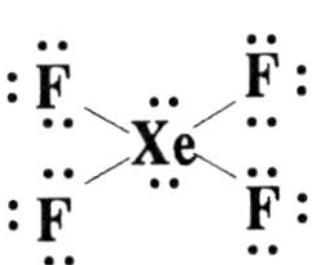

9.100

(a) (b)

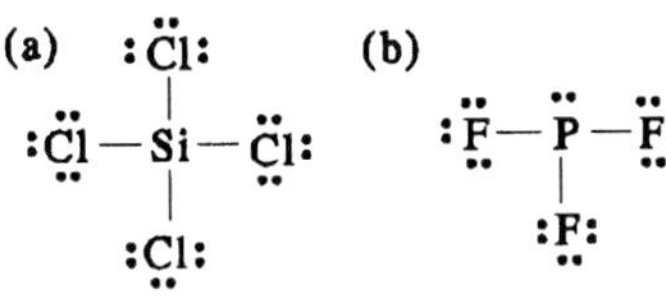

(c) H—P—H (with H below P)

(d) :Cl—S—Cl:

9.101

(a) (b)

(c) (d)

9.102 (a) S=C=S (b) [:C≡N:]⁻

9.103 (a) O—Se—O (with =O below Se) (b) O—Se=O

9.104 (a) H—As—H (with H below As) (b) O—Cl—O—H

(c) (d)

9.105

(a) [:N≡O:]+

(b) [:O–N=O]+

(c) $[SbCl_6]^-$

(d) $[IO_3]^-$

9.106

$H_2C=O$

$SOCl_2$

9.107

(a) $GeCl_4$

(b) $[CO_3]^{2-}$

(c) $[PO_4]^{3-}$

(d) $[O–O]^{2-}$

9.108

(a) H–O–Cl–O: formal charges 0, 0, 1+, 1-

(b) O=S–O, O: formal charges 0, 2+, 1-, 1-

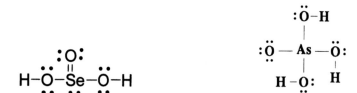

(c)

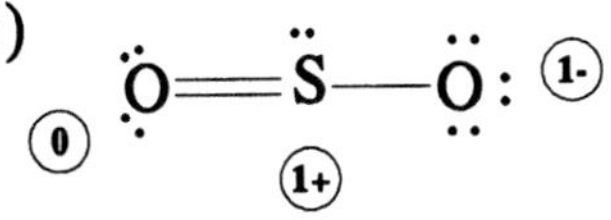

9.109 (a)

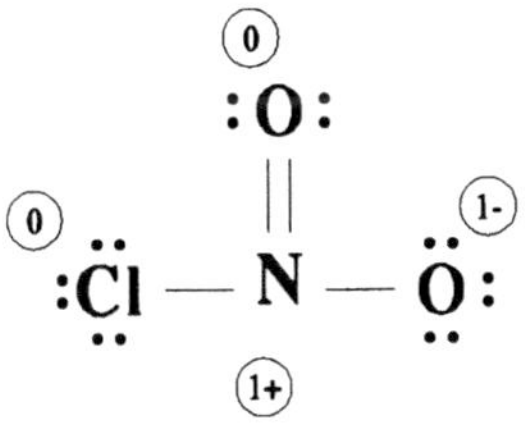

(b)

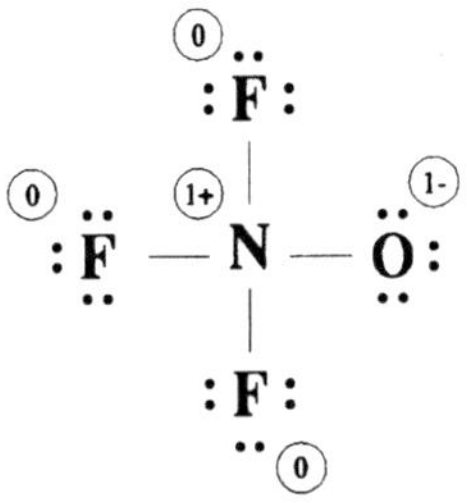

(c)

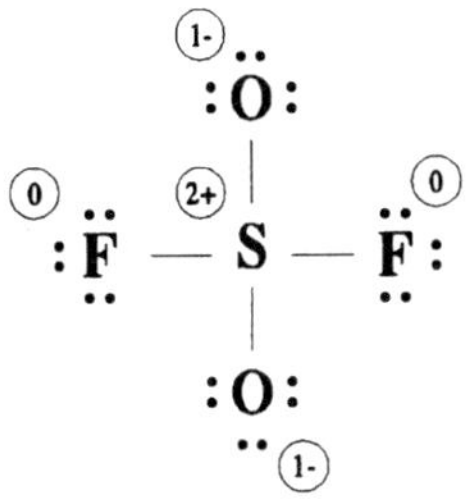

9.110

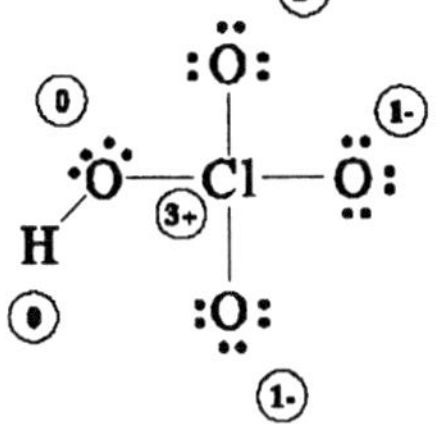

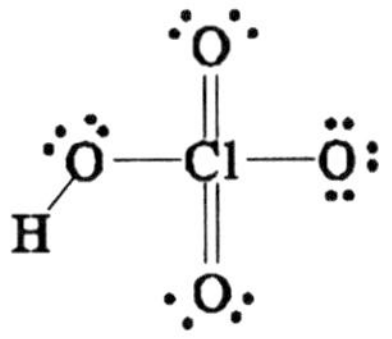

9.111

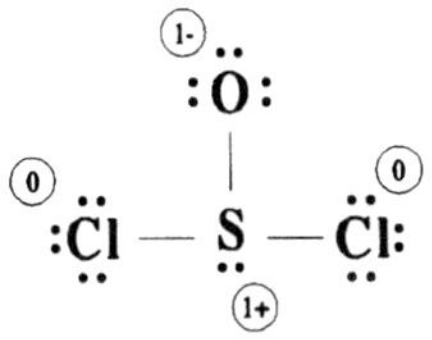

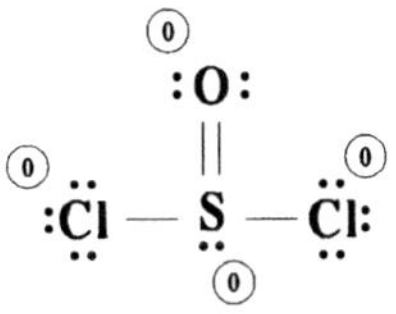

9.112 The formal charges on all of the atoms of the left structure are zero, therefore, the potential energy of this molecule is lower and it is more stable.

9.113 The one on the right is better. The one on the left gives one chlorine atom a formal charge of +1, which is not a likely situation for such a highly electronegative element.

9.114 The average bond order is 4/3

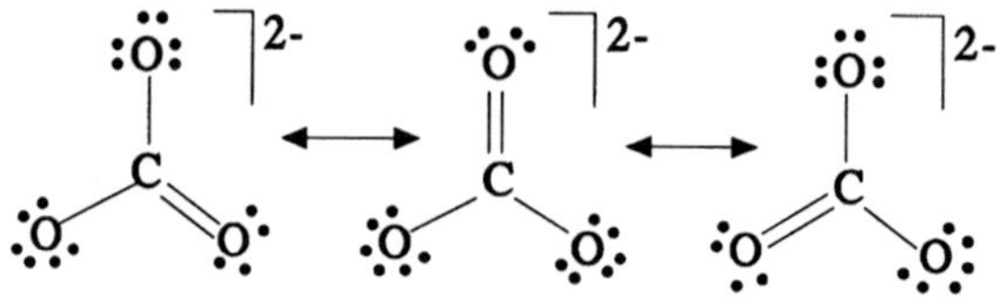

9.115

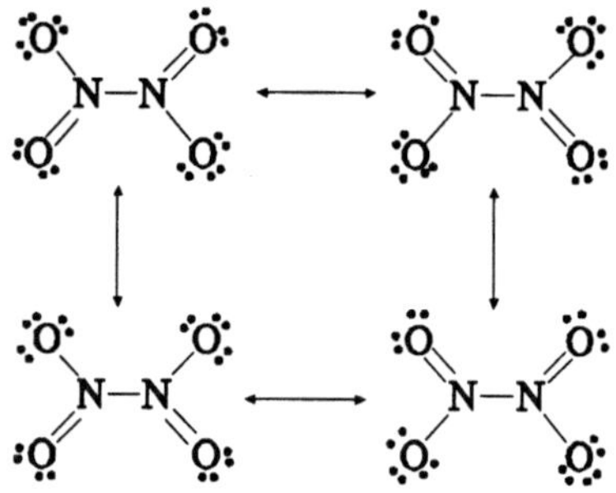

The average bond order is 1.5.

9.116 The Lewis structure for NO_3^- is given in the answer to practice exercise 9, and that for NO_2^- is given below.

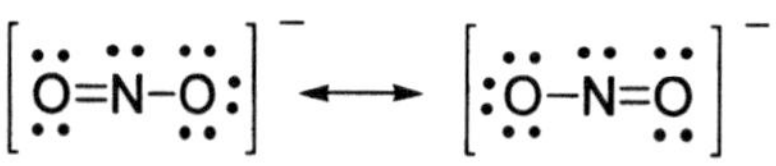

$[O=N-O:]^- \longleftrightarrow [:O-N=O]^-$

Resonance causes the average number of bonds in each N—O linkage of NO_3^- to be 1.33. Resonance causes the average number of electron pair bonds in each linkage of NO_2^- to be 1.5. We conclude that the N—O bond in NO_2^- should be shorter than that in NO_3^-.

9.117 The Lewis structures of the four must be compared. Carbon monoxide has a triple bond, :O≡C:; the formate ion (see text) has an average CO bond order of 1.5; CO_2 has double bonds; and CO_3^{2-} has a bond order of 1.33 (from review problem 9.107)
The order of increasing CO bond length is: $CO < CO_2 < HCO_2^- < CO_3^{2-}$

9.118

$:O\equiv C-O: \longleftrightarrow :O-C\equiv O:$

These are not preferred structures, because in each Lewis diagram, one oxygen bears a formal charge of +1 whereas the other bears a formal charge of −1. The structure with the formal charges of zero has a lower potential energy and is more stable.

9.119

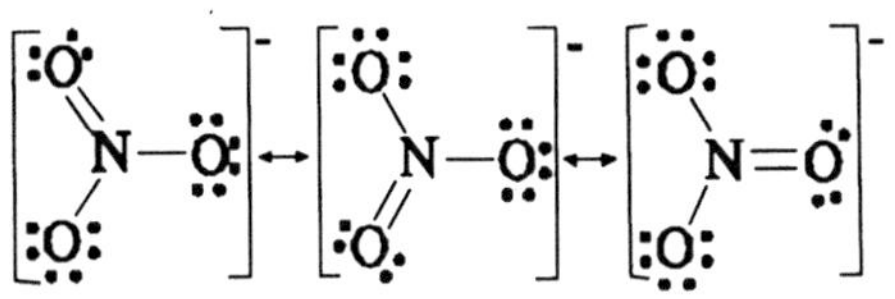

(The three oxygen atoms are equivalent in nitrate.)

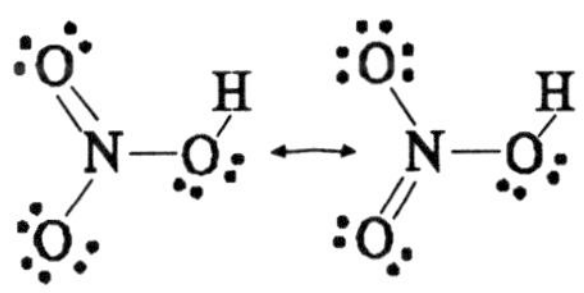

(Only two oxygen atoms are equivalent in the acid.)

In the nitrate anion, there are three equivalent O atoms, and therefore we can write three resonance forms. In nitric acid, there are only two equivalent O atoms (the ones not having an H atom attached), and we may, therefore, write only two resonance forms.

9.120 The Lewis structure that is obtained using the rules of Figure 9.8 is:

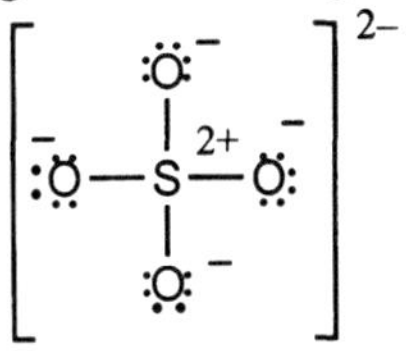

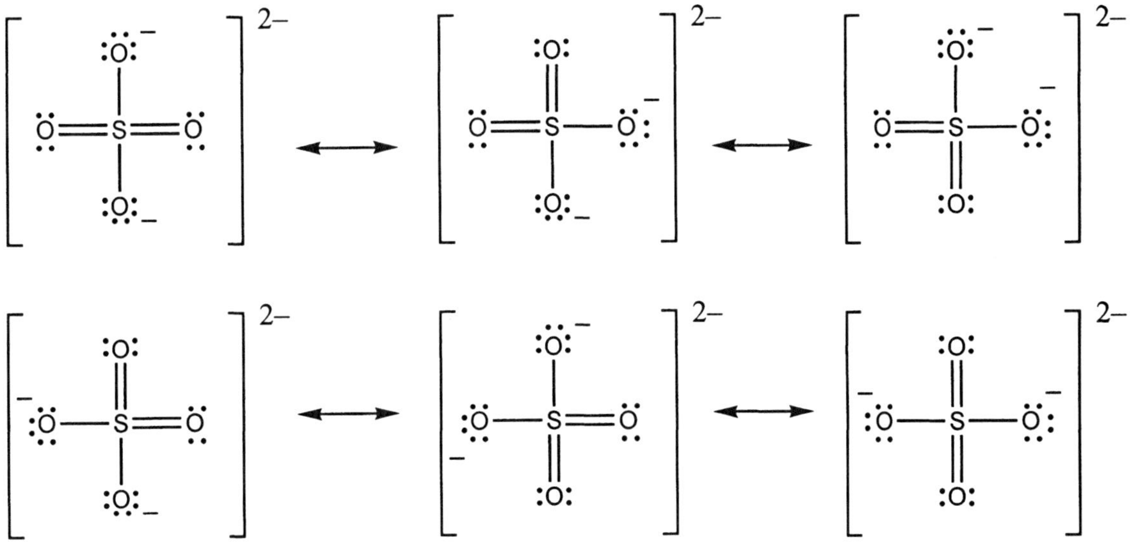

Formal charges are indicated. The average bond order is 1.5.

9.121

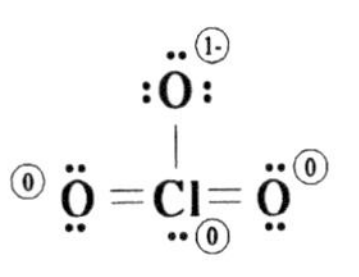

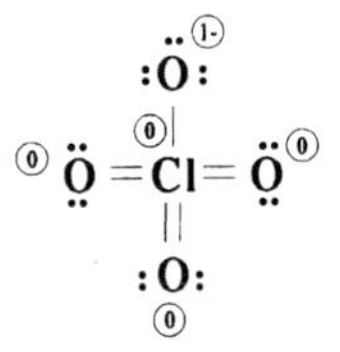

The average bond order in ClO_3^- is 1.67 and the average bond order in ClO_4^- is 1.75. Since ClO_4^- has the larger bond order, it will have the shorter bond length.

9.122

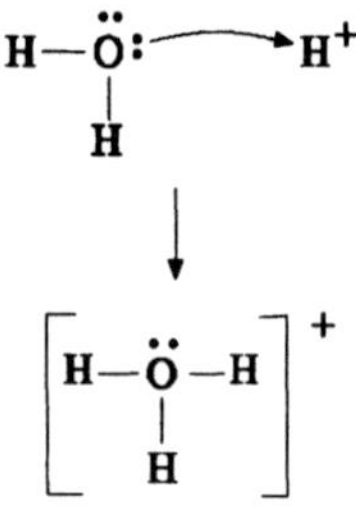

9.123

Additional Exercises

9.124 $Na(g) \rightarrow Na^+(g) + e^-$ IE = 496 kJ/mol
$Cl(g) + e^- \rightarrow Cl^-(g)$ EA = –348 kJ/mol
$Na(g) + Cl(g) \rightarrow Na^+(g) + Cl^-(g)$ ΔH = 148 kJ/mol

$Na(g) \rightarrow Na^+(g) + e^-$ IE = 496 kJ/mol
$Na^+(g) \rightarrow Na^{2+}(g) + e^-$ IE = 4563 kJ/mol
$2Cl(g) + 2e^- \rightarrow 2Cl^-(g)$ EA = 2(–348kJ/mol)
$Na(g) + 2Cl(g) \rightarrow Na^{2+}(g) + 2Cl^-(g)$ $\Delta H = 4.36 \times 10^3$ kJ/mol

In order for $NaCl_2$ to be more stable than NaCl, the lattice energy should be almost 30 times larger 436 kJ/148 kJ = 29.5.

9.125

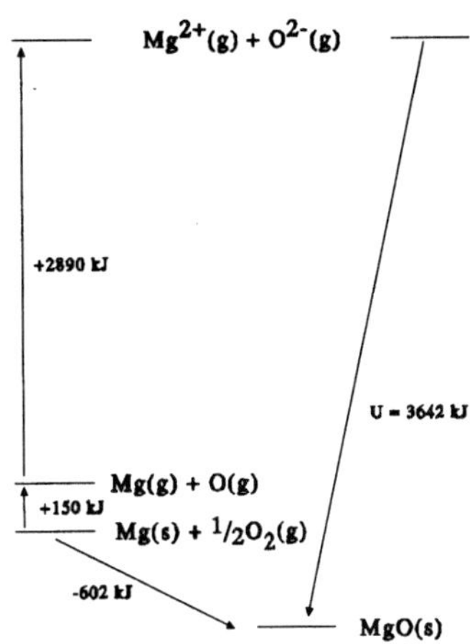

The lattice energy for NaCl is 787 kJ. The large difference in lattice energy is a result of charge (divalent versus univalent) and size of ions.

9.126

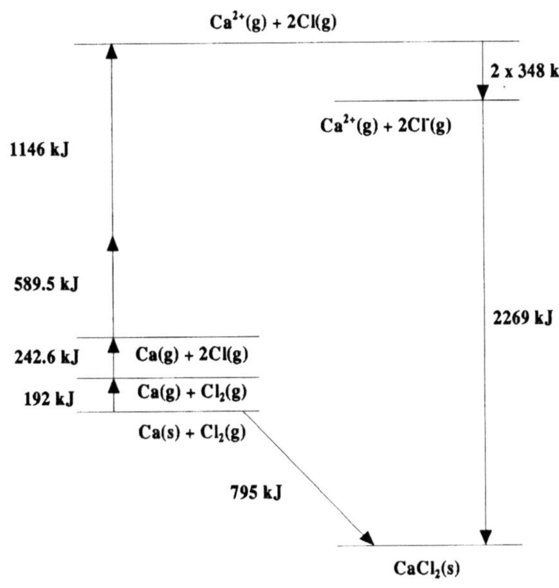

9.127

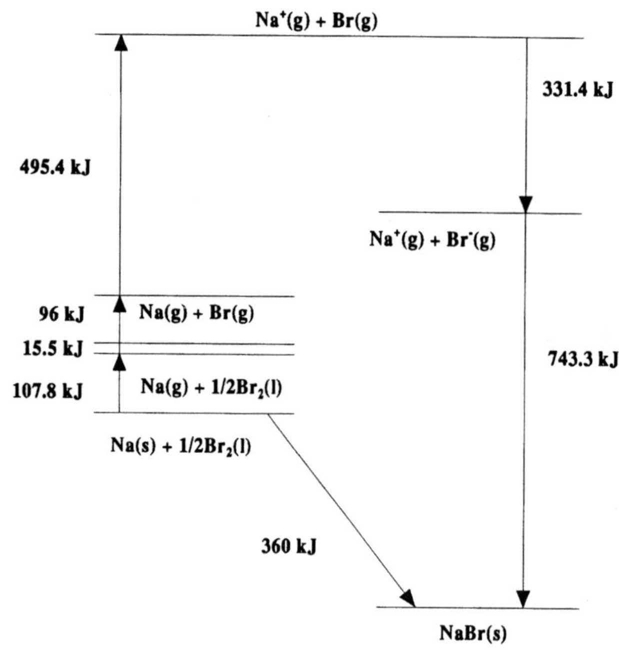

Electron affinity for Br = –331.4 kJ/mol

9.128

```
        ..
       :Cl:
        |
  ..    |    ..
 :Cl— Sn — Cl:
  ..    |    ..
       :Cl:
        ..
```

9.129 (a) PCl_3 (b) GeF_4 (c) $SiCl_4$ (d) SrO

9.130 Eighteen electrons: $3s^2$, $3p^6$, $3d^{10}$

9.131
(a) Carbon nearly always makes four bonds and this structure has only 3. Additionally, the formal charge on the carbon atom is –1.
(b) The formal charge on the oxygen atom bonded to the hydrogen is +1 which is unlikely since it is the most electronegative element in the molecule.
(c) Carbon never makes more than four bonds.

9.132 Given: Better (lower energy) structure:

Given: H(0)— N(1+)≡N(1+)—N̤̈:(2-)

Better (lower energy) structure: H(0)—N̈(0)= N(1+) =N̤̈(1-)

9.133

```
      (0)    (0)
       :Ö — H
(-1)  (+1)|     (0)
 :Ö — As — Ö — H
      |     (0)
    H — Ö:
 (0)       (0)
```

```
     :Ö——H
       |
 Ö=As—Ö——H
       |
     :Ö——H
```

9.134 Of the four resonance structures, the one in the upper left is the best possible structure. Each of the terminal nitrogen atoms has a –1 formal charge and the central atom has a formal charge of +1. The structure in the upper right is also a good structure except that the formal charge on the rightmost nitrogen atom is –2 and the central nitrogen is +1 while the leftmost nitrogen has a formal charge of zero. The bottom two resonance structures are both poor because each violates the octet rule. The left bottom structure has too many electrons around the central nitrogen atom while the right bottom structure has too few electrons on the central atom.

9.135 The average bond orders are: SO_3 (1.00), SO_2 (1.50), SO_3^{2-} (1.33), and SO_4^{2-} (1.50). Therefore the SO bond lengths should vary in the following way: $SO_3 > SO_3^{2-} > SO_2 \cong SO_4^{2-}$.

9.136

$$:\ddot{\underset{..}{Cl}} - \ddot{\underset{..}{S}} - \ddot{\underset{..}{S}} - \ddot{\underset{..}{Cl}}:$$

9.137 Because this is an acid it is probable that the hydrogen atom is bonded to the most electronegative atom. For this molecule, that is O. So, the structures HOCN and HONC are the most likely.

9.138 E = energy required to atomize H_2 (bond energy) + IE_H + EA_H

E = 435 kJ/mol + 1312 kJ/mol + (–73 kJ/mol)

$$E = 1674 \text{ kJ/mol}\left(\frac{1000 \text{ J}}{1 \text{ kJ}}\right)\left(\frac{1 \text{ mol}}{6.022 \times 10^{23} \text{ molecule}}\right) = 2.78 \times 10^{-18} \text{ J/molecule}$$

$$E = h\nu \qquad E = \frac{hc}{\lambda}$$

$$2.78 \times 10^{-18} \text{ J/molecule} = \frac{(6.626 \times 10^{-34} \text{ Js})(3.00 \times 10^{8} \text{ m/s})}{\lambda}$$

7.15×10^{-8} m = 71.5 nm

(These are very high energy photons, in the X-ray region of the electromagnetic spectrum.)

9.139 $\mu = q \cdot r$

$$8.59 \text{ D}\left(\frac{3.34 \times 10^{-30} \text{ C}\cdot\text{m}}{1 \text{ D}}\right) = q \cdot 217 \text{ pm}\left(\frac{1 \text{ m}}{10^{12} \text{ pm}}\right)$$

$q = 1.32 \times 10^{-19}$ C

The charge of an electron is 1.60×10^{-19} C.

The amount of charge is $\frac{1.32 \times 10^{-19} \text{ C}}{1.60 \times 10^{-19} \text{ C}}$

Therefore, the amount of charge on the K is $+0.826e^-$ and the amount of charge on the F is $-0.826e^-$.

9.140

$$C_6H_5-CH_2-C(CH_3)_2-NH_3^+$$

Practice Exercises

10.1 Each bond below corresponds to a bonding domain, including the double bond, which is considered a single bonding domain. (Carbon has 3 bonding domains.)

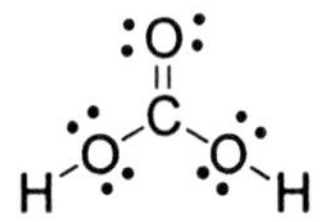

The lone pairs on the oxygens correspond to nonbonding domains.

10.2 $SbCl_5$ should have a trigonal bipyramidal shape (Figure 10.4) because, like PCl_5, it has five electron pairs around the central atom.

10.3 I_3^- should have a linear shape (Figure 10.4) because although it has five electron pairs around the central I atom, only two are being used for bonding.

10.4 HArF should have a linear shape (Figure 10.4) because although it has five electron pairs around the central Ar atom, only two are being used for bonding.

10.5 XeF_4 should have a square planar shape (Figure 10.5) because although it has six electron pairs around the central Xe atom, only four are being used for bonding.

10.6 In SO_3^{2-}, there are three bond pairs and one lone pair of electrons at the sulfur atom, and as shown in Figure 10.3, this ion has a trigonal pyramidal shape.

In CO_3^{2-}, there are three bonding domains (2 single bonds and 1 double bond) at the carbon atom, and as shown in Figure 10.2, this ion has a planar triangular shape.

In XeO_4, there are four bond pairs of electrons around the Xe atom, and as shown in Figure 10.3, this molecule is tetrahedral.

In OF_2, there are two bond pairs and two lone pairs of electrons around the oxygen atom, and as shown in Figure 10.3, this molecule is bent.

10.7 a) SF_6 is octahedral, and it is not polar.
b) SO_2 is bent, and it is polar.
c) BrCl is polar because there is a difference in electronegativity between Br and Cl.
d) AsH_3, like NH_3, is pyramidal, and it is polar.
e) CF_2Cl_2 is polar, because there is a difference in electronegativity between F and Cl.

10.8 The H–Cl bond is formed by the overlap of the half–filled 1s atomic orbital of a H atom with the half–filled 3*p* valence orbital of a Cl atom:

Cl atom in HCl (x = H electron):

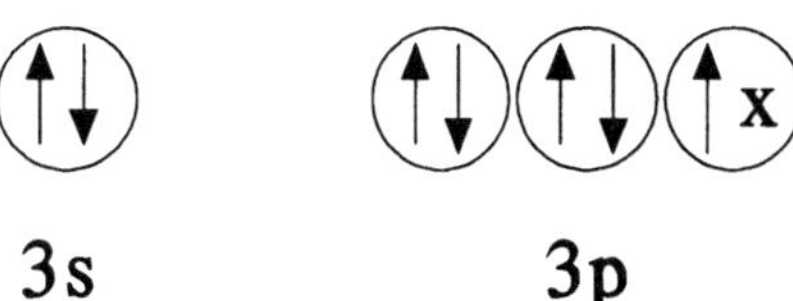

The overlap that gives rise to the H–Cl bond is that of a 1*s* orbital of H with a 3*p* orbital of Cl:

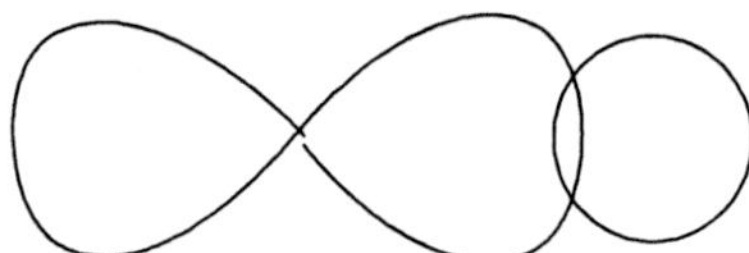

10.9 The half–filled 1*s* atomic orbital of each H atom overlaps with a half–filled 3*p* atomic orbital of the P atom, to give three P–H bonds. This should give a bond angle of 90°.

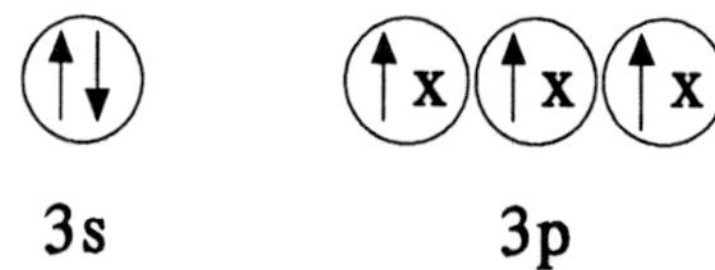

The orbital overlap that forms the P–H bond combines a 1*s* orbital of hydrogen with a 3*p* orbital of phosphorus (note: only half of each p orbital is shown):

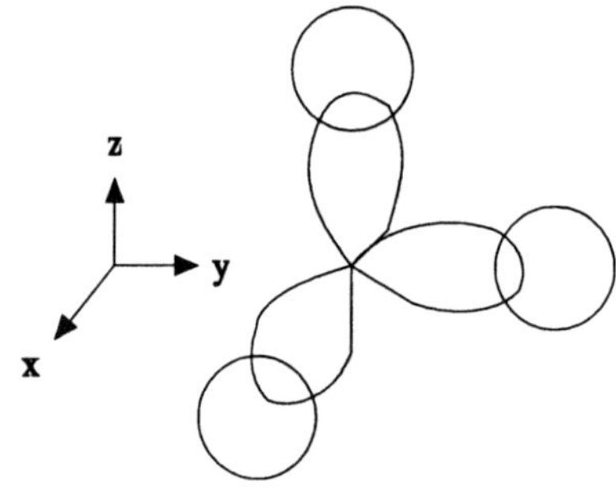

10.10 Since there are five bonding pairs of electrons on the central arsenic atom, we choose sp^3d hybridization for the As atom. Each of arsenic's five sp^3d hybrid orbitals overlaps with a 3*p* atomic orbital of a chlorine atom to form a total of five As–Cl single bonds. Four of the 3*d* atomic orbitals of As remain unhybridized.

10.11 a) sp^3 b) sp^3d

10.12 a) sp^3 b) sp^3d

10.13 sp^3d^2, since six atoms are bonded to the central atom.

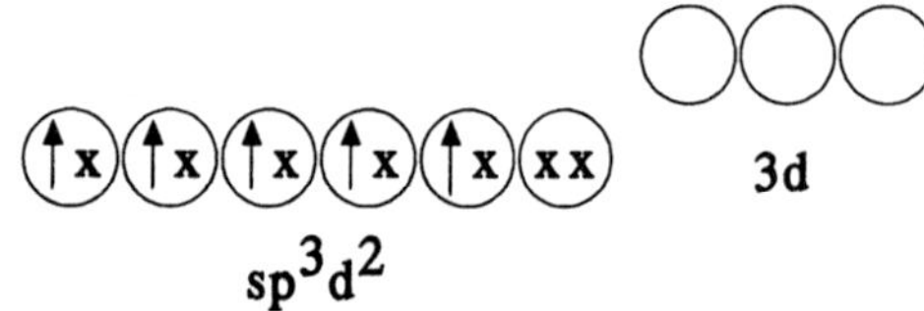

The ion is octahedral because six atoms and no lone pairs surround the central atom.

10.14 NO has 11 valence electrons, and the MO diagram is similar to that shown in Table 10.1 for O_2, except that one fewer electron is employed at the highest energy level

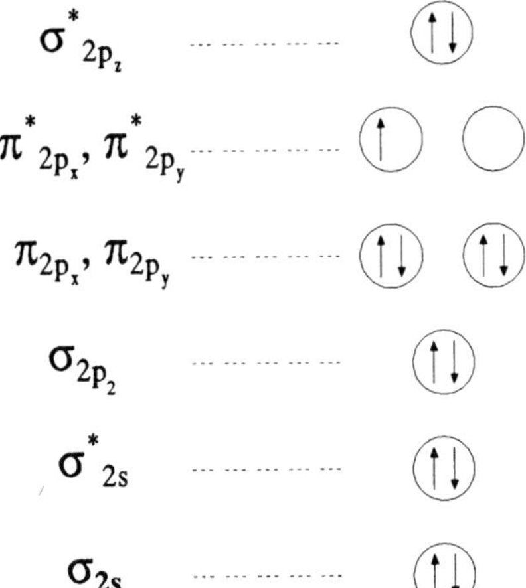

The bond order is calculated to be 5/2:

$$\text{Bond Order} = \frac{(8 \text{ bonding e}^-) - (3 \text{ antibonding e}^-)}{2} = \frac{5}{2}$$

Thinking It Through

T10.1 Like NH_3, all pyramidal molecules have a lone pair of electrons on the central atom.

T10.2 Each molecule has the same number of bonded atoms, and we conclude that it must be the number of lone pairs that governs geometry. After drawing the Lewis diagrams, we see that XeF_2 has three lone pairs, whereas SF_2 has only two lone pairs.

T10.3 Because there are six atoms attached to the central atom, we know that the molecule is octahedral. Since it is an unsymmetrical molecule, it is polar.

T10.4 Since the angles are nearer to 90° than to 109°, we must use atomic rather than hybrid orbitals on antimony.

T10.5 The Lewis structure indicates there are two pairs of bonding electrons and two pairs of nonbonding electrons on the central atom. Figure 10.3 indicates the molecule is bent and polar.

T10.6 NO has one less valence electron than O_2. The highest occupied molecular orbitals are π^* orbitals, so upon removing one of these electrons, the bond order increases. So, the bond order for NO is 2.5.

T10.7 NO is paramagnetic because it has one unpaired electron in a π^* orbital.

T10.8 Five valence electrons → Group VA

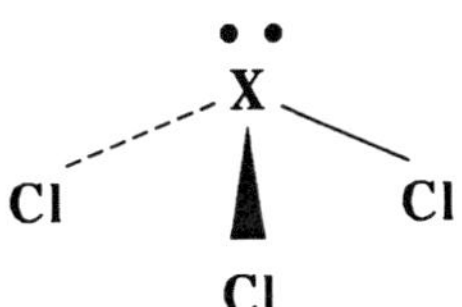

Three valence electrons → Group IIIA

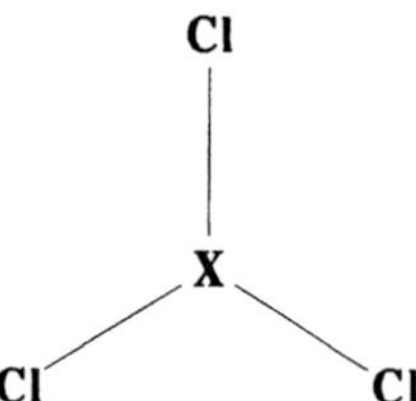

Seven valence electrons → Group VIIA

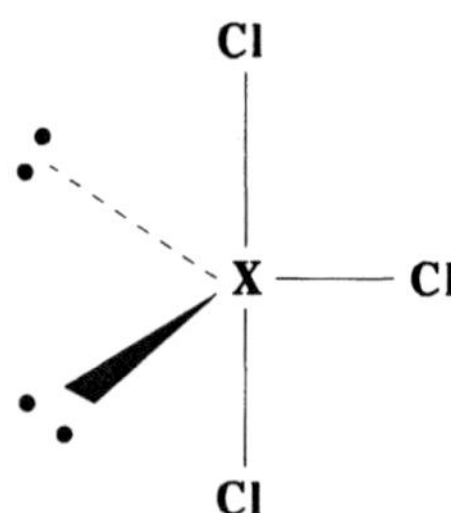

It is unlikely that X would be in Group VI because it would have an unpaired electron since X would have six valence electrons and three would be used for bonding with the Cl, and that would leave three remaining electrons.

T10.9 The electronegativity difference between O and Cl will force the Cl-C-Cl bond to become smaller. In addition, the short double bond brings the charge of the oxygen atom closer to the carbon atom.

Review Questions

10.1 (a) See Figure 10.2. The angles are 120°.
(b) See Figures 10.2 and 10.3. The bond angles are 109.5°.
(c) See Figures 10.2 and 10.5. The bond angles are 90°.

10.2 (a) See Figure 10.2. The bond angle is 180°.
(b) See Figures 10.2 and 10.4. The bond angles are 120° between the equatorial bonds and 180° between the two axial bonds.

10.3 VSEPR Theory is based on the principle that adjacent electron pairs repel one another, and that these destabilizing repulsions are reduced to a minimum when electron pairs stay as far apart as possible.

10.4 An electron domain is a region in space where electrons can be found.

10.5 In HCHO there are bonding domains around the C; one bonding domain around each H, and one bonding domain around O and two nonbonding domains around O.

10.6 (a) Planar triangular, otherwise known as trigonal planar
(b) Octahedral
(c) Tetrahedral
(d) Trigonal bipyramidal

10.7 (a) See Figure 10.3. The shape is pyramidal.

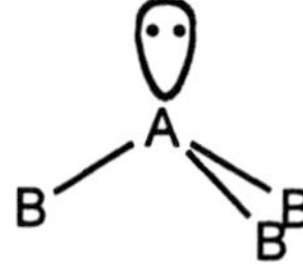

(b) See Figure 10.5. The shape is planar (square).

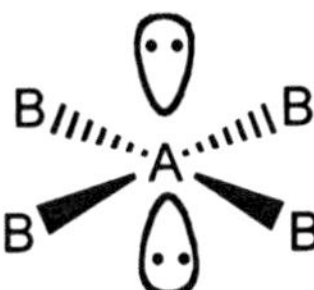

(c) This is linear.

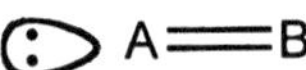

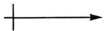

10.8 Polar molecules attract one another, and that influences the physical and chemical properties of substances.

10.9 A bond's dipole moment is depicted with an arrow having a + sign on one end, where the "barb" of the arrow is taken to represent the location of the opposing negative charge of the dipole:

10.10 Asymmetric molecules are polar only if there is a difference is the electronegativity of the atoms.

10.11 A molecule having polar bonds will be nonpolar only if the bond dipoles are arranged so as to cancel one another's effect.

10.12

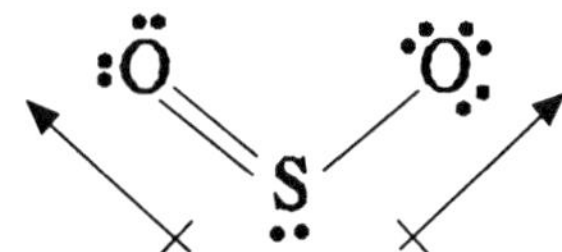

The individual bond dipoles do not cancel one another.

10.13 Both VB and MO theory have wave mechanics as their theoretical basis. In each theory, bonds are considered to arise from the overlap of orbitals.

10.14 Lewis structures do not explain how atoms share electrons, nor do they explain why molecules adopt particular shapes. Also, electrons are known to be delocalized, a fact which only MO theory addresses from the beginning. Also, odd-electron systems can be more effectively discussed with MO theory.

10.15 VB theory views atoms coming together with their orbitals already containing specific electrons. The bonds that are formed according to VB theory do so by the overlap of orbitals on neighboring atoms, and this is accompanied by the pairing (sharing) of the electrons that are contained in the orbitals.

MO theory, on the other hand, considers a molecule to be a collection of positive nuclei surrounded by a set of molecular orbitals, which, by definition, belong to the molecule as a whole, rather than to any specific atom. The electrons of the molecule are distributed among the molecular orbitals according to the same rules that govern the filling of atomic orbitals.

10.16 Orbital overlap occurs when orbitals from different atoms share the same space. This overlap provides a more stable region for the electrons, which find themselves under the stabilizing influence of the positive charge of two nuclei.

10.17 The expected bond angle would be 90°, the angle between the *p* orbitals.

10.18 According to VB theory, spin–paired electrons are shared between atoms via the overlap of orbitals from the different atoms.

10.19 This is shown in Figure 10.11.

10.20 This is the same as the HF molecule, shown in Figure 10.12. A half–filled valence p orbital of Br overlaps with the half–filled $1s$ orbital of the hydrogen atom.

10.21 Hybridization

10.22 Hybrid orbitals provide better overlap than do atomic orbitals, and this results in stronger bonds.

10.23 These are shown in Figures 10.17 and 10.22.

10.24 Elements in period two do not have a d subshell in the valence level.

10.25 Lewis structures are something like shorthand representations of VB descriptions of molecules.

10.26 The VSEPR model gives the shape of the electron domains around the central atom, this information is used to predict the hybridization of the atom.

10.27

(↑)(↑)(↑)(↑) sp^3 C atom

(↑↓)(↑)(↑)(↑) sp^3 N atom

(↑↓)(↑↓)(↑)(↑) sp^3 O atom

There are zero, one and two lone pairs, respectively, on these atoms when sp^3 hybridization is utilized to form bonds.

10.28 (a)

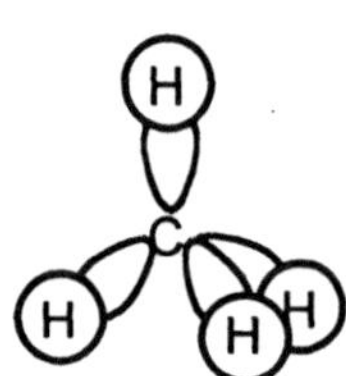

(b)

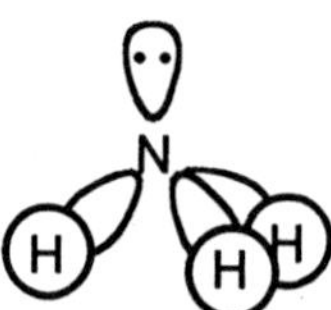

(c)

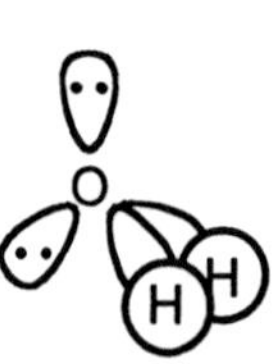

10.29 This angle would have had to be 90°, the angle between one atomic p orbital and another.

10.30 90°

10.31 See Figure 10.18 and Figure 10.25.

Valence shell for boron:

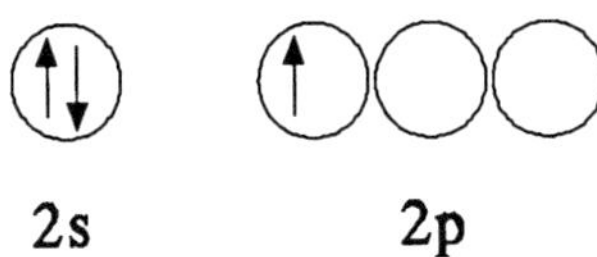

Valence shell for carbon:

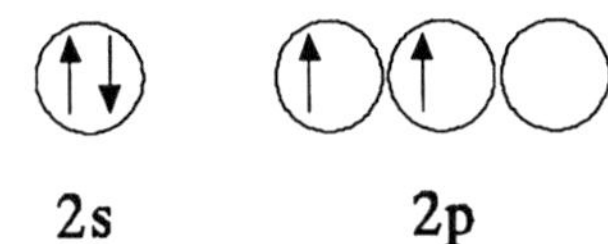

sp^2-hybridized boron atom: **sp^2-hybridized carbon atom:**

10.32 sp^3d – trigonal bipyramid sp^3d^2 – octahedral

10.33 The ammonium ion has a tetrahedral geometry, with bond angles of 109.5°.

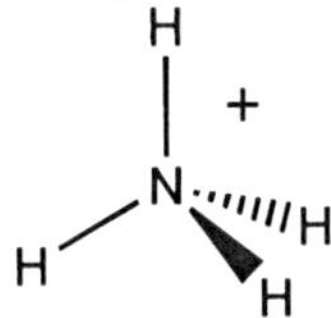

10.34 The oxygen atom in H_2O has two lone pairs of electrons, either one of which can form a coordinate covalent bond to a proton:

O in H_3O^+:

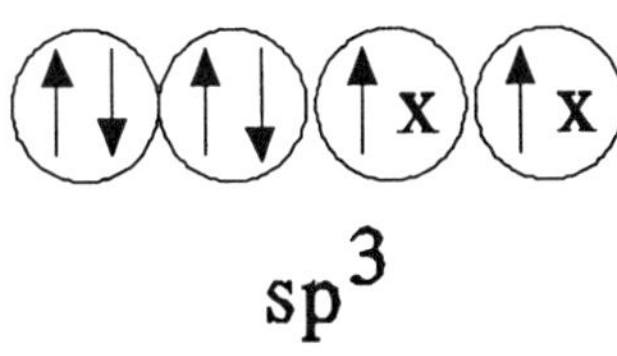

(x = H electron)

10.35 σ bond – The electron density is concentrated along an imaginary straight line joining the nuclei of the bonded atoms.

π bond – The electron density lies above and below an imaginary straight line joining the bonded nuclei.

10.36 The characteristic side–to–side overlap of *p* atomic orbitals that characterizes π bonds is destroyed upon rotation about the bond axis. This is not the case for a σ bond, because regardless of rotation, a σ bond is still effective at overlap.

10.37 See Figure 10.25.

10.38 See Figure 10.28.

10.39 Two resonance hybrids should be drawn.

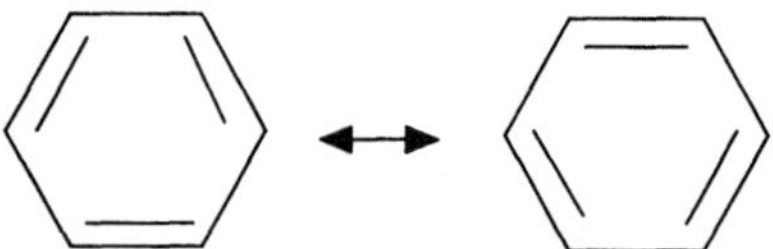

sp^2 hybrid orbitals are used for each carbon atom. Consequently, each bond angle is 120°. See Figure 10.37.

10.40 If an electron is forced to occupy this MO, the molecule loses stability and the bond is made weaker than if an electron, or a pair of electrons, is made to occupy the lower energy (bonding) MO.

10.41 See Figures 10.30 and 10.31. It is important to note that, relative to the atomic orbitals from which they are formed, the bonding orbital is lower in energy and the antibonding orbital is higher in energy.

10.42 In the hypothetical molecule He_2, both the bonding and the antibonding MO are doubly occupied, and the net bond order is zero, as shown in Figure 10.32. See Figure 10.31 for the MO diagram for H_2.

10.43 As shown in Table 10.1, the highest energy electrons in dioxygen occupy the doubly degenerate π–antibonding level:

$$\underset{\pi^*_{2p_x}}{\uparrow} \qquad \underset{\pi^*_{2p_y}}{\uparrow}$$

Since their spins are unpaired, the molecule is paramagnetic.

10.44 As shown in Table 10.1, the bond order of Li_2 is 1.0. The bond order of Be_2 would be zero. Yes, Be_2^+ could exist since the bond order would be 1/2.

10.45 (a) 2.5 (b) 1.5 (c) 1.5

10.46 As bond order increases, bond energy (and strength) increases.

10.47 See Figure 10.33.

10.48 A delocalized MO is one that extends over more than two nuclei.

10.49

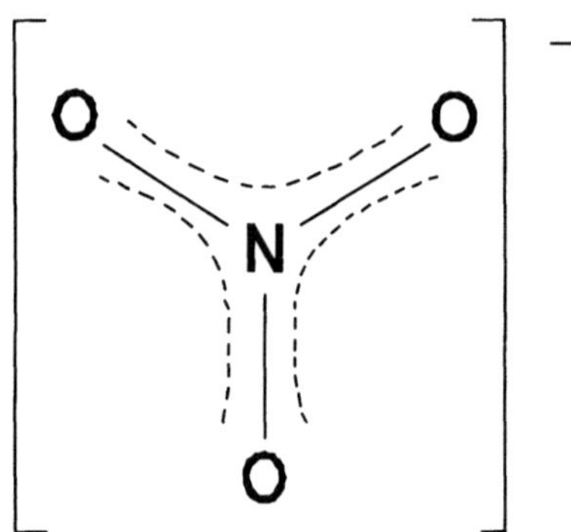

10.50 MO theory avoids the need in VB theory for the cumbersome and numerous resonance forms.

10.51

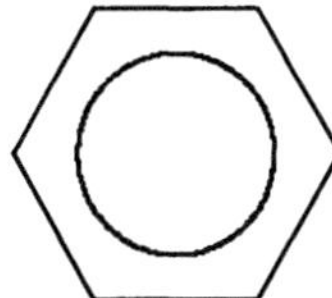

10.52 Delocalization increases stability.

10.53 Delocalization energy is a term used in MO theory to mean essentially the same thing as the term resonance energy, which derives from VB theory. They both represent the additional stability associated with a spreading out of electron density.

Review Problems

10.54
(a) bent (central N atom has two single bonds and two lone pairs)
(b) planar triangular (central C atom has three bonding domains—a double bond and a single bond)
(c) T-shaped (central I atom has three single bonds and two lone pairs)
(d) Linear (central Br atom has two single bonds and three lone pairs)
(e) planar triangular (central Ga atom has three single bonds and no lone pairs)

10.55
(a) trigonal pyramidal (b) planar triangular
(c) tetrahedral (d) nonlinear (bent)
(e) linear

10.56
(a) nonlinear (b) trigonal bipyramidal
(c) trigonal pyramidal (d) trigonal pyramidal
(e) nonlinear

10.57
(a) distorted tetrahedral (b) octahedral
(c) nonlinear (d) tetrahedral
(e) tetrahedral

10.58
(a) tetrahedral (b) square planar
(c) octahedral (d) tetrahedral
(e) linear

10.59
(a) linear (b) square planar
(c) T–shaped (d) pyramidal
(e) planar triangular

10.60 180°

10.61 all angles 120°

10.62 The ones that contain nonbonding domains are (a), (b), (c), and (e). Actually, all of these compounds contain nonbonding domains. Compound (d) contains no nonbonding domains on the central atom, but contains them on the outer atoms.

10.63 The ones that contain nonbonding domains are (a), (c), (d), and (e).

10.64
a) 109.5° b) 109.5°
c) 120° d) 180°
e) 109.5°

10.65 a) 109.5° b) 109.5°
c) 180° d) 120°
e) 120°

10.66 The ones that are polar are (a), (b), and (c). The last two have symmetrical structures, and although individual bonds in these substances are polar bonds, the geometry of the bonds serves to cause the individual dipole moments of the various bonds to cancel one another.

10.67 Two of these substances have planar triangular structures that are not polar because the individual bond dipole moments cancel one another: SeO_3 and BCl_3. Two of these molecules have pyramidal structures, and are, therefore, polar: PBr_3 and $AsCl_3$. ClF_3 is T–shaped and, therefore, polar.

10.68 All are polar. (a)-(c) and (e) have asymmetrical structures, and (d) only has one bond, which is polar.

10.69 BeH_2 is nonpolar since it is linear. H_2S is bent and, therefore, polar. SCN^- and CN^- are linear, but have dipoles that do not cancel out. $BrCl_3$ is T-shaped and, therefore, polar.

10.70 In SF_6, although the individual bonds in this substance are polar bonds, the geometry of the bonds is symmetrical which serves to cause the individual dipole moments of the various bonds to cancel one another. In SF_5Br, one of the six bonds has a different polarity so the individual dipole moments of the various bonds do not cancel one another.

10.71 In CCl_4, although the individual bonds in this substance are polar bonds, the geometry of the bonds is symmetrical which serves to cause the individual dipole moments of the various bonds to cancel one another. In CH_3Cl, one of the four bonds has a different polarity so the individual dipole moments of the various bonds do not cancel one another.

10.72 The 1*s* atomic orbitals of the hydrogen atoms overlap with the mutually perpendicular *p* atomic orbitals of the selenium atom.

Se atom in H_2Se (x = H electron):

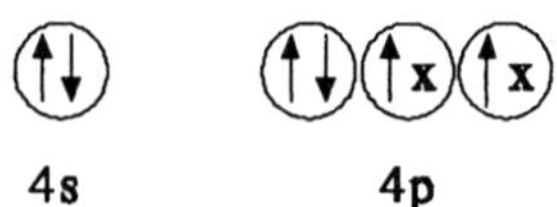

10.73 This is shown in Figure 10.14. Another way to diagram it would be to show the orbitals of one of the fluorine atoms:

Each F atom:

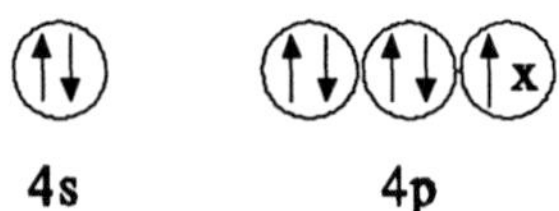

(x = an electron from the other F atom)

10.74 Atomic Be:

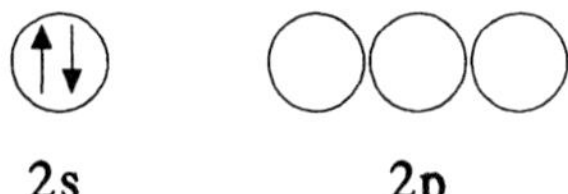

Hybridized Be:

(x = a Cl electron)

10.75 (a)

Sn atom in $SnCl_4$ (x = Cl electron):

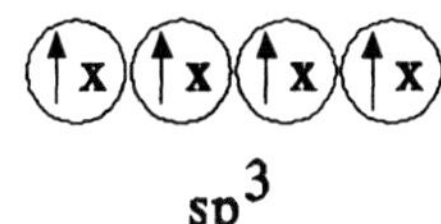

(b)

Sb atom in $SbCl_5$ (x = Cl electron):

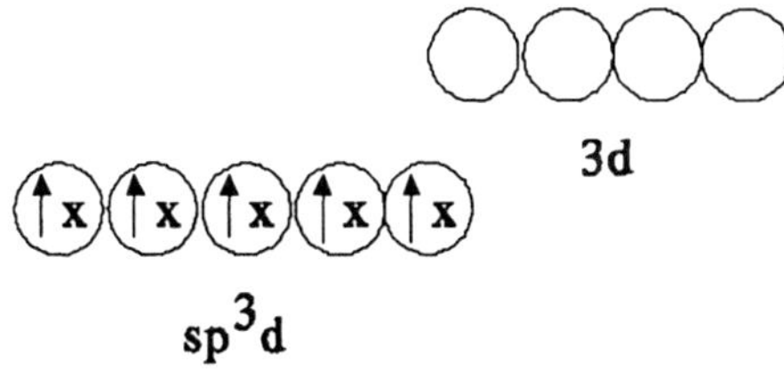

10.76 (a) There are three bonds to the central Cl atom, plus one lone pair of electrons. The geometry of the electron pairs is tetrahedral so the Cl atom is to be sp^3 hybridized:

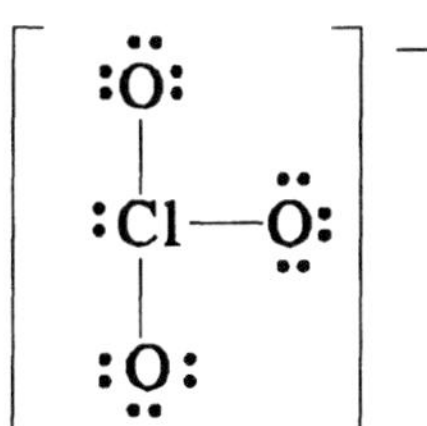

(b) There are three atoms bonded to the central sulfur atom, and no lone pairs on the central sulfur. The geometry of the electron pairs is that of a planar triangle, and the hybridization of the S atom is sp^2:

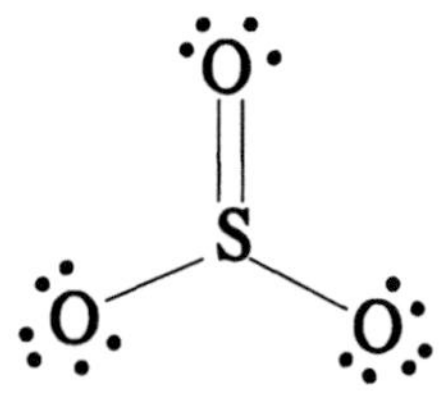

Two other resonance structures should also be drawn for SO_3.

(c) There are two bonds to the central O atom, as well as two lone pairs. The O atom is to be sp^3 hybridized, and the geometry of the electron pairs is tetrahedral.

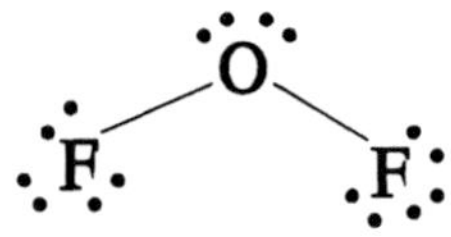

10.77 (a) Six Cl atoms surround the central Sb atom in an octahedral geometry, and the hybridization of Sb is sp^3d^2.

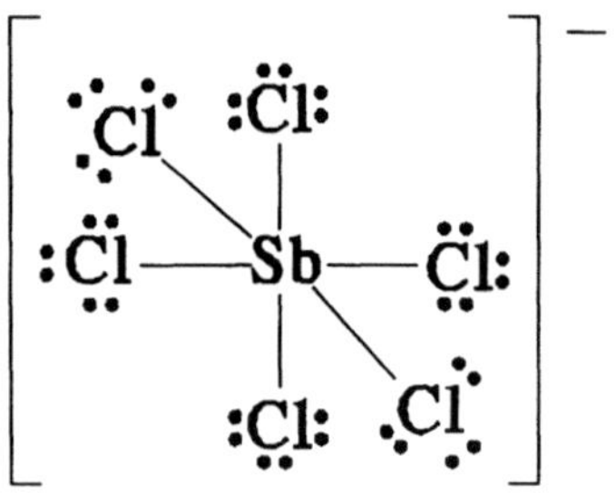

(b) Three Cl atoms are bonded to the Br atom, plus the Br atom has two lone pairs of electrons. This requires the Br atom to be sp^3d hybridized, and the geometry is T–shaped.

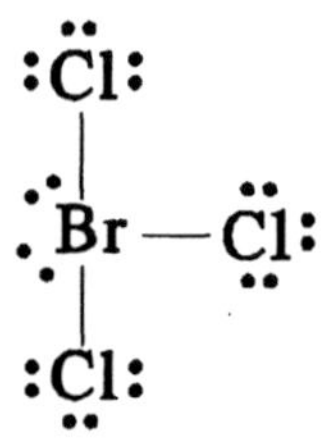

(c) The central Xe atom is bonded to four F atoms, plus it has two lone pairs of electrons. The molecule has an square planar geometry. This requires sp^3d^2 hybridization of Xe.

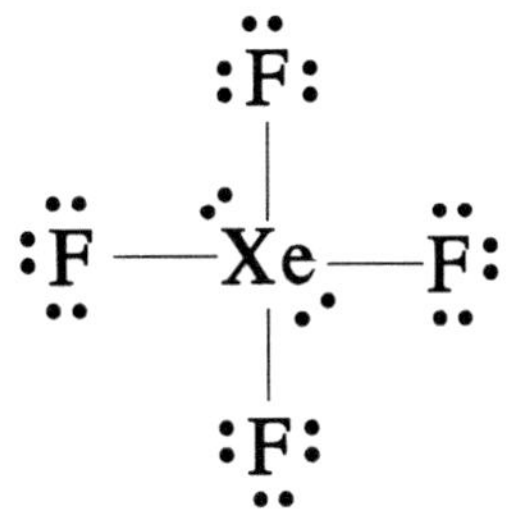

```
      ..
     :F:
      |
 ..  .|     ..
:F ——Xe——F:
 ..   |.    ..
      |
     :F:
      ..
```

10.78 (a) There are three bonds to As and one lone pair at As, requiring As to be sp^3 hybridized.

The Lewis diagram:

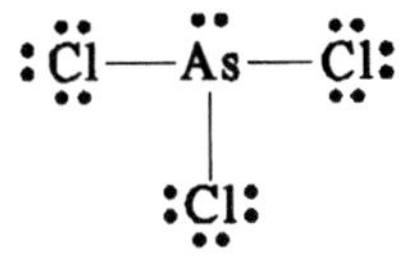

```
 ..   ..   ..
:Cl——As——Cl:
 ..   |    ..
      |
    :Cl:
      ..
```

The hybrid orbital diagram for As:

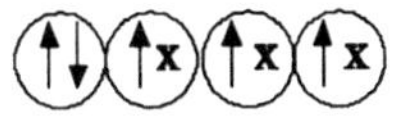

(↑↓)(↑x)(↑x)(↑x)

sp^3

(x = a Cl electron)

(b) There are three atoms bonded to the central Cl atom, and it also has two lone pairs of electrons. The hybridization of Cl is thus sp^3d.

The Lewis diagram:

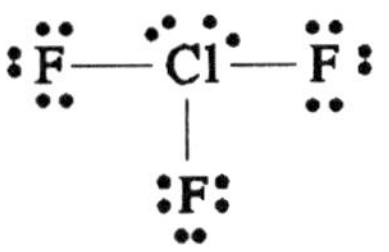

```
 ..   . .    ..
:F——Cl——F:
 ..   |      ..
      |
    :F:
      ..
```

The hybrid orbital diagram for Cl:

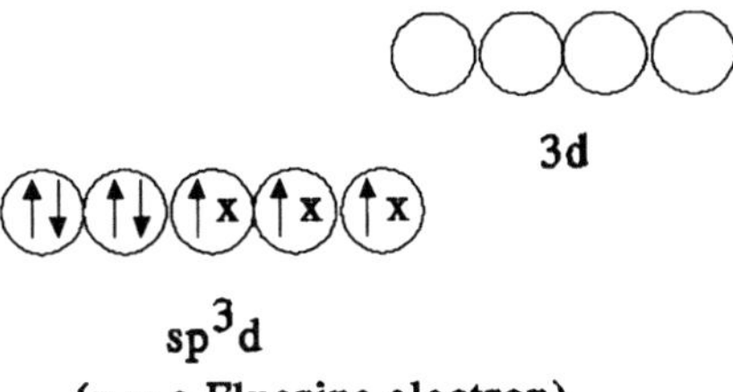

(x = a Fluorine electron)

10.79 (a) Sb has five bonds to Cl atoms and no lone pairs and is therefore sp^3d hybridized.

The Lewis diagram:

The hybrid orbital diagram for Sb:

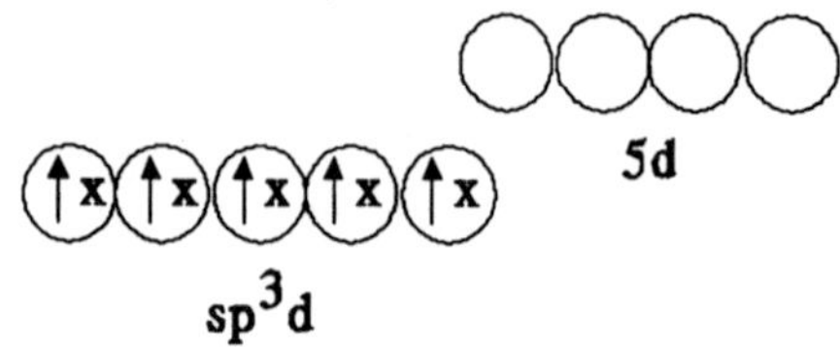

(x = a chlorine electron)

(b) Se has two bonds to Cl atoms and two lone pairs of electrons, requiring it to be sp^3 hybridized.

The Lewis diagram:

The hybrid orbital diagram for Se:

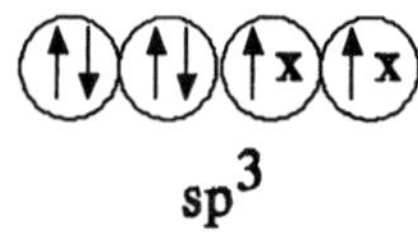

(x = a chlorine electron)

10.80 We can consider that this ion is formed by reaction of SbF_5 with F^-. The antimony atom accepts a pair of electrons from fluoride:

Sb in SbF_6^-:

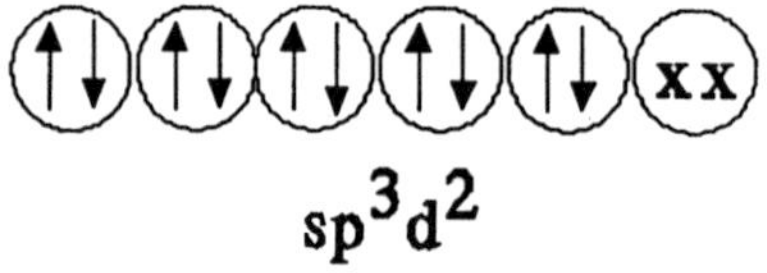

(xx = an electron pair from the donor F^-)

10.81 This is an octahedral ion with sp^3d^2 hybridized tin:

Sn atom in $SnCl_6^{2-}$ (x = Cl electron):

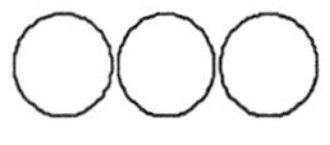

5d

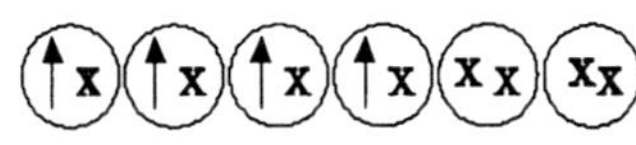

sp^3d^2

10.82 (a)

N in the C=N system:

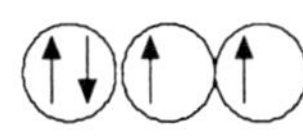

sp^2 2p

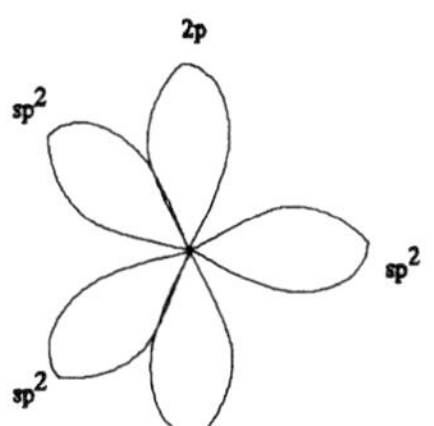

(b)

sigma bond pi bond

(c)

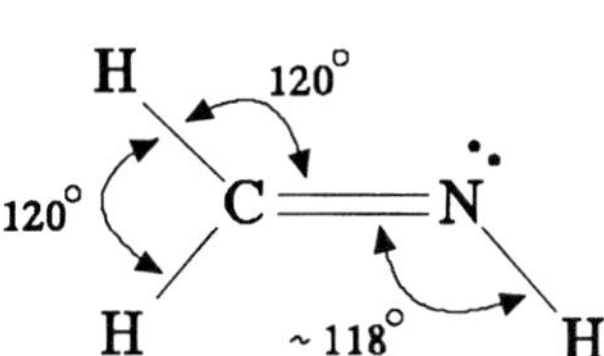

10.83 (a)

sp-hybridized N atom:

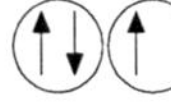

sp 2p

(b) The σ bonds:

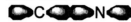

The π bonds:

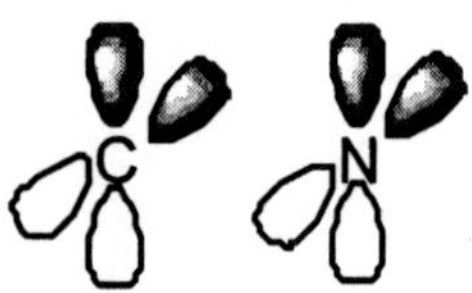

(c) For HCN, the only difference from (b) is the formation of another σ bond when an H 1*s* orbital overlaps the C *sp* hybrid orbital.

(d) The HCN bond angle should be 180°.

10.84 Each carbon atom is sp^2 hybridized, and each C–Cl bond is formed by the overlap of an sp^2 hybrid of carbon with a *p* atomic orbital of a chlorine atom. The C=C double bond consists first of a C–C σ bond formed by "head on" overlap of sp^2 hybrids from each C atom. Secondly, the C=C double bond consists of a side–to–side overlap of unhybridized *p* orbitals of each C atom, to give one π bond. The molecule is planar, and the expected bond angles are all 120°.

10.85 The bonding in phosgene is the same as that diagrammed for H_2CO in Figure 10.27, except that H atom 1*s* orbitals are replaced by Cl atom sp^3 hybrid orbitals.

10.86 1. sp^3 2. sp 3. sp^2 4. sp^2

10.87
1. one σ bond
2. one σ bond and two π bonds
3. one σ bond
4. one σ bond and one π bond

10.88 Here we pick the one with the higher bond order.

(a) O_2^+ (b) O_2 (c) N_2

10.89 NO has 11 valence electrons, and the MO diagram is similar to that shown in Table 10.1 for O_2, except that one fewer electron is employed at the highest energy level
The bond order for NO is calculated to be 5/2:

$$\text{Bond Order} = \frac{(8\ \text{bonding e}^-) - (3\ \text{antibonding e}^-)}{2} = \frac{5}{2}$$

If we remove one electron to form NO^+, the bond order becomes 3 (there are only two antibonding electrons). The larger bond order indicates a shorter bond length.

10.90 Using Fig. 10.34b, and filling in the appropriate number of valence electrons, we see that (a)-(d) all have unpaired electrons, and hence are paramagnetic. (e) does not have any unpaired electrons, and hence is not paramagnetic.

10.91 (a) $$\text{Bond Order} = \frac{(8\ \text{bonding e}^-) - (2\ \text{antibonding e}^-)}{2} = 3$$

(b) $\text{Bond Order} = \frac{(7 \text{ bonding } e^-) - (2 \text{ antibonding } e^-)}{2} = \frac{5}{2} = 2.5$

(c) $\text{Bond Order} = \frac{(5 \text{ bonding } e^-) - (2 \text{ antibonding } e^-)}{2} = \frac{3}{2} = 1.5$

(d) $\text{Bond Order} = \frac{(6 \text{ bonding } e^-) - (2 \text{ antibonding } e^-)}{2} = \frac{4}{2} = 2$

(e) $\text{Bond Order} = \frac{(8 \text{ bonding } e^-) - (3 \text{ antibonding } e^-)}{2} = \frac{5}{2} = 2.5$

Additional Exercises

10.92 planar triangular

10.93 (a) planar triangular → tetrahedral
(b) trigonal bipyramidal → octahedral
(c) T-shaped → square planar
(d) pyramidal → trigonal bipyramidal
(e) linear → planar

10.94 The normal C–C–C angle for an sp^3 hybridized carbon atom is 109.5°. The 60° bond angle in cyclopropane is much less than this optimum bond angle. This means that the bonding within the ring cannot be accomplished through the desirable "head on" overlap of hybrid orbitals from each C atom. As a result, the overlap of the hybrid orbitals in cyclopropane is less effective than that in the more normal, noncyclic propane molecule, and this makes the C–C bonds in cyclopropane comparatively weaker than those in the noncyclic molecule. We can also say that there is a severe "ring strain" in the molecule.

10.95 Boron typically forms substances such as BF_3, which have triangular geometries. Also, the valence shell of boron contains only three electrons. Since these three electrons of boron could half–fill only three hybrid orbitals, it is likely that only three hybrids should be formed. This leads to sp^2 hybridization.

10.96 (a) PF_3 is a pyramidal molecule and uses sp^3 hybrid orbitals. The expected bond angle is 109.5°.
(b) Using the unhybidized *p* orbitals, we would anticipate a bond angle of 90°.
(c) The observed bond angle is almost exactly the average of the bond angles listed in parts (a) and (b) above. So, neither hybid orbitals nor unhybridized atomic orbitals explain the observed bond angle.

10.97 Whereas the arrangement around an sp^2 hybridized atom may have 120° bond angles (which well accommodates a six–membered ring), the geometrical arrangement around an atom that is *sp* hybridized must be linear, as in $C - C \equiv C - C$ systems.

10.98 (a) The C–C single bonds are formed from head–to–head overlap of C atom sp^2 hybrids. This leaves one unhybridized atomic *p* orbital on each carbon atom, and each such atomic orbital is oriented perpendicular to the plane of the molecule.

(b) Sideways or π type overlap is expected between the first and the second carbon atoms, as well as between the third and the fourth carbon atoms. However, since all of these atomic *p* orbitals are properly aligned, there can be continuous π type overlap between all four carbon atoms.

(c) The situation described in part (b) is delocalized. We expect completely delocalized π type bonding among the carbon atoms.

(d) The bond is shorter because of the extra stability associated with the delocalization energy.

10.99 This molecule has a tetrahedral geometry and is polar. The polarity results because of the electronegativity difference between the fluorine atoms and the chlorine atoms.

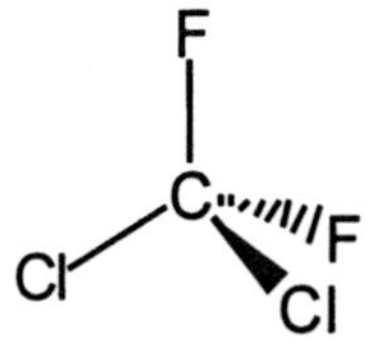

F
C
Cl F
Cl

10.100 The arrangement of the atoms is trigonal bipyramidal.

Recall that the bond angle between equatorial atoms is 120°. The bond angle from the equatorial position to the axial position is 90°. Due to the smaller bond angles, the atoms in the axial positions create more repulsions. The structure with the least amount of total repulsion is preferred; the statement implies that the more electronegative atoms create less repulsion, therefore the more electronegative atom should be placed in the axial position. Since fluorine is more electronegatove then chlorine, the F atoms will be in the axial positions and the Cl atoms will be in the equatorial positions. The molecule is non–polar.

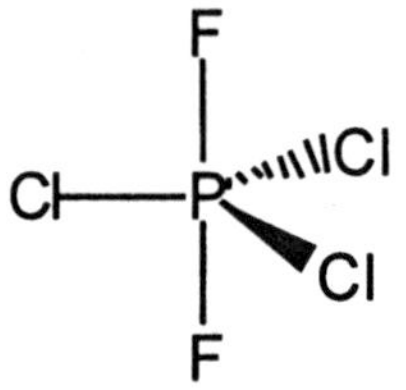

F
Cl—P Cl
Cl
F

10.101 The lone pair of electrons repels the Te–F bonding pairs, causing the F–Te–F angles to be smaller than in the ideal trigonal bipyramid.

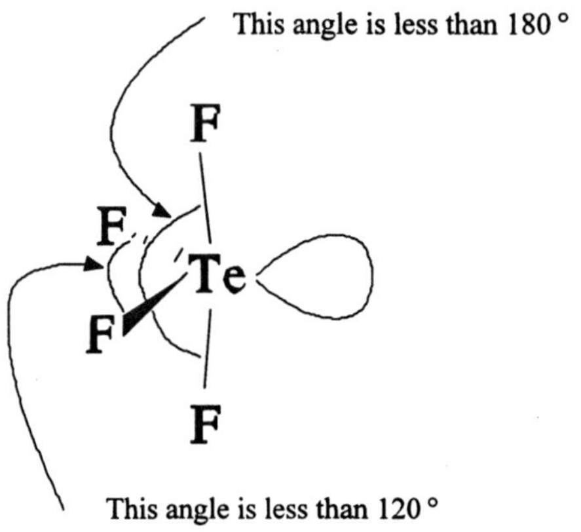

This angle is less than 180 °

F
F
Te
F
F

This angle is less than 120 °

In BrF_4^-, the two lone pairs are located as far as possible from one another, giving a square geometry to the ion:

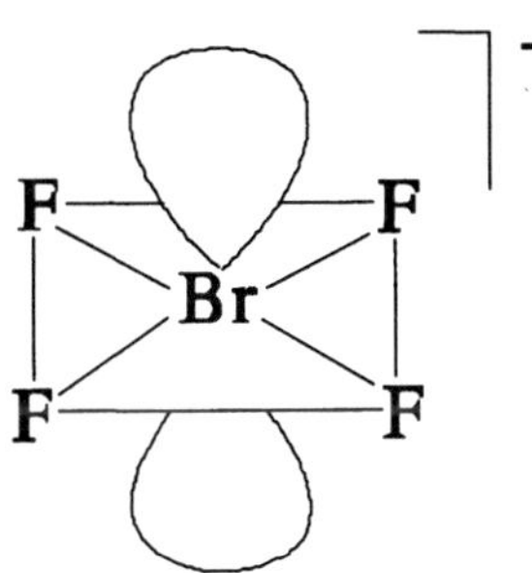

10.102 The double bonds are predicted to be between S and O atoms. Hence, the Cl–S–Cl angle diminishes under the influence of the S=O double bonds.

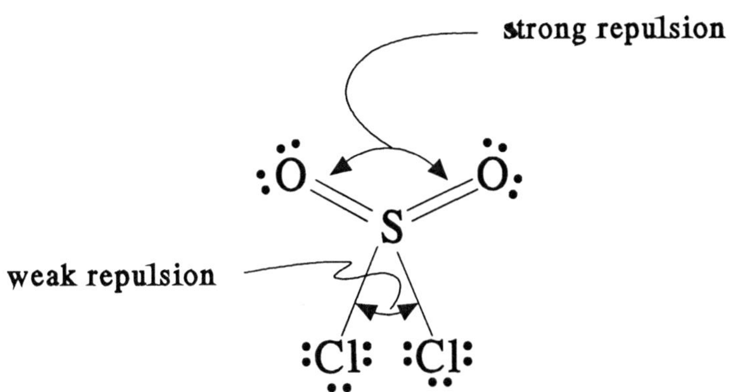

10.103 The polarity in the N–H bond, as measured by the difference in electronegativities between N(3.1) and H(2.1) adds to the polarity created by the sp^3 hybridized lone pair on the N atom. However, the higher electronegativity of fluorine (4.1) causes the N–F bond polarity to oppose the polarity associated with the hybridized lone pair on the N atom. Thus in the first case, a net dipole exists in the molecule, whereas in the second case, the polarity of the N–F bonds cancel the polarity of the hybrid lone pair.

10.104 This is a π bond, since overlap is *side to side* rather than the *end to end.* Also, consider that no bond rotation is possible here without breaking the bond since overlap occurs both above and below the bond axis.

10.105 Only the p_x orbital can form a π bond with d_{xz}, if the internuclear axis is the z axis.

10.106 (a) The O-O-N bond angle is about 109.5° and the O-N-O bond angle is around 120°.
(b) sp^2
(c) The nitrate ion benefits from three stable resonance structures (or electron delocalization) which stabilize it. The peroxynitrite ion does not.

10.107 (a) 109.5°
(b) 120°
(c) sp^3

Practice Exercises

11.1 $\# \text{ torr} = 887 \text{ mbar}\left(\frac{1 \text{ bar}}{1000 \text{ mbar}}\right)\left(\frac{0.9868 \text{ atm}}{1 \text{ bar}}\right)\left(\frac{760 \text{ torr}}{1 \text{ atm}}\right) = 665 \text{ torr}$

11.2 Assuming the pressure of the atmosphere in the room is 756 torr, the pressure in the container would be 756 – 87 = 669 torr.

11.3 $h_B = h_A \times \frac{d_A}{d_B}$

$$h_B = 760 \text{ mm} \times \frac{13.6 \text{ g/mL}}{1.00 \text{ g/mL}} = 10,300 \text{ mm}$$

$$\# \text{ ft} = 10,300 \text{ mm}\left(\frac{1 \text{ cm}}{10 \text{ mm}}\right)\left(\frac{1 \text{ inch}}{2.54 \text{ cm}}\right)\left(\frac{1 \text{ ft}}{12 \text{ inches}}\right) = 33.8 \text{ ft}$$

11.4 Since volume is to decrease, pressure must increase, and we multiply the starting pressure by a volume ratio that is larger than one. Also, since $P_1V_1 = P_2V_2$, we can solve for P_2:

$$P_2 = \frac{P_1V_1}{V_2} = \frac{(740 \text{ torr})(880 \text{ mL})}{(870 \text{ mL})} = 750 \text{ torr}$$

11.5 In general the combined gas law equation is: $\frac{P_1V_1}{T_1} = \frac{P_2V_2}{T_2}$, and in particular, for this problem, we have:

$$P_2 = \frac{P_1V_1T_2}{T_1V_2} = \frac{(745 \text{ torr})(950 \text{ m}^3)(333.2 \text{ K})}{(1150 \text{ m}^3)(298.2 \text{ K})} = 688 \text{ torr}$$

11.6 $n = \frac{PV}{RT} = \frac{(57.8 \text{ atm})(12.0 \text{ L})}{\left(0.0821 \frac{\text{L atm}}{\text{mol K}}\right)(298 \text{ K})} = 28.3 \text{ moles gas}$

28.3 mol Ar (39.95 g Ar/mol) = 1,130 g Ar = 1.13 kg Ar

11.7 Since PV = nRT, then n = PV/RT

$$n = \frac{PV}{RT} = \frac{(685 \text{ torr})\left(\frac{1 \text{ atm}}{760 \text{ torr}}\right)(0.300 \text{ L})}{\left(0.0821 \frac{\text{L atm}}{\text{mol K}}\right)(300.2 \text{ K})} = 0.0110 \text{ moles gas}$$

$$\text{molar mass} = \frac{1.45 \text{ g}}{0.0110 \text{ mol}} = 132 \text{ g mol}^{-1}$$

The gas must be Xenon.

11.8 d = m/V

Taking 1.00 mol SO2:

m = 64.1 g

$$V = \frac{nRT}{P} = \frac{(1.00\text{ mol})(0.0821 \frac{\text{L atm}}{\text{mol K}})(253.15\text{ K})}{\left(96.5\text{ kPa}\frac{1\text{ atm}}{101.325\text{ kPa}}\right)} = 21.8\text{ L} = 21{,}800\text{ mL}$$

$$\text{density} = \frac{64.1\text{ g}}{21{,}800\text{ mL}} = 2.94\times10^{-3}\text{ g/mL}$$

11.9 In general PV = nRT, where n = mass × formula mass. Thus

$$PV = \frac{\text{mass}}{\text{formula mass}}RT$$

We can rearrange this equation to get;

$$\text{formula mass} = \frac{(\text{mass/V})RT}{P} = \frac{dRT}{P}$$

$$\text{formula mass} = \frac{(5.60\text{ g L}^{-1})(0.0821 \frac{\text{L atm}}{\text{mol K}})(295.2\text{ K})}{(750\text{ torr})\left(\frac{1\text{ atm}}{760\text{ torr}}\right)} = 138\text{ g mol}^{-1}$$

The empirical mass is 69 g mol^{-1}. The ratio of the molecular mass to the empirical mass is 138 g mol^{-1}/69 g mol^{-1} = 2. Therefore, the molecular formula is 2 times the empirical formula, i.e., P_2F_4.

11.10 We can determine the pressure due to the oxygen since $P_{total} = P_{N_2} + P_{O_2}$.
$P_{O_2} = P_{total} - P_{N_2}$ = 30.0 atm – 15.0 atm = 15.0 atm. We can now use the ideal gas law to determine the number of moles of O_2:

$$n = \frac{PV}{RT} = \frac{(15.0\text{ atm})(5.00\text{ L})}{\left(0.0821\frac{\text{L atm}}{\text{mol K}}\right)(298\text{ K})} = 3.06\text{ moles }O_2$$

$$\#\text{ g }O_2 = (3.06\text{ moles }O_2)\left(\frac{32.0\text{ g }O_2}{1\text{ mol }O_2}\right) = 98.1\text{ g }O_2$$

11.11 First we find the partial pressure of nitrogen, using the vapor pressure of water at 15 °C:

$$P_{N_2} = P_{total} - P_{water} = 745\text{ torr} - 12.79\text{ torr} = 732\text{ torr.}$$

To calculate the volume of the nitrogen we can use the combined gas law

$$\frac{P_1V_1}{T_1} = \frac{P_2V_2}{T_2}$$

For this problem,

$$V_2 = \frac{P_1V_1T_2}{P_2T_1} = \frac{(732\text{ torr})(310\text{ mL})(273\text{ K})}{(760\text{ torr})(288\text{ K})} = 283\text{ mL}$$

11.12 The mole fraction is defined in Equation 11.8:

$$X_{O_2} = \frac{P_{O_2}}{P_{total}} = \frac{116\text{ torr}}{760\text{ torr}} = 0.153\text{ or }15.3\%$$

11.13 When gases are held at the same temperature and pressure, and dispensed in this fashion during chemical reactions, then they react in a ratio of volumes that is equal to the ratio of the coefficients (moles) in the balanced chemical equation for the given reaction. We can, therefore, directly use the stoichiometry of the balanced chemical equation to determine the combining ratio of the gas volumes:

$$\#\ \text{L } O_2 = (4.50 \text{ L } CH_4)\left(\frac{2 \text{ volume } O_2}{1 \text{ volume } CH_4}\right) = 9.00 \text{ L } O_2$$

11.14 First, we determine the number of moles of NO using the ideal gas law:

$$n = \frac{PV}{RT} = \frac{[720 \text{ torr}(1 \text{atm}/760 \text{ torr})][0.180 \text{ L } CH_4]}{(0.0821 \frac{\text{L atm}}{\text{mol K}})(318 \text{ K})} = 0.00653 \text{ mol NO}$$

Now we calculate the number of moles of O_2 which will be consumed, using the mole ratio from the balanced equation:

$$\#\ \text{mol } O_2 = 0.00653 \text{ mol NO}\frac{1 \text{ mol } O_2}{2 \text{ mol NO}} = 0.003265 \text{ mol } O_2$$

Finally, we use the ideal gas law to determine the volume of O_2 :

$$V = \frac{nRT}{P} = \frac{(0.003265 \text{ mol } O_2)(0.0821 \frac{\text{L atm}}{\text{mol K}})(293 \text{ K})}{755 \text{ torr}(1 \text{atm}/760 \text{ torr})} = 0.0791 \text{ L } O_2 = 79.1 \text{ mL } O_2$$

11.15 First lets determine the number of moles of CO_2 that are produced:

$$n = \frac{PV}{RT} = \frac{(738 \text{ torr})\left(\frac{1 \text{ atm}}{760 \text{ torr}}\right)(0.250 \text{ L})}{\left(0.0821 \frac{\text{L atm}}{\text{mol K}}\right)(296 \text{ K})} = 9.99 \times 10^{-3} \text{ moles } CO_2$$

The stoichiometry of the reaction indicates that one mole of $Na_2CO_3(s)$ will produce one mole $CO_2(g)$, so we will need to use 9.99×10^{-3} mol of $Na_2CO_3(s)$.

$$\#\ \text{g } Na_2CO_3 = (9.99 \times 10^{-3} \text{ mol } Na_2CO_3)\left(\frac{106 \text{ g } Na_2CO_3}{1 \text{ mol } Na_2CO_3}\right) = 1.06 \text{ g } Na_2CO_3$$

11.16 Use Equation 11.10;

$$\frac{\text{effusion rate (HX)}}{\text{effusion rate (HCl)}} = \sqrt{\frac{M_{HCl}}{M_{HX}}}$$

$$M_{HX} = M_{HCl} \times \left(\frac{\text{effusion rate (HX)}}{\text{effusion rate (HCl)}}\right)^2 = 36.46 \text{ g mol}^{-1} \times (1.88)^2 = 128.9 \text{ g mol}^{-1}$$

The unknown gas must be HI.

Thinking It Through

T11.1 Since only pressure and volume change, this is a Boyle's Law calculation. We need only the initial volume $V_1 = 175$ mL and the initial and final pressures, $P_1 = 734$ torr and $P_2 = 2.5$ atm. First convert P_1 in torr to atm, the calculate V_2.

$$P_1 = 734 \text{ torr}\left(\frac{1 \text{ atm}}{760 \text{ torr}}\right) = 0.966 \text{ atm}$$

$$V_2 = \frac{P_1V_1}{P_2} = \frac{(0.966 \text{ atm})(175 \text{ mL})}{(2.5 \text{ atm})}$$

T11.2 Since pressure does not change, this is a Charles' Law calculation. We need only the initial temperature, $T_1 = 70 + 273 = 343$ K, and the initial and final volumes, $V_1 = 550$ mL and $V_2 = 500$ mL. The temperature should decrease, because volume has decreased.

$$T_2 = \frac{T_1V_2}{V_1} = \frac{(343 \text{ K})(500 \text{ mL})}{(550 \text{ mL})}$$

T11.3 Since we are given both an initial and a final temperature, $T_1 = 25 + 273 = 298$ K and $T_2 = 50 + 273 = 323$ K, and an initial pressure, $P_1 = 1$ atm, we have all of the information that is needed to use Gay–Lussac's Law. Since the temperature has been increased, we expect the pressure also to increase:

$$P_2 = \frac{P_1T_2}{T_1} = \frac{(1 \text{ atm})(323 \text{ K})}{(298 \text{ K})}$$

T11.4 Start by determining the pressure of the N_2 gas: we know V = 1.50 L and T = 24 + 273 = 297 K. We can determine the number of moles of N_2 from the mass and the molar mass;

$$n_{N_2} = 1.68 \text{ g} \div 28.0 \text{ g/mol} = 0.0600 \text{ mol}$$

Using the ideal gas law we can determine that $P_{N_2} = 0.975$ atm.

The atmospheric pressure is determined by adding the pressure of the nitrogen to the height difference of the manometer, i.e, $P_{atm} = P_{N_2} + (18 \text{ mm Hg})\left(\frac{1 \text{ torr}}{1 \text{ mm Hg}}\right)\left(\frac{1 \text{ atm}}{760 \text{ torr}}\right) = 0.999 \text{ atm} = 759 \text{ torr}$

T11.5 The total pressure is equal to the pressure of the hydrogen plus the pressure of the water. Determine the vapor pressure of the water at 20 °C and subtract from total pressure to determine the pressure of the hydrogen. Then, use the ideal gas law to solve for the number of moles of hydrogen. From the balanced equation the number of moles of magnesium is related to the number of moles of hydrogen. Finally, convert the number of moles to a mass.

T11.6 Recall that $n = \frac{\text{mass}}{\text{molar mass}}$ so the ideal gas law may be written as $PV = \frac{mRT}{MM}$ and $MM = \frac{mRT}{PV}$. Be sure to convert T, V and P to units consistent with your value of R.

T11.7 First determine the pressure of the gas sample by subtracting the vapor pressure of water at 25 °C. We can write the ideal gas law as $PV = \frac{mRT}{MM}$, and we can determine the molecular mass of the gas from the first part of the question. Density is mass/volume and we can rearrange the equation above to get $D = \frac{m}{V} = \frac{P(MM)}{RT}$. Be sure to use the new conditions of P and T to calculate D.

T11.8 First, measure the volume of liquid water and convert this to a number of moles. Second, determine the vapor pressure of water at 25 °C and convert this to a number of moles $\frac{P_{H_2O}V}{RT}$. All of the water resulted from burning the hydrogen. The moles of hydrogen = moles of liquid and vapor water. Finally, convert the moles hydrogen to grams hydrogen.

T11.9 A correction for the vapor pressure of water is not necessary, since this pressure contributes equally to both the initial and the final pressure readings. In other words, this pressure remains constant throughout this constant temperature experiment. First, calculate the partial pressure of oxygen, using Dalton's Law.

$$P_{O_2} = (907 - 740 \text{ torr})\left(\frac{1 \text{ atm}}{760 \text{ torr}}\right) = 0.220 \text{ atm}$$

Before we use the ideal gas law to determine the value of n, we must realize that the net volume available to the oxygen is the initial volume less the volume occupied by the peroxide solution. This is very nearly equal to 20.0 mL, since the density of dilute aqueous solutions is 1.0 g/ml:

$$n = \frac{PV}{RT} = \frac{(0.220 \text{ atm})(1.00\text{L} - 0.0200 \text{ L})}{(0.0821 \frac{\text{L atm}}{\text{mol K}})(298 \text{ K})} = 0.00881 \text{ mol}$$

From the stoichiometry, twice this number of moles of hydrogen peroxide (0.0176 mol) was present. This number of moles of hydrogen peroxide may then be converted to grams, by multiplying by its molar mass.

T11.10 First, determine the gas pressure by subtracting the vapor pressure of water at 35 °C. Using Boyle's law we can write $P_1V_1 = P_2V_2$ and rearrange to get $P_1 = P_2\frac{V_2}{V_1}$ and $\frac{V_2}{V_1} = \frac{1}{2}$, and we determined P_2 above. Solve for P_1.

T11.11 The mass of N in the sample may be determined by using the ideal gas law $n = \frac{PV}{RT}$, and then multiplying this by the molar mass of N. We know the amount of C and H in the sample so we can subtract the masses of C, H and N from the total mass of the sample to get the mass of O. Convert all masses to a number of moles for each atom. Determine the simplest mole ratio of the four atoms.

T11.12 The substance with the larger molecular mass will effuse through pinhole leaks more slowly. Therefore we choose SF_6.

T11.13 As the weather balloon rises, the pressure on the outside of the balloon decreases and the external temperature decreases. If the only external change were a decrease in pressure, the constant internal pressure would cause the balloon to expand. If the only external change is a decrease in temperature, the particles inside the balloon would move more slowly and the balloon would decrease in size. Since we know that the balloon expands, the decreased external pressure has a greater impact than the decreased external temperature.

T11.14 The best representation is Figure D. Figures A, B and E were rejected because the probability that all the molecules are in a specific volume region or are separated by type of molecule is extremely low. Figure C was rejected because H_2 and O_2 do not readily react to produce water molecules.

T11.15 (a) The area under the curve represents the total number of particles in the system.

(b)

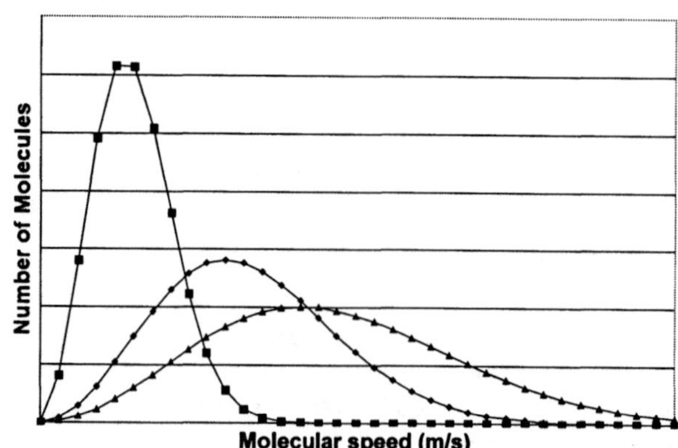

The lowest temperature curve is on the left and the highest temperature is to the right. Notice that the highest temperature curve has a maximum at higher energy and has a flatter curve.

T11.16 (a) We would expect to see some brown color forming at the stopcock. The NO and O_2 will react here initially and the color will spread throughout the two bulbs.

(b) The concept of diffusion is demonstrated by the spreading of the color between the flasks.

(c)

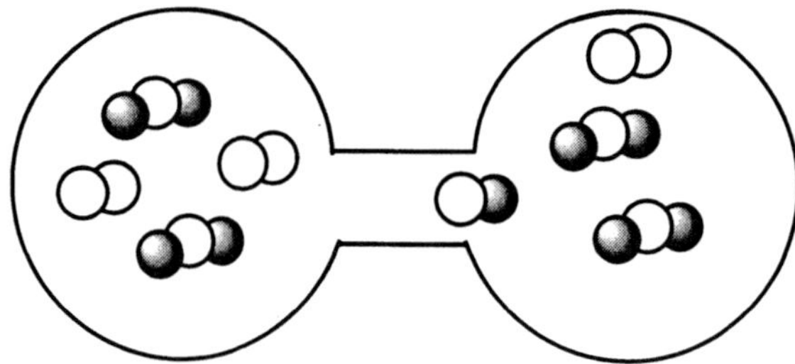

Review Questions

11.1 Pressure is defined as a force per unit area. P=F/A, as the area increases, the force also increases.

11.2 The reason it hurts more to be jabbed by a point of a pencil rather than the eraser, even though the force is the same, is because the area of the point is smaller than the area of the eraser, and therefore, the pressure is higher.

11.3 (a) $\frac{1 \text{ atm}}{101.325 \text{ kPa}}$ and $\frac{101.325 \text{ kPa}}{1 \text{ atm}}$

(b) $\frac{1 \text{ torr}}{1 \text{ mm Hg}}$ and $\frac{1 \text{ mm Hg}}{1 \text{ torr}}$

(c) $\frac{760 \text{ torr}}{1 \text{ atm}}$ and $\frac{1 \text{ atm}}{760 \text{ torr}}$

(d) $\frac{760 \text{ torr}}{101{,}325 \text{ Pa}}$ and $\frac{101{,}325 \text{ Pa}}{760 \text{ torr}}$

(e) $\frac{1.013 \text{ bar}}{101{,}325 \text{ Pa}}$ and $\frac{101{,}325 \text{ Pa}}{1.013 \text{ bar}}$

(f) $\frac{1 \text{ bar}}{0.9868 \text{ atm}}$ and $\frac{0.9868 \text{ atm}}{1 \text{ bar}}$

11.4 Since the density of water is approximately 13 times smaller than that of mercury, a barometer constructed with water as the moveable liquid would have to be some 13 times longer than one constructed using mercury. Also, the vapor pressure of water is large enough that the closed end of the barometer may fill with sufficient water vapor so as to affect atmospheric pressure readings. In fact, the measurement of atmospheric pressure at normal temperatures would be about 18 torr too low, due to the presence of water vapor in the closed end of the barometer.

11.5 A closed–end manometer reads pressure without the need to correct for atmospheric pressure.

11.6 Water has a relatively high vapor pressure and the water vapor would fill the closed end of the barometer enough to affect the pressure readings.

11.7 (a) Temperature–Volume Law: The volume of a given mass of a gas is directly proportional to the Kelvin temperature, provided the pressure is held constant:
$V \propto T$ or $V_1/T_1 = V_2/T_2$, at constant P. This is Charles' Law.

(b) Temperature–Pressure Law: The pressure of a gas is directly proportional to the Kelvin temperature, provided the volume is held constant:
$P \propto T$ or $P_1/T_1 = P_2/T_2$, at constant V. This is Gay-Lussac's Law.

(c) Pressure–Volume Law: The volume of a given mass of a gas is inversely proportional to the pressure, provided the temperature is held constant:
$V \propto 1/P$ or $P_1V_1 = P_2V_2$, at constant T. This is Boyle's Law.

(d) Combined Gas Law: The pressure and volume of a gas are directly proportional to the Kelvin temperature, provided the number of particles is held constant.
$PV \propto T$ or $P_1V_1/T_1 = P_2V_2/T_2$, at constant n.

11.8 (a) number of moles and temperature
(b) number of moles and pressure
(c) number of moles and volume
(d) number of moles

11.9 An ideal gas obeys the gas laws over all pressures and temperatures. A real gas behaves most like an ideal gas at low pressures and high temperatures.

11.10 $PV = nRT$; $R = 0.082057 \text{ L atm mol}^{-1} \text{ K}^{-1}$

11.11 Molecule: the smallest unit of a gas, except those gases such as argon that are monatomic. The oxygen molecule is O_2.

Mole: Avogadro's number of gas molecules (or of gaseous atoms if the gas is monatomic). For oxygen, this is 6.02×10^{23} molecules of O_2.

Molar mass: the mass in grams of one mole of the substance. For oxygen, this is 32.0 g/mole.

Molar volume: the volume occupied by one mole of a gas at STP. For oxygen and other ideal gasses, this is 22.4 L.

11.12 Avagadro's number; 6.022×10^{23} molecules of H_2

11.13 $P_{total} = P_a + P_b + P_c + \cdots$

11.14 Mole fraction is the ratio of the number of moles of one component of a mixture to the total number of moles of all components.

11.15 Drawing B

A: 0.500 atm C: 0.667 atm
A: 0.500 atm C: 0.333 atm

11.16 Diffusion is the spontaneous intermingling of one substance with another while effusion is the movement of a gas through a very tiny opening into a region of lower pressure.

$$\frac{\text{effusion rate (A)}}{\text{effusion rate (B)}} = \sqrt{\frac{d_B}{d_A}} = \sqrt{\frac{M_B}{M_A}}$$

11.17 $M = \frac{m}{n}$ M = molecular mass of a gas m = mass n = number of moles

$d = \frac{m}{V}$ d = density m = mass V = volume

rearrange: $m = d \times V$

$M = \frac{d \times V}{n}$ if V and n are constant, then $M \propto d$

11.18 A gas consists of hard, supersmall or volumeless particles in random motion, and the particles neither attract nor repel one another.

11.19 (a) $T \propto$ average kinetic energy of the particles
(b) $T \propto PV$

11.20 The randomness of the motion of the particles and their super-small nature guarantees mixing.

11.21 The temperature and pressure will decrease.

11.22 The increase in temperature requires an increase in kinetic energy. This can happen only if the gas velocities increase. Higher velocities cause the gas particles to strike the walls of the container with more force, and this in turn causes the container to expand if a constant pressure is to be maintained.

11.23 The increase in temperature causes an increase in the force with which the gas particles strike the container walls. If the container cannot expand, an increase in pressure must result.

11.24 The minimum temperature corresponds to zero kinetic energy, which is accomplished only when velocity is zero. In other words, the molecules have ceased all movement.

11.25 The root mean square speed represents the speed of a molecule that would have the average kinetic energy. The kinetic energies of the molecules of a gas are proportional to the root mean square speed of the molecules.

11.26 (c) NH_3 will have the largest $\overline{v_{rms}}$ since it has the lowest molecular mass.

11.27 At any given temperature, all of the gas molecules possess the same kinetic energy. A heavy molecule must therefore have a lower velocity than a light molecule, if their kinetic energies are to be the same. The molecules with lower formula mass travel more quickly, chances for their effusion through a small hole are increased.

11.28 (a) As the pressure of a gas increases, the rate of effusion should increase since the molecules will hit the walls of the container more frequently and with greater force. If the molecules hit more frequently, they are more likely to go through the small openings in the walls of the container.
(b) As the temperature of the gas increases, the rate of effusion will increase since temperature is proportional to kinetic energy which is dependent on the velocity of the particles. The faster the particles move, the more likely they are to hit the walls and pass through the small openings.

11.29 The fact that a simple sum of partial pressures gives the total pressure of a collection of different gases indicates that the individual gas molecules do not physically interact with one another, either to attract or to repel each other. This is one of the postulates of the Kinetic Theory of Gases.

11.30 It is not true that the gas particles occupy no volume themselves, apart from the volume between the gas particles. Also, it is not true that the gas particles exert no force on one another. In other words, real molecules occupy space and attract or repel one another. Because of short-range interactions, it is also not true that particles travel always in straight paths.

11.31 The van der Waal's *b* constant is related to the size of the atom or molecule in question. A larger value for this constant implies that the particles of gas A are larger than the particles of gas B.

11.32 A small value for the constant *a* suggests that the gas molecules have weak forces of attraction among themselves.

11.33 (b) has a larger value of the van der Waals constant *b*, since it is a larger molecule.

11.34 Under the same conditions of *T* and *V*, the pressure of a real gas is less than the pressure of an ideal gas because real gases do not have perfectly elastic collisions and may clump together and stick to the walls of the container, thus decreasing the number of collisions the gas makes. The volume of a real gas is greater than the volume of an ideal gas because the atoms and molecules take up space.

11.35 The helium atoms are moving faster than the argon atoms because they have less mass and the rate of effusion is inversely proportional to mass.

Review Problems

11.36
(a) # torr = (1.26 atm)(760 torr / 1 atm) = 958 torr
(b) # atm = (740 torr)(1 atm / 760 torr) = 0.974 atm
(c) 738 torr = 738 mm Hg
(d) # torr = $(1.45 \times 10^3$ Pa)(760 torr / 1.01325×10^5 Pa) = 10.9 torr

11.37
(a) # torr = (0.625 atm)(760 torr / 1 atm) = 475 torr
(b) # atm = (825 torr)(1 atm / 760 torr) = 1.09 atm
(c) 62 mm Hg = 62 torr
(d) # bar = 1.22 kPa(1.013 bar / 101.32 kPa) = 0.0122 bar

11.38
(a) # torr = (0.329 atm)(760 torr / 1 atm) = 250 torr
(b) # torr = (0.460 atm)(760 torr / 1 atm) = 350 torr

11.39
(a) # atm = (595 torr)(1.00 atm / 760 torr) = 0.783 atm
(b) # atm = (160 torr)(1.00 atm / 760 torr) = 0.211 atm
(c) # atm = (0.300 torr)(1.00 atm / 760 torr) = 3.95×10^{-4} atm

11.40 765 torr – 720 torr = 45 torr 45 torr = 45 mm Hg

$$\# \text{ cm Hg} = 45 \text{ mm Hg}\left(\frac{1 \text{ cm}}{10 \text{ mm}}\right) = 4.5 \text{ cm Hg}$$

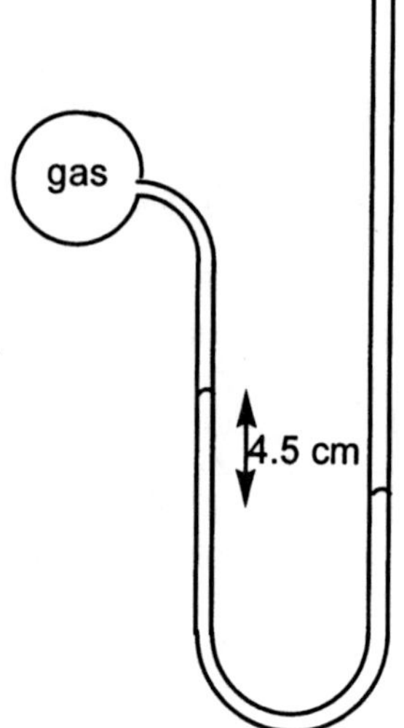

11.41 820 torr – 750 torr = 70 torr 70 torr = 70 mm Hg

$$\# \text{ cm Hg} = 70 \text{ mm Hg}\left(\frac{1 \text{ cm}}{10 \text{ mm}}\right) = 7.0 \text{ cm Hg}$$

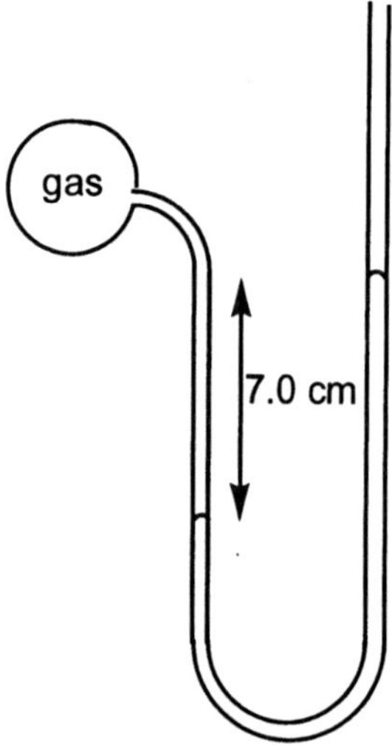

11.42 65 mm Hg = 65 torr 748 torr + 65 torr = 813 torr

11.43 82 mm Hg = 82 torr 752 torr – 82 torr = 670 torr

11.44 P_{gas} = P_{atm} + (height difference)
= 746 mm + 80.0 mm = 826 mm = 826 torr

The reaction produced gases. When the mercury level in the manometer arm nearest the bulb goes down by 4.00 cm (12.50 to 8.50) the mercury in the other arm goes up by 4.00 cm. Hence, the difference in the heights of the two arms is 8.00 cm = 80.0 mm.

11.45 P_{gas} = P_{atm} + (height difference)
= 741 mm + 46.0 mm = 787 mm = 787 torr

When the mercury level in the manometer arm nearest the bulb goes down by 2.30 cm (12.50 to 10.20) the mercury in the other arm goes up by 2.30 cm. Hence, the difference in the heights of the two arms is 4.60 cm = 46.0 mm. The temperature is immaterial in this problem since the temperature was constant for the two measurements. Since the pressure in the bulb increased, the reaction must have produced gases.

11.46 In a closed-end manometer the difference in height of the mercury levels in the two arms corresponds to the pressure of the gas.

$$\#\,\text{torr} = (12.5\text{ cm Hg})\left(\frac{10\text{ mm}}{1\text{ cm}}\right)\left(\frac{1\text{ torr}}{1\text{ mm Hg}}\right) = 125\text{ torr}$$

11.47 The closed-end manometer data indicates that the pressure inside the flask is 236 mm Hg. The open-end manometer data indicate that P_{atm} = 512 mm Hg + 236 mm Hg = 748 mm Hg.

11.48 First convert the pressure in torr to mm Hg:

$$\#\,\text{mm Hg} = (755\text{ torr})\left(\frac{1\text{ mm Hg}}{1\text{ torr}}\right) = 755\text{ mm Hg}$$

Then use Equation 11.3 to convert height of the mercury column (liquid A) to height of the fluid column with a density of 1.22 g/mL (liquid B).

$$h_B = h_A\left(\frac{d_A}{d_B}\right) = 755\text{ mm} \times \frac{13.6\text{ g mL}^{-1}}{1.22\text{ g mL}^{-1}} = 8.42 \times 10^3\text{ mm}$$

11.49 The most perfect vacuum pump can, at most, cause a column of mercury to rise to a height of that day's atmospheric pressure, since that is the maximum height of a column of mercury that the atmosphere can support. That is to say, a vacuum pump cannot induce a mercury column to rise any farther than what is forced up by atmospheric pressure. Correspondingly, the highest a pump could cause a column of water to rise will be only that amount of water that will be supported by atmospheric pressure, that is the weight of water that is equivalent to the weight of 760 mm of mercury. Since the density of mercury is 13.6 g/cm^3 whereas that of water is only 1.00 g/cm^3, the heights of the two columns are also related by the proportion 13.6 – to – 1.00. Thus, the height of an equivalent column of water would be: 760 mm Hg × 13.6 = 1.03 × 10^4 mm. This is next converted to a value in feet:

$$\#\text{ ft} = (1.03 \times 10^4\text{ mm})\left(\frac{1\text{ cm}}{10\text{ mm}}\right)\left(\frac{1\text{ in.}}{2.54\text{ cm}}\right)\left(\frac{1\text{ ft}}{12\text{ in.}}\right) = 33.9\text{ ft}$$

This means that the best conceivable vacuum pump (one capable of producing a perfect vacuum on the column of water being pulled up) could cause the water in the pipe attached to the pump to rise only 33.9 ft above the surface of the water in the pit. The water at the bottom of a 35 ft pit could not be removed by use of a vacuum pump alone.

11.50 Use Boyle's Law to solve for the second volume:

$$V_2 = \frac{P_1V_1}{P_2} = \frac{255\text{ mL}(725\text{ torr})}{365\text{ torr}} = 507\text{ mL}$$

11.51 $P_1V_1 = P_2V_2$

$V_2 = \dfrac{P_1V_1}{P_2}$ since the pump has a fixed diameter, the length of the tube is proportional to its volume

$$P_1l_1 = P_2l_2$$

$$l_2 = \frac{P_1l_1}{P_2} = \frac{1\text{ atm}(75.0\text{ cm})}{5.50\text{ atm}} = 13.6\text{ cm}$$

11.52 Use Charles's Law to solve the second volume:

$$V_2 = \frac{V_1T_2}{T_1} = \frac{3.86\ \text{L}\ (353\ \text{K})}{318\ \text{K}} = 4.28\ \text{L}$$

11.53 Use Charles's Law to solve for the second volume:

$$V_2 = \frac{V_1T_2}{T_1} = \frac{2.50\ \text{L}\ (258\ \text{K})}{295\ \text{K}} = 2.19\ \text{L}$$

11.54 Compare pressure change to temperature to solve for temperature change:

$$T_2 = \frac{P_2T_1}{P_1} = \frac{1700\ \text{torr}(558\ \text{K})}{850\ \text{torr}} = 1120\ \text{K} \qquad 1120\text{K} - 273\text{K} = 843\ °\text{C}$$

11.55 $P_2 = \frac{P_1T_2}{T_1} = \frac{45\ \text{lb in}^{-2}(316\ \text{K})}{283\ \text{K}} = 50.2\ \text{lb in}^{-2}$

11.56 In general the combined gas law equation is: $\frac{P_1V_1}{T_1} = \frac{P_2V_2}{T_2}$, and in particular, for this problem, we have:

$$P_2 = \frac{P_1V_1T_2}{T_1V_2} = \frac{(740\ \text{torr})(2.58\ \text{L})(348.2\ \text{K})}{(297.2\ \text{K})(2.81\ \text{L})} = 796\ \text{torr}$$

11.57 In general the combined gas law equation is: $\frac{P_1V_1}{T_1} = \frac{P_2V_2}{T_2}$, and in particular, for this problem, we have:

$$P_2 = \frac{P_1V_1T_2}{T_1V_2} = \frac{(0.985\ \text{atm})(648\ \text{mL})(336.2\ \text{K})}{(289.2\ \text{K})(689\ \text{mL})} = 1.08\ \text{atm}$$

11.58 In general the combined gas law equation is $\frac{P_1V_1}{T_1} = \frac{P_2V_2}{T_2}$, and in particular, for this problem, we have:

$$V_2 = \frac{P_1V_1T_2}{T_1P_2} = \frac{(745\ \text{torr})(2.68\ \text{L})(648.2\ \text{K})}{(297.2\ \text{K})(760\ \text{torr})} = 5.73\ \text{L}$$

11.59 In general the combined gas law equation is: $\frac{P_1V_1}{T_1} = \frac{P_2V_2}{T_2}$, and in particular, for this problem, we have:

$$V_2 = \frac{P_1V_1T_2}{T_1P_2} = \frac{(741\ \text{torr})(280\ \text{mL})(306.2\ \text{K})}{(291.2\ \text{K})(760\ \text{torr})} = 287\ \text{mL}$$

11.60 In general the combined gas law equation is: $\frac{P_1V_1}{T_1} = \frac{P_2V_2}{T_2}$, and in particular, for this problem, we have:

$$T_2 = \frac{P_2V_2T_1}{P_1V_1} = \frac{(373\ \text{torr})(9.45\ \text{L})(293.2\ \text{K})}{(761\ \text{torr})(6.18\ \text{L})} = 220\ \text{K} = -53.4°\text{C}$$

11.61 In general the combined gas law equation is: $\frac{P_1V_1}{T_1} = \frac{P_2V_2}{T_2}$, and in particular, for this problem, we have:

$$T_2 = \frac{P_2V_2T_1}{P_1V_1} = \frac{(2.00\ \text{atm})(220\ \text{mL})(298.2\ \text{K})}{(1.51\ \text{atm})(455\ \text{mL})} = 191\ \text{K} = -82.2°\text{C}$$

11.62 $R = \left(0.0821\ \frac{\text{L atm}}{\text{mol K}}\right)\left(\frac{1000\text{ mL}}{1\text{ L}}\right)\left(\frac{760\text{ torr}}{1\text{ atm}}\right) = 6.24 \times 10^4\ \frac{\text{mL torr}}{\text{mol K}}$

11.63 If PV = nRT, then R = PV/nT.

Let P = 1 atm = 101,325 Pa, T = 273 K, and n = 1.
Next, express the volume of the standard mole using the units m^3, instead of L, remembering that 22.4 L = 22,400 cm^3:

$$\#\,m^3 = 22{,}400\text{ cm}^3 \times \left(\frac{1\text{ m}}{100\text{ cm}}\right)^3 = 0.0224\text{ m}^3$$

$$R = \left(\frac{(101{,}325\text{ Pa})(0.0224\text{ m}^3)}{(1\text{ mole})(273\text{ K})}\right) = 8.31\text{ m}^3\text{ Pa mol}^{-1}\text{ K}^{-1}$$

11.64 $V = \frac{nRT}{P} = \frac{\left(0.136\text{g}\left(\frac{1\text{ mol}}{32.0\text{ g}}\right)\right)\left(0.0821\frac{\text{L atm}}{\text{mol K}}\right)(298\text{ K})}{\left(748\text{ torr}\left(\frac{1\text{ atm}}{760\text{ torr}}\right)\right)} = 0.106\text{ L}$

11.65 $V = \frac{nRT}{P} = \frac{\left(1.67\text{ g}\left(\frac{1\text{ mol}}{28.0\text{ g}}\right)\right)\left(0.0821\frac{\text{L atm}}{\text{mol K}}\right)(295\text{ K})}{\left(756\text{ torr}\left(\frac{1\text{ atm}}{760\text{ torr}}\right)\right)} = 1.45\text{ L}$

11.66 $P = \frac{nRT}{V} = \frac{\left(10.0\text{g}\left(\frac{1\text{ mol}}{32.0\text{ g}}\right)\right)\left(0.0821\frac{\text{L atm}}{\text{mol K}}\right)(300\text{ K})}{(2.50\text{ L})} = 3.08\text{ atm}\,\frac{760\text{ torr}}{1\text{ atm}} = 2340\text{ torr}$

11.67 $P = \frac{nRT}{V} = \frac{\left(12.0\text{ g}\left(\frac{1\text{ mol}}{18.0\text{ g}}\right)\right)\left(0.0821\frac{\text{L atm}}{\text{mol K}}\right)(381\text{ K})}{(3.60\text{ L})} = 5.80\text{ atm}$

11.68 $n = \frac{PV}{RT} = \frac{(0.821\text{ atm})(0.0265\text{ L})}{\left(0.0821\frac{\text{L atm}}{\text{mol K}}\right)(293\text{ K})} = (9.04\times10^{-4}\text{ mol})\left(\frac{44.0\text{ g}}{1\text{ mol}}\right) = 0.0398\text{ g}$

11.69 $n = \frac{PV}{RT} = \frac{\left((750\text{ torr})\left(\frac{1\text{ atm}}{760\text{ torr}}\right)\right)(0.250\text{ L})}{\left(0.0821\frac{\text{L atm}}{\text{mol K}}\right)(300\text{ K})} = (1.00\times10^{-2}\text{ mol})\left(\frac{16\text{ g}}{1\text{ mol}}\right) = 0.160\text{ g}$

11.70 Since PV = nRT, then;

$$P = \frac{nRT}{V} = \frac{(4.18\text{ mol})\left(0.0821\frac{\text{L atm}}{\text{mol K}}\right)(291.2\text{ K})}{(24.0\text{ L})} = 4.16\text{ atm}$$

11.71 Since PV = nRT, then;

$$P = \frac{nRT}{V} = \frac{(10.0\text{ mol})(0.0821\,\frac{\text{L atm}}{\text{mol K}})(298.2\text{ K})}{(1.60\text{ L})} = 153\text{ atm}$$

11.72 (a) density C_2H_6 $= \left(\frac{30.1\text{ g } C_2H_6}{1\text{ mol } C_2H_6}\right)\left(\frac{1\text{ mol}}{22.4\text{ L}}\right) = 1.34\text{ g L}^{-1}$

(b) density N_2 $= \left(\frac{28.0\text{ g } N_2}{1\text{ mol } N_2}\right)\left(\frac{1\text{ mol}}{22.4\text{ L}}\right) = 1.25\text{ g L}^{-1}$

(c) density Cl_2 $= \left(\frac{70.9\text{ g } Cl_2}{1\text{ mol } Cl_2}\right)\left(\frac{1\text{ mol}}{22.4\text{ L}}\right) = 3.17\text{ g L}^{-1}$

(d) density Ar $= \left(\frac{39.9\text{ g Ar}}{1\text{ mol Ar}}\right)\left(\frac{1\text{ mol}}{22.4\text{ L}}\right) = 1.78\text{ g L}^{-1}$

11.73 (a) density Ne $= \left(\frac{20.2\text{ g Ne}}{1\text{ mol Ne}}\right)\left(\frac{1\text{ mol}}{22.4\text{ L}}\right) = 0.902\text{ g L}^{-1}$

(b) density O_2 $= \left(\frac{32.0\text{ g } O_2}{1\text{ mol } O_2}\right)\left(\frac{1\text{ mol}}{22.4\text{ L}}\right) = 1.43\text{ g L}^{-1}$

(c) density CH_4 $= \left(\frac{16.0\text{ g } CH_4}{1\text{ mol } CH_4}\right)\left(\frac{1\text{ mol}}{22.4\text{ L}}\right) = 0.714\text{ g L}^{-1}$

(d) density CF_4 $= \left(\frac{88.0\text{ g } CF_4}{1\text{ mol } CF_4}\right)\left(\frac{1\text{ mol}}{22.4\text{ L}}\right) = 3.93\text{ g L}^{-1}$

11.74 In general PV = nRT, where n = mass ÷ formula mass. Thus

$$PV = \frac{\text{mass}}{(\text{formula mass})}RT$$

and we arrive at the formula for the density (mass divided by volume) of a gas:

$$D = \frac{P \times (\text{formula mass})}{RT}$$

$$D = \frac{(742\text{ torr})(\frac{1\text{ atm}}{760\text{ torr}})(32.0\text{ g/mol})}{(0.0821\,\frac{\text{L atm}}{\text{mol K}})(297.2\text{ K})}$$

$$D = 1.28\text{ g/L for } O_2$$

11.75 In general PV = nRT, where n = mass ÷ formula mass. Thus

$$PV = \frac{\text{mass}}{(\text{formula mass})}RT$$

and we arrive at the formula for the density (mass divided by volume) of a gas:

$$d = \frac{P \times (\text{formula mass})}{RT}$$

$$d = \frac{(748.0\text{ torr})(\frac{1\text{ atm}}{760\text{ torr}})(39.95\text{ g/mol})}{(0.0821\,\frac{\text{L atm}}{\text{mol K}})(293.80\text{ K})}$$

$$d = 1.63\text{ g/L for Ar}$$

11.76 First determine the number of moles:

$$n = \frac{PV}{RT} = \frac{(10.0\ \text{torr})\left(\frac{1\ \text{atm}}{760\ \text{torr}}\right)(255\ \text{mL})\left(\frac{1\ \text{L}}{1000\ \text{mL}}\right)}{\left(0.0821\ \frac{\text{L atm}}{\text{mol K}}\right)(298.2\ \text{K})} = 1.37 \times 10^{-4}\ \text{mol}$$

Now calculate the molecular mass:

$$\text{molecular mass} = \frac{\text{mass}}{\#\ \text{of moles}} = \frac{(12.1\ \text{mg})\left(\frac{1\ \text{g}}{1000\ \text{mg}}\right)}{1.37 \times 10^{-4}\ \text{mol}} = 88.3\ \text{g/mol}$$

11.77

$$\text{formula mass} = \frac{dRT}{P} = \frac{(\text{mass})RT}{PV}$$

$$\text{formula mass} = \frac{(6.3 \times 10^{-3}\ \text{g})\left(0.0821\ \frac{\text{L atm}}{\text{mol K}}\right)(298.2\ \text{K})}{(11\ \text{torr})\left(\frac{1\ \text{atm}}{760\ \text{torr}}\right)(385\ \text{mL})\left(\frac{1\ \text{L}}{1000\ \text{mL}}\right)}$$

$$\text{formula mass} = 28\ \text{g mol}^{-1}$$

(b) The formula weights of the boron hydrides are:
BH_3, 13.8
B_2H_6, 27.7
B_4H_{10}, 53.3
And we conclude that the sample must have been B_2H_6.

11.78 $$\text{molecular mass} = \frac{dRT}{P} = \frac{\text{mass}\ RT}{PV} = \frac{(1.13\ \text{g/L})\left(0.0821\ \frac{\text{L atm}}{\text{mol K}}\right)(295\ \text{K})}{(0.995\ \text{atm})} = 27.5\ \text{g/mol}$$

11.79 Note: 0.08747 mg/mL = 0.08747 g/L

$$\text{molecular mass} = \frac{dRT}{P} = \frac{(0.08747\ \text{g/L})\left(0.0821\ \frac{\text{L atm}}{\text{mol K}}\right)(290.2\ \text{K})}{(760\ \text{torr})\left(\frac{1\ \text{atm}}{760\ \text{torr}}\right)} = 2.08\ \text{g/mol}$$

This gas is most likely H_2.

11.80 When gases are held at the same temperature and pressure, and dispensed in this fashion during chemical reactions, then they react in a ratio of volumes that is equal to the ratio of the coefficients (moles) in the balanced chemical equation for the given reaction. We can, therefore, directly use the stoichiometry of the balanced chemical equation to determine the combining ratio of the gas volumes:

$$\#\ \text{L}\ F_2 = (4.00\ \text{L}\ H_2)\left(\frac{1\ \text{volume}\ F_2}{1\ \text{volume}\ H_2}\right) = 4.00\ \text{L}\ F_2$$

11.81 When gases are held at the same temperature and pressure, and dispensed in this fashion during chemical reactions, then they react in a ratio of volumes that is equal to the ratio of the coefficients (moles) in the balanced chemical equation for the given reaction. We can, therefore, directly use the stoichiometry of the balanced chemical equation to determine the combining ratio of the gas volumes:

$$\#\ \text{L}\ N_2 = (45.0\ \text{L}\ H_2)\left(\frac{1\ \text{volume}\ N_2}{3\ \text{volume}\ H_2}\right) = 15.0\ \text{L}\ N_2$$

11.82 $\# \text{ mL } O_2 = (175 \text{ mL } C_4H_{10})\left(\frac{13 \text{ mL } O_2}{2 \text{ mL } C_4H_{10}}\right) = 1.14 \times 10^3 \text{ mL}$

11.83 $\# \text{ mL } O_2 = (855 \text{ mL } CO_2)\left(\frac{19 \text{ mL } O_2}{12 \text{ mL } CO_2}\right) = 1.35 \times 10^3 \text{ mL}$

11.84

$$\# \text{mol } C_3H_6 = (18.0 \text{ g } C_3H_6)\left(\frac{1 \text{ mol } C_3H_6}{42.08 \text{ g } C_3H_6}\right) = 0.428 \text{ mol } C_3H_6$$

$$\# \text{mol } H_2 = (0.428 \text{ mol } C_3H_6)\left(\frac{1 \text{ mol } H_2}{1 \text{ mol } C_3H_6}\right) = 0.428 \text{ mol } H_2$$

$$V = \frac{nRT}{P} = \frac{(0.428 \text{ mol } H_2)(0.0821 \frac{\text{L atm}}{\text{mol K}})(297.2 \text{ K})}{(740 \text{ torr})(\frac{1 \text{ atm}}{760 \text{ torr}})} = 10.7 \text{ L } H_2$$

11.85

$$\# \text{moles } HNO_3 = (12.0 \text{ g } HNO_3)\left(\frac{1 \text{ mole } HNO_3}{63.01 \text{ g } HNO_3}\right) = 0.190 \text{ moles } HNO_3$$

$$\# \text{moles } NO_2 = (0.190 \text{ moles } HNO_3)\left(\frac{3 \text{ moles } NO_2}{2 \text{ moles } HNO_3}\right) = 0.286 \text{ moles } NO_2$$

$$V = \frac{nRT}{P} = \frac{(0.286 \text{ moles } NO_2)\left(0.0821 \frac{\text{L atm}}{\text{mol K}}\right)(298 \text{ K})}{(752 \text{ torr})\left(\frac{1 \text{ atm}}{760 \text{ torr}}\right)} = 7.06 \text{ L or } 7.06 \times 10^3 \text{ mL}$$

11.86 $CH_4 + 2O_2 \rightarrow CO_2 + 2H_2O$

$$V_{O_2} = \frac{nRT}{P} = \frac{(1.27 \times 10^{-3} \text{ moles})\left(0.0821 \frac{\text{L atm}}{\text{mol K}}\right)(300 \text{ K})}{(654 \text{ torr})\left(\frac{1 \text{ atm}}{760 \text{ torr}}\right)} = 3.63 \times 10^{-2} \text{ L} = 36.3 \text{ mL } O_2$$

11.87 $$n_{CH_4} = \frac{PV}{RT} = \frac{(725 \text{ torr})\left(\frac{1 \text{ atm}}{760 \text{ torr}}\right)(16.8 \times 10^{-3} \text{ L})}{\left(0.0821 \frac{\text{L atm}}{\text{mol K}}\right)(308 \text{ K})} = 6.34 \times 10^{-4} \text{ moles} \Rightarrow 1.27 \times 10^{-3} \text{ moles } O_2$$

$$n_{NH_3} = \frac{PV}{RT} = \frac{(825 \text{ torr})\left(\frac{1 \text{ atm}}{760 \text{ torr}}\right)(33.6 \times 10^{-3} \text{ L})}{\left(0.0821 \frac{\text{L atm}}{\text{mol K}}\right)(400 \text{ K})} = 1.11 \times 10^{-3} \text{ moles } NH_3 = 1.67 \times 10^{-3} \text{ moles } H_2O$$

$$V_{H_2O} = \frac{nRT}{P} = \frac{(1.67 \times 10^{-3} \text{ moles})\left(0.0821 \frac{\text{L atm}}{\text{mol K}}\right)(591 \text{ K})}{(735 \text{ torr})\left(\frac{1 \text{ atm}}{760 \text{ torr}}\right)} = 8.36 \times 10^{-2} \text{ L} = 83.6 \text{ mL}$$

11.88 $2CO + O_2 \rightarrow 2CO_2$

$$\# \text{ moles CO} = \frac{(683 \text{ torr})\left(\frac{1 \text{ atm}}{760 \text{ torr}}\right)(0.300 \text{ L})}{\left(0.0821 \frac{\text{L atm}}{\text{mol K}}\right)(298 \text{ K})} = 1.10\times10^{-2} \text{ moles}$$

$$\# \text{ moles } O_2 = \frac{(715 \text{ torr})\left(\frac{1 \text{ atm}}{760 \text{ torr}}\right)(0.150 \text{ L})}{\left(0.0821 \frac{\text{L atm}}{\text{mol K}}\right)(398 \text{ K})} \quad 4.32\times10^{-3} \text{ moles}$$

$\therefore O_2$ is the limiting reactant

$$\# \text{ moles } CO_2 = (4.32\times10^{-3} \text{ moles } O_2)\left(\frac{2 \text{ mol } CO_2}{1 \text{ mol } O_2}\right) = 8.64\times10^{-3} \text{ moles } CO_2$$

$$V = \frac{(8.64\times10^{-3} \text{ mol})\left(0.0821 \frac{\text{L atm}}{\text{mol K}}\right)(300 \text{ K})}{(745 \text{ torr})\left(\frac{1 \text{ atm}}{760 \text{ torr}}\right)} \quad 2.17\times10^{-1} \text{ L} = 217 \text{ mL}$$

11.89 $$\# \text{ moles } NH_3 = \frac{(750 \text{ torr})\left(\frac{1 \text{ atm}}{760 \text{ torr}}\right)(0.300 \text{ L})}{\left(0.0821 \frac{\text{L atm}}{\text{mol K}}\right)(301 \text{ K})} = 1.20\times10^{-2} \text{ moles}$$

$$\# \text{ moles } O_2 = \frac{(780 \text{ torr})\left(\frac{1 \text{ atm}}{760 \text{ torr}}\right)(0.220 \text{ L})}{\left(0.0821 \frac{\text{L atm}}{\text{mol K}}\right)(323 \text{ K})} = 8.51\times10^{-3} \text{ moles}$$

Assume NH_3 is the limiting reagent.

$$\# \text{ moles } N_2 = (1.20 \times 10^{-2} \text{ moles } NH_3)\left(\frac{2 \text{ mol } N_2}{4 \text{ moles } NH_3}\right) = 6.00 \times 10^{-3} \text{ moles } N_2$$

Assume O_2 is the limiting reagent:

$$\# \text{ moles } N_2 = (8.51 \times 10^{-3} \text{ moles } O_2)\left(\frac{2 \text{ mol } N_2}{3 \text{ mol } O_2}\right) = 5.68 \times 10^{-3} \text{ mol } N_2 \;\therefore\; O_2 \text{ is limiting reactant}$$

$$\# \text{ mL } N_2 = \frac{(5.68\times10^{-3} \text{ mol})\left(0.0821 \frac{\text{L atm}}{\text{mol K}}\right)(373 \text{ K})\left(\frac{1000 \text{ mL}}{1 \text{ L}}\right)}{(740 \text{ torr})\left(\frac{1 \text{ atm}}{760 \text{ torr}}\right)} = 179 \text{ mL}$$

11.90 $P_{Tot} = 200 \text{ torr} + 150 \text{ torr} + 300 \text{ torr} = 650 \text{ torr}$

11.91 $P_{Tot} = P_{N_2} + P_{O_2} + P_{CO_2}$
$P_{CO_2} = P_{Tot} - P_{N_2} - P_{O_2}$
$P_{CO_2} = 740 \text{ torr} - 120 \text{ torr} - 400 \text{ torr} = 220 \text{ torr}$

11.92 Assume all gases behave ideally and recall that 1 mole of an ideal gas at 0 °C and 1 atm occupies a volume of 22.4 L. So,

$P_{N_2} = 0.30 \text{ atm} = 228 \text{ torr}$
$P_{O_2} = 0.20 \text{ atm} = 152 \text{ torr}$
$P_{He} = 0.40 \text{ atm} = 304 \text{ torr}$
$P_{CO_2} = 0.10 \text{ atm} = 76 \text{ torr}$

11.93 $P_{CO_2} = 840 \text{ torr} - 320 \text{ torr} = 520 \text{ torr}$

$$n_{CO_2} = (0.200 \text{ mol})\left(\frac{520 \text{ torr}}{840 \text{ torr}}\right) = 0.124 \text{ moles}$$

11.94 $P_{total} = (P_{CO} + P_{H_2O})$
$P_{H_2O} = 17.54$ torr at 20 °C, from Table 11.2.
$P_{CO} = 754 - 17.54 \text{ torr} = 736 \text{ torr}$
The temperature stays constant so, $P_1V_1 = P_2V_2$, and

$$V_2 = \frac{P_1V_1}{P_2} = \frac{(736 \text{ torr})(268 \text{ mL})}{(760 \text{ torr})} = 260 \text{ mL}$$

11.95 $P_{total} = P_{H_2} + P_{H_2O}$
$P_{H_2O} = 23.76$ torr at 25 °C, from Table 11.2.
$P_{H_2} = P_{total} - P_{H_2O} = 742 - 23.76 = 718 \text{ torr}$
The temperature stays constant so, $P_1V_1 = P_2V_2$, and

$$V_2 = \frac{P_1V_1}{P_2} = \frac{(718 \text{ torr})(288 \text{ mL})}{(760 \text{ torr})} = 272 \text{ mL}$$

11.96 From Table 11.2, the vapor pressure of water at 20 °C is 17.54 torr. Thus only (742 – 17.54) = 724 torr is due to "dry" methane. In other words, the fraction of the wet methane sample that is pure methane is 724/742 = 0.976. The question can now be phrased: What volume of wet methane, when multiplied by 0.976, equals 244 mL?

Volume"wet" methane × 0.976 = 244 mL

Volume"wet" methane = 244 mL/0.976 = 250 mL

In other words, one must collect 250 total mL of "wet methane" gas in order to have collected the equivalent of 244 mL of pure methane.

11.97 First convert the needed amount of oxygen at 760 torr to the volume that would correspond to the laboratory conditions of 746 torr: $P_1V_1 = P_2V_2$ or $V_2 = P_1V_1/P_2$
$V_2 = 275 \text{ mL} \times 760 \text{ torr}/746 \text{ torr} = 280$ mL of dry oxygen gas

The wet sample of oxygen gas will also be collected at atmospheric pressure, 746 torr. The vapor pressure of water at 15 °C is equal to 12.8 torr (from Table 11.2), and the wet sample will have the following partial pressure of oxygen, once it is collected:
$P_{O_2} = P_{total} - P_{H_2O} = 746 - 12.8 = 733$ torr of oxygen in the wet sample. Thus the wet sample of oxygen is composed of the following % oxygen:

% oxygen in the wet sample = 733/746 × 100 = 98.3 %

The question now becomes what amount of a wet sample of oxygen will contain the equivalent of 280 mL of pure oxygen, if the wet sample is only 98.3 % oxygen (and 1.7 % water). $0.983 \times V_{wet} = 280$ mL, hence

V_{wet} = 285 mL. This means that 285 mL of a wet sample of oxygen must be collected in order to obtain as much oxygen as would be present in 280 mL of a pure sample of oxygen.

11.98 Use equation 11.8 to convert all partial pressures to mole fraction. (P_{total} = 760 torr)

$$X_{N_2} = \frac{P_{N_2}}{P_{total}} = \frac{570 \text{ torr}}{760 \text{ torr}} = 0.750;\ 75.0\ \%$$

$$X_{O_2} = \frac{P_{O_2}}{P_{total}} = \frac{103 \text{ torr}}{760 \text{ torr}} = 0.136;\ 13.6\ \%$$

$$X_{CO_2} = \frac{P_{CO_2}}{P_{total}} = \frac{40 \text{ torr}}{760 \text{ torr}} = 0.053;\ 5.3\ \%$$

$$X_{H_2O} = \frac{P_{H_2O}}{P_{total}} = \frac{47 \text{ torr}}{760 \text{ torr}} = 0.062;\ 6.2\ \%$$

11.99 Use equation 11.8 to convert all partial pressures to mole fraction.

(a)

$$X_{N_2} = \frac{P_{N_2}}{P_{total}} = \frac{197 \text{ torr}}{250 \text{ torr}} = 0.788;\ 78.8\ \%$$

$$X_{O_2} = \frac{P_{O_2}}{P_{total}} = \frac{53 \text{ torr}}{250 \text{ torr}} = 0.21;\ 21\ \%$$

(b)

$$X_{N_2} = \frac{P_{N_2}}{P_{total}} = \frac{600 \text{ torr}}{760 \text{ torr}} = 0.789;\ 78.9\ \%$$

$$X_{O_2} = \frac{P_{O_2}}{P_{total}} = \frac{160 \text{ torr}}{760 \text{ torr}} = 0.211;\ 21.1\ \%$$

(c) Since the relative amount of oxygen is the same at both altitudes, the difficulty in breathing experienced at the higher altitude must result from the lower total pressure, which in turn provides a lower amount of oxygen.

11.100 Effusion rates for gases are inversely proportional to the square root of the gas density, and the gas with the lower density ought to effuse more rapidly. Nitrogen in this problem has the higher effusion rate because it has the lower density:

$$\frac{\text{rate}(N_2)}{\text{rate}(CO_2)} = \sqrt{\frac{1.96 \text{ g L}^{-1}}{1.25 \text{ g L}^{-1}}} = 1.25$$

11.101 Ethylene, C_2H_4, the lightest of these three, diffuses the most rapidly.

10.102 The relative rates are inversely proportional to the square roots of their molecular masses:

$$\frac{\text{rate}(^{235}UF_6)}{\text{rate}(^{238}UF_6)} = \sqrt{\frac{\text{molar mass }(^{238}UF_6)}{\text{molar mass }(^{235}UF_6)}} = \sqrt{\frac{352 \text{ g mol}^{-1}}{349 \text{ g mol}^{-1}}} = 1.0043$$

Meaning that the rate of effusion of the $^{235}UF_6$ is only 1.0043 times faster than the $^{238}UF_6$ isotope.

11.103 Use equation 11.10

$$\frac{\text{effusion rate x}}{\text{effusion rate } C_3H_8} = \sqrt{\frac{M_{C_3H_8}}{M_x}}$$

$$M_x = M_{C_3H_8}\left(\frac{\text{effusion rate } C_3H_8}{\text{effusion rate x}}\right)^2$$

$$= 44.1 \text{ g/mol}\left(\frac{1}{1.65}\right)^2 = 16.2 \text{ g/mol}$$

Additional Exercises

11.104 We found that 1 atm = 33.9 ft of water. This is equivalent to 33.9 ft × 12 in./ft = 407 in. of water, which in this problem is equal to the height of a water column that is uniformly 1.00 $in.^2$ in diameter. Next, we convert the given density of water from the units g/mL to the units lb/$in.^3$:

$$\#\frac{\text{lb}}{\text{in.}^3} = \left(\frac{1.00 \text{ g}}{1.00 \text{ mL}}\right)\left(\frac{1 \text{ lb}}{454 \text{ g}}\right)\left(\frac{1 \text{ mL}}{1 \text{ cm}^3}\right)\left(\frac{2.54 \text{ cm}}{1 \text{ in.}}\right)^3 = 0.0361 \frac{\text{lb}}{\text{in.}^3}$$

The area of the total column of water is now calculated: 1.00 $in.^2$ × 407 in. = 407 $in.^3$, along with the mass of the total column of water: 407 $in.^3$ × 0.0361 lb/$in.^3$ = 14.7 lb. Finally, we can determine the pressure (force/unit area) that corresponds to one atm: 1 atm = 14.7 lb ÷ 1.00 $in.^2$ = 14.7 lb/$in.^2$

11.105

$$\text{Total footprint} = (4 \text{ tires})\left(\frac{6.0 \text{ in} \times 3.2 \text{ in}}{\text{tire}}\right) = 76.8 \text{ in}^2$$

$$\text{Total pressure} = \frac{3500 \text{ lb}}{76.8 \text{ in}^2} = 45.6 \text{ lb/in}^2$$

$$\text{Gauge pressure} = 45.6 \text{ lb/in}^2 - 14.7 \text{ lb/in}^2 = 30.9 \text{ lb/in}^2$$

11.106

$$\text{Total weight} = (45.6 \text{ tons} + 8.3 \text{ tons})\left(\frac{2000 \text{ lb}}{1 \text{ ton}}\right) = 1.08 \times 10^5 \text{ lbs}$$

$$\text{Total pressure} = 85 \text{ psi} + 14.7 \text{ psi} = 99.7 \text{ psi/tire}$$

$$\# \text{ tires} = \frac{1.08 \times 10^5 \text{ lbs}}{(99.7 \text{ lbs in}^{-2}\text{/tire})(100 \text{ in}^2)} = 10.8 \text{ tires}$$

The minimal number of tires is 12 since tires are mounted in multiples of 2

11.107 Assume a 1 sq in. cylinder of water

$$V = (12{,}468 \text{ ft})\left(\frac{12 \text{ in.}}{1 \text{ ft}}\right)\left(1 \text{ in.}^2\right) = (149616 \text{ in.}^3)\left(\frac{2.54 \text{ cm}}{1 \text{ in.}}\right)^3\left(\frac{1 \text{ mL}}{1 \text{ cm}^3}\right) = 2.4518 \times 10^6 \text{ mL}$$

$$\text{mass} = (2.4518 \times 10^6 \text{ mL})\left(\frac{1.025 \text{ g}}{1 \text{ mL}}\right)$$

$$= (2.51306 \times 10^6 \text{ g})\left(\frac{1 \text{ lb}}{453.6 \text{ g}}\right) = 5.54026 \times 10^3 \text{ lb}$$

$$\text{Pressure} = (5.54026 \times 10^3 \text{ lb in}^{-2})\left(\frac{1 \text{ atm}}{14.7 \text{ lb in}^{-2}}\right) = 376.89 \text{ atm}$$

11.108 From the data we know that the pressure in flask 1 is greater than atmospheric pressure, and greater than the pressure in flask 2. The pressure in flask 1 can be determined from the manometer data. The pressure in flask 1 is:

$$(0.827 \text{ atm})\left(\frac{760 \text{ mm Hg}}{1 \text{ atm}}\right)\left(\frac{1 \text{ cm}}{10 \text{ mm}}\right) + 12.26 \text{ cm} = 75.11 \text{ cm Hg}$$

The pressure in flask 2 is lower than flask 1

$$P = 75.11 \text{ cm Hg} - (16.24 \text{ cm oil})\left(\frac{0.826 \text{ g mL}^{-1}}{13.6 \text{ g mL}^{-1}}\right) = 74.12 \text{ cm Hg} = 741.2 \text{ torr}$$

11.109 To calculate the pressure at 100 ft assume a cylinder of water 100 ft long and 1 in^2.

$$\text{mass} = (100 \text{ ft})\left(\frac{12 \text{ in}}{1 \text{ ft}}\right)(1 \text{ in})^2\left(\frac{2.54 \text{ cm}}{1 \text{ in}}\right)^3\left(\frac{1 \text{ mL}}{1 \text{ cm}^3}\right)\left(\frac{1.025 \text{ g}}{1 \text{ mL}}\right)\left(\frac{1 \text{ lb}}{453.6 \text{ g}}\right) = 44.4 \text{ lb}$$

$$P = (44.4 \text{ lb in}^{-2})\left(\frac{1 \text{ atm}}{14.7 \text{ lb in}^{-2}}\right) = 3.02 \text{ atm}$$

Since the pressure decreases by a factor of 3, the volume must increase by a factor of 3. Divers exhale to decrease the amount of gas in their lungs, so it does not expand to a volume larger than the divers lungs.

11.110 First calculate the initial volume (V_1) and the final volume (V_2) of the cylinder, using the given geometrical data, noting that the radius is half the diameter (10.7/2 = 5.35 cm): $V_1 = \pi \times (5.35 \text{ cm})^2 \times 13.4 \text{ cm} = 1.20 \times 10^3 \text{ cm}^3$

$V_2 = \pi \times (5.35 \text{ cm})^2 \times (13.4 \text{ cm} - 12.7 \text{ cm}) = 62.9 \text{ cm}^3$

In general the combined gas law equation is: $\frac{P_1V_1}{T_1} = \frac{P_2V_2}{T_2}$, and in particular, for this problem, we have:

$$T_2 = \frac{P_2V_2T_1}{P_1V_1} = \frac{(34.0 \text{ atm})(62.9 \text{ cm}^3)(364 \text{ K})}{(1.00 \text{ atm})(1.20 \times 10^3 \text{ cm}^3)} = 649 \text{ K} = 376 \text{ °C}$$

11.111 First convert the temperature data to the Kelvin scale: 273 + 5/9(60.0 – 32.0) = 289 K and 273 + 5/9(104 – 32) = 313 K. Next, calculate the final pressure at the gauge, taking into account the temperature change only:

$$P_2 = \frac{P_1T_2}{T_1} = \frac{(64.7 \text{ lb in.}^{-2})(313 \text{ K})}{(289 \text{ K})} = 70.1 \text{ lb in.}^{-2}$$

This represents the actual pressure inside the tire. The pressure gauge measures only the difference between the pressure inside the tire and the pressure outside the tire (atmospheric pressure). Hence the gauge reading is equal to the internal pressure of the tire less atmospheric pressure: (70.1 – 14.7) lb/in^2 = 55.4 lb/in^2

11.112 The temperatures must first be converted to Kelvin:

$$\text{°C} = \frac{5}{9} \times (\text{°F} - 32) = \frac{5}{9} \times (-50 - 32) = -46 \text{ °C}$$

$$\text{°C} = \frac{5}{9} \times (\text{°F} - 32) = \frac{5}{9} \times (120 - 32) = 49 \text{ °C}$$

Next, the pressure calculation is done using the following equation:

$$P_2 = \frac{P_1T_2}{T_1} = \frac{(35 \text{ lb in.}^{-2})(322 \text{ K})}{(227 \text{ K})} = 50 \text{ lb in.}^{-2}$$

11.113 $Cl_2 + SO_3^{2-} + H_2O \rightarrow 2Cl^- + SO_4^{2-} + 2H^+$

$$\#\,\text{moles Cl}_2 = (50.0\text{ mL Na}_2\text{SO}_3)\left(\frac{0.200\text{ moles Na}_2\text{SO}_3}{1000\text{ mL Na}_2\text{SO}_3}\right)\left(\frac{1\text{ mole SO}_3{}^{2-}}{1\text{ mole Na}_2\text{SO}_3}\right)\left(\frac{1\text{ mole Cl}_2}{1\text{ mole SO}_3{}^{2-}}\right)$$

$$= 1.00\times10^{-2}\text{ moles Cl}_2$$

$$V_{Cl_2} = \frac{(1.00\times10^{-2}\text{ moles})\left(0.0821\,\frac{\text{L atm}}{\text{mol K}}\right)(298\text{ K})}{(734\text{ torr})\left(\frac{1\text{ atm}}{760\text{ torr}}\right)} = 0.253\text{ L} = 253\text{ mL}$$

11.114 The relative rates are inversely proportional to the square roots of their molecular masses:

$$\frac{\text{rate(NH}_3)}{\text{rate(unknown gas)}} = \sqrt{\frac{\text{molar mass (unknown gas)}}{\text{molar mass (NH}_3)}} = 2.93$$

$$\frac{\text{molar mass (unknown gas)}}{\text{molar mass (NH}_3)} = (2.93)^2$$

$$\text{molar mass (unknown gas)} = \text{molar mass (NH}_3) \times (2.93)^2$$

$$\text{molar mass (unknown gas)} = 17.30\text{ g/mol} \times 8.58 = 146\text{ g/mol}$$

11.115 (a) $Zn(s) + 2HCl(aq) \rightarrow H_2(g) + ZnCl_2(aq)$

(b) Calculate the number of moles of hydrogen:

$$n = \frac{PV}{RT} = \frac{(760\text{ torr})\left(\frac{1\text{ atm}}{760\text{ torr}}\right)(12.0\text{ L})}{\left(0.0821\,\frac{\text{L atm}}{\text{mol K}}\right)(293.2\text{ K})} = 0.499\text{ mol H}_2$$

and the number of moles of zinc:

$$\#\text{ mol Zn} = (0.499\text{ mol H}_2)\left(\frac{1\text{ mol Zn}}{1\text{ mol H}_2}\right) = 0.499\text{ mol Zn}$$

The number of grams of zinc needed is, therefore:

$$\#\text{ g Zn} = (0.499\text{ mol Zn})\left(\frac{65.39\text{ g Zn}}{1\text{ mol Zn}}\right) = 32.6\text{ g Zn}$$

(c)

$$\#\text{ mol HCl} = (0.499\text{ mol Zn})\left(\frac{2\text{ mol HCl}}{1\text{ mol Zn}}\right) = 0.998\text{ mol HCl}$$

$$\#\text{ mL HCl} = (0.998\text{ mol HCl})\left(\frac{1000\text{ mL HCl}}{8.00\text{ mol HCl}}\right) = 125\text{ mL HCl}$$

11.116 $P_{total} = 740\text{ torr} = P_{H_2} + P_{water}$

The vapor pressure of water at 25 °C is available in Table 11.2: 23.76 torr. Hence:

$P_{H_2} = (740 - 24)\text{ torr} = 716\text{ torr}$

Next, we calculate the number of moles of hydrogen gas that this represents:

$$n = \frac{PV}{RT} = \frac{(716 \text{ torr})\left(\frac{1 \text{ atm}}{760 \text{ torr}}\right)(0.335 \text{ L})}{\left(0.0821 \frac{\text{L atm}}{\text{mol K}}\right)(298.2 \text{ K})} = 0.0129 \text{ mol } H_2$$

The balanced chemical equation is: $Zn(s) + 2HCl(aq) \rightarrow H_2(g) + ZnCl_2(aq)$
and the quantities of the reagents that are needed are:

$$\# \text{ g Zn} = (0.0129 \text{ mol } H_2)\left(\frac{1 \text{ mol Zn}}{1 \text{ mol } H_2}\right)\left(\frac{65.39 \text{ g Zn}}{1 \text{ mol Zn}}\right) = 0.844 \text{ g Zn}$$

$$\# \text{ mL HCl} = (0.0129 \text{ mol } H_2)\left(\frac{2 \text{ mol HCl}}{1 \text{ mol } H_2}\right)\left(\frac{1000 \text{ mL HCl}}{6.00 \text{ mol HCl}}\right) = 4.30 \text{ mL HCl}$$

11.117 This is a limiting reactant problem. First we need to calculate the moles of dry CO_2 that can be produced from the given quantities of $CaCO_3$ and HCl:

$$\# \text{ mol } CaCO_3 = (12.3 \text{ g } CaCO_3)\left(\frac{1 \text{ mol } CaCO_3}{100.09 \text{ g } CaCO_3}\right) = 0.123 \text{ mol } CaCO_3$$

$$\# \text{ mol HCl} = (185 \text{ mL HCl})\left(\frac{0.250 \text{ mol HCl}}{1000 \text{ mL HCl}}\right) = 0.0463 \text{ mol HCl}$$

Thus, HCl is limiting and we use this to determine the moles of CO_2 that can be produced:

$$\# \text{ mol } CO_2 = (0.0463 \text{ mol HCl})\left(\frac{1 \text{ mol } CO_2}{2 \text{ mol HCl}}\right) = 0.0231 \text{ mol } CO_2$$

Next we realize that CO_2 is collected over water at a total pressure of 745 torr. Thus, the pressure of "dry" CO_2 is calculated as follows: $P_{total} = 745 \text{ torr} = P_{CO_2} + P_{water}$. The vapor pressure of water at 20°C is available in Table 11.2, 17.54 torr. Hence, $P_{CO_2} = 745 \text{ torr} - 18 \text{ torr} = 727 \text{ torr}$. Finally, the volume of this "dry" CO_2 is calculated using the ideal gas equation:

$$V = \frac{nRT}{P} = \frac{(0.0231 \text{ mol})\left(0.0821 \frac{\text{L atm}}{\text{mol K}}\right)(293.2 \text{ K})}{(727 \text{ torr})\left(\frac{1 \text{ atm}}{760 \text{ torr}}\right)} = 0.582 \text{ L } CO_2 = 582 \text{ mL}$$

11.118 Using the ideal gas law, determine the number of moles of H_2 and O_2 gas initially present:
For hydrogen:

$$n = \frac{PV}{RT} = \frac{(1250 \text{ torr})\left(\frac{1 \text{ atm}}{760 \text{ torr}}\right)(400 \text{ mL})\left(\frac{1 \text{ L}}{1000 \text{ mL}}\right)}{\left(0.0821 \frac{\text{L atm}}{\text{mol K}}\right)(318 \text{ K})} = 2.52\times10^{-2} \text{ mol } H_2$$

for oxygen:

$$n = \frac{PV}{RT} = \frac{(740 \text{ torr})\left(\frac{1 \text{ atm}}{760 \text{ torr}}\right)(300 \text{ mL})\left(\frac{1 \text{ L}}{1000 \text{ mL}}\right)}{\left(0.0821 \frac{\text{L atm}}{\text{mol K}}\right)(298 \text{ K})} = 1.19\times10^{-2} \text{ mol } O_2$$

This problem is an example of a limiting reactant problem in that we know the amounts of H_2 and O_2 initially present. Since 1 mol of O_2 reacts completely with 2 mol of H_2, we can see, by inspection, that there is excess H_2 present. Using the amounts calculated above, we can make 2.38×10^{-2} mol of H_2O and have an excess of 1.4×10^{-3} mol of H_2. Thus, the total amount of gas present after complete reaction is 2.52×10^{-2} mol. Using this value for n, we can calculate the final pressure in the reaction vessel:

$$P = \frac{nRT}{V} = \frac{\left(2.52\times10^{-2} \text{ mol}\right)\left(0.0821 \frac{\text{L atm}}{\text{mol K}}\right)(393 \text{ K})}{(500 \text{ mL})\left(\frac{1 \text{ L}}{1000 \text{ mL}}\right)} = 1.63 \text{ atm} = 1.24\times10^{3} \text{ torr}$$

11.119 We first need to determine the pressure inside the apparatus. Since the water level is 8.5 cm higher inside than outside, the pressure inside the container is lower than the pressure outside. To determine the inside pressure, we first need to convert 8.5 cm of water to an equivalent dimension for mercury. This is done using the density of mercury: P_{Hg} = 85 mm/13.6 = 6.25 mm (where the density of mercury, 13.6 g/mL, has been used.) $P_{inside} = P_{outside} - P_{Hg}$ = 746 torr – 6 torr = 740 torr. In order to determine the P_{H_2}, we need to subtract the vapor pressure of water at 24 °C. This value may be found in Appendix E4 and is equal to 22.4 torr. The $P_{H_2} = P_{inside} - P_{H_2O}$ = 740 torr – 22.4 torr = 717 torr. Now, we can use the ideal gas law in order to determine the number of moles of H_2 present;

$$n = \frac{PV}{RT} = \frac{(717\ \text{torr})\left(\frac{1\ \text{atm}}{760\ \text{torr}}\right)(18.45\ \text{mL})\left(\frac{1\ \text{L}}{1000\ \text{mL}}\right)}{\left(0.0821\ \frac{\text{L atm}}{\text{mol K}}\right)(297\ \text{K})} = 7.14\times10^{-4}\ \text{mol}\ H_2$$

The balanced equation described in this problem is:

$$Zn(s) + 2HCl(aq) \rightarrow ZnCl_2(aq) + H_2(g)$$

By inspection we can see that 1 mole of $Zn(s)$ reacts to form 1mole of $H_2(g)$ and we must have reacted 7.14 × 10^{-4} mol Zn in this reaction.

$$\#\,\text{g Zn} = (7.14\times10^{-4}\ \text{mol Zn})\left(\frac{65.39\ \text{g Zn}}{1\ \text{mol Zn}}\right) = 4.67\times10^{-2}\ \text{g Zn}$$

11.120 $2H_2 + O_2 \rightarrow 2H_2O$

$$\#\ \text{moles}\ H_2 = (12.7\ \text{g}\ H_2)\left(\frac{1\ \text{mol}\ H_2}{2.02\ \text{g}\ H_2}\right) = 6.30\ \text{moles}\ H_2$$

$$\#\ \text{moles}\ O_2 = (87.5\ \text{g}\ O_2)\left(\frac{1\ \text{mol}\ O_2}{32.0\ \text{g}\ O_2}\right) = 2.73\ \text{moles}\ O_2 \quad \therefore\ O_2\ \text{is the limiting reactant}$$

$$\#\ \text{moles}\ H_2O = (2.73\ \text{mol}\ O_2)\left(\frac{2\ \text{mol}\ H_2O}{1\ \text{mol}\ O_2}\right) = 5.47\ \text{moles}\ H_2O$$

$$\#\ \text{moles}\ H_2\ \text{needed} = (2.73\ \text{mol}\ O_2)\left(\frac{2\ \text{mol}\ H_2}{1\ \text{mol}\ O_2}\right) = 5.47\ \text{moles}\ H_2$$

$$\#\ \text{moles}\ H_2\ \text{needed} = (2.73\ \text{mol}\ O_2)\left(\frac{2\ \text{mol}\ H_2}{1\ \text{mol}\ O_2}\right) = 5.47\ \text{moles}\ H_2$$

$$P_{H_2O} = \frac{(5.47\ \text{mol})\left(0.0821\ \frac{\text{L atm}}{\text{mol K}}\right)(433\ \text{K})}{12.0\ \text{L}} = 16.2\ \text{atm}$$

$$P_{H_2} = \frac{(0.83\ \text{mol})\left(0.0821\ \frac{\text{L atm}}{\text{mol K}}\right)(433\ \text{K})}{12.0\ \text{L}} = 2.5\ \text{atm}$$

$$P_{Tot} = 16.2\ \text{atm} + 2.5\ \text{atm} = 18.7\ \text{atm}$$

11.121 (a) $$\#\ \text{L}\ NH_3 = (28.3 \times 10^3\ \text{L}\ CH_4)\left(\frac{16\ \text{volume}\ NH_3}{7\ \text{volume}\ CH_4}\right) = 6.47 \times 10^4\ \text{L}\ NH_3$$

(b) $$\#\ \text{mol}\ NH_3 = (6.47 \times 10^4\ \text{L}\ NH_3)\left(\frac{1\ \text{mol}\ NH_3}{22.4\ \text{L}\ NH_3}\right) = 2.89 \times 10^3\ \text{mol}\ NH_3$$

$$\#\ \text{kg}\ NH_3 = (2.89 \times 10^3\ \text{mol}\ NH_3)\left(\frac{17.0\ \text{g}\ NH_3}{1\ \text{mol}\ NH_3}\right)\left(\frac{1\ \text{kg}}{1000\ \text{g}}\right) = 49.1\ \text{kg}\ NH_3$$

11.122 (a) First determine the % by mass S and O in the sample:

% S = 1.448 g/3.620 g × 100 = 40.00 % S

% O = 2.172 g/3.620 g × 100 = 60.00 % O

(b) Next, determine the number of moles of S and O in a sample of the material weighing 100 g exactly, in order to make the conversion from % by mass to grams straightforward: In 100 g of the material, there are 40.00 g S and 60.00 g O:

40.00 g S ÷ 32.07 g/mol = 1.247 mol S

60.00 g O ÷ 16.00 g/mol = 3.750 mol O

Dividing each of these mole amounts by the smaller of the two gives the relative mole amounts of S and O in the material: for S, 1.247 mol ÷ 1.247 mol = 1.000 relative moles, for O, 3.750 mol ÷ 1.247 mol= 3.007 relative moles, and the empirical formula is, therefore, SO_3.

(c) We determine the formula mass of the material by use of the ideal gas law:

$$n = \frac{PV}{RT} = \frac{(750\text{ torr})\left(\frac{1\text{ atm}}{760\text{ torr}}\right)(1.120\text{ L})}{\left(0.0821\,\frac{\text{L atm}}{\text{mol K}}\right)(298.2\text{ K})} = 0.0451\text{ mol}$$

The formula mass is given by the mass in grams (given in the problem) divided by the moles determined here: formula mass = 3.620 g ÷ 0.0451 mol = 80.2 g mol^{-1}. Since this is equal to the formula mass of the empirical unit determined in step (b) above, namely SO_3, then the molecular formula is also SO_3.

11.123 (a) $P_{total} = 746.0\text{ torr} = P_{H_2O} + P_{N_2}$

$P_{N_2} = 746.0\text{ torr} - 22.1\text{ torr} = 723.9\text{ torr}$

Now, use the ideal gas equation to determine the moles of N_2 that have been collected:

$$n = \frac{PV}{RT} = \frac{(723.9\text{ torr})\left(\frac{1\text{ atm}}{760\text{ torr}}\right)(18.90\text{ mL})\left(\frac{1\text{ L}}{1000\text{ mL}}\right)}{\left(0.0821\,\frac{\text{L atm}}{\text{mol K}}\right)(296.95\text{ K})} = 7.384\times10^{-4}\text{ mol N}_2$$

Then the mass of nitrogen that has been collected is determined: 7.384×10^{-4} mol N_2 × 28.0 g/mol = 2.068×10^{-2} g N_2. Next, the % by mass nitrogen in the material is calculated: % N = (0.02068 g)/(0.2394 g) × 100 = 8.638 % N

(b) mass of C in the sample:

$$\#\text{ g C} = (17.57 \times 10^{-3}\text{ g CO}_2)\left(\frac{1\text{ mole CO}_2}{44.01\text{ g CO}_2}\right)\left(\frac{1\text{ mol C}}{1\text{ mol CO}_2}\right)\left(\frac{12.01\text{ g C}}{1\text{ mol C}}\right)$$

$$= 4.795 \times 10^{-3}\text{ g C}$$

mass of H in the sample:

$$\#\text{ g H} = (4.319 \times 10^{-3}\text{ g H}_2\text{O})\left(\frac{1\text{ mole H}_2\text{O}}{18.02\text{ g H}_2\text{O}}\right)\left(\frac{2\text{ mol H}}{1\text{ mol H}_2\text{O}}\right)\left(\frac{1.008\text{ g H}}{1\text{ mol H}}\right)$$

$$= 4.832 \times 10^{-4}\text{ g H}$$

mass of N in the sample:

$$\#\text{ g N} = (6.478 \times 10^{-3}\text{ g sample})\left(\frac{8.638\text{ g N}}{100\text{ g sample}}\right) = 5.596 \times 10^{-4}\text{ g N}$$

mass of O in the sample = total mass – (mass C + H + N)

$$\# \text{ mg O} = 6.478 \text{ mg sample} - (4.795 \text{ mg C} + 0.4832 \text{ mg H} + 0.5596 \text{ mg N}) = 0.640 \text{ mg O}$$

Next we convert each of these mass amounts into the corresponding mole values:

for C, 4.795×10^{-3} g ÷ 12.01 g/mol = 3.993×10^{-4} mol C
for H, 4.832×10^{-4} g ÷ 1.008 g/mol = 4.794×10^{-4} mol H
for N, 5.596×10^{-4} g ÷ 14.01 g/mol = 3.994×10^{-5} mol N
for O, 6.40×10^{-4} g ÷ 16.00 g/mol = 4.00×10^{-5} mol O

Last, we convert these mole amounts into relative mole amounts by dividing each by the smallest of the four:

for C, 3.993×10^{-4} mol/ 3.994×10^{-5} mol = 9.998
for H, 4.794×10^{-4} mol/ 3.994×10^{-5} mol = 12.00
for N, 3.994×10^{-5} mol/ 3.994×10^{-5} mol = 1.000
for O, 4.00×10^{-5} mol/ 3.994×10^{-5} mol = 1.00

The empirical formula is therefore $C_{10}H_{12}NO$

(c) The formula mass of the empirical unit is 162. Since this is half the value of the known molecular mass, the molecular formula must be twice the empirical formula, $C_{20}H_{24}N_2O_2$.

11.124 (a) The equation can be rearranged to give:

$$0.04489 \times \frac{V(P - P_{H_2O})}{273 + t_{°C}} = \%N \times W$$

This means that the left side of the above equation should be obtainable simply from the ideal gas law, applied to the nitrogen case. If PV = nRT, then for nitrogen: PV = (mass nitrogen)/(28.01 g/mol) × RT, and the mass of nitrogen that is collected is given by: (mass nitrogen) = PV(28.01)/RT, where R = 82.1 mL atm/K mol × 760 torr/atm = 6.24×10^4 mL torr/K mol. Using this value for R in the above equation, we have the following result for the mass of nitrogen, remembering that the pressure of nitrogen is less than the total pressure, by an amount equal to the vapor pressure of water:

$$(\text{mass nitrogen}) = \frac{28.01 \times V \times (P_{total} - P_{H_2O})}{\left(6.24 \times 10^4 \frac{\text{mL torr}}{\text{mol K}}\right)(273 + °C)}$$

Finally, it is only necessary to realize that the value

$$\frac{28.01}{6.24 \times 10^4} \times 100 = 0.04489$$

is exactly the value given in the problem.

(b) $$\%N = 0.04489 \times \frac{(18.90 \text{ mL})(746 \text{ torr} - 22.1 \text{ torr})}{(0.2394 \text{ g})(273.15 + 23.80)} = 8.639\ \%$$

11.125 (a) We begin by converting the dimensions of the room into cm: 40 ft × 30.48 cm/ft = 1.2×10^3 cm, 20 ft × 30.48 cm/ft = 6.1×10^2 cm, 8 ft × 30.48 cm/ft = 2.4×10^2 cm. Next, the volume of the room is determined: V = $(1.2 \times 10^3 \text{ cm})(6.1 \times 10^2 \text{ cm})(2.4 \times 10^2 \text{ cm}) = 1.8 \times 10^8 \text{ cm}^3$. Since there are 1000 cm^3 in a liter, volume is: V = 1.8×10^5 L

The calculation of the amount of H_2S goes as follows:

$$\#\,\text{L}\,\text{H}_2\text{S} = \left(1.8\times10^5\ \text{L space}\right)\left(\frac{0.15\,\text{L}\,\text{H}_2\text{S}}{1\times10^9\ \text{L space}}\right) = 2.7\times10^{-5}\ \text{L}\,\text{H}_2\text{S}$$

(b) Convert volume (in liters) to moles at STP:

$$\#\,\text{mol}\,\text{H}_2\text{S} = \left(2.7\times10^{-5}\ \text{L}\,\text{H}_2\text{S}\right)\left(\frac{1\,\text{mol}\,\text{H}_2\text{S}}{22.4\,\text{L}\,\text{H}_2\text{S}}\right) = 1.2\times10^{-6}\ \text{mol}\,\text{H}_2\text{S}$$

Since the stoichiometry is 1:1, we require the same number of moles of Na_2S:

$$\#\,\text{mL Na}_2\text{S} = \left(1.2\times10^{-6}\ \text{mol Na}_2\text{S}\right)\left(\frac{1000\,\text{mL Na}_2\text{S}}{0.100\,\text{mol Na}_2\text{S}}\right)$$

$$= 1.2\times10^{-2}\ \text{mL Na}_2\text{S}$$

Practice Exercises

12.1 Propylamine would have a substantially higher boiling point because of its ability to form hydrogen bonds (there are N-H bonds in propylamine, but not in trimethylamine.)

12.2 The number of molecules in the vapor will increase, and the number of molecules in the liquid will decrease, but the sum of the molecules in the vapor and the liquid remains the same.

12.3 We use the curve for water, and find that at 330 torr, the boiling point is approximately 75 °C.

12.4 Adding heat will shift the equilibrium to the right, producing more vapor. This increase in the amount of vapor causes a corresponding increase in the pressure, such that the vapor pressure generally increases with increasing temperature.

12.5 Refer to the phase diagram for water, Figure 12.28. We "move" along a horizontal line marked for a pressure of 2.15 torr. At –20 °C, the sample is a solid. If we bring the temperature from –20 °C to 50 °C, keeping the pressure constant at 2.15 torr, the sample becomes a gas. The process is thus solid $\rightarrow$ gas, i.e. sublimation.

12.6 As diagramed in Figure 12.28, this falls in the liquid region.

Thinking It Through

T12.1 It is necessary only to compare the densities of hot and cool water in order to answer this question. The cool water, being more dense, sinks to the bottom.

T12.2 The rate of evaporation on these two types of days must be compared. The relative humidity is the key to the question, because evaporation is more efficient on a dry day.

T12.3 The reason for the hiss must be the presence of vapors, which appear when gasoline vapors establish an equilibrium with liquid gasoline in the container.

T12.4 The molecular mass of the wax is much larger than that of water. If the comparison of boiling points is to be used in order to determine the relative strengths of intermolecular forces, it is important to compare substances having approximately the same molecular masses. Also, London forces have been shown to increase with increased molecular masses.

T12.5 First we look for the possibility for hydrogen bonding in either of the two substances, since this is the strongest of the intermolecular forces. The second compound, $(CH_3CH_2)_2NH$, is an amine and can have hydrogen bonds. The first compound is an ether, and cannot have hydrogen bonds, since it does not possess an OH group. Other intermolecular forces may exist, but having found this difference in hydrogen–bonding, we can proceed to answer the question. The first compound (the ether) has the weaker intermolecular forces and, therefore, the higher vapor pressure.

T12.6 Since the boiling point depends on the pressure, we would need to know the pressure inside the flask in order to answer this question. Obviously the pressure inside the flask at the temperature of the ice water is lower than that of the vapor pressure of the trapped water.

T12.7 Air with 100% humidity is saturated with water vapor, meaning that the partial pressure of water in the air has become equal to the vapor pressure of water, at whatever temperature we may have. Since vapor pressure decreases with decreasing temperature, the total amount of water vapor in saturated air should decrease with decreasing temperature. When cold air is brought inside, the low water content constitutes only a small amount of the water content that is possible in the warm room.

T12.8 It is necessary to compare the water content of the two air masses. A cloud of tiny liquid water dropletswill form if there is enough water in the humid air mass to cause the relative humidity to exceed 100 % at the lower temperature.

T12.9 Here we have two quantities that differ in one important way. Vapor pressure does not depend on the amount of a sample, whereas the rate of evaporation does depend on the amount of surface area.

T12.10 Water will vaporize into such a container until the vapor pressure becomes as large as the equilibrium value, provided there is some amount of liquid water after equilibrium is reached. Evidently all of the liquid evaporated before this equilibrium was reached.

T12.11 We must compare two enthalpy values, ΔH_{vap} and ΔH_{fus}. The process of vaporization (and of condensation) generally involves more energy than the process of melting (and of crystallization). This is because a greater disruption of intermolecular forces is required when a substance is vaporized than when it is melted.

T12.12 A large molar heat of vaporization will correspond with the substance having the stronger intermolecular forces of attraction. We can make this comparison straightforwardly, since the chain lengths are the same for the two substances. In this case, we choose the O–containing analog, because of the high strength of its hydrogen bonds.

T12.13 Hydrogen bonding is not an issue here. The chain lengths are different, and we expect a difference in polarity between the two. The greater chain length of the first compound causes it to have the greater heat of vaporization.

T12.14 Since room temperature is already above the critical temperature, we conclude that condensation is not possible, regardless of the pressure.

T12.15 In order to answer this question, we must compare the strengths only of two types of forces, dipole–dipole and London forces, since these are the only two possibilities. We expect a difference in the permanent dipole moment of these two substances as well as a difference in the London forces, since chlorine is a polarizable atom. Based on this, $CHCl_3$ has a higher vapor pressure.

T12.16 The fact that the sample undergoes a cooling tells us that some energy must have been "spent" in the process of expansion into a vacuum. This loss of energy is only necessary if intermolecular forces have to be overcome in order to move the gas molecules apart.

T12.17 The metal cation with the larger charge is more able to polarize the electron cloud of the anion towards itself. As the electron cloud of the anion becomes more polarized towards the cation, we have, by definition, a more covalent situation. The bond between cation and anion becomes more covalent because the cation is able to attract and share the anion's electrons more completely.

Review Exercises

12.1 This would be the condition that minimizes intermolecular interactions, high temperature and low pressure.

12.2 Liquids have strong intermolecular interactions and these interactions are affected by the composition of the liquids and in turn, affect the properties of the liquids.

12.3 Gases are able to behave ideally (that is behave in ways that are unrelated to chemical composition) because the molecules of an ideal gas sample are far apart from one another, and intermolecular interactions are negligible.

12.4 Intermolecular forces in liquids and solids are more important than in gases because the molecules and atoms of liquid and solid samples are so much closer together than they are in a gas.

12.5 The transfer of a gas from one container to another may be accompanied by either a change in shape or volume, or both. Only the shape of a liquid may change when its container is altered; the volume of a liquid does not change when the liquid is transferred to a new container. A solid changes neither its shape nor its volume when it is transferred into a new container.

12.6 At a molecular level, gases have minimal intermolecular interactions, they are in constant random motion. Liquids, have stronger intermolecular attractions, they move, but are in contact with the other molecules. For solids, the molecules vibrate in place, but are not free to move about due to the stronger intermolecular attractions.

12.7 The intermolecular attractive forces are strongest for a solid and weakest for a gas with the liquid state in between.

12.8 The strengths of intermolecular attractions are much weaker than the strengths of chemical bonds.

12.9 It is the intramolecular forces, the bonds, that are responsible for the chemical properties, not the intermolecular forces. For the physical properties, the intermolecular forces are: dipole-dipole attractions, hydrogen bonds, London forces, and ion-dipole attractions.

12.10 Dipole–dipole interactions arise from the attraction of the permanent dipole moment of one molecule with that of an adjacent molecule, the positive end of one dipole being drawn to the negative end of the other dipole. This is diagrammed in Figure 12.3 of the text.

12.11 London forces are diagrammed in Figure 12.5. These weak forces of attraction are caused by instantaneous dipoles that attract induced dipoles in neighboring molecules. London forces increase in strength with increasing molecular size, as illustrated in Figure 12.6 and as shown by the data of Table 12.1 and Table 12.2. London forces increase in strength as the number of atoms in a molecule increases, as illustrated in Figure 12.7. London forces decrease the more compact the molecule is compared to a more chainlike molecule with the same number and types of atom as illustrated in Figure 12.8.

12.12 Polarizability is a measure of the ease with which the electron cloud is distorted. If an electron cloud is easily polarizable, that is large and easily deformed, then instantaneous dipoles and induced dipoles form without much difficulty and stronger London forces are experienced by that molecule.

12.13 Hydrogen bonds are special types of dipole–dipole attractions, and they have greater strength than other types of intermolecular forces. A hydrogen bond arises when a hydrogen atom is covalently bonded to fluorine, oxygen, or nitrogen. Compounds in which hydrogen bonding is important have boiling points that are higher than might be otherwise expected.

12.14 These are fluorine, oxygen and nitrogen, which have small atomic size and high electronegativities.

12.15 Since these are both nonpolar molecular substances, the only type of intermolecular force that we need to consider is London forces. The larger molecule has the greater London force of attraction and hence the higher boiling point: C_8H_{18}.

12.16 Whereas the ether has no O–H linkage, ethanol does. Therefore, ethanol can have hydrogen bonding between molecules and ether cannot. Ethanol thus has stronger intermolecular forces, and its boiling point is consequently higher.

12.17 Covalent bonds are normally about 100 times stronger than normal dipole–dipole attractions; hydrogen bonds are about 5–10 times stronger than dipole–dipole attractions.

12.18 The instantaneous dipole moment in an otherwise nonpolar substance arises from a momentary imbalance in the electron distribution within the molecule. This creates an induced dipole in a neighboring molecule. As the polarizability of an atom increases, it becomes easier to distort electron clouds and generate London forces.

12.19 These are the attractive forces that are created when an ion induces a dipole in a neighboring molecule. The strength of these forces increases with increasing charge if the ion size is constant.

12.20 The ion induced dipole attractions would be stronger for CaO because it is more ionic than CaS and because the oxide ion is smaller and, consequently, the charge is more densely packed for CaO.

12.21 Physical properties that depend on tightness of packing: compressibility and diffusion. Physical properties that depend on the strengths of intermolecular interactions: retention of volume and shape, surface tension, wetting of a surface by a liquid, viscosity, evaporation, and sublimation.

12.22 The particles of a gas are free to move randomly, and thus to diffuse readily. Diffusion in liquids is comparatively slower because of the more numerous collisions that a molecule in a liquid sample must undergo in traveling from place to place. The particles of a solid are not free to move from place to place in a solid sample.

12.23 The average kinetic energies of the molecules of both the ice and the liquid water, at 0 °C, are the same.

12.24 Liquids and solids are difficult to compress because there is not much extra space between the particles of the solid or liquid. The compressibility of a gas, however, arises because there may be a great deal of room between the particles, and compression reduces the amount of this empty space.

12.25 The rate of diffusion should increase because the molecules move faster at the higher temperature.

12.26 Surface tension is related to the energy needed to decrease the surface area of a liquid. Molecules at the surface of a liquid have no other molecules above them, and they consequently are attracted only to those molecules that are next to them – namely, those in the interior of the liquid. This is illustrated in Figure 12.12.

12.27 Surface tension is a property that causes a liquid to form rounded droplets and allows liquids to rise above the rim of containers such as a glass (see Figure 12.13).

12.28 The greater the intermolecular attractions, the greater the surface tension.

12.29 Water should have the greater surface tension because it has the stronger intermolecular forces, i.e., hydrogen bonding.

12.30 Wetting – spreading a liquid across a surface to form a thin film.
Surfactant – a substance that lowers surface tension in a liquid and thereby promotes wetting.

12.31 There is no intermolecular force common to both polyethylene and water that can allow for wetting. The surface tension of water, which is high, is not disrupted by any effective interaction between water and polyethylene.

12.32 Glycerol ought to wet the surface of glass quite nicely, because the dipolar bonds at the surface of glass can interact strongly with the polar O–H groups of glycerol.

12.33 Water has a very high surface tension and it is more attracted to itself than to the nonpolar surface, whereas a nonpolar liquid does not have a high surface tension, therefore it will not form rounded droplets.

12.34 Since it is the high energy molecules in a sample that are the first to evaporate, the remaining molecules have a lower average kinetic energy. A reduction in kinetic energy corresponds to a decrease in temperature.

12.35 Raising the temperature of the sample increases the fraction of molecules in the sample that have enough kinetic energy to escape by evaporation.

12.36 An increase in surface area causes an increase in the rate of evaporation because, when the surface area is increased, there are more molecules in position at the surface of the liquid sample, where they are capable of evaporation. The stronger the intermolecular forces, the less readily a substance can evaporate.

12.37 The snow dissipates by sublimation. Even at low temperatures (by human standard) many compounds have measurable vapor pressures.

12.38 Freeze drying is accomplished by sublimation. It offers the advantage that the sample does not have to be subjected to high temperatures.

12.39 The rate of evaporation of a liquid increases with increasing temperature.

12.40 (a) sublimation
(b) vaporization
(c) condensation
(d) melting
(e) freezing

12.41 After the molecule is in the vapor phase for a while its kinetic energy will be less because it collided with other molecules in the vapor phase and transferred some of its energy to the other molecules. This molecule is not likely to bounce out of the surface of the liquid unless it regains some kinetic energy from the liquid.

12.42 This happens because of the loss in kinetic energy that the colliding molecule experiences when it hits the surface molecules. The colliding molecule has less kinetic energy after striking the surface, and its ability to escape subsequently from the liquid is momentarily diminished by the presence of intermolecular forces of attraction generated upon mixing.

12.43 A dynamic equilibrium is established if the liquid evaporates into a sealed container. It is termed a dynamic equilibrium because opposing processes (evaporation and condensation) continue to take place, once the condition of equilibrium has been achieved. At equilibrium, the rate of condensation is equal to the rate of evaporation, and there is consequently no net change in the number of molecules in the vapor or in the liquid.

12.44 A dynamic equilibrium is achieved when a solid is held at its melting temperature. At this point, particles are melting and freezing at an equivalent rate. This is the melting or freezing point of a substance.

12.45 Yes. This is the sublimation process.

12.46 When moist air contacts a surface that has a temperature that is below the freezing point of water, then the water in the moist air will freeze onto the surface.

12.47 Equilibrium vapor pressure is the pressure exerted by a vapor that is in equilibrium with its liquid. It is a dynamic equilibrium because events have not ceased. Liquid continues to evaporate once the state of equilibrium has been reached, but the rate of evaporation is equal to the rate of condensation. These two opposing processes occur at equal rates, such that there is no further change in the amount of either the liquid or the gas.

12.48 Changing the volume only upsets the equilibrium for a moment, provided the volume is not increased to a point that all the liquid evaporates at which no equilibrium would exist. After sufficient time has elapsed, the rates of evaporation and condensation again become equal to one another, and the same condition of equilibrium is achieved. The vapor pressure (or the ease of evaporation) only depends on the strength of intermolecular forces in the liquid sample.

12.49 The equilibrium vapor pressure is governed only by the strength of the attractive forces within the liquid and by the temperature.

12.50 Raising the temperature increases the vapor pressure by imparting enough kinetic energy (for evaporation) to more of the liquid molecules.

12.51 Air with 100% humidity is saturated with water vapor, meaning that the partial pressure of water in the air has become equal to the vapor pressure of water at that temperature. Since vapor pressure increases with increasing temperature, so too should the total amount of water vapor in humid (saturated) air increase with increasing temperature.

12.52 As the air rises, it becomes cooler, and eventually the amount of moisture in the air becomes greater than is required for equilibrium with the liquid. The air is less able to hold a given amount of vapor at the lower temperature, and condensation occurs.

12.53 At the temperature of the cool glass, the equilibrium vapor pressure of the water is lower than the partial pressure of water in the air. The air in contact with the cool glass is induced to relinquish some of its water, and condensation occurs.

12.54 In humid air, the rate of condensation on the skin is more nearly equal to the rate of evaporation from the skin, and the net rate of evaporation of perspiration from the skin is low. The cooling effect of the evaporation of perspiration is low, and our bodies are cooled only slowly under such conditions. In dry air, however, perspiration evaporates more rapidly, and the cooling effect is high.

12.55 Although the cold air outside the building may be nearly saturated at the low temperature, the same air circulated inside to a higher temperature becomes unsaturated. In fact the water content of the air at the higher temperature may be only a fraction of the maximum % humidity for that temperature. The % humidity of the air at the indoor temperature is, therefore, comparatively low.

12.56 Boiling point – the temperature at which the vapor pressure of a liquid is equal to the prevailing atmospheric pressure.

Normal boiling point – the temperature at which the vapor pressure of a liquid is equal to 1 atm, or the boiling point when the atmospheric pressure is 1 atm.

12.57 This happens because boiling is a process that is a function of pressure. Since the vapor pressure varies with temperature, the boiling point must also change as the pressure changes.

12.58 At about 77 °C

12.59 Boiling point is an easily measured property that varies considerably from one substance to another. It is thus characteristic of a substance, and useful as an identifying property of various substances.

12.60 In a sealed container, the water reaches the boiling point only at a temperature that is higher than the normal boiling point. This is because the vaporization of the liquid into a closed container causes an increase in pressure and a corresponding increase in the boiling point. This means that the food is cooked at a higher temperature.

12.61 Ethanol vapor is present inside the bubbles of boiling ethanol.

12.62 Even at higher temperatures, the contents of the radiator do not boil, because the pressure in the system increases with temperature, since the system is closed. The boiling point of the liquid is higher because the pressure is higher.

12.63 Inside the lighter, the liquid butane is in equilibrium with its vapor, which exerts a pressure somewhat above normal atmospheric pressure. This keeps the butane as a liquid.

12.64 Since H_2Se is larger than H_2S, its London forces are stronger than in H_2S. Because water is capable of hydrogen bonding, whereas H_2S is not, its boiling point is higher than that of H_2S.

12.65 The hydrogen bond network in HF is less extensive than in water, because it is a monohydride not a dihydride.

12.66
(a) 1, 3, and 5
(b) 2 and 4
(c) 2
(d) 4
(e) The heat of vaporization is larger.
(f) This is the temperature of line 4.

(g) This is the temperature of line 2.
(h) Line 3 would descend lower in temperature than line 4, before rising to the temperature of line 4.

12.67 (a) molar heat of vaporization, ΔH_{vap}
(b) molar heat of sublimation, ΔH_{sub}
(c) molar heat of fusion, ΔH_{fus}

12.68 The heat of vaporization of a molecular substance is generally larger than the heat of fusion, because, in vaporization, the molecules undergo much larger changes in their distance of separation (and require the disruption of much stronger intermolecular forces) than is true of melting. The heat of sublimation is typically larger than the heat of vaporization of a liquid because sublimation involves a greater change in intermolecular separation, a larger disruption of intermolecular forces of attraction, and hence a larger change in potential energy.

12.69 The heat of condensation is exothermic, and it is equal in magnitude, but opposite in sign to the heat of vaporization (which is endothermic).

12.70 The moisture which powers the hurricane, condenses when the hurricane travels over cold water. As the amount of vapor in the storm decreases, its energy decreases.

12.71 The condensation process is exothermic. The energy that is released warms the air, which then rises. This warming and rising of the air creates wind currents.

12.72 The substance with the larger molar heat of vaporization has the stronger intermolecular forces. This is ethanol, which has hydrogen bonding, whereas ethyl acetate does not.

12.73 Steam releases a considerable amount of energy in the form of condensation energy as opposed to simply cooling liquid water.

12.74 $CH_4 < CF_4 < HCl < HF$

12.75 When a system at equilibrium is modified so as to upset the equilibrium, the system will respond in a manner which enables the equilibrium to be reestablished. See also Section 12.8.

12.76 By "position of equilibrium" we mean the relative amounts of the various reactants and products that exist in the equilibrium mixture.

12.77 This is an endothermic system, and adding heat to the system will shift the position of the equilibrium to the right, producing a new equilibrium mixture having more liquid and less solid. Some of the solid melts when heat is added to the system.

12.78 Changing the temperature disrupts the sublimation, deposition equilibrium that was established. By lowering the temperature, the kinetic energy of the molecules will decrease and more deposition will occur, giving rise to a lower vapor pressure. This will continue until a new equilibrium is established in which the rate of sublimation and deposition are again equal.

12.79 An increase in pressure should favor the system with the lower volume, i.e. the solid. Therefore, if the substance is at its melting point at a pressure of one atm, and then if the pressure were to be increased, more solid would form at the expense of liquid – that is, more of the substance would freeze. If melting were to be accomplished at the higher pressure, it would require a temperature that is higher than the normal melting temperature.

12.80 Critical temperature – the temperature above which the substance can not exist as a liquid, regardless of the applied pressure. It is, therefore, the temperature above which a gas cannot be made to liquefy, regardless of the amount of pressure that is applied.

Critical pressure – the vapor pressure of a liquid at the liquid's critical temperature.

A critical temperature and critical pressure together constitute a substance's critical point.

12.81 A supercritical fluid is a substance at a temperature above its critical temperature. Supercritical CO_2 is used to decaffeinate coffee because it replaces organic solvents such as methylene chloride and ethyl acetate which cannot be completely removed from the coffee bean. Supercritical CO_2 can remove as much as 97% of the caffeine.

12.82 In order for a refrigeration system to work, the gas must be condensed into a liquid, if the critical temperature is too low, then the gas will not condense. The boiling point of the substance must be low because the refrigerant needs to be expanded into a gas, and if the boiling point is too high, then the substance will not form a gas.

12.83 Solid, liquid and gas are all in equilibrium at the triple point.

12.84 Carbon dioxide does not have a normal boiling point because its triple point lies above one atmosphere. Thus, the liquid–vapor equilibrium that is taken to represent the boiling point does not exist at the pressure (1 atm) conventionally used to designate the "normal" boiling point.

12.85 The critical temperature of hydrogen is below room temperature because, at room temperature, it cannot be liquefied by the application of pressure. The critical temperature of butane is above room temperature, because butane can be liquefied by the application of pressure.

Review Problems

12.86 Diethyl ether has the faster rate of vaporization, since it does not have hydrogen bonds, as does butanol.

12.87 Diethyl ether should have a higher vapor pressure since it has weaker intermolecular forces. Butanol has a higher boiling point since it has stronger intermolecular forces of attraction.

12.88 London forces are possible in them all. Where another intermolecular force can operate, it is generally stronger than London forces, and this other type of interaction overshadows the importance of the London force. The substances in the list that can have dipole–dipole attractions are those with permanent dipole moments: (a), (b), and (d). SF_6, (c), is a non–polar molecular substance. HF, (a), has hydrogen bonding.

12.89
(a) London forces, dipole-diploe, H-bonding
(b) London forces, dipole-dipole
(c) London forces
(d) London forces, dipole-dipole, H-bonding

12.90 Chloroform would be expected to display larger dipole-dipole attractions because it has a larger dipole moment than bromoform. (Chlorine has a higher electronegativity which results in each C-Cl bond having a larger dipole than each C-Br bond.) On the other hand, bromoform would be expected to show stronger London forces due to having larger electron clouds which are more polarizable than those of chlorine.

Since bromoform in fact has a higher boiling point that chloroform, we must conclude that it experiences stronger intermolecular attractions than chloroform, which can only be due to London forces. Therefore, London forces are more important in determining the boiling points of these two compounds.

12.91 NO_2 is capable of forming a liquid at atmospheric pressure while CO_2 does not, this suggests that NO_2 has stronger intermolecular attractions than CO_2. Since CO_2 and NO_2 cannot form hydrogen bonds, the next strongest intermolecular interactions are dipole-dipole interactions; therefore, NO_2 probably has dipole-dipole interactions. This would require NO_2 to have a dipole which is only possible if it is bent, while CO_2 is linear.

12.92 Ethanol, because it has H-bonding.

12.93 The London forces are stronger in CS_2 because the larger S atoms are more easily polarized than O atoms. Consequently, CS_2 has a higher boiling point than CO_2.

12.94 ether < acetone < benzene < water < acetic acid

12.95 diethyl ether < ethanol < water < ethylene glycol

12.96 Relative humidity is determined by the formula given in Chapter 12, shown below. The equilibrium vapor pressure of water at 25 °C may be found in the table "Vapor Pressure of Water…" in the Appendix of the text.

$$\% \text{ relative humidity} = \left(\frac{\text{actual } P_{H_2O}}{\text{equilibrium } P_{H_2O}}\right)\times 100$$

$$= \left(\frac{18.42 \text{ torr}}{23.8 \text{ torr}}\right)\times 100 = 77.4\%$$

12.97 The vapor pressure for water at 30 °C is 31.8 torr. If the relative humidity is 65%, then the pressure exerted by the water vapor is:

$$31.8 \text{ torr} \times 0.65 = 20.7 \text{ torr}$$

12.98 $$\# \text{ kJ} = (125 \text{ g } H_2O)\left(\frac{1 \text{ mol } H_2O}{18.015 \text{ g } H_2O}\right)\left(\frac{43.9 \text{ kJ}}{1 \text{ mol } H_2O}\right) = 305 \text{ kJ}$$

12.99 $$\# \text{ kJ} = (5.00 \text{ g } C_3H_6O)\left(\frac{1 \text{ mol } C_3H_6O}{58.1 \text{ g } C_3H_6O}\right)\left(\frac{-30.3 \text{ kJ}}{1 \text{ mol } C_3H_6O}\right) = -2.61 \text{ kJ}$$

12.100 We can approach this problem by first asking either of two equivalent questions about the system: how much heat energy (q) is needed in order to melt the entire sample of solid water (105 g), or how much energy is lost when the liquid water (45.0 g) is cooled to the freezing point? Regardless, there is only one final temperature for the combined (150.0 g) sample, and we need to know if this temperature is at the melting point (0 °C, at which temperature some solid water remains in equilibrium with a certain amount of liquid water) or above the melting point (at which temperature all of the solid water will have melted).

Heat flow supposing that all of the solid water is melted:

$$q = 6.01 \text{ kJ/mole} \times 105 \text{ g} \times 1 \text{ mol}/18.0 \text{ g} = 35.1 \text{ kJ}$$

Heat flow on cooling the liquid water to the freezing point:

$$q = 45.0 \text{ g} \times 4.18 \text{ J/g °C} \times 85 \text{ °C} = 1.60 \times 10^4 \text{ J} = 16.0 \text{ kJ}$$

The lesser of these two values is the correct one, and we conclude that 16.0 kJ of heat energy will be transferred from the liquid to the solid, and that the final temperature of the mixture will be 0 °C. The system will be an equilibrium mixture weighing 150 g and having some solid and some liquid in equilibrium with one another. The amount of solid that must melt in order to decrease the temperature of 45.0 g of water from 85 °C to 0 °C is: 16.0 kJ ÷ 6.01 kJ/mol = 2.66 mol of solid water. 2.66 mol × 18.0 g/mol = 47.9 g of water must melt.

(a) The final temperature will be 0 °C.
(b) 47.9 g of water must melt.

12.101 The amount of heat gained by melting all the benzene is:

$$\# \text{ kJ} = (10.0\text{g benzene})\left(\frac{1 \text{ mol benzene}}{78.11 \text{ g benzene}}\right)\left(\frac{9.92 \text{ kJ}}{1 \text{ mol}}\right) = 1.27 \text{ kJ}$$

Assuming all of this heat is removed from the water and using:
Heat = mass × specific heat × ΔT

$$\Delta T = \frac{\text{heat}}{\text{mass} \times \text{specific heat}} = \frac{1.27 \times 10^3 \text{ J}}{(10.0\text{g})(4.184\text{J/g}^\circ\text{C})} = 30.4^\circ\text{C}$$

Since the final temperature cannot be lower than the initial temperature, the final temperature of the water will be 5.45 °C and there will still be some solid benzene remaining.

12.102 Water boils at a higher temperature than ethanol. This reveals large intermolecular attractions in water, which would cause us to expect a larger molar heat of vaporization for water than ethanol.

12.103 Ethylene glycol would be expected to have a higher molar heat of fusion since it has a higher boiling point which indicates stronger intermolecular forces.

12.104

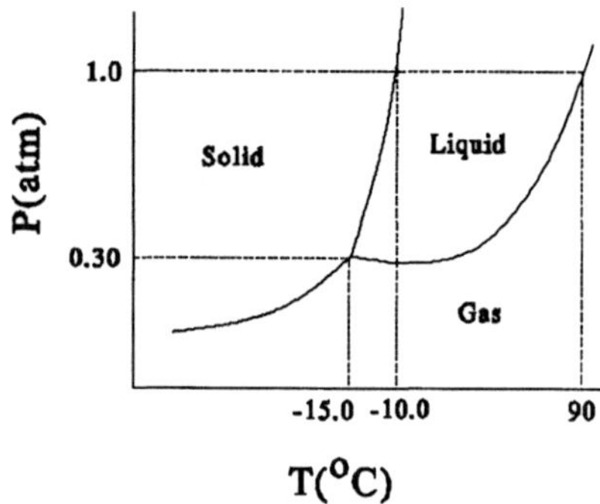

12.105 Sublimation is possible only below a pressure of 0.30 atm, as marked on the phase diagram. The density of the solid is higher than that of the liquid. Notice that the line separating the solid from the liquid slopes to the right, in contrast to the diagram for water, Figure 12.28 of the text.

12.106 (a) solid (b) gas (c) liquid (d) solid, liquid, and gas

12.107 The solid–liquid line slants toward the right.

12.108 At –56 °C, the vapor is compressed until the liquid–vapor line is reached, at which point the vapor condenses to a liquid. As the pressure is increased further, the solid–liquid line is reached, and the liquid freezes.
At –58 °C, the gas is compressed until the solid–vapor line is reached, at which point the vapor condenses directly to a solid.

12.109 The solid's temperature increases until the solid–gas line is reached, at which point the solid sublimes. After it has completely vaporized, the vapor's temperature increases to 0 °C.

Additional Exercises

12.110 Intermolecular (between different particles): The principal attractive forces are ion–ion forces and ion–dipole forces. These are of overwhelming strength compared to London forces, which do technically exist. Intramolecular (within certain particles): The sulfite ion (SO_3^{2-}) is a polyatomic ion whose atoms are held together by *covalent bonds.* Although a covalent bond is not an "attraction" in itself, the attractive forces which make up these bonds are valence electron's attractions to the nuclei of neighboring atoms in the polyatomic ion.

12.111 The pressure of the water is:

$$P_{H_2O} = (0.75)(17.54 \text{ torr}) = 13.16 \text{ torr} = 13.16 \text{ mm Hg}$$

The number of moles of water is calculated from the ideal gas equation:

$$PV = nRT$$

$$n = \frac{PV}{RT}$$

$$n = \frac{(13.16 \text{ torr})\left(\frac{1 \text{ atm}}{760 \text{ torr}}\right)(10.0 \text{ L})}{(0.0821 \text{ L atm/mol K})(293 \text{ K})} = 7.20 \times 10^{-3} \text{ mol } H_2O$$

The number of grams of water is calculated from the molecular mass of water:

$$\# \text{ g } H_2O = 7.20 \times 10^{-3} \text{ mol } H_2O\left(\frac{18.02 \text{ g } H_2O}{1 \text{ mol } H_2O}\right) = 0.130 \text{ g } H_2O$$

12.112 Yes, because of the possibility of a weak hydrogen bond between the carbonyl oxygen of acetone and an OH group of water.

12.113 The comparatively weak intermolecular forces in acetone allow for very rapid evaporation. This provides for significantly more cooling than does ethylene glycol. The intermolecular forces of ethylene glycol, plus its ability to adhere to the skin through OH groups, causes it to evaporate slowly.

12.114 Using Hess's Law, sublimation may be considered equivalent to melting followed by vaporization.
$\Delta H_{sublimation} = \Delta H_{fusion} + \Delta H_{vaporization} = 10.8 \text{ kJ/mol} + 24.3 \text{ kJ/mol} = 35.1 \text{ kJ}$

12.115 As the cooling takes place, the average kinetic energy of the gas molecules decreases, and the attractive forces that can operate among the gas molecules become able to bind the various molecules together in a process that leads to condensation. The air thus loses much of its moisture on the ascending side of the mountain. On descending the other side of the mountain, the air is compressed, and the temperature rises according to Charles' Law. The relative humidity of the air is now very low for two reasons: much of the moisture was released on the other side of the mountain range, and now that the temperature is higher, there is far less than the maximum allowable water content. The coast of California receives the rain since the air releases its moisture to the west of the mountains and the valleys east of the mountains do not receive any rain.

12.116 At the higher temperature, all of the carbon dioxide exists in the gas phase, because the critical temperature has been exceeded.

12.117 Acetone is polar and has dipole–dipole forces. Propane is nonpolar and has only London forces. Therefore, we need only reach 97 °C before the propane cannot be liquified, regardless of pressure. Above 97 °C, the weak intermolecular forces of propane cannot cause condensation. Accordingly, the stronger intermolecular forces of attraction in acetone produce a significantly higher T_{crit}.

12.118

$$\ln\left(\frac{P_1}{P_2}\right) = \frac{\Delta H_{vap}}{8.314 \text{ J mol}^{-1}\text{K}^{-1}}\left(\frac{1}{T_2} - \frac{1}{T_1}\right)$$

$$\ln\left(\frac{72.8}{186.2}\right) = \frac{\Delta H_{vap}}{8.314 \text{ J mol}^{-1}\text{K}^{-1}}\left(\frac{1}{313 \text{ K}} - \frac{1}{293 \text{ K}}\right)$$

$$-0.939 \times 8.314 \text{ J mol}^{-1}\text{K}^{-1} = \Delta H_{vap} \times (-0.000218 \text{ K}^{-1})$$

$$\Delta H_{vap} = 3.58 \times 10^4 \text{ J mol}^{-1} = 35.8 \text{ kJ/mol}$$

12.119

$$\ln\left(\frac{P_1}{P_2}\right) = \frac{\Delta H_{vap}}{8.314\ \text{J mol}^{-1}\text{K}^{-1}}\left(\frac{1}{T_2} - \frac{1}{T_1}\right)$$

$$\ln\left(\frac{31.6}{P_2}\right) = \frac{42.09 \times 10^3\ \text{J mol}^{-1}}{8.314\ \text{J mol}^{-1}\text{K}^{-1}}\left(\frac{1}{333\ \text{K}} - \frac{1}{293\ \text{K}}\right)$$

Take the antilog of both sides

$$\left(\frac{31.6}{P_2}\right) = e^{-2.08} = 0.125$$

$$P_2 = 31.6\ \text{torr}/0.125 = 253\ \text{torr}$$

12.120

$$\ln\left(\frac{P_1}{P_2}\right) = \frac{\Delta H_{vap}}{8.314\ \text{J mol}^{-1}\text{K}^{-1}}\left(\frac{1}{T_2} - \frac{1}{T_1}\right)$$

$$\ln\left(\frac{28.1}{47.0}\right) = \frac{\Delta H_{vap}}{8.314\ \text{J mol}^{-1}\text{K}^{-1}}\left(\frac{1}{60\ \text{K}} - \frac{1}{57\ \text{K}}\right)$$

$$-0.514 \times 8.314\ \text{J mol}^{-1}\text{K}^{-1} = \Delta H_{vap} \times \left(-8.77 \times 10^{-4}\ \text{K}^{-1}\right)$$

$$\Delta H_{vap} = 4.87 \times 10^3\ \text{J mol}^{-1} = 4.87\ \text{kJ/mol}$$

The normal boiling point is where the vapor pressure equals 760 torr. So,

$$\ln\left(\frac{P_1}{P_2}\right) = \frac{\Delta H_{vap}}{8.314\ \text{J mol}^{-1}\text{K}^{-1}}\left(\frac{1}{T_2} - \frac{1}{T_1}\right)$$

$$\ln\left(\frac{28.1}{760}\right) = \frac{\Delta H_{vap}}{8.314\ \text{J mol}^{-1}\text{K}^{-1}}\left(\frac{1}{T_2} - \frac{1}{57\ \text{K}}\right)$$

$$-3.298 \times 8.314\ \text{J mol}^{-1}\text{K}^{-1}/\ 4.87 \times 10^3\ \text{J/mol} = \frac{1}{T_2} - \frac{1}{57\ \text{K}}$$

$$-5.64 \times 10^{-3}\ \text{K}^{-1} = \frac{1}{T_2} - 1.75 \times 10^{-2}$$

$$\frac{1}{T_2} = 1.19 \times 10^{-2}$$

$$T_2 = 84.0\ \text{K} = -189.2\ ^\circ\text{C}$$

12.121

$$\ln\left(\frac{P_1}{P_2}\right) = \frac{\Delta H_{vap}}{8.314\ \text{J mol}^{-1}\text{K}^{-1}}\left(\frac{1}{T_2} - \frac{1}{T_1}\right)$$

$$\ln\left(\frac{185}{292}\right) = \frac{\Delta H_{vap}}{8.314\ \text{J mol}^{-1}\text{K}^{-1}}\left(\frac{1}{283\ \text{K}} - \frac{1}{273\ \text{K}}\right)$$

$$-0.456 \times 8.314\ \text{J mol}^{-1}\text{K}^{-1} = \Delta H_{vap} \times \left(-1.29 \times 10^{-4}\ \text{K}^{-1}\right)$$

$$\Delta H_{vap} = 2.93 \times 10^4\ \text{J mol}^{-1} = 29.3\ \text{kJ/mol}$$

12.122 The "two point" form of the Clausius-Clapeyron equation is:

$$\ln\left(\frac{P_1}{P_2}\right) = \frac{\Delta H_{vap}}{8.314\ \text{J mol}^{-1}\text{K}^{-1}}\left(\frac{1}{T_2} - \frac{1}{T_1}\right)$$

Inserting the given values from the problem, we have:

$$\ln\left(\frac{P_1}{P_2}\right) = \frac{\Delta H_{vap}}{8.314 \text{ J mol}^{-1}\text{K}^{-1}}\left(\frac{1}{T_2} - \frac{1}{T_1}\right)$$

$$\ln\left(\frac{20.0}{P_2}\right) = \frac{38{,}000 \text{ J/mol}}{8.314 \text{ J mol}^{-1}\text{K}^{-1}}\left(\frac{1}{298} - \frac{1}{291.6 \text{ K}}\right)$$

$$\ln\left(\frac{20.0}{P_2}\right) = 4{,}570 \text{ K} \times (-0.000{,}0737\text{K})$$

$$\ln\left(\frac{20.0}{P_2}\right) = -0.337$$

$$\frac{20.0}{P_2} = e^{-0.337}$$

$$\frac{20.0}{P_2} = 0.714$$

$$P_2 = 28.0 \text{ torr}$$

12.123 The "two point" form of the Clausius-Clapeyron equation is:

$$\ln\left(\frac{P_1}{P_2}\right) = \frac{\Delta H_{vap}}{8.314 \text{ J mol}^{-1}\text{K}^{-1}}\left(\frac{1}{T_2} - \frac{1}{T_1}\right)$$

Inserting the given values from the problem, we have:

$$\ln\left(\frac{P_1}{P_2}\right) = \frac{\Delta H_{vap}}{8.314 \text{ J mol}^{-1}\text{K}^{-1}}\left(\frac{1}{T_2} - \frac{1}{T_1}\right)$$

$$\ln\left(\frac{40}{20.0}\right) = \frac{38{,}000 \text{ J/mol}}{8.314 \text{ J mol}^{-1}\text{K}^{-1}}\left(\frac{1}{291.6} - \frac{1}{T_1}\right)$$

$$0.693 = 4{,}570 \text{ K} \times \left(3.43\times10^{-3} - \frac{1}{T_1}\right)$$

$$1.52\times10^{-4} = \left(3.43\times10^{-3} - \frac{1}{T_1}\right)$$

$$3.28\times10^{-3} = \frac{1}{T_1}$$

$$T_1 = 305 \text{ K}$$

12.124 Here we need only solve for the temperature T_2 at which the vapor pressure P_2 has reached 760 torr:

$$\ln\left(\frac{P_1}{P_2}\right) = \frac{\Delta H_{vap}}{8.314 \text{ J mol}^{-1}\text{K}^{-1}}\left(\frac{1}{T_2} - \frac{1}{T_1}\right)$$

$$\ln\left(\frac{31.6}{760}\right) = \frac{42.09 \times 10^3 \text{ J/mol}}{8.314 \text{ J mol}^{-1}\text{K}^{-1}}\left(\frac{1}{T_2} - \frac{1}{293 \text{ K}}\right)$$

$$-3.18 = 5063 \text{ K} \times \left(\frac{1}{T_2} - \frac{1}{293 \text{ K}}\right)$$

$$-6.28 \times 10^{-4} \text{ K}^{-1} = \frac{1}{T_2} - \frac{1}{293 \text{ K}}$$

$$\frac{1}{T_2} = 2.78 \times 10^{-3} \text{ K}^{-1}; \quad T_2 = 359 \text{ K} = 86 \text{ °C}$$

12.125 From equation (1) in Facets of Chemistry 12.2, we have the relationship between ln P and 1/T:

$$\ln P = \left(\frac{-\Delta H_{vap}}{R}\right) \times \frac{1}{T} + c$$

This equation is in the form of a straight line, with the dependent variable being ln P and the independent variable being 1/T. We must therefore graph the quantity ln P vs. 1/T, and the slope will be equal to the value: $(-\Delta H_{vap}/R)$. The data are converted as follows, taking temperature in K and pressure in torr:

$1/T \times 10^3$ (K^{-1})	ln P
5.52	3.69
5.38	4.09
5.18	4.61
4.72	5.70
4.42	6.40

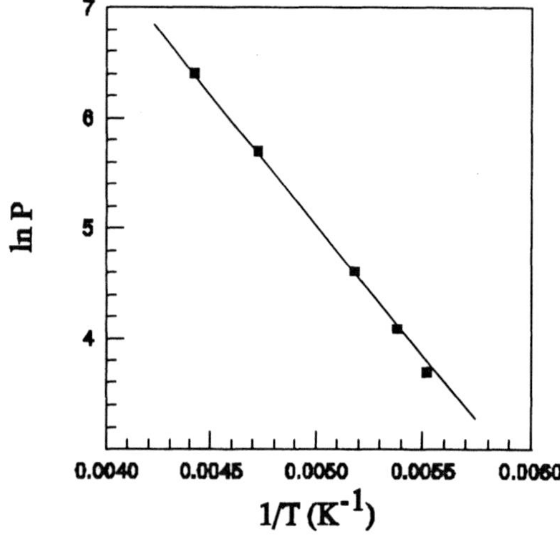

The slope is found to be: $-2440\ K = -\Delta H_{vap}/R$
The value of ΔH is thus: 20.3×10^3 J/mol

12.126 Here we use Equation (2) in Facets of Chemistry 12.2 simultaneously for liquids A and B.

Liquid A: $$\ln\frac{100}{P_2} = \frac{32{,}000\ \text{J/mol}}{8.314\ \text{J/mol K}}\left(\frac{1}{T_2} - \frac{1}{273\ \text{K}}\right)$$

Liquid B: $$\ln\frac{200}{P_2} = \frac{18{,}000\ \text{J/mol}}{8.314\ \text{J/mol K}}\left(\frac{1}{T_2} - \frac{1}{273\ \text{K}}\right)$$

Since the problem asks for the case where $P_2(A) = P_2(B) = P_2$ and $T_2(A) = T_2(B) = T_2$, we solve the two equations for P_2 and T_2. Subtracting the second from the first we have:

$$\ln(0.5) = 1.683 \times 10^3\ K^{-1}\left(\frac{1}{T_2} - \frac{1}{273\ K}\right)$$

$$\frac{1}{T_2} = 3.25 \times 10^{-3}\ K^{-1}$$

$$T_2 = 307.6\ K = 34.4\ ^\circ C$$

Practice Exercises

13.1 For cesium:

8 corners × 1/8 Cs^+ per corner = 1 Cs^+

For chloride:

1 Cl^- in center, Total: 1 Cl^-

Thus, the ratio is 1 to 1.

13.2 Because this is a high melting, hard material, it must be a covalent or network solid. Covalent bonds link the various atoms of the crystal.

13.3 Since the melt does not conduct electricity, it is not an ionic substance. The softness and the low melting point suggest that this is a molecular solid, and indeed the formula is most properly written S_8.

13.4 -(CF_2-CF_2)-(CF_2-CF_2)-(CF_2-CF_2)-

Thinking It Through

T13.1 First, it is necessary to determine the number of chromium atoms in the unit cell, a body–centered cubic unit cell, then determine the number of atoms in 26.0 g of chromium using the molecular mass of chromium and Avogadro's number. Finally, the number of unit cells would be the number of atoms in 26.0 g of chromium divided by the number of atoms in the unit cell.

T13.2 If the unit cell contains four atoms of A then there must be six atoms of B in the unit cell. The formula for the compound is A_2B_3, giving a stoichiometric ratio of 2 A atoms to 3 B atoms. The number of atoms in the unit cell must have the same ratio.

T13.3 The pentagon (a) cannot be a unit cell in a two-dimensional lattice. When it is moved in one direction parallel to one edge, this does not align with the unit cell that was move in the direction of the adjacent edge. The unit cells do not build up a two-dimensional lattice.

T13.4 The structure of (b) is more likely to form a condensation polymer since it has both an acid and a base functionality. These can react to form a bond an eliminate water. Structure (a) is more likely to form an addition polymer.

T13.5 The light that has passed through a polarizing filter is less intense than light passed through ordinary glass because the polarizing filter has removed all of the vibrations except those in one plane.

T13.6 Liquid crystals have a rigid region, are long and thin, and have a strong dipole. The hydrocarbon does not have the rigid region and it does not have a strong dipole even though it is long and thin.

T13.7 In general, ceramics are made from metals with large positive oxidation states combined with small nonmetals with large negative oxidation states. For $TiBr_3$, Ti has only a +3 oxidation state and Br^- is a large nonmetal ion with a low negative oxidation state.

T13.8 First, it is necessary to determine the length of the unit cell edge. For NaCl type structures, the cell edge is always twice the radius of the cation plus twice the radius of the anion. Next, it is necessary to calculate the volume of the unit cell, which is simply the cube of the cell edge. Finally, the mass of the atoms contained in one unit cell is divided by the volume of the unit cell in order to calculate density.

T13.9 The unit cell for gold, a face centered cubic structure, using the Pythagorem theorem, $a^2 + b^2 = c^2$, the length of the diagonal across the face to the cube can be determined. This length is 4 × radius of the gold atom.

T13.10 To calculate the percentage of unoccupied space for a simple cubic structure, first determine the volume of the unit cell. Second, there is the equivalent of one atom in the unit cell, and the radius for the atom is 1/2 the length of the side of the unit cell. Determine the volume of the atom using the equation:

$$\text{Volume of the atom} = \frac{4}{3}\pi r^3$$

Finally subtract the volume of the atom from the volume of the cell; that will be the volume of empty space. The percentage of empty space is:

$$\text{Percentage of empty space} = \frac{\text{volume of empty space}}{\text{volume of the unit cell}} \times 100\%$$

T13.11 Convert 200 μm to nm

$$200\ \mu\text{m}\ \frac{1000\ \text{nm}}{1\ \mu\text{m}} = 2 \times 10^5\ \text{nm}$$

Then, assume that if the nanotubes will bundle together such that they will line up across the diameter of the human hair, and the number of nanotubes is:

$$\text{Number of nanotubes per hair} = \frac{200 \times 10^5\ \text{nm/hair}}{1.4\ \text{nm/nanotube}}$$

T13.12 To determine the largest value that the wavelength of the X rays, use equation 13.1

$$n\lambda = 2\,d\sin\theta$$

$$\lambda = \frac{2\,d\sin\theta}{n}$$

$\theta = 18.5°$ d = distance between the gold layers = 4.07×10^5 m n = integer = 1

Review Exercises

13.1 Crystalline solids have an ordered internal structure while amorphous solids do not have the long-range repetitive order of crystalline solids.

13.2 The surface features that suggest a high degree of order among the particles within them are the flat faces and the regular angles.

13.3 A lattice is a set of points that have the same repeat distances and are arranged along lines oriented at the same angles. A unit cell is the smallest repeating unit of the lattice that can be used to define the lattice.

13.4 The entire crystal lattice of the solid can be generated by repeated use of the unit cell only.

13.5 a)

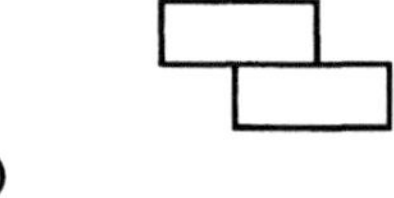

b)

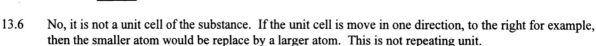

13.6 No, it is not a unit cell of the substance. If the unit cell is move in one direction, to the right for example, then the smaller atom would be replace by a larger atom. This is not repeating unit.

13.7 Simple cubic – possesses lattice points only at the eight corners of a cube.

Face–centered cubic – possesses the eight lattice points of the simple cubic cell, plus one in the center of each of the six faces of the cube.

Body–centered cubic – possesses the eight lattice points of the simple cubic cell, plus one at the center of the cube.

For sample sketches see Figures 13.4, 13.5 and 13.7.

13.8 See Figure 13.8 and 13.11.

13.9 These structures are both of the face–centered cubic variety. They differ only in the length of an edge for a unit cell, that is the length of the cube edge is different in the two metals. Silver might be expected to have a face–centered unit cell, also.

13.10 Zinc sulfide has a face centered cubic lattice.

13.11 Calcium fluoride also has a face centered cubic lattice.

13.12 Although there are only fourteen different kinds of lattice geometries that can fill space, there are essentially an infinite variety of cell dimensions that can be adopted by substances.

13.13 $n\lambda = 2d \sin\theta$

n = an integer (1, 2, 3, . . .)
λ = wavelength of the X–rays
d = the interplane spacing in the crystal
θ = the angle of incidence and the angle of reflectance of X–rays to the various crystal planes.

13.14 There are 100 pm in 1 Å.

$$\frac{100 \text{ pm}}{1 \text{ Å}} \qquad \frac{1 \text{ Å}}{100 \text{ pm}}$$

13.15 By measuring the various angles θ, one can compute the *d*–spacings between planes of atoms in the crystal lattice. This, plus the intensities of the reflected X–rays, is used to deduce the locations of the atoms in the unit cell. Some chemical intuition is then needed in order to decide which atoms are bonded together.

13.16 These are not 1:1 ionic substances. The cubic unit cell of NaCl contains the same number of sodium and chloride ions.

13.17 The lattice positions are occupied by metal cations, which are then surrounded by the core electrons of the metal. A sea of valence electrons encompasses the entire metallic solid.

13.18 (a) dipole–dipole, London forces or hydrogen bonds
(b) electrostatic forces
(c) covalent bonds

13.19 Covalent crystals are also termed network solids because they are constructed of atoms that are covalently bonded to one another, giving a giant interlocking network.

13.20 We expect ionic substances in general to be hard, brittle, high melting, and nonconducting in the solid but conducting when melted.

13.21 Amorphous means, literally, without form. It is taken here to represent a solid that does not have the regular, repeating geometrical form normally associated with a crystal lattice.

13.22 An amorphous solid is a noncrystalline solid. It is a solid that lacks the long–range order that characterizes a crystalline substance. When cooled, a liquid that will form an amorphous solid gradually becomes viscous and slowly hardens to give a glass, or a supercooled liquid. When crystalline solids are broken, the angles are regular and the faces are flat. When amorphous solids are broken, the faces are smooth and flat.

13.23 In a conductor, the valence band is partially filled, making the valence band a conduction band. In an insulator, the valence band is filled, preventing the electrons from moving. Also, in insulators, there is a large band gap between the valence band and the nearest vacant band. In semiconductors, although the valence band is filled (as in the case of insulators), there is a relatively small energy gap between the filled valence band and the empty conduction band. Consequently, electrons are readily promoted from the valence band into the conduction band.

13.24 The valence band is that set of energy levels containing the valence (outer) electrons. Any uninterrupted energy band that is partially filled is called a conduction band, because electrons are able to travel through the sample continuously.

13.25 As the temperature increases, so does the number of electrons that are promoted into the conduction band. See Figure 13.23.

13.26 The 2*s* level in calcium is filled, but the vacant 4*p* conduction band overlaps with the filled 4*s* level and provides for conductivity.

13.27 In a p–type semiconductor, a sample (for instance silicon) is doped with an "impurity" that provides a positive (hole) charge carrier. This impurity characteristically has fewer electrons than the host material. In n–type semiconductors, the "impurity" has more electrons than the sample, and the charge carriers that enter the conduction band are negative.

13.28 Since germanium is in Group IV, an element from Group III must be added to make a p–type semiconductor. Here we could use boron, aluminum, or gallium.

13.29 We must choose elements that have one more electron in the valence shell than does silicon. Phosphorus and arsenic are reasonable choices, since both are in Group V.

13.30 This is diagrammed in Figure 13.24. A p–type semiconductor is layed down over an n–type semiconductor. The energy supplied by light causes the equilibrium distribution of electrons and holes between the layers to be disturbed. As a result, excess numbers of electrons jump the so–called p–n junction, and enter the n layer from the p layer. This causes current to flow from the n layer into the external circuit.

13.31 A macromolecule is a molecule whose molecular mass is very large.

13.32 A polymer is a substance formed by linking together many simpler units called monomers. Not all macromolecules are polymers because they are not made of monomers.

13.33 A monomer is a substance of relatively low formula mass that is used to make a polymer.

a)

```
H   CH3
|   |
C ═ C
|   |
H   H
```

b)

```
F   F
|   |
C ═ C
|   |
F   F
```

c)

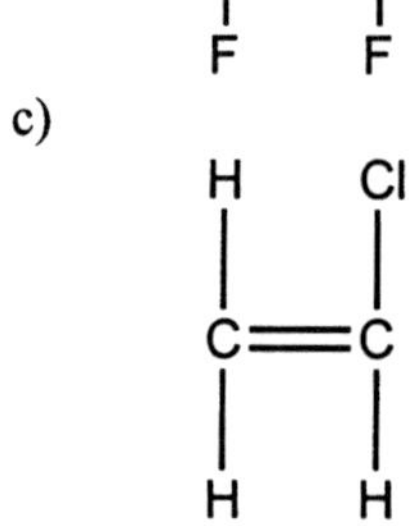

13.34 The polymer backbone is the long chain of atoms that have substituents at regular intervals.

13.35 The repeat unit for polypropylene is

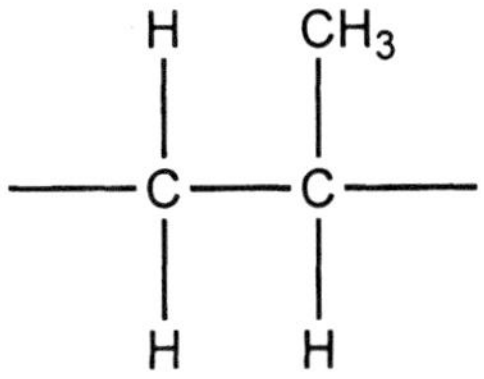

$$\left[-\mathrm{CH_2}-\mathrm{CH(CH_3)}- \right]$$

The formula for the polypropylene polymer is $[CH_2CH(CH_3)]_n$.
Uses of polypropylene are dishwasher-safe food containers, indoor-outdoor carpeting, and artificial turf.

13.36 Propylene has a double bond between the two carbons, while polypropylene has a single bond.

13.37 An addition polymer is the simple addition of one monomer unit to another. The condensation polymer eliminates a small molecule when the two monomer units become joined.

13.38

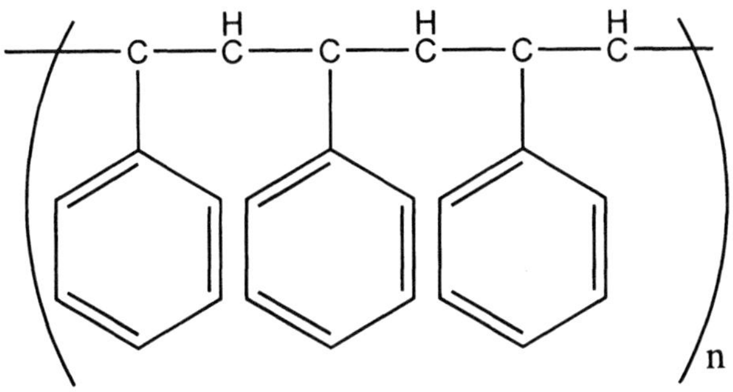

$$\left(-C-\overset{H}{C}-C-\overset{H}{C}-C-\overset{H}{C}- \right)_n \text{ (each unlabeled C bearing a phenyl ring)}$$

Three uses of polystyrene are clear plastic drinking glasses, molded car parts, housing of computers and kitchen appliances, Styrofoam cups and insulation materials.

13.39 A copolymer is formed by combing two different compounds.

13.40 a) Nylon 6,6

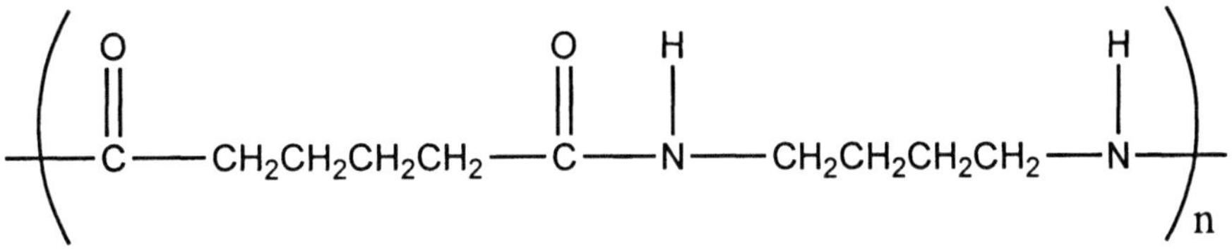

$$\left(-\overset{O}{\overset{\|}{C}}-CH_2CH_2CH_2CH_2-\overset{O}{\overset{\|}{C}}-\overset{H}{\overset{|}{N}}-CH_2CH_2CH_2CH_2-\overset{H}{\overset{|}{N}}- \right)_n$$

b)

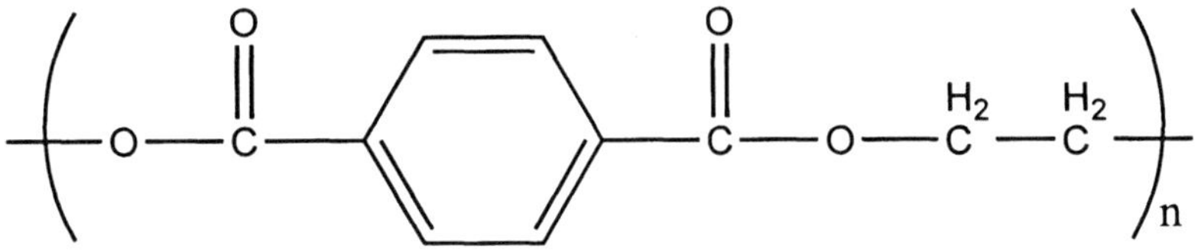

$$\left(-O-\overset{O}{\overset{\|}{C}}-C_6H_4-\overset{O}{\overset{\|}{C}}-O-\overset{H_2}{C}-\overset{H_2}{C}- \right)_n$$

13.41

$$\left(-\overset{O}{\overset{\|}{C}}-C_6H_4-\overset{O}{\overset{\|}{C}}-\underset{H}{\underset{|}{N}}-C_6H_4-\underset{H}{\underset{|}{N}}- \right)_n$$

amide bond

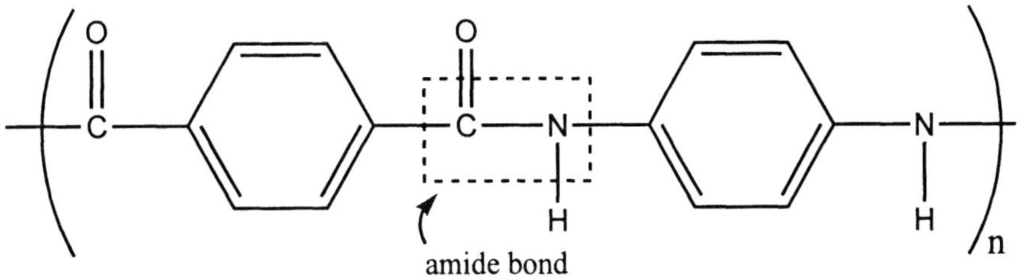

13.42

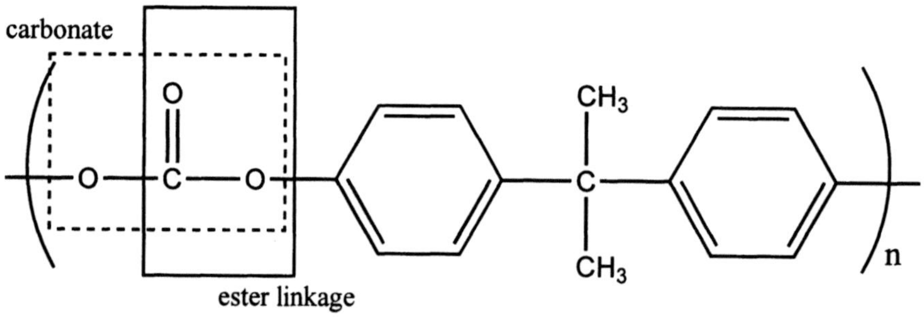

carbonate

ester linkage

13.43 Branching is when a polymer chain grows off the main backbone.

13.44 Cross-linking is occurs between two strands of a polymer in which the strands are chemically bonded to each other. The polymer becomes stronger, prevents it from becoming brittle in the cold, and gives the polymer a "memory."

13.45 Vulcanized rubber is natural rubber that has been reacted with sulfur in order for the polymer strands to become cross-linked.

13.46 The higher the polymer crystallinity, the stronger the polymer

13.47 Low-density polyethylene has short polymer chains and extensive branching which prevents the polymer form lining up in an orderly fashion.

13.48 Nylon forms strong fibers since the substituents on the chains can form hydrogen bonds between the polymer chains.

13.49

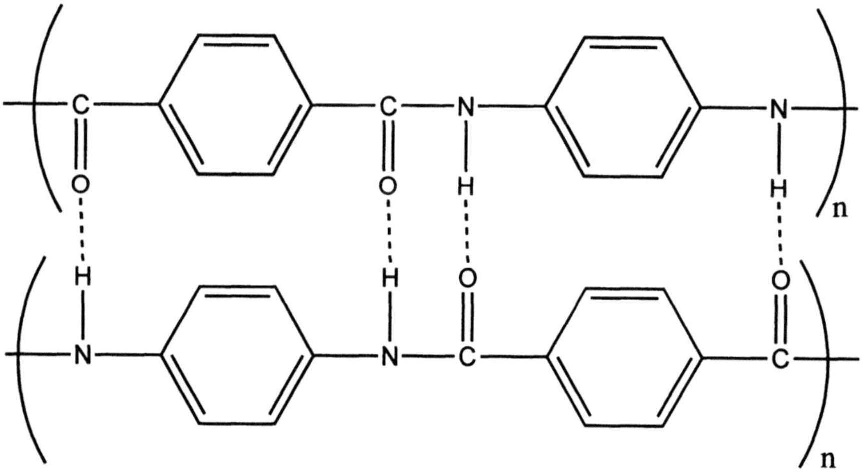

13.50 HDPE – lightweight, strong, tear and water resistant material like paper

UHDPE	bulletproof vests, cut-resistant fabric, surgical gloves
Kevlar	Bulletproof vests, thin yet strong hulls of racing boats.

13.51 Inorganic polymers do not use carbon for the backbone, but use other atoms such as silicon.

13.52

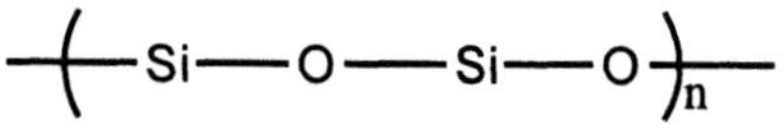

$$\left(\text{Si—O—Si—O}\right)_n$$

13.53 The group $(CH_3)_3SiCl$ is used to terminate a polymer chain. If CH_3SiCl_3 were added to the reaction, then the chain would branch.

13.54 The polymer found in the antigas medication simethicone is a silicone polymer:

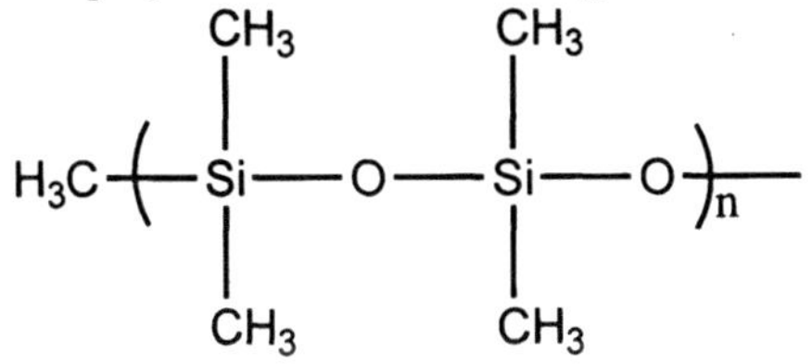

$$H_3C\left(-\underset{CH_3}{\overset{CH_3}{|}}{Si}-O-\underset{CH_3}{\overset{CH_3}{|}}{Si}-O-\right)_n-$$

13.55 A liquid crystal is a substance that has liquid properties but the molecules are not arranges altogether randomly.

13.56 (b) is most likely to form a liquid crystal since it is long and thin and has a rigid center as well as having a strong dipole.
(a) and (c) are not long and thin and do not have strong dipoles.

13.57 (a) In a nematic phase, the molecules are lined up parallel to each other in one direction, but are still able to move past each other up, down and sideways.
(b) In a smectic phase, the molecules are arranged parallel to each other and approximately in layers which are able to slide over each other.
(c) In a cholesteric phase, the molecules align as in the nematic phase, but in thin layers, and in the layers the orientations of the molecules rotate.

13.58 The color of reflected light from a cholesteric liquid crystal depend on the temperature because the wavelength of light that is reflected is equal to the distance require for the twist to make one full turn. This distance changes with temperature.

13.59 A smectic liquid crystal might have a slippery feel since the layers may slide past each other.

13.60 Plane polarized light has the oscillations of the electric and magnetic fields of the photons vibrate in the same plane. For unpolarized light, these vibrations are oriented randomly around the direction of the light beam.

13.61 Crossed polarizers do not permit the passage of light since one polarizer removes all the light except the vibrations in one direction. The other polarizer, at 90°, then removes those photons of light.

13.62 The electric field forces the liquid crystal molecules to leave their helical orientation and become aligned with the electric field. Now the liquid crystal molecules do not rotate the plane polarized light and the light is block by the second polarizer and the image looks black.

13.63 Student dependent answer.

13.64 Ceramic materials have high melting points and are very hard.

13.65 A refractory is a heat-resistant material. They are used to line furnaces and rocket engine exhausts.

13.66 Many ceramic materials, at room temperature, are used to make supports for high-voltage electrical transmission cables since they do not conduct electricity.

13.67 (a) and (d) have small none metals with large negative oxidation states and metals with large positive oxidation states.

13.68 Ceramic materials are formed into ceramic products by pulverizing the ceramic and molding it into the desired shape, then heating the object to a high temperature.

13.69 Sintering is heating the ceramic to a high temperature (>1000 °C) and allowing the fine particles to stick together. This can have a negative effect on the physical properties such as strength, since sintered materials often contain small cracks and voids.

13.70 The sol-gel process resembles the formation of a condensation polymer because a small molecule is eliminated in the formation of the network of atoms.

13.71 An alkoxide is an anion formed by removing a hydrogen ion from an alcohol. A hydrolysis reaction is one that involves a reaction with water.

13.72

$$Si(OC_2H_5)_4 + 4\,H_2O \longrightarrow Si(OH)_4 + 4\,C_2H_5OH$$

13.73 A xerogel is a porous solid formed when the solvent is evaporated from a wet gel. If a xerogel is heated to high temperatures, then the porous structure of the xerogel collapses and forms a dense ceramic of glass with a uniform structure.

13.74 An areogel is a very porous and extremely low density solid. If is made when the solvent is removed from the wet gel under supercritical conditions. It is an excellent insulator.

13.75

$$\left(-Si(CH_3)(H)-Si(CH_3)(H)-Si(CH_3)(H)-Si(CH_3)(H)- \right)_n \xrightarrow{\text{heat}} \left(\text{Si–C–Si network} \right)_n + H_2$$

13.76 Thin ceramic films make antireflecitvie coatings and filters for lighting, and TiN coatings for drill bits to make them more wear resistant. Golf spikes for golf shoes, hip-joint replacements, and knives are made from ZrO_2. Boron nitride is used to make silky textured cosmetics, and boron carbide is used with Kevlar for bulletproof vests.

13.77 A superconductor is a material in a state in which it offers no resistance of the flow of electricity. Typical metal superconductors work on at a few degrees above absolute zero and must be immersed in liquid helium which is very expensive.

13.78 If a substance that was an excellent superconductor at –175 °C, it would be of economic interest since liquid nitrogen could be used as a refrigerant, and liquid nitrogen is relatively inexpensive compared to liquid helium

13.79 $YBa_2Cu_3O_7$

13.80 It is possible to levitate a strong magnet above a substance that is in a superconducting state because a substance in its superconducting state permits no magnetic field within itself. A weak magnetic field is actually repelled by a superconductor.

13.81 Nano- means 10^{-9}. Nanotechnology looks at substances on the level of tens to hundreds of atoms. Femtotechnology would be smaller than an atom and millitechnology is closer to a scale of objects we can see.

13.82 The ultimate goal of nanotechnology is to be able to build materials from the atom up.

13.83 A scanning tunneling microscopy works on conducting samples by brining the tip of a sharp metal probe to the surface or the material and an electric current bridging the gap is begun. The flow of current is extremely sensitive to the distance between the tip of the probe and the sample. As the tip is moved across the surface, the height of the tip is continually adjusted to keep the current flow constant. By accurately recording the height fluctuations of the tip, a map of the hills and valleys on the surface is obtained. Ceramic objects do not conduct electricity, therefore, STM cannot be used with them.

13.84 Atomic force microscopy uses a very sharp stylus, moving it across the surface of the sample. Intermolecular forces between the tip of the probe and the surface molecules causes the probe to flex as it follows the ups and downs of the bumps that are the individual molecules and atoms. A mirrored surface attached to the probe reflects a laser beam at angles proportional to the amount of deflection of the probe. A sensor picks up the signal from the laser and translates it into data that can be analyzed by a computer to give three-dimensional images of the sample's surface. This could be used to investigate the surface of a ceramic object.

13.85 Graphite consists of layers of sp^2 hybridized carbon atoms, each of which consists of many hexagonal "benzene-like" rings fused together. Diamond is a three dimensional lattice in which each sp^3 hybridized carbon is at the center of a tetrahedron and is bonded to another carbon. Buckminsterfullerene is composed of sp^2 hybridized carbon atoms in a pattern of five- and six- member rings arranged like the seams in a soccer ball.

13.86 A single-walled carbon nanotube is a tube of carbon atoms in the form of a rolled-up sheet of graphite. Single-walled tubes have high strengths, and low densities, and they can be conductors or semiconductors depending on their structure.

13.87 Graphite is a flat structure while nanotubes are a tube of the same structure.

13.88 The diameter of a carbon nanotube is 1/50,000th the thickness of a human hair.

13.89 MoS_2 and NB

13.90 $BaTiO_3$ and $SrTiO_3$ nanorods are attractive candidates for data storage devices because they can be electrically polarized in either of two directions.

13.91 Al_2O_3

Review Problems

13.92 For zinc:

4 surrounding center = **4 Zn^{2+}**

For sulfide:

8 corners × 1/8 S^{2-} per corner = 1 S^{2-}

6 faces × 1/2 S^{2-} per face = 3 S^{2-}

Total = **4 S^{2-}**

13.93 A cube has six faces and eight corners. Each of the six face atoms is shared by two adjacent unit cells: 6 × 1/2 = 3 atoms. The eight corner atoms are each shared by eight unit cells: 8 × 1/8 = 1 atom. The total number atoms to be assigned to any one cell is thus 3 + 1 = 4.

(6 × 1/2 copper atoms) + (8 × 1/8 copper atoms) = 4 copper atoms

13.94 From figure 13.6, we can see that the length of the diagonal of the cell = 4r, where r = radius of the atom. According to the Pythagorean theorem,

$$a^2 + b^2 = c^2$$

for a right triangle. Since a = b here, we may re-write this as

$$2l^2 = c^2,$$

where l = length of the edge of the unit cell. As mentioned above, the diagonal of the unit cell = 4r, so we may say that

$$2l^2 = (4r)^2$$
$$l^2 = (4r)^2/2$$
$$l^2 = 16r^2/2$$
$$l^2 = 8r^2$$
$$l = \sqrt{(8r^2)}$$

Finally, substituting the value provided for r in the problem, $l = \sqrt{[8(1.24Å)^2]} = 3.51$ Å. Using the conversion factor 1pm = 100 Å, this is 351 pm.

13.95 The following diagram is appropriate:

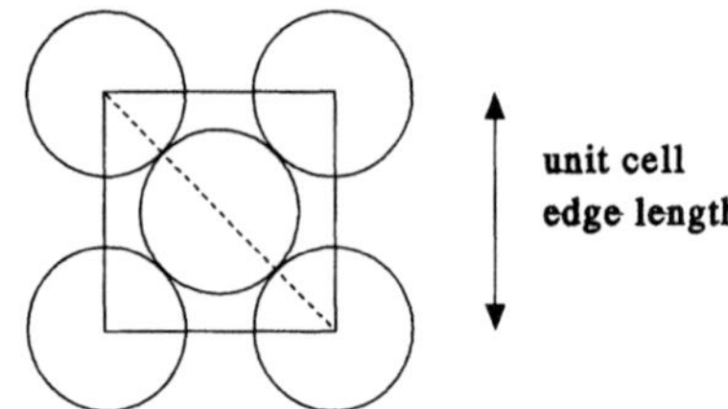

The face diagonal is 4 times the radius of the atom. The Pythagorean theorem is: diagonal^2 = edge^2 + edge^2. Hence we have: $[4(144 \text{ pm})]^2 = 2 \times \text{edge}^2$. Solving for the edge length we get 407 pm.

13.96 Each edge is composed of 2 × radius of the cation plus 2 × radius of the anion. The edge is therefore 2 × 133 + 2 × 195 = 656 pm.

13.97 $2 \times r_{Na} + d_{Cl} = 564.0$ pm
$2 \times 95 \text{ pm} + d_{Cl} = 564.0$ pm
$d_{Cl} = 374.0$ pm

13.98 Using the Bragg equation (eqn. 13.1), $n\lambda = 2d\sin\theta$

a)

$$n(229\text{pm}) = 2(1{,}000)\sin\theta$$
$$0.1145n = \sin\theta$$
$$\theta = 6.57°$$

b)

$$n(229\text{pm}) = 2(250)\sin\theta$$
$$0.458n = \sin\theta$$
$$\theta = 27.3°$$

13.99 $n\lambda = 2d \sin \theta$

$$d = \frac{n\lambda}{2\sin\theta}$$

At 20.0°

$$d = \frac{1 \times 141 \text{ pm}}{2 \times \sin 20.0°}$$

$$d = 206 \text{ pm}$$

At 27.4°

$$d = \frac{1 \times 141 \text{ pm}}{2 \times \sin 27.4°}$$

$$d = 153 \text{ pm}$$

At 35.8°

$$d = \frac{1 \times 141 \text{ pm}}{2 \times \sin 35.8°}$$

$$d = 121 \text{ pm}$$

13.100 From figure 13.6, we can see that the length of the diagonal of the cell = 4r, where r = radius of the atom. According to the Pythagorean theorem,

$$a^2 + b^2 = c^2$$

for a right triangle. Since a = b here, we may re-write this as

$$2l^2 = c^2,$$

where l = length of the edge of the unit cell. As mentioned above, the diagonal of the unit cell = 4r, so we may say that

$$2l^2 = (4r)^2$$
$$l^2 = (4r)^2/2$$
$$l^2 = 16r^2/2$$
$$l^2 = 8r^2$$
$$l = \sqrt{(8r^2)}$$

Finally, substituting the value provided for r in the problem, $l = \sqrt{[8(1.43\text{Å})^2]} = 4.04$ Å.

13.101 The body diagonal can be calculated using the Pythagorean theorem: $\text{diagonal}^2 = \text{edge}_1^2 + \text{edge}_2^2$, but in this case, one of the edges is the diagonal of the face of the cube ($\text{edge}_1^2 = \text{diagonal}_f^2$). This face diagonal can also be calculated using the Pythagorean theorem: $\text{diagonal}_f^2 = \text{edge}_2^2 + \text{edge}_2^2$. Putting it all together,

$\text{diagonal}^2 = \text{edge}_2^2 + \text{edge}_2^2 + \text{edge}_2^2$ or

$d^2 = 3a^2$

$a = 2.884$ Å

$d^2 = 24.95$ Å

$d = 4.995$ Å

13.102 First, let us convert the given density into units of amu/pm^{-3}:

$$\frac{\text{amu}}{\text{pm}^3} = \left(\frac{3.99 \text{ g}}{1 \text{ cm}^3}\right)\left(\frac{1 \text{ amu}}{1.66 \times 10^{-24} \text{ g}}\right)\left(\frac{1 \times 10^{-10} \text{ cm}}{1 \text{ pm}}\right)^3 = 2.40 \times 10^{-6} \text{ amu/pm}^3$$

Thus, the density of CsCl is $\underline{2.40 \times 10^{-6} \text{ amu/pm}^3}$.

If CsCl were *body-centered cubic* (see Figure 13.7)…

Its mass would be:

(2 Cs ions × 132.91) + (2 chloride ions × 35.45) = 336.72 amu

and its volume would be:

$(412.3 \text{ pm})^3 = 7.009 \times 10^7 \text{ pm}^3$

Therefore, its density would be:

$d = m/V = \underline{4.804 \times 10^{-6} \text{ amu/pm}^3}$

If CsCl were *face-centered cubic* (see Figure 13.10)…

Its mass would be:

(4 Cs ions × 132.91) + (4 chloride ions × 35.45) = 673.44 amu

and its volume would be:

$(412.3 \text{ pm})^3 = 7.009 \times 10^7 \text{ pm}^3$

Therefore, its density would be:

$d = m/V = \underline{9.609 \times 10^{-6} \text{ amu/pm}^3}$

Thus, CsCl is neither face-centered cubic, nor body-centered cubic.

13.103 First determine the number of unit cells in 1 cm^{-3} of sodium:

$$\text{number of unit cells} = \left(\frac{0.97 \text{ g Na}}{1 \text{ cm}^{-3}}\right)\left(\frac{1 \text{ mol Na}}{22.99 \text{ g Na}}\right)\left(\frac{6.022\times10^{23} \text{ atoms Na}}{1 \text{ mol Na}}\right)\left(\frac{1 \text{ unit cell}}{2 \text{ atoms Na}}\right)$$

$$= 1.27 \times 10^{22} \text{ unit cell in 1 cm}^3 \text{ of sodium}$$

Taking the inverse will give the size of the unit cell in cm^3:

7.87×10^{-23} cm^3 per unit cell.

The volume of a cube is: $v_{cube} = edge_{cube}{}^3$

$7.87 \times 10^{-23}\ cm^3 = edge^3$

$edge = 4.29 \times 10^{-8}$ cm

The body diagonal for a cube can be determined from the Pythagorean theorem:

$diagonal^2 = edge^2 + edge^2 + edge^2$

$diagonal^2 = 3(4.29 \times 10^{-8}\ cm)^2$

$diagonal^2 = 5.52 \times 10^{-15}\ cm^2$

$diagonal = 7.43 \times 10^{-8}$ cm

The radius of the sodium atom comes from:

diagonal = 4r

$r = 1.86 \times 10^{-8}$ cm = 1.86 Å = 186 pm

13.104 According to the Pythagorean theorem,

$$a^2 + b^2 = c^2$$

for a right triangle. First, we need to find the length of a diagonal on a face of the unit cell. Since a = b here, we may re-write this as

$$2l^2 = c^2,$$

where l = length of the edge of the unit cell and c = the diagonal length. Using the given 412.3 pm as the length of the edge, c = 583.1 pm. The diagonal length inside the cell from corner to opposite corner may now be found by the same theorem:

$$a^2 + b^2 = c^2$$
$$(412.3)^2 + (583.1)^2 = c^2$$
$$c = 714.1 \text{ pm}$$

This diagonal length inside the cell from corner to opposite corner is due to 1 Cs^+ ion and 1 Cl^- ion (see Figure 13.9). Therefore:

$$2r_{Cs+} + 2r_{Cl-} = 714.1\text{pm}$$
$$2r_{Cs+} + 2(181\text{pm}) = 714.1\text{pm}$$
$$2r_{Cs+} = 352 \text{ pm}$$
$$r_{Cs+} = 176 \text{ pm}$$

13.105 $(2 \times r_{Rb}) + (2 \times r_{Cl}) = 658$ pm

$(2 \times r_{Rb}) + (2 \times 181\ pm) = 658$ pm

$r_{Rb} = 148.0$ pm

13.106 This must be a molecular solid, because if it were ionic it would be high–melting, and the melt would conduct.

13.107 This is a covalent network solid.

13.108 This is a metallic solid.

13.109 This is a covalent, molecular solid.

13.110 This is a metallic solid.

13.111 This is a molecular, or covalent solid.

13.112 (a) molecular (b) ionic (c) ionic (d) metallic (e) covalent (f) molecular (g) ionic

13.113 (a) molecular (b) molecular (c) metallic (d) ionic (e) covalent (f) molecular (g) molecular

13.114

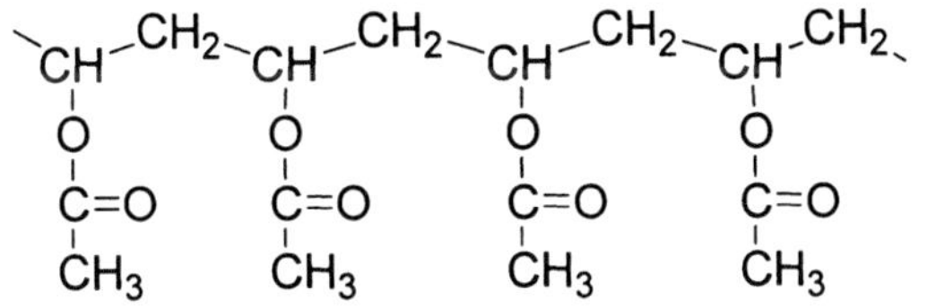

$$\text{—CH(O–C(=O)CH}_3\text{)–CH}_2\text{–CH(O–C(=O)CH}_3\text{)–CH}_2\text{–CH(O–C(=O)CH}_3\text{)–CH}_2\text{–CH(O–C(=O)CH}_3\text{)–CH}_2\text{—}$$

13.115

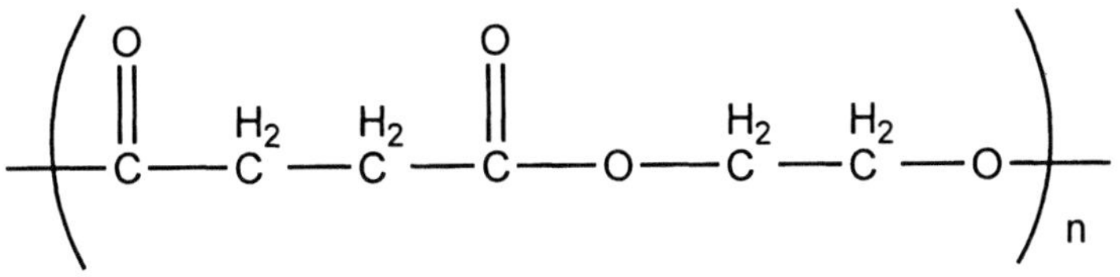

$$\left(\text{–C(=O)–CH}_2\text{–CH}_2\text{–C(=O)–O–CH}_2\text{–CH}_2\text{–O–}\right)_n$$

13.116

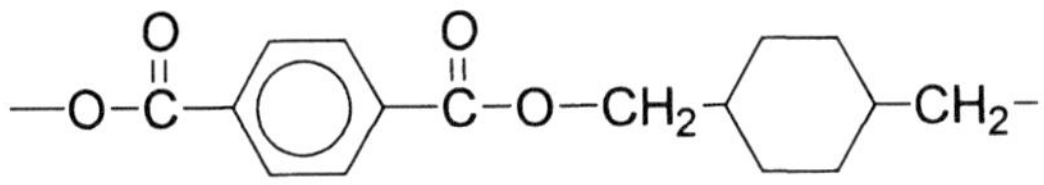

$$\text{–O–C(=O)–C}_6\text{H}_4\text{–C(=O)–O–CH}_2\text{–C}_6\text{H}_{10}\text{–CH}_2\text{–}$$

13.117

$$\left(\text{–C(=O)–C}_6\text{H}_4\text{–C(=O)–N(H)–CH}_2\text{–CH}_2\text{–CH}_2\text{–CH}_2\text{–N(H)–}\right)_n$$

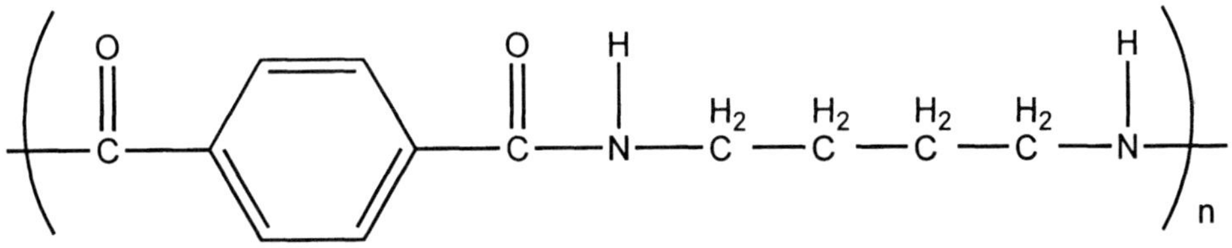

13.118

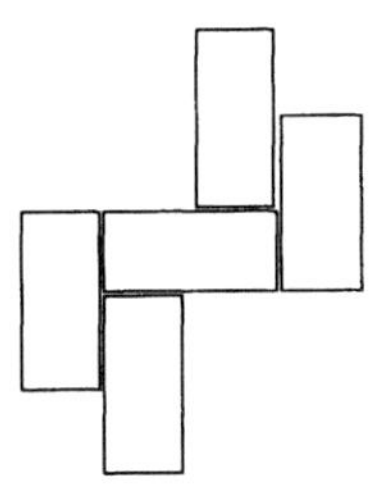

13.119 We start by determining the volume of one unit cell:
$407.86\ \text{pm} = 407.86 \times 10^{-12}\ \text{m} = 407.86 \times 10^{-10}\ \text{cm}$
$\text{Vol} = (407.86 \times 10^{-10}\ \text{cm})^3 = 6.7847 \times 10^{-23}\ \text{cm}^3$
Next, we calculate the volume per atom, remembering that each unit cell contains a total of 4 gold atoms:
$6.7847 \times 10^{-23}\ \text{cm}^3 / 4\ \text{atoms} = 1.6962 \times 10^{-23}\ \text{cm}^3$
This value is multiplied by the density:
$1.6962 \times 10^{-23}\ \text{cm}^3/\text{atom} \times 19.31\ \text{g/cm}^3 = 3.275 \times 10^{-22}\ \text{g/atom}$
Finally, it is necessary to divide this value into the atomic mass of gold:

$$\frac{196.97\ \text{g/mol}}{3.275 \times 10^{-22}\ \text{g/atom}} = 6.014 \times 10^{23}\ \text{atoms/mol}$$

13.120 From figure 13.6, we can see that the length of the diagonal of the cell = 4r, where r = radius of the atom. According to the Pythagorean theorem,

$$a^2 + b^2 = c^2$$

for a right triangle. Since a = b here, we may re-write this as

$$2l^2 = c^2,$$

where l = length of the edge of the unit cell. As mentioned above, the diagonal of the unit cell = 4r, so we may say that

$$2l^2 = (4r)^2$$
$$2l^2 = 16r^2$$
$$(0.125)l^2 = r^2$$
$$\sqrt{[(0.125)l^2]} = r$$

Finally, substituting the value provided for l in the problem, $r = \sqrt{[(0.125)(407.86)^2]} = 144.20$ pm.

13.121 Radius = 0.5diameter = 0.50 pm = r

Simple Cubic:

Volume of cube: $(2r)^3 = (2 \times 0.50\text{ pm})^3 = 1\text{ pm}^3$

Volume of atoms: 1 atom in unit cell

Volume of atoms = $4/3\pi r^3 = 4/3\pi(0.5\text{ pm})^3 = 0.524\text{ pm}^3$

Volume of empty space = Volume of cube – Volume of atoms = $1\text{ pm}^3 - 0.524\text{ pm}^3 = 0.458\text{ pm}^3$

Body-Centered Cubic:

Volume of cube: $(2r)^3$

$(4r)^2 = 3(\text{edge})^2$ edge = a

$(4 \times 0.50\text{ pm})^2 = 3(a)^2$

a = 1.15 pm

Volume of cube = $a^3 = (1.15\text{ pm})^3 = 1.52\text{ pm}^3$

Volume of atoms: 2 atoms in unit cell

Volume of atoms = $2(4/3\pi r^3) = 2(4/3\pi(0.5\text{ pm})^3) = 1.05\text{ pm}^3$

Volume of empty space = Volume of cube – Volume of atoms

= $1.52\text{ pm}^3 - 1.05\text{ pm}^3 = 0.470\text{ pm}^3$

Face-Centered Cubic:

Volume of cube: $(2r)^3$

$(4r)^2 = 2(\text{edge})^2$ edge = a

$(4 \times 0.50\text{ pm})^2 = 2(a)^2$

a = 1.41 pm

Volume of cube = $a^3 = (1.41\text{ pm})^3 = 2.83\text{ pm}^3$

Volume of atoms: 4 atoms in unit cell

Volume of atoms = $4(4/3\pi r^3) = 4(4/3\pi(0.5\text{ pm})^3) = 2.09\text{ pm}^3$

Volume of empty space = Volume of cube – Volume of atoms

= $2.83\text{ pm}^3 - 2.09\text{ pm}^3 = 0.74\text{ pm}^3$

The efficiency of packing is determined by dividing the volume occupied by the atoms divided by the volume occupied by the cube:

Simple Cubic:	$0.524\text{ pm}^3 \div 1\text{ pm}^3 = 0.524$
Body-Centered Cubic	$1.05\text{ pm}^3 \div 1.52\text{ pm}^3 = 0.691$
Face-Centered Cubic	$2.09\text{ pm}^3 \div 2.83\text{ pm}^3 = 0.739$

13.122 From Figure 13.6, we can see that the length of the diagonal of the cell = 4r, where r = radius of the particle. In our case, we will consider r as the radius of the bromide ion. According to the Pythagorean theorem,

$$a^2 + b^2 = c^2$$

for a right triangle. Since a = b here, we may re-write this as

$$2l^2 = c^2,$$

where l = length of the edge of the unit cell. As mentioned above, the diagonal of the unit cell = 4r, so we may say that

$$2l^2 = (4r)^2$$
$$2l^2 = 16r^2$$
$$(0.125)l^2 = r^2$$
$$\sqrt{[(0.125)l^2]} = r$$

Finally, substituting the value provided for l in the problem, $r = \sqrt{[(0.125)(550)^2]} = 194$ pm.

By inspection of Figure 13.6, we see that the length of the unit cell is equal to 2r + d, where r = radius of bromide ion, and d = the space in-between, which corresponds to the diameter of the lithium ion:

$$550 = 2r + d$$
$$550 = 2(194 \text{ pm}) + d$$
$$d = 162 \text{ pm}$$

Therefore the radius of the lithium ion is half of this, or 81 pm.

In this case, the value we have calculated is larger than that given in the table because the bromide ions are so large that they leave a bigger space to fill than the actual diameter of the lithium ion. (81 pm is simply the radius of that space.)

13.123 a = edge of a unit cell

First determine the mass of 1 atom of silver in grams:

$$\text{mass of 1 atom of silver} = \left(\frac{107.87 \text{ g Ag}}{1 \text{ mol Ag}}\right)\left(\frac{1 \text{ mol Ag}}{6.022\times10^{23} \text{ atoms Ag}}\right) = 1.791 \times 10^{-22} \text{ g /atom Ag}$$

Then determine the volume of each unit cell and the number of atoms in each unit cell.

(a) Simple Cubic Lattice:

2r = a

volume of the unit cell = $a^3 = (2r)^3 = (2 \times 144 \text{ pm})^3 = 2.39 \times 10^7 \text{ pm}^3$

$$2.39 \times 10^7 \text{ pm}^3 \left(\frac{1 \text{ cm}}{10^{10} \text{ pm}}\right)^3 = 2.39 \times 10^{-23} \text{ cm}^3\text{/unit cell}$$

number of atoms in the unit cell = 1

1 atom/unit cell

$$\text{Density} = \text{mass/volume} = \left(\frac{1.791\times10^{-22} \text{ g}}{1 \text{ atom Ag}}\right)\left(\frac{1 \text{ atom Ag}}{1 \text{ unit cell}}\right)\left(\frac{1 \text{ unit cell}}{2.39\times10^{-23} \text{ cm}^3}\right) = 7.49 \text{ g cm}^{-3}$$

(b) Body-Centered Cubic Lattice:

$(4r)^2 = 3a^2$

$(4 \times 144 \text{ pm})^2 = 3a^2$

a = 333 pm

Volume of the unit cell = $a^3 = (333 \text{ pm})^3 = (333 \text{ pm})^3 = 3.69 \times 10^7 \text{ pm}^3$

$$3.69 \times 10^7 \text{ pm}^3 \left(\frac{1 \text{ cm}}{10^{10} \text{ pm}}\right)^3 = 3.69 \times 10^{-23} \text{ cm}^3\text{/unit cell}$$

number of atoms in the unit cell = 2

2 atom/unit cell

$$\text{Density} = \text{mass/volume} = \left(\frac{1.791\times10^{-22} \text{ g}}{1 \text{ atom Ag}}\right)\left(\frac{2 \text{ atom Ag}}{1 \text{ unit cell}}\right)\left(\frac{1 \text{ unit cell}}{3.69\times10^{-23} \text{ cm}^3}\right) = 9.70 \text{ g cm}^{-3}$$

(c) Face-Centered Cubic Lattice:

$(4r)^2 = 2a^2$

$(4 \times 144 \text{ pm})^2 = 2a^2$

a = 407 pm

Volume of the unit cell = $a^3 = (407 \text{ pm})^3 = (407 \text{ pm})^3 = 6.76 \times 10^7 \text{ pm}^3$

$$6.76 \times 10^7\ \text{pm}^3 \left(\frac{1\ \text{cm}}{10^{10}\ \text{pm}}\right)^3 = 6.76\times 10^{-23}\ \text{cm}^3/\text{unit cell}$$

number of atoms in the unit cell = 4
4 atom/unit cell

$$\text{Density} = \text{mass/volume} = \left(\frac{1.791\times10^{-22}\ \text{g}}{1\ \text{atom Ag}}\right)\left(\frac{4\ \text{atom Ag}}{1\ \text{unit cell}}\right)\left(\frac{1\ \text{unit cell}}{6.76\times10^{-23}\ \text{cm}^3}\right) = 10.6\ \text{g cm}^{-3}$$

Silver has a face-centered cubic lattic.

*13.124 Using the Bragg equation (eqn. 13.1), $n\lambda = 2d \sin \theta$

$$154 = 2d\sin(6.97°)$$
$$77 = d(0.124)$$
$$d = 635\ \text{pm}$$

The mass of the unit cell would be:

(4 potassium ions × 39.10) + (4 chloride ions × 35.45) = 298.20 amu

and its volume would be:

$(635\ \text{pm})^3 = 2.56 \times 10^8\ \text{pm}^3$

Therefore, its density would be:

$d = m/V = \underline{1.16 \times 10^{-6}\ \text{amu/pm}^3}$

Practice Exercises

14.1 Since these solutions are saturated, the maximum amount of each gas is dissolved in the solution. Hence, 0.00430 g of O_2 and 0.00190 g of N_2 are dissolved in the water.

14.2 The total mass of the solution is to be 250 g. If the solution is to be 1.00 % (w/w) NaOH, then the mass of NaOH will be: 250 g × 1.00 g NaOH/100 g solution = 2.50 g NaOH. We therefore need 2.50 g of NaOH and (250 – 2.50) = 248 g H_2O. The volume of water that is needed is: 248 g × 0.988 g/mL = 251 mL H_2O.

14.3 An HCl solution that is 37 % (w/w) has 37 grams of HCl for every 1.0×10^2 grams of solution.

$$\#\text{ g solution} = (7.5\text{ g HCl})\left(\frac{1.0 \times 10^2\text{ g solution}}{37\text{ g HCl}}\right) = 2.0 \times 10^1\text{ g solution}$$

14.4 $$\#\text{ g CH}_3\text{OH} = (2000\text{ g H}_2\text{O})\left(\frac{0.250\text{ mol CH}_3\text{OH}}{1000\text{ g H}_2\text{O}}\right)\left(\frac{32.0\text{ g CH}_3\text{OH}}{1\text{ mol CH}_3\text{OH}}\right)$$

$$= 16.0\text{ g CH}_3\text{OH} \Rightarrow 20\text{ g CH}_3\text{OH (rounded to 1 sig. fig.)}$$

14.5 We need to know the number of moles of NaOH and the number of kg of water.
4.00 g NaOH × 40.0 g/mol = 0.100 mol NaOH
250 g H_2O × 1 kg/1000 g = 0.250 kg H_2O

The molality is thus given by:
m = 0.100 mol/0.25 kg = 0.40 mol NaOH/kg H_2O = 0.40 m

14.6 If a solution is 37.0 % (w/w) HCl, then 37.0 % of the mass of any sample of such a solution is HCl and (100.0 – 37.0) = 63.0 % of the mass is water. In order to determine the molality of the solution, we can conveniently choose 100.0 g of the solution as a starting point. Then 37.0 g of this solution are HCl and 63.0 g are H_2O. For molality, we need to know the number of moles of HCl and the mass in kg of the solvent:

37.0 g HCl ÷ 36.46 g/mol = 1.01 mol HCl
63.0 g H_2O × 1 kg/1000 g = 0.0630 kg H_2O
molality = mol HCl/kg H_2O = 1.01 mol/0.0630 kg = 16.1 m

14.7 We first determine the mass of one L (1000 mL) of this solution, using the density:

1000 mL × 1.38 g/mL = 1.38×10^3 g

Next, we use the fact that 40.0 % of this total mass is due to HBr, and calculate the mass of HBr in the 1000 mL of solution:

$0.400 \times (1.38 \times 10^3)$ = 552 g HBr

This is converted to the number of moles of HBr in 552 g:

552 g HBr ÷ 80.91 g/mol = 6.82 mol HBr

Last, the molarity is the number of moles of HBr per liter of solution

6.82 mol/1 L = 6.82 M

14.8 First determine the number of moles of each component of the solution:

For $C_{16}H_{22}O_4$, 20.0 g/278 g/mol = 0.0719 mol
For C_8H_{18}, 50.0 g/114 g/mol = 0.439 mol

The mole fraction of solvent is:

0.439 mol/(0.439 mol + 0.0719 mol) = 0.859

Using Raoult's Law, we next find the vapor pressure to expect for the solution, which arises only from the solvent (since the solute is known to be nonvolatile):

$P_{solvent} = X_{solvent} \times P°_{solvent}$ = 0.859 × 10.5 torr = 9.02 torr

14.9 $P_{cyclohexane} = X_{cyclohexane} \times P°_{cyclohexane} = 0.500 \times 66.9$ torr = 33.5 torr
$P_{toluene} = X_{toluene} \times P°_{toluene} = 0.500 \times 21.1$ torr = 10.6 torr
$P_{total} = P_{cyclohexane} + P_{toluene} = 33.5$ torr + 10.6 torr = 44.1 torr

14.10 A 10 % solution contains 10 g sugar and 90 g water.
$10 \text{ g } C_{12}H_{22}O_{11} \div 342 \text{ g/mol} = 0.029 \text{ mol } C_{12}H_{22}O_{11}$
$90 \text{ g } H_2O \times 1 \text{ kg}/1000 \text{ g} = 0.090 \text{ kg } H_2O$
m = 0.029 mol/0.090 kg = 0.32 mol/kg
$\Delta T_b = K_b \times m = 0.51\ °C\ m^{-1} \times 0.32\ m = 0.16\ °C$
$T_b = 100.16\ °C$

14.11 It is first necessary to obtain the values of the freezing point of pure benzene and the value of K_f for benzene from Table 12.4 of the text. We proceed to determine the number of moles of solute that are present and that have caused this depression in the freezing point: $\Delta T = K_f m$
$\therefore m = \Delta T/K_f = (5.45\ °C - 4.13\ °C)/(5.07\ °C \text{ kg mol}^{-1}) = 0.260\ m$
Next, use this molality to determine the number of moles of solute that must be present:
0.260 mol solute/kg solvent × 0.0850 kg solvent = 0.0221 mol solute
Last, determine the formula mass of the solute:
3.46 g/0.0221 mol = 157 g/mol

14.12 We can use the equation $\Pi = MRT$:
$\Pi = (0.0115 \text{ mol/L})(0.0821 \text{ L atm/K mol})(310 \text{ K})$
$\Pi = 0.293$ atm
Converting to torr, we have:

$$\#\text{ torr} = (0.293 \text{ atm})\left(\frac{760 \text{ torr}}{1 \text{ atm}}\right) = 223 \text{ torr}$$

14.13 We can use the equation $\Pi = MRT$, remembering to convert pressure to atm:

$$\#\text{ atm} = (25.0 \text{ torr})\left(\frac{1 \text{ atm}}{760 \text{ torr}}\right) = 0.0329 \text{ atm}$$

$\Pi = 0.0329 \text{ atm} = M \times (0.0821 \text{ L atm/K mol})(298 \text{ K})$
$M = 1.34 \times 10^{-3} \text{ mol L}^{-1}$
$\#\text{ mol} = 1.34 \times 10^{-3} \text{ mol L}^{-1} \times 0.100 \text{ L} = 1.34 \times 10^{-4} \text{ mol}$

$$\text{formula mass} = \frac{72.4 \times 10^{-3} \text{ g}}{1.34 \times 10^{-4} \text{ mol}} = 5.38 \times 10^{2} \text{ g mol}^{-1}$$

14.14 For the solution as if the solute were 100 % dissociated:
$\Delta T = (1.86\ °C\ m^{-1})(2 \times 0.237\ m) = 0.882\ °C$ and the freezing point should be $-0.882\ °C$.
For the solution as if the solute were 0 % dissociated:
$\Delta T = (1.86\ °C\ m^{-1})(1 \times 0.237\ m) = 0.441\ °C$ and the freezing point should be $-0.441\ °C$.

Thinking It Through

T14.1 This is a Henry's Law calculation. Each pressure must be in the same units, either both in atm or both in torr, before use of the following equation:

$$\frac{C_1}{P_1} = \frac{C_2}{P_2}$$

where C_1, P_1, and P_2 are given. The temperature is constant.

T14.2 The solution is 10.0 % sulfuric acid, meaning that there are 10.0 g of sulfuric acid in every 100 g of solution. This mass is first converted to a mole amount, in order to get the necessary conversion factor:

$$\#\ \text{mol}\ H_2SO_4 = (10.0\ \text{g}\ H_2SO_4)\left(\frac{1\ \text{mol}\ H_2SO_4}{98.0\ \text{g}\ H_2SO_4}\right)$$

There are thus 0.102 mol H_2SO_4 per 100 g of solution, and this conversion factor can now be used to solve the problem:

$$\#\ \text{g solution} = (0.100\ \text{mol}\ H_2SO_4)\left(\frac{100\ \text{g solution}}{0.102\ \text{mol}\ H_2SO_4}\right)$$

To determine the number of milliliters we would need to know the density of the solution.

T14.3 The mass percent tells us that there are 12.5 g of sugar per 100 g of solution. Thus, we have 100.0 – 12.5 = 87.5 g of water. The moles of sugar are determined by dividing mass (12.5 g) by molecular mass. The kilograms of water are found by dividing 87.5 g by 1000. Molality is then moles of sugar divided by kilograms of water.

T14.4 The mole fraction tells us that there is the ratio of 0.0100 mole of NaCl for every mole of substance in the solution. The mass of the NaCl is determined by multiplying the moles of NaCl (0.0100 mol) by the formula mass of NaCl (58.44 g mol^{-1}). The mass of the water is determined from the moles of water (1 mol of solution – 0.0100 mol NaCl = 0.99 mol H_2O) and the formula mass of water (18.02 g mol^{-1}). Multiply the moles of water by its formula mass. Then, using the mass of the NaCl and the total mass of the solution (mass of NaCl + mass of water) determine the mass percent of the NaCl in solution.

$$\text{Mass percent} = \left(\frac{\text{mass of NaCl}}{\text{mass of water} + \text{mass of NaCl}}\right)\times 100\%$$

T14.5 The molality indicates that there are 0.750 mole of solute per 1000 g of water in this solution. The number of moles of water in 1000 g is:

$$\#\ \text{mol}\ H_2O = (1000\ \text{g}\ H_2O)\left(\frac{1\ \text{mol}\ H_2O}{18.02\ \text{g}\ H_2O}\right) = 55.49\ \text{mol}\ H_2O$$

The mole fraction of solute is then 0.750 mol divided by the total number of moles (55.49 + 0.750 = 56.24).

T14.6 The molality is 0.125, meaning 0.125 moles of solute per 1000 g of solvent, not per 1000 g of solution. We therefore cannot simply weigh out the solution. First, it is necessary to calculate the number of grams of solute by multiplying 0.125 moles by the formula mass:

$$0.125\ \text{mol} \times 53.5\ \text{g/mol} = 6.69\ \text{g}$$

Therefore, an equivalent value to describe this solution would be to state the concentration in units of (g solute) ÷ (g solution). In this solution, we have

$$\frac{6.69\ \text{g}\ NH_4Cl}{1006.69\ \text{g solution}}$$

We can use this conversion factor, along with the molar mass in order to determine the number of grams needed to obtain the required amount of NH_4Cl.

In order to calculate volume in mL, we first need to know the density of the solution.

T14.7 In order to calculate molarity, we need to know the number of moles of H_2SO_4 in each liter of solution. There are 50.0 g of H_2SO_4 in every 100 g of solution. We find moles of H_2SO_4 by dividing 50.0 g by the molecular mass (98.0 g/mol). The volume of the solution is found by dividing the mass (100 g) by the density, which is the same as the specific gravity, at this temperature. Molarity is finally determined by dividing moles of H_2SO_4 by the volume (in L) of the solution.

T14.8 We can use Raoult's Law, $P_{solution} = X_{solvent} P^{\circ}_{solvent}$. We know $P_{solution}$ and $P^{\circ}_{solvent}$ so we can determine $X_{solvent}$. We also know the mass of solvent (100 g) so we can determine the total number of moles. Subtract the moles of solvent from the total number of moles of solute. To get mass of solute, multiply the number of moles solute by the molar mass.

T14.9 We need the molality, to get the number of moles of the solution. For simplicity, assume you have 1000 mL of water (the solvent) (1 kg) and 1000 mL of ethylene glycol. We know the density of ethylene glycol, so we can determine the mass and the number of moles of ethylene glycol, which are then used to calculate m. Then using $\Delta T_f = k_f m$ and $\Delta T_b = k_b m$ we can calculate the change in freezing and boiling points.

T14.10 We must determine the Raoult's Law pressure for each volatile component separately, and then add the two to get the total vapor pressure, since each component is volatile. Multiply the mole fraction of tetrachloroethylene (0.450) by its pure vapor pressure (40.0 torr), multiply the mole fraction of methyl acetate (1 – 0.450 = 0.550) by its pure vapor pressure (400 torr), and add the two partial pressures together to get the total vapor pressure above the mixture.

T14.11 Multiply the molarity by the gas constant (0.0821 L atm/K mol) and by the temperature (293 K), as in the equation for osmotic pressure of a solution: $\Pi = MRT$

T14.12 The initial solution consisted of 15.0 g of NaCl and 135 g of water. Adding 75.0 g of water increases the total mass to 225 g. The new mass percent is 15.0 g/225 g = 6.67 %.

T14.13 Start by assuming there is 1 kg solvent present. Use the equation $\Delta T_f = i k_f m$. We know that $\Delta T_f = 4.50°C$, $i = 4$ ($Al^{3+} + 3Cl^-$) and k_f for water. We can solve for m and determine the number of moles $AlCl_3$ and H_2O. From these we can determine X_{AlCl_3} and X_{H_2O}. The vapor pressure of the solution can be determined from Raoult's Law. We need to know the vapor pressure of water at 35°C to complete the calculation.

T14.14 Using Raoult's Law, determine the mole fraction of H_2O. We know the mass of water, 150 g, so we can determine the number of moles of water, and from the mole fraction, determine the number of moles of glycerol. To raise the vapor pressure, we have to dilute the solution.
Determine the $X_{solvent}$ for the higher vapor pressure using Raoult's Law. Determine the total number of moles in the new solution using the number of moles solute from the first part and the new mole fraction. Using the total number of moles and the mole fraction of solvent, determine the number of moles solvent. Convert this to a mass and subtract the original 150 g of water to determine the amount of water to add.

T14.15 The easiest way to solve this problem is to set up a ratio of the two experiments:

$$\frac{\Delta T_f(2)}{\Delta T_f(1)} = \frac{m(2)}{m(1)}$$

since $m = \dfrac{\text{moles solute}}{\text{\# kg solvent}}$ and the amount of solvent is constant we can simplify to

$$\frac{\Delta T_f(2)}{\Delta T_f(1)} = \frac{\text{moles (2)}}{\text{moles (1)}}$$

$$\text{moles(1)} = \text{moles(2)}\frac{\Delta T_f(1)}{\Delta T_f(2)}$$

and since $\text{moles} = \dfrac{\text{mass}}{\text{molar mass}}$, $\text{mass(1)} = \text{mass(2)}\dfrac{\Delta T_f(1)}{\Delta T_f(2)}$

If we let x = mass of solute present in the initial solution, we have:

$$x = (x + 1.20)\left(\frac{-0.835\ ^\circ C}{-1.045\ ^\circ C}\right) = (x + 1.20)(0.799)$$

$x = 4.77$ g

T14.16 (a) No. Vapor pressure is a fixed value (at a constant T) irrespective of container size or surface area (so long as sufficient liquid is present to provide enough vapor). The surface area simply controls the rate of evaporation.

(b) The total energy of the two beakers is the same if the temperature is the same. The molecules that compose the beaker on the right will exhibit the same energy distribution as the molecules that compose the beaker on the left. However, the beaker on the right will have fewer water molecules possessing sufficient energy to overcome the intermolecular forces of attraction. (This results because some of the high-energy particles will be non-volatile solute particles.) As a result, the observed vapor pressure is lower in the beaker on the right than it is in the beaker on the left.

T14.17 (a) The van't Hoff factor for acetic acid should be slightly greater than one. Acetic acid is a weak acid and some of the acid will dissociate in aqueous solution.

(b) The van't Hoff factor of 0.5 indicates that there is an association of acetic acid molecules in this non-polar solvent. Two acetic acid molecules are attracted to each other through hydrogen bonding. The apparent molality of the solution is reduced due to the formation of these dimers.

T14.18 See Figures 14.5 and 14.7

Review Questions

14.1 Three types of homogeneous mixtures are solutions, colloidal dispersions, and suspensions. The particles are smallest in solutions.

14.2 This event, diagrammed in Figure 14.1, is due to the tendency for all systems to proceed spontaneously towards a state with a higher degree of randomness (disorder).

14.3 First, the tendency towards randomness drives the solution process, and second, the new forces of attraction between solute and solvent molecules drive the process. Thus the relative degree of solute–solute, solvent–solvent and solute–solvent interactions will determine if a solute is soluble in a solvent.

14.4 Since water and methyl alcohol both have OH groups, there can be hydrogen bonding between a water molecule and a methyl alcohol molecule. This allows any proportion of methyl alcohol in water to be nearly as stable as either separate water samples or separate methyl alcohol samples.

14.5 Water molecules are tightly linked to one another by hydrogen bonding. In gasoline, however, which is a nonpolar organic substance, we have only weak London forces of attraction. This means that gasoline as a solute in water offers no advantage in attraction to individual water molecules, and the solvent is therefore not disrupted to allow the solute to dissolve.

14.6 The dipole moments of water molecules can be oriented so as to stabilize both the dissolved cation and the dissolved anion.

14.7 Hydration of electrolytes is the stabilization of ions in water solution by interaction of the ions with the dipole moments of water molecules. Hydration is most effective in bringing ions (electrolytes) into solution.

14.8 Hydrophilic compounds are water "loving" and, thus, are more soluble in water.

14.9 There is no solvating force provided by carbon tetrachloride that can overcome and offset the very strong ion–ion forces of the solid KCl sample.

14.10 Since iodine is nonpolar, it dissolves more readily in the solvent that is also nonpolar, in this case carbon tetrachloride.

14.11 Ethyl alcohol molecules are more nonpolar (less polar) than those of water.

14.12 The enthalpy of solution is a molar quantity, the enthalpy change associated with dissolving a mole of a solute in a solvent. The net enthalpy change arises from a combination of the energy necessary to dissociate both the solute and the solvent, and the energy associated with solvating the solute.

Heat of solution = lattice energy – solvation energy

14.13 This analysis works because it employs Hess's Law. It is convenient to view the process in this two–step fashion because it divides the enthalpy changes into two readily understandable and distinguishable processes, one involving the disruption of the solute's forces (lattice energy), and one involving the solvent–to–solute type forces (solvation energy) that exist once the solution is obtained.

14.14 Since the enthalpy of solution is positive, the process is endothermic. The system thus requires heat for the dissolving process, and the heat flow should cause the temperature to decrease.

14.15 It must be the endothermic step (disruption of the lattice) that is larger.

14.16 The Al^{3+} ion, having the greater positive charge, should have the larger hydration energy.

14.17 When a gas dissolves in a liquid, there is no endothermic step analogous to the lattice energy of a solid. The only enthalpy change is the one associated with hydration, and this is always negative.

14.18 This is case (a) in Figure 14.12. There is a greater attraction between water and acetone molecules in the resulting solution than there is among acetone molecules in the starting pure solute or water molecules in the starting pure solvent.

14.19 This is case (b) in Figure 14.12. The disruption of ethyl alcohol and the disruption of hexane together cost more energy than is gained on formation of the solution. This is because the two liquids are not alike; ethyl alcohol is a polar substance with hydrogen bonding, whereas hexane is a nonpolar liquid having only London forces.

14.20 As shown in Figure 14.15, most substances dissolve in water more extensively at the higher temperatures; in other words, solubility increases as temperature increases. For substances of this class, heat may be regarded as a reactant, and it is written into the left side of the thermochemical equation, meaning that dissolution is endothermic. By the principle of Le Châtelier, the stress that is caused by adding heat to such a system is relieved by the equilibrium's shifting to the right, the result being that more material dissolves.

14.21 The fact that the ΔH_{soln} value for the formation of a mixture of A and B is zero, implies that the relative strengths of A-A, B-B, and A-B intermolecular attractions are similar.

14.22 We can estimate from Figure 14.15 that the solubility of NH_4NO_3 in 100 g of H_2O is 500 g at 70 °C and 150 g at 10 °C. The amount of solid that will crystallize is the difference between these two solubilities, namely 500 – 150 = 350 g.

14.23 NaBr is the least soluble, and it should crystallize first.

14.24 Oxygen solubility increases as the temperature decreases. The larger fish will need more oxygen and will be found in the colder areas of lake bottoms.

14.25 Henry's Law is the statement, applied to the dissolving of a gas in a solvent, that at a given temperature, the concentration (C_g) of the gas in a solution is directly proportional to the partial pressure (P_g) of the gas on the solution, where k in the following equation is the constant of proportionality: $C_g = k \times P_g$. As discussed in the text, an alternate statement expresses the relationship of concentration at one pressure P_1 to the concentration that would exist at some new pressure P_2: $C_1/P_1 = C_2/P_2$

14.26 The atmospheric pressure on a mountain is less than the atmospheric pressure at sea level. From Henry's Law, as the partial pressure of oxygen decreases, the concentration of the oxygen also decreases. Therefore, there is less oxygen to sustain life in mountain streams.

14.27 Ammonia is more soluble in water than nitrogen because ammonia is able to hydrogen bond with solvent molecules, whereas nitrogen cannot. Nitrogen is a nonpolar molecular substance, whereas ammonia is a polar substance capable of hydrogen bonding. Also, ammonia reacts with water to form nonvolatile ions:

$$NH_3(g) + H_2O(l) \rightarrow NH_4^+(aq) + OH^-(aq)$$

Carbon dioxide is more soluble in water than nitrogen because it reacts with the water forming carbonic acid:

$$CO_2(g) + H_2O(l) \rightarrow H_2CO_3(aq)$$

14.28 Sulfur dioxide reacts with water to give an aqueous solution of the weak acid H_2SO_3:

$$SO_2(g) + H_2O(l) \rightarrow H^+(aq) + HSO_3^-(aq)$$

14.29 When the cap is removed from a bottle of carbonated beverage, the liquid fizzes because CO_2 is being released from the liquid. When the cap is on, the CO_2 fills the space above the liquid until an equilibrium is established between the gas and the liquid. After the cap is removed, the equilibrium is disrupted and more of the gas leaves the solution. This is the fizzing.

14.30 mole fraction = moles component/total number of moles
mole percent = mole fraction × 100%
molality = moles solute/kg solvent
percent by mass = (mass component/total mass) × 100%

14.31 Molality is independent of temperature. Molarity decreases with increasing temperature because the volume of the solvent increases with increasing temperature.

14.32 The molarity will be greater than 1.0. Since the density of the solution is greater than one, the mass of the solution in kg will be greater than its volume in liters.

14.33 A colligative property of a solution is one that depends only on the molal concentration of the solute particles, and not on the identity of the solute.

14.34 The total energy of a pure liquid and a solution are identical at the same temperature. The energy distribution of particles in the liquid and solution was described in a previous section. In the pure liquid, a certain number of particles possess enough energy to break the intermolecular forces of attraction and contribute to the vapor pressure. In the solution, some of the solute particles have high energies so fewer solvent particles have high energies. Consequently, the vapor pressure is reduced.

14.35 We use Raoult's Law to calculate the individual partial pressures for such a solution. The sum of the partial pressures is then taken to be the total vapor pressure of the solution.

14.36 A solution is ideal if the sum of the partial pressures of the components of the solution equals the observed vapor pressure of the solution, i.e., if the solution obeys Raoult's Law. Also, it should be true that the heat of solution is nearly zero.

14.37 A positive deviation indicates that the vapor pressure of the real solution is greater than expected if the solution behaved ideally. Positive deviations result when mixtures with weaker intermolecular forces of attraction between the two substances compared to the intermolecular forces of the pure substances are formed.

14.38 The crystal structure of ice accommodates only water molecules, not ions of a solute. In other words, for thermodynamic reasons, the presence of salt ions would destabilize the crystal lattice of pure water.

14.39 When a solute is dissolved in a solvent, the vapor pressure is lowered. As a result, the boiling point is increased to a temperature where the vapor pressure is high enough to once again allow boiling to occur. This affect also reduces the triple point and the entire solid-liquid equilibrium curve on a phase diagram shifts to lower temperatures. The net result is a lowering of the freezing point.

14.40 These are semipermeable because only certain substances are able to pass through the membrane. A nonpermeable material would allow nothing to pass through.

14.41 An osmotic membrane allows only solvent to pass, whereas a dialyzing membrane allows solvated ions of a certain minimum size to pass as well as solvent molecules. A dialyzing membrane prevents the passage of only certain solute particles, usually those of large size, such as colloid particles.

14.42 Dialysis is the use of a membrane, which allows the passage of solvent molecules and certain small solute molecules, to separate large solute molecules from a solution.

14.43 The side of the membrane less concentrated in solute will be more concentrated in solvent. Therefore, the escaping tendency of the solvent will be greater than on the side of the membrane less concentrated in solute. The solvent will shift through the membrane from the side less concentrated in solute to the side more concentrated in solute.

14.44 The solution that loses solvent into the other solution is the one with the lower molarity.

14.45 In each case, the osmotic pressure Π is given by the equation: $\Pi = M \times R \times T$. Since we do not know either the density of the solution or the volume of the solution, we cannot convert values for % by mass into molarities. However, we do know that glucose, having the smaller molecular mass, has the higher molarity, and we conclude that it will have the larger osmotic pressure.

14.46 By the "association of solute particles" we mean that some particles are attracted to others, or that solvent does not perfectly insulate solute particles from attachment to one another. This is another way of saying that there is less than 100 % dissociation or dissolution of solute in such a solution.

14.47 If a cell is placed in a solution, the concentration of salts in the solution will affect the cell. If the solution is hypertonic, the concentration of salts is higher than the concentration of salts in the cell. If the solution is hypotonic, then the concentration of salts is lower than the concentration of salts in the cell.

14.48 Ionic compounds dissociate in solution. The dissociation results in an increase in the number of particles in the solution, i.e., one NaCl "molecule" will dissociate creating two ions; Na^+ and Cl^-. Colligative properties depend upon the concentration of particles so any compound that dissociates into multiple particles will have pronounced effects on colligative properties.

14.49 The van't Hoff factor is the ratio of the value for a colligative property as actually measured to that value of the colligative property that is expected in the complete absence of any solute dissociation. A van't Hoff factor of one is expected for all nondissociating molecular solutes. A van't Hoff factor greater than one indicates a dissociation of the solute. A van't Hoff factor less than one indicates association of the solute. If the van't Hoff factor is 0.5 this indicates the formation of dimers.

14.50 The solute NaI has the larger formula mass, and it thus has the solution with the smaller number of moles per kg of solvent. Thus, the NaI solution should have the smaller depression in the freezing point. It is, therefore, the NaCl solution that should have the lower freezing point.

14.51 The solute that dissolves to produce the greater number of ions, Na_2CO_3, gives the solution with the larger boiling point elevation and, thus, the higher boiling point.

14.52 A homogeneous mixture is one that has only one phase and that has uniform properties throughout.

14.53 It is the size of the particles that distinguishes a solution, a colloidal dispersion and a suspension.

14.54 Blood cells would be considered a suspension since the blood cells are larger than 1 μm.

14.55 The stability of a colloidal dispersion can be influenced by Brownian movement, the attachment of charged particles to the outer surface of the particles, and the use of an emulsifying agent.

14.56 A dispersion of one liquid in another is an emulsion.

14.57 Emulsifying agents stabilize emulsions. Some useful emulsifying agents are egg yolk protein, natural products from soybeans, and the milk protein casein.

14.58 The Tyndall effect is the light scattering that is caused by the relatively large particles of a colloidal dispersion. The particles of a solution are not large enough to cause this effect.

14.59 Brownian movement by colloidal particles is caused by the continuous, random movement of molecules of solvent, which collide with the colloidal particles and cause them to move erratically.

14.60 Sols are dispersions of solid particles in a liquid medium. They are often stabilized by attracting ions of one charge type to their particle surfaces.

14.61 A Tyndall effect should not be exhibited by a solution, but it should be evident in a colloidal dispersion. Also, solutions should not exhibit Brownian movement.

14.62 The formation of micelles is diagramed in Figure 14.30. The driving force for this process is the tendency of the hydrophobic hydrocarbon groups to avoid interaction with water.

Review Problems

14.63 This is to be very much like that shown in Figures 14.8 and 14.9:

(a) $KCl(s) \rightarrow K^+(g) + Cl^-(g)$, $\Delta H° = +690 \text{ kJ mol}^{-1}$

(b) $K^+(g) + Cl^-(g) \rightarrow K^+(aq) + Cl^-(aq)$, $\Delta H° = -686 \text{ kJ mol}^{-1}$

$KCl(s) \rightarrow K^+(aq) + Cl^-(aq)$, $\Delta H° = +4 \text{ kJ mol}^{-1}$

14.64 New lattice energy for KCl: $\Delta H° = 690 \text{ kJ mol}^{-1} + (0.02 \times 690 \text{ kJ mol}^{-1}) = 704 \text{ kJ mol}^{-1}$

$\Delta H_{soln} = \Delta H_{lattice\ energy} + \Delta H_{hydration}$

$\Delta H_{soln} = 704 \text{ kJ mol}^{-1} + (-686 \text{ kJ mol}^{-1}) = 18 \text{ kJ mol}^{-1}$

The new value is larger than the value given in the table. The new value is 350% larger than the old value.

14.65 lattice energy + hydration energy = enthalpy of solution

Note: It is sometimes conventional to list lattice energies as negative values, however it always *requires* energy to separate oppositely-charged ions, therefore the lattice energy should be a positive value in the equation below. (Energy is put in to the system.)

$\Delta H_{soln} = \Delta H_{lattice\ energy} + \Delta H_{hydration}$

$\Delta H_{hydration} = \Delta H_{soln} - \Delta H_{lattice\ energy}$

$\Delta H_{hydration} = 14 \text{ kJ mol}^{-1} - (630 \text{ kJ mol}^{-1}) = -616 \text{ kJ mol}^{-1}$

14.66 $\Delta H_{soln} = \Delta H_{lattice\ energy} + \Delta H_{hydration}$

$\Delta H_{lattice\ energy} = \Delta H_{soln} - \Delta H_{hydration}$

$\Delta H_{lattice\ energy} = -50 \text{ kJ mol}^{-1} - (890 \text{ kJ mol}^{-1}) = 940 \text{ kJ mol}^{-1}$

14.67 $C_1/P_1 = C_2/P_2$, $C_2 = (C_1 \times P_2)/P_1 = (0.025\text{ g/L} \times 1.5\text{ atm})/1.0\text{ atm} = 0.038\text{ g/L}$.

14.68 We can compare the solubility that is actually observed with the predicted solubility based on Henry's Law. If the actual and the predicted solubilities are the same, we conclude that the gas obeys Henry's Law. We proceed as in Review Problem 14.67: $C_1/P_1 = C_2/P_2$
$C_2 = (C_1 \times P_2)/P_1$. $C_2 = (0.018\text{ g/L} \times 620\text{ torr})/740\text{ torr} = 0.015\text{ g/L}$. The calculated value of C_2 is the same as the observed value, and we conclude that over this pressure range, nitrogen does obey Henry's Law.

14.69 $C_1/P_1 = C_2/P_2$, $C_2 = (C_1 \times P_2)/P_1$
$C_2 = (0.010\text{ gL}^{-1} \times 2.0\text{ atm})/1.0\text{ atm} = 0.020\text{ gL}^{-1}$

14.70 $C_{gas} = k_H \times P_{gas}$
$C_{O_2} = 0.0039\text{ g } O_2/100\text{ mL solution}$ $\qquad P_{O_2} = 1.0\text{ atm}$

$$k_H = \frac{3.9\times10^5\text{ g mL}^{\text{Ğ1}}}{1.0\text{ atm}} = 3.9\times10^5\text{ g mL}^{\text{Ğ1}}\text{atm}^{\text{Ğ1}}$$

14.71 One liter of solution has a mass of:

$$\#\text{ g solution} = 1\text{ L solution}\left(\frac{1{,}000\text{ mL solution}}{1\text{ L solution}}\right)\left(\frac{1.07\text{ g solution}}{1\text{ mL solution}}\right) = 1{,}070\text{ g}$$

According to the given molarity, it contains 3.000 mol NaCl. This has a mass of:

$$\#\text{ g NaCl} = 3.000\text{ mol NaCl}\left(\frac{58.45\text{ g NaCl}}{1\text{ mol NaCl}}\right) = 175.4\text{ g NaCl}$$

Thus, the mass of water in 1 L solution must be:
1,070 g – 175.4 g = 895 g water

$$m = \frac{\#\text{ mol solute}}{\text{kg solvent}} = \left(\frac{3.000\text{ mol NaCl}}{0.895\text{ kg solvent}}\right) = 3.35\text{ m}$$

14.72 Since the density of the solution is 1.00 g mL^{-1}, the molarity and molality are the same

$$\text{molality of the acetic acid solution} = \left(\frac{0.143\text{ mol }CH_3O_2H}{1\text{ L soln}}\right)\left(\frac{1\text{ L soln}}{1000\text{ mL soln}}\right)\left(\frac{1\text{ mL soln}}{1.00\text{ g soln}}\right) \times\left(\frac{1000\text{ g soln}}{1\text{ kg soln}}\right) = 0.143\ m$$

14.73 24.0 g glucose ÷ 180 g/mol = 0.133 mol glucose
molality = 0.133 mol glucose/1.00 kg solvent = 0.133 molal
mole fraction = moles glucose/total moles
moles glucose = 0.133

$$\text{moles } H_2O = (1.00\times10^3\text{ g } H_2O)\left(\frac{1\text{ mole } H_2O}{18\text{ g } H_2O}\right) = 55.5\text{ moles } H_2O$$

$$X_{glucose} = \frac{1.33}{55.5 + 0.133} = 2.39 \times 10^{-3}$$

mass % = 24.0/1024 × 100% = 2.34%

14.74 $$\text{mol of NaCl} = 11.5\text{ g NaCl}\left(\frac{1\text{ mol NaCl}}{58.44\text{ g NaCl}}\right) = 0.197\text{ mol NaCl}$$

molality = 0.197 mol NaCl/1.00 kg H_2O = 0.197 molal

mass % = 11.5/1011.5 g × 100% = 1.14%
mole % = (moles NaCl/total moles) × 100%
moles NaCl = 0.197 moles
mole % = (0.197 moles/(55.5 moles + 0.197 moles)) × 100% = 0.354 %
Since the density of water is 1.00 g/mL, the volume of 1 kg is 1 L. Thus, the molarity is: 0.197 mol/1.00 L = 0.197 M. A solvent must have a density close to 1 g/mL for this to happen. Also, the volume of the solvent must not change appreciably on addition of the solute.

14.75 We need to know the mole amounts of both components of the mixture. It is convenient to work from an amount of solution that contains 1.25 mol of ethyl alcohol and, therefore, 1.00 kg of solvent. Convert the number of moles into mass amounts as follows:

For CH_3CH_2OH, $\frac{\text{\# g ethanol}}{\text{g solution}} = \left(\frac{1.25\text{ mol ethanol}}{1\text{ kg water}}\right)\left(\frac{46.08\text{ g ethanol}}{1\text{ mol ethanol}}\right) = 57.6\text{ g ethanol}$

Mass % ethanol = (mass ethanol/(total solution mass) × 100%
Mass % ethanol = (57.6 g ethanol/(1,000 g water + 57.5 g ethanol) × 100%
= 5.45 %

14.76 If we have 100.0 g of the solution, then 19.5 g is NaCl and the remainder, 80.5 g, is water. We need to know the number of moles of NaCl and the number of kg of water: 19.5 g NaCl ÷ 58.5 g/mol = 0.333 mol NaCl; 80.5 g ÷ 1000 g/kg = 8.05×10^{-2} kg H_2O. Molality = 0.333 mol/8.05×10^{-2} kg = 4.14 *m* NaCl.

14.77 If we assume 100 g of solution we have 5 g NH_3 and 95 g H_2O.

$$\text{\# moles NH}_3 = (5.00\text{ g NH}_3)\left(\frac{1\text{ mole NH}_3}{17.03\text{ g NH}_3}\right) = 0.294\text{ moles NH}_3$$

$$\text{\# kg H}_2\text{O} = (95.0\text{ g})\left(\frac{1\text{ kg}}{1000\text{ g}}\right) = 0.095\text{ kg water}$$

$$\text{m} = \frac{0.294\text{ moles}}{0.095\text{ kg}} = 3.09\text{ m}$$

$$\text{\# moles H}_2\text{O} = (95.0\text{ g})\left(\frac{1\text{ mole H}_2\text{O}}{18.0\text{ g H}_2\text{O}}\right) = 5.28\text{ moles H}_2\text{O}$$

mole percent = 0.294 moles/(5.27 moles+ 0.294 moles) × 100% = 5.28%

14.78 Assume 1 mole total.
\# moles C_3H_8O = 0.250 mol

$$\text{\# g C}_3\text{H}_8\text{O} = (0.250\text{ mol})\left(\frac{60.10\text{ g}}{1\text{ mol}}\right) = 15.02\text{ g C}_3\text{H}_8\text{O}$$

$$\text{\# moles H}_2\text{O} = (0.750\text{ moles})$$

$$\text{\# g H}_2\text{O} = (0.750\text{ mol})\left(\frac{18.02\text{ g}}{1\text{ mol}}\right) = 13.5\text{ g H}_2\text{O}$$

$$\text{C}_3\text{H}_8\text{O mass \%} = \frac{15.02\text{ g}}{13.5\text{ g} + 15.02\text{ g}} \times 100\% = 52.7\ \%$$

$$\text{molality} = \frac{0.250\text{ mol}}{(13.5\text{ g})\left(\frac{1\text{kg}}{1000\text{ g}}\right)} = 18.5\text{ m}$$

14.79 If we choose, for convenience, an amount of solution that contains 1 kg of solvent, then it also contains 0.363 moles of $NaNO_3$. The number of moles of solvent is:

1.00×10^3 g ÷ 18.0 g/mol = 55.6 mol H_2O

Now, convert the number of moles to a number of grams: for $NaNO_3$, 0.363 mol × 85.0 g/mol = 30.9 g; for H_2O, 1000 g was assumed and the percent (w/w) values are:

% $NaNO_3$ = 30.9 g/1031 g × 100 = 3.00 %

% H_2O = 1000 g/1031 g × 100 = 97.0 %

To determine the molar concentration of $NaNO_3$ assume 1 kg of solvent which would then contain 0.363 mole of $NaNO_3$ or 30.9 g $NaNO_3$. The total mass of the solution would be 1000 g + 30.9 g = 1031 g of solution. Now, the ratio of moles of solute to grams of solution is 0.363 mol $NaNO_3$/1031 g solution. From this calculate the molarity of the solution

$$\text{M of solution} = \left(\frac{0.363 \text{ mol NaNO}_3}{1031 \text{ g solution}}\right)\left(\frac{1.0185 \text{ g soln}}{1 \text{ mL soln}}\right)\left(\frac{1000 \text{ mL soln}}{1 \text{ L soln}}\right) = 0.359 \text{ M NaNO}_3$$

$$\#\text{ mol H}_2\text{O} = (1000 \text{ g})\left(\frac{1 \text{ mol}}{18.02 \text{ g}}\right) = 55.6 \text{ moles H}_2\text{O}$$

$$X_{NaNO_3} = \frac{0.363}{55.6 + 0.363} = 6.49 \times 10^{-3}$$

14.80 (a) If the sample is 1.89 mol % H_2SO_4, then an amount of the solution that contains 1.89 mol of H_2SO_4 also contains (100 – 1.89) = 98.11 mol water. We can calculate the molality if we know the number of moles of H_2SO_4 and the number of kg of solvent. The latter is determined as follows: 98.11 mol H_2O × 18.02 g/mol × 1 kg/1000 g = 1.768 kg H_2O. Molality = 1.89 mol H_2SO_4/1.768 kg H_2O = 1.07 *m* H_2SO_4.

(b) The mass of H_2SO_4 in the above sample is: 1.89 mol × 98.1 g/mol = 185 g H_2SO_4. The total mass of the solution is then equal to [185 g + (1.768×10^3 g] = 1.953×10^3 g, and the % (w/w) values are: for H_2SO_4, [185 g/(1.953×10^3 g)] × 100 = 9.47 %; for H_2O, [(1.768×10^3 g)/(1.953×10^3 g)] × 100 = 90.53 %.

(c) If we have on hand 100 mL (0.100 L) of this solution, it will have a mass that can be determined using its known density: mass = 1.0645 g/mL × 100.0 mL = 106.4 g of solution. Since this solution has 9.49 % (w/w) H_2SO_4, the mass of H_2SO_4 in 0.100 L of the solution is: 106.4 g × 0.0949 = 10.1 g H_2SO_4. The number of moles of H_2SO_4 is thus: 10.1 g ÷ 98.1 g/mol = 0.103 mol H_2SO_4. The molarity is the number of moles of H_2SO_4 divided by the volume of solution: 0.103 mol/0.100 L = 1.03 M H_2SO_4.

14.81 $P_{solution} = P°_{solvent} \times X_{solvent}$

We need to determine $X_{solvent}$:

$$\#\text{ mol glucose} = (65.0 \text{ g})\left(\frac{1 \text{ mol}}{180.2 \text{ g}}\right) = 0.361 \text{ moles}$$

$$\#\text{ mol H}_2\text{O} = (150 \text{ g H}_2\text{O})\left(\frac{1 \text{ mol H}_2\text{O}}{18.02 \text{ g H}_2\text{O}}\right) = 8.32 \text{ mol H}_2\text{O}$$

The total number of moles is thus: 8.32 mol + 0.361 mol = 8.69 mol and the mole fraction of the solvent is:

$$X_{solvent} = \left(\frac{8.32 \text{ mol solvent}}{8.69 \text{ mol solution}}\right) = 0.957. \text{ Therefore,}$$

$P_{solution}$ = 23.8 torr × 0.957 = 22.8 torr.

14.82 In 100 g of the mixture we have the following mole amounts:

80.0 g H_2O ÷ 18.02 g/mol = 4.44 mol H_2O

20.0 g $C_2H_6O_2$ ÷ 62.07 g/mol = 0.322 mol ethylene glycol

X_{H_2O} = 4.44/(4.44 + 0.322) = 0.932

$P_{solution} = P°_{solvent} \times X_{solvent}$ = 17.5 torr × 0.932 = 16.3 torr

14.83 $P_{benzene} = X_{benzene} \times P°_{benzene}$
$P_{toluene} = X_{toluene} \times P°_{toluene}$
$P_{Tot} = P_{benzene} + P_{Toluene}$

$$\# \text{ mol benzene} = (60.0 \text{ g})\left(\frac{1 \text{ mol}}{78.11 \text{ g}}\right) = 0.768 \text{ mol benzene}$$

$$\# \text{ mol toluene} = (40.0 \text{ g})\left(\frac{1 \text{ mol}}{92.14 \text{ g}}\right) 0.434 \text{ mol toluene}$$

$$X_{benzene} = \frac{0.768}{0.768 + 0.434} = 0.639$$

$$X_{toluene} = \frac{0.434}{0.768 + 0.434} = 0.361$$

$$P_{benzene} = (0.639)(93.4 \text{ torr}) = 59.7 \text{ torr}$$

$$P_{toluene} = (0361)(26.9 \text{ torr}) = 9.71 \text{ torr}$$

$$P_{Tot} = 59.7 \text{ torr} + 9.71 \text{ torr} = 69.4 \text{ torr}$$

14.84 Assume 50 g of each

$$\# \text{ mol pentane} = (50 \text{ g})\left(\frac{1 \text{ mol}}{72.15 \text{ g}}\right) = 0.693 \text{ mol}$$

$$\# \text{ mol heptane} = (50 \text{ g})\left(\frac{1 \text{ mol}}{100.21 \text{ g}}\right) = 0.499 \text{ mol}$$

$$X_{pentane} = \frac{0.693}{0.693 + 0.499} = 0.581$$

$$X_{heptane} = \frac{0.499}{0.693 + 0.499} = 0.419$$

$P_{pentane} = X_{pentane} \times P°_{pentane} = 0.581 \times 420 \text{ torr} = 244 \text{ torr}$
$P_{heptane} = X_{heptane} \times P°_{heptane} = 0.419 \times 36 \text{ torr} = 15.1 \text{ torr}$
$P_{Total} = P_{pentane} + P_{heptane} = (244 + 15.1) \text{ torr} = 259 \text{ torr}$

14.85 The following relationships are to be established: $P_{Total} = 96 \text{ torr} = (P°_{benzene} \times X_{benzene}) + (P°_{toluene} \times X_{toluene})$. The relationship between the two mole fractions is: $X_{benzene} = 1 - X_{toluene}$, since the sum of the two mole fractions is one. Substituting this expression for $X_{benzene}$ into the first equation gives:
$96 \text{ torr} = [P°_{benzene} \times (1 - X_{Toluene})] + [P°_{Toluene} \times X_{Toluene}]$,
$96 \text{ torr} = [180 \text{ torr} \times (1 - X_{Toluene})] + [60 \text{ torr} \times X_{Toluene}]$.
Solving for $X_{Toluene}$ we get: $120 \times X_{Toluene} = 84$,
$X_{Toluene} = 0.70$ and $X_{benzene} = 0.30$. The mole % values are to be 70 % toluene and 30 % benzene.

14.86 $X_{CH_3OH} = P/P° = 140 \text{ torr}/160 \text{ torr} = 0.875$

$$\# \text{ mol } CH_3OH = (100 \text{ g})\left(\frac{1 \text{ mol}}{32 \text{ g}}\right) = 3.12 \text{ mol}$$

$$0.875 = \frac{3.12 \text{ mol}}{3.12 \text{ mol} + x \text{ mol}}$$

$$x = \frac{3.12 \text{ mol}(1 - 0.875)}{0.875} = 0.446 \text{ mol}$$

$$\# \text{ g } C_3H_5(OH)_3 = (0.446 \text{ mol})\left(\frac{92.1 \text{ g}}{1 \text{ mol}}\right) = 41.1 \text{ g}$$

14.87 (a) $X_{solvent} = P/P° = 511 \text{ torr}/526 \text{ torr} = 0.971$
$X_{solute} = 1 - X_{solvent} = 0.029$

(b) We know 0.971 = 1 mol/1 mol + x moles $x = 2.99 \times 10^{-2}$ moles

(c) molar mass = 8.3 g/2.99×10^{-2} moles = 278 g/mol

14.88 $P_{solvent} = X_{solvent} \times P°_{solvent}$

336.0 torr = $X_{solvent} \times$ 400.0 torr

$X_{solvent} = 0.8400$

$X_{solute} = 1 - 0.8400 = 0.1600$

The number of moles of solvent is: 33.25 g ÷ 109.0 g/mol = 0.3050 mol and the following expression for mole fraction of solvent can be solved to determine the number of moles of solute

We know 0.8400 = 0.3050 mol/(0.3050 mol + x), x = 5.81×10^{-2} moles

Molar mass = 18.26 g/5.81×10^{-2} mol = 314.3 g/mol

14.89 $\Delta T = K_b m = 0.51$ °C kg mol^{-1} × 2.0 mol kg^{-1} = 1.0 °C

T_b = 100.0 + 1.0 = 101 °C

$\Delta T = K_f m = 1.86$ °C kg mol^{-1} × 2.00 mol kg^{-1} = 3.72 °C

T_f = 0.0 – 3.72 = –3.72 °C

14.90 The number of moles of glycerol is: 46.0 g ÷ 92 g/mol = 0.50 mol and the molality of this solution is 0.50 mol/0.250 kg = 2.0 *m*.

$\Delta T_b = K_b \times m$ = 0.51 °C kg mol^{-1} × 2.0 mol kg^{-1} = 1.0 °C, and the boiling point is 100.0 + 1.0 = 101.0 °C

$\Delta T_f = K_f \times m$ = 1.86 °C kg mol^{-1} × 2.0 mol kg^{-1} = 3.7 °C, and the freezing point is 0.0 – 3.7 = –3.7 °C

14.91 $\Delta T_f = K_f m$

$m = \Delta T_f/K_f$ = 3.00 °C/1.86 °C kg/mol = 1.61 mol/kg

$$\#\ \text{kg} = (100\ \text{g})\left(\frac{1\ \text{kg}}{1000\ \text{g}}\right) = 0.1\ \text{kg}$$

mol = (1.61 mol/kg)(0.1 kg) = 0.161 mol

$$\#\ \text{g} = (0.161\ \text{mol})\left(\frac{342.3\ \text{g}}{1\ \text{mol}}\right) = 55.1\ \text{g}$$

14.92 $\Delta T_b = K_b m$ = (0.51 °Ckg/mol)(1.61 mol/kg) = 0.82 °C, and the boiling point is 100.0 + 0.82 = 100.8 °C.

14.93 ΔT = (5.45 – 3.45) = 2.00 °C = $K_f \times m$ = 5.07 °C kg mol^{-1} × m

m = 0.394 mol solute/kg solvent

0.394 mol/kg benzene × 0.200 kg benzene = 0.0788 mol solute and the molecular mass is: 12.00 g/0.0788 mol = 152 g/mol

14.94 $\Delta T_b = K_b m$

ΔT_b = 81.7 °C – 80.2 °C = 1.5 °C

$m = \Delta T_b/K_b$ = 1.5 °C/2.53 °C kg/mol = 0.593 mol/kg

mol solute = (0.593 mol/kg)(1kg) = 0.593 mol

molar mass = (14 g)/0.593 mol = 24 g/mol

14.95 $\Delta T_f = K_f m$

$m = \Delta T_f/K_f$ = 0.307 °C/5.07 °C kg/mol = 0.0606 mol/kg

mol = (0.0606 mol/kg)(0.5 kg) = 0.0303 mol

$$\text{molar mass} = \frac{3.84\ \text{g}}{0.0303\ \text{mol}} = 127\ \text{g/mol}$$

The empirical formula has a mass of 64.1 g/mol. So the molecular formula is

$C_8H_4N_2$

14.96 (a) For convenience we choose to work with 100 g of the compound, and then to convert the mass amounts of each element found in this compound into mole amounts:

for C, 42.86 g ÷ 12.01 g/mol = 3.569 mol C
for H, 2.40 g ÷ 1.01 g/mol = 2.38 mol H
for N, 16.67 g ÷ 14.01 g/mol = 1.190 mol N
for O, 38.07 g ÷ 16.00 g/mol = 2.379 mol O

The relative mole amounts that represent the empirical formula are determined by dividing the above mole amounts each by the smallest mole amount:

for C: 3.569 mol ÷ 1.190 mol = 2.999
for H: 2.37 mol ÷ 1.190 mol = 1.99
for N: 1.190 mol ÷ 1.190 mol = 1.000
for O: 2.379 mol ÷ 1.190 mol = 1.999

and the empirical formula is $C_3H_2NO_2$.

(b) $\Delta T_b = 1.84\ °C = K_b \times m = 2.53\ °C\ kg\ mol^{-1} \times m$
m = 0.727 mol solute/kg benzene.
The number of moles of solute is: 0.727 mol/kg benzene × 0.045 kg benzene = 0.0327 mol, and the formula mass is: 5.5g/0.0327 mol = 168 g/mol. Since the mass of the empirical unit is 84, the molecular formula must be twice the empirical formula, namely $C_6H_4N_2O_4$.

14.97 (a) If the equation is correct, the units on both sides of the equation should be g/mol. The units on the right side of this equation are:

$$\frac{(g) \times (L\ atm\ mol^{-1}\ K^{-1}) \times (K)}{L \times atm} = g/mol$$

which is correct.

(b) $\Pi = MRT = (n/V)RT$, $n = \Pi V/RT$
This means that we can calculate the number of moles of solute in one L of solution, as follows:

$$n = \frac{(0.021\ torr)(1\ atm/760\ torr)(1.0\ L)}{(0.0821\ L\ atm\ mol^{-1}\ K^{-1})(298\ K)} = 1.1 \times 10^{-6}\ mol$$

The molecular mass is the mass in 1 L divided by the number of moles in 1 L:
2.0 g/1.1 × 10^{-6} mol = 1.8 × 10^{6} g/mol

14.98 $\Pi = MRT = (n/V)RT$, $n = \Pi V/RT$

$$n = \frac{(3.74\ torr)\left(\frac{1\ atm}{760\ torr}\right)(1\ L)}{(0.0821\ \frac{Latm}{mol\ K})(300\ K)} = 2.00 \times 10^{-4}\ mol$$

$$\text{molar mass} = \frac{0.400\ g}{2.00 \times 10^{-4}\ mol} = 2.00 \times 10^{3}\ g/mol$$

14.99 The equation for the vapor pressure is:

$$P_{solution} = P°_{H_2O} \times X_{H_2O}$$

Where $P°_{H_2O}$ is 17.5 torr. To calculate the vapor pressure we need to find the mole fraction of water first.

$$X_{H_2O} = \text{moles } H_2O/(\text{moles } H_2O + \text{moles NaCl})$$

Calculate the moles of NaCl in 10.0 g

$$\#\ mol\ NaCl = (10.0\ g\ NaCl)\left(\frac{1\ mol\ NaCl}{58.44\ g\ NaCl}\right) = 0.171\ moles\ NaCl$$

When NaCl dissolves in water, Na^+ and Cl^- are formed. So, for every mole of NaCl that dissolves, two moles of ions are formed. For this solution, the number of moles of ions is 0.342.
The number of moles of solvent (water) is:

$$\#\text{ mol } H_2O = (100\text{ g } H_2O)\left(\frac{1\text{ mol } H_2O}{18.02\text{ g } H_2O}\right) = 5.55\text{ moles } H_2O$$

Calculate the mole fraction as

$$X_{H_2O} = \frac{(\text{moles } H_2O)}{(\text{moles } H_2O + \text{moles NaCl})} = \frac{5.55\text{ mol}}{(5.55\text{ mol} + 0.342\text{ mol})} = 0.942$$

The vapor pressure is then $P_{solution} = P°_{H_2O} \times X_{H_2O} = 17.5\text{ torr} \times 0.942 = 16.5\text{ torr}$

14.100 $X_{H_2O} = P/P° = 38.7\text{ torr}/42.2\text{ torr} = 0.917$

$$\#\text{ mol } H_2O = (150\text{ mL})\left(\frac{1\text{ g}}{1\text{ mL}}\right)\left(\frac{1\text{ mol}}{18.02\text{ g}}\right) = 8.32\text{ mol}$$

$$0.917 = \frac{8.32\text{ mol}}{8.32\text{ mol} + x\text{ mol}}$$

$x = 0.753$ mol

Since the van't Hoff factor for $AlCl_3$ is 4, we need:

$0.753/4 = 0.188$ mol $AlCl_3$

$$\#\text{ g } AlCl_3 = (0.188\text{ mol})\left(\frac{133.3\text{ g}}{1\text{ mol}}\right) = 25.1\text{ g}$$

14.101 $\Pi = MRT$

$$M = \frac{(2.0\text{ g NaCl})\left(\frac{1\text{ mol NaCl}}{58.45\text{ g NaCl}}\right)}{0.100\text{ L}} = 0.34\text{ M}$$

For every NaCl there are two ions produced so M = 0.68 M

$$\Pi = (0.68\text{ M})(0.0821\text{ Latm/molK})(298\text{ K})\left(\frac{760\text{ torr}}{1\text{ atm}}\right) = 1.3 \times 10^3\text{ torr}$$

14.102 $\Pi = MRT$ For each ion, multiply the concentration by 24.47 L atm/mol

Ion	Molality (mol/L)	Π(atm)
Cl^-	0.566	13.9
Na^+	0.486	11.9
Mg^{2+}	0.055	1.35
SO_4^{2-}	0.029	0.710
Ca^{2+}	0.011	0.269
K^+	0.011	0.269
HCO_3^-	0.002	0.0489

Adding these together we get $\Pi = 28.4$ atm. Thus, a pressure greater than 28.4 atm is needed to desalinate seawater by reverse osmosis.

14.103 $CaCl_2 \rightarrow Ca^{2+} + 2Cl^-$; van't Hoff factor, i = 3
$\Delta T_f = i \times k_f \times m = (3)(1.86\text{ °C } m^{-1})(0.20\ m) = 1.1\text{ °C}$
The freezing point is –1.1 °C.

14.104 If we assume that mercury(I) nitrate has the formula $HgNO_3$, we predict a freezing point of –0.37 °C, $\Delta T_f = i \times k_f \times m = 2 \times 1.86\ °C/m \times 0.10, m = 0.37\ °C$. However, the observed freezing point depression is lower than this. So, assume that the correct formula of the compound is $Hg_2(NO_3)_2$ where the mercury ion is dimeric and divalent, i.e., Hg_2^{2+}. Now, the concentration of the solution is correctly stated as 0.050 *m* and $\Delta T_f = i \times k_f \times m = 3 \times 1.86\ °C/m \times 0.050\ m = 0.28\ °C$.
Therefore, the dissociation produces three ions and the equation is:
$Hg_2(NO_3)_2 \rightarrow Hg_2^{2+} + 2NO_3^-$

14.105 The freezing point depression that is expected from this solution if HF behaves as a nonelectrolyte is:

$$\Delta T_f = 1.86\ °C/m \times 1.00\ m = 1.86\ °C.$$

The freezing point that is expected upon complete dissociation of HF is:

$$\Delta T_f = K_f \times (2 \times m) = 3.72\ °C.$$

The observed freezing point depression is 1.91 ∞C, and the apparent molality is:

$$m = \Delta T_f/K_f$$

$$m = (1.91\ °C)/(1.86\ °C/m) = 1.03\ m,$$

or 1.03 mol solute particles per kg of solvent. This represents an excess of 3 solute particles per mol of HF (1.03 *m* – 1.00 *m* = 0.03 *m*), and we conclude that the percent ionization is 3 %.

4.106 $\Delta T_f = 0.261\ °C = K_f \times$ (apparent molality). Thus, the apparent molality of solute particles is: $m = 0.261\ °C/(1.86\ °C\ molal^{-1}) = 0.140$ molal. If the solute were dissolved as a nonelectrolyte, the apparent molality would be 0.125. The excess apparent molality arises from dissociation of the solute, and the amount (0.140 – 0.125) = 0.015 is the excess molality due to dissociation in this case. That is, 0.015 mol of solute per kg of solvent have been generated by dissociation of some certain % of the solute.

% ionization = 0.015/0.125 × 100 = 12%

14.107 Any electrolyte such as $NiSO_4$, that dissociated to give 2 ions, if fully dissociated should have a van't Hoff factor of 2.

14.108 $K_2SO_4 \rightarrow 2K^+ + SO_4^{2-}$, the van't Hoff factor is expected to be three.

14.109 $\Delta T_f = i \times k_f \times m$

$i = \Delta T_f/k_f \times m = 0.415°C/(1.86\ °C\ m^{1-})(0.118\ m) = 1.89$

14.110 To solve this problem, we need to assume the density of the solution is 1 g/mL. From problem 14.109 we know that most of the LiCl has dissociated. As a result, the affect of the dissociated ions will increase the osmotic pressure.
$\Pi = MRT$ If we consider the dissociation

$$\Pi = iMRT$$

$=(1.87)(0.118\ mol\ L^{-1})(0.0821\ L\ atm/mol\ K)(283\ K)(760\ torr/1atm) = 3.90 \times 10^3$ torr

Additional Exercises

14.111 The partial pressure of N_2 in air is:

$$P_{N2} = 1.00atm(0.78\ mol\%) = 0.78\ atm$$

Therefore, according to Henry's Law, the amount of N_2 dissolved per liter of blood at 1.00 atm is:

$$(1\ L)(0.015\ g/L)(0.78/1.00) = 0.012\ g\ N_2$$

$$0.012\ g\ N_2\ (1\ mol\ N_2/28.0\ g\ N_2) = 0.00043\ mol\ N_2$$

The amount of N_2 dissolved per liter of blood at 4.00 atm would be four times that, or: 0.0017 mol N_2

The amount of nitrogen released per liter of blood upon quickly surfacing is the difference between the two, or (0.0017 mol – 0.00043 mol) = 0.0013 mol N_2. The volume of that gas at 1.00 atm and 37 °C would be given by the ideal gas law:

$PV = nRT$

$V = nRT/P$

$V = (0.0013 \text{ mol } N_2)(0.0821 \text{ L·atm/mol·K})[(273+37)\text{K}]/1 \text{ atm}$

$V = 0.033 \text{ L} = 33 \text{ mL } N_2$ per liter of blood

14.112 In the simple molecular model of gas solubility, the solvent must expand slightly to allow the gas to enter, this is an endothermic process; then the gas enters the empty spaces and intermolecular attractions between the gas and the solvent hold the gas in the solution, this is an exothermic process. Since helium is a very small, nonpolar atom, the interaction between helium and the blood is not enough to overcome the energy required to expand the solvent. Helium is not very soluble in blood.

14.113 Around 70 °C, water molecules begin to acquire enough energy on average so that the gaps created between them can accommodate a molecule of nitrogen.

14.114 Let A = CCl_4 and B = unknown

$P_{Tot} = P_A + P_B$

$P_{Tot} = X_A P_A^\circ + X_B P_B^\circ$

We also know that $X_A + X_B = 1$

So, $P_{Tot} = (1 - X_B)P_A^\circ + X_B P_B^\circ$

$P_{Tot} - P_A^\circ = X_B(P_B^\circ - P_A^\circ)$

$X_B = (P_{Tot} - P_A^\circ)/(P_B^\circ - P_A^\circ) = (137 \text{ torr} - 143 \text{ torr})/(85 \text{ torr} - 143 \text{ torr}) = 0.103$

$X_A = 0.897$

$$X_{CCL_4} = \frac{\text{mol } CCl_4}{\text{mol } CCl_4 + \text{mol unknown}}, \text{ rearranging we get}$$

$$\text{moles unknown} = \text{mole } CCl_4\left(\frac{1}{X_{CCl_4}} - 1\right)$$

$$\text{moles } CCl_4 = (400 \text{ g})\left(\frac{1 \text{ mol}}{153.8 \text{ g}}\right) = 2.60 \text{ mol}$$

$$\text{moles unknown} = (2.60 \text{ moles})\left(\frac{1}{0.897} - 1\right) = 0.299 \text{ moles}$$

$$\text{molar mass} = \frac{43.3 \text{ g}}{0.299 \text{ moles}} = 145 \text{ g/mol}$$

14.115 (a) The formula masses are $Na_2Cr_2O_7{\bullet}2H_2O$: 298 g/mol, C_3H_8O: 60.1 g/mol, and C_3H_6O: 58.1 g/mol.

$$\# \text{ g } Na_2Cr_2O_7{\bullet}2H_2O = (21.4 \text{ g } C_3H_8O)\left(\frac{1 \text{ mol } C_3H_8O}{60.1 \text{ g } C_3H_8O}\right)$$

$$\times \left(\frac{1 \text{ mol } Na_2Cr_2O_7{\bullet}2H_2O}{3 \text{ mol } C_3H_8O}\right)\left(\frac{298 \text{ g } Na_2Cr_2O_7{\bullet}2H_2O}{1 \text{ mol } Na_2Cr_2O_7{\bullet}2H_2O}\right)$$

$$= 35.4 \text{ g } Na_2Cr_2O_7{\bullet}2H_2O$$

(b) The theoretical yield is:

$$\# \text{ g } C_3H_6O = (21.4 \text{ g } C_3H_8O)\left(\frac{1 \text{ mol } C_3H_8O}{60.1 \text{ g } C_3H_8O}\right)\left(\frac{3 \text{ mol } C_3H_6O}{3 \text{ mol } C_3H_8O}\right)\left(\frac{58.1 \text{ g } C_3H_6O}{1 \text{ mol } C_3H_6O}\right)$$

$$= 20.7 \text{ g } C_3H_6O$$

The percent yield is therefore: $12.4/20.7 \times 100\% = 59.9\ \%$

(c) First, we determine the number of grams of C, H, and O that are found in the products, and then the % by mass of C, H, and O that were present in the sample that was analyzed by combustion, i.e. the by–product:

$$\# \text{ g C} = (22.368 \times 10^{-3} \text{ g } CO_2)\left(\frac{12.011 \text{ g C}}{44.010 \text{ g } CO_2}\right) = 6.1046 \times 10^{-3} \text{ g C}$$

and the % C is: $(6.1046 \times 10^{-3}$ g/8.654 $\times 10^{-3}$ g) $\times$ 100% = 70.54% C

$$\# \text{ g H} = (10.655 \times 10^{-3} \text{ g } H_2O)\left(\frac{2.0159 \text{ g H}}{18.015 \text{ g } H_2O}\right) = 1.1923 \times 10^{-3} \text{ g H}$$

and the % H is: $(1.1923 \times 10^{-3}$ g H/8.654 $\times 10^{-3}$ g) $\times$ 100% = 13.78% H

For O, the mass is the total mass minus that of C and H in the sample that was analyzed:

8.654×10^{-3} g total – $(6.1046 \times 10^{-3}$ g C + 1.1923×10^{-3} g H) = 1.357×10^{-3} g O
and the % O is: $(1.357 \times 10^{-3}$ g)/(8.654 $\times 10^{-3}$ g) $\times$ 100% = 15.68% O.
Alternatively, we could have determined the amount of oxygen by using the mass % values, realizing that the sum of the mass percent values should be 100.
Next, we convert these mass amounts for C, H, and O into mole amounts by dividing the amount of each element by the atomic mass of each element:

For C, 6.1046×10^{-3} g C $\div$ 12.011 g/mol = 0.50825×10^{-3} mol C
For H, 1.1923×10^{-3} g H $\div$ 1.0079 g/mol = 1.1829×10^{-3} mol H
For O, 1.357×10^{-3} g O $\div$ 16.00 g/mol = 0.08481×10^{-3} mol O

Lastly, these are converted to relative mole amounts by dividing each of the above mole amounts by the smallest of the three (We can ignore the 10^{-3} term since it is common to all three components):

For C, 0.50825 mol/0.08481 mol = 5.993
For H, 1.1829 mol/0.08481 mol = 13.95
For O, 0.08481 mol/0.08481 mol = 1.000

and the empirical formula is given by this ratio of relative mole amounts, namely $C_6H_{14}O$.

(d) $\Delta T_f = K_f m$, (5.45 °C – 4.87 °C) = (5.07 °C/m) $\times$ m, m = 0.11 molal, and there are 0.11 moles of solute dissolved in each kg of solvent. Thus, the number of moles of solute that have been used here is:

0.11 mol/kg $\times$ 0.1150 kg = 1.3×10^{-2} mol solute.

The formula mass is thus: 1.338 g/0.013 mol = 102 g/mol. Since the empirical formula has this same mass, we conclude that the molecular formula is the same as the empirical formula, i.e. $C_6H_{14}O$.

14.116 Π = MRT = (0.0100 mol/L)(0.0821 L atm K^{-1} mol^{-1})(298 K) = 0.245 atm
0.245 atm $\times$ 760 torr/atm = 186 torr.

14.117 (a) Since –40 °F is also equal to –40 °C, the following expression applies: $\Delta T = K_f m$, so
40 °C = (1.86 °C kg mol^{-1}) $\times$ m, m = 40/1.86 mol/kg = 22 molal
Therefore, 22 moles must be added to 1 kg of water.

(b)

$$\# \text{ mL} = (22 \text{ moles})\left(\frac{62.1 \text{ g}}{1 \text{ mol}}\right)\left(\frac{1.00 \text{ mL}}{1.11 \text{ g}}\right) = 1.2 \times 10^3 \text{ mL}$$

(c) There are 946 mL in one quart. Thus, for 1 qt of water we are to have 946 mL, and the required number of quarts of ethylene glycol is:

$$\frac{\#\ \text{qt } C_2H_6O_2}{1\ \text{qt } H_2O} = \left(\frac{1.2 \times 10^3\ \text{mL } C_2H_6O_2}{1000\ \text{g } H_2O}\right)\left(\frac{1\ \text{g } H_2O}{1\ \text{mL } H_2O}\right) \times \left(\frac{946\ \text{mL } H_2O}{1\ \text{qt } H_2O}\right)\left(\frac{1\ \text{qt } C_2H_6O_2}{946\ \text{mL } C_2H_6O_2}\right)$$

$$= 1.2\ \text{qt } C_2H_6O_2$$

The proper ratio of ethylene glycol to water is 1.2 qt to 1 qt.

14.118 (a) The height difference is proportional to the osmotic pressure, therefore Π may be calculated by converting the height difference to the height of a mercury column in mm, which is equal to the pressure in torr (1 mm Hg = 1 torr):

$h_{Hg} = h_{solution} \times (d_{solution}/d_{Hg}) = (12.6\text{ mm}) \times (1.00\text{ g/mL}/13.6\text{ g/mL}) = 0.926\text{ mm Hg}$

P = 0.926 torr

(b) $\Pi = MRT$

$$M = \Pi/RT = \frac{(0.926\ \text{torr})\left(\frac{1\,\text{atm}}{760\ \text{torr}}\right)}{\left(0.0821\ \frac{\text{L atm}}{\text{mol K}}\right)(298\ \text{K})} = 4.98 \times 10^{-5}\ \text{M}$$

(c) Since this is a dilute solution and the solute does not dissociate, we can assume that the molarity and molality are equivalent. So,

$\Delta T_f = k_f m = (1.86\ °\text{C}m^{-1})(4.98 \times 10^{-5}\ \text{m})$

$= 9.26 \times 10^{-5}\ °\text{C}$

Freezing point will be -9.26×10^{-5} °C

(d) The magnitude of the temperature change is too small to measure.

14.119 (a) Since the molarity of the solution is 4.613 mol/L, then one L of this solution contains:

4.613 mol × 46.07 g/mol = 212.5 g C_2H_5OH.

The mass of the total 1 L of solution is:

1000 mL × 0.9677 g/mL = 967.7 g.

The mass of water is thus 967.7 g – 212.5 g = 755.2 g H_2O, and the molality is:

4.613 mol C_2H_5OH/0.7552 kg H_2O = 6.108 *m*.

(b) % (w/w) C_2H_5OH = (212.5 g/967.7 g) × 100% = 21.96%

14.120 (a) $\Delta T_b = k_b m = (0.51\ °\text{C}\ m^{-1})(1.00\ m)$

= 0.51°C

ΔT_b = 100 °C + 0.51 °C = 100.51 °C

(b) $\Delta T_b = ik_b m = (4)(0.51\ °\text{C}\ m^{-1})(1.00\ m)$

= 2.04 °C

ΔT_b = 100 °C + 2.04 °C = 102.04 °C

(c) i = 0.183°C/0.51°C = 0.36

14.121 (a) Assume 100 mol of solution, so there are 0.9159 mol solute and 100–0.9159 = 99.0841 mol of water. Converting this we get 1.785 kg water. So the molality is 0.5131 molal.

(b) $$\%(\text{w/w})\ KNO_3 = \frac{\text{mass } KNO_3}{(\text{mass } KNO_3 + \text{mass } H_2O)} \times 100\%$$

$$\#\ \text{g } KNO_3 = (0.9159\ \text{mol } KNO_3)\left(\frac{101.1\ \text{g } KNO_3}{1\ \text{mol } KNO_3}\right) = 92.6\ \text{g } KNO_3$$

$$\%(w/w)\ KNO_3 = \frac{92.6\ g\ KNO_3}{(92.6\ g + 1785\ g)} \times 100\% = 4.93\%$$

(c) Add the mass of water and the mass of KNO_3 then divide by the density to determine a volume:

92.6 g KNO_3 + 1785 g H_2O = 1878 g solution

$$d = \left(\frac{1878\ g\ solution}{1.0849\ {}^{g}\!/\!_{mL}}\right)\left(\frac{1\ L}{1000\ mL}\right) = 1.790\ L$$

$$M = \left(\frac{0.9159\ mol\ KNO_3}{1.790\ L\ solution}\right) = 0.5117\ M$$

Practice Exercises

15.1 From the coefficients in the balanced equation we see that, for every two moles of SO_2 that is produced, 2 moles of H_2S are consumed, three moles of O_2 are consumed, and two moles of H_2O are produced.

Rate of disappearance of O_2 = $\left(\frac{3\text{ mol }O_2}{2\text{ molS }O_2}\right)\left(\frac{0.30\text{ mol}}{\text{L s}}\right) = 0.45\text{ mol L}^{-1}\text{ s}^{-1}$

Rate of disappearance of H_2S = $\left(\frac{2\text{ mol }H_2S}{2\text{ mol }SO_2}\right)\left(\frac{0.30\text{ mol}}{\text{L s}}\right) = 0.30\text{ mol L}^{-1}\text{ s}^{-1}$

15.2 The rate of the reaction after 250 seconds have elapsed is equal to the slope of the tangent to the curve at 250 seconds. First draw the tangent, and then estimate its slope as follows, where A is taken to represent one point on the tangent, and B is taken to represent another point on the tangent:

$$\text{rate} = \left(\frac{\text{A (mol/L)} - \text{B (mol/L)}}{\text{A (s)} - \text{B (s)}}\right) = \frac{\text{change in concentration}}{\text{change in time}}$$

A value near 1×10^{-4} mol L^{-1} s^{-1} is correct.

15.3 (a) First use the given data in the rate law:

Rate = $k[\text{HI}]^2$
$2.5 \times 10^{-4}\text{ mol L}^{-1}\text{ s}^{-1} = k[5.58 \times 10^{-2}\text{ mol/L}]^2$
$k = 8.0 \times 10^{-2}\text{ L mol}^{-1}\text{ s}^{-1}$

(b) L mol^{-1} s^{-1}

15.4 The order of the reaction with respect to a given substance is the exponent to which that substance is raised in the rate law:
order of the reaction with respect to $[BrO_3^-] = 1$
order of the reaction with respect to $[SO_3^{2-}] = 1$
overall order of the reaction = 1 + 1 = 2

15.5 In each case, $k = \text{rate}/[A][B]^2$, and the units of k are $L^2\text{ mol}^{-2}\text{ s}^{-1}$.

Each calculation is performed as follows, using the second data set as the example:

$$k = \frac{0.40\text{ mol L}^{-1}\text{ s}^{-1}}{(0.20\text{ mol L}^{-1})(0.10\text{ mol L}^{-1})^2} = 2.0\times10^2\text{ L}^2\text{ mol}^{-2}\text{ s}^{-1}$$

Each of the other data sets also gives the same value:
$k = 2.0 \times 10^2\text{ L}^2\text{ mol}^{-2}\text{ s}^{-1}$.

15.6 When the concentration of sucrose is doubled, the rate doubles.
When the concentration of sucrose is raised five times, the rate goes up five times.
The concentration and rate are directly proportional; therefore the reaction is first order with respect to sucrose.

15.7 The rate law will likely take the form rate = $k[A]^n[B]^{n'}$, where n and n' are the order of the reaction with respect to A and B, respectively. On comparing the first two lines of data, in which the concentration of B is held constant, we note that increasing the concentration of A by a factor of 2 (from 0.40 to 0.80) causes an increase in the rate by a factor of 4 (from 1.0×10^{-4} to 4.0×10^{-4}). Thus, we have a rate increase by 2^2, caused by a concentration increase by a factor of 2. This corresponds to the case in Table 15.3 for which n = 2, and we conclude that the reaction is second–order with respect to A.

On comparing the second and third lines of data (wherein the concentration of A is held constant), we note that increasing the concentration of B by a factor of 2 (from 0.30 to 0.60) causes an increase in the rate by a factor of 4 (from 4.0×10^{-4} to 16.0×10^{-4}). This is an increase in rate by a factor of 2^2, and it forces us to conclude, using the information of Table 15.3, that the value of n' is 2. Thus the reaction is also second–order with respect to B. The rate law is then written:

Rate = $k[A]^2[B]^2$

15.8 a) We substitute into equation 15.3, first converting the time to seconds:

t = 2 hr × 3600 s/hr = 7200 s

$$\ln \frac{[A]_0}{[A]_t} = kt$$

$$\text{antiln}\left[\ln \frac{[A]_0}{[A]_t}\right] = \text{antiln}\,[kt] = \text{antiln}\left[\left(6.2 \times 10^{-5}\ s^{-1}\right)\left(7200\ s\right)\right]$$

$$\frac{[A]_0}{[A]_t} = 1.56 = \frac{0.40\ M}{[A]_t}$$

$$[A]_t = \frac{0.40\ M}{1.56} = 0.26\ M$$

b) Again, we use equation 15.3, this time solving for time:

$$\ln \frac{[A]_0}{[A]_t} = kt$$

$$t = \frac{1}{k} \times \ln \frac{[A]_0}{[A]_t} = \frac{1}{6.2 \times 10^{-5}\ s^{-1}} \times \ln \frac{0.40\ M}{0.30\ M} = 4600\ s$$

4.6×10^3 s × 1 min/60 s = 77 min

15.9 This is a second–order reaction, and we use equation 15.4:

$$\frac{1}{[NOCl]_t} - \frac{1}{[NOCl]_0} = kt$$

$$\frac{1}{[0.010\ M]} - \frac{1}{[0.040\ M]} = \left(0.020\ L\ mol^{-1}\ s^{-1}\right) \times t$$

t = 3.8×10^3 s

t = 3.8×10^3 s × 1 min/60 s = 63 min

15.10 For a first–order reaction:

$$t_{1/2} = \frac{0.693}{k} = \frac{0.693}{6.17 \times 10^{-4}\ s^{-1}} = 1.12 \times 10^3\ s$$

$$t_{1/2} = 1.12 \times 10^3\ s \times \frac{1\ min}{60\ s}$$

$$= 18.7\ min$$

If we refer to the chart given in the text in example 15.10, we see that two half lives will have passed if there is to be only one quarter of the original amount of material remaining. This corresponds to:

18.7 min per half–life × 2 half lives = 37.4 min

15.11 The reaction is first–order. A second–order reaction should have a half–life that depends on the initial concentration according to equation 15.6.

15.12 a) Use equation 15.9:

$$\ln \frac{k_2}{k_1} = \frac{-E_a}{R}\left[\frac{1}{T_2} - \frac{1}{T_1}\right]$$

$$\ln\left[\frac{23 \text{ L mol}^{-1} \text{ s}^{-1}}{3.2 \text{ L mol}^{-1} \text{ s}^{-1}}\right] = \frac{-E_a}{8.314 \text{ J mol}^{-1} \text{ K}^{-1}}\left[\frac{1}{673 \text{ K}} - \frac{1}{623 \text{ K}}\right]$$

Solving for E_a gives 1.4×10^5 J/mol = 1.4×10^2 kJ/mol

b) We again use equation 15.9, substituting the values:
$k_1 = 3.2 \text{ L mol}^{-1} \text{ s}^{-1}$ at $T_1 = 623$ K
$k_2 = ?$ at $T_2 = 573$ K

$$\ln \frac{k_2}{k_1} = \frac{-E_a}{R}\left[\frac{1}{T_2} - \frac{1}{T_1}\right]$$

$$\ln\left[\frac{k_2}{3.2 \text{ L mol}^{-1} \text{ s}^{-1}}\right] = \frac{-1.4 \times 10^5 \text{ J mol}^{-1}}{8.314 \text{ J mol}^{-1} \text{ K}^{-1}}\left[\frac{1}{573 \text{ K}} - \frac{1}{623 \text{ K}}\right]$$

Solving for k_2 gives $0.30 \text{ L mol}^{-1} \text{ s}^{-1}$.

15.13 If the reaction occurs in a single step, one molecule of each reactant must be involved, according to the balanced equation. Therefore, the rate law is expected to be: Rate = $k[NO][O_3]$.

15.14 The slow step (second step) of the mechanism determines the rate law:

$$\text{Rate} = k[NO_2Cl]^1[Cl]^1$$

However, Cl is an intermediate and cannot be part of the rate law expression. We need to solve for the concentration of Cl by using the first step of the mechanism. Assuming that the first step is an equilibrium, the rates of the forward and reverse reactions are equal:

$$\text{Rate} = k_{forward}[NO_2Cl] = k_{reverse}[Cl][NO_2]$$

Solving for [Cl] we get

$$[Cl] = \frac{k_f}{k_r}\frac{[NO_2Cl]}{[NO_2]}$$

Substituting into the rate law expression for the second step yields:

$$\text{Rate} = \frac{k[NO_2Cl]^2}{[NO_2]}$$, where all the constants have been combined into one new constant.

Thinking It Through

T15.1 The coefficients in the balanced chemical equation are used to construct conversion factors which represent the number of moles of a given reactant (or product) that appear (or disappear) for each mole of C_3H_8 that reacts. The rate for C_3H_8 can then be multiplied by these rate conversion factors in order to obtain the new rate. Remember that rates of reaction for reactants are given a negative sign, since these materials have concentrations that decrease with time. An example is:

$$\text{rate for } O_2 = \left(\frac{-0.400 \text{ mol } C_3H_8}{\text{L s}}\right)\left(\frac{5 \text{ mol } O_2}{1 \text{ mol } C_3H_8}\right) = -2.00 \text{ mol L}^{-1} \text{ s}^{-1}$$

T15.2 A graph of concentration vs. time must be constructed. The slope of the tangent to the curve at time = 25 min gives us the rate of disappearance of A. Conversion factors based on the stoichiometry of the equation are then used to convert to rates for B and C.

T15.3 The second–order rate equation (equation 15.2, where n = m = 1) can be used directly here, substituting values for the two reactant concentrations, since a value for the rate constant has been given. The concentration of hydroxide ion is not needed in order to work the problem. We must assume that the rate equation is first–order with respect to I^- and OCl^-.

T15.4 The rate constant can be calculated from the half–life, using equation 15.5. This rate constant can then be used in equation 15.3 in order to calculate time. The value for [A] should be set to 99, and the value for $[A]_0$ should be set to 100.

T15.5 The units given with this rate constant indicate that the reaction is second–order. It is therefore necessary to use equation 15.4 for this calculation. All of the necessary data are provided.

T15.6 The number of reactions per second constitutes a frequency, and has units s^{-1}. This can be obtained by multiplying the rate constant (L mol^{-1} s^{-1}) by the concentration (mol L^{-1}).

T15.7 We must examine the influence on the rate that a change in concentration has had. On comparing the data of experiments 1 and 2, we see that multiplying the concentration by 1.5 has produced an increase of the measured rate by 1.5. This indicates a first–order rate expression. The data from any of the experiments can then be used to calculate a value for the rate constant, by simple substitution into the first–order rate law.

T15.8 The stoichiometry of the reaction indicates that 5 moles of gas are produced from 2 moles of reactants. The net change in pressure will therefore be 5 – 2 = 3 mol of gas per unit time. This means that for every two moles of reactant that disappear, there will be an increase in pressure due to the net appearance of three moles of gas. The following conversion factor can be constructed:

$$\frac{2 \text{ mol } N_2O_5}{3 \text{ mol increase}}$$

and used in the necessary calculations:

$$\text{rate for } N_2O_5 = 20.0 \text{ torr s}^{-1} \times \frac{2 \text{ mol } N_2O_5}{3 \text{ mol increase}}$$

T15.9 Equation 15.9 can be used directly. Remember to convert temperature to Kelvins.

T15.10 We can start by deducing what the rate law should be if the first step is rate determining. Each reactant contributes to the rate law for such a mechanistic step.

(a) Since the first step involves a direct collision between F_2 and NO_2, it should be first–order in both reactants.

(b) If the second step is rate determining, the observed rate law will not be simply first–order in both F_2 and NO_2.

(c) We conclude that the first step is rate determining.

T15.11 (a) Figure C with the largest amount of A atoms will have the highest rate and Figure A with the smallest number of A atoms will have the slowest rate.

(b) Figure A and D will have the slowest rate and Figure C will have the fastest rate. To calculate an approximate rate, assume each sphere represents one concentration unit. Using the rate law, the rate for Figure A is Rate = k[1][4] = 4k.

(c) Figure E will have the fastest rate and Figure D will have the slowest.

T15.12 (a) The effect of a catalyst

(b) $A + B \rightarrow C + D$

(c) $A + B \rightarrow C + D$

(d) $A + E \rightarrow F + C$

$B + F \rightarrow D + E$

Overall: $A + B \rightarrow D + C$

(e) The activation energy of the higher energy path is represented by the height from the line labeled "A+B+E" to the point labeled "A–B". The activation energy of the lower energy path is represented by the height from the line labeled "A+B+E" to the point labeled "A–E". ΔH for the reaction is represented by the height from the line labeled "A+B+E" to the line labeled "C+D+E".

(f) A – reactant
B – reactant
C – product
D – product
E – catalyst
F – intermediate

Review Questions

15.1 Rate of reaction constitutes the speed with which a reactant or product changes concentration during a chemical reaction.

15.2 Student responses will vary.
(a) combustion of gasoline
(b) cooking an egg in boiling water
(c) curing of cement

15.3 An explosion is an extremely rapid reaction in which the energy that is released cannot be dissipated, causing the products to expand, pushing everything outward.

15.4 Industrialists need to be able to adjust the conditions so as to make products as rapidly as possible.

15.5 A collision between only two molecules is much more probable than the simultaneous collision of three molecules. We therefore conclude that a reaction involving a two–body collision is faster (i.e. will occur more frequently) than one requiring a three–body collision.

15.6 Five factors that affect reaction rate are:
- the nature of the reactants
- the ability of the reactants to collide effectively
- the concentration of the reactants
- the temperature
- the presence of a catalyst

15.7 The instantaneous rate of reaction is the rate of the reaction at a particular moment. The average rate of reaction is the average rate for the reaction over the time of the whole reaction. This includes the very rapid rates when the concentration of reactants is high and the very slow rates when the concentration of reactants is low.

15.8 A tangent line to the curve of concentration as a function of time at time zero is drawn. The slope of this line is determined which is the initial instantaneous rate of reaction.

15.9 A homogeneous reaction is one in which all reactants and products are in the same phase. An example would be:

$$2H_2(g) + O_2(g) \rightarrow 2H_2O(g)$$

15.10 A heterogeneous reaction is one in which all reactants and products are not in the same phase. An example would be:

$$25O_2(g) + 2C_8H_{18}(\ell) \rightarrow 16CO_2(g) + 18H_2O(g)$$

15.11 Chemical reactions that are carried out in solution take place smoothly because the reactants can mingle effectively at the molecular level.

15.12 Heterogeneous reactions are most affected by the extent of surface contact between the reactant phases.

15.13 In a heterogeneous reaction, the smaller the particle size, the faster the rate. This is because decreasing the particle size increases the surface area of the material, thereby increasing contact with another reactant phase.

15.14 This illustrates the effect of concentration.

15.15 Reaction rate generally increases with increasing temperature.

15.16 A catalyst is a substance that affects the rate of a reaction without being used up by the reaction.

15.17 In cool weather, the rates of metabolic reactions of cold–blooded insects decrease, because of the affect of temperature on rate.

15.18 Water boils at a higher temperature under pressure, and foods cook faster at the higher temperature.

15.19 The low temperature causes the rate of metabolism to be very low.

15.20 Reaction rate has the units mol L^{-1} s^{-1}, or molar per second (M s^{-1}).

15.21 A rate law is an equation that states the experimentally determined relationship between the rate of a reaction (for which the units are mol L^{-1} s^{-1}, or M s^{-1}) to the concentrations of reactants and products that are found to influence reaction rate. The proportionality constant is k, the rate constant.

15.22 The overall order of a reaction is the sum of the exponents for the concentration terms of the rate law.

15.23 The units are, in each case, whatever is required to give the units of rate (mol L^{-1} s^{-1}) to the overall rate law:

(a) s^{-1} (b) L mol^{-1} s^{-1} (c) L^2 mol^{-2} s^{-1}

15.24 The exponents in a rate law (i.e. the order of reaction with respect to each of the reactants) must be determined experimentally.

15.25 A zero–order reaction has no dependence on concentration while the first–order reaction is linearly dependent on concentration.

15.26 No, the coefficients of a balanced chemical reaction do not predict with any certainty the exponents in a rate law. These exponents must be determined through experiment.

15.27 This is the first case in Table 15.3 of the text, that is a zero-order reaction.

15.28 This is the fourth case Table 15.3 of the text, that is a first–order reaction.

15.29 This is the seventh case in Table 15.3, and the rate increases by a factor of $2^2 = 4$.

15.30 This is the eleventh case in Table 15.3, and the order of the reaction with respect to the reactant is 3.

15.31 Since the substrate concentration does not influence rate, its concentration does not appear in the rate law, and the order of the reaction with respect to substrate is zero.

15.32 $L^3\ mol^{-3}\ s^{-1}$

15.33 First–order: $\ln\dfrac{[A]_0}{[A]_t} = kt$

Second–order: $\dfrac{1}{[B]_t} - \dfrac{1}{[B]_0} = kt$

15.34 The half–life of a reaction is the time required for the concentration of a reactant to be reduced to half of its initial value.

15.35 The half–life of a first–order reaction is unaffected by the initial concentration.

15.36 The half–life of a second–order reaction is inversely proportional to the initial concentration, as expressed in the equation.

15.37 We use Equation 15.3, substituting exactly one–half the value of $[A]_0$ for the value of $[A]_t$:

$$\ln\frac{[A]_0}{[A]_t} = kt$$

$$\ln\frac{1.0}{0.50} = kt_{1/2}$$

This simplifies to give: $kt_{1/2} = \ln(2) = 0.693$. Thus $t_{1/2} = 0.693/k$

For the second–order reaction, we also substitute half the value of $[B]_0$ for the value of $[B]_t$ in equation 15.4:

$$\frac{1}{[B]_t} - \frac{1}{[B]_0} = kt$$

$$\frac{1}{\frac{1}{2}\times[B]_0} - \frac{1}{[B]_0} = \frac{2}{[B]_0} - \frac{1}{[B]_0} = kt_{1/2}$$

$$kt_{1/2} = \frac{1}{[B]_0} \quad \text{and} \quad t_{1/2} = \frac{1}{k\times[B]_0}$$

15.38 For the zero–order reaction, substitute half the value of $[A]_0$ for the value of $[A]_t$ in the equation

$[A]_t - [A]_0 = kt$

$1/2[A]_0 - [A]_0 = kt$

$-1/2[A]_0 = kt$

$$t_{1/2} = \frac{[A]_0}{2t}$$

15.39 According to collision theory, the rate is proportional to the number of collisions per second among the reactants.

15.40 The effectiveness of collisions is influenced by the orientation of the reactants and by the activation energy.

15.41 This happens because a larger fraction of the reactant molecules possess the minimum energy necessary to surpass E_a.

15.42

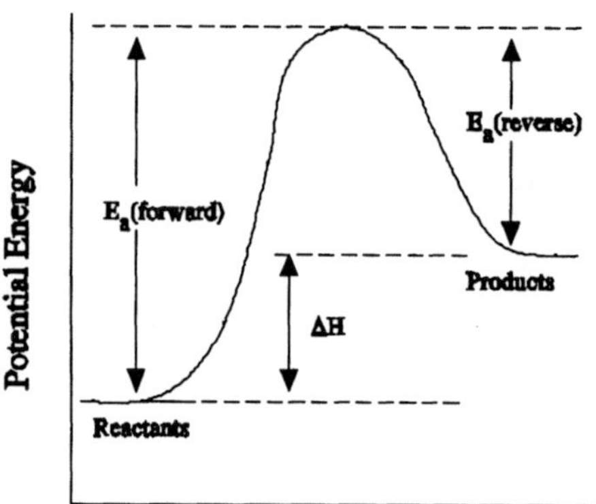

15.43 The energy that is necessary for the reaction is obtained at the expense of the total kinetic energy of the system. Thus, kinetic energy is lost, being converted into potential energy, and the temperature goes down.

15.44 Transition state – the high point on the potential energy diagram of the reaction.

Activated complex – the chemical arrangement that exists at the transition state.

15.45

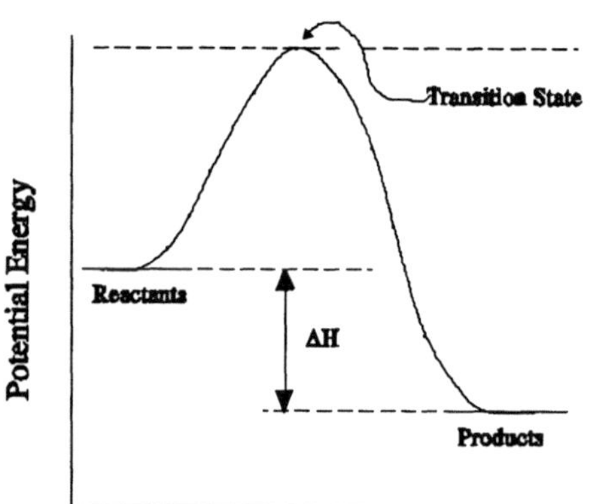

15.46 The reaction is slow due to the requirement for a very particular orientation in the activated complex (Arrhenius frequency factor A), rather than due to a high activation energy.

15.47 Breaking a strong bond requires a large input of energy, hence a large E_a.

15.48

$$k = Ae^{-E_a/RT}$$

where k = the rate constant
A = a frequency factor
e = the base for natural logarithms

E_a = the activation energy in units of J
R = the gas constant in units of J mol^{-1} K^{-1}
T = absolute temperature (Kelvin)

15.49 An elementary process is an actual collision event that occurs during the reaction. It is one of the key events that moves the reaction along in the stepwise process that leads to the overall reaction that is observed. It is thus one step in a potentially multi–step mechanism.

15.50 The rate–determining step in a mechanism is the slowest step.

15.51 The rate law for a reaction is based on the rate–determining step.

15.52 Adding all of the steps gives:

$2NO + 2H_2 \rightarrow N_2 + 2H_2O$

15.53 The rate-determining step would be step 2, $N_2O_2 + H_2 \rightarrow N_2O + H_2O$. Since N_2O_2 is not one of the reactants, but is made from 2NO, this second step is the rate-determining step.

15.54 For such a mechanism, the rate law should be:
rate = $k[NO_2][CO]$

Since this is not the same as the observed rate law, this is not a reasonable mechanism to propose.

15.55 Add all of the steps together:
$CO + O_2 + NO \rightarrow CO_2 + NO_2$

15.56 On adding together all of the steps in each separate mechanism, we get, in each case:

$OCl^- + I^- \rightarrow Cl^- + OI^-$

15.57 The predicted rate law is based on the rate–determining step:

rate = $k[NO_2]^2$

15.58 Step 1 is the slower step: rate = k[E][A]
Step 2 is the slower step: rate = k[EA] but EA is not a reactant, therefore, it must stated as rate = k[E][A].

15.59 A catalyst changes the mechanism of a reaction, and provides a reaction path having a smaller activation energy.

15.60 A homogeneous catalyst is one that is present in the same phase as the reactants. It is used in one step of a cycle, but regenerated in a subsequent step, so that in a net sense, it is not consumed.

15.61 Heterogeneous catalysts are present in a phase that is different from that of the reactants. They function by an adsorption of reactants, where the catalyst causes the activation energy to become smaller than would otherwise be true.

15.62 Adsorption is a clinging to a surface. Absorption involves a penetration below the surface, as in the action of a sponge. Heterogeneous catalysis involves adsorption.

15.63 The catalytic converter promotes oxidation of unburned hydrocarbons, as well as the decomposition of nitrogen oxide pollutants. Lead poisons the catalyst and renders it ineffective.

Review Problems

15.64

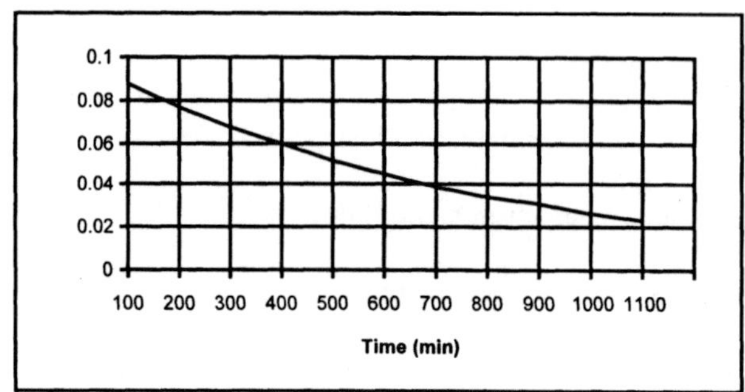

Since they are in a 1-to-1 mol ratio, the rate of formation of SO_2 is *equal* and *opposite* to the rate of consumption of SO_2Cl_2. This is equal to the slope of the curve at any point on the graph (see below). At 200 min, we obtain a value of about 1×10^{-4} M/s. At 600 minutes, this has decreased to about 7×10^{-5} M/s.

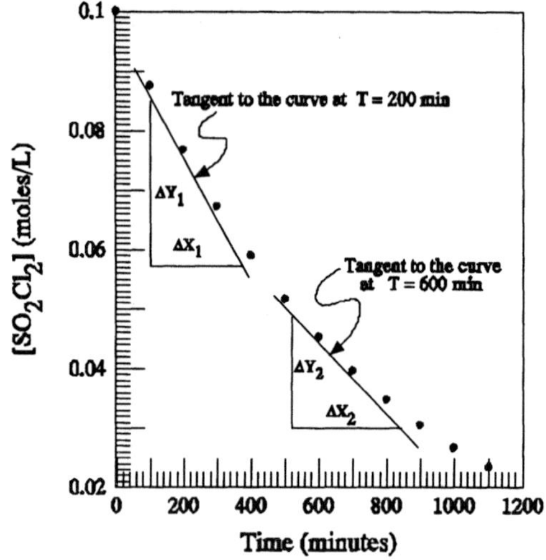

15.65 The slope of the tangent to the curve at each time is the negative of the rate at each time:
$Rate_{60} = 8.5 \times 10^{-4}$ mol L^{-1} s^{-1}
$Rate_{120} = 4.0 \times 10^{-4}$ mol L^{-1} s^{-1}

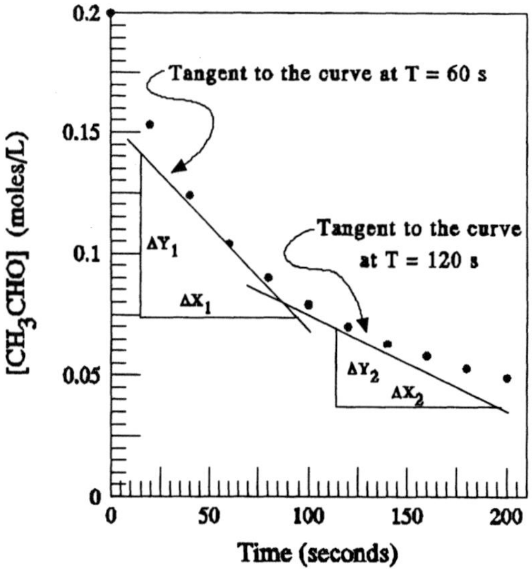

15.66 This is determined by the coefficients of the balanced chemical equation. For every mole of N_2 that reacts, 3 mol of H_2 will react. Thus the rate of disappearance of hydrogen is three times the rate of disappearance of nitrogen. Similarly, the rate of disappearance of N_2 is half the rate of appearance of NH_3, or NH_3 appears twice as fast as N_2 disappears.

15.67 From the coefficients in the balanced equation we see that, for every mole of B that reacts, 2 mol of A are consumed, and three mol of C are produced. This means that A will be consumed twice as fast as B, and C will be produced three times faster than B is consumed.

rate of disappearance of A = $2(-0.30) = -0.60$ mol L^{-1} s^{-1}
rate of appearance of C = $3(0.30) = 0.90$ mol L^{-1} s^{-1}

15.68 (a) rate for $O_2 = -1.20$ mol L^{-1} $s^{-1} \times 19/2 = -11.4$ mol L^{-1} s^{-1}
By convention, this is reported as a positive number: 11.4 mol L^{-1} s^{-1}
(b) rate for $CO_2 = +1.20$ mol L^{-1} $s^{-1} \times 12/2 = 7.20$ mol L^{-1} s^{-1}
(c) rate for $H_2O = +1.20$ mol L^{-1} $s^{-1} \times 14/2 = 8.40$ mol L^{-1} s^{-1}

15.69 We rewrite the balanced chemical equation to make the problem easier to answer: $N_2O_5 \rightarrow 2NO_2 + 1/2O_2$. Thus, the rates of formation of NO_2 and O_2 will be, respectively, twice and one half the rate of disappearance of N_2O_5.

rate of formation of $NO_2 = 2(2.5 \times 10^{-6}) = 5.0 \times 10^{-6}$ mol L^{-1} s^{-1}
rate of formation of $O_2 = 1/2(2.5 \times 10^{-6}) = 1.2 \times 10^{-6}$ mol L^{-1} s^{-1}

15.70 The rate can be found by simply inserting the given concentration values:

Rate = $(5.0 \times 10^5\ L^5\ mol^{-5}\ s^{-1})[H_2SeO_3][I^-]^3[H^+]^2$
rate = $(5.0 \times 10^5\ L^5\ mol^{-5}\ s^{-1})(2.0 \times 10^{-2}\ mol/L)(2.0 \times 10^{-3}\ mol/L)^3(1.0 \times 10^{-3}\ mol/L)^2$
rate = 8.0×10^{-11} mol L^{-1} s^{-1}

15.71 rate = $(1.3 \times 10^{11}\ L\ mol^{-1}\ s^{-1})(1.0 \times 10^{-7}\ mol/L)(1.0 \times 10^{-7}\ mol/L)$
rate = 1.3×10^{-3} mol L^{-1} s^{-1}

15.72 rate = $(7.1 \times 10^9\ L^2\ mol^{-2}\ s^{-1})(1.0 \times 10^{-3}\ mol/L)^2(3.4 \times 10^{-2}\ mol/L)$
rate = 2.4×10^2 mol L^{-1} s^{-1}

15.73 rate = $(1.0 \times 10^{-5}\ s^{-1})(1.0 \times 10^{-3}\ mol/L) = 1.0 \times 10^{-8}$ mol L^{-1} s^{-1}

15.74 In each case, the order with respect to a reactant is the exponent to which that reactant's concentration is raised in the rate law.
(a) For $HCrO_4^-$, the order is 1.
For HSO_3^-, the order is 2.
For H^+, the order is 1.
(b) The overall order is 1 + 2 + 1 = 4.

15.75 In each case, the order with respect to a reactant is the exponent to which that reactant's concentration is raised in the rate law. The order of reaction with respect to each reactant is 1. The overall order is 1 + 1 = 2.

15.76 On comparing the data of the first and second experiments, we find that, whereas the concentration of N is unchanged, the concentration of M has been doubled, causing a doubling of the rate. This corresponds to the fourth case in Table 15.3, and we conclude that the order of the reaction with respect to M is 1. In the second and third experiments, we have a different result. When the concentration of M is held constant, the concentration of N is tripled, causing an increase in the rate by a factor of nine. This constitutes the eighth case in Table 15.3, and we conclude that the order of the reaction with respect to N is 2. This means that the overall rate expression is: rate = $k[M][N]^2$ and we can solve for the value of k by substituting the appropriate data:

5.0×10^{-3} mol L^{-1} s^{-1} = k × [0.020 mol/L][0.010 mol/L]^2
k = 2.5×10^3 L^2 mol^{-2} s^{-1}

15.77 First, compare the first and second experiments, in which there has been an increase in concentration by a factor of 2. This has caused an increase in rate of:

$$\frac{5.90 \times 10^{-5} \text{ mol L}^{-1} \text{ s}^{-1}}{2.95 \times 10^{-5} \text{ mol L}^{-1} \text{ s}^{-1}} = 2.00 = 2^1$$

This is the fourth case in Table 15.3, and the order of the reaction is found to be 1. Similarly, on comparing the first and third experiments, an increase in concentration by a factor of 3 has caused an increase in rate by a factor of:

$$\frac{8.85 \times 10^{-5} \text{ mol L}^{-1} \text{ s}^{-1}}{2.95 \times 10^{-5} \text{ mol L}^{-1} \text{ s}^{-1}} = 3.00 = 3^1$$

This is the fifth case in Table 13.5, and we again conclude that the order is 1.
rate = k[C_3H_6]
We can use any of the three sets of data to solve for k. Using the first data set gives:
2.95×10^{-5} mol L^{-1} s^{-1} = k(0.050 mol L^{-1})
k = 5.9×10^{-4} s^{-1}

15.78 The reaction is first–order in OCl^-, because an increase in concentration by a factor of two, while holding the concentration of I^- constant (compare the first and second experiments of the table), has caused an increase in rate by a factor of $2^1 = 2$. The order of reaction with respect to I^- is also 1, as is demonstrated by a comparison of the first and third experiments.

rate = k[OCl^-][I^-]
Using the last data set:
3.5×10^4 mol L^{-1} s^{-1} = k[1.7×10^{-3} mol/L][3.4×10^{-3} mol/L]
k = 6.1×10^9 L mol^{-1} s^{-1}

15.79 Compare the data of the first and second experiments, in which the concentration of NO is held constant and the concentration of O_2 is increased by a factor of 4. Since this caused a rate increase by a factor of $28.4/7.10 = 4^1$, we conclude that the order of the reaction with respect to O_2 is one (case number six in Table 15.3). In the second and third experiments, an increase in the concentration of NO by a factor of 3 (while holding the concentration of O_2 constant) caused a rate increase by a factor of 255.6/28.4 = 9. This is the eighth case in Table 15.3, and the order is seen to be two.

rate = k[O_2][NO]^2

We can use any of the three sets of data to solve for k. Using the first data set gives:
7.10 mol L^{-1} s^{-1} = k[1.0×10^{-3} mol/L][1.0×10^{-3} mol/L]^2
k = 7.10×10^9 L^2 mol^{-2} s^{-1}

15.80 Compare the first and second experiments. On doubling the ICl concentration, the rate is found to increase by a factor of $2 = 2^1$, and the order of the reaction with respect to ICl is 1 (case number four in Table 15.3). In the first and third experiments, the concentration of ICl is constant, whereas the concentration of H_2 in the first experiment is twice that in the third. This causes a change in the rate by a factor of 2 also, and the rate law is found to be: rate = k[ICl][H_2]. Using the data of the first experiment:
1.5×10^{-3} mol L^{-1} s^{-1} = k[0.10 mol L^{-1}][0.10 mol L^{-1}]
k = 1.5×10^{-1} L mol^{-1} s^{-1}

15.81 In the first, fourth, and fifth experiments, the concentration of OH^- has been made to increase while the [$(CH_3)_3CBr$] is left unchanged. In each of these experiments, there is no change in rate. This means that the rate is independent of [OH^-], and the order of reaction with respect to OH^- is zero. The concentration of $(CH_3)_3CBr$ doubles from the first to the second experiment, and triples from the first to the third experiment, as the OH^- concentration is held constant. There is a corresponding 2–fold (i.e. 2^1) increase in rate from the first to the second experiment, and there is a 3–fold (i.e. 3^1) increase in rate from the first to the third experiment. In both cases, we conclude that the order with respect to $(CH_3)_3CBr$ is one.

rate = k[$(CH_3)_3CBr$]
Using the third set of data gives:
3.0×10^{-3} mol L^{-1} s^{-1} = k[3.0×10^{-1} mol/L]
k = 1.0×10^{-2} s^{-1}

15.82 A graph of ln [SO_2Cl_2]$_t$ versus t will yield a straight line if the data obeys a first–order rate law.

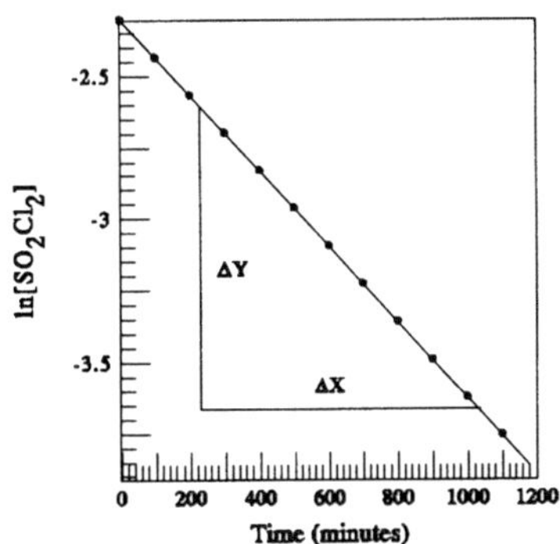

These data do yield a straight line when ln [SO_2Cl_2]$_t$ is plotted against the time, t. The slope of this line equals –k. Plotting the data provided and using linear regression to fit the data to a straight line yields a value of 1.32×10^{-3} min^{-1} for k.

15.83 Since it is the plot of 1/conc. that gives a straight line, the order of the reaction with respect to CH_3CHO is two. The rate constant is given by the slope directly:

k = 0.0771 M^{-1} s^{-1}

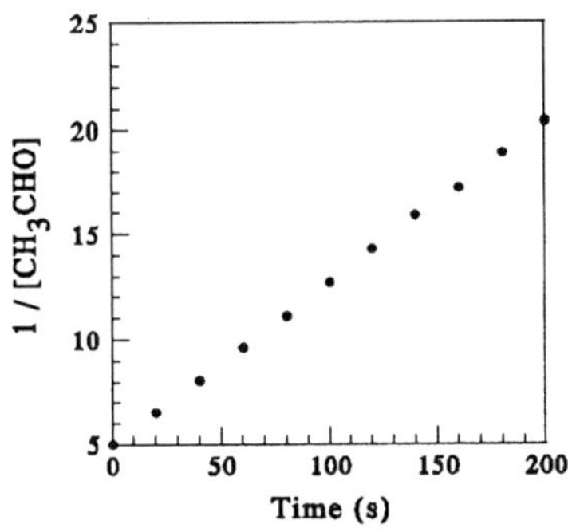

15.84 (a) The time involved must be converted to a value in seconds:
1 hr × 3600 s/hr = 3.6×10^3 s, and then we make use of equation 15.3, where x is taken to represent the desired SO_2Cl_2 concentration:

$$\ln \frac{0.0040 \text{ M}}{x} = (2.2 \times 10^{-5} \text{ s}^{-1})(3.6 \times 10^3 \text{ s})$$

x = 3.7×10^{-3} M

(b) The time is converted to a value having the units seconds

24 hr × 3600 s/hr = 8.64×10^4 s, and then we use equation 15.3, where x is taken to represent the desired SO_2Cl_2 concentration:

$$\ln \frac{0.0040\text{ M}}{x} = (2.2 \times 10^{-5}\text{ s}^{-1})(8.64 \times 10^4\text{ s})$$

$x = 6.0 \times 10^{-4}$ M

15.85 Use equation 15.3 and assume the initial concentration is 100 and the final concentration is, therefore, 20. The units are immaterial in this case since they will cancel out in the calculation:

$$\ln \frac{[A]_0}{[A]_t} = kt$$

$$\ln \frac{100}{20} = k(75.0\text{ min})$$

Solving for k we get 2.15×10^{-2} min^{-1}

15.86 Any consistent set of units for expressing concentration may be used in equation 15.3, where we let A represent the drug that is involved:

$$\ln \frac{[A]_0}{[A]_t} = kt$$

$$\ln \frac{25.0\ \text{mg/kg}}{15.0\ \text{mg/kg}} = k(120\text{ min})$$

Solving for k we get 4.26×10^{-3} min^{-1}

15.87 Use the equation, taking time in minutes; 3 hr = 180 min.

$$\ln \frac{x}{5.0\ \text{mg/kg}} = (4.26 \times 10^{-3}\text{ min}^{-1})(180\text{ min})$$

x = 11 mg/kg

15.88 We use the equation:

$$\frac{1}{[HI]_t} - \frac{1}{[HI]_0} = kt$$

$$\frac{1}{[8.0 \times 10^{-4}\text{ M}]} - \frac{1}{[3.4 \times 10^{-2}\text{ M}]} = (1.6 \times 10^{-3}\text{ L mol}^{-1}\text{ s}^{-1}) \times t$$

Solving for t gives:
$t = 7.6 \times 10^5$ s or $t = (7.6 \times 10^5\text{ s}) \times 1$ min/60 s = 1.3×10^4 min

15.89 Use equation 15.4, where the time is: $(2.5 \times 10^3\text{ min}) \times 60$ s/min = 1.5×10^5 s

$$\frac{1}{[B]_t} - \frac{1}{[B]_0} = kt$$

$$\frac{1}{[4.5 \times 10^{-4}\text{ M}]} - \frac{1}{[HI]_0} = (1.6 \times 10^{-3}\text{ L mol}^{-1}\text{ s}^{-1})(1.5 \times 10^5\text{ s})$$

$$\frac{1}{[HI]_0} = 2.0 \times 10^3\text{ L mol}^{-1}$$

$$[HI]_0 = 5.0 \times 10^{-4}\text{ M}$$

15.90 $\#\text{ half lives} = (2.0\text{ hrs})\left(\frac{60\text{ min}}{1\text{ hr}}\right)\left(\frac{1\text{ half life}}{15\text{ min}}\right) = 8.0\text{ half lives}$

Eight half lives correspond to the following fraction of original material remaining:

Number of half lives	Fraction remaining
1	1/2
2	1/4
3	1/8
4	1/16
5	1/32
6	1/64
7	1/128
8	1/256

15.91 Since 1/2 of the Sr–90 decays every half–life, it will take 5 half–lives, or 5 × 28 yrs = 140 yrs, for the Sr–90 to decay to 1/32 of its present amount. (See 15.90)

15.92 It requires approximately 500 min (as determined from the graph) for the concentration of SO_2Cl_2 to decrease from 0.100 M to 0.050 M, i.e., to decrease to half its initial concentration. Likewise, in another 500 minutes, the concentration decreases by half again, i.e. from 0.050 M to 0.025 M. This means that the half–life of the reaction is independent of the initial concentration, and we conclude that the reaction is first–order in SO_2Cl_2.

15.93 From the graph, we see that the first half–life is achieved in about 65 seconds, because this is the amount of time it requires for the concentration to decrease from its initial value (0.200 M) to half its initial value (0.100 M). The second half–life period, i.e., the time required for the concentration to decrease from 0.100 M to 0.050 M, is longer, namely about 125 seconds (this is determined by estimating the time when the concentration is 0.050 M and subtracting the time when the concentration is 0.100 M). Since the half–life does depend on concentration; we conclude that the reaction is not first–order. In fact, the data are consistent with second–order kinetics, because the value of the half–life decreases in proportion to the inverse of the initial concentration. That is, as the initial concentration for any half–life period becomes smaller, the half–life becomes larger.

15.94 $t = 0.693/k = 0.693/1.6 \times 10^{-3}\text{ s}^{-1} = 4.3 \times 10^{2}$ seconds

15.95 Using equation 15.6,

$$t_{1/2} = \frac{1}{k \times [B]_0} = \frac{1}{(6.7\times10^{-4}\text{ L mol}^{-1}\text{ s}^{-1})(0.20\text{ mol L}^{-1})} = 7.5\times10^{3}\text{ s}$$

15.96 The graph is prepared exactly as in example 15.12 of the text. The slope is found using linear regression, to be: -9.5×10^3 K. Thus $-9.5 \times 10^3\text{ K} = -E_a/R$

$$E_a = -(-9.5 \times 10^3\text{ K})(8.314\text{ J K}^{-1}\text{ mol}^{-1}) = 7.9 \times 10^4\text{ J/mol} = 79\text{ kJ/mol}$$

Using the equation, we proceed as follows:

$$\ln\frac{k_2}{k_1} = \frac{-E_a}{R}\left[\frac{1}{T_2} - \frac{1}{T_1}\right]$$

$$\ln\left[\frac{1.94 \times 10^{-3}\text{ L mol}^{-1}\text{ s}^{-1}}{2.88 \times 10^{-4}\text{ L mol}^{-1}\text{ s}^{-1}}\right] = \frac{-E_a}{8.314\text{ J mol}^{-1}\text{ K}^{-1}}\left[\frac{1}{673\text{ K}} - \frac{1}{593\text{ K}}\right]$$

$$1.907 = \frac{2.00 \times 10^{-4}\text{ K}^{-1}}{8.314\text{ J mol}^{-1}\text{ K}^{-1}} \times E_a$$

$E_a = 7.93 \times 10^4$ J/mol = 79.3 kJ/mol

15.97 The graph is prepared exactly as in example 15.12 of the text. The slope is found using linear regression, to be:
-1.67×10^4 K. Thus $-1.67 \times 10^4 \text{ K} = -E_a/R$
$E_a = -(-1.67 \times 10^4 \text{ K})(8.314 \text{ J K}^{-1} \text{ mol}^{-1}) = 1.39 \times 10^5$ J/mol = 139 kJ/mol

Using equation 15.9 we have:

$$\ln \frac{k_2}{k_1} = \frac{-E_a}{R}\left[\frac{1}{T_2} - \frac{1}{T_1}\right]$$

$$\ln\left[\frac{1.08 \times 10^{-1} \text{ L mol}^{-1}\text{ s}^{-1}}{1.91 \times 10^{-2} \text{ L mol}^{-1}\text{ s}^{-1}}\right] = \frac{-E_a}{8.314 \text{ J mol}^{-1}\text{ K}^{-1}}\left[\frac{1}{503 \text{ K}} - \frac{1}{478 \text{ K}}\right]$$

$$1.732 = \frac{1.04 \times 10^{-4} \text{ K}^{-1}}{8.314 \text{ J mol}^{-1}\text{ K}^{-1}} \times E_a$$

$E_a = 1.38 \times 10^5$ J/mol = 138 kJ/mol

15.98 Using the equation we have:

$$\ln \frac{k_2}{k_1} = \frac{-E_a}{R}\left[\frac{1}{T_2} - \frac{1}{T_1}\right]$$

$$\ln\left[\frac{1.0 \times 10^{-3} \text{ L mol}^{-1}\text{ s}^{-1}}{9.3 \times 10^{-5} \text{ L mol}^{-1}\text{ s}^{-1}}\right] = \frac{-E_a}{8.314 \text{ J mol}^{-1}\text{ K}^{-1}}\left[\frac{1}{403 \text{ K}} - \frac{1}{373 \text{ K}}\right]$$

$$2.37 = \frac{2.00 \times 10^{-4} \text{ K}^{-1}}{8.314 \text{ J mol}^{-1}\text{ K}^{-1}} \times E_a$$

$E_a = 9.89 \times 10^4$ J/mol = 99 kJ/mol

Equation states $k = A \exp\left(\frac{-E_a}{RT}\right)$

$$A = \frac{k}{\exp\left(\frac{-E_a}{RT}\right)}$$

$$= \frac{9.3 \times 10^{-5} \text{ L mol}^{-1}\text{ s}^{-1}}{\exp\left(\frac{-9.89 \times 10^4 \text{ J/mol}}{(8.314 \text{ J/mol K})(373 \text{ K})}\right)}$$

$$= 6.6 \times 10^9 \text{ L mol}^{-1}\text{ s}^{-1}$$

15.99 (a) Substituting into the equation:

$$\ln \frac{k_2}{k_1} = \frac{-E_a}{R}\left[\frac{1}{T_2} - \frac{1}{T_1}\right]$$

$$\ln\left[\frac{1.1 \times 10^{-5} \text{ L mol}^{-1}\text{ s}^{-1}}{1.3 \times 10^{-6} \text{ L mol}^{-1}\text{ s}^{-1}}\right] = \frac{-E_a}{8.314 \text{ J mol}^{-1}\text{ K}^{-1}}\left[\frac{1}{703 \text{ K}} - \frac{1}{673 \text{ K}}\right]$$

$$2.1 = \frac{6.34 \times 10^{-5}\ K^{-1}}{8.314\ J\ mol^{-1}\ K^{-1}} \times E_a$$

$E_a = 2.8 \times 10^5$ J/mol $= 2.8 \times 10^2$ kJ/mol

(b) Equation states $k = A \exp\left(\frac{-E_a}{RT}\right)$

$$A = \frac{k}{\exp\left(\frac{-E_a}{RT}\right)}$$

$$= \frac{1.3\times10^{-6}\ L\ mol^{-1}\ s^{-1}}{\exp\left(\frac{-2.8\times10^5\ J/mol}{(8.314\ J/mol\ K)(673\ K)}\right)}$$

$$= 7.0\times10^{15}\ L\ mol^{-1}\ s^{-1}$$

(c) The equation can be used directly now that the value of A is known:

$$k = A \exp\left(\frac{-E_a}{RT}\right)$$

$$= \left(7.0\times10^{15}\ L\ mol^{-1}\ s^{-1}\right)\exp\left(\frac{-2.8\times10^5\ J\ mol^{-1}}{(8.314\ J/mol\ K)(623\ K)}\right)$$

$$= 2.3\times10^{-8}\ L\ mol^{-1}\ s^{-1}$$

15.100 Substituting into the equation:

$$\ln\frac{k_2}{k_1} = \frac{-E_a}{R}\left[\frac{1}{T_2} - \frac{1}{T_1}\right]$$

$$\ln\left[\frac{3.75 \times 10^{-2}\ s^{-1}}{2.1 \times 10^{-3}\ s^{-1}}\right] = \frac{-E_a}{8.314\ J\ mol^{-1}\ K^{-1}}\left[\frac{1}{298\ K} - \frac{1}{273\ K}\right]$$

$(3.70 \times 10^{-5}$ mol/J$)(E_a) = 2.88$

$E_a = 7.8 \times 10^4$ J/mol $= 78$ kJ/mol

15.101 Substituting into the equation gives:

$$\ln\frac{k_2}{k_1} = \frac{-E_a}{R}\left[\frac{1}{T_2} - \frac{1}{T_1}\right]$$

$$\ln\left[\frac{k_2}{3.0 \times 10^{-4}\ s^{-1}}\right] = \frac{-100.0 \times 10^3\ J\ mol^{-1}}{8.314\ J\ mol^{-1}\ K^{-1}}\left[\frac{1}{323.2\ K} - \frac{1}{298.2\ K}\right]$$

$$\ln\left[\frac{k_2}{3.0 \times 10^{-4}\ s^{-1}}\right] = 3.120$$

Taking the exponential of both sides of this equation gives:

$k_2/3.0 \times 10^{-4}\ s^{-1} = \exp(3.120) = 22.65$

$k_2 = 6.8 \times 10^{-3}\ s^{-1}$

15.102 We can use the equation:

(a)

$$k = A\exp\left(\frac{-E_a}{RT}\right)$$

$$= \left(4.3\times10^{13}\ s^{-1}\right)\exp\left(\frac{-103\times10^{3}\ J\ mol^{-1}}{\left(8.314\ J/_{mol\,K}\right)\left(293\ K\right)}\right)$$

$$= 1.9\times10^{-5}\ s^{-1}$$

(b)

$$k = A\exp\left(\frac{-E_a}{RT}\right)$$

$$= \left(4.3\times10^{13}\ s^{-1}\right)\exp\left(\frac{-103\times10^{3}\ J\ mol^{-1}}{\left(8.314\ J/_{mol\,K}\right)\left(373\ K\right)}\right)$$

$$= 1.6\times10^{-1}\ s^{-1}$$

15.103 Substitute into the equation:

$$\ln\frac{k_2}{k_1} = \frac{-E_a}{R}\left[\frac{1}{T_2} - \frac{1}{T_1}\right]$$

$$\ln\left[\frac{k_2}{6.2\times10^{-5}\ s^{-1}}\right] = \frac{-108\times10^{3}\ J\ mol^{-1}}{8.314\ J\ mol^{-1}\ K^{-1}}\left[\frac{1}{318\ K} - \frac{1}{308\ K}\right]$$

$\ln(k_2/6.2\times10^{-5}\ s^{-1}) = 1.33$

$k_2 = 6.2\times10^{-5}\ s^{-1} \times \exp(1.33) = 2.3\times10^{-4}\ s^{-1}$

Additional Exercises

15.104 The concentration of CH_4 at any time is equal to the starting concentration of CH_3CHO (0.200 M) minus the remaining concentration of CH_3CHO, because for every mole of CH_3CHO that disappears, one mole of CH_4 appears. The rates of reaction at t = 40 s and at t = 100 s are given by the slopes of tangents to the curve at t = 40 s and t = 100 s respectively.

$rate_{40} = 1.2\times10^{-3}\ mol\ L^{-1}\ s^{-1}$

$rate_{100} = 5.1\times10^{-4}\ mol\ L^{-1}\ s^{-1}$

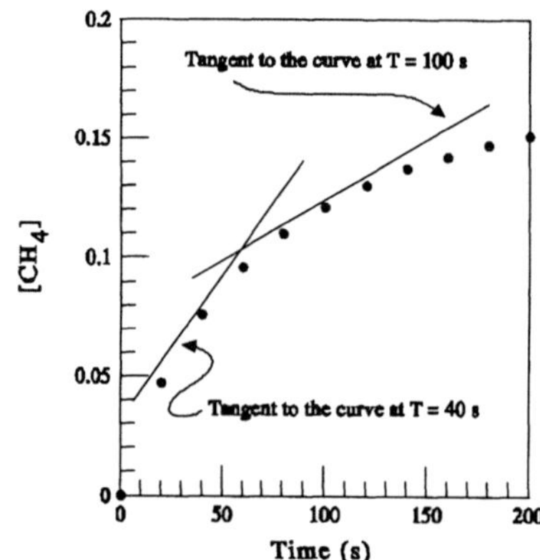

15.105 $k = 0.693/12.5\ y = 5.54 \times 10^{-2}\ y^{-1}$

$$\ln \frac{1}{0.1} = \left(5.54 \times 10^{-2}\ y^{-1}\right) \times t$$

$t = 41.6$ y

15.106 From Section 15.5, the fraction remaining after n half-lives is $1/2^n$. Therefore, we have:

$$0.810 = 1/2^n$$

Inverting both sides of the equation gives:

$$1.23 = 2^n$$

We can solve this for now using trial and error: We know that $2^1 = 2$, and that $2^0 = 1$. Therefore n must be between 0 and 1. We start by trying n = 0.50:

$$2^{0.50} = 1.41$$

This is too high, so we might try 0.30:

$$2^{0.30} = 1.23$$

This is just what we were looking for, therefore n = 0.30. Since one half life of C-14 is 5730 years, one might estimate the age of the mummy as (0.30)(5730) = about 1,700 years old.

15.107 $NO_2 + O_3 \rightarrow NO_3 + O_2$ (slow)
$NO_3 + NO_2 \rightarrow N_2O_5$ (fast)

15.108 (a) rate = $k_1[A]^2$
(b) rate = $k_{-1}[A_2]^1$
(c) rate = $k_2[A_2]^1[E]^1$
(d) $2A + E \rightarrow B + C$
(e) The rates for the forward and reverse directions of step one are set equal to each other in order to arrive at an expression for the intermediate $[A_2]$ in terms of the reactant [A]: $k_1[A]^2 = k_{-1}[A_2]$

$$[A_2] = \frac{k_1}{k_{-1}}[A]^2$$

This is substituted into the rate law for question (c) above, giving a rate expression that is written using only observable reactants: rate = $k_2 \frac{k_1}{k_{-1}}[A]^2[E]^1$

15.109 (a) The order with respect to [B] is:

$$\left(\frac{0.040\ \text{mol L}^{-1}}{0.030\ \text{mol L}^{-1}}\right)^n = \left(\frac{-0.0334\ \text{mol L}^{-1}\ \text{s}^{-1}}{-0.0188\ \text{mol L}^{-1}\ \text{s}^{-1}}\right) = 1.78 \qquad \therefore n = 2$$

The order with respect to [A] is:

$$\left(\frac{0.025\ \text{mol L}^{-1}}{0.020\ \text{mol L}^{-1}}\right)^n = \left(\frac{-0.0188\ \text{mol L}^{-1}\ \text{s}^{-1}}{-0.0150\ \text{mol L}^{-1}\ \text{s}^{-1}}\right) = 1.25 \qquad \therefore n = 1$$

rate = $k[A]^1[B]^2$

(b) One fact that is not immediately obvious, but that you may have noticed throughout this chapter, is that the value of a rate constant is always positive. Hence, we will take the absolute value of the initial rate for this calculation. Using the first rate value, we substitute into the rate expression from part (a).

$$0.0150\ \text{mol L}^{-1}\ \text{s}^{-1} = k(0.020\ \text{mol L}^{-1})^1(0.030\ \text{mol L}^{-1})^2$$

$$k = 8.33 \times 10^2\ \text{L}^2\ \text{mol}^{-2}\ \text{s}^{-1}$$

15.110 First, determine a value for E_a using the equation:

$$\ln\frac{k_2}{k_1} = \frac{-E_a}{R}\left[\frac{1}{T_2} - \frac{1}{T_1}\right]$$

$$\ln\left[\frac{2.25 \times 10^{-5}\ \text{min}^{-1}}{5.84 \times 10^{-6}\ \text{min}^{-1}}\right] = \frac{-E_a}{8.314\ \text{J mol}^{-1}\ \text{K}^{-1}}\left[\frac{1}{343\ \text{K}} - \frac{1}{333\ \text{K}}\right]$$

$$1.35 = \frac{8.76 \times 10^{-5}\ \text{K}^{-1}}{8.314\ \text{J mol}^{-1}\ \text{K}^{-1}} \times E_a$$

$E_a = 1.28 \times 10^5$ J/mol = 128 kJ/mol

Next, use this value of E_a and the data at 70 °C to calculate a rate constant at 80 °C:

$$\ln\frac{k_2}{k_1} = \frac{-E_a}{R}\left[\frac{1}{T_2} - \frac{1}{T_1}\right]$$

$$\ln\left[\frac{k_2}{2.25 \times 10^{-5}\ \text{min}^{-1}}\right] = \frac{-1.28 \times 10^5\ \text{J mol}^{-1}}{8.314\ \text{J mol}^{-1}\ \text{K}^{-1}}\left[\frac{1}{353\ \text{K}} - \frac{1}{343\ \text{K}}\right]$$

$$\ln\left[\frac{k_2}{2.25 \times 10^{-5}\ \text{min}^{-1}}\right] = 1.27$$

$k_2 = (2.25 \times 10^{-5}\ \text{min}^{-1}) \times \exp(1.27) = 8.02 \times 10^{-5}\ \text{min}^{-1}$.

Finally, use the first–order rate expression to determine time:

$$\ln\frac{0.0020\ \text{M}}{0.0012\ \text{M}} = (8.02 \times 10^{-5}\ \text{min}^{-1}) \times t$$

Solving for t we get 6.37×10^3 min.

15.111 Taking the log of both sides of this equation, we have:

log(rate) = log k + n × log[A]

Thus, if we plot log(rate) vs. log[A], the slope should be equal to the value for n, and the intercept is equal to log k.

A plot of the data is used to determine slopes of the tangents to the curve at time = 150, 300, 450, and 600 min. The negatives of these slopes are equal to the value of the rate constants at these particular times during the reaction. (Notice that the rate decreases with time.)

The four rates are found to be approximately:
1.1×10^{-4} mol L^{-1} min^{-1}, at t = 150 min
8.9×10^{-5} mol L^{-1} min^{-1}, at t = 300 min
7.2×10^{-5} mol L^{-1} min^{-1}, at t = 450 min
6.0×10^{-5} mol L^{-1} min^{-1}, at t = 600 min

Suitable values for $[SO_2Cl_2]$ at each of these times can be read from the graph of concentration vs. time. Once each of these concentration values has been converted to $\log[SO_2Cl_2]$, the plot of log rate vs. log $[SO_2Cl_2]$ can be constructed:

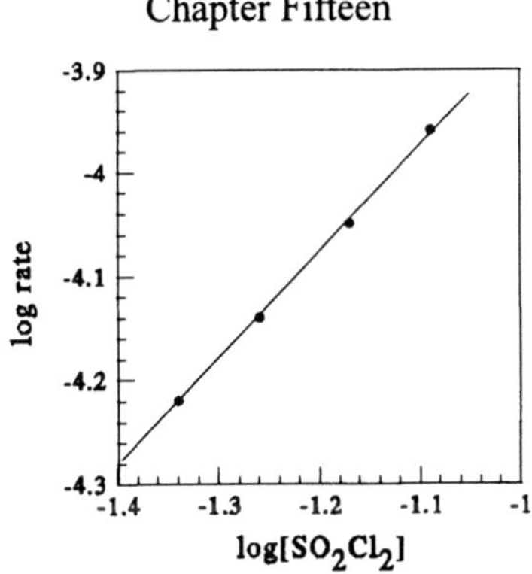

The slope of this straight line has the value 1.0, which is the order of the reaction with respect to $[SO_2Cl_2]$.

15.112 Molecular oxygen exists highly in the diradical state. That is, the two electrons in the HOMOs exist in separate, degenerate orbitals (see discussion in section 10.7). This means that the "double bond" of O_2 does not have the stability of most other double bonds. Therefore diatomic oxygen is at a higher energy level relative to the transition state energy for its reactions, and the activation energy is smaller.

Molecular nitrogen, on the other hand, contains two π-bonds. Neither of these are diradical in character and lend exceptional stability to the molecule. Therefore diatomic nitrogen is at a lower energy level relative to the transition state energy for its reactions, and the activation energy is greater.

15.113

$$\ln \frac{k_2}{k_1} = \frac{-E_a}{R}\left[\frac{1}{T_2} - \frac{1}{T_1}\right]$$

$$\ln\left[\frac{2}{1}\right] = \frac{-E_a}{8.314\ \text{J mol}^{-1}\ \text{K}^{-1}}\left[\frac{1}{308\ \text{K}} - \frac{1}{298\ \text{K}}\right]$$

$(1.31 \times 10^{-5}\ \text{mol/J})(E_a) = 0.693$

$E_a = 5.29 \times 10^4\ \text{J/mol} = 52.9\ \text{kJ/mol}$

15.114 To solve this problem, plot the data provided as 1/T vs 1/t where T is the absolute temperature and 1/t is proportional to the rate constant.

t (min)	T (K)	1/T	ln(1/t)
10	291	0.003436	-2.302585
9	293	0.003412	-2.197224
8	294	0.003401	-2.079441
7	295	0.003389	-1.945910
6	297	0.003367	-1.791759
13.6	288	0.003472	-2.608

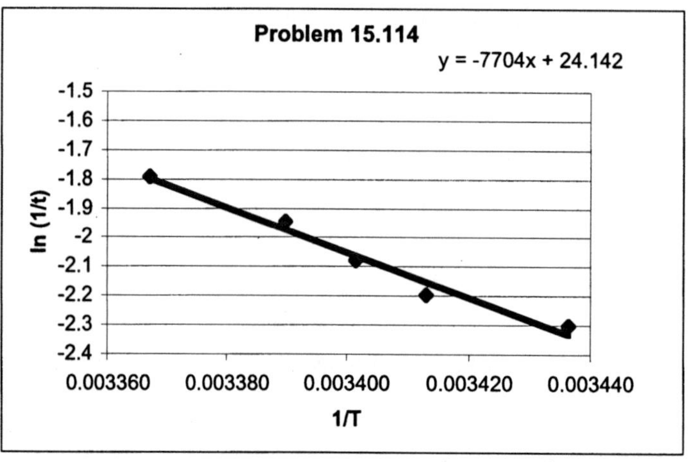

The slope of the graph is equal to $-E_a/R$, therefore:

$$-7{,}704 = -E_a/R$$
$$7{,}704R = E_a$$
$$7{,}704\text{ K}(8.314\text{ J/mol}\cdot\text{K}) = E_a$$
$$E_a = 64{,}050\text{ J/mol}$$
$$E_a = 64\text{ kJ/mol}$$

From the straight–line equation, we can determine the time needed to develop the film at 15 °C is 14 min.

15.115 (a) The number of chirps in eight seconds is simply the temperature minus 4. So, in order, there will be 16, 21, 26 and 31 chirps at these temperatures.

(b) The data are plotted below.

# of chirps	T	1/T	ln (chirps)
16	293	0.003412	2.772588
21	298	0.003355	3.044522
26	303	0.003300	3.258096
31	308	0.003246	3.433987

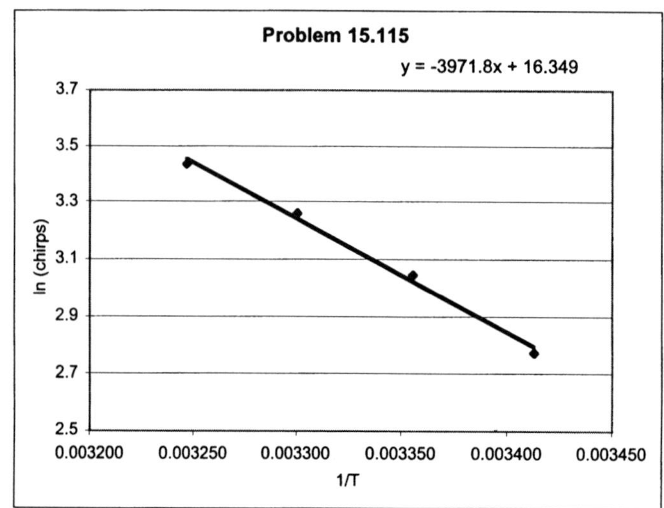

The slope of the line is equal to the Activation energy divided by R so $E_a = 477.4$ J/mol.

(c) At 40 °C the cricket would make 36 chirps.

15.116 Taking note of the inverse relationship between the reaction rate constant, k, and the cooking time, t, we set up equation 15.9 in the following manner:

$$\ln \frac{k_2}{k_1} = \frac{-E_a}{R}\left[\frac{1}{T_2} - \frac{1}{T_1}\right]$$

$$\ln\left[\frac{1/t_2}{1/t_1}\right] = \frac{-E_a}{R}\left[\frac{1}{T_2} - \frac{1}{T_1}\right]$$

We are provided with some subtle but key information about the physical conditions. For instance, the 3–minute traditional egg provides a cooking time of 3 minutes at the normal boiling point of water, 100 °C or 373 K. We are also given the atmospheric pressure (355 torr) on Mt. McKinley where the cooking is to be carried out at a lower temperature. At 355 torr, H_2O will boil when its vapor pressure equals 355 torr. The temperature corresponding to this pressure is 80 °C or 323 K (See Appendix). Thus, with the given value of the activation energy, i.e., E_a = 418 kJ/mol, we can proceed with the calculation to obtain t_2:

$$\ln\left[\frac{1/t_2}{1/3\text{ min}}\right] = \frac{-418 \times 10^3 \text{ J/mol}}{8.314 \text{ J/mol K}}\left[\frac{1}{353 \text{ K}} - \frac{1}{373 \text{ K}}\right]$$

$$\ln\left[\frac{3 \text{ min}}{t_2}\right] = -7.64$$

$$\frac{3 \text{ min}}{t_2} = \exp(-7.64)$$

$$t_2 = 3 \text{ min}/\exp(-7.64) = 6.2 \times 10^3 \text{ min} = 104 \text{ hrs}$$

Thus, to get the same degree of protein denaturization, it would take roughly 4 days to cook the egg at an atmospheric pressure of 355 torr as opposed to cooking the egg at normal atmospheric pressure.

15.117 In step 2, the reaction of the $Cl^\bullet$ radical with O_3 is very fast, producing the $ClO^\bullet$ radical which then reacts in step 3 rapidly with O atoms. In this step the reactive $Cl^\bullet$ radical is regenerated to react again with O_3. This, in effect, is a chain reaction.

15.118 (a) The first step, in which a free radical is produced, is the initiation step.

(b) Both the second and third steps are propagating steps since HBr, the desired product, and an additional free radical are produced.

(c) The final step in which two bromine free radicals recombine to give a bromine molecule is the termination step.

The presence of the additional reaction step serves to decrease the concentration of HBr.

Practice Exercises

16.1 (a) $$\frac{[H_2O]^2}{[H_2]^2[O_2]} = K_c$$ (b) $$\frac{[CO_2][H_2O]^2}{[CH_4][O_2]^2} = K_c$$

16.2 Since the starting equation has been reversed and divided by two, we must invert the equilibrium constant, and then take the square root: $K_c = 1.2 \times 10^{-13}$

16.3 If we divide both equations by 2 and reverse the second we get:

$CO(g) + 1/2O_2(g) \rightarrow CO_2(g)$ $K_c = 5.7 \times 10^{45}$
$H_2O(g) \rightarrow H_2(g) + 1/2O_2(g)$ $K_c = 3.3 \times 10^{-41}$

Note that when we divide the equation by two, we need to take the square root of the rate constant. When we reverse the reaction, we need to take the inverse.

Adding these equations we get the desired equation so we need simply multiply the values for K_c in order to obtain the new value: $K_c = 1.9 \times 10^5$

16.4 $$K_c = \frac{(P_{HI})^2}{(P_{H_2})(P_{I_2})}$$

16.5 Reaction (b) will proceed farthest to completion since it has the largest value for K_c.

16.6 We would expect K_P to be smaller than K_c since Δn_g is negative.

Use the equation:

$$K_p = K_c(RT)^{\Delta n_g}$$
$$K_c = \frac{K_p}{(RT)^{\Delta n_g}}$$

In this case, $\Delta n_g = (1 - 3) = -2$, and we have:

$$K_c = \frac{K_p}{(RT)^{\Delta n_g}} = \frac{3.8\times10^{-2}}{\left(\left(0.0821\,\frac{L\,atm}{mol\,K}\right)(473\,K)\right)^{-2}} = 57$$

16.7 Use the equation $K_p = K_c(RT)^{\Delta n_g}$. In this reaction, $\Delta n_g = 3 - 2 = 1$, so

$$K_p = K_c(RT)^{\Delta n_g} = (7.3\times10^{34})\left(\left(0.0821\,\frac{L\,atm}{mol\,K}\right)(298\,K)\right)^1 = 1.8\times10^{36}$$

16.8 a) $$K_c = \frac{1}{[Cl_2\,(g)]}$$

b) $$K_c = \frac{1}{[NH_3\,(g)][HCl(g)]}$$

c) $K_c = [Na^+ (aq)][OH^- (aq)][H_2 (g)]$

d) $K_c = [Ag^+]^2[CrO_4^{2-}]$

e) $K_c = \dfrac{[Ca^{+2} (aq)][HCO_3^- (aq)]^2}{[CO_2 (aq)]}$

16.9 a) The equilibrium will shift to the right, decreasing the concentration of Cl_2 at equilibrium, and consuming some of the added PCl_3. The value of K_p will be unchanged.

b) The equilibrium will shift to the left, consuming some of the added PCl_5 and increasing the amount of Cl_2 at equilibrium. The value of K_p will be unchanged.

c) For any exothermic equilibrium, an increase in temperature causes the equilibrium to shift to the left, in order to remove energy in response to the stress. This equilibrium is shifted to the left, making more Cl_2 and more PCl_3 at the new equilibrium. The value of K_p is given by the following:

$$K_p = \frac{P_{PCl_5}}{P_{PCl_3} \times P_{Cl_2}}$$

In this system, an increase in temperature (which causes an increase in the equilibrium concentrations of both PCl_3 and Cl_2 and a decrease in the equilibrium concentration of PCl_5) causes an increase in the denominator of the above expression as well as a decrease in the numerator of the above expression. Both of these changes serve to decrease the value of K_p.

d) Decreasing the container volume for a gaseous system will produce an increase in partial pressures for all gaseous reactants and products. In order to lower the increase in partial pressures, the equilibrium will shift so as to favor the reaction side having the smaller number of gaseous molecules, in this case to the right. This shift will decrease the amount of Cl_2 and PCl_3 at equilibrium, and it will increase the amount of PCl_5 at equilibrium. This increases the size of the numerator and decreases the size of the denominator in the above expression for K_p, causing the value of K_p to increase.

16.10 $K_c = \dfrac{[CO_2][H_2]}{[CO][H_2O]} = \dfrac{(0.150)(0.200)}{(0.180)(0.0411)} = 4.06$

16.11 $2CO(g) + O_2(g) \leftrightarrows 2CO_2(g)$

Using the stoichiometry of the reaction we can see that for every mol of O_2 that is used, twice as much CO will react and twice as much CO_2 will be produced. Consequently, if the $[O_2]$ decreases by 0.030 mol/L, the [CO] decreases by 0.060 mol/L and $[CO_2]$ increases by 0.060 mol/L.

16.12 a) The initial concentrations were:
$[PCl_3]$ = 0.200 mol/1.00 L = 0.200 M
$[Cl_2]$ = 0.100 mol/1.00 L = 0.100 M
$[PCl_5]$ = 0.00 mol/1.00 L = 0.000 M

b) The change in concentration of PCl_3 was (0.200 – 0.120) M = 0.080 mol/L. The other materials must have undergone changes in concentration that are dictated by the coefficients of the balanced chemical equation, namely: $PCl_3 + Cl_2 \rightarrow PCl_5$ or both PCl_3 and Cl_2 have decreased by 0.080 M and PCl_5 has increased by 0.080 M.

c) As stated in the problem, the equilibrium concentration of PCl_3 is 0.120 M. The equilibrium concentration of PCl_5 is 0.080 M since initially there was no PCl_5. The equilibrium concentration of Cl_2 equals the initial concentration minus the amount that reacted, 0.100 M - 0.080 M = 0.020 M.

d) $K_c = \dfrac{[PCl_5]}{[PCl_3][Cl_2]} = \dfrac{(0.080)}{(0.120)(0.020)} = 33$

16.13 $K_c = \dfrac{[CH_3CO_2C_2H_5][H_2O]}{[CH_3CO_2H][C_2H_5OH]} = \dfrac{(0.910)(0.00850)}{(0.210)[C_2H_5OH]} = 4.10$

$[C_2H_5OH] = 8.98 \times 10^{-3}$ M

16.14 Initially we have $[H_2] = [I_2] = 0.200$ M.

	$[H_2]$	$[I_2]$	[HI]
I	0.200	0.200	–
C	–x	–x	+2x
E	0.200–x	0.200–x	+2x

Substituting the above values for equilibrium concentrations into the mass action expression gives:

$$K_c = \frac{[HI]^2}{[H_2][I_2]} = \frac{(2x)^2}{(0.200-x)(0.200-x)} = 49.5$$

Take the square root of both sides of this equation to get; $\dfrac{2x}{(0.200-x)} = 7.04$. This equation is easily solved giving x = 0.156. The substances then have the following concentrations at equilibrium: $[H_2] = [I_2] = 0.200 - 0.156 = 0.044$ M, [HI] = 2(0.156) = 0.312 M.

16.15 $$N_2(g) + O_2(g) \rightleftharpoons 2NO(g)$$

	$[N_2]$	$[O_2]$	[NO]
I	0.033	0.00810	–
C	–x	–x	+2x
E	0.033–x	0.00810–x	+2x

Substituting the above values for equilibrium concentrations into the mass action expression gives:

$$K_c = \frac{[NO]^2}{[N_2][O_2]} = \frac{(2x)^2}{(0.033-x)(0.00810-x)} = 4.8 \times 10^{-31}$$

If we assume that x << 0.033 and x << 0.00810, we can simplify this equation. (Because the value of K_c is so low, this assumption should be valid.) The equation simplifies as:

$$K_c = \frac{(2x)^2}{(0.033)(0.00810)} = 4.8 \times 10^{-31}$$

This equation is easily solved to give $x = 5.7 \times 10^{-18}$ M. The equilibrium concentration of NO is 2x according to the ICE table so, $[NO] = 1.1 \times 10^{-17}$ M.

Thinking It Through

T16.1 After writing the equilibrium law for the reaction:

$$K = \frac{[N_2O_3]}{[NO_2][NO]}$$

we should find that substitution of the equilibrium concentrations for any of the three experiments listed in the problem gives the same value for K.

T16.2 We must decide how the first equation has been manipulated in arriving at the second equation. By inspection, it appears that the first equation has been reversed and multiplied by the factor two. Thus, the first equilibrium constant must be inverted and squared in order to determine a value for the second equilibrium constant.

T16.3 We can always use equation 16.4 to convert from K_c to K_p values. Here the value for Δn is equal to zero, so K_c and K_p are equal to each other.

T16.4 A change in the initial concentration of any reactant or product would change the position of the final equilibrium, as would a change in the temperature of the system. Since the value for Δn is zero, changes in pressure and in volume will not affect the position of the equilibrium. A catalyst will similarly not affect the position of the equilibrium.

T16.5 Once the equilibrium law expression for this system has been written, we can simply substitute values for K, $[NH_3]$ and $[N_2]$ into the equation, and solve for $[H_2]$.

$$K_c = \frac{[NH_3]^2}{[H_2]^3[N_2]}$$

T16.6 The equilibrium concentrations for each of the reactants and products can be converted to pressures, using the ideal gas equation, remembering that a molar concentration for a gas corresponds to the quantity n/V: $P = (n/V) \times RT$

T16.7 Once partial pressures have been determined for each reactant and product, they can be substituted into the equilibrium law expression:

$$K_P = \frac{P^2_{NH_3}}{P_{N_2}P^2_{H_2}}$$

T16.8 The equilibrium law expression can be set up using values from the following typical table:

	$[N_2O_3]$	[NO]	$[NO_2]$
I	–	0.40	0.60
C	+x	–x	–x
E	+x	0.40–x	0.60–x

Based on the balanced equation provided, we know that the NO and NO_2 will react to form N_2O_3. The balanced equation tells us that for every mole of NO and NO_2 that react, one mole of N_2O_3 is produced. We also know that the equilibrium concentration of N_2O_3 is 0.13 M. Hence, we know x in the equilibrium table and we can determine the equilibrium concentrations of all compounds.

T16.9 A catalyst should not change the position of an equilibrium. Rather it changes the speed at which equilibrium is achieved. We conclude that a catalyst does not change the equilibrium position, but does increase the rate of attaining equilibrium. It does this by providing an alternative pathway to the formation of products. Also, noting the endothermic nature of the reaction, application of Le Châtelier's Principle would lead one to argue that the reaction is driven further to the right (formation of NO(*g*)) as the converter is warmed up.

T16.10 The mass action expression for this equilibrium is needed first:

$$K = \frac{[NH_3]^2}{[H_2]^3[N_2]}$$

This indicates that a decrease in the value for K (as temperature is increased) corresponds to a decrease in the equilibrium concentration of ammonia, since ammonia appears in the numerator for the equilibrium law.

T16.11 The value for the equilibrium constant dictates to us how far the system will move towards products upon achieving equilibrium. Here the equilibrium constant is an extremely large number, signifying that practically all of the oxygen and hydrogen are consumed by the reaction. Obviously x is not a small number at all.

T16.12 The system can be "allowed" to react completely so as to use up the initial concentration of H_2. This produces a corresponding new initial amount of O_2 and H_2O:

	$[H_2]$	$[O_2]$	$[H_2O]$
I	–	0.020	0.020
C	+2x	+x	–x
E	+2x	0.020+x	0.020–x

Substituting the above values for equilibrium concentrations into the mass action expression gives an equivalent, but readily simplified result:

$$K_c = \frac{[H_2O]^2}{[H_2]^2[O_2]} = \frac{(0.020 - x)^2}{(2x)^2(0.020 + x)} = 1.4 \times 10^{17}$$

We know the equilibrium will not shift far to the left so we may assume that x << 0.020. The mass action expression becomes:

$$K_c = \frac{(0.020)^2}{(2x)^2(0.020)} = 1.4 \times 10^{17}$$

This equation is easily solved for x.

T16.13 We start by determining the initial concentration of NO_2: $[NO_2]$ = 0.20 mol/4.00 L = 0.050 M.

Next, this value is used in an equilibrium law type calculation using the equation:

$$3NO_2 \rightarrow N_2O_5 + NO$$

	$[NO_2]$	$[N_2O_5]$	$[NO]$
I	0.050	–	–
C	–3x	+x	+x
E	0.050–3x	+x	+x

Substituting the above values for equilibrium concentrations into the mass action expression gives:

$$K_c = \frac{[N_2O_5][NO]}{[NO_2]^3} = \frac{x^2}{(0.050-3x)^3} = 1.0 \times 10^{-11}$$

Upon solving the above expression for the value of x, we will have determined the concentration of N_2O_5 (and of NO) at equilibrium. In order to determine number of moles, we then multiply the concentration (x) by volume (4.00 L).

T16.14 First, the partial pressure of water should be converted into a molar concentration using the ideal gas law. Next, since the initial concentration of HNO_2 is zero, we realize that the system must shift to the right to reach equilibrium. Furthermore, the final concentration of HNO_2 is equal to twice the change in molar concentration that NO, NO_2, and H_2O each undergo. This allows us to calculate the equilibrium concentrations for each of the three reactants. Values of the equilibrium concentrations for all four substances are then substituted directly into the equilibrium law in order to arrive at a numerical value for K.

T16.15 Performing an equilibrium analysis

	$[Br_2]$	$[Cl_2]$	$[BrCl]$
I	0.6	0.4	–
C	–x	–x	+2x
E	0.6–x	0.4–x	+2x

$$K_c = 2 = \frac{[BrCl]^2}{[Br_2][Cl_2]} = \frac{(2x)^2}{(0.6-x)(0.4-x)}$$

This leads to a quadratic equation whose solution gives x = 0.2. Thus, at equilibrium, $[Br_2]$ = 0.4 M, $[Cl_2]$ = 0.2 M and [BrCl] = 0.4 M. These initial conditions and equilibrium results are best represented in Figure D. Figures A and E are rejected because in each case, one reactant completely reacts, i.e., no equilibrium is achieved. Figure C is rejected because the $[BrCl]_{equil}$ is too high. Finally, Figure B represents the nature of an oscillatory reaction and not a dynamic equilibrium.

T16.16 (a)

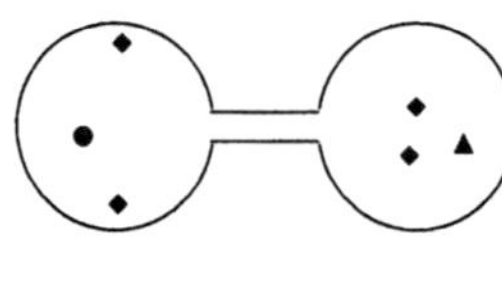

◆ = PF_5
▲ = PF_3
● = F_2

(b) $F_2(g) + PF_3(g) \rightarrow PF_5(g)$
$PF_5(g) \rightarrow F_2(g) + PF_3(g)$

(c)

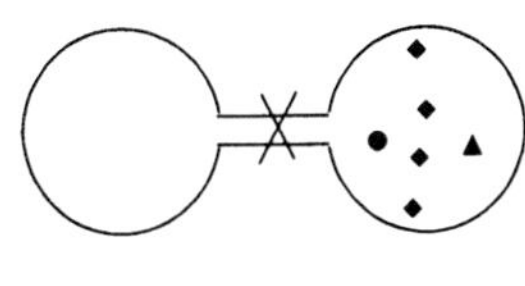

◆ = PF_5
▲ = PF_3
● = F_2

Because the volume is smaller, LeChateliers Principle states that the equilibrium will favor the products. Consequently, equilibrium will be shifted to the right in a smaller volume. As can be seen here, a higher proportion of the reactants have reacted to form more PF_5 as compared to the case in (a).

Review Questions

16.1 See Figure 16.1.

16.2 By reversibility we mean, first, that the reaction can proceed in either of the two possible directions, and, secondly, that for a given overall composition, the same equilibrium mixture can be attained from either the forward or the reverse directions.

16.3 An equilibrium law is the statement that the reaction quotient, i.e., the value of K_c must be equal to the numerical value of the mass action expression once equilibrium is attained.

16.4 The reaction quotient is the numerical value of the mass action expression.

16.5 The reaction quotient becomes equal to the value of K_c once equilibrium has been attained.

16.6 By convention, the products are always written into the numerator and the reactants are written into the denominator of the mass action expression.

16.7 This equilibrium constant is small, and we do not expect the equilibrium to favor products.

16.8 The increasing tendency to go to completion is: (a) < (c) < (b), based on the relative magnitudes of K_c.

16.9 $$K_p = K_c \times (RT)^{\Delta n_g}$$

where Δn_g is equal to the change in the number of moles of gaseous material on going from reactants to products. The value to be used for the gas constant, R, is 0.0821 L atm K^{-1} mol^{-1}.

16.10 For the ith component of a gas mixture, the partial pressure is given by the ideal gas law:

$$P_i = \frac{n_i RT}{V_i} = M_i RT$$, since n_i/V_i is equal to molarity, M_i.

The proportionality constant between P_i and the molarity of the ith gas, M_i, is the product RT.

16.11 In a homogeneous equilibrium, all of the reactants and products are in the same phase. In heterogeneous equilibria, at least two different phases are found among the reactants and products.

16.12 This is possible because their concentrations are constants that are incorporated into the numerical values of equilibrium constants. The concentrations may be considered constant because as the solid or liquid loses (or gains) mass, it loses (or gains) a volume proportional to the mass.

16.13 When a system at equilibrium is disturbed so that the equilibrium is upset, the system changes in a way that opposes the disturbance and returns the system to a state of equilibrium.

16.14 (a) The system shifts to the right to consume some of the added methane.
(b) The system shifts to the left to consume some of the added hydrogen.
(c) The system shifts to the right to make some more carbon disulfide.
(d) The system shifts to the left to decrease the amount of gaseous moles.
(e) The system shifts to the right to absorb some of the added heat.

16.15 (a) increase; we are adding a reactant
(b) decrease; we are removing a reactant
(c) increase; decreasing the volume favors the side with fewer gas molecules
(d) no change; a catalyst increases the rate but does not affect the concentration
(e) decrease; this is an exothermic reaction so heat may be considered a product

16.16 (a) right (b) left (c) left
(d) right (e) no effect (f) left

16.17 The value of K_c can be changed only by a change in temperature.

16.18 (a) increase (b) increase
(c) increase (d) decrease

Review Problems

16.19 (a) $K_c = \dfrac{[POCl_3]^2}{[PCl_3]^2[O_2]}$ (d) $K_c = \dfrac{[NO_2]^2[H_2O]^8}{[N_2H_4][H_2O_2]^6}$

(b) $K_c = \dfrac{[SO_2]^2[O_2]}{[SO_3]^2}$ (e) $K_c = \dfrac{[SO_2][HCl]^2}{[SOCl_2][H_2O]}$

(c) $K_c = \dfrac{[NO]^2[H_2O]^2}{[N_2H_4][O_2]^2}$

16.20 (a) $K_c = \dfrac{[NCl_3][HCl]^3}{[Cl_2]^3[NH_3]}$ (d) $K_c = \dfrac{[HOCl]^2}{[H_2O][Cl_2O]}$

(b) $K_c = \dfrac{[PCl_2Br][PClBr_2]}{[PCl_3][PBr_3]}$ (e) $K_c = \dfrac{[BrF_5]^2}{[Br_2][F_2]^5}$

(c) $K_c = \dfrac{[HNO_2]^2}{[NO][NO_2][H_2O]}$

16.21 (a) $K_p = \frac{(P_{POCl_3})^2}{(P_{PCl_3})^2(P_{O_2})}$

(d) $K_p = \frac{(P_{NO_2})^2(P_{H_2O})^8}{(P_{N_2H_4})(P_{H_2O_2})^6}$

(b) $K_p = \frac{(P_{SO_2})^2(P_{O_2})}{(P_{SO_3})^2}$

(e) $K_p = \frac{(P_{SO_2})(P_{HCl})^2}{(P_{SOCl_2})(P_{H_2O})}$

(c) $K_p = \frac{(P_{NO})^2(P_{H_2O})^2}{(P_{N_2H_4})(P_{O_2})^2}$

16.22 (a) $K_p = \frac{(P_{NCl_3})(P_{HCl})^3}{(P_{Cl_2})^3(P_{NH_3})}$

(d) $K_p = \frac{(P_{HOCl})^2}{(P_{H_2O})(P_{Cl_2O})}$

(b) $K_c = \frac{(P_{PCl_2Br})(P_{PClBr_2})}{(P_{PCl_3})(P_{PBr_3})}$

(e) $K_c = \frac{(P_{BrF_5})^2}{(P_{Br_2})(P_{F_2})^5}$

(c) $K_c = \frac{(P_{HNO_2})^2}{(P_{NO})(P_{NO_2})(P_{H_2O})}$

16.23 (a) $K_c = \frac{[Ag(NH_3)_2^+]}{[Ag^+][NH_3]^2}$

(b) $K_c = \frac{[Cd(SCN)_4^{2-}]}{[Cd^{2+}][SCN^-]^4}$

16.24 (a) $K_c = \frac{[H_3O^+][ClO^-]}{[HClO]}$

(b) $K_c = \frac{[HCO_3^-][SO_4^{2-}]}{[CO_3^{2-}][HSO_4^-]}$

16.25 The first equation has been reversed in making the second equation. We therefore take the inverse of the value of the first equilibrium constant in order to determine a value for the second equilibrium constant: K = 1×10^{85}

16.26 If we reverse the second reaction and double it, we can then add the first and second reactions together to obtain the desired reaction. When we double the second reaction we square the rate constant. When we reverse it, we invert the rate constant. Consequently, we get the following:

$2CH_4(g) \leftrightarrows C_2H_6(g) + H_2(g)$ $K_c = 9.5 \times 10^{-13}$
$2CH_3OH(g) + 2H_2(g) \leftrightarrows 2CH_4(g) + 2H_2O(g)$ $K_c = 1.3 \times 10^{41}$

Adding these equations we get:

$2CH_3OH(g) + H_2(g) \leftrightarrows C_2H_6(g) + 2H_2O(g)$ $K_c = 1.2 \times 10^{29}$

Where the K_c for the final reaction is the product of the K_c for the first reaction and the K_c for the "modified" second reaction.

16.27 (a) $K_c = \dfrac{[HCl]^2}{[H_2][Cl_2]}$ (b) $K_c = \dfrac{[HCl]}{[H_2]^{1/2}[Cl_2]^{1/2}}$

K_c for reaction (b) is the square root of K_c for reaction (a).

16.28 $K_c = \dfrac{[H_2][Cl_2]}{[HCl]^2}$

This is equal to $1/K_c$ for Review Exercise 16.27(a).

16.29 M = P/RT

$$M = \frac{(745\ \text{torr})\left(\frac{1\ \text{atm}}{760\ \text{torr}}\right)}{\left(0.0821\ \frac{\text{L atm}}{\text{mol K}}\right)(318\ \text{K})} = 0.0375\ \text{M}$$

16.30 P = M × RT = (0.0200 mol/L)(0.0821 L atm/K mol)(418 K)
P = 0.686 atm

16.31 b, since $\Delta n_g = 0$

16.32 a, since $\Delta n_g = 0$

16.33 $K_p = K_c \times (RT)^{\Delta n_g}$
$6.3 \times 10^{-3} = K_c[(0.0821\ \text{L atm K}^{-1}\ \text{mol}^{-1})(498\ \text{K})]^{-2} = (5.98 \times 10^{-4}) \times K_c$
$K_c = 11$

16.34 $K_p = K_c \times (RT)^{\Delta n_g}$

$1.6 \times 10^{6} = K_c[(0.0821\ \text{L atm K}^{-1}\ \text{mol}^{-1})(673\ \text{K})]^{1}$
$K_c = 2.9 \times 10^{4}$

16.35 $K_p = K_c \times (RT)^{\Delta n_g}$
$K_p = 4.2 \times 10^{-4}[(0.0821\ \text{L atm K}^{-1}\ \text{mol}^{-1})(773\ \text{K})]^{1} = 2.7 \times 10^{-2}$

16.36 $K_p = K_c \times (RT)^{\Delta n_g}$
$K_p = (2.2 \times 10^{59}) \times [(0.0821\ \text{L atm K}^{-1}\ \text{mol}^{-1})(573\ \text{K})]^{-1} = 4.7 \times 10^{57}$

16.37 $K_p = K_c \times (RT)^{\Delta n_g}$
$K_p = (0.40)[(0.0821\ \text{L atm K}^{-1}\ \text{mol}^{-1})(1046\ \text{K})]^{-2} = 5.4 \times 10^{-5}$

16.38 $K_p = K_c \times (RT)^{\Delta n_g}$

$$K_c = \frac{K_p}{(RT)^{\Delta n_g}} = \frac{4.6 \times 10^{-2}}{\left[\left(0.0821\ \frac{\text{L atm}}{\text{mol K}}\right)(668\ \text{K})\right]} = 8.4 \times 10^{-4}$$

16.39 In each case we get approximately 55.5 M:

(a)

$$\# \text{ mol } H_2O = (18.0 \text{ mL } H_2O)\left(\frac{1 \text{ g}}{1 \text{ mL}}\right)\left(\frac{1 \text{ mol } H_2O}{18.02 \text{ g } H_2O}\right) = 0.999 \text{ mol } H_2O$$

$$M = \left(\frac{0.999 \text{ mol } H_2O}{18.0 \text{ mL } H_2O}\right)\left(\frac{1000 \text{ mL}}{1 \text{ L}}\right) = 55.5 \text{ M}$$

(b)

$$\# \text{ mol } H_2O = (100.0 \text{ mL } H_2O)\left(\frac{1 \text{ g}}{1 \text{ mL}}\right)\left(\frac{1 \text{ mol } H_2O}{18.02 \text{ g } H_2O}\right) = 5.549 \text{ mol } H_2O$$

$$M = \left(\frac{5.549 \text{ mol } H_2O}{100.0 \text{ mL } H_2O}\right)\left(\frac{1000 \text{ mL}}{1 \text{ L}}\right) = 55.49 \text{ M}$$

(c)

$$\# \text{ mol } H_2O = (1.00 \text{ L } H_2O)\left(\frac{1000 \text{ mL}}{1 \text{ L}}\right)\left(\frac{1 \text{ g}}{1 \text{ mL}}\right)\left(\frac{1 \text{ mol } H_2O}{18.02 \text{ g } H_2O}\right)$$

$$= 55.5 \text{ mol } H_2O$$

$$M = \left(\frac{55.5 \text{ mol } H_2O}{1.00 \text{ L } H_2O}\right) = 55.5 \text{ M}$$

16.40 The density is 2.164 g/mL.

$$\# \text{ mol NaCl} = (12.0 \text{ cm}^3 \text{ NaCl})\left(\frac{2.164 \text{ g}}{1 \text{ cm}^3}\right)\left(\frac{1 \text{ mol NaCl}}{58.44 \text{ g NaCl}}\right)$$

$$= 0.4444 \text{ mol NaCl}$$

$$M = \left(\frac{0.4444 \text{ mol NaCl}}{0.0120 \text{ L NaCl}}\right) = 37.03 \text{ M}$$

$$\# \text{ mol NaCl} = (25.0 \text{ g NaCl})\left(\frac{1 \text{ mol NaCl}}{58.44 \text{ g NaCl}}\right)$$

$$= 0.4278 \text{ mol NaCl}$$

$$\# \text{ L NaCl} = 25.0 \text{ g NaCl}\left(\frac{1 \text{ cm}^3 \text{ NaCl}}{2.164 \text{ g NaCl}}\right)\left(\frac{1 \text{ L NaCl}}{1000 \text{ cm}^3 \text{ NaCl}}\right) = 1.155\times10^{-2} \text{ L NaCl}$$

$$M = \left(\frac{0.4278 \text{ mol NaCl}}{1.155\times10^{-2} \text{ L NaCl}}\right) = 37.04 \text{ M}$$

16.41 (a) $K_c = \dfrac{[CO]^2}{[O_2]}$ (d) $K_c = \dfrac{[H_2O][CO_2]}{[HF]^2}$

(b) $K_c = [H_2O][SO_2]$ (e) $K_c = [H_2O]^5$

(c) $K_c = \dfrac{[CH_4][CO_2]}{[H_2O]^2}$

16.42 a) $K_c = \frac{[CO_2]}{[SO_2]}$ (d) $K_c = [H_2O]$

(b) $K_c = \frac{[Cl^-]}{[Br^-]}$ (e) $K_c = \frac{[N_2][H_2O]^3}{[NH_3]^2}$

(c) $K_c = [Cu^{2+}][OH^-]^2$

16.43

	[HCl]	[HI]	[Cl_2]
I	0.100	–	–
C	–2x	+2x	+x
E	0.100–2x	+2x	+x

Note: Since the $I_2(s)$ has a constant concentration, it may be neglected.

$$K_c = \frac{[HI]^2[Cl_2]}{[HCl]^2} = 1.6 \times 10^{-34}$$

$$K_c = \frac{(2x)^2(x)}{(0.100-2x)^2} = 1.6 \times 10^{-34}$$

Because the value of K_c is so small, we make the simplifying assumption that $(0.100 - 2x) \approx 0.100$, and the above equation becomes:

$$K_c = \frac{[HI]^2[Cl_2]}{[HCl]^2} = 1.6 \times 10^{-34}$$

$$K_c = \frac{(2x)^2(x)}{(0.100)^2} = 1.6 \times 10^{-34}$$

$4x^3 = 1.6 \times 10^{-36}$; $\therefore$ $x = 7.37 \times 10^{-13}$, and the above assumption is seen to have been valid.

$[HI] = 2x = 1.47 \times 10^{-12}$ M
$[Cl_2] = x = 7.37 \times 10^{-13}$ M
$[HCl] = (0.100 - 2x) \approx 0.100$ M

16.44 $AgCl(s) + Br^-(aq) \leftrightarrows AgBr(s) + Cl^-(aq)$

	[Br^-]	[Cl^-]
I	0.10	–
C	–x	+x
E	0.10–x	+x

Substituting the above values for equilibrium concentrations into the mass action expression gives:

$$K_c = \frac{[Cl^-]}{[Br^-]} = \frac{x}{0.10 - x} = 360$$

Solving for the value of x: x = $[Cl^-]$ = 0.0997 M and $[Br^-]$ = 0.10 – 0.0997 = 0.00030 M.
If we concern ourselves with significant figures, the $[Br^-] \approx 0.00$ M.

16.45 The mass action expression for this equilibrium is:

$$K_c = \frac{[PCl_5]}{[PCl_3][Cl_2]} = 0.18$$

and the value for the reaction quotient for this system is:

$$Q = \frac{(0.00500)}{(0.0420)(0.0240)} = 4.96$$

(a) No. This is not the value of the equilibrium constant, and we conclude that the system is not at equilibrium.

(b) Since the value of the reaction quotient for this system is larger than that of the equilibrium constant, the system must shift to the left to reach equilibrium.

16.46 The mass action expression for the system is:

$$K_c = \frac{[NO][SO_3]}{[SO_2][NO_2]} = 85.0$$

and the reaction quotient for the system is:

$$Q = \frac{[NO][SO_3]}{[SO_2][NO_2]} = \frac{(0.0250)(0.0400)}{(0.00250)(0.00350)} = 114$$

(a) The system is not at equilibrium because $Q \neq K_c$.

(b) Since the value of the reaction quotient is larger than the value of the equilibrium constant, the system must shift to the left in order to reach equilibrium.

16.47 $$K_c = \frac{[CH_3OH]}{[CO][H_2]^2} = \frac{[CH_3OH]}{(0.180)(0.220)^2} = 0.500$$

$[CH_3OH] = 4.36 \times 10^{-3}$ M.

16.48 The mass action expression is:

$$K_c = \frac{[NH_3]^2}{[N_2][H_2]^3} = 64$$

Solving for $[H_2]$ gives:

$$[H_2] = \sqrt[3]{\frac{[NH_3]^2}{[N_2]K_c}} = \sqrt[3]{\frac{(0.360)^2}{(0.0192)(64)}} = 0.47\text{ M}$$

16.49 $$K_c = \frac{[CH_3OH]}{[CO][H_2]^2} = \frac{(0.00261)}{(0.105)(0.250)^2} = 0.398$$

16.50 $$K_c = \frac{[C_2H_5OH]}{[C_2H_4][H_2O]} = \frac{(0.180)}{(0.0148)(0.0336)} = 362$$

16.51

	[HBr]	$[H_2]$	$[Br_2]$
I	0.500	–	–
C	–2x	+x	+x
E	0.500–2x	+x	+x

The problem tell us that $[Br_2]$ = 0.0955 M = x at equilibrium. Using the ICE table as a guide we see that the equilibrium concentrations are; $[H_2]$ = $[Br_2]$ = 0.0955 M and [HBr] = 0.500–2(0.0955) = 0.309 M.

$$K_c = \frac{[H_2][Br_2]}{[HBr]^2} = \frac{(0.0955)(0.0955)}{(0.309)^2} = 0.0955$$

16.52 $CH_2O(g) \leftrightarrows H_2(g) + CO(g)$ $\qquad K_c = \frac{[H_2][CO]}{[CH_2O]}$

	$[CH_2O]$	$[H_2]$	[CO]
I	0.100	–	–
C	–x	+x	+x
E	0.100–x	+x	+x

The problem states that the equilibrium concentration of CH_2O is 0.066 M. Therefore, x = 0.034 M = $[H_2]$ = [CO] at equilibrium. Substituting these values into the mass action expression we can determine K_c:

$$K_c = \frac{[H_2][CO]}{[CH_2O]} = \frac{(0.034)(0.034)}{0.066} = 1.8 \times 10^{-2}$$

16.53 According to the problem, the concentration of NO_2 increases in the course of this reaction. This means our ICE table will look like the following:

	$[NO_2]$	[NO]	$[N_2O]$	$[O_2]$
I	0.0560	0.294	0.184	0.377
C	+x	+x	–x	–x
E	0.0560 + x	0.294 + x	0.184 – x	0.377 – x

The problem tell us that $[NO_2]$ = 0.118 M = 0.0560 + x at equilibrium. Solving we get; x = 0.062 M. Using the ICE table as a guide we see that the equilibrium concentrations are; [NO] = 0.356 M, $[N_2O]$ = 0.122 M and $[O_2]$ = 0.315 M.

$$K_c = \frac{[N_2O][O_2]}{[NO_2][NO]} = \frac{(0.122)(0.315)}{(0.118)(0.356)} = 0.915$$

16.54 $K_c = \frac{[NO_2]^4}{[N_2O]^2[O_2]^3}$

	$[N_2O]$	$[O_2]$	$[NO_2]$
I	0.020	0.0560	–
C	–2x	–3x	+4x
E	0.020 – 2x	0.0560 – 3x	+4x

At equilibrium $[NO_2]$ = 4x = 0.020 M
x = 0.0050M
So, $[N_2O]$ = 0.020 – 2(0.0050) = 0.010 M
$[O_2]$ = 0.0560 – 3(0.0050) = 0.041M

$$K_c = \frac{[0.020]^4}{[0.010]^2[0.041]^3} = 23$$

16.55 $2BrCl \leftrightarrows Br_2 + Cl_2$

	[BrCl]	$[Br_2]$	$[Cl_2]$
I	0.050	–	–
C	–2x	+x	+x
E	0.050–2x	+x	+x

Substituting the above values for equilibrium concentrations into the mass action expression gives:

$$K_c = \frac{[Br_2][Cl_2]}{[BrCl]^2} = \frac{(x)(x)}{(0.050-2x)^2} = 0.145$$

Take the square root of both sides to get

$$K_c = \frac{x}{0.050-2x} = 0.381$$

Solving for x gives: x = 0.011 M = $[Br_2]$ = $[Cl_2]$

16.56 $2BrCl \leftrightarrows Br_2 + Cl_2$
In this problem, the reaction will shift from right to left as written above. This is due to the presence of products and the absence of reactants.

	[BrCl]	$[Br_2]$	$[Cl_2]$
I	–	0.0250	0.0250
C	+2x	–x	–x
E	+2x	0.0250–x	0.0250–x

Substituting the above values for equilibrium concentrations into the mass action expression gives:

$$K_c = \frac{[Br_2][Cl_2]}{[BrCl]^2} = \frac{(0.0250 - x)(0.0250 - x)}{(2x)^2} = 0.145$$

Take the square root of both sides to get

$$K_c = \frac{0.0250 - x}{2x} = 0.381$$

Solving for x gives: x = 0.0142 M. The individual concentrations are then: $[Br_2] = [Cl_2] = 0.0250 - 0.0142 = 0.0108$ M, [BrCl] = 2(0.0142) = 0.0284 M.

16.57 The initial concentrations are each 0.240 mol/2.00 L = 0.120 M.

	$[SO_3]$	[NO]	$[NO_2]$	$[SO_2]$
I	0.120	0.120	–	–
C	–x	–x	+x	+x
E	0.120–x	0.120–x	+x	+x

Substituting the above values for equilibrium concentrations into the mass action expression gives:

$$K_c = \frac{[NO_2][SO_2]}{[SO_3][NO]} = \frac{(x)(x)}{(0.120 - x)(0.120 - x)} = 0.500$$

Taking the square root of both sides of this equation gives: 0.707 = x/(0.120 – x)
Solving for x we have: 1.707(x) = 0.0848,
x = 0.0497 mol/L = $[NO_2] = [SO_2]$,
$[NO] = [SO_3]$ = 0.120 – x = 0.0703 mol/L

16.58 The initial concentrations are each 0.120 M.

	$[SO_3]$	[NO]	$[NO_2]$	$[SO_2]$
I	–	–	0.120	0.120
C	+x	+x	–x	–x
E	+x	+x	0.120–x	0.120–x

Substituting the above values for equilibrium concentrations into the mass action expression gives:

$$K_c = \frac{[NO_2][SO_2]}{[SO_3][NO]} = \frac{(0.120 - x)(0.120 - x)}{(x)(x)} = 0.500$$

Taking the square root of both sides of this equation gives: 0.707 = (0.120 – x)/x. Solving for x we have:
1.707(x) = 0.120
∴ x = 0.0703 mol/L = $[NO] = [SO_3]$,
$[NO_2] = [SO_2]$ = 0.120 – x = 0.050 mol/L

Although these two systems reach different equilibrium positions, the equilibrium concentrations of the four substances in each experiment give the same value for the equilibrium constant when substituted into the mass action expression.

An alternative method for solving problems where the reaction proceeds from right to left, such as this problem, is to reverse the reaction written and use this "new" reaction as the basis for the problem. In doing this, the mass action expression and the equilibrium constant must change to reflect the "new" reaction. As outlined in Section 16.3, "when the direction of an equation is reversed, the new equilibrium constant is the reciprocal of the original" and the "new" mass action expression is also the reciprocal of the original. Try solving this problem using this method. Your answer should, of course, remain the same as stated above.

16.59 The initial concentrations are all 1.00 mol/100 L = 0.0100 M. Since the initial concentrations are all the same, the reaction quotient is equal to 1.0, and we conclude that the system must shift to the left to reach equilibrium since $Q > K_c$.

	$[CO]$	$[H_2O]$	$[CO_2]$	$[H_2]$
I	0.0100	0.0100	0.0100	0.0100
C	+x	+x	−x	−x
E	0.0100+x	0.0100+x	0.0100−x	0.0100−x

Substituting the above values for equilibrium concentrations into the mass action expression gives:

$$K_c = \frac{[CO_2][H_2]}{[CO][H_2O]} = \frac{(0.0100 - x)(0.0100 - x)}{(0.0100 + x)(0.0100 + x)} = 0.400$$

We take the square root of both sides of the above equation:

$$\frac{(0.0100 - x)}{(0.0100 + x)} = 0.632$$

and $(0.632)(0.0100 + x) = 0.0100 - x$
$(1.632)x = 3.68 \times 10^{-3}$, or
$x = 2.25 \times 10^{-3}$ mol/L

The equilibrium concentrations are then:
$[H_2] = [CO_2] = (0.0100 - 2.25 \times 10^{-3}) = 7.7 \times 10^{-3}$ M,
$[CO] = [H_2O] = (0.0100 + 2.25 \times 10^{-3}) = 0.0123$ M.

16.60 If we substitute the initial concentrations of 0.0400 M for each component of this mixture into the reaction quotient expression that results from the balanced equation in the text, we see that the reaction quotient equals one. Since $Q > K_c$, the reaction will proceed from right to left as written, i.e., $[Br_2]$ and $[Cl_2]$ will decrease and $[BrCl]$ will increase.

	$[BrCl]$	$[Br_2]$	$[Cl_2]$
I	0.0400	0.0400	0.0400
C	+ 2x	− x	− x
E	0.0400 + 2x	0.0400 − x	0.0400 − x

Substituting the above values for equilibrium concentrations into the mass action expression gives:

$$K_c = \frac{[Br_2][Cl_2]}{[BrCl]^2} = \frac{(0.0400 - x)(0.0400 - x)}{(0.0400 + 2x)^2} = 0.145$$

Take the square root of both sides of this equation to get; $\frac{(0.0400 - x)}{(0.0400 + 2x)} = 0.381$.

This equation is easily solved giving x = 0.0141 M.
The substances then have the following concentrations at equilibrium:
$[Cl_2] = [Br_2] = 0.0400 - 0.0141 = 0.0259$ M,
$[BrCl] = 0.0400 + 2(0.0141) = 0.0682$ M.

16.61

	[HCl]	$[H_2]$	$[Cl_2]$
I	0.0500	–	–
C	–2x	+x	+x
E	0.0500–2x	+x	+x

$$K_c = \frac{[H_2][Cl_2]}{[HCl]^2} = \frac{(x)(x)}{(0.0500 - 2x)^2} = 3.2 \times 10^{-34}$$

Because K_c is so exceedingly small, we can make the simplifying assumption that x is also small enough to make $(0.0500 - 2x) \approx 0.0500$. Thus we have: $3.2 \times 10^{-34} = (x)^2/(0.0500)^2$

Taking the square root of both sides, and solving for the value of x gives:
$x = 8.9 \times 10^{-19}$ M = $[H_2]$ = $[Cl_2]$
$[HCl] = (0.0500 - x) \approx 0.0500$ mol/L

16.62 The initial concentrations are :
0.200 mol/4.00 L = 0.0500 M N_2O
0.400 mol/4.00 L = 0.100 M NO_2

	$[N_2O]$	$[NO_2]$	[NO]
I	0.0500	0.100	–
C	–x	–x	+3x
E	0.0500–x	0.100–x	+3x

$$K_c = \frac{[NO]^3}{[N_2O][NO_2]} = 1.4 \times 10^{-10}$$

$$K_c = \frac{(3x)^3}{(0.0500)(0.100)}$$

Where we have assumed that x << 0.0500. Solving we get $x = 2.96 \times 10^{-5}$.

16.63 $$K_c = \frac{[CO]^2[O_2]}{[CO_2]^2} = 6.4 \times 10^{-7}$$

	$[CO_2]$	[CO]	$[O_2]$
I	1.0×10^{-2}	–	–
C	–2x	+2x	+x
E	$1.0 \times 10^{-2} - 2x$	+2x	+x

$$K_c = \frac{[2x]^2[x]}{[1.0 \times 10^{-2} - 2x]^2} = 6.4 \times 10^{-7}$$

Assume $x << 1.0 \times 10^{-2}$.

$$\frac{4x^3}{(1.0 \times 10^{-2})^2} = 6.4 \times 10^{-7} \qquad x = 2.5 \times 10^{-4}$$

$$[CO] = 2x = 5.0 \times 10^{-4} \text{ M}$$

16.64 $K_c = \frac{[H_2]^2[O_2]}{[H_2O]^2} = 6.0 \times 10^{-28}$

	$[H_2O]$	$[H_2]$	$[O_2]$
I	0.015 mol/5.00L	–	–
C	–2x	+2x	+x
E	$3.0 \times 10^{-3} - 2x$	+2x	+x

$$K_c = \frac{[2x]^2[x]}{[3.0 \times 10^{-3} - 2x]^2} = 6.0 \times 10^{-28}$$

Assume $2x << 3.0 \times 10^{-3}$.

$$\frac{4x^3}{(3.0 \times 10^{-3})^2} = 6.0 \times 10^{-28} \qquad x = 1.1 \times 10^{-11}$$

$[H_2] = 2.2 \times 10^{-11}$ M, $[O_2] = 1.1 \times 10^{-11}$ M

16.65 We first approach the problem in the normal fashion with an initial concentration of $PCl_5 = 0.013$ M.

	$[PCl_3]$	$[Cl_2]$	$[PCl_5]$
I	–	–	0.013
C	+x	+x	–x
E	+x	+x	0.013–x

Substituting the above values for equilibrium concentrations into the mass action expression gives:

$$K_c = \frac{[PCl_5]}{[PCl_3][Cl_2]} = \frac{(0.013 - x)}{(x)(x)} = 0.18$$

rearranging; $0.18x^2 + x - 0.013 = 0$

We next attempt to use the quadratic equation to solve for the value of x, setting a = 0.18; b = 1; c = –0.013.

However, we find that unless we carry one more significant figure than is allowed, the quadratic formula for this problem gives us a concentration of zero for PCl_5. A better solution is obtained by "allowing" the initial equilibrium to shift **completely** to the left, giving us a new initial situation from which to work:

	$[PCl_3]$	$[Cl_2]$	$[PCl_5]$
I	0.013	0.013	–
C	–x	–x	+x
E	0.013–x	0.013–x	+x

Substituting the above values for equilibrium concentrations into the mass action expression gives:

$$K_c = \frac{[PCl_5]}{[PCl_3][Cl_2]} = \frac{(+x)}{(0.013-x)(0.013-x)} = 0.18$$

Now, we may assume that x << 0.013. The equation is simplified and we solve for x
$x = [PCl_5] = 3.0 \times 10^{-5}$ M.

16.66 All concentrations are a factor of ten less than the stated value due to the container size. Since $Q < K_c$ for this reaction, the reaction proceeds as written, i.e., from left to right.

	$[SO_2]$	$[NO_2]$	$[NO]$	$[SO_3]$
I	0.0100	0.00600	0.00800	0.0120
C	–x	–x	+x	+x
E	0.0100–x	0.00600–x	0.00800+x	0.0120+x

Substituting the above values for equilibrium concentrations into the mass action expression gives:

$$K_c = \frac{[NO][SO_3]}{[SO_2][NO_2]} = \frac{(0.00800+x)(0.0120+x)}{(0.0100-x)(0.00600-x)} = 85.0$$

$0.00510 - 1.36(x) + 85.0(x)^2 = 9.60 \times 10^{-5} + 0.0200(x) + (x)^2$
$84.0(x)^2 - 1.38(x) + 0.00500 = 0$

The quadratic equation is used, with the following values: $a = 84.0$; $b = -1.38$;
$c = 0.00500$. Upon solving for x, we find that only the negative root is sensible:
$x = 0.00542$ M.

$[SO_2] = 0.0100 - 0.00542 = 0.0046$ M
$[NO_2] = 0.00600 - 0.00542 = 0.00058$ M
$[NO] = 0.00800 + 0.00542 = 0.0134$ M
$[SO_3] = 0.0120 + 0.00542 = 0.0174$ M

16.67

	$[SO_3]$	$[NO]$	$[NO_2]$	$[SO_2]$
I	0.0500	0.100	–	–
C	–x	–x	+x	+x
E	0.0500–x	0.100–x	+x	+x

Substituting the above values for equilibrium concentrations into the mass action expression gives:

$$K_c = \frac{[NO_2][SO_2]}{[SO_3][NO]} = \frac{(x)(x)}{(0.0500-x)(0.100-x)} = 0.500$$

Since the equilibrium constant is not much larger than either of the values 0.0500 or 0.100, we cannot neglect the size of x in the above expression. A simplifying assumption is not therefore possible, and we must solve for the value of x using the quadratic equation. Multiplying out the above denominator, collecting like terms, and putting the result into the standard quadratic form gives:

$$0.500x^2 + (7.50 \times 10^{-2})x - (2.50 \times 10^{-3}) = 0$$

$$x = \frac{-7.50 \times 10^{-2} \pm \sqrt{\left(7.50 \times 10^{-2}\right)^2 - 4(0.500)\left(-2.50 \times 10^{-3}\right)}}{2(0.500)} = 0.0281 \text{ M}$$

using the (+) root. So, $[NO_2] = [SO_2] = 0.0281$ M

16.68 If we substitute the initial concentrations for each component of this mixture into the reaction quotient expression that results from the balanced equation in the text, we see that the reaction quotient equals 0.375. Since $Q > K_c$ for this situation, the reaction will proceed from right to left, i.e., $[Br_2]$ and $[Cl_2]$ will decrease and [BrCl] will increase.

	[BrCl]	**[Br_2]**	**[Cl_2]**
I	0.0400	0.0300	0.0200
C	+2x	–x	–x
E	0.0400+2x	0.0300–x	0.0200–x

Substituting the above values for equilibrium concentrations into the mass action expression gives:

$$K_c = \frac{[Br_2][Cl_2]}{[BrCl]^2} = \frac{(0.0300 - x)(0.0200 - x)}{(0.0400 + 2x)^2} = 0.145$$

Rearranging we get the following quadratic expression: $0 = 0.42x^2 - 0.0732x + 0.000368$. Solving and using the negative root gives x = 0.0052 M. The individual concentrations are: $[Cl_2] = 0.0200 - 0.0052 = 0.0148$ M, $[Br_2] = 0.0300 - 0.0052 = 0.0248$ M, $[BrCl] = 0.0400 + 2(0.0052) = 0.0504$ M.

16.69 $$K_c = \frac{[CO][H_2O]}{[HCHO_2]^2} = 4.3 \times 10^5$$

Since K_c is large, start by assuming all of the $HCHO_2$ decomposes to give CO and H_2O

	[$HCHO_2$]	**[CO]**	**[O_2]**
I	–	0.200	0.200
C	+x	–x	–x
E	+x	0.200 –x	0.200 –x

$$Kc = \frac{[0.200][0.200]}{[x]} = 4.3 \times 10^5$$

$x = 9.3 \times 10^{-8}$
so, at equilibrium
$[CO] = [H_2O] = 0.200 - x = 0.200$

16.70 Because of the very large value of K_c, we start by realizing that nearly all of the 0.100 moles of H_2 will react with 0.100 moles of Br_2 to form 0.200 moles of HBr. This brings us to a new "initial" condition that is more realistically close to the true equilibrium condition, i.e., [HBr] = (0.200 moles/10.0 L) = 0.0200 M, $[Br_2]$ = (0.100 moles/10.0 L) = 0.0100 M, $[H_2]$ = 0 M. Next, we proceed in the normal fashion, allowing 2x mol/L of HBr to disappear, and making x mol/L each of H_2 and Br_2 in order to reach equilibrium:

	[H_2]	**[Br_2]**	**[HBr]**
I	–	0.0100	0.0200
C	+x	+x	–2x
E	+x	0.0100+x	0.0200–2x

Substituting the above values for equilibrium concentrations into the mass action expression gives:

$$K_c = \frac{[HBr]^2}{[H_2][Br_2]} = \frac{(0.0200-2x)^2}{(+x)(0.0100+x)} = 7.9 \times 10^{18}$$

We next make the assumptions that (0.0200 – 2x) ≈ 0.0200, and (0.0100 + x) ≈ 0.0100, giving: 7.9×10^{18} = $(0.0200)^2/(0.0100)(x)$
$x = 5.1 \times 10^{-21}$ M = $[H_2]$.
The small size of x demonstrates that the assumptions made above were justified.

[HBr] = 0.0200 – 2x = 0.0200 M
$[Br_2]$ = 0.0100 + x = 0.0100 M

Additional Exercises

16.71 (a) The mass action expression is:

$$K_p = \frac{(P_{NO_2})^2}{(P_{N_2O_4})} = 0.140 \text{ atm}$$

Solving the above expression for the partial pressure of NO_2, we get:

$$P_{NO_2} = \sqrt{P_{N_2O_4} \times K_p} = \sqrt{(0.250 \text{ atm})(0.140 \text{ atm})} = 0.187 \text{ atm}$$

(b) $P_{total} = P_{NO_2} + P_{N_2O_4} = 0.187 + 0.250 = 0.437$ atm

16.72 $$K_c = \frac{[H_5IO_6]^5}{[IO_3^-]^7[H^+]^7} = 1 \times 10^{-85}$$

H_5IO_6 is a weak acid and must be part of the equilibrium expression. As ions, both H^+ and IO_3^- are part of the expression. H_2O is ignored since it is a pure liquid and I_2 is ignored since it is a solid at 25°C.

16.73 The initial concentrations are:
$[NO_2]$ = (0.200 mol/4.00 L) = 0.0500 M
[NO] = (0.300 mol/4.00 L) = 0.0750 M
$[N_2O]$ = (0.150 mol/4.00 L) = 0.0375 M
$[O_2]$ = (0.250 mol/4.00 L) = 0.0625 M

Substituting these values into the mass action expression we determine:

$$Q = \frac{[N_2O][O_2]}{[NO_2][NO]} = \frac{(0.0375)(0.0625)}{(0.0500)(0.0750)} = 0.625$$

Since $Q<K_c$, this reaction will proceed from left to right as written. The ICE table becomes:

	$[NO_2]$	[NO]	$[N_2O]$	$[O_2]$
I	0.0500	0.0750	0.0375	0.0625
C	–x	–x	+x	+x
E	0.0500 – x	0.0750 – x	0.0375 + x	0.0625 + x

Substituting the above values for equilibrium concentrations into the mass action expression gives:

$$K_c = \frac{[N_2O][O_2]}{[NO_2][NO]} = \frac{(0.0375 + x)(0.0625 + x)}{(0.0500 - x)(0.0750 - x)} = 0.914$$

To solve we need to use the quadratic equation. Expanding the above calculation we get:

$0.086x^2 + 0.214x - 0.00109 = 0$

Solving we get x = 0.00508. So,

$[NO_2] = 0.0500 - x = 0.0449M$

$[NO] = 0.0750 - x = 0.0699$ M

$[N_2O] = 0.0375 + x = 0.0426$ M

$[O_2] = 0.0625 + x = 0.0676$ M

16.74 $HCHO_2(g) \leftrightarrows CO(g) + H_2O(g)$

Since there is no $HCHO_2(g)$ present initially, we know this reaction will move from right to left as written.

	$[HCHO_2]$	$[CO]$	$[H_2O]$
I	–	0.20	0.30
C	+x	–x	–x
E	+x	0.20–x	0.30–x

Substituting the above values for equilibrium concentrations into the mass action expression gives:

$$K_c = \frac{[CO][H_2O]}{[HCHO_2]} = \frac{(0.20 - x)(0.30 - x)}{(x)} = 2.9 \times 10^4$$

Since K_c is so large for the forward direction, we do not anticipate that the reverse reaction will proceed to any appreciable extent. Therefore, we may assume x << 0.20 and x << 0.30. This simplifies the problem immensely and we may solve for x.

$x = [HCHO_2] = 2.1 \times 10^{-6}$ M.

16.75 First, calculate a value for K_c using a rearranged form of equation 16.4,and setting the value for Δn to -1:

$$K_c = \frac{K_P}{(RT)^{\Delta n_g}} = \frac{1.5 \times 10^{18}}{\left[\left(0.0821 \frac{\text{L atm}}{\text{mol K}}\right)(300\text{ K})\right]^{-1}} = 3.7 \times 10^{19}$$

The value for the equilibrium constant is very large, indicating that the equilibrium lies far to the right. We therefore anticipate that the initial conditions are unrealistic. The system is "allowed" to come to a more realistic new initial set of concentrations, by reaction of all of the starting amount of NO, to give a stoichiometric amount of N_2O and NO_2. Only then can we solve for the equilibrium concentrations in the usual manner:

	$[NO]$	$[N_2O]$	$[NO_2]$
I	–	0.010	0.010
C	+3x	–x	–x
E	+3x	0.010 – x	0.010 – x

Substituting the above values for equilibrium concentrations into the mass action expression gives:

$$K_c = \frac{[N_2O][NO_2]}{[NO]^3} = \frac{(0.010-x)(0.010-x)}{(3x)^3} = 3.7 \times 10^{19}$$

The simplifying assumption can be made that the value for x is much smaller than the number 0.010. Upon solving for x we get: $27x^3 = 2.7 \times 10^{-24}$ $\quad x = 4.6 \times 10^{-9}$ M
$[NO] = 3x = 1.4 \times 10^{-8}$ M,
$[N_2O] = [NO_2] = 0.010 - x = 0.010$ M.

16.76 (a) Equilibrium is unaffected by the addition of a solid.
(b) Equilibrium is unaffected by the removal of a solid (provided it is not completely removed.)
(c) The equilibrium will shift to the left.
(d) The equilibrium will shift to the right.

16.77 First, calculate the number of moles of CO used in the experiment, using the ideal gas law:

$$n = \frac{PV}{RT} = \frac{(0.177\text{ atm})(2.00\text{ L})}{(0.0821\,\frac{\text{L atm}}{\text{mol K}})(298\text{ K})} = 0.0145\text{ mol}$$

Next, calculate the partial pressure of CO at 400 °C:

$$P = \frac{nRT}{V} = \frac{(0.0145\text{ mol})(0.0821\,\frac{\text{L atm}}{\text{mol K}})(673\text{ K})}{(2.00\text{ L})} = 0.401\text{ atm}$$

Next, we calculate the number of moles of water that are supplied to the reaction, and convert to partial pressure for water, using the ideal gas equation:

$$\#\text{ mol mol } H_2O = (0.391\text{ g } H_2O)\left(\frac{1\text{ mol } H_2O}{18.02\text{ g } H_2O}\right) = 0.0217\text{ mol } H_2O$$

$$P = \frac{nRT}{V} = \frac{(0.0217\text{ mol})(0.0821\,\frac{\text{L atm}}{\text{mol K}})(673\text{ K})}{(2.00\text{ L})} = 0.600\text{ atm}$$

Finally, we solve for the equilibrium partial pressure in the usual manner:

	P_{HCHO_2}	P_{CO}	P_{H_2O}
I	–	0.401	0.600
C	+x	–x	–x
E	+x	0.401–x	0.600–x

Substituting the above values for equilibrium partial pressures into the mass action expression gives:

$$K_p = \frac{(P_{CO})(P_{H_2O})}{(P_{HCHO_2})} = \frac{(0.401-x)(0.600-x)}{x} = 1.6 \times 10^6$$

Because K_p is so large, we assume x << 0.401 and x << 0.600. We then solve for $P_{HCHO_2} = x = 1.5 \times 10^{-7}$ atm.

16.78 The initial concentrations are: $[NO_2] = (0.200\text{ mol}/5.00\text{ L}) = 0.0400$ M,
$[NO] = (0.200\text{ mol}/5.00\text{ L}) = 0.0400$ M, $[N_2O] = (0.200\text{ mol}/5.00\text{ L}) = 0.0400$ M,
$[O_2] = (0.200\text{ mol}/5.00\text{ L}) = 0.0400$ M.

Substituting these values into the mass action expression we determine:

$$Q = \frac{[N_2O][O_2]}{[NO_2][NO]} = \frac{(0.0400)(0.0400)}{(0.0400)(0.0400)} = 1.00$$

Since Q > K_c, this reaction will proceed from right to left as written. The ICE table becomes:

	$[NO_2]$	$[NO]$	$[N_2O]$	$[O_2]$
I	0.0400	0.0400	0.0400	0.0400
C	+x	+x	−x	−x
E	0.0400 + x	0.0400 + x	0.0400 − x	0.0400 − x

Substituting the above values for equilibrium concentrations into the mass action expression gives:

$$K_c = \frac{[N_2O][O_2]}{[NO_2][NO]} = \frac{(0.0400-x)(0.0400-x)}{(0.0400+x)(0.0400+x)} = 0.914$$

Taking the square root of both sides we get $K_c = \frac{(0.0400-x)}{(0.0400+x)} = 0.956$

Solving for x we get: $x = 9.0 \times 10^{-4}$. Thus, at equilibrium we have:
$[N_2O] = [O_2] = 0.0391$ M, $[NO_2] = [NO] = 0.0409$ M.

If 0.050 moles of NO_2 are added (note: 0.050 mol/5.00 L = 0.01 M NO_2 added), the ICE table becomes:

	$[NO_2]$	$[NO]$	$[N_2O]$	$[O_2]$
I	0.0409 + 0.01	0.0409	0.0391	0.0391
C	−x	−x	+x	+x
E	0.0509 − x	0.0409 − x	0.0391 + x	0.0391 + x

Substituting these values into the mass action expression gives:

$$K_c = \frac{[N_2O][O_2]}{[NO_2][NO]} = \frac{(0.0391+x)(0.0391+x)}{(0.0509-x)(0.0409-x)} = 0.914$$

Solving the quadratic expression, $0.086x^2 + 0.162\,x - 3.7 \times 10^{-4} = 0$, we get x = 0.00226. Thus, at equilibrium, we now have:

$[N_2O] = 0.0414$ M
$[O_2] = 0.0414$ M
$[NO_2] = 0.0486$ M
$[NO] = 0.0386$ M

16.79 $$Kc = \frac{[N_2O][O_2]}{[NO_2][NO]} = 0.914$$

Let z = the initial concentration.

	$[NO_2]$	$[NO]$	$[N_2O]$	$[O_2]$
I	z	z	−	−
C	−x	−x	+x	+x
E	z − x	z −x	+x	+x

$$Kc = \frac{[x][x]}{[z - x][z - x]} = 0.914$$

Take the square roots of both sides to get

$\frac{x}{z - x} = 0.956$ $x = 0.050$ from the data in the problem so

$\frac{0.0500}{z - 0.0500} = 0.956$ solving for z we get $z = 0.10$

moles NO = # moles NO_2 = $\left(\frac{0.10 \text{ mol}}{L}\right)(5.00 \text{ L}) = 0.50$ moles

Practice Exercises

17.1 In each case the conjugate base is obtained by removing a proton from the acid:

(a) OH^- (b) I^- (c) NO_2^- (d) $H_2PO_4^-$
(e) HPO_4^{2-} (f) PO_4^{3-} (g) H^- (h) NH_3

17.2 In each case the conjugate acid is obtained by adding a proton to the base:

(a) H_2O_2 (b) HSO_4^- (c) HCO_3^- (d) HCN
(e) NH_3 (f) NH_4^+ (g) H_3PO_4 (h) $H_2PO_4^-$

17.3 The Brønsted acids are $H_2PO_4^-(aq)$ and $H_2CO_3(aq)$
The Brønsted bases are $HCO_3^-(aq)$ and $HPO_4^{2-}(aq)$

conjugate pair

$$\underset{\text{base}}{PO_4^{3-}(aq)} + \underset{\text{acid}}{HC_2H_3O_2(aq)} \rightleftharpoons \underset{\text{acid}}{HPO_4^{2-}(aq)} + \underset{\text{base}}{C_2H_3O_2^-(aq)}$$

conjugate pair

17.4 conjugate pair

$$\underset{\text{base}}{PO_4^{3-}(aq)} + \underset{\text{acid}}{HC_2H_3O_2(aq)} \rightleftharpoons \underset{\text{acid}}{HPO_4^{2-}(aq)} + \underset{\text{base}}{C_2H_3O_2^-(aq)}$$

conjugate pair

17.5 $HPO_4^{2-}(aq) + OH^-(aq) \rightarrow PO_4^{3-}(aq) + H_2O$; HPO_4^{2-} acting as an acid
$HPO_4^{2-}(aq) + H_3O^+(aq) \rightarrow H_2PO_4^- + H_2O$; HPO_4^{2-} acting as a base

17.6 The substances on the right because they are the weaker acid and base.

17.7 (a) HBr is the stronger acid since binary acid strength increases from left to right within a period.
(b) H_2Te is the stronger acid since binary acid strength increases from top to bottom within a group.
(c) CH_3SH since acid strength increases from top to bottom within a group.

17.8 $HClO_3$ because Cl is more electronegative than Br.

17.9 (a) HIO_4 (b) H_2TeO_4 (c) H_3AsO_4

17.10 (a) Fluoride ions have a filled octet of electrons and are likely to behave as Lewis bases, i.e., electron pair donors.
(b) $BeCl_2$ is a likely Lewis acid since it has an incomplete shell. The Be atom has only two valence electrons and it can easily accept a pair of electrons.
(c) It could reasonably be considered a potential Lewis base since it contains three oxygens, each with lone pairs and partial negative charges. However, it is more effective as a Lewis acid, since the central sulfur bears a significant positive charge.

17.11 $K_w = 1.0 \times 10^{-14} = [H^+][OH^-]$

$$[H^+] = \frac{1.0 \times 10^{-14}}{[OH^-]} = \frac{1.0 \times 10^{-14}}{7.8 \times 10^{-6}} = 1.3 \times 10^{-9} \text{ M}$$

Since $[OH^-] > [H^+]$, the solution is basic.

17.12 $pH = -\log[H^+] = -\log[3.67 \times 10^{-4}] = 3.44$
$pOH = 14.00 - pH = 14.00 - 3.44 = 10.56$

The solution is acidic since pH is below 7.0.

17.13 $pOH = -\log[OH^-] = -\log[1.47 \times 10^{-9}] = 8.83$
$pH = 14.00 - pOH = 14.00 - 8.83 = 5.17$

17.14 In general, we have the following relationships between pH, $[H^+]$, and $[OH^-]$:

$$[H^+] = 10^{-pH}$$
$$[H^+][OH^-] = 1.00 \times 10^{-14}$$

(a) $[H^+] = 10^{-2.90} = 1.3 \times 10^{-3}$ M
$[OH^-] = 1.00 \times 10^{-14}/1.3 \times 10^{-3}$ M $= 7.7 \times 10^{-12}$ M
The solution is acidic.

(b) $[H^+] = 10^{-3.85} = 1.4 \times 10^{-4}$ M
$[OH^-] = 1.00 \times 10^{-14}/1.4 \times 10^{-4}$ M $= 7.1 \times 10^{-11}$ M
The solution is acidic.

(c) $[H^+] = 10^{-10.81} = 1.5 \times 10^{-11}$ M
$[OH^-] = 1.00 \times 10^{-14}/1.5 \times 10^{-11}$ M $= 6.7 \times 10^{-4}$ M
The solution is basic.

(d) $[H^+] = 10^{-4.11} = 7.8 \times 10^{-5}$ M
$[OH^-] = 1.00 \times 10^{-14}/7.8 \times 10^{-5}$ M $= 1.3 \times 10^{-10}$ M
The solution is acidic.

(e) $[H^+] = 10^{-11.61} = 2.5 \times 10^{-12}$ M
$[OH^-] = 1.00 \times 10^{-14}/2.5 \times 10^{-12}$ M $= 4.0 \times 10^{-3}$ M
The solution is basic.

17.15 $[OH^-] = 0.0050$ M
$pOH = -\log[OH^-] = -\log[0.0050] = 2.30$
$pH = 14.0 - pOH = 14.00 - 2.30 = 11.70$
$[H^+] = 10^{-11.70} = 2.0 \times 10^{-12}$ M

17.16 $[H^+] = 10^{-5.5} = 3.2 \times 10^{-6}$ M

Thinking It Through

T17.1 A conjugate base is formed by removal of a proton, here giving us CH_3^- ion. This is an exceedingly strong base because methane, CH_4, is not at all acidic.

T17.2 The amphoteric ones are those that can react by either addition or loss of a proton: H_2O (which forms OH^- and H_3O^+). We might be tempted to list HSO_4^- as well, which forms H_2SO_4 when protonated. H_2SO_4 is a strong acid and would immediately dissociate to give HSO_4^-. Consequently, HSO_4^- is not amphoteric.

T17.3 Ammonia is a Lewis base because it has a lone pair of electrons it can donate to form a coordinate covalent bond. It can't act as a Lewis acid because its octet is completely filled.

T17.4 The Al^{3+} ion accepts a pair of electrons when the ion is hydrated. Water acts as the Lewis base in the formation of the coordinate covalent bond.

T17.5 When the charge on the metal ion is small, the metal hydroxide easily dissociates to give M^{n+} and OH^-. The metal hydroxide acts as an Arrhenius base. When the charge on the metal ion is large, the ion acts as a Lewis acid. The metal ion forms coordinate covalent bonds with water.

T17.6 NH_3 is the electron pair donor and is, therefore, the Lewis base. Thus, $N_2H_5^+$ is the Lewis acid.

T17.7 The solute is a weak acid. If it were a strong acid, the pH would be calculated by taking the negative log of the concentration of the acid, assuming the acid is monoprotic. Using the concentration listed, a pH of 1.46 would be expected. The actual pH is higher than this so the solute must be a weak acid.

T17.8 $HCl + H_2O \rightarrow H_3O^+ + Cl^-$
$2H_2O \rightarrow H_3O^+ + OH^-$
From the first reaction, $[H_3O^+] = 1.0 \times 10^{-6}$ M, from the second reaction, $[H_3O^+] = 1.0 \times 10^{-7}$ M. So, $[H_3O^+] = 1.0 \times 10^{-6} + 1.0 \times 10^{-7} = 1.1 \times 10^{-6}$.

T17 .9 The equilibrium for this problem is $HCHO_2 \leftrightarrows H^+ + CHO_2^-$ and the equilibrium table is:

	$[HCHO_2]$	$[H^+]$	$[CHO_2^-]$
I	0.50	–	–
C	– x	+ x	+ x
E	0.50 – x	+ x	+ x

We can use the pH to determine x and solve for the percent ionization.

T17.10 The initial HCl solution has a $[H^+]$ of 0.0100 M. The final concentration will be 0.00100 M from:
$pH = -\log[H^+]$
$3.00 = -\log[H^+]$
$[H^+] = 1.0 \times 10^{-3}$
First calculate the number of moles of H^+ are in the solution from the volume of the solution, 0.150 L and the concentration, 0.0100 M; this gives 1.5×10^{-3} mol H^+. Since solid NaOH is being added, we can assume the volume of the solution will remain at 150 mL. Thus the amount of H^+ that will be in solution after the addition of NaOH will be determined from M × V = mol, or 1.5×10^{-4} mol H^+. Therefore, the amount of NaOH added will be (1.5×10^{-3} mol H^+) – (1.5×10^{-4} mol H^+) = 1.35×10^{-3} mol NaOH. Calculate the number of grams of NaOH by multiplying the mol of NaOH by the formula mass of NaOH

Review Questions

17.1 A Brønsted acid is a proton donor, whereas a Brønsted base is a proton acceptor.

17.2 In a conjugate acid–base pair, the acid has one more hydrogen ion than does the base.

17.3 H_2SO_4 is not the conjugate acid of SO_4^{2-} because H_2SO_4 has two more hydrogen ions than does SO_4^{2-}.

17.4 An amphoteric substance can act either as an acid or as a base.
$HCl(aq) + H_2O(l) \rightarrow H_3O^+(aq) + Cl^-(aq)$, in which water serves as a base
$NH_3(g) + H_2O(l) \rightarrow OH^-(aq) + NH_4^+(aq)$, in which water serves as an acid

17.5 A compound that can act either as a proton donor or a proton acceptor.

17.6 Acid strength increases from left to right in the same period and from top to bottom in the same family.

17.7 (a) The relative strength of the binary acids increases from top to bottom in a group of the periodic table, so we expect HAt to be a stronger acid than HI.

(b) The relative strength of oxoacids increases from bottom to top in a group of the periodic table, so we expect $HAtO_4$ to be a weaker acid that $HBrO_3$.

17.8 In nitric acid, there are more oxygen atoms bound to the nitrogen atom than in nitrous acid. As the number of oxygen atoms increases, the "pull" on the electrons in the OH bond increases withdrawing electrons away from the hydrogen atom. This makes it easier to lose a hydrogen ion.

17.9 The relative strengths of the binary acids increases from top to bottom in a group of the periodic table, so we expect H_2S to be a stronger acid than H_2O. This is because the H–S bond strength is lower than the H–O bond strength.

17.10 $CH_3CH_2O^-$ is a stronger Brønsted base than $CH_3CH_2S^-$. The S–H bond is weaker than the O–H bond, therefore, CH_3CH_2SH is a stronger acid than CH_3CH_2OH this in turn makes the conjugate base of the stronger acid a weaker base.

17.11 There are more oxygen atoms not attached to protons in $HClO_4$ than in H_2SeO_4. In addition, the chlorine atom is more electronegative than selenium which makes the OH bond more polarized in $HClO_4$.

17.12 NO_2^- is a stronger base than SO_3^{2-} because its conjugate acid, HNO_2, is a weaker acid than the conjugate acid of SO_3^{2-}, i.e., HSO_3^-. HSO_3^- is a stronger acid than HNO_2 as evidenced by the trend presented in the text for oxoacids regarding the number of oxygens attached to the central atom.

17.13 conjugate pair

$HOCl(aq) + H_2O \leftrightarrows H_3O^+(aq) + OCl^-(aq)$

acid base acid base

conjugate pair

OCl^- is a stronger base than water and H_3O^+ is a stronger acid than HOCl

17.14 The equilibrium would lie to the left, more so than to the right, so as to lie in the direction which favors formation of the weaker acid and base.

17.15 $C_2H_3O_2^-$ is a stronger base than NO_2^-.

17.16 This equilibrium should lie to the left, because if HNO_3 is a strong acid, then, by definition, NO_3^- must be a weak base.

17.17 Both $HClO_4$ and HNO_3 are 100% ionized in water and, therefore, appear to be the same strength. To distinguish between these two, we would need a solvent that is a weaker proton acceptor than H_2O.

17.18 Water, formic acid, and acetic acid have similar but different dissociation constants. Hence, it is possible to distinguish between them. In NH_3 solution, formic acid and acetic acid are fully dissociated and appear the same strength.

17.19 The molecule on the right will be the stronger acid because the very electronegative F atoms will stabilize the resulting ion.

17.20 The hydrogen on the tri-chloro compound is more acidic because the high electronegativity of Cl will stabilize the ion.

17.21 Acid: Electron pair acceptor
Base: Electron pair donor

17.22 The H^+ ion accepts a pair of electrons from the oxygen atom of the water molecule. This makes the H^+ ion the Lewis acid and the H_2O molecule is the Lewis base.

17.23

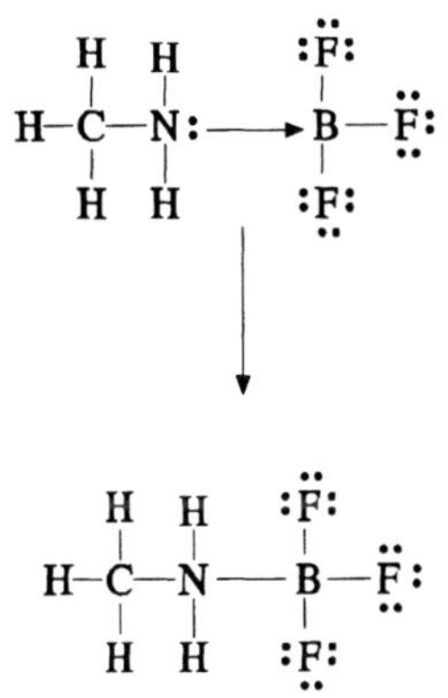

17.24

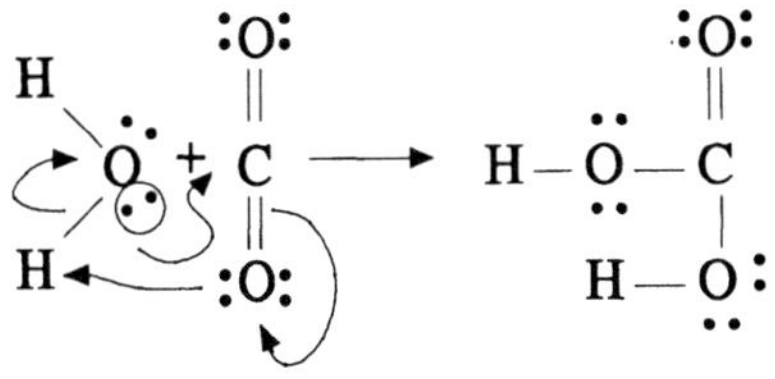

The Lewis base is the water molecule and the Lewis acid is the CO_2.

17.25 The O^{2-} ion has a complete valence shell. It can donate electron pairs functioning as a Lewis base. It cannot accept additional electrons and, therefore, cannot serve as a Lewis acid.

17.26 SbF_5 has an empty orbital, it can accept a pair of electrons to form a Lewis acid-base adduct.

17.27 In the reaction of calcium with oxygen, there is a complete transfer of the electron pair from the calcium to the oxygen, then the electrostatic attractions bind the two ions together. In a Lewis acid–base reaction, a covalent bond is formed between the two atoms and the electrons are shared.

17.28

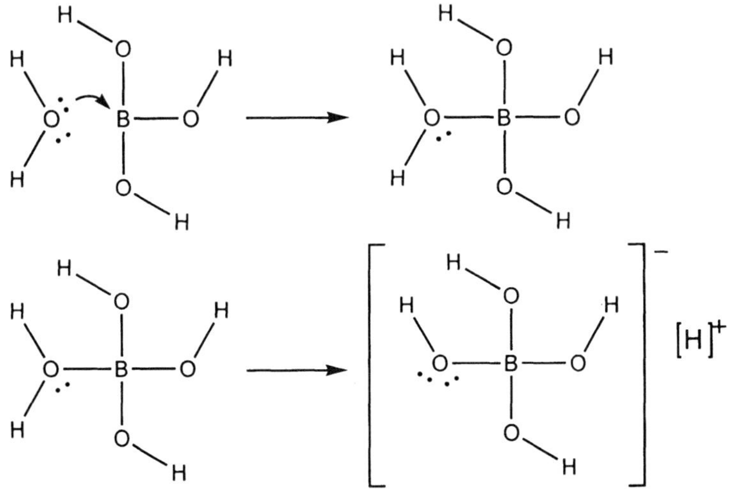

17.29 The element would be classified as a metal if it had a characteristic metallic luster, was a good conductor of heat and electricity, was ductile and malleable, formed ionic compounds with nonmetals and formed an oxide that was a basic anhydride.

17.30 Metal oxides are typically basic so this must be a non-metal oxide.

17.31 This is due to the equilibrium: $Cr(H_2O)_6^{3+} + H_2O \leftrightharpoons Cr(H_2O)_5OH^{2+} + H_3O^+$

17.32 The charge density on the Fe^{3+} is higher, so it is more acidic.

17.33 Because they have such a small charge density, the metal ions have little impact and essentially zero interaction with H_2O in an acid–base sense.

17.34 (a) H_2SO_4
(b) H_2CO_3
(c) H_3PO_4

17.35 (a) CrO_3
(b) CrO
(c) Cr_2O_3

17.36 (a) $Al_2O_3 + 6H^+ \rightarrow 2Al^{3+} + 3H_2O$
(b) $Al_2O_3 + 2OH^- \rightarrow 2AlO^{2-} + H_2O$

17.37 $H_2O + H_2O \leftrightharpoons H_3O^+ + OH^-$
$K_w = [H_3O^+][OH^-]$

17.38 (a)

neutral solution	$[H_3O^+] = [OH^-] = 1.0 \times 10^{-7}$ M
acidic solution	$[H_3O^+] > [OH^-]$
basic solution	$[H_3O^+] < [OH^-]$

(b)

neutral solution	pH = 7.00
acidic solution	pH < 7.00
basic solution	pH > 7.00

17.39 pH + pOH = 14.00

Review Problems

17.40 (a) HF (b) $N_2H_5^+$ (c) $C_5H_5NH^+$
(d) HO_2^- (e) H_2CrO_4

17.41 (a) NH_2O^- (b) SO_3^{2-} (c) CN^-
(d) $H_4IO_6^-$ (e) NO_2^-

17.42 (a)

conjugate pair

$$\underset{\text{acid}}{HNO_3} + \underset{\text{base}}{N_2H_4} \leftrightharpoons \underset{\text{acid}}{N_2H_5^+} + \underset{\text{base}}{NO_3^-}$$

conjugate pair

(b)

conjugate pair

$$\underset{\text{acid}}{N_2H_5^+} + \underset{\text{base}}{NH_3} \leftrightharpoons \underset{\text{acid}}{NH_4^+} + \underset{\text{base}}{N_2H_4}$$

conjugate pair

(c) conjugate pair

$H_2PO_4^- + CO_3^{2-} \leftrightarrows HCO_3^- + HPO_4^{2-}$

acid base acid base

conjugate pair

(d) conjugate pair

$HIO_3 + HC_2O_4^- \leftrightarrows H_2C_2O_4 + IO_3^-$

acid base acid base

conjugate pair

17.43 (a) conjugate pair

$HSO_4^- + SO_3^{2-} \leftrightarrows HSO_3^- + SO_4^{2-}$

acid base acid base

conjugate pair

(b) conjugate pair

$H_2O + S^{2-} \leftrightarrows HS^- + OH^-$

acid base acid base

conjugate pair

(c) conjugate pair

$H_3O^+ + CN^- \leftrightarrows HCN + H_2O$

acid base acid base

conjugate pair

(d) conjugate pair

$H_2Se + H_2O \leftrightarrows H_3O^+ + HSe^-$

acid base acid base

conjugate pair

17.44 (a) H_2Se, larger central atom
(b) HI, more electronegative atom
(c) PH_3, larger central atom

17.45 (a) HBr, HBr bond is weaker
(b) HF, more electronegative F polarizes and weakens the bond
(c) HBr, larger Br forms a weaker bond

17.46 (a) HIO_4, because it has more oxygen atoms
(b) H_3AsO_4, because it has more oxygen atoms

17.47 (a) $HClO_2$, because it has more oxygen atoms
(b) H_2SeO_4, because it has more lone oxygen atoms

17.48 (a) H_3PO_4, since P is more electronegative
(b) HNO_3, because N is more electronegative (HNO_3 is a strong acid)
(c) $HClO_4$, because Cl is more electronegative

17.49 (a) $HClO_3$, because Cl is more electronegative
(b) $HClO_3$, because the charge is more evenly distributed
(c) $HBrO_4$, because the negative charge is more evenly distributed

17.50

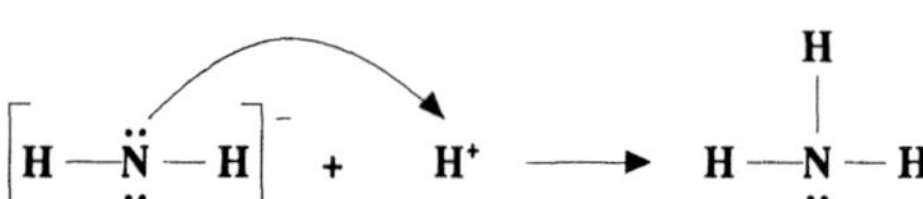

Lewis acid: H^+ Lewis base: NH_2^-

17.51 Lewis acid: BF_3 Lewis Base: F^-

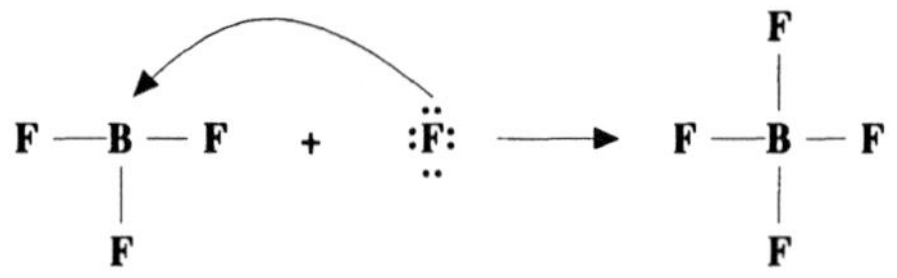

F—B—F (with F below B) + :F: → F—B—F (with F above and below B)

17.52

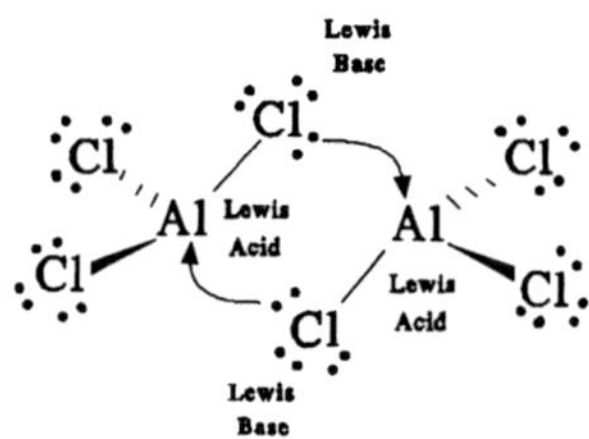

17.53

:Cl—Be—Cl: + :Cl—Be—Cl: + :Cl—Be—Cl:

↓

:Cl—Be—Cl:
:Cl—Be—Cl:
:Cl—Be—Cl:

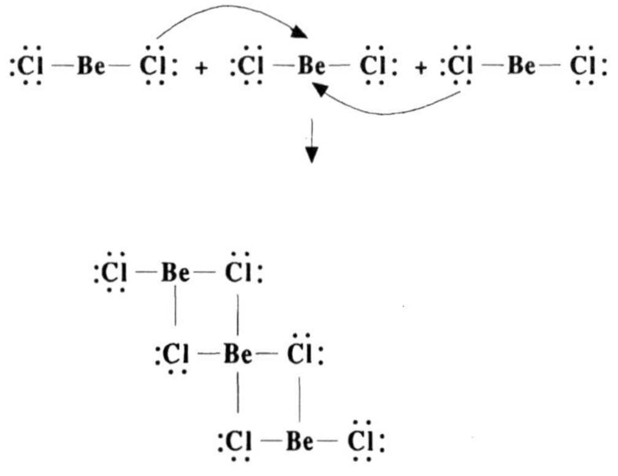

17.54 Lewis base Lewis acid

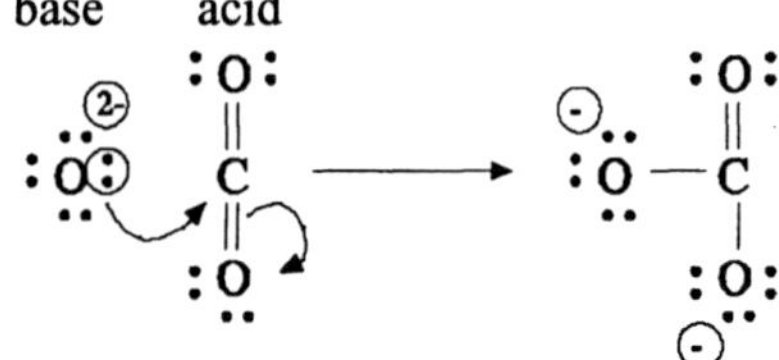

→ O—C(=O)—O (each O with a negative charge)

17.55 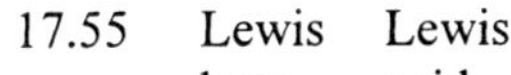

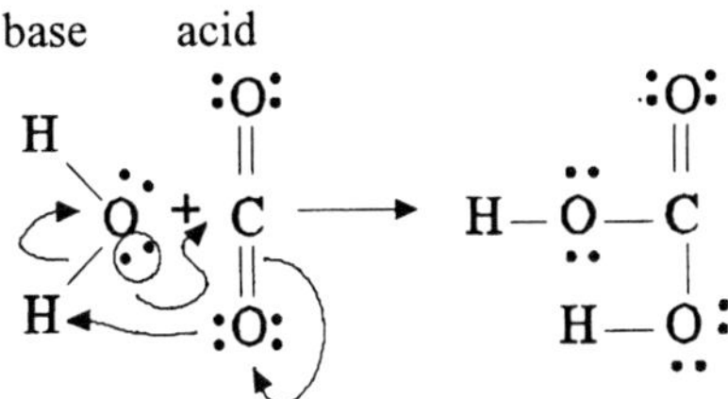

17.56

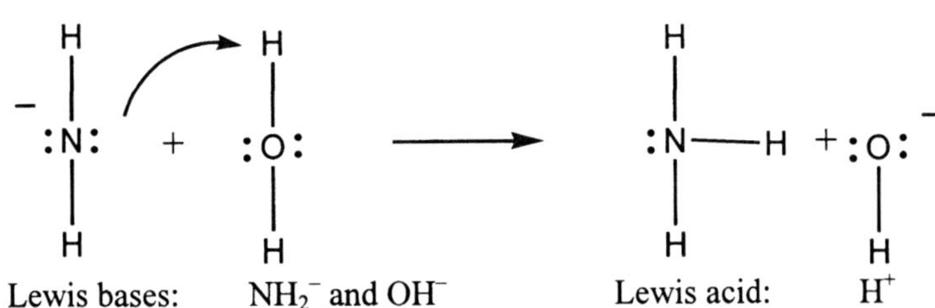

Lewis bases: NH_2^- and OH^- Lewis acid: H^+

17.57

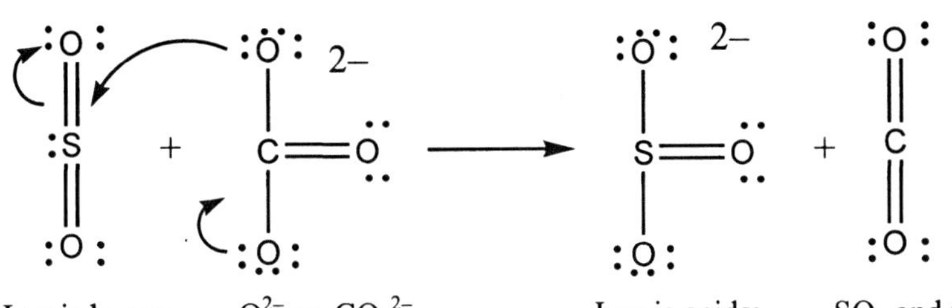

Lewis bases: O^{2-} on CO_3^{2-} Lewis acids: SO_2 and CO_2

17.58 It should be stated from the outset that water at this temperature is neutral by definition, since $[H^+] = [OH^-]$. In other words, the self–ionization of water still occurs on a one–to–one mole basis:

$$H_2O \leftrightarrows H^+ + OH^-.$$

$$K_w = 2.4 \times 10^{-14} = [H^+] \times [OH^-]$$

Since $[H^+] = [OH^-]$, we can rewrite the above relationship:

$$2.4 \times 10^{-14} = ([H^+])^2,$$

$$\therefore\ [H^+] = [OH^-] = 1.5 \times 10^{-7}\ M$$

$pH = -\log[H^+] = -\log(1.5 \times 10^{-7}) = 6.82$
$pOH = -\log[OH^-] = -\log(1.5 \times 10^{-7}) = 6.82$
$pK_w = pH + pOH = 6.82 + 6.82 = 13.64$

Alternatively, for the last calculation we can write:
$pK_w = -\log(K_w) = -\log(2.4 \times 10^{-14}) = 13.62$

Water is neutral at this temperature because the concentration of the hydrogen ion is the same as the concentration of the hydroxide ion.

17.59 $D_2O \leftrightarrows D^+ + OD^-$, $K_w = [D^+] \times [OD^-] = 8.9 \times 10^{-16}$
Since $[D^+] = [OD^-]$, we can rewrite the above expression to give:
$8.9 \times 10^{-16} = ([D^+])^2$, $[D^+] = 3.0 \times 10^{-8}\ M = [OD^-]$
$pD = -\log[D^+] = -\log(3.0 \times 10^{-8}) = 7.52$
$pOD = -\log[OD^-] = -\log(3.0 \times 10^{-8}) = 7.52$
$pK_w = pD + pOD = 15.04$

Alternatively, we can calculate:
$pK_w = -\log(K_w) = -\log(8.9 \times 10^{-16}) = 15.05$

17.60 At 25 °C, $K_w = 1.0 \times 10^{-14} = [H^+] \times [OH^-]$. Let $x = [H^+]$, for each of the following:

(a) $x(0.0024) = 1.0 \times 10^{-14}$
$[H^+] = (1.0 \times 10^{-14}) \div (0.0024) = 4.2 \times 10^{-12}$ M

(b) $x(1.4 \times 10^{-5}) = 1.0 \times 10^{-14}$
$[H^+] = (1.0 \times 10^{-14}) \div (1.4 \times 10^{-5}) = 7.1 \times 10^{-10}$ M

(c) $x(5.6 \times 10^{-9}) = 1.0 \times 10^{-14}$
$[H^+] = (1.0 \times 10^{-14}) \div (5.6 \times 10^{-9}) = 1.8 \times 10^{-6}$ M

(d) $x(4.2 \times 10^{-13}) = 1.0 \text{ x } 10^{-14}$
$[H^+] = (1.0 \times 10^{-14}) \div (4.2 \times 10^{-13}) = 2.4 \times 10^{-2}$ M

17.61 At 25 °C, $K_w = 1.0 \times 10^{-14} = [H^+] \times [OH^-]$. Let $x = [OH^-]$, for each of the following:

(a) $(3.5 \times 10^{-8})x = 1.0 \times 10^{-14}$
$[OH^-] = (1.0 \times 10^{-14}) \div (3.5 \times 10^{-8}) = 2.9 \times 10^{-7}$ M

(b) $(0.0065)x = 1.0 \times 10^{-14}$
$[OH^-] = (1.0 \times 10^{-14}) \div (0.0065) = 1.5 \times 10^{-12}$ M

(c) $(2.5 \times 10^{-13})x = 1.0 \times 10^{-14}$
$[OH^-] = (1.0 \times 10^{-14}) \div (2.5 \times 10^{-13}) = 4.0 \times 10^{-2}$ M

(d) $(7.5 \times 10^{-5})x = 1.0 \times 10^{-14}$
$[OH^-] = (1.0 \times 10^{-14}) \div (7.5 \times 10^{-5}) = 1.3 \times 10^{-10}$ M

17.62 $pH = -\log[H^+]$

$[H^+] = 4.2 \times 10^{-12}$ M	pH = 11.38
$[H^+] = 7.1 \times 10^{-10}$ M	pH = 9.15
$[H^+] = 1.8 \times 10^{-6}$ M	pH = 5.74
$[H^+] = 2.4 \times 10^{-2}$ M	pH = 1.62

17.63 $pH = -\log[H^+]$

$[H^+] = 3.5 \times 10^{-8}$ M	pH = 7.46
$[H^+] = 0.0065$ M	pH = 2.19
$[H^+] = 2.5 \times 10^{-13}$ M	pH = 12.60
$[H^+] = 7.5 \times 10^{-5}$ M	pH = 4.12

17.64 $pH = -\log[H^+] = -\log(1.9 \times 10^{-5}) = 4.72$

17.65 $pH = -\log[H^+] = -\log(1.4 \times 10^{-5}) = 4.85$

17.66 $[H^+] = 10^{-pH}$ and $[OH^-] = 10^{-pOH}$
At 25 °C, pH + pOH = 14.00

(a) $[H^+] = 10^{-pH} = 10^{-3.14} = 7.2 \times 10^{-4}$ M
pOH = 14.00 – pH = 14.00 – 3.14 = 10.86
$[OH^-] = 10^{-pOH} = 10^{-10.86} = 1.4 \times 10^{-11}$ M

(b) $[H^+] = 10^{-pH} = 10^{-2.78} = 1.7 \times 10^{-3}$ M
pOH = 14.00 – pH = 14.00 – 2.78 = 11.22
$[OH^-] = 10^{-pOH} = 10^{-11.22} = 6.0 \times 10^{-12}$ M

(c) $[H^+] = 10^{-pH} = 10^{-9.25} = 5.6 \times 10^{-10}$ M
pOH = 14.00 – pH = 14.00 – 9.25 = 4.75
$[OH^-] = 10^{-pOH} = 10^{-4.75} = 1.8 \times 10^{-5}$ M

(d) $[H^+] = 10^{-pH} = 10^{-13.24} = 5.8 \times 10^{-14}$ M
$pOH = 14.00 - pH = 14.00 - 13.24 = 0.76$
$[OH^-] = 10^{-pOH} = 10^{-0.76} = 1.7 \times 10^{-1}$ M

(e) $[H^+] = 10^{-pH} = 10^{-5.70} = 2.0 \times 10^{-6}$ M
$pOH = 14.00 - pH = 14.00 - 5.70 = 8.30$
$[OH^-] = 10^{-pOH} = 10^{-8.30} = 5.0 \times 10^{-9}$ M

17.67 (a) $[OH^-] = 10^{-pOH} = 10^{-8.26} = 5.5 \times 10^{-9}$ M
$pH = 14.00 - pOH = 14.00 - 8.26 = 5.74$
$[H^+] = 10^{-pH} = 10^{-5.74} = 1.8 \times 10^{-6}$ M

(b) $[OH^-] = 10^{-pOH} = 10^{-10.25} = 5.6 \times 10^{-11}$ M
$pH = 14.00 - pOH = 14.00 - 10.25 = 3.75$
$[H^+] = 10^{-pH} = 10^{-3.75} = 1.8 \times 10^{-4}$ M

(c) $[OH^-] = 10^{-pOH} = 10^{-4.65} = 2.2 \times 10^{-5}$ M
$pH = 14.00 - pOH = 14.00 - 4.65 = 9.35$
$[H^+] = 10^{-pH} = 10^{-9.35} = 4.5 \times 10^{-10}$ M

(d) $[OH^-] = 10^{-pOH} = 10^{-6.18} = 6.6 \times 10^{-7}$ M
$pH = 14.00 - pOH = 14.00 - 6.18 = 7.82$
$[H^+] = 10^{-pH} = 10^{-7.82} = 1.5 \times 10^{-8}$ M

(e) $[OH^-] = 10^{-pOH} = 10^{-9.70} = 2.0 \times 10^{-10}$ M
$pH = 14.00 - pOH = 14.00 - 9.70 = 4.30$
$[H^+] = 10^{-pH} = 10^{-4.30} = 5.0 \times 10^{-5}$ M

17.68 $[H^+] = 10^{-pH} = 10^{-5.6} = 2.5 \times 10^{-6}$ M
$pOH = 14.00 - pH = 14.00 - 5.6 = 8.4$
$[OH^-] = 10^{-pOH} = 10^{-10} = 4.0 \times 10^{-9}$ M

17.69 $[H^+] = 10^{-pH} = 10^{-3.42} = 3.80 \times 10^{-4}$ M
$pOH = 14.00 - pH = 14.00 - 3.42 = 10.58$
$[OH^-] = 10^{-pOH} = 10^{-10.58} = 2.63 \times 10^{-11}$ M

17.70 HNO_3 is a strong acid so $[H^+] = [HNO_3] = 0.030$ M
$pH = -\log[H^+] = -\log(0.030) = 1.52$
$pOH = 14.00 - pH = 14.00 - 1.52 = 12.48$
$[OH^-] = 10^{-pOH} = 10^{-12.48} = 3.31 \times 10^{-13}$ M

? No

17.71 $HClO_4$ is a strong acid so $[H^+] = [HClO_4] = 0.0025$ M
$pH = -\log[H^+] = -\log(0.0050) = 2.60$
$pOH = 14.00 - pH = 14.00 - 2.60 = 11.40$
$[OH^-] = 10^{-pOH} = 10^{-11.40} = 3.98 \times 10^{-12}$ M

17.72

$$\text{M OH}^- = \frac{\#\text{ moles OH}^-}{\#\text{ L solution}} = \left(\frac{6.0\text{ g NaOH}}{1.00\text{ L solution}}\right)\left(\frac{1\text{ mole NaOH}}{40.0\text{ g NaOH}}\right)\left(\frac{1\text{ mole OH}^-}{1\text{ mole NaOH}}\right)$$

$$= 0.15\text{ M OH}^-$$

$pOH = -\log[OH^-] = -\log(0.15) = 0.82$
$pH = 14.00 - pOH = 14.00 - 0.82 = 13.18$
$[H^+] = 10^{-pH} = 10^{-13.18} = 6.61 \times 10^{-14}$ M

17.73

$$\text{M OH}^- = \frac{\text{\# moles OH}^-}{\text{\# L solution}}$$

$$= \left(\frac{0.837 \text{ g Ba(OH)}_2}{0.100 \text{ L solution}}\right)\left(\frac{1 \text{ mole Ba(OH)}_2}{171.3 \text{ g Ba(OH)}_2}\right)\left(\frac{2 \text{ mole OH}^-}{1 \text{ mole Ba(OH)}_2}\right)$$

$$= 0.0977 \text{ M OH}^-$$

$pOH = -\log[OH^-] = -\log(9.77 \times 10^{-2}) = 1.01$
$pH = 14.00 - pOH = 14.00 - 1.01 = 12.99$
$[H^+] = 10^{-pH} = 10^{-12.99} = 1.02 \times 10^{-13}$ M

17.74 $pOH = 14.00 - pH = 14.00 - 11.60 = 2.40$
$[OH^-] = 10^{-pOH} = 10^{-2.40} = 4.0 \times 10^{-3}$ M

$$[\text{Ca(OH)}_2] = \left(\frac{4.0 \times 10^{-3} \text{ mol OH}^-}{1 \text{ L solution}}\right)\left(\frac{1 \text{ mol Ca(OH)}_2}{2 \text{ mol OH}^-}\right)$$

$$= 2.0 \times 10^{-3} \text{ M Ca(OH)}_2$$

17.75 $[H^+] = 10^{-pH} = 10^{-2.50} = 3.2 \times 10^{-3}$ M

$$\text{\# g HCl} = (0.250 \text{ L})\left(\frac{3.2 \times 10^{-3} \text{ mol H}^+}{1 \text{ L solution}}\right)\left(\frac{1 \text{ mol HCl}}{1 \text{ mol H}^+}\right)\left(\frac{36.46 \text{ g HCl}}{1 \text{ mol HCl}}\right)$$

$$= 2.9 \times 10^{-2} \text{ g HCl}$$

17.76 First, we must find the concentration of the HCl solution. Since HCl is a strong acid, we know that [HCl] = $[H^+]$. We can find $[H^+]$ from the pH:
$[H^+] = 10^{-pH} = 10^{-2.25} = 5.6 \times 10^{-3}$ M H^+

Now we can solve the problem using the given conversion factors:

$$\text{\# mL 0.100 M KOH} = 300 \text{ mL}\left(\frac{5.6\times10^{-3} \text{ mol H}^+}{1000 \text{ mL solution}}\right)\left(\frac{1 \text{ mol OH}^-}{1 \text{ mol H}^+}\right)\left(\frac{1000 \text{ mL OH}^-}{0.100 \text{ mol OH}^-}\right)$$

$$= 16.8 \text{ mL 0.100 M KOH}$$

17.77 $pOH = 14 - pH = 14 - 12.05 = 1.95$
$[OH^-] = 10^{-pOH} = 10^{-1.95} = 1.1 \times 10^{-2}$ M OH^-

$$\text{mol of H}^+ = 300 \text{ mL LiOH solution}\left(\frac{1.1\times10^{-2} \text{ mol OH}^-}{1000 \text{ mL LiOH solution}}\right)\left(\frac{1 \text{ mol H}^+}{1 \text{ mol OH}^-}\right) = 3.3 \times 10^{-3} \text{ mol H}^+$$

$$\text{M HNO}_3 = \left(\frac{3.4\times10^{-3} \text{ mol H}^+}{0.02020 \text{ L solution}}\right) = 1.6 \times 10^{-1} \text{ M HNO}_3$$

17.78 Since NaOH is a strong base, $[NaOH] = [OH^-] = 0.0020$ M OH^-. (Next, we make the simplifying assumption that the amount of hydroxide ion formed from the dissociation of water is so small that we can neglect it in calculating pOH for the solution.)

$[H^+] = 1 \times 10^{-14} \div 0.0020 = 5 \times 10^{-12}$ M H^+
The only source of H^+ is the autoionization of water. Therefore, the molarity of OH^- from the ionization water is 5×10^{-12} M.

17.79 The H^+ concentration from the ionization of water also gives the same OH^- concentration. Since the only source of OH^- is the water, the concentration of OH^- is 3.4×10^{-11} M.

The total H^+ concentration can be determined from:
$1 \times 10^{-14} = [H^+] \times [OH^-] = [H^+] \times (3.4 \times 10^{-11})$
$pH = 2.9 \times 10^{-4}$ M H^+

Additional Exercises

17.80 Acid: $(CH_3)_2NH_2^+$ Base: $(CH_3)_2N^-$

17.81 The moles of HCl is determined from the ideal gas law:

$$n = \frac{PV}{RT} = \frac{(734\text{ torr})\left(\frac{1\text{ atm}}{760\text{ torr}}\right)(0.250\text{ L})}{0.0821\,\frac{\text{L atm}}{\text{mol K}}\,298\text{ K}} = 9.87\times10^{-3}\text{ mol HCl}$$

The molarity of the solution is:

$$\text{M of solution} = \frac{9.87\times10^{-3}\text{ mol HCl}}{0.250\text{ L solution}} = 3.95\times10^{-2}\text{ M HCl}$$

$PH = -\log[H^+] = -\log(3.95 \times 10^{-2}\text{ M}) = 1.404$

17.82 From problem 17.81, we find there are 3.95×10^{-4} mol HCl present. We need to know how many mol NaOH are in the NaOH solution to react with the HCl. We can find $[OH^-]$ from the given pH:

$pOH = 14.00 - pH = 14.00 - 10.50 = 3.50$
$[OH^-] = 10^{-pOH} = 10^{-3.50} = 3.16 \times 10^{-4}$ M

$$\text{mol NaOH} = 200\text{ mL NaOH soln}\left(\frac{1\text{ L NaOH soln}}{1000\text{ mL NaOH soln}}\right)\left(\frac{3.16\times10^{-4}\text{ mol NaOH}}{1\text{ L NaOH soln}}\right)$$
$$= 6.32\times10^{-5}\text{ mol NaOH}$$

Since HCl and NaOH react in a 1–to–1 ratio, 6.32×10^{-5} mol of each will neutralize one another. This leaves (3.95×10^{-4} mol HCl – 6.32×10^{-5} mol HCl) = 3.32×10^{-4} mol HCl remaining.

The concentration of H^+ is now:

$[H^+]$ = mol H^+/L solution
= 3.32×10^{-4} mol/0.200 L solution
= 1.66×10^{-3} M

Therefore $pH = -\log [H^+] = -\log [1.66 \times 10^{-3}] = 2.78$

17.83 $HCO_3^- + H_2O \leftrightarrows H_2CO_3 + OH^-$
$HCO_3^- + H_2O \leftrightarrows H_3O^+ + CO_3^{2-}$

17.84 (a) In H_2O_2, the oxygen atom helps to stabilize the HO_2^- ion making it easier for H_2O_2 to lose a proton than can H_2O.
(b) acidic

17.85 $HClO_3$ will ionize more in the more weakly acidic solvent, H_2O. H_2F^+ is a very strong acid and the second equilibrium will be to the left.

17.86 The equilibrium lies to the right since reactions favor the weaker acid and base.

17.87 acids: $CH_3NH_3^+$, NH_3OH^+
bases: NH_2OH, CH_3NH_2

17.88 NH_3OH^+ is the strongest acid

17.89

$$\# \text{ moles } H^+ = (38.0 \text{ mL})\left(\frac{0.00200 \text{ moles}}{1000 \text{ mL}}\right) = 7.60 \times 10^{-5} \text{ moles } H^+$$

$$\# \text{ moles } OH^- = (40.0 \text{ mL})\left(\frac{0.0018 \text{ moles}}{1000 \text{ mL}}\right) = 7.20 \times 10^{-5} \text{ moles } OH^-$$

excess $H^+ = 0.40 \times 10^{-5}$ moles H^+

$$[H^+] = \frac{0.40 \times 10^{-5} \text{ moles}}{(38 \text{ mL} + 40 \text{ mL})\left(\frac{1 \text{ L}}{1000 \text{ mL}}\right)} = 5.12 \times 10^{-5} \text{ M}$$

pH = 4.29

17.90 The total $[H^+]$ is from the HCl and from the dissociation of H_2O. Since HCl is a strong acid, it will contribute 1.0×10^{-7} mol of H^+ per liter of solution. We need to use the equilibrium expression to determine the amount of H^+ contributed by the water.

$$K_w = [H^+][OH^-] = (1.0 \times 10^{-7} + x)(x) = 1.0 \times 10^{-14}$$

Solving a quadratic equation we see that $x = 6 \times 10^{-8} = [OH^-] = [H^+]$ from water dissociation.
So, $[H^+]_{total} = 1.0 \times 10^{-7} + 6 \times 10^{-8} = 1.6 \times 10^{-7}$ and pH = 6.80.

17.91 $K_w = 2.42 \times 10^{-14} = [H^+][OH^-]$
In a neutral solution, $[H^+] = [OH^-]$; let $x = [H^+]$
$2.42 \times 10^{-14} = x^2$ $\quad x = 1.56 \times 10^{-7}$
$pH = -\log[H^+] = -\log[1.56 \times 10^{-7}] = 6.81$

17.92 A solution with pH = 3.00 would have:

$[H^+] = 10^{-pH} = 10^{-3.00}$ $= 1.00 \times 10^{-3}$ M
$= 1.00 \times 10^{-3}$ mol $[H^+]$/L solution
$= 0.00100$ mol $[H^+]$/L solution

We begin with:

$$\text{mol HCl} = 200 \text{ mL HCl soln}\left(\frac{1 \text{ L HCl soln}}{1{,}000 \text{ mL HCl soln}}\right)\left(\frac{0.010 \text{ mol HCl}}{1 \text{ L HCl soln}}\right)$$

= 0.002 mol HCl
= 0.002 mol H^+

Our final molarity will be: M = mol $[H^+]$ /L solution. This can be written as:

$$M = (0.002 - n_{NaOH})/(0.200 + x),$$

where n_{NaOH} = moles of NaOH added and, x = volume of 0.10 M NaOH added.

But n_{NaOH} is simply $M_{NaOH} \cdot V_{NaOH}$, so

$$M = (0.002 - 0.10x)/(0.200 + x)$$
$$0.00100 = (0.002 - 0.10x)/(0.200 + x)$$
$$(0.200 + x)0.00100 = (0.002 - 0.10x)$$
$$0.000200 + 0.00100\,x = 0.002 - 0.10x$$
$$0.1001\,x = 0.0018$$
$$x = 0.018, \text{ or } 18 \text{ mL}$$

17.93 $pOH = 14.00 - pH = 3.92$
$pOH = -\log [OH^-]$
$[OH^-] = 10^{-pOH} = 1.2 \times 10^{-4}$ M

$$\# \text{ grams } Mg(OH)_2 = (100 \text{ mL})\left(\frac{1.2 \times 10^{-4} \text{ mol}}{1000 \text{ mL}}\right)\left(\frac{1 \text{ mol } Mg(OH)_2}{2 \text{ mol } OH^-}\right)$$

$$\left(\frac{58.32 \text{ g } Mg(OH)_2}{1 \text{ mol } Mg(OH)_2}\right) = 3.5 \times 10^{-4} \text{ g } Mg(OH)_2$$

17.94 $[H^+] = 10^{-pH} = 4.27 \times 10^{-3}$ M

$$\% \text{ ionization} = \frac{4.27 \times 10^{-3}}{1.0} \times 100\% = 0.43\ \%$$

Practice Exercises

18.1 a) $HCHO_2 + H_2O \rightleftharpoons H_3O^+ + CHO_2^-$

$$K_a = \frac{[H_3O^+][CHO_2^-]}{[HCHO_2]}$$

b) $(CH_3)_2NH_2^+ + H_2O \rightleftharpoons H_3O^+ + (CH_3)_2NH$

$$K_a = \frac{[H_3O^+][CH_3NH]}{\left[(CH_3)_2NH_2^+\right]}$$

c) $H_2PO_4^- + H_2O \rightleftharpoons H_3O^+ + HPO_4^{2-}$

$$K_a = \frac{[H_3O^+][HPO_4^{2-}]}{\left[H_2PO_4^-\right]}$$

18.2 The acid with the smaller pK_a (H_A) is the strongest acid.
Since $pK_a = -\log K_a$, $K_a = 10^{-pKa}$
For H_A: $K_a = 10^{-3.16} = 6.9 \times 10^{-4}$
For H_B: $K_a = 10^{-4.14} = 7.2 \times 10^{-5}$

18.3 a) $(CH_3)_3N + H_2O \rightleftharpoons (CH_3)_3NH^+ + OH^-$

$$K_b = \frac{[(CH_3)_3NH^+][OH^-]}{[(CH_3)_3N]}$$

b) $SO_3^{2-} + H_2O \rightleftharpoons HSO_3^- + OH^-$

$$K_b = \frac{[HSO_3^-][OH^-]}{\left[SO_3^{2-}\right]}$$

c) $NH_2OH + H_2O \rightleftharpoons NH_3OH^+ + OH^-$

$$K_b = \frac{[NH_3OH^+][OH^-]}{[NH_2OH]}$$

18.4 For conjugate acid base pairs, $K_a \times K_b = K_w$:
$K_b = K_w \div K_a = 1.0 \times 10^{-14} \div 1.8 \times 10^{-4} = 5.6 \times 10^{-11}$

18.5 $HBu \rightleftharpoons H^+ + Bu^-$

$$K_a = \frac{[H^+][Bu^-]}{[HBu]}$$

	[HBu]	[H⁺]	[Bu⁻]
I	0.01000	–	–
C	–x	+x	+x
E	0.01000–x	+x	+x

We know that the acid is 4.0% ionized so x = 0.01000 M × 0.040 = 0.00040 M. Therefore, our equilibrium concentrations are $[H^+] = [Bu^-] = 0.00040$ M, and [HBu] = 0.01000 M – 0.00040 M = 0.00960 M.

Substituting these values into the mass action expression gives:

$$K_a = \frac{(0.00040)(0.00040)}{0.00960} = 1.7 \times 10^{-5}$$

$pK_a = -\log(K_a) = -\log(1.7 \times 10^{-5}) = 4.78$

18.6 We will use the symbol Mor and $HMor^+$ for the base and its conjugate acid respectively:

$$Mor + H_2O \rightleftharpoons HMor^+ + OH^-$$

$$K_b = \frac{[HMor^+][OH^-]}{[Mor]}$$

	[Mor]	[HMor$^+$]	[OH$^-$]
I	0.010	–	–
C	–x	+x	+x
E	0.010–x	+x	+x

At equilibrium, the pH = 10.10 and the pOH = 14.00 – 10.10 = 3.90.
The $[OH^-] = 10^{-pOH} = 10^{-3.90} = 1.3 \times 10^{-4}$ M = x.

Substituting these values into the mass action expression gives:

$$K_b = \frac{(1.3 \times 10^{-4})(1.3 \times 10^{-4})}{0.010 - 1.3 \times 10^{-4}} = 1.7 \times 10^{-6}$$

$pK_b = -\log(K_b) = -\log(1.7 \times 10^{-6}) = 5.77$

18.7 $HC_2H_6NO_2 \rightleftharpoons H^+ + C_2H_6NO_2^-$

$$K_a = \frac{[H^+][C_2H_6NO_2^-]}{[HC_2H_6NO_2]} = 1.4 \times 10^{-5}$$

	[$HC_2H_6NO_2$]	[H^+]	[$C_2H_6NO_2^-$]
I	0.050	–	–
C	–x	+x	+x
E	0.050–x	+x	+x

Assume that x << 0.050 and substitute the equilibrium values into the mass action expression to get:

$$K_a = \frac{[x][x]}{[0.050]} = 1.4 \times 10^{-5}$$

Solving for x we determine that x = 8.4×10^{-4} M = [H^+].
pH = $-\log[H^+] = -\log(8.4 \times 10^{-4}) = 3.08$

18.8 $C_5H_5N + H_2O \rightleftharpoons C_5H_5NH^+ + OH^-$

$$K_b = \frac{[C_5H_5NH^+][OH^-]}{[C_5H_5N]} = 1.5 \times 10^{-9}$$

	[C_5H_5N]	[$C_5H_5NH^+$]	[OH^-]
I	0.010	–	–
C	–x	+x	+x
E	0.010–x	+x	+x

Assume that x << 0.010 and substitute the equilibrium values into the mass action expression to get:

$$K_b = \frac{[x][x]}{[0.010]} = 1.7 \times 10^{-9}$$

Solving for x we determine that x = 4.1×10^{-6} M = $[OH^-]$.
pOH = $-\log[OH^-] = -\log(4.1 \times 10^{-6}) = 5.38$
pH = 14.00 – pOH = 8.62

18.9 We will use the notation Hphenol and phenol for the acid and its conjugate base:

$$\text{Hphenol} \rightleftharpoons H^+ + \text{phenol}$$

$$K_a = \frac{[H^+][\text{phenol}]}{[\text{Hphenol}]} = 1.3 \times 10^{-10}$$

	[Hphenol]	$[H^+]$	[phenol]
I	0.15	–	–
C	–x	+x	+x
E	0.15–x	+x	+x

If we assume that x << 0.15, a good assumption based upon the size of K_a, we can substitute the equilibrium values in to the mass action expression to get:

$$K_a = \frac{(x)(x)}{0.15} = 1.3 \times 10^{-10}$$

Solving gives x = 4.4×10^{-6} M = $[H^+]$
pH = $-\log[H^+] = -\log(4.4 \times 10^{-6}) = 5.36$

18.10 We examine the ions in solution one at a time: The cation for acidity, and the anion for basicity. Na^+ is an ion of a Group IA metal and is *not* acidic. NO_3^- is the conjugate base of HNO_3 (a strong acid), therefore it is *not* a strong base. This solution should be neutral.

18.11 We examine the ions in solution one at a time: The cation for acidity, and the anion for basicity. K^+ is an ion of a Group IA metal and is *not* acidic. Cl^- is the conjugate base of HCl (a strong acid), therefore it is *not* a strong base. This solution should be neutral.

18.12 We examine the ions in solution one at a time: The cation for acidity, and the anion for basicity. NH_4^+ is the conjugate acid of NH_3, a weak, molecular base. Therefore it is slightly *acidic.* Br^- is the conjugate base of HBr (a strong acid), therefore it is *not* a strong base. This solution should be acidic.

18.13 The sodium ion is neutral since it is the salt of the strong base, NaOH. The nitrite ion is basic since it is the salt of nitrous acid, HNO_2, a weak acid. The equilibrium we are interested in for this problem is:

$$NO_2^- + H_2O \rightleftharpoons HNO_2 + OH^-.$$

$$K_b = \frac{[HNO_2][OH^-]}{[NO_2^-]}$$

In order to determine the value for K_b recall that $K_a - K_b = K_w$. We can look for the value of K_a for HNO_2
$K_b = 1.0 \times 10^{-14} \div 7.1 \times 10^{-4} = 1.4 \times 10^{-11}$.

	$[NO_2^-]$	$[HNO_2]$	$[OH^-]$
I	0.10	–	–
C	–x	+x	+x
E	0.10–x	+x	+x

Assume that x << 0.10 and substitute the equilibrium values into the mass action expression to get:

$$K_b = \frac{[x][x]}{[0.10]} = 1.4 \times 10^{-11}$$

Solving we determine that $x = 1.2 \times 10^{-6}$ M = $[OH^-]$.
$pOH = -\log[OH^-] = -\log(1.2 \times 10^{-6}) = 5.93$
$pH = 14.00 - pOH = 8.07$

18.14 As previously determined, a solution of NH_4Br will be acidic since NH_4^+ is the salt of a weak base and Br^- is the salt of a strong acid. As in the previous Practice Exercise, we need to determine the value for the dissociation constant using the relationship $K_a \times K_b = K_w$ and the value of K_b for NH_3 as listed in Table 16.5.
$K_a = 1.0 \times 10^{-14} \div 1.8 \times 10^{-5} = 5.6 \times 10^{-10}$. The equilibrium reaction is:
$NH_4^+ \rightleftharpoons NH_3 + H^+$

	$[NH_4^+]$	$[NH_3]$	$[H^+]$
I	0.10	–	–
C	–x	+x	+x
E	0.10–x	+x	+x

Assume that x << 0.10 and substitute the equilibrium values into the mass action expression to get:

$$K_a = \frac{[x][x]}{[0.10]} = 5.6 \times 10^{-10}$$

Solving we determine that $x = 7.5 \times 10^{-6}$ M = $[H^+]$.
$pH = -\log[H^+] = -\log(7.5 \times 10^{-6}) = 5.13$

18.15 Since the ammonium ion is the salt of a weak base, NH_3, it is acidic. The cyanide ion is the salt of a weak acid, HCN, so it is basic. In order to determine if the solution is acidic or basic, we need to determine the relative strength of the two components. Use the relationship $K_a \times K_b = K_w$ in order to determine the dissociation constants for the cyanide ion and the ammonium ion.

$K_a(NH_4^+) = K_w \div K_b(NH_3) = 1.0 \times 10^{-14} \div 1.8 \times 10^{-5} = 5.6 \times 10^{-10}$
$K_b(CN^-) = K_w \div K_a(HCN) = 1.0 \times 10^{-14} \div 6.2 \times 10^{-14} = 1.6 \times 10^{-5}$

Since the $K_b(CN^-)$ is larger than the $K_a(NH_4^+)$ the NH_4CN solution will be basic.

18.16 $(CH_3)_2NH + H_2O \rightleftharpoons (CH_3)_2NH_2^+ + OH^-$

$$K_b = \frac{[(CH_3)_2NH_2^+][OH^-]}{[(CH_3)_2NH]} = 9.6 \times 10^{-4}$$

	$[(CH_3)_2NH]$	$[(CH_3)_2NH_2^+]$	$[OH^-]$
I	0.0010	–	–
C	–x	+x	+x
E	0.0010–x	+x	+x

We cannot neglect x in this calculation due to the large size of the dissociation constant. Consequently, we will have to solve the quadratic equation. Substituting the equilibrium values into the mass action expression gives:

$$K_b = \frac{[x][x]}{[0.0010 - x]} = 9.6 \times 10^{-4}$$

Rearranging and collecting terms on one side of the equal sign gives:

$x^2 + (9.6 \times 10^{-4})x - (9.6 \times 10^{-7}) = 0$

Using the quadratic equation

$x = 6.1 \times 10^{-4}$ M = [OH^-], pOH = 3.21 and the pH = 10.79

18.17 The equation is: $C_2H_3O_2^- + H_2O \rightleftharpoons HC_2H_3O_2 + OH^-$

Start by determining K_b for acetate ion using $K_w = K_aK_b$

$$K_b = K_w/K_a = 1.0 \times 10^{-14}/1.8 \times 10^{-5} = 5.6 \times 10^{-10}$$

$$K_b = \frac{[HC_2H_3O_2][OH^-]}{[C_2H_3O_2^-]}$$

	[$C_2H_3O_2^-$]	[$HC_2H_3O_2$]	[OH^-]
I	0.11	0.090	–
C	–x	+x	+x
E	0.11 – x	0.090 +x	+x

$$K_b = \frac{[x][0.090 + x]}{[0.11 - x]}$$

assume x << 0.090 and solve for x

$x = [OH^-] = 6.8 \times 10^{-10}$

pOH = 9.16

pH = 14.00 – 9.16 = 4.84

18.18 We will use the relationship $K_a \times K_b = K_w$.

$$K_a(NH_4^+) = K_w \div K_b(NH_3) = 1.0 \times 10^{-14} \div 1.8 \times 10^{-5} = 5.6 \times 10^{-10}$$

The equation for the acid dissociation is $NH_4^+ \rightleftharpoons H^+ + NH_3$. The dissociation constant is written:

$$K_a = \frac{[H^+][NH_3]}{[NH_4^+]} = 5.6 \times 10^{-10} = \frac{[H^+](0.12)}{(0.095)}$$

Solving gives $[H^+] = 4.4 \times 10^{-10}$ M

$pH = -\log[H^+] = -\log(4.4 \times 10^{-10}) = 9.36$

18.19 Yes, formic acid and sodium formate would make a good buffer solution since $pK_a = 3.74$ and the desired pH is within one pH unit of this value.

Using Equation 18.12; $[H^+] = K_a \times \frac{\text{mol } HCHO_{2\,initial}}{\text{mol } CHO_2^-{}_{initial}}$

Rearranging and substituting the known values we get;

$$\frac{\text{mol } HCHO_{2\,initial}}{\text{mol } CHO_2^-{}_{initial}} = \frac{[H^+]}{K_a} = \frac{1.3 \times 10^{-4}}{1.8 \times 10^{-4}} = 0.72 \text{ mol } HCHO_2 \text{ for every mol } CHO_2^-.$$

For 0.10 mol $HCHO_2$, 0.072 mol $NaCHO_2$ would be needed. Converting to grams, this is (0.072 mol $NaCHO_2$)(68.02 g/1 mol) = 4.9 g $NaCHO_2$.

18.20 Using the data provided in Example 18.17, the initial pH of this buffer is 4.74. When OH^- is added to the solution, it will react with the H^+ present from the dissociation of the acetic acid. The acetic acid in solution dissociates further to maintain the equilibrium and any unreacted hydroxide will react with H^+ as it is produced. Eventually, the hydroxide will be completely reacted. The net change that occurs is a reduction in the amount of acetic acid in solution and an equivalent increase in the amount of acetate ion in solution. We started with 1 mol of acetic acid and 1 mol of acetate ion. The addition of 0.11 mol OH^- will reduce the amount of acetic acid to 0.89 mol and increase the amount of acetate ion to 1.11 mol. Substituting these amounts into the mass action expression gives:

$$K_a = \frac{[H^+](1.11)}{(0.89)} = 1.8 \times 10^{-5}$$

$$[H^+] = 1.4 \times 10^{-5}\ M$$

$pH = -\log[H^+] = 4.84$

18.21 $[H^+]$ is determined by the first protic equilibrium:

$H_2C_6H_6O_6 \rightleftharpoons H^+ + HC_6H_6O_6^-$

The mass action expression is: $K_{a_1} = 6.8 \times 10^{-5} = x^2/0.10$

$x = [H^+] = 2.6 \times 10^{-3}\ M$

$pH = -\log(2.6 \times 10^{-3}) = 2.6$

The concentration of the anion, $[HC_6H_6O_6^-]$, is given almost entirely by the second ionization equilibrium: $HC_6H_6O_6^- \rightleftharpoons H^+ + C_6H_6O_6^{2-}$ for which the mass action expression is:

$$K_{a2} = \frac{[H^+][C_6H_6O_6^{2-}]}{[HC_6H_6O_6^-]} = 2.7 \times 10^{-12}$$

We have used the value for K_{a_2} from Table 18.3. Using the value of x from the first step above gives:

$$2.7 \times 10^{-12} = \frac{(2.6 \times 10^{-3})[C_6H_6O_6^{2-}]}{(2.6 \times 10^{-3})}$$

$$[HC_6H_6O_6^-] = 2.7 \times 10^{-12}$$

18.22 The equilibrium we are interested in for this problem is:

$$SO_3^{2-}(aq) + H_2O(l) \rightleftharpoons HSO_3^-(aq) + OH^-(aq)$$

$$K_b = K_w/K_{a_2} = 1.0 \times 10^{-14} / 6.6 \times 10^{-8} = 1.5 \times 10^{-7}$$

$$K_b = 1.5 \times 10^{-7} = \frac{[HSO_3^-][OH^-]}{[SO_3^{2-}]}$$

	$[SO_3^{2-}]$	$[HSO_3^-]$	$[OH^-]$
I	0.20	–	–
C	–x	+x	+x
E	0.20–x	+x	+x

Substituting these values into the mass action expression gives:

$$K_b = 1.5 \times 10^{-7} = \frac{(x)(x)}{0.20 - x}$$

Assume that x << 0.20 and solving gives $x = 1.7 \times 10^{-4}$.
$x = [OH^-] = 1.7 \times 10^{-4}$ M
$pOH = -\log(1.7 \times 10^{-4}) = 3.76$
$pH = 14.00 - pOH = 14.00 - 3.76 = 10.24$

18.23 For weak polyprotic acids, the concentration of the polyvalent ions is equal to the volume of K_{an} where n is the valency. By analogy, the concentration of H_2SO_3 in 0.010 M Na_2SO_3 will be equal to K_{b2} for SO_3^{2-}.

18.24 $HCHO_2 + H_2O \rightleftharpoons H_3O^+ + CHO_2^-$

$$Ka = \frac{[H_3O^+][CHO_2^-]}{[HCHO_2]} = 1.8 \times 10^{-4}$$

a)

	$[HCHO_2]$	$[H_3O^+]$	$[CHO_2^-]$
I	0.100	–	–
C	–x	+x	+x
E	0.100–x	+x	+x

Assume x << 0.100. Solving we get $x = [H_3O^+] = 4.2 \times 10^{-3}$. The pH is 2.37.

b) $[HCHO_2] = [CHO_2^-]$ so $[H_3O^+] = Ka = 1.8 \times 10^{-4}$ and the pH = 3.74.

c)

$$\text{\# moles base added} = (15.0 \text{ mL})\left(\frac{0.100 \text{ mol}}{1000 \text{ mL}}\right) = 1.50 \times 10^{-3}$$

$$\text{\# moles acid initially} = (20.0 \text{ mL})\left(\frac{0.10 \text{ mol}}{1000 \text{ mL}}\right) = 2.00 \times 10^{-3}$$

$$\text{excess acid} = 2.00 \times 10^{-3} - 1.50 \times 10^{-3} = 0.50 \times 10^{-3} \text{ moles acid}$$

$$[\text{acid}] = \frac{0.50 \times 10^{-3} \text{ moles}}{(35 \text{ mL})\left(\frac{1 \text{ L}}{1000 \text{ mL}}\right)} = 1.43 \times 10^{-2} \text{ M}$$

$$[\text{base}] = \frac{1.50 \times 10^{-3} \text{ moles}}{(35 \text{ mL})\left(\frac{1 \text{ L}}{1000 \text{ mL}}\right)} = 4.29 \times 10^{-2} \text{ M}$$

Substituting into the equilibrium expression and solving we get $[H_3O^+] = 6.00 \times 10^{-5}$ and the pH = 4.22.

d) We now have a solution of formate ion with a concentration of 0.0500 M. We need K_b for formate ion: $K_b = K_w/K_a = 5.6 \times 10^{-11}$. If we set up the equilibrium problem and solve we get: $[OH^-] = 1.7 \times 10^{-6}$. The pOH = 5.78 and the pH = 8.22.

18.25

$$\text{\# moles base added} = (30.0 \text{ mL})\left(\frac{0.15 \text{ mol}}{1000 \text{ mL}}\right) = 4.50 \times 10^{-3}$$

$$\text{\# moles acid initially} = (50.0 \text{ mL})\left(\frac{0.20 \text{ mol}}{1000 \text{ mL}}\right) = 1.00 \times 10^{-2}$$

$$\text{excess acid} = 1.00 \times 10^{-2} - 4.50 \times 10^{-3} = 5.50 \times 10^{-3} \text{ moles acid}$$

$$[\text{acid}] = \frac{5.50 \times 10^{-3}\ \text{moles}}{(80\ \text{mL})\left(\frac{1\ \text{L}}{1000\ \text{mL}}\right)} = 6.88 \times 10^{-2}\ \text{M}$$

$$[\text{base}] = \frac{4.50 \times 10^{-3}\ \text{moles}}{(80\ \text{mL})\left(\frac{1\ \text{L}}{1000\ \text{mL}}\right)} = 5.63 \times 10^{-2}\ \text{M}$$

$$HCHO_2 + H_2O \rightleftharpoons H_3O^+ + CHO_2^-$$

$$Ka = \frac{[H_3O^+][CHO_2^-]}{[HCHO_2]}$$

	$[HCHO_2]$	$[H_3O^+]$	$[CHO_2^-]$
I	6.88×10^{-2}	–	5.63×10^{-2}
C	–x	+x	+x
E	6.88×10^{-2}–x	+x	5.63×10^{-2}+x

Assume x << 5.63×10^{-2}

$$Ka = \frac{[x][5.63 \times 10^{-2}]}{[6.88 \times 10^{-2}]}$$

$$x = 2.20 \times 10^{-2} = [H_3O^+]$$

pH = 3.66

Thinking It Through

T18.1 We need to know the pH of the solution at equilibrium in order to determine K_a. From the pH we can get $[H^+]$ which is then used directly in the K_a expression.

T18.2 In this problem there is sufficient information to determine the pH of the solution at equilibrium. The initial base concentration is 0.10 M. At equilibrium the concentration of conjugate acid and hydroxide ion will be equal. Call this concentration x. Since K_b is so small, we can neglect any change in the initial concentration of the base. The $[OH^-]$ is, thus, the square root of K_b multiplied by 0.10 M, the initial base concentration. From this value we may calculate a pOH and, subsequently, a pH.

We should check to insure that the % dissociation is less than 5% but, due to the small size of K_b, this should be true.

T18.3 To answer the question we need to ask, what is the pH of a 0.10 M solution of a weak acid with a K_a of 2.5 × 10^{-6}? Start by looking at the equilibrium:

$$HA + H_2O \leftrightarrows H_3O^+ + A^- \qquad K_a = \frac{[H_3O^+][A^-]}{[HA]}$$

Set up the following equilibrium table:

	[HA]	$[H_3O^+]$	$[A^-]$
I	0.10	–	–
C	–x	+ x	+ x
E	0.10 – x	+ x	+ x

Substitute the equilibrium values into the mass action expression and solve for x:

$$K_a = \frac{(x)(x)}{0.10 - x} = 2.5 \times 10^{-6}$$

Assume that x << 0.10. The mass action expression simplifies to give:

$$K_a = \frac{(x)(x)}{0.10} = 2.5 \times 10^{-6}$$

Solving gives $x = 5.0 \times 10^{-4}$ M = $[H^+]$

pH = $-\log[H^+] = -\log(5.0 \times 10^{-4}) = 3.30$

The pH of the resulting solution is almost the same as that reported in the problem. It can safely be concluded that, within experimental uncertainty, the weak acid is the only solute present in the solution which can affect the acidity/basicity of the solution.

T18.4 To prepare a buffer of a known pH you need to know the pK_a (or the K_a) of the buffer system. Further, we need to determine the ratio of acid and conjugate base needed to establish the buffer at a specific pH. This ratio is easily determined using a rearranged mass action expression:

$$K_a = \frac{[H_3O^+][A^-]}{[HA]}$$

$$\frac{[A^-]}{[HA]} = \frac{K_a}{[H_3O^+]}$$

Finally, we need to have some idea as to the buffer capacity which is desired. This will enable us to establish the concentration of either the weak acid or its conjugate base. Once this is accomplished, the concentration of the remaining component of the buffer may be determined.

T18.5 As in any other equilibrium calculation, a table listing the initial concentrations of all components of the mixture should be constructed. The change that occurs in establishing equilibrium may be determined from the balanced equation and the equilibrium concentrations may be determined by adding the initial concentration to the change that is expected.
For this situation:

	$[HC_2H_3O_2]$	$[H^+]$	$[C_2H_3O_2^-]$
I	0.10	0.0010	–
C	–x	+x	+x
E	0.10 – x	0.0010 + x	+x

The equilibrium amounts may be substituted into the mass action expression. Our normal approximation will probably not hold for this problem. While the amount of acetic acid which dissociates, x, may be much smaller than 0.10, the initial concentration of acetic acid, it will probably not be much smaller than the initial concentration of H^+. Consequently, we will need to solve a quadratic equation to determine x. The value we calculate for x is the equilibrium concentration of acetate ion as is indicated in the equilibrium table.

T18.6 The solute is a weak acid. If it were a strong acid, the pH would be calculated by taking the negative log of the concentration of the acid, assuming the acid is monoprotic. Using the concentration listed, a pH of 1.46 would be expected. The actual pH is higher than this so the solute must be a weak acid.

T18.7 Yes, this solution would function as a buffer. The acetic acid will serve to resist changes in pH due to the addition of base and the salt of propionic acid will resist changes in pH due to the addition of acid.

$$HC_2H_3O_2(aq) + OH^-(aq) \rightarrow H_2O(l) + C_2H_3O_2^-(aq)$$
$$C_3H_5O_2^-(aq) + H^+(aq) \rightarrow HC_3H_5O_2(aq)$$

T18.8 The phosphate ion is the conjugate base of the HPO_4^{2-} ion. We know that $K_a \times K_b = K_w$. If we know K_a for HPO_4^{2-}, we may calculate K_b for PO_4^{3-}.

T18.9 The three sources of OH^- are the sulfite ion, the bisulfite ion and water.

$$SO_3^{2-}(aq) + H_2O(l) \leftrightarrows HSO_3^-(aq) + OH^-(aq)$$
$$HSO_3^-(aq) + H_2O(l) \leftrightarrows H_2SO_3(aq) + OH^-(aq)$$
$$2H_2O(l) \leftrightarrows H_3O^+(aq) + OH^-(aq)$$

T18.10 If we mix an equal number of moles of a weak acid with a strong base, all of the acid is neutralized forming the conjugate base. If half the amount of acid is neutralized, an equivalent amount of conjugate base is produced and there remains an amount of acid equal to the amount of base. In this situation, the equilibrium expression may be rearranged to show that $pK_a = pH$. Consequently, measuring the pH when half the acid is neutralized gives a direct measure of the pK_a.

T18.11 In this mixture, we have neutralized exactly half of the base initially present. Consequently, there are equal amounts of weak base and its conjugate acid present. In this situation, the pOH will equal the pK_b and we can easily calculate the pH.

T18.12 (a) If K_w increases, the concentrations of H^+ and OH^- will also increase. Consequently, the pH will decrease as the temperature increases.

(b) The water will remain neutral. In part (a) it was stated that the concentrations of both H^+ and OH^- increase as the temperature increases. As the concentrations increase, they remain the same. And if a neutral solution is one in which $[H^+] = [OH^-]$, the solution must remain neutral.

T18.13 In order to determine the freezing point, we need to determine the percent dissociation of this weak acid. Using standard equilibrium methods, we can determine the amount of weak acid, conjugate base and hydrogen ion present in an equilibrium mixture. Adding these together gives the total concentration of solute particles and we can then use the freezing point depression expression to calculate the freezing point of the mixture.

T18.14 Using the ideal gas law you may calculate the number of moles of NH_3 present in the mixture. The number of moles of HCl is easily calculated from the information provided. The HCl will neutralize the NH_3 producing NH_4^+. When the limiting reactant, HCl, is completely consumed, use the concentrations of NH_3 and NH_4^+ that remain as the initial concentrations of a normal equilibrium buffer problem.

T18.15 To solve this problem, start by setting-up the normal equilibrium calculation. We know the concentration of acetate ion desired upon completion of the reaction. The complicating factor in this problem, however, is that addition of a volume of acid will cause a reaction and dilute the sample. It turns out that the dilution is not a significant problem and may be neglected completely since all components of the mixture will be affected identically. Consequently, we need to solve a problem that asks what initial amount of H^+ is needed to insure a concentration of acetate ion equal to 1.0×10^{-5} M at equilibrium. This is easily solved using normal equilibrium techniques.

T18.16 Carbonate ion acts as a base in water solutions. In order to calculate the $[H_2CO_3]$ we need to consider two reactions; the first produces HCO_3^- and the second produces H_2CO_3.

In a previous example (example 16.19 in the text), we calculated the $[HCO_3^-]$ in a solution of Na_2CO_3. In that example it was stated that in a solution that contains a polyfunctional base **as the only solute**, the concentration of the ion formed in the first step of the ionization equals K_{b_1}.

Using the same logic we see that in a solution that contains a polyfunctional base **as the only solute**, the concentration of the diprotic acid formed in the second step of the hydrolysis equals K_{b_2}.

So, $[H_2CO_3]$ equals K_{b_2} for CO_3^{2-}.

T18.17 The $[H_3O^+]$ may be determined from K_{a_1} for H_3PO_4

$$H_3PO_4(aq) + H_2O(l) \rightleftarrows H_3O^+(aq) + H_2PO_4^-(aq)$$

$$K_{a_1} = \frac{[H_3O^+][H_2PO_4^-]}{[H_3PO_4]} = 7.1 \times 10^{-3}$$

Solving the resulting quadratic equation gives $[H_3O^+] = 0.023$ M. All of the hydrogen ion results from the first dissociation since the other dissociation constants are so small.

In order to determine the $[PO_4^{3-}]$ we need to look at the third dissociation expression:

$$HPO_4^{2-}(aq) + H_2O(l) \rightleftarrows H_3O^+(aq) + PO_4^{3-}(aq)$$

$$K_{a_3} = \frac{[H_3O^+][PO_4^{3-}]}{[HPO_4^{2-}]} = 4.5 \times 10^{-13}$$

The $[H_3O^+] = 0.023$ M, the result of the first calculation. The $[HPO_4^{2-}] = K_{a_2}$ since it is such a small value. Since $K_{a_2} \neq [H_3O^+]$, the $[PO_4^{3-}] \neq K_{a_3}$.

T18.18 We need to consider two dissociations
$H_2SO_4 + H_2O \rightarrow H_3O^+ + HSO_4^-$
$HSO_4^- + H_2O \rightleftarrows H_3O^+ + SO_4^{2-}$

While the first dissociation is nearly 100%, the second requires consideration normally used for weak acids. Because of the presence of H_3O^+ from the first dissociation, then the second dissociation will be reduced. To solve this problem, approach it as though it were a buffer solution with a known concentration of H_3O^+ and HSO_4^-.

T18.19 For each OH^- ion added an additional F^- will form and an HF molecule will dissociate:
$HF + H_2O \rightleftarrows H_3O^+ + F^-$
As OH^- is added it reacts with the H_3O^+ causing additional HF to dissociate. For each H^+ ion added an F^- ion will react with it forming additional HF. Using the equation above, the added H^+ reacts with F^- producing HF.

T18.20 The pair of electrons that are donated are on the nitrogen atom and not the carbon atom.

Review Questions

18.1 $HA + H_2O \rightleftarrows H_3O^+ + A^-$

$$K_a = \frac{[H_3O^+][A^-]}{[HA]}$$

18.2
(a) $HNO_2 \rightleftarrows H^+ + NO_2^-$
(b) $H_3PO_4 \rightleftarrows H^+ + H_2PO_4^-$
(c) $HAsO_4^{2-} \rightleftarrows H^+ + AsO_4^{3-}$
(d) $(CH_3)_3NH^+ \rightleftarrows H^+ + (CH_3)_3N$

18.3 (a) $K_a = \frac{[H^+][NO_2^-]}{[HNO_2]}$ (b) $K_a = \frac{[H^+][H_2PO_4^-]}{[H_3PO_4]}$

(c) $K_a = \dfrac{[H^+][AsO_4^{3-}]}{[HAsO_4^{2-}]}$ (d) $K_a = \dfrac{[H^+][(CH_3)_3N]}{[(CH_3)_3NH^+]}$

18.4 $B + H_2O \leftrightharpoons HB^+ + OH^-$

$$K_b = \frac{[HB^+][OH^-]}{[B]}$$

18.5 (a) $(CH_3)_3N + H_2O \leftrightharpoons (CH_3)_3NH^+ + OH^-$
(b) $AsO_4^{3-} + H_2O \leftrightharpoons HAsO_4^{2-} + OH^-$
(c) $NO_2^- + H_2O \leftrightharpoons HNO_2 + OH^-$
(d) $(CH_3)_2N_2H_2 + H_2O \leftrightharpoons (CH_3)_2N_2H_3^+ + OH^-$

18.6 (a) $K_b = \dfrac{[(CH_3)_3NH^+][OH^-]}{[(CH_3)_3N]}$ (b) $K_b = \dfrac{[HAsO_4^{2-}][OH^-]}{[AsO_4^{3-}]}$

(c) $K_a = \dfrac{[HNO_2][OH^-]}{[NO_2^-]}$ (d) $K_b = \dfrac{[(CH_3)_2N_2H_3^+][OH^-]}{[(CH_3)_2N_2H_2]}$

18.7 $C_6H_5CO_2H \leftrightharpoons H^+ + C_6H_5CO_2^-$

$$K_a = \frac{[H^+][C_6H_5CO_2^-]}{[C_6H_5CO_2H]}$$

18.8 $C_6H_5CO_2^- + H_2O \leftrightharpoons C_6H_5CO_2H + OH^-$

$$K_b = \frac{[C_6H_5CO_2H][OH^-]}{[C_6H_5CO_2^-]}$$

18.9 In general, there is an inverse relationship between the strength of an acid and its conjugate base; the stronger the acid, the weaker its conjugate base. Since HCN has the larger value for pK_a, we know that it is a weaker acid than HF. Accordingly, CN^- is a stronger base than F^-.

18.10 (a) (b)

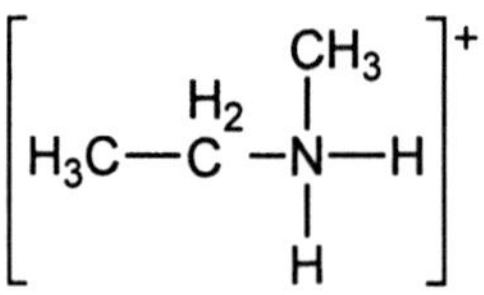

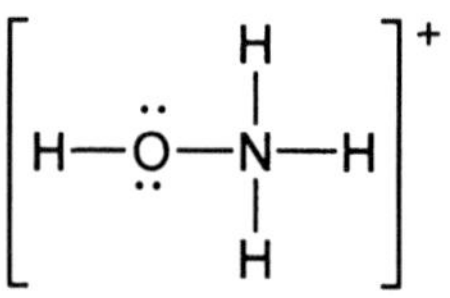

```
(a)
            CH3   ]+
       H2   |
[ H3C—C —N—H ]
            |
            H

(b)
            H    ]+
       ..   |
[ H—O—N—H ]
       ..   |
            H

(c)
[ (pyridine ring)N+—N ]+
```

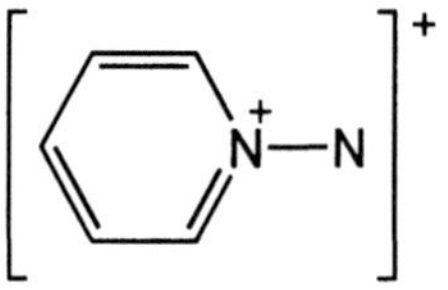

18.11 (a)

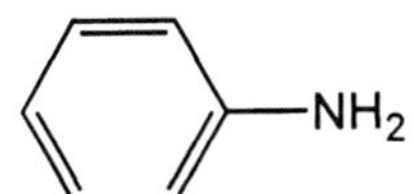

(b)

```
    H   H
    |   |
H — N — N — H
    ..  ..
```

(c)

```
       CH3
       |
H3C — N
       |
       CH3
```

18.12 Percentage ionization is the number of moles ionized per liter of solution divided by the moles available per liter of solution times 100.

$$\text{Percentage ionization} = \frac{\text{moles ionized per liter}}{\text{moles available per liter}} \times 100\%$$

18.13 First write out the chemical equation:

$$HA \leftrightarrows H^+ + A^-$$

$$\text{Percentage ionization} = \frac{\text{moles ionized per liter}}{\text{moles available per liter}} \times 100\%$$

$$1.2\% = = \frac{x\ M}{0.22\ M} \times 100\% \qquad x = 0.00264\ M\ H^+$$

18.14 The acetylsalicylate ion is the anion of the salt of a weak acid and is therefore a weak base. The solution of the sodium salt will be basic since the sodium ion is neutral.

18.15 The oxalate ion hydrolyzes in water, to give a basic solution, according to the equation:

$$C_2O_4^{2-} + H_2O \leftrightarrows HC_2O_4^- + OH^-$$

18.16 Recall that the salt of a weak acid is basic, the cation of a salt of a weak base is acidic and the salt of a strong acid or base is neutral.

(a) acidic: $(NH_4)_2SO_4$: NH_4^+ is the salt of NH_3, a weak base.

(b) basic: KF: F^- is the anion of a salt of HF, a weak acid.
KCN: CN^- is the anion of a salt of HCN, a weak acid.
$KC_2H_3O_2$: $C_2H_3O_2^-$ is the anion of a salt of $HC_2H_3O_2$, a weak acid.

(c) neutral: NaI: This is a salt of a strong acid, HI, and a strong base, NaOH.
$CsNO_3$: Metal ions with small charges are neutral and NO_3^- is the anion of a salt of HNO_3, a strong acid.
KBr: This is the salt of a strong acid, HBr, and a strong base, KOH.

18.17 The litmus paper will turn red indicating the presence of an acid. In section 17.4 it was stated that small, highly charged cations are acidic in water because the water molecules that surround the metal ion easily lose H^+ ions. Al^{3+} was a specific example.

18.18 The ratio of charge to size in Be^{2+} is greater than in Ca^{2+}. Consequently, Be^{2+} may become hydrated like other small, highly charged metals. These hydrates behave as weak acids.

18.19 Since ammonium ion is the cation of a salt of a weak base, it is acidic. The anion in the fertilizer, NO_3^-, is the anion of a salt of a strong acid and is, therefore, neutral. Consequently, adding ammonium nitrate should lower the pH of the soil.

18.20 K_b for hydrazine is larger than K_a for acetic acid since the solution is acidic. The hydrazine reacts with water to form a weak acid. Since the solution is acidic, the more of the hydrazine has dissociated than the acetate ion has reacted, this occurs if K_b is larger than K_a.

18.21 $Mg(OH)_2 + 2NH_4^+ \rightarrow Mg^{2+} + 2NH_3 + 2H_2O$

18.22 If $[HA]_{initial} < 400 \times K_a$ or if the % ionization ≥ 5%, the initial concentration of the acid is not equal to the equilibrium concentration. The same argument holds true for bases.

18.23 Simplification works when $[HA]_{initial} \geq 400 \times K_a$

(a) $400 \times (1.8 \times 10^{-5}) = 0.0072$
$0.0072\text{ M} < 0.020\text{ M}$ Use initial concentration

(b) $400 \times (4.4 \times 10^{-4}) = 0.176$
$0.176\text{ M} > 0.10\text{ M}$ May not use initial concentration

(c) $400 \times (1.7 \times 10^{-6}) = 6.8 \times 10^{-4}$
$6.8 \times 10^{-4}\text{ M} < 0.002\text{ M}$ Use initial concentration

(d) $400 \times (1.8 \times 10^{-4}) = 7.2 \times 10^{-2}$
$7.2 \times 10^{-2}\text{ M} > 0.050\text{ M}$ May not use initial concentration

18.24 In the method of successive approximations we assume that the unknown quantity is much less than the known quantity and solve for the unknown. The result is the first approximation. The determined value is substituted into the original equation in any location where an addition or subtraction occurs. Again, solve for the unknown quantity. This is the second approximation. Compare the values determined in the first approximation and the second approximation. If they are the same the problem is finished. If they are different, repeat the process using the newly determined value. Repeat until successive approximations are equivalent.

18.25 (a) $H_2CO_3 + OH^- \rightleftarrows H_2O + HCO_3^-$
$H^+ + HCO_3^- \rightleftarrows H_2CO_3$

(b) $H_2PO_4^- + OH^- \rightleftarrows H_2O + HPO_4^{2-}$
$H^+ + HPO_4^{2-} \rightleftarrows H_2PO_4^-$

(c) $NH_4^+ + OH^- \rightleftarrows H_2O + NH_3$
$H^+ + NH_3 \rightleftarrows NH_4^+$

18.26 Buffer 1 is made with an excess of NH_3 and will, therefore, be better able to resist changes in pH upon addition of acid. Buffer 2 is made with an excess of NH_4^+ and will, therefore, be better able to resist changes in pH upon addition of base. Consequently, buffer 1 would hold the pH steadier upon addition of strong acid.

18.27 $HCO_3^- + H^+ \rightleftarrows H_2CO_3$
$HCO_3^- + OH^- \rightleftarrows CO_3^{2-} + H_2O$

18.28 (a) $H_3AsO_4 \rightleftarrows H_2AsO_4^- + H^+$
$H_2AsO_4^- \rightleftarrows HAsO_4^{2-} + H^+$
$HAsO_4^{2-} \rightleftarrows AsO_4^{3-} + H^+$

(b) $K_1 = 10^{-2.2} = 6.3 \times 10^{-3}$
$K_2 = 10^{-6.9} = 1.3 \times 10^{-7}$
$K_3 = 10^{-11.5} = 3.2 \times 10^{-12}$

(c) $HAsO_4^{2-}$ is the stronger acid because it has the larger K_a (See Table 18.3).

18.29 $H_2SO_3 \leftrightarrows HSO_3^- + H^+$ $HSO_3^- \leftrightarrows SO_3^{2-} + H^+$

$$K_{a_1} = \frac{[HSO_3^-][H^+]}{[H_2SO_3]} \qquad K_{a_2} = \frac{[SO_3^{2-}][H^+]}{[HSO_3^-]}$$

18.30 $H_3C_6H_5O_7 + H_2O \leftrightarrows H_3O^+ + H_2C_6H_5O_7^-$

$$K_{a_1} = \frac{[H_3O^+][H_2C_6H_5O_7^-]}{[H_3C_6H_5O_7]}$$

$H_2C_6H_5O_7^- + H_2O \leftrightarrows H_3O^+ + HC_6H_5O_7^{2-}$

$$K_{a_2} = \frac{[H_3O^+][HC_6H_5O_7^{2-}]}{[H_2C_6H_5O_7^-]}$$

$HC_6H_5O_7^{2-} + H_2O \leftrightarrows H_3O^+ + C_6H_5O_7^{3-}$

$$K_{a_3} = \frac{[H_3O^+][C_6H_5O_7^{3-}]}{[HC_6H_5O_7^{2-}]}$$

18.31 $H_2O \leftrightarrows H^+ + OH^-$ $K_{a_1} = [H^+][OH^-] = K_w$ They are equal.

$OH^- \leftrightarrows H^+ + O^{2-}$ $K_{a_2} = \dfrac{[H^+][O^{2-}]}{[OH^-]}$

18.32 Nearly all of the H^+ in solution comes from the first ionization. The concentration of the conjugate base of a weak polyprotic acid once deprotonated is approximately equal to the value of the equilibrium constant. These approximations are usually valid because the value of the first ionization constant is much larger than the second ionization constant. If the two ionization constants are not very different, the approximation stated above will fail.

18.33 (a) $SO_3^{2-} + H_2O \leftrightarrows HSO_3^- + OH^-$
$HSO_3^- + H_2O \leftrightarrows H_2SO_3 + OH^-$
(b) $PO_4^{3-} + H_2O \leftrightarrows HPO_4^{2-} + OH^-$
$HPO_4^{2-} + H_2O \leftrightarrows H_2PO_4^- + OH^-$
$H_2PO_4^- + H_2O \leftrightarrows H_3PO_4 + OH^-$
(c) $C_4H_4O_6^{2-} + H_2O \leftrightarrows HC_4H_4O_6^- + OH^-$
$HC_4H_4O_6^- + H_2O \leftrightarrows H_2C_4H_4O_6 + OH^-$

18.34 Simplifying assumptions are essentially the same for polyprotic acids (see 18.32 above). The assumptions are valid because the first dissociation constant is so much larger than the second.

18.35 At the equivalence point, the titrated formic acid solution will be basic.

18.36 At the equivalence point, the titrated hydrazine solution will be acidic.

18.37 The equivalence point is when the moles of acid exactly equal the moles of base. The end point is a measured value that approximates the equivalence point.

18.38 An acid-base indicator is a weak acid that changes color when it is converted from its acidic form to its basic form. Because indicators are weak acids, they will react with the titrant. Consequently, small amounts of indicator are used so the results will not be impaired excessively by the indicator.

18.39 This is an example of a titration of a weak base using a strong acid. At the equivalence point, all of the NH_3 will be converted to NH_4^+. This cation is a weak acid and solutions of NH_4^+ will have a $pH < 7$. Ethyl red changes color at a $pH < 7$ while phenolphthalein changes color at a $pH > 7$. Ethyl red is the better indicator since its color change is in the pH region we expect for the equivalence point.

18.40 Both potassium hydroxide and hydrobromic acid are strongly dissociating. Consequently, the pH at the equivalence point is 7. Any indicator that changes color around pH 7 would be appropriate. Phenolphthalein would also be a good indicator since the pH changes so rapidly in titrations of a strong acid with a strong base. Only a small volume of added acid, frequently less than one drop, will cause a large change in pH. In addition, since the color change for phenolphthalein is from a colorless solution to a pink solution, it is easily noticed when the color change occurs.

18.41 At the equivalence point, all of the acetic acid will have been converted to acetate ions. Since this would be a solution of a weak base having a $pH > 7$, it would be argued that the use of methyl orange would be inappropriate since the color change occurs in the range $pH = 3.2$ to $pH = 4.4$.

Review Problems

18.42 At 25 °C, $K_a \times K_b = K_w$

$K_b = K_w/K_a = 1.0 \times 10^{-14} \div 6.8 \times 10^{-4} = 1.5 \times 10^{-11}$

18.43 At 25 °C, $K_a \times K_b = K_w$

$K_a = K_w/K_b = 1.0 \times 10^{-14} \div 1.0 \times 10^{-10} = 1.0 \times 10^{-4}$

18.44 At 25 °C, $K_a \times K_b = K_w$

$K_b = K_w/K_a = 1.0 \times 10^{-14} \div 1.8 \times 10^{-12} = 5.6 \times 10^{-3}$

18.45 $K_a \times K_b = K_w$

$K_a = K_w/K_b = 1.0 \times 10^{-14} \div 4.4 \times 10^{-4} = 2.3 \times 10^{-11}$

18.46 At 25 °C, $K_a \times K_b = K_w$

$K_b = K_w/K_a = 1.0 \times 10^{-14} \div 1.4 \times 10^{-4} = 7.1 \times 10^{-11}$

18.47 (a) The conjugate base is IO_3^-

$pK_b = 14.00 - pK_a = 14.00 - 0.23 = 13.77$

$K_b = 10^{-pK_b} = 10^{-13.77} = 1.7 \times 10^{-14}$

(b) IO_3^- is a weaker base than an acetate anion, because its K_b value is smaller than that of an acetate anion.

18.48 pH = 3.22 $[H^+] = 10^{-3.22} = 6.03 \times 10^{-4}$ M

$$\text{Percentage ionization} = \frac{\text{moles ionized per liter}}{\text{moles available per liter}} \times 100\%$$

$$\text{Percentage ionization} = \frac{6.03 \times 10^{-4}\ \text{M}}{0.20\ \text{M}} \times 100\% = 0.30\%$$

18.49 Percentage ionization = $\dfrac{\text{moles ionized per liter}}{\text{moles available per liter}} \times 100\%$

Let x = moles of H^+ in solution

$0.030\% = \dfrac{x\ M}{0.030\ M} \times 100\%$ $\qquad x = 9.0 \times 10^{-6}$ M

pH = -log[H^+] = -log[9.0×10^{-6}] = 5.05

18.50 $HA \leftrightarrows H^+ + A^-$

	[HA]	[H^+]	[A^-]
I	0.20	–	–
C	-6.03×10^{-4}	$+6.03 \times 10^{-4}$	$+6.03 \times 10^{-4}$
E	0.20	6.03×10^{-4}	6.03×10^{-4}

$$K_a = \frac{[H^+][A^-]}{[HA]} = \frac{[6.03\times10^{-4}][6.03\times10^{-4}]}{[0.20]} = 1.82 \times 10^{-6}$$

18.51 $HA \leftrightarrows H^+ + A^-$

	[HA]	[H^+]	[A^-]
I	0.030	–	–
C	-9.0×10^{-6}	$+9.0 \times 10^{-6}$	$+9.0 \times 10^{-6}$
E	0.030	9.0×10^{-6}	9.0×10^{-6}

$$K_a = \frac{[H^+][A^-]}{[HA]} = \frac{[9.0\times10^{-6}][9.0\times10^{-6}]}{[0.030]} = 2.7 \times 10^{-9}$$

18.52 $HIO_4 \leftrightarrows H^+ + IO_4^-$

$$K_a = \frac{[H^+][IO_4^-]}{[HIO_4]}$$

	[HIO_4]	[H^+]	[IO_4^-]
I	0.10	–	–
C	–x	+x	+x
E	0.10–x	+x	+x

We know that at equilibrium [H^+] = 0.038 M = x. The equilibrium concentrations of the other components of the mixture are:

[HIO_4] = 0.10 – x = 0.06 M and [IO_4^-] = x = 0.038 M.

Substituting the above values for equilibrium concentrations into the mass action expression gives:

$$K_a = \frac{(0.038)(0.038)}{0.06} = 2 \times 10^{-2}$$

$pK_a = -\log(K_a) = -\log(2 \times 10^{-2}) = 1.7$

18.53 $HC_2H_2ClO_2 \rightleftharpoons H^+ + C_2H_2ClO_2^-$

$$K_a = \frac{[H^+][C_2H_2ClO_2^-]}{[HC_2H_2ClO_2]}$$

	$[HC_2H_2ClO_2]$	$[H^+]$	$[C_2H_2ClO_2^-]$
I	0.10	–	–
C	–x	+x	+x
E	0.10–x	+x	+x

pH = 1.96 $\quad [H^+] = 10^{-1.96} = 0.011$ M = x

The equilibrium concentrations of the other components of the mixture are:

$[HC_2H_2ClO_2] = 0.10 - x = 0.089$ M and $[C_2H_2ClO_2^-] = x = 0.011$ M.

Substituting the above values for equilibrium concentrations into the mass action expression gives:

$$K_a = \frac{(0.011)(0.011)}{0.089} = 1.4 \times 10^{-3}$$

$pK_a = -\log(K_a) = -\log(1.4 \times 10^{-3}) = 2.87$

18.54 pOH = 14.00 – pH = 14.00 – 11.86 = 2.14

$[OH^-] = 10^{-pOH} = 10^{-2.14} = 7.2 \times 10^{-3}$ M

$CH_3CH_2NH_2 + H_2O \leftrightarrows CH_3CH_2NH_3^+ + OH^-$

$$K_b = \frac{[CH_3CH_2NH_3^+][OH^-]}{[CH_3CH_2NH_2]}$$

	$[CH_3CH_2NH_2]$	$[CH_3CH_2NH_3^+]$	$[OH^-]$
I	0.10	–	–
C	–x	+x	+x
E	0.10–x	+x	+x

In the equilibrium analysis, the value of x is, therefore, equal to 7.2×10^{-3} M. Therefore, our equilibrium concentrations are $[CH_3CH_2NH_3^+] = [OH^-] = 7.2 \times 10^{-3}$ M, and $[CH_3CH_2NH_2] = 0.10$ M $- 7.2 \times 10^{-3}$ M = 0.09 M.

Substituting these values into the mass action expression gives:

$$K_b = \frac{(7.2 \times 10^{-3})(7.2 \times 10^{-3})}{0.09} = 6 \times 10^{-4}$$

$pK_b = -\log(K_b) = -\log(6 \times 10^{-4}) = 3.2$

$$\text{Percentage ionization} = \frac{\text{moles ionized per liter}}{\text{moles available per liter}} \times 100\%$$

$$\text{Percentage ionization} = \frac{7.2 \times 10^{-3}}{0.10} \times 100\% = 7.2\%$$

18.55 $HONH_2 + H_2O \leftrightarrows HONH_3^+ + OH^-$

$$K_b = \frac{[HONH_3^+][OH^-]}{[HONH_2]}$$

pOH = 14.00 – pH = 14.00 – 10.12 = 3.88

$[OH^-] = 10^{-3.88} = 1.3 \times 10^{-4}$ M

In the equilibrium analysis, the value of x is, therefore, equal to 1.3×10^{-4} M:

	$[HONH_2]$	$[HONH_3^+]$	$[OH^-]$
I	0.15	–	–
C	–x	+x	+x
E	0.15–x	+x	+x

Therefore, our equilibrium concentrations are $[HONH_3^+] = [OH^-] = 1.3 \times 10^{-4}$ M, and $[HONH_2] = 0.15$ M – 1.3×10^{-4} M = 0.15 M.

Substituting these values into the mass action expression gives:

$$K_b = \frac{(1.3 \times 10^{-4})(1.3 \times 10^{-4})}{0.15} = 1.1 \times 10^{-7}$$

$pK_b = -\log(K_b) = -\log(1.1 \times 10^{-7}) = 6.96$

$$\text{Percentage ionization} = \frac{\text{moles ionized per liter}}{\text{moles available per liter}} \times 100\%$$

$$\text{Percentage ionization} = \frac{1.3 \times 10^{-4}}{0.15} \times 100\% = 0.087\%$$

18.56 $HC_3H_5O_2 + H_2O \leftrightarrows H_3O^+ + C_3H_5O_2^-$

$$K_a = \frac{[H_3O^+][C_3H_5O_2^-]}{[HC_3H_5O_2]} = 1.4 \times 10^{-4}$$

	$[HC_3H_5O_2]$	$[H_3O^+]$	$[C_3H_5O_2^-]$
I	0.15	–	–
C	–x	+x	+x
E	0.150 – x	+x	+x

Assume x << 0.15

$$K_a = \frac{[x][x]}{[0.150]} = 1.4 \times 10^{-4} \quad x = 4.6 \times 10^{-3} = [H_3O^+]$$

pH = 2.34

$[HC_3H_5O_2] = 0.145$
$[H^+] = 4.5 \times 10^{-3}$
$[C_3H_5O_2^-] = 4.5 \times 10^{-3}$

18.57 $H_2O_2 \leftrightarrows H^+ + HO_2^-$

$$K_a = \frac{[H^+][HO_2^-]}{[H_2O_2]} = 1.8 \times 10^{-12}$$

	$[H_2O_2]$	$[H^+]$	$[HO_2^-]$
I	1.0	–	–
C	–x	+x	+x
E	1.0–x	+x	+x

Substituting these values into the mass action expression gives:

$$K_a = \frac{(x)(x)}{1.0 - x} = 1.8 \times 10^{-12}$$

Assuming x << 1.0 we determine that $x = 1.3 \times 10^{-6}$ M = $[H^+]$
pH = $-\log[H^+]$ = 5.89

18.58 $HN_3 + H_2O \leftrightarrows H_3O^+ + N_3^-$

$$K_a = \frac{[H_3O^+][N_3^-]}{[HN_3]} = 1.8 \times 10^{-5}$$

	$[HN_3]$	$[H_3O^+]$	$[N_3^-]$
I	0.15	–	–
C	–x	+x	+x
E	0.15 –x	+x	+x

Assume x << 0.15

$$K_a = \frac{[x][x]}{[0.15]} = 1.8 \times 10^{-5} \quad x = 1.6 \times 10^{-3} = [H_3O^+]$$

pH = 2.78

$$\text{Percentage ionization} = \frac{\text{moles ionized per liter}}{\text{moles available per liter}} \times 100\%$$

$$\text{Percentage ionization} = \frac{1.6 \times 10^{-3}}{0.15} \times 100\% = 1.1\%$$

18.59 $HC_6H_5O \leftrightarrows H^+ + C_6H_5O^-$

$$K_a = \frac{[H^+][C_6H_5O^-]}{[HC_6H_5O]} = 1.3 \times 10^{-10}$$

	$[HC_6H_5O]$	$[H^+]$	$[C_6H_5O^-]$
I	0.050	–	–
C	–x	+x	+x
E	0.050–x	+x	+x

Substituting these values into the mass action expression gives:

$$K_a = \frac{(x)(x)}{0.050 - x} = 1.3 \times 10^{-10}$$

Assuming x << 0.050 we determine that $x = 2.5 \times 10^{-6}$ M = $[H^+] = [C_6H_5O^-]$, $[HC_6H_5O]$ = 0.050 – x = 0.050 M. pH = $-\log[H^+]$ = 5.59.

$$\% \text{ ionization} = \frac{\text{amount ionized}}{\text{amount available}} \times 100\%$$

$$= \frac{2.5 \times 10^{-6}\ \text{M}}{0.050\ \text{M}} \times 100\%$$

$$= 5.0 \times 10^{-3}\ \%$$

18.60 $K_b = 10^{-pKb} = 10^{-5.79} = 1.6 \times 10^{-6}$
$Cod + H_2O \leftrightarrows HCod^+ + OH^-$

$$K_b = \frac{[HCod^+][OH^-]}{[Cod]} = 1.6 \times 10^{-6}$$

	[Cod]	[HCod$^+$]	[OH$^-$]
I	0.020	–	–
C	–x	+x	+x
E	0.020–x	+x	+x

Substituting these values into the mass action expression gives:

$$K_b = \frac{(x)(x)}{0.020 - x} = 1.6 \times 10^{-6}$$

If we assume that x << 0.020 we get; $x^2 = 3.2 \times 10^{-8}$,
$x = 1.8 \times 10^{-4}$ M = [OH$^-$]
pOH = –log[OH$^-$] = –log(1.8×10^{-4}) = 3.74
pH = 14.00 – pOH = 14.00 – 3.74 = 10.26

18.61 $C_6H_5N + H_2O \leftrightarrows HC_6H_5N^+ + OH^-$

$$K_b = \frac{[HC_6H_5N^+][OH^-]}{[C_6H_5N]} = 1.7 \times 10^{-9}$$

	[C_6H_5N]	[$HC_6H_5N^+$]	[OH$^-$]
I	0.20	–	–
C	–x	+x	+x
E	0.20–x	+x	+x

Assume x << 0.20

$$K_b = \frac{(x)(x)}{0.20} = 1.7 \times 10^{-9} \qquad x = 1.8 \times 10^{-5} = [OH^-]$$

pOH = 4.73 pH = 9.27

18.62 [H$^+$] = $10^{-pH} = 10^{-2.54} = 2.9 \times 10^{-3}$ M
$HC_2H_3O_2 \leftrightarrows H^+ + C_2H_3O_2^-$

$$K_a = \frac{[H^+][C_2H_3O_2^-]}{[HC_2H_3O_2]} = 1.8 \times 10^{-5}$$

	[$HC_2H_3O_2$]	[H$^+$]	[$C_2H_3O_2^-$]
I	Z	–	–
C	–x	+x	+x
E	Z–x	+x	+x

Substituting these values into the mass action expression gives:

$$K_a = \frac{(x)(x)}{Z - x} = 1.8 \times 10^{-5}$$

Assuming x << Z and knowing that $x = 2.9 \times 10^{-3}$ M, we can solve for Z and find Z = 0.47. The initial concentration of $HC_2H_3O_2$ is 0.47 M.

18.63 Let Z = initial concentration of NH_3.
$NH_3 + H_2O \leftrightarrows NH_4^+ + OH^-$

$$K_b = \frac{[NH_4^+][OH^-]}{[NH_3]} = 1.8 \times 10^{-5}$$

	$[NH_3]$	$[NH_4^+]$	$[OH^-]$
I	Z	–	–
C	–x	+x	+x
E	Z–x	+x	+x

Assume x << Z

$$K_b = \frac{(x)(x)}{Z} = 1.8 \times 10^{-5} \quad Z = \frac{(x)(x)}{1.8 \times 10^{-5}}$$

$x = [OH^-] = 10^{-pOH} = 1.7 \times 10^{-3}$

$Z = 0.15$ M

$$\#\text{ moles } NH_3 = (500\text{ mL})\left(\frac{0.15\text{ mol}}{1000\text{ mL}}\right) = 0.075\text{ moles}$$

18.64 NaCN will be basic in solution since CN^- is a basic ion and Na^+ is a neutral ion.

$CN^- + H_2O \leftrightarrows HCN + OH^-$

For HCN, $K_a = 6.2 \times 10^{-10}$, we need K_b for CN^-;

$K_b = K_w/K_a = (1.0 \times 10^{-14}) \div (6.2 \times 10^{-10}) = 1.6 \times 10^{-5}$

$$K_b = \frac{[HCN][OH^-]}{[CN^-]} = 1.6 \times 10^{-5}$$

	$[CN^-]$	$[HCN]$	$[OH^-]$
I	0.20	–	–
C	–x	+x	+x
E	0.20–x	+x	+x

Substituting these values into the mass action expression gives:

$$K_b = \frac{(x)(x)}{0.20 - x} = 1.6 \times 10^{-5}$$

Assuming that x << 0.20 we can solve for x an determine;

$x = 1.8 \times 10^{-3}$ M $= [OH^-]$

$pOH = -\log[OH^-] = -\log(1.8 \times 10^{-3}) = 2.74$

$pH = 14.00 - pOH = 14.00 - 2.74 = 11.26$

Concentration of HCN is equal to that of hydroxide ion: 1.8×10^{-3} M

18.65 KNO_2 will be basic in solution since NO_2^- is a basic ion and K^+ is neutral.

$NO_2^- + H_2O \leftrightarrows HNO_2 + OH^-$

For HNO_2, $K_a = 7.1 \times 10^{-4}$, we need K_b for NO_2^-;

$K_b = K_w/K_a = (1.0 \times 10^{-14}) \div (7.1 \times 10^{-4}) = 1.4 \times 10^{-11}$

$$K_b = \frac{[HNO_2][OH^-]}{[NO_2^-]} = 1.4 \times 10^{-11}$$

	$[NO_2^-]$	$[HNO_2]$	$[OH^-]$
I	0.40	–	–
C	–x	+x	+x
E	0.40–x	+x	+x

Substituting these values into the mass action expression gives:

$$K_b = \frac{(x)(x)}{0.40 - x} = 1.4 \times 10^{-11}$$

Assuming that x << 0.40 we can solve for x and determine the equilibrium concentrations;
$x = 2.4 \times 10^{-6}$ M = $[OH^-]$ = $[HNO_2]$
pOH = $-\log[OH^-] = -\log(2.4 \times 10^{-6}) = 5.63$
pH = 14.00 – pOH = 14.00 – 5.63 = 8.37

18.66 A solution of CH_3NH_3Cl will be acidic since the Cl^- ion is neutral and the $CH_3NH_3^+$ ion is acidic.
$CH_3NH_3^+ \leftrightarrows H^+ + CH_3NH_2$
For CH_3NH_2, $K_b = 4.4 \times 10^{-4}$. We need K_a for $CH_3NH_3^+$;
$K_a = K_w/K_b = (1.0 \times 10^{-14}) \div (4.4 \times 10^{-4}) = 2.3 \times 10^{-11}$

$$K_a = \frac{[H^+][CH_3NH_2]}{[CH_3NH_3^+]} = 2.3 \times 10^{-11}$$

	$[CH_3NH_3^+]$	$[H^+]$	$[CH_3NH_2]$
I	0.15	–	–
C	–x	+x	+x
E	0.15–x	+x	+x

Substituting these values into the mass action expression gives:

$$K_a = \frac{(x)(x)}{0.15 - x} = 2.3 \times 10^{-11}$$

Assuming that x << 0.15 we can solve for x an determine;
$x = 1.9 \times 10^{-6}$ M = $[H_3O^+]$
pH = $-\log[H_3O^+] = -\log(1.9 \times 10^{-6}) = 5.72$

18.67 Hydrazinium chloride is the acidic salt of hydrazine N_2H_4.
$N_2H_5^+ + H_2O \leftrightarrows H_3O^+ + N_2H_4$

$$K_a = \frac{[H_3O^+][N_2H_4]}{[N_2H_5^+]}$$ K_b for hydrazine is 1.7×10^{-6} so $K_a = \frac{K_w}{K_b} = 5.9 \times 10^{-9}$

	$[N_2H_5^+]$	$[H_3O^+]$	$[N_2H_4]$
I	0.10	–	–
C	–x	+x	+x
E	0.10–x	+x	+x

Assume x << 0.10

$$K_a = \frac{(x)(x)}{0.10} = 5.9 \times 10^{-9} \qquad x = 2.4 \times 10^{-5} = [H_3O^+]$$

pH = 4.62

18.68 The reaction for this problem is:
$H\text{–}Mor^+ \leftrightarrows H^+ + Mor$
We know that $pK_a + pK_b = pK_w = 14.00$.
So, $pK_a = 14.00 - pK_b = 14.00 - 6.13 = 7.87$, and $K_a = 10^{-pKa} = 1.3 \times 10^{-8}$.

$$K_a = \frac{[H^+][Mor]}{[H-Mor^+]} = 1.3 \times 10^{-8}$$

	$[H\text{–}Mor^+]$	$[H^+]$	$[Mor]$
I	0.20	–	–
C	–x	+x	+x
E	0.20–x	+x	+x

Substituting these values into the mass action expression gives:

$$K_a = \frac{(x)(x)}{0.20 - x} = 1.3 \times 10^{-8}$$

Assuming that x << 0.20 we can solve for x an determine;
$x = 5.1 \times 10^{-5}$ M = $[H^+]$
pH = $-\log[H^+] = -\log(5.1 \times 10^{-5}) = 4.29$

18.69 The reaction for this problem is:
$H\text{–}Qu^+ \leftrightarrows H^+ + Qu$

$K_a = 10^{-pKa} = 3.0 \times 10^{-9}$.

$$K_a = \frac{[H^+][Qu]}{[H-Qu^+]} = 3.0 \times 10^{-9}$$

	$[H\text{–}Qu^+]$	$[H^+]$	$[Qu]$
I	0.15	–	–
C	–x	+x	+x
E	0.15–x	+x	+x

Substituting these values into the mass action expression gives:

$$K_a = \frac{(x)(x)}{0.15 - x} = 3.0 \times 10^{-9}$$

Assuming that x << 0.15 we can solve for x and determine the equilibrium concentrations;
$x = 2.1 \times 10^{-5}$ M = $[H^+]$
pH = $-\log[H^+] = -\log(2.1 \times 10^{-5}) = 4.67$

18.70 Let HNic symbolize the nicotinic acid
$Nic^- + H_2O \leftrightarrows HNic + OH^-$

$$K_b = \frac{[HNic][OH^-]}{[Nic^-]}$$

	$[Nic^-]$	$[HNic]$	$[OH^-]$
I	0.18	–	–
C	–x	+x	+x
E	0.18–x	+x	+x

Assume x << 0.18

$$K_b = \frac{(x)(x)}{0.18} \qquad \text{we know } x = [OH^-]$$

we know pH = 9.05 so pOH = 4.95

and $[OH^-] = 10^{-pOH} = 1.1 \times 10^{-5} = x$

$$\text{Now } K_b = \frac{(1.1 \times 10^{-5})^2}{0.18} = 7.0 \times 10^{-10}$$

$$K_a = \frac{1\times10^{-14}}{7.0\times10^{-10}} = 1.4 \times 10^{-5}$$

18.71 The reaction of the weak base in water is: $B + H_2O \leftrightarrows BH^+ + OH^-$
The salt will react as: $BH^+ \leftrightarrows H^+ + B$

$$K_a = \frac{[H^+][B]}{[BH^+]}$$

	$[BH^+]$	$[H^+]$	[B]
I	0.15	–	–
C	–x	+x	+x
E	0.15–x	+x	+x

At equilibrium, the pH = 4.28 and $[H^+] = 5.2 \times 10^{-5}$ M = x. Substituting these values into the mass action expression gives:

$$K_a = \frac{(x)(x)}{0.15 - x} = \frac{(5.2 \times 10^{-5})(5.2 \times 10^{-5})}{0.15 - (5.2 \times 10^{-5})} = 1.8 \times 10^{-8}$$

Recall that $K_a \times K_b = K_w$, so $K_b = K_w \div K_a = 1.0 \times 10^{-14} \div 1.8 \times 10^{-8} = 5.6 \times 10^{-7}$.

18.72 $C_5H_5NH^+$ is the conjugate acid of pyridine, C_5H_5N, which has a K_b listed in Table 18.2; $K_b = 1.7 \times 10^{-9}$. For a conjugate acid-base pair:

$$K_a \times K_b = 1.00 \times 10^{-14}$$
$$K_a \times 1.7 \times 10^{-9} = 1.00 \times 10^{-14}$$
$$K_a = 5.9 \times 10^{-6}$$

$$C_5H_5NH^+ + H_2O \rightleftharpoons H_3O^+ + C_5H_5N$$

$$K_a = \frac{[H_3O^+][C_5H_5N]}{[C_5H_5N_3H^+]} = 5.9 \times 10^{-6}$$

	$[C_5H_5NH^+]$	$[H_3O^+]$	$[C_5H_5N]$
I	0.10	–	–
C	–x	+x	+x
E	0.10 –x	+x	+x

Assume x << 0.10

$$K_a = \frac{[x][x]}{[0.10]} = 5.9 \times 10^{-6} \qquad x = 7.7 \times 10^{-4} = [C_5H_5N]$$

% reacting = $([C_5H_5N]/[\text{total acid}]) \times 100 = (7.7 \times 10^{-4}/0.10) \times 100 = 0.77$ %

18.73 The only dissociation that we need to consider is: $NH_4^+ \leftrightharpoons NH_3 + H^+$,
if pH = 5.16, then $[H^+] = 10^{-pH} = 10^{-5.16} = 6.9 \times 10^{-6}$ M
This is also the required value for $[NH_3]$, since they are formed in a one-to-one mole ratio.

$K_a = [H^+][NH_3]/[NH_4^+] = 5.6 \times 10^{-10}$ and we arrive at:
$5.6 \times 10^{-10} = (6.9 \times 10^{-6})(6.9 \times 10^{-6})/[NH_4^+]$
$[NH_4^+] = 8.5 \times 10^{-2}$ M
8.5×10^{-2} mol/L × 97.9 g/mol = 8.3 g NH_4Br are needed per liter.

18.74 $OCl^- + H_2O \leftrightharpoons HOCl + OH^-$

$$K_b = \frac{[HOCl][OH^-]}{[OCl^-]} = \frac{K_w}{K_a} = \frac{1.0 \times 10^{-14}}{3.0 \times 10^{-8}} = 3.3 \times 10^{-7}$$

$$[OCl^-] = \left(\frac{5.0 \text{ g NaOCl}}{100 \text{ g solution}}\right)\left(\frac{1 \text{ mol NaOCl}}{74.5 \text{ g NaOCl}}\right)\left(\frac{1.0 \text{ g}}{1 \text{ mL}}\right)\left(\frac{1000 \text{ mL}}{1 \text{ L}}\right)$$

$= 0.67$ M

	$[OCl^-]$	$[HOCl]$	$[OH^-]$
I	0.67	–	–
C	–x	+x	+x
E	0.67 –x	+x	+x

Assume that x << 0.67

$$K_b = \frac{(x)(x)}{0.67} = 3.3 \times 10^{-7} \qquad x = 4.7 \times 10^{-4} = [OH^-]$$

pOH = 3.33

pH = 10.67

18.75 The conjugate acid BH^+ is acidic due to the following equilibrium:
$BH^+ \leftrightharpoons H^+ + B$

The base Y^- has the following equilibrium:
$Y^- + H_2O \leftrightharpoons HY + OH^-$

If a solution of the salt BHY is basic, it can only mean that Y^- is a stronger base than B:
$BH^+ + Y^- \rightarrow HY + B$

This can happen only if HY is a weaker acid than BH^+. We conclude that pK_a for HY is greater than 5.

18.76 $HF \leftrightharpoons H^+ + F^-$

$$K_a = \frac{[H^+][F^-]}{[HF]} = 6.8 \times 10^{-4}$$

	$[HF]$	$[H^+]$	$[F^-]$
I	0.15	–	–
C	–x	+x	+x
E	0.15–x	+x	+x

Substituting the above values for equilibrium concentrations into the mass action expression and assuming that x << 0.15 gives:

$$K_a = \frac{(x)(x)}{0.15} = 6.8 \times 10^{-4}$$

$x^2 = 1.0 \times 10^{-4}$, $x = 1.0 \times 10^{-2}$ M = $[H^+]$
pH = –log$[H^+]$ = –log(0.010) = 2.00
% ionization = 0.010/0.15 × 100 = 6.7 %

Since the % ionization is > 5%, we cannot make the simplifying assumption. Consequently, we need to solve for x using a quadratic equation. The mass action expression is:

$$K_a = \frac{(x)(x)}{0.15 - x} = 6.8 \times 10^{-4}$$

We may rearrange this expression to obtain the following quadratic equation:

$$x^2 + 6.8 \times 10^{-4}x - 1.0 \times 10^{-4} = 0$$

where a = 1, b = 6.8×10^{-4}, and c = -1.0×10^{-4}. If we substitute these values into the quadratic equation (see Example 18.13) and take the positive root we get x = 0.0097 M.
$[H^+]$ = x = 0.0097 M.
pH = –log$[H^+]$ = –log(0.0097) = 2.01
% ionization = 0.0097/0.15 × 100 = 6.5 %

In this problem, the change in pH between assuming x to be small and not assuming x to be small is very slight.

18.77 $HC_2H_3O_2 \rightleftarrows H^+ + C_2H_3O_2^-$

$$K_a = \frac{[H^+][C_2H_3O_2^-]}{[HC_2H_3O_2]} = 1.8 \times 10^{-5}$$

	$[HC_2H_3O_2]$	$[H^+]$	$[C_2H_3O_2^-]$
I	0.0010	–	–
C	–x	+x	+x
E	0.0010–x	+x	+x

Substituting the above values for equilibrium concentrations into the mass action expression and assuming that x << 0.0010 gives:

$$K_a = \frac{(x)(x)}{0.0010} = 1.8 \times 10^{-5}$$

x = 1.3×10^{-4} M = $[H^+]$
pH = –log$[H^+]$ = –log(1.3×10^{-4}) = 3.87
% ionization = (1.3×10^{-4})/0.0010 × 100% = 13 %

Clearly, the % ionization > 5%, and our simplification is not justified.

So, without making any assumptions, we obtain the following mass action expression:

$$K_a = \frac{(x)(x)}{0.0010 - x} = 1.8 \times 10^{-5}$$

Rearranging we get: $x^2 + 1.8 \times 10^{-5}x - 1.8 \times 10^{-8} = 0$, where a =1, b = 1.8×10^{-5}, and c = -1.8×10^{-5}.
Substituting these values into the quadratic equation and taking the positive root gives x = 1.3×10^{-4} M = $[H^+]$
pH = –log$[H^+]$ = –log(1.3×10^{-4}) = 3.87
% ionization = (1.3×10^{-4})/0.0010 × 100% = 13 %

18.78 $CN^- + H_2O \leftrightarrows HCN + OH^-$

$$K_b = \frac{[HCN][OH^-]}{[CN^-]} = \frac{K_w}{K_a} = \frac{1.0 \times 10^{-14}}{6.2 \times 10^{-10}} = 1.6 \times 10^{-5}$$

	$[CN^-]$	[HCN]	$[OH^-]$
I	0.0050	–	–
C	–x	+x	+x
E	0.0050 –x	+x	+x

Assume that x << 0.0050

$$K_b = \frac{(x)(x)}{0.0050} = 1.6 \times 10^{-5} \qquad x = 2.8 \times 10^{-4} = [OH^-]$$

Wait!!!! 2.8×10^{-4} is not << 0.0050

So, use the method of successive approximations.

$$K_b = \frac{(x)(x)}{0.0050 - 0.00028} = 1.6 \times 10^{-5} \qquad x = 2.7 \times 10^{-4}$$

$$Kb = \frac{(x)(x)}{0.0050 - 0.00027} = 1.6 \times 10^{-5} \qquad x = 2.7 \times 10^{-4}$$

$x = 2.7 \times 10^{-4} = [OH^-]$

pOH = 3.56

pH = 10.44

18.79 $HA + H_2O \leftrightarrows H_3O^+ + A^-$

$$K_a = \frac{[H_3O^+][A^-]}{[HA]} = 1.36 \times 10^{-3}$$

	[HA]	$[H_3O^+]$	$[A^-]$
I	0.020	–	–
C	–x	+x	+x
E	0.020 –x	+x	+x

Substituting the above values for equilibrium concentrations into the mass action expression and assuming that x << 0.020 gives:

$$K_a = \frac{(x)(x)}{0.020} = 1.36 \times 10^{-3} \qquad x = 5.2 \times 10^{-3}$$ Wait!!!! 5.2×10^{-3} is not << 0.020

Use method of successive approximations

$$K_a = \frac{(x)(x)}{0.020 - 0.0052} = 1.36 \times 10^{-3} \qquad x = 4.5 \times 10^{-3}$$

$$K_a = \frac{(x)(x)}{0.020 - 0.0045} = 1.36 \times 10^{-3} \qquad x = 4.6 \times 10^{-3}$$

So, $x = 4.6 \times 10^{-3} = [H_3O^+]$

pH = 2.34

18.80 $K_a = 10^{-pKa} = 10^{-4.92} = 1.2 \times 10^{-5}$

H–Paba $\leftrightarrows$ H^+ + $Paba^-$

$$K_a = \frac{[H^+][Paba^-]}{[H-Paba]} = 1.2 \times 10^{-5}$$

	[H–Paba]	$[H^+]$	$[Paba^-]$
I	0.030	–	–
C	–x	+x	+x
E	0.030–x	+x	+x

Substituting the above values for equilibrium concentrations into the mass action expression and assuming that x << 0.030 gives:

$$K_a = \frac{[x][x]}{[0.030]} = 1.2 \times 10^{-5}$$

$x^2 = 3.6 \times 10^{-7}$, $x = 6.0 \times 10^{-4}$ M = $[H^+]$

pH = $-\log[H^+] = -\log(6.0 \times 10^{-4}) = 3.22$

18.81 $K_a = 10^{-pKa} = 10^{-4.01} = 9.8 \times 10^{-5}$

H–Bar $\leftrightarrows$ H^+ + Bar^-

$$K_a = \frac{[H^+][Bar^-]}{[H-Bar]} = 9.8 \times 10^{-5}$$

	[H–Bar]	$[H^+]$	$[Bar^-]$
I	0.020	–	–
C	–x	+x	+x
E	0.020–x	+x	+x

Substituting the above values for equilibrium concentrations into the mass action expression and assuming that x << 0.020 gives:

$K_a = \frac{(x)(x)}{0.020} = 9.8 \times 10^{-5}$ $\quad x = 1.4 \times 10^{-3}$ Wait!!!! 1.4×10^{-3} is not << 0.020

Use method of successive approximations

$K_a = \frac{(x)(x)}{0.020 - 0.0014} = 9.8 \times 10^{-5}$ $\quad x = 1.4 \times 10^{-3}$

So, $x = 1.4 \times 10^{-3} = [H_3O^+]$

pH = 2.85

18.82 $HC_2H_3O_2 \leftrightarrows H^+ + C_2H_3O_2^-$

$$K_a = \frac{[H^+][C_2H_3O_2^-]}{[HC_2H_3O_2]} = 1.8 \times 10^{-5}$$

	$[HC_2H_3O_2]$	$[H^+]$	$[C_2H_3O_2^-]$
I	0.15	–	0.25
C	–x	+x	+x
E	0.15–x	+x	0.25+x

Substituting these values into the mass action expression gives:

$$K_a = \frac{(x)(0.25+x)}{0.15-x} = 1.8 \times 10^{-5}$$

Assume that x << 0.15 M and x << 0.25 M, then;

$$x \times \left(\frac{0.25}{0.15}\right) \approx 1.8 \times 10^{-5}$$

$$x \approx \left(\frac{0.15}{0.25}\right) \times 1.8 \times 10^{-5}$$

$$x \approx 1.1 \times 10^{-5}\ M = [H^+]$$

pH = –log [H^+] = 4.97

18.83 $C_2H_3O_2^- + H_2O \leftrightarrows HC_2H_3O_2 + OH^-$

$$K_b = \frac{[HC_2H_3O_2][OH^-]}{[C_2H_3O_2^-]} = \frac{K_w}{K_a} = \frac{1.0 \times 10^{-14}}{1.8 \times 10^{-5}} = 5.6 \times 10^{-10}$$

	$[C_2H_3O_2^-]$	$[HC_2H_3O_2]$	$[OH^-]$
I	0.25	0.15	–
C	–x	+x	+x
E	0.25–x	0.15+x	+x

Substituting these values into the mass action expression gives:

$$K_b = \frac{(0.15+x)(x)}{0.25-x} = 5.6 \times 10^{-10}$$

Assume that x << 0.15 M and x << 0.25 M, then;

$$x \times \left(\frac{0.15}{0.25}\right) \approx 5.6 \times 10^{-10}$$

$$x \approx \left(\frac{0.25}{0.15}\right) \times 5.6 \times 10^{-10}$$

$$x \approx 9.3 \times 10^{-10}\ M = [OH^-]$$

pOH = –log [OH^-] = 9.03
pH = 14.00 – pOH = 4.97

The answer is identical no matter which direction we choose to solve it.

18.84 The equilibrium we will consider in this problem is: $NH_3 + H_2O \leftrightarrows NH_4^+ + OH^-$

$$K_b = \frac{[NH_4^+][OH^-]}{[NH_3]} = 1.8 \times 10^{-5}$$

$$= \frac{(0.45)[OH^-]}{0.25} = 1.8 \times 10^{-5}$$

$[OH^-] = 1.0 \times 10^{-5}$ M
$pOH = -\log[OH^-] = -\log(1.0 \times 10^{-5}) = 5.00$
$pH = 14.00 - pOH = 14.00 - 5.00 = 9.00$

18.85 Now we consider the equilibrium; $NH_4^+ \leftrightarrows H^+ + NH_3$
We know that $K_aK_b = K_w$, so $K_a = K_w/K_b = 5.6 \times 10^{-10}$.

$$K_a = \frac{[H^+][NH_3]}{[NH_4^+]} = 5.6 \times 10^{-10}$$

$$= \frac{[H^+](0.25)}{0.45} = 5.6 \times 10^{-10}$$

$[H^+] = 1.0 \times 10^{-9}$ M and pH = 9.00

Notice that the two values calculated are essentially the same.

18.86 For a conjugate acid-base pair, $K_a \times K_b = 1.00 \times 10^{-14}$. The K_b of ammonia is given in Table 18.2, so we can find the K_a for the ammonium ion:

$$K_a \times K_b = 1.00 \times 10^{-14}$$
$$K_a \times 1.8 \times 10^{-5} = 1.00 \times 10^{-14}$$
$$K_a = 1.4 \times 10^{-5}$$

Using equation 18.10:

$$[H^+] = K_a \frac{[HA]}{[A^-]}$$

$$[H^+] = K_a \frac{[NH_4^+]}{[NH_3]}$$

$$[H^+] = 1.4 \times 10^{-5} \frac{[0.20]}{[0.25]} = 1.1 \times 10^{-5}$$

Therefore, the initial pH of the buffer = $-\log(1.1 \times 10^{-5}) = 5.0$

Initial amounts in the solution are:
mol NH_3 = (.25 mol/L)(.25 L) = 0.063 mol NH_3
mol NH_4^+ = (.20 mol/L)(.25 L) = 0.050 mol NH_4^+

The added acid (0.0250 L(0.10 mol/L) = 0.00250 mol HCl) will react with the ammonia present in the buffer solution. Assume the added acid reacts completely. For each mole of acid added, one mole of NH_3 is converted to NH_4^+. Since 0.00250 mol of acid is added;

mol $NH_4^+{}_{final}$ = (0.050 + 0.00250) mol = 0.053 mol
mol NH_{3final} = (0.063 – 0.00250) mol = 0.061 mol

The final volume of solution is 250 mL + 25.0 mL = 275 mL.

The final concentrations are:

$[NH_4^+]_{final}$ = 0.053 mol/0.275 L = 0.19 M NH_4^+

$[NH_3]_{final}$ = 0.061 mol/0.275 L = 0.22 M NH_3

So the *changes* in concentrations are:

$\Delta[NH_4^+] = [NH_4^+]_{final} - [NH_4^+]_{initial}$ = 0.19 M – 0.20 M = –0.01 M
$\Delta[NH_3] = [NH_3]_{final} - [NH_3]_{initial}$ = 0.22 M – 0.25 M = –0.03 M

18.87 $HC_2H_3O_2 \leftrightarrows H^+ + C_2H_3O_2^-$

$K_a = 1.8 \times 10^{-5}$

Using equation 18.10:

$$[H^+] = K_a \frac{[HA]}{[A^-]}$$

$$[H^+] = K_a \frac{[HC_2H_3O_2]}{[C_2H_3O_2^-]}$$

$$[H^+] = 1.8 \times 10^{-5} \frac{[0.15]}{[0.25]} = 1.1 \times 10^{-5}$$

Therefore, the initial pH of the buffer = $-\log(1.1 \times 10^{-5})$ = 5.0
Initial amounts in the solution are:
mol $C_2H_3O_2^-$ = (0.25 mol/L)(0.25 L) = 0.063 mol $C_2H_3O_2^-$
mol $HC_2H_3O_2$ = (0.15 mol/L)(0.25 L) = 0.038 mol $HC_2H_3O_2$

The added base (0.100 L(0.10 mol/L) = 0.010 mol NaOH) will react with the acetic acid present in the buffer solution. Assume the added base reacts completely. For each mole of base added, one mole of $HC_2H_3O_2$ is converted to $C_2H_3O_2^-$. Since 0.010 mol of base is added;

mol $HC_2H_3O_{2\,final}$ = (0.038 – 0.010) mol = 0.028 mol
mol $C_2H_3O_2^-{}_{final}$ = (0.063 + 0.010) mol = 0.073 mol

The final volume of solution is 250 mL + 100 mL = 350 mL.

The final concentrations are:

$[HC_2H_3O_2]_{final}$ = 0.028 mol/0.350 L = 0.08 M $HC_2H_3O_2$
$[C_2H_3O_2^-]_{final}$ = 0.073 mol/0.350 L = 0.21 M $C_2H_3O_2^-$

So the *changes* in concentrations are:

$\Delta[HC_2H_3O_2] = [HC_2H_3O_2]_{final} - [HC_2H_3O_2]_{initial}$ = 0.08 M – 0.15 M = –0.07 M
$\Delta[C_2H_3O_2^-] = [C_2H_3O_2^-]_{final} - [C_2H_3O_2^-]_{initial}$ = 0.21 M – 0.25 M = –0.04 M

18.88 The initial pH of the buffer is 4.97 as determined in Exercise 18.82. The added acid, 0.050 mol, will react with the acetate ion present in the buffer solution. Assume the added acid reacts completely. For each mole of acid added, one mole of $C_2H_3O_2^-$ is converted to $HC_2H_3O_2$. Since 0.050 mol of acid is added;

$[HC_2H_3O_2]_{final}$ = (0.15 + 0.050) M = 0.20 M
$[C_2H_3O_2^-]_{final}$ = (0.25 – 0.050) M = 0.20 M

Now, substitute these values into the mass action expression to calculate the final $[H^+]$ in solution;

$$\frac{[H^+](0.20)}{(0.20)} = 1.8 \times 10^{-5}$$

$[H^+] = 1.8 \times 10^{-5}$ mol L^{-1} and the pH = 4.74

The pH of the solution changes by 4.74 – 4.93 = –0.23 pH units upon addition of the acid.

18.89 The initial pH of the buffer is 4.97 as determined in Exercise 18.82. The added base, 0.0500 L × 0.10 mol L^{-1} = 0.0050 mol, will react with the acetic acid present in the buffer solution. Assume the added base reacts completely. For each mole of base added, one mole of $HC_2H_3O_2$ is converted to $C_2H_3O_2^-$. Since 0.0050 mol of base is added;

$[HC_2H_3O_2]_{final}$ = (0.075 – 0.0050) mol = 0.070 mol
$[C_2H_3O_2^-]_{final}$ = (0.125 + 0.0050) mol = 0.130 mol

(Note: the amount of acid and base present in the solution is half the stated molarity since we only have 500 mL. Additionally, we have carried one additional sig. fig. when calculating the number of moles of acetate ion in order to make the calculation more meaningful.)

Now, substitute these values into the mass action expression to calculate the final $[H^+]$ in solution (remember to convert back to molarity accounting for the new volume of 550 mL);

$$\frac{[H^+](0.24)}{(0.13)} = 1.8 \times 10^{-5}$$

$[H^+] = 9.8 \times 10^{-6}$ mol L^{-1} and the pH = 5.01

The pH of the solution rises but does not change significantly upon addition of the base.

18.90 The initial pH is 9.00 as calculated in Exercise 18.84. For every mole of H^+ added, one mol of NH_3 will be changed to one mol of NH_4^+. Since we added 0.020 mol H^+:

$$[NH_4^+]_{final} = 0.45 \text{ M} + 0.020 \text{ M} = 0.47 \text{ M}$$
$$[NH_3]_{final} = 0.25 \text{ M} - 0.020 \text{ M} = 0.23 \text{ M}$$

Using these new concentrations, we can calculate a new pH:

$$K_b = \frac{[NH_4^+][OH^-]}{[NH_3]} = \frac{(0.47)[OH^-]}{0.23} = 1.8 \times 10^{-5}$$

$[OH^-] = 8.8 \times 10^{-6}$ M, the pOH = 5.06 and the pH = 8.94.

As expected, when an acid is added, the pH decreases. In this problem, the pH decreases by 0.06 pH units from 9.00 to 8.94.

18.91 The initial pH is 9.00 as calculate in Exercise 18.84. First, we will determine the number of moles of OH^- added:

$$\# \text{ mol OH}^- = (75 \text{ mL KOH})\left(\frac{0.10 \text{ mol KOH}}{1000 \text{ mL KOH}}\right)\left(\frac{1 \text{ mol OH}^-}{1 \text{ mol KOH}}\right) = 7.5 \times 10^{-3} \text{ mol OH}^-$$

Then, determine the number of moles of NH_4^+ and NH_3 initially present:

$$\# \text{ mol } NH_4^+ = (200 \text{ mL solution})\left(\frac{0.45 \text{ mol } NH_4^+}{1000 \text{ mL solution}}\right)$$

$$= 9.0 \times 10^{-2} \text{ mol } NH_4^+$$

$$\# \text{ mol } NH_3 = (200 \text{ mL solution})\left(\frac{0.25 \text{ mol } NH_3}{1000 \text{ mL solution}}\right)$$

$$= 5.0 \times 10^{-2} \text{ mol } NH_3$$

For every mole of OH^- added, one mol of NH_4^+ will be changed to one mol of NH_3. Since we added 7.5×10^{-3} mol OH^-;

$$\left[NH_4^+\right]_{final} = \frac{(0.090 \text{ mol} - 0.0075 \text{ mol})}{0.275 \text{ L}} = 0.300 \text{ M}$$

$$\left[NH_3\right]_{final} = \frac{(0.050 \text{ mol} + 0.0075 \text{ mol})}{0.275 \text{ L}} = 0.21 \text{ M}$$

Note: the volume of the solution changes as a result of the addition of KOH.

Using these new concentrations, we can calculate a new pH;

$$K_b = \frac{\left[NH_4^+\right]\left[OH^-\right]}{\left[NH_3\right]} = \frac{(0.300)\left[OH^-\right]}{0.21} = 1.8 \times 10^{-5}$$

$[OH^-] = 1.3 \times 10^{-5}$ M, the pOH = 4.90 and the pH = 9.10.

As expected, when a base is added, the pH increases. In this problem, the pH increases by 0.10 pH units, from 9.00 to 9.10.

18.92 $$\text{pOH} = \text{p}K_b + \log\frac{[\text{cation}]}{[\text{base}]}$$

Now pOH = 14.00 – pH = 14.00 – 9.25 = 4.75
$4.75 = 4.74 + \log([NH_4^+]/[NH_3])$
$\log([NH_4^+]/[NH_3]) = 0.01$
Taking the antilog of both sides of this equation gives:
$[NH_4^+]/[NH_3] = 10^{0.01} = 1$
The ratio is 1 to 1.

18.93 $HC_3H_5O_2 + H_2O \leftrightarrows H_3O^+ + C_3H_5O_2^-$

$$K_a = \frac{\left[H_3O^+\right]\left[C_3H_5O_2^-\right]}{\left[HC_3H_5O_2\right]} = 1.34\times10^{-5} \quad \text{or} \quad \text{p}Ka = \text{pH} - \log\frac{C_3H_5O_2^-}{HC_3H_5O_2}$$

rearranging we get $$\frac{[C_3H_5O_2^-]}{[HC_3H_5O_2]} = 10^{(\text{pH - p}K)} = 10^{(4.15 - 4.87)} = 0.19$$

18.94 $$\text{pH} = \text{p}K_a + \log\frac{[\text{anion}]}{[\text{acid}]} = \text{p}K_a + \log\frac{\left[A^-\right]}{[HA]}$$

$5.00 = 4.74 + \log([NaC_2H_3O_2]/[HC_2H_3O_2])$
$[NaC_2H_3O_2]/[HC_2H_3O_2] = 1.8$
$[NaC_2H_3O_2] = 1.8 \times [HC_2H_3O_2] = 1.8 \times 0.15 = 0.27$ M
Thus to the 1 L of acetic acid solution we add: 0.27 mol $NaC_2H_3O_2 \times 82.0$ g/mol = 22 g $NaCHO_2$.

18.95 $$pH = pK_a + \log\frac{[\text{anion}]}{[\text{acid}]} = pK_a + \log\frac{[A^-]}{[HA]}$$

$3.80 = 3.74 + \log([NaCHO_2]/[HCHO_2])$
$[NaCHO_2]/[HCHO_2] = 1.15$
$[NaCHO_2] = 1.15 \times [HCHO_2] = 1.15 \times 0.12 = 0.14$ M
Thus, to the 1 L of formic acid solution we add: 0.14 mol $NaCHO_2 \times 68.0$ g/mol = 9.5 g $NaCHO_2$.

18.96 $$pOH = pK_b + \log\frac{[\text{cation}]}{[\text{base}]}$$

Now pOH = 14.00 – pH = 14.00 – 10.00 = 4.00
$4.00 = 4.74 + \log([NH_4^+]/[NH_3])$
$\log([NH_4^+]/[NH_3]) = -0.74$
Taking the antilog of both sides of this equation gives:
$[NH_4^+]/[NH_3] = 10^{-0.74} = 0.18$
Thus, $[NH_4^+] = 0.18 \times [NH_3] = 0.18 \times 0.20\text{ M} = 3.6 \times 10^{-2}\text{ M} = [NH_4Cl]$

Finally, 0.500 L of buffer solution would require:
$0.500\text{ L} \times 3.6 \times 10^{-2}\text{ mol/L} \times 53.5\text{ g/mol} = 0.96$ g NH_4Cl

18.97 $$pOH = pK_b + \log\frac{[\text{cation}]}{[\text{base}]}$$

Now, pOH = 14.00 – pH = 14.00 – 9.15 = 4.85
$4.85 = 4.74 + \log([NH_4^+]/[NH_3])$
$\log([NH_4^+]/[NH_3]) = 0.11$
Taking the antilog of both sides of this equation gives:
$[NH_4^+]/[NH_3] = 10^{0.11} = 1.3$
Thus $[NH_4^+] = 1.3 \times [NH_3] = 1.3 \times 0.10\text{ M} = 0.13\text{ M} = [NH_4Cl]$

Finally, 0.125 L of buffer solution would require:
$0.125\text{ L} \times 0.13\text{ mol/L} \times 53.5\text{ g/mol} = 0.86$ g NH_4Cl

18.98 The equilibrium is; $HC_2H_3O_2 \leftrightarrows H^+ + C_2H_3O_2^-$

$$K_a = \frac{[H^+][C_2H_3O_2^-]}{[HC_2H_3O_2]} = 1.8 \times 10^{-5}$$

The initial pH is; $\dfrac{[H^+](0.110)}{0.100} = 1.8 \times 10^{-5}$, $[H^+] = 1.64 \times 10^{-5}$ M and pH = 4.786. In this calculation we are able to use either the molar concentration or the number of moles since the volume is constant in this portion of the problem.

In order to calculate the change in pH, we need to determine the concentrations of $HC_2H_3O_2$ and $C_2H_3O_2^-$ after the complete reaction of the added acid. One mole of $C_2H_3O_2^-$ will be consumed for every mole of acid added and one mole of $HC_2H_3O_2$ will be produced. The number of moles of acid added is:

$$\# \text{ mol H}^+ = (25.00 \text{ mL HCl})\left(\frac{0.100 \text{ mol HCl}}{1000 \text{ mL HCl}}\right)\left(\frac{1 \text{ mol H}^+}{1 \text{ mol HCl}}\right)$$

$$= 2.50 \times 10^{-3} \text{ mol H}^+$$

Since the volume of the new concentrations of $HC_2H_3O_2$ and $C_2H_3O_2^-$ are:

$$[HC_2H_3O_2]_{final} = \frac{(0.100 \text{ mol} - 0.00250 \text{ mol})}{0.525 \text{ L}} = 0.195 \text{ M}$$

$$[C_2H_3O_2^-]_{final} = \frac{(0.110 \text{ mol} + 0.00250 \text{ mol})}{0.525 \text{ L}} = 0.205 \text{ M}$$

Note: The new volume has been used in these calculations.

$$K_a = \frac{[H^+][C_2H_3O_2^-]}{[HC_2H_3O_2]} = \frac{[H^+](0.205)}{(0.195)} = 1.8 \times 10^{-5}$$

$[H^+] = 1.71 \times 10^{-5}$ and pH = 4.766.

Notice that the change in pH is very small in spite of adding a strong acid. If the same amount of HCl were added to water, a completely different effect would be observed.

Since HCl is a strong acid, the $[H^+]$ in a water solution will be the result of the strong acid dissociation. We do, of course, need to account for the dilution. Using the dilution equation, $M_1V_1 = M_2V_2$, we determine the $[H^+] = 0.0167$ mol L^{-1} and the pH = 1.78. The change in pH in this case is 7.00 – 1.78 = 5.22 pH units. A significantly larger change!!

18.99 The initial pH is 4.786. We want to lower the pH to 4.736, a decrease of 0.050 pH units. First we will calculate the ratio of acid and its conjugate base:

$$pH = pK_a + \log\frac{[\text{anion}]}{[\text{acid}]} = pK_a + \log\frac{[A^-]}{[HA]}$$

$4.736 = 4.745 + \log([NaC_2H_3O_2]/[HC_2H_3O_2])$

$[NaC_2H_3O_2]/[HC_2H_3O_2] = 0.979$

We need a final ratio of conjugate base to acid of 0.979. Initially there are 0.110 moles of acetate ion and 0.100 moles of acetic acid. We will add a strong acid using up some of the acetate ion and forming additional acetic acid. For every mole of acetate that reacts, one mole of acetic acid will be formed. We can solve the following equation:

$$\frac{0.110 \text{ moles acetate} - x \text{ moles H}^+}{0.100 \text{ moles acetic acid} + x \text{ moles H}^+} = 0.979$$

Solving we determine $x = 6.11 \times 10^{-3}$ mole H^+. This is the number of moles of HCl needed to change the pH by 0.05 pH units.

$$\# \text{ mL} = (6.11 \times 10^{-3} \text{ mol H}^+)\left(\frac{1 \text{ mol HCl}}{1 \text{ mol H}^+}\right)\left(\frac{1000 \text{ mL}}{0.15 \text{ mol HCl}}\right) = 41 \text{ mL}.$$ Thus, we need to add only 41 mL to the 100 mL of buffer.

If we added 41 mL of 0.15 M HCl directly to 100 mL of pure water, we would find the pH = 1.36. Thus, the pH of pure water would change by 5.64 pH units as opposed to the 0.05 pH unit change in the buffer solution.

18.100 $H_2C_6H_6O_6 + H_2O \leftrightarrows H_3O^+ + HC_6H_6O_6^-$ $K_{a1} = 6.7 \times 10^{-5}$
$HC_6H_6O_6^- + H_2O \leftrightarrows H_3O^+ + C_6H_6O_6^{2-}$ $K_{a2} = 2.7 \times 10^{-12}$

	$[H_2C_6H_6O_6]$	$[H_3O^+]$	$[HC_6H_6O_6^-]$
I	0.15	–	–
C	–x	+x	+x
E	0.15–x	+x	+x

Assume x << 0.15

$$K_a = \frac{[x][x]}{[0.15]} = 6.7 \times 10^{-5} \qquad x = 3.2 \times 10^{-3}$$

$[H_2C_6H_6O_6] \cong 0.15$ M

$[H_3O^+] = [HC_6H_6O_6^-] = 3.2 \times 10^{-3}$ M

$[C_6H_6O_6^{2-}] = 2.7 \times 10^{-12}$ M

pH = 2.50

pOH = 11.5

$[OH^-] = 3.2 \times 10^{-12}$ M

18.101 As we do with most diprotic weak acids, assume that all of the H^+ results from the first dissociation and all of the divalent anion results from the second dissociation. And, since the second dissociation constant is much smaller than the first dissociation constant, the concentration of the divalent anion is equal to the second dissociation constant.

$H_6TeO_6 \leftrightarrows H_5TeO_6^- + H^+$ $\qquad K_{a_1} = \frac{[H_5TeO_6^-][H^+]}{[H_6TeO_6]} = 2 \times 10^{-8}$

$H_5TeO_6^- \leftrightarrows H_4TeO_6^{2-} + H^+$ $\qquad K_{a_2} = \frac{[H_4TeO_6^{2-}][H^+]}{[H_5TeO_6^-]} = 1 \times 10^{-11}$

	$[H_6TeO_6]$	$[H_5TeO_6^-]$	$[H^+]$
I	0.25	–	–
C	–x	+x	+x
E	0.25–x	+x	+x

$$K_{a_1} = \frac{[H_5TeO_6^-][H^+]}{[H_6TeO_6]} = \frac{(x)(x)}{(0.25 - x)} = 2 \times 10^{-8}$$

Assume that x << 0.25 then; $x^2 = (0.25)(2 \times 10^{-8})$ and $x = 7.1 \times 10^{-5}$ M

$pH = -\log[H^+] = -\log(7.1 \times 10^{-5}) = 4.2$
$[H_4TeO_6^{2-}] = K_{a_2} = 1 \times 10^{-11}$ M

18.102 $H_3PO_4 \leftrightarrows H_2PO_4^- + H^+$ $\qquad K_{a1} = \dfrac{[H_2PO_4^-][H^+]}{[H_3PO_4]} = 7.1 \times 10^{-3}$

$H_2PO_4^- \leftrightarrows HPO_4^{2-} + H^+$ $\qquad K_{a2} = \dfrac{[HPO_4^{2-}][H^+]}{[H_2PO_4^-]} = 6.3 \times 10^{-8}$

$HPO_4^{2-} \leftrightarrows PO_4^{3-} + H^+$ $\qquad K_{a3} = \dfrac{[PO_4^{3-}][H^+]}{[HPO_4^{2-}]} = 4.5 \times 10^{-13}$

The following assumptions are made:

$[H^+]_{total} \approx [H^+]_{first\ step}$
$[H_2PO_4^-]_{total} \approx [H_2PO_4^-]_{first\ step}$
$[HPO_4^{2-}]_{total} \approx [HPO_4^{2-}]_{second\ step}$

The first dissociation:

	$[H_3PO_4]$	$[H_2PO_4^-]$	$[H^+]$
I	3.0	–	–
C	–x	+x	+x
E	3.0–x	+x	+x

$$K_{a1} = \frac{[H_2PO_4^-][H^+]}{[H_3PO_4]} = \frac{(x)(x)}{(3.0 - x)} = 7.1 \times 10^{-3}$$

Solve for x by successive approximations or by the quadratic equation:

$x = 0.14\ M = [H^+] = [H_2PO_4^-]$
$[H_3PO_4] = 3.0 - 0.14 = 2.9\ M$
$pH = -\log(0.14) = 0.85$

The second dissociation:

	$[H_2PO_4^-]$	$[HPO_4^{2-}]$	$[H^+]$
I	0.14	–	0.14
C	–x	+x	+x
E	0.14–x	+x	0.14+x

Assume that $x << 0.14$, therefore $0.14 - x \approx 0.14$ and $0.14 + x \approx 0.14$

$$K_{a2} = \frac{(x)(0.14)}{(0.14)} = 6.3 \times 10^{-8}$$

$x = 6.3 \times 10^{-8} = [HPO_4^{2-}]$

The third dissociation:

	$[HPO_4^{2-}]$	$[PO_4^{3-}]$	$[H^+]$
I	6.3×10^{-8}	–	0.14
C	–x	+x	+x
E	6.3×10^{-8}–x	+x	0.14+x

Assume that $x << 6.3 \times 10^{-8}$, therefore $6.3 \times 10^{-8} - x \approx 6.3 \times 10^{-8}$ and $0.14 + x \approx 0.14$

$$K_{a3} = \frac{(x)(0.14)}{(6.3 \times 10^{-8})} = 4.5 \times 10^{-13}$$

Solving for x we get, $x = 2.0 \times 10^{-19} = [PO_4^{3-}]$

18.103 $H_3AsO_4 + H_2O \leftrightarrows H_3O^+ + H_2AsO_4^-$ $K_{a_1} = 5.6 \times 10^{-3}$

$H_2AsO_4^- + H_2O \leftrightarrows H_3O^+ + HAsO_4^{2-}$ $K_{a_2} = 1.7 \times 10^{-7}$

$HAsO_4^{2-} + H_2O \leftrightarrows H_3O^+ + AsO_4^{3-}$ $K_{a_3} = 4.0 \times 10^{-12}$

$[HAsO_4^{2-}] = 1.7 \times 10^{-7}$
$[AsO_4^{3-}] = 4.0 \times 10^{-12}$

Now solve using $K_{a_1} = 5.6 \times 10^{-3}$ for the other concentrations:

$[H_3O^+] = [H_2AsO_4^-] = 5.3 \times 10^{-2}$, pH = 1.28.

18.104 The equilibrium is:

$$H_2Tar \leftrightarrows HTar^- + H^+ \quad K_{a_1} = \frac{[HTar^-][H^+]}{[H_2Tar]} = 9.2 \times 10^{-4}$$

$$HTar^- \leftrightarrows Tar^{2-} + H^+ \quad K_{a_2} = \frac{[Tar^{2-}][H^+]}{[HTar^-]} = 4.3 \times 10^{-5}$$

The $[H^+]$ and $[HTar^-]$ are determined by the first dissociation equation.

	$[H_2Tar]$	$[HTar^-]$	$[H^+]$
I	0.50	–	–
C	–x	+x	+x
E	0.50–x	+x	+x

$$K_{a_1} = \frac{[HTar^-][H^+]}{[H_2Tar]} = \frac{(x)(x)}{(0.50\text{-}x)} = 9.2 \times 10^{-4}$$

Solving this equation requires a quadratic equation or the method of successive approximations:

$x = 0.021\ M = [H^+] = [HTar^-]$
$[H_2Tar] = [H^+] = 0.50 - 0.021 = 0.48\ M$
$[Tar^{2-}] = K_{a2} = 4.3 \times 10^{-5}\ M$

18.105 The equilibrium is:

$$H_2Tar \leftrightarrows HTar^- + H^+ \quad K_{a_1} = \frac{[HTar^-][H^+]}{[H_2Tar]} = 9.2 \times 10^{-4}$$

$$HTar^- \leftrightarrows Tar^{2-} + H^+ \quad K_{a_2} = \frac{[Tar^{2-}][H^+]}{[HTar^-]} = 4.3 \times 10^{-5}$$

The $[H^+]$ and $[HTar^-]$ are determined by the first dissociation equation.

	$[H_2Tar]$	$[HTar^-]$	$[H^+]$
I	0.20	–	0.030
C	–x	+x	+x
E	0.20–x	+x	0.030

$$K_{a_1} = \frac{[HTar^-][H^+]}{[H_2Tar]} = \frac{(x)(0.030)}{(0.20 - x)} = 9.2 \times 10^{-4}$$

Solving this equation requires a quadratic equation or the method of successive approximations:

x = 0.0060 M = [$HTar^-$]
[H_2Tar] = 0.20 – 0.0060 = 0.19 M
[H^+] = 0.030 M

We need to use the second equilibrium to determine the [Tar^{2-}]

$$K_{a_2} = \frac{[Tar^{2-}][H^+]}{[HTar^-]} = \frac{(x)(0.030)}{(0.0060)} = 4.3 \times 10^{-5}$$

Solving for x we get, x = [Tar^{2-}] = 8.6 × 10^{-6} M

18.106 $H_3PO_3 \leftrightarrows H_2PO_3^- + H^+$ $\qquad K_{a_1} = \frac{[H_2PO_3^-][H^+]}{[H_3PO_3]} = 1.0 \times 10^{-2}$

$H_2PO_3^- \leftrightarrows HPO_3^{2-} + H^+$ $\qquad K_{a_2} = \frac{[HPO_3^{2-}][H^+]}{[H_2PO_3^-]} = 2.6 \times 10^{-7}$

To simplify the calculation, assume that the second dissociation does not contribute a significant amount of H^+ to the final solution. Solving the equilibrium problem for the first dissociation gives:

	[H_3PO_3]	**[$H_2PO_3^-$]**	**[H^+]**
I	1.0	–	–
C	–x	+x	+x
E	1.0–x	+x	+x

$$K_{a_1} = \frac{[H_2PO_3^-][H^+]}{[H_3PO_3]} = \frac{(x)\,(x)}{(1.0 - x)} = 1.0 \times 10^{-2}$$

Because K_{a_1} is so large, a quadratic equation must be solved. On doing so we learn that
x = 0.095 M = [H^+] = [$H_2PO_3^-$].
pH = – log[H^+] = – log(0.095) = 1.022

The [HPO_3^{2-}] may be determined from the second ionization constant.

	[$H_2PO_3^-$]	**[HPO_3^{2-}]**	**[H^+]**
I	0.095	–	0.095
C	–x	+x	+x
E	0.095–x	+x	0.095+x

$$K_{a_2} = \frac{[HPO_3^{2-}][H^+]}{[H_2PO_3^-]} = \frac{(x)\,(0.95 + x)}{(0.095 - x)} = 2.6 \times 10^{-7}$$

If we assume that x is small then 0.095 ± x ≈ 0.095. Then, x = [HPO_3^{2-}] = 2.6 × 10^{-7} M.

18.107 $H_2C_2O_4 + H_2O \leftrightarrows H_3O^+ + HC_2O_4^-$ $\quad Ka = \dfrac{[H_3O^+][HC_2O_4^-]}{[H_2C_2O_4]} = 6.5 \times 10^{-2}$

	$[H_2C_2O_4]$	$[H_3O^+]$	$[HC_2O_4^-]$
I	0.20	–	–
C	–x	+x	+x
E	0.20–x	+x	+x

Because K_a is large, use a quadratic equation to determine $x = 8.6 \times 10^{-2} = [H_3O^+]$, pH = 1.06.

18.108 The hydrolysis equation is:

$SO_3^{2-} + H_2O \leftrightarrows HSO_3^- + OH^-$ $\quad K_b = \dfrac{[HSO_3^-][OH^-]}{[SO_3^{2-}]}$

In order to obtain K_b we will use the relationship $K_w = K_a \times K_b$

$$K_b = \frac{K_w}{K_a} = \frac{1.0 \times 10^{-14}}{6.6 \times 10^{-8}} = 1.5 \times 10^{-7}$$

	$[SO_3^{2-}]$	$[HSO_3^-]$	$[OH^-]$
I	0.12	–	–
C	–x	+x	+x
E	0.12–x	+x	+x

Since K_b is so small, assume that x << 0.12 and we determine $x = 1.3 \times 10^{-4}$ M = $[OH^-]$
pOH = $-\log(1.3 \times 10^{-4}) = 3.87$, pH = 14.00 – pOH = 10.13

For the concentration of H_2SO_3

$HSO_3^- + H_2O \leftrightarrows H_2SO_3 + OH^-$ $\quad K_{b_2} = \dfrac{[H_2SO_3][OH^-]}{[HSO_3^-]}$

$$K_{b_2} = \frac{K_w}{K_{a_1}} = \frac{1.0 \times 10^{-14}}{1.2 \times 10^{-2}} = 8.3 \times 10^{-13}$$

$$K_{b_2} = \frac{[H_2SO_3][1.3 \times 10^{-4}]}{[1.3 \times 10^{-4}]} = 8.3 \times 10^{-13}$$

$[H_2SO_3] = 8.3 \times 10^{-13}$

18.109 The hydrolysis equation is:

$CO_3^{2-} + H_2O \leftrightarrows HCO_3^- + OH^-$ $\quad K_b = \dfrac{[HCO_3^-][OH^-]}{[CO_3^{2-}]}$

In order to obtain K_b we will use the relationship $K_w = K_a \times K_b$

$$K_b = \frac{K_w}{K_a} = \frac{1.0 \times 10^{-14}}{4.7 \times 10^{-11}} = 2.1 \times 10^{-4}$$

	$[CO_3^{2-}]$	$[HCO_3^-]$	$[OH^-]$
I	0.15	–	–
C	–x	+x	+x
E	0.15–x	+x	+x

Since K_b is so small, assume that $x << 0.15$ and we determine $x = 5.6 \times 10^{-3}$ M = $[OH^-]$
pOH = $-\log(5.6 \times 10^{-3}) = 2.25$, pH = 14.00 – pOH = 11.75

$HCO_3^- + H_2O \leftrightarrows H_2CO_3 + OH^-$ $\qquad K_{b_2} = \frac{[H_2CO_3][OH^-]}{[HCO_3^-]}$

For the concentration of H_2CO_3

$$K_{b_2} = \frac{K_w}{K_{a_1}} = \frac{1.0 \times 10^{-14}}{4.5 \times 10^{-7}} = 2.2 \times 10^{-8}$$

$$K_{b_2} = \frac{[H_2SO_3][5.6\times10^{-3}]}{[5.6\times10^{-3}]} = 2.2\times10^{-8}$$

$[H_2SO_3] = 2.2 \times 10^{-8}$

18.110 Only the first hydrolysis needs to be examined. The difficulty is that we are solving for the initial concentration in this problem.

$CO_3^{2-} + H_2O \leftrightarrows HCO_3^- + OH^-$ $\qquad K_b = \frac{[HCO_3^-][OH^-]}{[CO_3^{2-}]}$

In order to obtain K_b we will use the relationship $K_w = K_a \times K_b$

$$K_b = \frac{K_w}{K_a} = \frac{1.0 \times 10^{-14}}{4.7 \times 10^{-11}} = 2.1 \times 10^{-4}$$

	$[CO_3^{2-}]$	$[HCO_3^-]$	$[OH^-]$
I	**Z**	–	–
C	–x	+x	+x
E	**Z**–x	+x	+x

In this problem, we know that the pH at equilibrium is 11.62. Using this fact we calculate a pOH = 2.38 and a $[OH^-] = 10^{-pOH} = 4.2 \times 10^{-3}$ M = x. Substitute this into the equilibrium expression and solve for **Z**, our initial concentration of carbonate ion.

$$K_b = 2.1 \times 10^{-4} = \frac{[HCO_3^-][OH^-]}{[CO_3^{2-}]} = \frac{(x)(x)}{(\mathbf{Z} - x)} = \frac{(4.2 \times 10^{-3})^2}{\mathbf{Z} - 4.2 \times 10^{-3}}$$

Solving this equation we determine that Z = 8.8×10^{-2} M = $[CO_3^{2-}]$. (Note: We could have assumed that the change upon dissociation is small in this example, i.e., x << **Z**. It is unnecessary to do this for this problem. If you had made this assumption, a value of Z = 8.4×10^{-2} M would be obtained.)

The problem asks for the number of grams of $Na_2CO_3 \bullet 10H_2O$ so convert the concentration to a number of grams: 25.2 grams.

18.111 The equilibrium we are studying is:

$$SO_3^{2-} + H_2O \leftrightarrows HSO_3^- + OH^- \qquad K_b = \frac{[HSO_3^-][OH^-]}{[SO_3^{2-}]}$$

In order to obtain K_b we will use the relationship $K_w = K_a \times K_b$

$$K_b = \frac{K_w}{K_a} = \frac{1.0 \times 10^{-14}}{6.6. \times 10^{-8}} = 1.5 \times 10^{-7}$$

	$[SO_3^{2-}]$	$[HSO_3^-]$	$[OH^-]$
I	Z	–	–
C	–x	+x	+x
E	Z–x	+x	+x

In this problem, we know that the pH at equilibrium is 10.15. Using this fact we calculate a pOH = 3.85 and a $[OH^-] = 10^{-pOH} = 1.4 \times 10^{-4}$ M = x. Substitute this into the equilibrium expression and solve for **Z**, our initial concentration of sulfite ion.

$$K_b = 1.5 \times 10^{-7} = \frac{[HSO_3^-][OH^-]}{[SO_3^{2-}]} = \frac{(x)(x)}{(\mathbf{Z}-x)} = \frac{(1.4 \times 10^{-4})^2}{\mathbf{Z}-1.4 \times 10^{-4}}$$

Solving this equation we determine that Z = 0.13 M = $[SO_3^{2-}]$.

The problem asks for the number of grams of $Na_2SO_3{\bullet}7H_2O$ so;

$$\#\ g\ Na_2SO_3 \cdot 7H_2O = (0.50\ L)\left(\frac{0.13\ moles\ SO_3^{2-}}{L}\right)\left(\frac{1\ mole\ Na_2SO_3 \cdot 7H_2O}{1\ mole\ SO_3^{2-}}\right) \times \left(\frac{252\ g\ Na_2SO_3 \cdot 7H_2O}{1\ mole\ Na_2SO_3 \cdot 7H_2O}\right)$$

$$= 16\ g\ Na_2SO_3 \cdot 7H_2O$$

18.112 $C_6H_5O_7^{3-} + H_2O \leftrightarrows HC_6H_5O_7^{2-} + OH^-$

$$K_b = \frac{[HC_6H_5O_7^{2-}][OH^-]}{[C_6H_5O_7^{3-}]} = \frac{K_w}{K_a} = \frac{1.0 \times 10^{-14}}{4.0 \times 10^{-7}} = 2.5 \times 10^{-8}$$

	$[C_6H_5O_7^{3-}]$	$[HC_6H_5O_7^{2-}]$	$[OH^-]$
I	0.10	–	–
C	–x	+x	+x
E	0.10–x	+x	+x

Assume x << 0.10

$$K_b = \frac{(x)(x)}{0.10} = 2.5 \times 10^{-8} \qquad x = 5.0 \times 10^{-5} = [OH^-]$$

pOH = 4.30 pH = 9.70

18.113 $C_2O_4^{2-} + H_2O \leftrightarrows HC_2O_4^- + OH^-$

$$K_b = \frac{[HC_2O_4^-][OH^-]}{[C_2O_4^{2-}]} = \frac{K_w}{K_a} = \frac{1.0 \times 10^{-14}}{6.1 \times 10^{-5}} = 1.6 \times 10^{-10}$$

	$[C_2O_4^{2-}]$	$[HC_2O_4^-]$	$[OH^-]$
I	0.25	–	–
C	–x	+x	+x
E	0.25–x	+x	+x

Assume x << 0.25

$$K_b = \frac{(x)(x)}{0.25} = 1.6 \times 10^{-10} \qquad x = 6.3 \times 10^{-6} = [OH^-]$$

pOH = 5.20 pH = 8.80

18.114 $PO_4^{3-} + H_2O \leftrightarrows HPO_4^{2-} + OH^-$ $\qquad K_{b1} = \frac{K_w}{K_{a3}} = 2.2 \times 10^{-2}$

$HPO_4^{2-} + H_2O \leftrightarrows H_2PO_4^- + OH^-$ $\qquad K_{b2} = \frac{K_w}{K_{a2}} = 1.6 \times 10^{-7}$

$H_2PO_4^- + H_2O \leftrightarrows H_3PO_4 + OH^-$ $\qquad K_{b3} = \frac{K_w}{K_{a1}} = 1.4 \times 10^{-12}$

By analogy with polyprotic acids, we know that
$[H_2PO_4^-] = 1.6 \times 10^{-7}$
We need to solve the first equilibrium expression to determine $[HPO_4^{2-}]$.

$$K_{b1} = \frac{[HPO_4^{2-}][OH^-]}{[PO_4^{3-}]} = 2.2 \times 10^{-2}$$

	$[PO_4^{3-}]$	$[HPO_4^{2-}]$	$[OH^-]$
I	0.50	–	–
C	–x	+x	+x
E	0.50–x	+x	+x

Assume x << 0.50

$$K_{b1} = \frac{(x)(x)}{0.50 - x} = 2.2 \times 10^{-2}$$

Use the quadratic equation to solve for x since K_b is so large

$x = 9.4 \times 10^{-2}$ M $= [OH^-] = [HPO_4^{2-}]$

pOH = 1.027

pH = 12.97

$[PO_4^{3-}] = 0.50 - 9.4 \times 10^{-2} = 0.41$ M
Solve the third equilibrium expression to determine $[H_3PO_4]$:

$$K_{b3} = \frac{[H_3PO_4][OH^-]}{[H_2PO_4^-]} = 1.4 \times 10^{-12}$$

Substitute the calculated values of $\left[H_2PO_4^-\right]$ and $\left[OH^-\right]$ and solve for x :

$$K_{b3} = \frac{(9.4 \times 10^{-2})x}{(1.6 \times 10^{-7})} = 1.4 \times 10^{-12}$$

$$x = [H_3PO_4] = 2.4 \times 10^{-18}$$

18.115 Because the pH is so high and is greater than pK_a, the $\left[CO_3^{2-}\right] = 0.10$ M. The added OH^- reacts with most of the HCO_3^- to reform CO_3^{2-}.

18.116 Since HCO_2H and NaOH react in a 1:1 ratio:

$$HCO_2H + NaOH \rightarrow NaHCO_2 + H_2O$$

we can use the equation $V_a \times M_a = V_b \times M_b$ to determine the volume of NaOH that is required to reach the equivalence point, i.e. the point at which the number of moles of NaOH is equal to the number of moles of HCO_2H:

$V_{NaOH} = 50$ mL $\times$ 0.10/0.10 = 50 mL

Thus the final volume at the equivalence point will be 50 + 50 = 100 mL.

The concentration of $NaHCO_2$ would then be:

0.10 mol/L $\times$ 0.050 L = 5.0×10^{-3} mol HCO_2H = 5.0×10^{-3} mol $NaHCO_2$

5.0×10^{-3} mol/0.100 L = 5.0×10^{-2} M $NaHCO_2$

The hydrolysis of this salt at the equivalence point proceeds according to the following equilibrium:

$HCO_2^- + H_2O \leftrightarrows HCO_2H + OH^-$

$$K_b = \frac{[HCO_2H][OH^-]}{[HCO_2^-]} = 5.6 \times 10^{-11}$$

	$[HCO_2^-]$	$[HCO_2H]$	$[OH^-]$
I	0.050	–	–
C	–x	+x	+x
E	0.050–x	+x	+x

Substituting the above values for equilibrium concentrations into the mass action expression and assuming x << 0.050 gives:

$$K_b = \frac{(x)(x)}{0.050} = 5.6 \times 10^{-11}$$

$x^2 = 2.8 \times 10^{-12}$ $x = 1.7 \times 10^{-6}$ M = $[OH^-] = [HCO_2H]$

pOH = $-\log[OH^-] = -\log(1.7 \times 10^{-6}) = 5.77$

pH = 14.00 – pOH = 14.00 – 5.77 = 8.23

Cresol red would be a good indicator, since it has a color change near the pH at the equivalence point.

18.117 Since HBr and NH_3 react in a 1:1 ratio:

$$HBr + NH_3 \rightarrow NH_4^+ + Br^-$$

we can use the equation $V_a \times M_a = V_b \times M_b$ to determine the volume of HBr that is required to reach the equivalence point, i.e. the point at which the number of moles of HBr is equal to the number of moles of NH_3:

$V_{HBr} = 25$ mL $\times$ 0.10/0.10 = 25 mL

Thus the final volume at the equivalence point will be 25 + 25 = 50 mL.
The concentration of NH_4^+ would then be:
$0.10 \text{ mol/L} \times 0.025 \text{ L} = 2.5 \times 10^{-3} \text{ mol } NH_3 = 2.5 \times 10^{-3} \text{ mol } NH_4^+$
$2.5 \times 10^{-3} \text{ mol}/0.050 \text{ L} = 5.0 \times 10^{-2} \text{ M } NH_4^+$

At the equivalence point NH_4^+ dissociate according to the following equilibrium:

$$NH_4^+ \leftrightarrows H^+ + NH_3$$

$$K_a = \frac{[NH_3][H^+]}{[NH_4^+]} = 5.6 \times 10^{-10}$$

	$[NH_4^+]$	$[H^+]$	$[NH_3]$
I	0.050	–	–
C	–x	+x	+x
E	0.050–x	+x	+x

Substituting the above values for equilibrium concentrations into the mass action expression and assuming $x \ll 0.050$ gives:

$$K_a = \frac{(x)(x)}{0.050} = 5.6 \times 10^{-10}$$

$x = 5.3 \times 10^{-6} \text{ M} = [H^+]$
$pH = -\log[H^+] = -\log(5.3 \times 10^{-6}) = 5.28$

Bromocresol purple would be a good indicator, since it has a color change near the pH at the equivalence point.

18.118

$$\# \text{ moles } HC_2H_3O_2 = (25.0 \text{ mL } HC_2H_3O_2)\left(\frac{0.180 \text{ mol } HC_2H_3O_2}{1000 \text{ mL } HC_2H_3O_2}\right)$$

$$= 4.50 \times 10^{-3} \text{ moles } HC_2H_3O_2$$

$$\# \text{ moles } OH^- = (35.0 \text{ mL } OH^-)\left(\frac{0.250 \text{ mol } OH^-}{1000 \text{ mL } OH^-}\right) = 8.75 \times 10^{-3} \text{ moles } OH^-$$

$$\text{excess } OH^- = 8.75 \times 10^{-3} - 4.50 \times 10^{-3} = 4.25 \times 10^{-3} \text{ moles}$$

$$[OH^-] = \frac{4.25 \times 10^{-3} \text{ moles}}{(25.0 + 35.0 \text{ mL})\left(\frac{1 \text{ L}}{1000 \text{ mL}}\right)} = 7.08 \text{ x } 10^{-2} \text{ M}$$

$pOH = 1.150$

$pH = 12.850$

18.119

$$\# \text{ moles } HC_2H_3O_2 = (30.0 \text{ mL } HC_2H_3O_2)\left(\frac{0.200 \text{ mol } HC_2H_3O_2}{1000 \text{ mL } HC_2H_3O_2}\right)$$

$$= 6.00 \times 10^{-3} \text{ moles } HC_2H_3O_2$$

$$\# \text{ moles } OH^- = (15.0 \text{ mL } OH^-)\left(\frac{0.400 \text{ mol } OH^-}{1000 \text{ mL } OH^-}\right) = 6.00 \times 10^{-3} \text{ moles } OH^-$$

Since the moles are equal, we are at the equivalence point and the equilibrium is now:

$C_2H_3O_2^- + H_2O \leftrightarrows HC_2H_3O_2 + OH^-$

$$K_b = \frac{[HC_2H_3O_2][OH^-]}{[C_2H_3O_2^-]} = \frac{K_w}{K_a} = 5.56 \times 10^{-10}$$

$$[C_2H_3O_2^-] = \frac{6.00 \times 10^{-3}}{(15.0\text{ mL} + 30.0\text{ mL})\left(\frac{1\text{ L}}{1000\text{ mL}}\right)} = 0.133\text{ M}$$

	$[C_2H_3O_2^-]$	$[HC_2H_3O_2]$	$[OH^-]$
I	0.133	–	–
C	–x	+x	+x
E	0.133 –x	+x	+x

Assume x << 0.133

$$K_b = \frac{(x)(x)}{0.133} = \frac{K_w}{K_a} = 5.56 \times 10^{-10} \qquad x = 8.6 \times 10^{-6} = [OH^-]$$

pOH = 5.066

pH = 8.934

18.120 a) $HC_2H_3O_2 \leftrightarrows H^+ + C_2H_3O_2^-$ $\quad K_a = \frac{[H^+][C_2H_3O_2^-]}{[HC_2H_3O_2]} = 1.8 \times 10^{-5}$

	$[HC_2H_3O_2]$	$[H^+]$	$[C_2H_3O_2^-]$
I	0.1000	–	–
C	–x	+x	+x
E	0.1000–x	+x	+x

Substituting the above values for equilibrium concentrations into the mass action expression and assuming that x << 0.1000 gives:

$x = [H^+] = 1.342 \times 10^{-3}$ M.

$pH = -\log [H^+] = -\log (1.342 \times 10^{-3}) = 2.8724$.

(b) When NaOH is added, it will react with the acetic acid present decreasing the amount in solution and producing additional acetate ion. Since this a one–to–one reaction, the number of moles of acetic acid will decrease by the same amount as the number of moles of NaOH added and the number of moles of acetate ion will increase by an identical amount. We must determine the number of moles of all ions present and calculate new concentrations accounting for dilution.

$$\#\text{ moles } HC_2H_3O_2 = (0.02500\text{ L solution})\left(\frac{0.1000\text{ moles } HC_2H_3O_2}{1\text{ L solution}}\right)$$

$$= 2.500 \times 10^{-3}\text{ moles } HC_2H_3O_2$$

$$\#\text{ moles } OH^- = (0.01000\text{ L solution})\left(\frac{0.1000\text{ moles } OH^-}{1\text{ L solution}}\right)$$

$$= 1.000 \times 10^{-3}\text{ moles } OH^-$$

$$[HC_2H_3O_2] = \frac{2.500 \times 10^{-3}\text{ moles} - 1.000 \times 10^{-3}\text{ moles}}{0.02500\text{ L} + 0.01000\text{ L}}$$

$$= 4.286 \times 10^{-2}\text{ M } HC_2H_3O_2$$

$$\left[C_2H_3O_2^-\right] = \frac{0 \text{ moles} + 1.000 \times 10^{-3} \text{ moles}}{0.02500 \text{ L} + 0.01000 \text{ L}}$$

$$= 2.857 \times 10^{-2} \text{ M } C_2H_3O_2^-$$

$$pH = pK_a + \log \frac{\left[C_2H_3O_2^-\right]}{\left[HC_2H_3O_2\right]}$$

$$= 4.7447 + \log \frac{\left(2.857 \times 10^{-2}\right)}{\left(4.286 \times 10^{-2}\right)} = 4.5686$$

(c) When half the acetic acid has been neutralized, there will be equal amounts of acetic acid and acetate ion present in the solution. At this point, $pH = pK_a = 4.7447$.

(d) At the equivalence point, all of the acetic acid will have been converted to acetate ion. The concentration of the acetate ion will be half the original concentration of acetic acid since we have doubled the volume of the solution. We then need to solve the equilibrium problem that results when we have a solution that possesses a $[C_2H_3O_2^-] = 0.05000$ M.

$C_2H_3O_2^- + H_2O \leftrightarrows HC_2H_3O_2 + OH^-$

$$K_b = \frac{\left[HC_2H_3O_2\right]\left[OH^-\right]}{\left[C_2H_3O_2^-\right]} = 5.6 \times 10^{-10}$$

	$[C_2H_3O_2^-]$	$[HC_2H_3O_2]$	$[OH^-]$
I	0.05000	–	–
C	–x	+x	+x
E	0.05000–x	+x	+x

Substituting the above values for equilibrium concentrations into the mass action expression and assuming that $x \ll 0.05000$ gives: $x = [OH^-] = 5.292 \times 10^{-6}$ M.
$pOH = -\log [OH^-] = -\log (5.292 \times 10^{-6}) = 5.2764$.
$pH = 14.0000 - pOH = 14.0000 - 5.2764 = 8.7236$.

18.121 a) $NH_3 + H_2O \leftrightarrows NH_4^+ + OH^-$

$$K_b = \frac{\left[NH_4^+\right]\left[OH^-\right]}{\left[NH_3\right]} = 1.8 \times 10^{-5}$$

	$[NH_3]$	$[NH_4^+]$	$[OH^-]$
I	0.1000	–	–
C	–x	+x	+x
E	0.1000–x	+x	+x

Substituting the above values for equilibrium concentrations into the mass action expression and assuming that $x \ll 0.1000$ gives:

$x = [OH^-] = 1.342 \times 10^{-3}$ M.
$pOH = -\log [OH^-] = -\log (1.342 \times 10^{-3}) = 2.8724$.
$pH = 14.00 - pOH = 11.1276$

(b) When HCl is added, it will react with the ammonia present decreasing the amount in solution and producing additional ammonium ion. Since this a one–to–one reaction, the number of moles of ammonia will decrease by the same amount as the number of moles of HCl added and the number of moles of ammonium ion will increase by an identical amount. We must determine the number of moles of all ions present and calculate new concentrations accounting for dilution.

$$\#\text{ moles NH}_3 = (0.02500\text{ L solution})\left(\frac{0.1000\text{ moles NH}_3}{1\text{ L solution}}\right)$$

$$= 2.500 \times 10^{-3}\text{ moles NH}_3$$

$$\#\text{ moles H}^+ = (0.01000\text{ L solution})\left(\frac{0.1000\text{ moles H}^+}{1\text{ L solution}}\right)$$

$$= 1.000 \times 10^{-3}\text{ moles H}^+$$

$$[\text{NH}_3] = \frac{2.500 \times 10^{-3}\text{ moles} - 1.000 \times 10^{-3}\text{ moles}}{0.02500\text{ L} + 0.01000\text{ L}}$$

$$= 4.286 \times 10^{-2}\text{ M NH}_3$$

$$\left[\text{NH}_4^{\ +}\right] = \frac{0\text{ moles} + 1.000 \times 10^{-3}\text{ moles}}{0.02500\text{ L} + 0.01000\text{ L}}$$

$$= 2.857 \times 10^{-2}\text{ M C}_2\text{H}_3\text{O}_2^{\ -}$$

$$\text{pOH} = \text{pK}_b + \log\frac{\left[\text{NH}_4^{\ +}\right]}{[\text{NH}_3]}$$

$$= 4.7447 + \log\frac{(2.857 \times 10^{-2})}{(4.286 \times 10^{-2})} = 4.5686$$

$\text{pH} = 14.0000 - \text{pOH} = 9.4314$

(c) When half the ammonia has been neutralized, there will be equal amounts of ammonia and ammonium ion present in the solution. At this point, $\text{pH} = \text{pK}_b = 4.7447$.

(d) At the equivalence point, all of the ammonia will have been converted to ammonium ion. The concentration of the ammonium ion will be half the original concentration of ammonia since we have doubled the volume of the solution. We then need to solve the equilibrium problem that results when we have a solution that possesses a $[\text{NH}_4^+] = 0.05000$ M.

$\text{NH}_4^+ \leftrightarrows \text{H}^+ + \text{NH}_3$

$$K_a = \frac{[\text{H}^+][\text{NH}_3]}{\left[\text{NH}_4^{\ +}\right]} = 5.6 \times 10^{-10}$$

	$[\text{NH}_4^+]$	$[\text{H}^+]$	$[\text{NH}_3]$
I	0.05000	–	–
C	–x	+x	+x
E	0.05000–x	+x	+x

Substituting the above values for equilibrium concentrations into the mass action expression and assuming that x << 0.05000 gives:
$x = [H^+] = 5.292 \times 10^{-6}$ M.
$pH = -\log [H^+] = -\log (5.292 \times 10^{-6}) = 5.2764$.

Additional Exercises

18.122 $HC_2H_3O_2 + H_2O \rightleftharpoons H_3O^+ + C_2H_3O_2^-$

$$K_a = \frac{[H_3O^+][C_2H_3O_2^-]}{[HC_2H_3O_2]} = 1.8 \times 10^{-5}$$

	$[HC_2H_3O_2]$	$[H_3O^+]$	$[C_2H_3O_2^-]$
I	1.0	–	–
C	1.0–x	+x	+x
E	1.0 –x	+x	+x

Assume x << 1.0

$$K_a = \frac{[x][x]}{[1.0]} = 1.8 \times 10^{-5} \quad x = 4.2 \times 10^{-3} = [H_3O^+]$$

% ionization = $([H^+]/[\text{total acid}]) \times 100\% = (4.2 \times 10^{-3}/1.0) \times 100\% = 0.42\%$

At lower concentrations, the only change is the denominator in the final expression for K_a above. Calculating the answers, we have:

1.0 M = 0.42% ionization
0.10 M = 1.3 % ionization
0.010 M = 4.2 % ionization

As a weak acid becomes more dilute, its % ionization goes up.

18.123 From the complete ionization of HCl: $[H^+] = 0.100$ M
We must solve an equilibrium problem in order to determine the contribution of H^+ from the acetic acid.

$HC_2H_3O_2 \leftrightharpoons H^+ + C_2H_3O_2^-$ $\qquad K_a = \frac{[H^+][C_2H_3O_2^-]}{[HC_2H_3O_2]} = 1.8 \times 10^{-5}$

	$[HC_2H_3O_2]$	$[H^+]$	$[C_2H_3O_2^-]$
I	0.125	0.100	–
C	–x	+x	+x
E	0.125–x	0.100+x	+x

Substituting the above values for equilibrium concentrations into the mass action expression and assuming that x << 0.125 M and x << 0.100 M gives:
$x = [C_2H_3O_2^-] = 2.25 \times 10^{-5}$ M.
The total $[H^+]$ is therefore $0.100 + 2.25 \times 10^{-5} = 0.100$ M, and pH is 1.000. The weak acid contributes very little H^+ to the final solution.

18.124 When these two solutions are mixed, the HCl will react with the ammonia producing NH_4^+. Since HCl is the limiting reactant, we will have an excess of ammonia. The resulting solution will be a buffer as it consists of ammonia, a weak base, and ammonium ion, its conjugate acid.

We need to determine the amount of NH_4^+ that is produced and the amount of NH_3 that remains:

$$\#\text{ mol H}^+ = (100\text{ mL HCl})\left(\frac{0.500\text{ mol HCl}}{1000\text{ mL HCl}}\right)\left(\frac{1\text{ mol H}^+}{1\text{ mol HCl}}\right) = 0.0500\text{ mol H}^+$$

$$\#\text{ mol NH}_3 = \#\text{ mol initially} - \text{mol H}^+\text{ added}$$

$$= (300\text{ mL HCl})\left(\frac{0.500\text{ mol HCl}}{1000\text{ mL HCl}}\right)\left(\frac{1\text{ mol H}^+}{1\text{ mol HCl}}\right) - 0.0500\text{ mol H}^+$$

$$= 0.100\text{ mol NH}_3$$

$$\#\text{ mol NH}_4^+ = \#\text{ mol H}^+ = 0.0500\text{ mol NH}_4^+$$

$$\text{pOH} = \text{pK}_b + \log\frac{[\text{NH}_4^+]}{[\text{NH}_3]} = 4.74 - 0.30 = 4.44$$

pH = 14.00 – pOH = 9.56

18.125 Only the cation is hydrolyzed:

$$\text{NH}_4^+ + \text{H}_2\text{O} \leftrightarrows \text{H}_3\text{O}^+ + \text{NH}_3 \qquad K_a = \frac{[\text{H}_3\text{O}^+][\text{NH}_3]}{[\text{NH}_4^+]} = 5.6 \times 10^{-10}$$

	$[NH_4^+]$	$[H_3O^+]$	$[NH_3]$
I	0.120	–	–
C	–x	+x	+x
E	0.120–x	+x	+x

$$K_a = \frac{(x)(x)}{(0.120 - x)} = 5.6 \times 10^{-10}$$

Assuming x to be smaller than 0.120, we have: $x = 8.2 \times 10^{-6}$ M

$\text{pH} = -\log(8.2 \times 10^{-6}) = 5.09$

18.126 (a) We first need to determine the concentration of the components of the solution.

$$[\text{HC}_2\text{H}_3\text{O}_2] = \frac{\#\text{ mol HC}_2\text{H}_3\text{O}_2}{\#\text{ L soln}}$$

$$= \left(\frac{15.0\text{ g HC}_2\text{H}_3\text{O}_2}{0.750\text{ L}}\right)\left(\frac{1\text{ mol HC}_2\text{H}_3\text{O}_2}{60.05\text{ g HC}_2\text{H}_3\text{O}_2}\right)$$

$$= 0.333\text{ M}$$

$$[\text{NaC}_2\text{H}_3\text{O}_2] = \frac{\#\text{ mol NaC}_2\text{H}_3\text{O}_2}{\#\text{ L soln}}$$

$$= \left(\frac{25.0\text{ g NaC}_2\text{H}_3\text{O}_2}{0.750\text{ L}}\right)\left(\frac{1\text{ mol NaC}_2\text{H}_3\text{O}_2}{82.03\text{ g NaC}_2\text{H}_3\text{O}_2}\right)$$

$$= 0.406\text{ M}$$

$$pH = pK_a + \log \frac{[\text{anion}]}{[\text{acid}]} = 4.745 + \log \frac{0.406}{0.333} = 4.831$$

(b) The added base will react with the acid present using it up and producing additional acetate ion. To solve this problem, determine the number of moles of base added and subtract this amount from the amount of acid present. Add the same amount to the amount of acetate ion present. Be sure to account for dilution when you determine the new concentrations.

$$\#\text{ mol OH}^- = (25.00 \text{ mL NaOH})\left(\frac{0.25 \text{ mol NaOH}}{1000 \text{ mL NaOH}}\right)\left(\frac{1 \text{ mol OH}^-}{1 \text{ mol NaOH}}\right)$$

$$= 0.0063 \text{ mol OH}^-$$

Consequently, the # mol of acetic acid will decrease by 0.0063 and the # of moles of acetate ion will increase by 0.0063. The new concentrations are:

$$[HC_2H_3O_2] = \frac{\#\text{ mol } HC_2H_3O_2}{\#\text{ L soln}}$$

$$= \left(\frac{(0.250 - 0.0063)\text{ mol } HC_2H_3O_2}{0.775 \text{ L}}\right)$$

$$= 0.314 \text{ M}$$

$$[C_2H_3O_2^-] = \frac{\#\text{ mol } C_2H_3O_2^-}{\#\text{ L soln}}$$

$$= \left(\frac{(0.305 + 0.0063)\text{ mol } C_2H_3O_2^-}{0.775 \text{ L}}\right)$$

$$= 0.402 \text{ M}$$

$$pH = pK_a + \log \frac{[\text{anion}]}{[\text{acid}]} = 4.745 + \log \frac{0.402}{0.314} = 4.852$$

(c) This problem is identical to the previous except that we are adding acid to the original solution. Consequently, the acid concentration will increase and the acetate ion concentration will decrease.

$$\#\text{ mol H}^+ = (25.0 \text{ mL HCl})\left(\frac{0.40 \text{ mol HCl}}{1000 \text{ mL HCl}}\right)\left(\frac{1 \text{ mol H}^+}{1 \text{ mol HCl}}\right)$$

$$= 0.010 \text{ mol H}^+$$

$$[HC_2H_3O_2] = \frac{\#\text{ mol } HC_2H_3O_2}{\#\text{ L soln}}$$

$$= \left(\frac{(0.250 + 0.010)\text{ mol } HC_2H_3O_2}{0.775 \text{ L}}\right)$$

$$= 0.335 \text{ M}$$

$$\left[C_2H_3O_2^-\right] = \frac{\#\text{ mol } C_2H_3O_2^-}{\#\text{ L soln}}$$

$$= \left(\frac{(0.305 - 0.010)\text{mol } C_2H_3O_2^-}{0.775\text{ L}}\right)$$

$$= 0.381\text{ M}$$

$$pH = pK_a + \log\frac{[\text{anion}]}{[\text{acid}]} = 4.745 + \log\frac{0.381}{0.335} = 4.801$$

18.127 (a) The equation that will be used is the normal Henderson–Hasselbach equation, namely:

$$pH = pK_a + \log\frac{[\text{anion}]}{[\text{acid}]} = pK_a + \log\frac{[A^-]}{[HA]}$$

where $A^- = C_2H_3O_2^-$ and $HA = HC_2H_3O_2$. We note further that the log term involves a ratio of concentrations, but that the volume remains constant in a process such as that to be analyzed here. Thus the log term may be replaced by a ratio of mole amounts, since volumes cancel:

$$pH = pK_a + \log\frac{(\text{moles } C_2H_3O_2^-)}{(\text{moles } HC_2H_3O_2)}$$

Thus we need only determine the number of moles of acid and conjugate base that remain in the buffer after the addition of a certain amount of H^+ or OH^-, in order to determine the pH of the buffer mixture after that addition.

The buffer is changed in the following way by the addition of OH^-:

$$HC_2H_3O_2 + OH^- \rightleftharpoons H_2O + C_2H_3O_2^-$$

In other words, if 0.0100 moles of OH^- are added to the buffer, the amount of $HC_2H_3O_2$ goes down by 0.0100 moles, whereas the amount of $C_2H_3O_2^-$ goes up by 0.0100 moles. In addition, the pH of the solution will increase upon the addition of base.

The buffer is changed in the following way by the addition of H^+:

$$C_2H_3O_2^- + H^+ \rightleftharpoons HC_2H_3O_2$$

If 0.0100 mol of H^+ are added, then the amount of $C_2H_3O_2^-$ goes down by 0.0100 moles and the amount of $HC_2H_3O_2$ goes up by 0.0100 moles. As before, the addition of acid will decrease the pH of the buffer system.

For the general buffer mixture that contains x mol of $C_2H_3O_2^-$ and y mol of $HC_2H_3O_2$, we can apply the Henderson–Hasselbach equation, noting the maximum amount by which we want the pH to change (namely 0.10 units):

For the case of added base:

$$5.12 + 0.10 = 4.74 + \log\frac{(x + 0.0100)\text{moles } C_2H_3O_2^-}{(x - 0.0100)\text{moles } HC_2H_3O_2}$$

Simplifying and taking the antilog of both sides of the above equation gives: [x + 0.0100]/[y – 0.0100] = 3.0, which is now designated eq. 1.

For the case of added acid:

$$5.12 - 0.10 = 4.74 + \log \frac{(x-0.0100)\text{moles } C_2H_3O_2^-}{(x+0.0100)\text{moles } HC_2H_3O_2}$$

Simplifying and taking the antilog of both sides of the equation gives:
$[x - 0.0100]/[y + 0.0100] = 1.9$, which is now designated eq. 2.

The equations 1 and 2 as designated above are solved simultaneously, since they are two equations containing two unknowns:

x = initial mol $C_2H_3O_2^-$ = 0.15 mol
y = initial mol $HC_2H_3O_2$ = 6.3×10^{-2} mol

These values are converted to grams as follows:
6.3×10^{-2} mol × 60.1 g/mol = 3.8 g $HC_2H_3O_2$
0.15 mol $NaC_2H_3O_2$ × 118 g/mol = 18 g $NaC_2H_3O_2{\cdot}2H_2O$

These are the minimum amounts of acid and conjugate base that would be required in order to prepare a buffer that would change pH by only 0.10 units on addition of either 0.0100 mol of OH^- or 0.0100 mol of H^+.

(b) 6.3×10^{-2} mol/0.250 L = 0.25 M $HC_2H_3O_2$
0.15 mol $C_2H_3O_2^-$/0.250 L = 0.60 M $NaC_2H_3O_2$

(c) 0.0100 mol OH^-/0.250 L = 4.00×10^{-2} M OH^-
pOH = $-\log[OH^-] = -\log(4.00 \times 10^{-2}) = 1.398$
pH = 14.000 – pOH = 14.000 – 1.398 = 12.602

(d) 1.00×10^{-2} mol H^+/0.250 L = 4.00×10^{-2} M H^+
pH = $-\log[H^+] = -\log(4.0 \times 10^{-2}) = 1.398$

18.128 (a) Formic acid $pK_a = 3.74$
(b) Start by determining the ratio of acid and base in the buffer solution.

$$pH = pK_a + \log \frac{[NaCHO_2]}{[HCHO_2]}$$

$$\frac{[NaCHO_2]}{[HCHO_2]} = 10^{\,pH - pKa} = 10^{(3.80 - 3.74)}$$

$$= 1.15$$

Since the volume is constant, the ratio of moles $NaCHO_2$ to moles $HCHO_2$ = 1.15. We need a minimum of 5.0×10^{-3} mol of $HCHO_2$. So,

$$\#\text{ g } HCHO_2 = (5.0 \times 10^{-3}\text{ mol } HCHO_2)\left(\frac{46.0\text{ g } HCHO_2}{1\text{ mol } HCHO_2}\right) = 0.230\text{ g } HCHO_2$$

$$\#\text{ g } NaCHO_2 = (5.7 \times 10^{-3}\text{ mol } NaCHO_2)\left(\frac{68.0\text{ g } NaCHO_2}{1\text{ mol } NaCHO_2}\right) = 0.390\text{ g } NaCHO_2$$

(c)

$$[HCHO_2] = \frac{5.0 \times 10^{-3}\text{ mol}}{0.750\text{ L}} = 6.7 \times 10^{-3}\text{ M}$$

$$[NaCHO_2] = \frac{5.7 \times 10^{-3}\text{ mol}}{0.750\text{ L}} = 7.6 \times 10^{-3}\text{ M}$$

(d)

$$[H_3O^+] = 10^{-pH} = 1.6 \times 10^{-4}$$

$$\text{\# moles } H_3O^+ = (0.750 \text{ L})\left(\frac{1.6 \times 10^{-4} \text{ mol}}{1 \text{ L}}\right) = 1.2 \times 10^{-4} \text{ moles}$$

add 5.0×10^{-3} moles # moles $H_3O^+ = 5.1 \times 10^{-3}$ moles

$$[H_3O^+] = \frac{5.1 \times 10^{-3} \text{ mol}}{0.750 \text{ L}} = 6.8 \times 10^{-3} \text{ M}$$

$pH = 2.17$

(e) From part (d) we know there are 1.2×10^{-4} moles H_3O^+ before adding the base. After adding 5.0×10^{-3} moles NaOH the acid will be neutralized and there will be 4.9×10^{-3} moles excess base.

$$[OH^-] = \frac{4.9 \times 10^{-3} \text{ mol}}{0.750 \text{ L}} = 6.5 \times 10^{-3} \text{ M}$$

$pOH = 2.18$

$pH = 11.82$

18.129 Since the ammonium ion is the salt of a weak base, NH_3, it is acidic. The cyanide ion is the salt of a weak acid, HCN, so it is basic. In order to determine if the solution is acidic or basic, we need to determine the relative strength of the two components. Use the relationship $K_a \times K_b = K_w$ in order to determine the dissociation constants for the cyanide ion and the ammonium ion.

$K_a(NH_4^+) = K_w \div K_b(NH_3) = 1.0 \times 10^{-14} \div 1.8 \times 10^{-5} = 5.6 \times 10^{-10}$
$K_b(CN^-) = K_w \div K_a(HCN) = 1.0 \times 10^{-14} \div 6.2 \times 10^{-9} = 1.6 \times 10^{-5}$

Since the $K_b(CN^-)$ is larger than the $K_a(NH_4^+)$ the NH_4CN solution will be basic.

18.130 Recall the $\Delta t_f = K_f m$, where K_f is the molal freezing point depression constant for the solvent and m is the molality of the solution. We need to determine the molality of this solution. In order to do this, we need to determine the concentrations of the ions which result from the dissociation of the weak acid. Recall that each ion in the solution helps to lower the freezing point.

$$HC_2HO_2Cl_2 \leftrightarrows H^+ + C_2HO_2Cl_2^-$$

$$K_a = \frac{[H^+][C_2HO_2Cl_2^-]}{[HC_2HO_2Cl_2]} = 5.0 \times 10^{-2}$$

	$[HC_2HO_2Cl_2]$	$[H^+]$	$[C_2HO_2Cl_2^-]$
I	0.50	–	–
C	–x	+x	+x
E	0.50–x	+x	+x

$$K_a = \frac{(x)(x)}{(0.50 - x)} = 5.0 \times 10^{-2}$$

Because the value of K_a is relatively large, we must solve a quadratic equation in order to determine the H^+ concentration. Doing so gives:

$x = 0.14 \text{ M} = [H^+] = [C_2HO_2Cl_2^-]$
$[HC_2HO_2Cl_2] = 0.50 - x = 0.36 \text{ M}$

We now calculate the molality of the components of the solution. Because this is an aqueous solution with a density of 1.0 g/mL, the molality and the molarity have the same value: $m_{H^+} = 0.14$, $m_{C_2HO_2Cl_2^-} = 0.14$ and $m_{HC_2HO_2Cl_2} = 0.36$.

The total molality is 0.14 + 0.14 + 0.36 = 0.64 *m*.

The freezing point depression constant for water equals –1.86 °C/*m*.

$$\Delta t_F = (-1.86\ ^\circ\text{C/m})(0.64\ \text{m}) = -1.19\ ^\circ\text{C}.$$

The freezing point of the solution is –1.19 °C.

18.131 In order to solve this problem we must first neutralize the HNO_3 that is present in the solution. Since HNO_3 is a monoprotic acid and NH_3 is a monobasic base, they will react in a one to one stoichiometry. For every mole of NH_3 added, one mole of HNO_3 will be neutralized and one mole of NH_4^+ will be produced. There are initially 0.0125 moles HNO_3 = (0.250 L)(0.050 mol L^{-1}) present in the solution. Consequently, we must first add 0.0125 moles of NH_3 to neutralize all of the HNO_3.

Now the question may be rephrased to ask, how much ammonia should be added to a solution that is 0.050 M in NH_4^+ so that the final pH is 9.26?

	$[NH_3]$	$[NH_4^+]$	$[OH^-]$
I	Z	0.050	–
C	Z–x	0.050+x	+x
E	Z–x	0.050+x	+x

$$K_b = \frac{(0.050 + x)(x)}{(Z - x)} = 1.8 \times 10^{-5}$$

The problem states that the equilibrium pH = 9.26 and, therefore, the pOH = 4.74 which implies that $[OH^-]$ = 1.8×10^{-5} M = x. Substituting this into the equilibrium expression and assuming that Z >> x we get

$$\frac{(0.050)(1.8 \times 10^{-5})}{Z} = 1.8 \times 10^{-5}$$

and Z = 0.050 M. Since the volume of the solution is 250 mL, this concentration corresponds to 0.0125 moles NH_3 = (0.250 L)(0.050 mol L^{-1}).

The total amount of NH_3 that must be added to the original solution is thus 0.0125 moles + 0.0125 moles = 0.0250 mols. The first addition neutralizes the HNO_3 and the second amount adjusts the pH.

The volume of NH_3 that must be added may be calculated using the ideal gas law;

$$V = \frac{nRT}{P} = \frac{(0.0250\ \text{mol})\left(0.0821\ \frac{\text{L atm}}{\text{mol K}}\right)(298\ \text{K})}{(740\ \text{torr})\left(\frac{1\ \text{atm}}{760\ \text{torr}}\right)} = 0.63\ \text{L} = 630\ \text{mL}$$

18.132 (a) $HSO_4^- + H_2O \leftrightarrows H_3O^+ + SO_4^{2-}$

$$K_a = \frac{[SO_4^{2-}][H_3O^+]}{[HSO_4^-]} = 1.0 \times 10^{-2}$$

(b)

	$[HSO_4^-]$	$[H_3O^+]$	$[SO_4^{2-}]$
I	0.010	–	–
C	–x	+x	+x
E	0.010–x	+x	+x

$$K_a = \frac{[x][x]}{0.010 - x} = 1.0 \times 10^{-2}$$

Solve the quadratic equation to get $x = 6.2 \times 10^{-3} = [H_3O^+]$

(c) $K_a = \frac{[x][x]}{0.010} = 1.0 \times 10^{-2}$ $x = 1.0 \times 10^{-2} = [H_3O^+]$

(d) The exact answer (obtained in part b) is 38% smaller than the approximate solution.

18.133 Ascorbic acid has the formula $H_2C_6H_6O_6$ and is a diprotic weak acid. We only need look at the first dissociation to determine the pH.

$H_2C_6H_6O_6 \leftrightarrows HC_6H_6O_6^- + H^+$ $K_{a_1} = \frac{[HC_6H_6O_6^-][H^+]}{[H_2C_6H_6O_6]} = 6.8 \times 10^{-5}$

$$[H_2C_6H_6O_6] = \left(\frac{6.0 \text{ g } H_2C_6H_6O_6}{250 \text{ mL}}\right)\left(\frac{1 \text{ mole } H_2C_6H_6O_6}{176 \text{ g } H_2C_6H_6O_6}\right)\left(\frac{1000 \text{ mL}}{1 \text{ L}}\right) = 0.14 \text{ M}$$

	$[H_2C_6H_6O_6]$	$[HC_6H_6O_6^-]$	$[H^+]$
I	0.14	–	–
C	–x	+x	+x
E	0.14–x	+x	+x

$$K_{a_1} = \frac{[HC_6H_6O_6^-][H^+]}{[H_2C_6H_6O_6]} = \frac{(x)(x)}{(0.14 - x)} = 6.8 \times 10^{-5}$$

Assuming that $x << 0.14$ and solving we find;

$x = [H^+] = 3.1 \times 10^{-3}$ M $pH = -\log[H^+] = -\log(3.1 \times 10^{-3}) = 2.51$

18.134

$$pH = pK_a - \log\frac{[CO_2]}{[HCO_3^-]}$$

$$[HCO_3^-] = [CO_2]10^{pH - pK_a}$$

$$[HCO_3^-] = (0.022)10^{(7.35-6.35)} = 0.22 \text{ M}$$

18.135 The first dissociation gives the concentrations of H_3PO_4, H^+ and $H_2PO_4^-$:

$H_3PO_4(aq) \leftrightarrows H^+(aq) + H_2PO_4^-(aq)$

$$K_{a_1} = \frac{[H^+][H_2PO_4^-]}{[H_3PO_4]} = 7.1 \times 10^{-3}$$

	$[H_3PO_4]$	$[H_2PO_4^-]$	$[H^+]$
I	0.30	–	–
C	–x	+x	+x
E	0.30–x	+x	+x

$$K_{a_1} = \frac{[H^+][H_2PO_4^-]}{[H_3PO_4]} = \frac{(x)(x)}{(0.30 - x)} = 7.1 \times 10^{-3}$$

Assuming that x << 0.30 and solving we find;

$x = 4.6 \times 10^{-2}$ M

A second iteration gives: $x = 4.2 \times 10^{-2}$ M
A third iteration gives: $x = 4.3 \times 10^{-2}$ M
A fourth iteration gives: $x = 4.3 \times 10^{-2}$ M

$[H_3PO_4] = 0.30 - x = 0.26$ M.
$[H^+] = [H_2PO_4^-] = x = 0.043$ M

The second dissociation gives the concentration of HPO_4^{2-}. As was shown in Example 18.18, $[HPO_4^{2-}] = K_{a_2} = 6.3 \times 10^{-8}$ M.

The third dissociation gives the $[PO_4^{3-}]$

$HPO_4^{2-}(aq) \leftrightarrows H^+(aq) + PO_4^{3-}(aq)$

$$K_{a_3} = \frac{[H^+][PO_4^{3-}]}{[HPO_4^{2-}]} = 4.5 \times 10^{-13}$$

	$[HPO_4^-]$	$[PO_4^{3-}]$	$[H^+]$
I	6.3×10^{-8}	–	0.043
C	–x	+x	+x
E	6.3×10^{-8}–x	+x	0.043+x

$$K_{a_3} = \frac{[H^+][H_2PO_4^-]}{[H_3PO_4]} = \frac{(0.043 + x)(x)}{(6.3 \times 10^{-8} - x)} = 4.5 \times 10^{-13}$$

Assuming that $x << 6.3 \times 10^{-8}$ and solving we find;

$x = 6.6 \times 10^{-19}$ M = $[PO_4^{3-}]$

18.136 This exercise is an example of a titration using a strong acid and a strong base. Prior to reaching the equivalence point, there will be an excess of acid present. The amount of acid present may be determined by subtracting the number of moles of base added from the amount of acid initially present. We may then use the dilution equation, $M_1 \times V_1 = M_2 \times V_2$, to determine the new concentration of acid and then calculate the pH. After the equivalence point, there is an excess of base. The amount in excess is determined by subtracting the inital amount of acid present from the amount of base added. Again, the dilution equation may be used to determine the resulting concentration.

Start by determining the amount of H^+ initially present:

$$\text{\# moles H}^+ = (25.00 \text{ mL solution})\left(\frac{0.1000 \text{ moles H}^+}{1000 \text{ mL solution}}\right)$$

$$= 2.500 \times 10^{-3} \text{ moles H}^+$$

(a) Initially, there is only 0.1000 M HCl present: $[H^+] = 0.1000$ M, $pH = -\log[H^+] = 1.0000$.

(b) 10.00 mL added base

$$\text{\# moles OH}^- = (10.00 \text{ mL solution})\left(\frac{0.1000 \text{ moles OH}^-}{1000 \text{ mL solution}}\right)$$

$$= 1.000 \times 10^{-3} \text{ moles OH}^-$$

$$[H^+] = \frac{2.500 \times 10^{-3} \text{ moles} - 1.000 \times 10^{-3} \text{ moles}}{0.02500 \text{ L} + 0.01000 \text{ L}}$$

$$= 4.286 \times 10^{-2} \text{ M H}^+$$

$$pH = -\log[H^+] = 1.3680$$

(c) 24.90 mL added base

$$\text{\# moles OH}^- = (24.90 \text{ mL solution})\left(\frac{0.1000 \text{ moles OH}^-}{1000 \text{ mL solution}}\right)$$

$$= 2.490 \times 10^{-3} \text{ moles OH}^-$$

$$[H^+] = \frac{2.500 \times 10^{-3} \text{ moles} - 2.490 \times 10^{-3} \text{ moles}}{0.02500 \text{ L} + 0.02490 \text{ L}}$$

$$= 2.004 \times 10^{-4} \text{ M H}^+$$

$$pH = -\log[H^+] = 3.6981$$

(d) 24.99 mL added base

$$\text{\# moles OH}^- = (24.99 \text{ mL solution})\left(\frac{0.1000 \text{ moles OH}^-}{1000 \text{ mL solution}}\right)$$

$$= 2.499 \times 10^{-3} \text{ moles OH}^-$$

$$[H^+] = \frac{2.500 \times 10^{-3} \text{ moles} - 2.499 \times 10^{-3} \text{ moles}}{0.02500 \text{ L} + 0.02499 \text{ L}}$$

$$= 2.000 \times 10^{-5} \text{ M H}^+$$

$$pH = -\log[H^+] = 4.6989$$

(e) 25.00 mL added base

This is the equivalence point and $[H^+] = 1.000 \times 10^{-7}$ M
pH = 7.000

(f) 25.01 mL added base

We have now added excess base so we will subtract the # moles H^+ initially present from the # moles OH^- added.

$$\text{\# moles } OH^- = (25.01 \text{ mL solution})\left(\frac{0.1000 \text{ moles } OH^-}{1000 \text{ mL solution}}\right)$$

$$= 2.501 \times 10^{-3} \text{ moles } OH^-$$

$$[OH^-] = \frac{2.501 \times 10^{-3} \text{ moles} - 2.500 \times 10^{-3} \text{ moles}}{0.02501 \text{ L} + 0.02500 \text{ L}}$$

$$= 2.000 \times 10^{-5} \text{ M } OH^-$$

$$pOH = -\log[OH^-] = 4.6991$$

$$pH = 14.0000 - pOH = 9.3009$$

(g) 25.10 mL added base

$$\text{\# moles } OH^- = (25.10 \text{ mL solution})\left(\frac{0.1000 \text{ moles } OH^-}{1000 \text{ mL solution}}\right)$$

$$= 2.510 \times 10^{-3} \text{ moles } OH^-$$

$$[OH^-] = \frac{2.510 \times 10^{-3} \text{ moles} - 2.500 \times 10^{-3} \text{ moles}}{0.02510 \text{ L} + 0.02500 \text{ L}}$$

$$= 1.996 \times 10^{-4} \text{ M } OH^-$$

$$pOH = -\log[OH^-] = 3.6998$$

$$pH = 14.0000 - pOH = 10.3000$$

(h) 26.00 mL added base

$$\text{\# moles } OH^- = (26.00 \text{ mL solution})\left(\frac{0.1000 \text{ moles } OH^-}{1000 \text{ mL solution}}\right)$$

$$= 2.600 \times 10^{-3} \text{ moles } OH^-$$

$$[OH^-] = \frac{2.600 \times 10^{-3} \text{ moles} - 2.500 \times 10^{-3} \text{ moles}}{0.02600 \text{ L} + 0.02500 \text{ L}}$$

$$= 1.961 \times 10^{-3} \text{ M } OH^-$$

$$pOH = -\log[OH^-] = 2.7076$$

$$pH = 14.0000 - pOH = 11.2924$$

(i) 50.00 mL added base

$$\# \text{ moles OH}^- = (50.00 \text{ mL solution})\left(\frac{0.1000 \text{ moles OH}^-}{1000 \text{ mL solution}}\right)$$

$$= 5.000 \times 10^{-3} \text{ moles OH}^-$$

$$[\text{OH}^-] = \frac{5.000 \times 10^{-3} \text{ moles} - 2.500 \times 10^{-3} \text{ moles}}{0.05000 \text{ L} + 0.02500 \text{ L}}$$

$$= 3.333 \times 10^{-2} \text{ M OH}^-$$

$$\text{pOH} = -\log[\text{OH}^-] = 1.4771$$

$$\text{pH} = 14.0000 - \text{pOH} = 12.5229$$

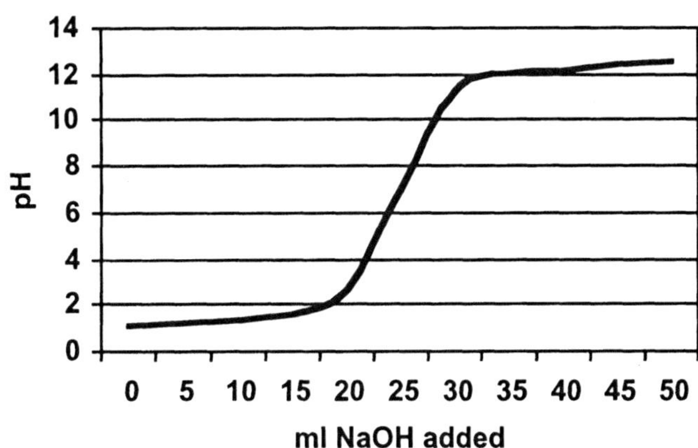

Practice Exercises

19.1 a) $K_{sp} = [Ba^{2+}][C_2O_4^{2-}]$ b) $K_{sp} = [Ag^+]^3[PO_4^{3-}]$

19.2 $TlI \leftrightarrows Tl^+ + I^-$

	$[Tl^+]$	$[I^-]$
I	–	–
C	+x	+x
E	+x	+x

$K_{sp} = x^2 = (1.8 \times 10^{-5})^2 = 3.2 \times 10^{-10}$

19.3 $PbF_2(s) \leftrightarrows Pb^{2+}(aq) + 2F^-(aq)$ $K_{sp} = [Pb^{2+}][F^-]^2$

$K_{sp} = (2.15 \times 10^{-3})(2(2.15 \times 10^{-3}))^2 = 3.98 \times 10^{-8}$

19.4 $CoCO_3(s) \leftrightarrows Co^{2+}(aq) + CO_3^{2-}(aq)$

	$[Co^{2+}]$	$[CO_3^{2-}]$
I	–	0.10
C	+1.0 × 10–9	+ 1.0 × 10–9
E	+1.0 × 10–9	0.10 + 1.0 × 10–9

Substituting the above values for equilibrium concentrations into the expression for K_{sp} gives:

$K_{sp} = [Co^{2+}][CO_3^{2-}] = (1.0 \times 10^{-9})(0.10 + 1.0 \times 10^{-9}) = 1.0 \times 10^{-10}$

19.5 $PbF_2(s) \leftrightarrows Pb^{2+}(aq) + 2F^-(aq)$

	$[Pb^{2+}]$	$[F^-]$
I	0.10	–
C	+ 3.1 × 10–4	+ 2(3.1 × 10–4)
E	0.10 + 3.1 × 10–4	+ 6.2 × 10–4

Substituting the above values for equilibrium concentrations into the expression for K_{sp} gives:

$K_{sp} = [Pb^{2+}][F^-]^2 = [0.10 + 3.1 \times 10^{-4}][6.2 \times 10^{-4}]^2$
Now $(0.10 + 3.1 \times 10^{-4})$ is also ≈ 0.10:
Hence, $K_{sp} = (0.10)(6.2 \times 10^{-4})^2 = 3.9 \times 10^{-8}$

19.6 a) $AgBr(s) \leftrightarrows Ag^+(aq) + Br^-(aq)$ $K_{sp} = [Ag^+][Br^-] = 5.0 \times 10^{-13}$

	$[Ag^+]$	$[Br^-]$
I	–	–
C	+ x	+ x
E	+ x	+ x

Substituting the above values for equilibrium concentrations into the expression for K_{sp} gives:

$K_{sp} = 5.0 \times 10^{-13} = [Ag^+][Br^-] = (x)(x)$

$x = \sqrt{5.0 \times 10^{-13}} = 7.1 \times 10^{-7}$
Thus the solubility is 7.1×10^{-7} M AgBr.

b) $Ag_2CO_3(s) \leftrightarrows 2Ag^+(aq) + CO_3^{2-}(aq)$ $K_{sp} = [Ag^+]^2[CO_3^{2-}] = 8.1 \times 10^{-12}$

	$[Ag^+]$	$[CO_3^{2-}]$
I	–	–
C	+ 2x	+ x
E	+ 2x	+ x

Substituting the above values for equilibrium concentrations into the expression for K_{sp} gives:

$K_{sp} = 8.1 \text{ x } 10^{-12} = [Ag^+]^2[CO_3^{2-}] = (2x)^2(x)$ and $4x^3 = 8.1 \times 10^{-12}$

$x = \sqrt[3]{\left(8.1 \times 10^{-12}\right)/4} = 1.3 \times 10^{-4}$

Thus the molar solubility is 1.3×10^{-4} M Ag_2CO_3.

19.7 $AgI(s) \leftrightarrows Ag^+(aq) + I^-(aq)$ $K_{sp} = [Ag^+][I^-] = 8.3 \times 10^{-17}$

	$[Ag^+]$	$[I^-]$
I	–	0.20
C	+ x	+ x
E	+ x	0.20 + x

Substituting the above values for equilibrium concentrations into the expression for K_{sp} gives:

$K_{sp} = 8.3 \times 10^{-17} = [Ag^+][I^-] = (x)(0.20 + x)$

We know that the value of K_{sp} is very small, and it suggests the simplifying assumption that $(0.20 + x) \approx 0.20$:

Hence, $8.3 \times 10^{-17} = (0.20)x$, and $x = 4.2 \times 10^{-16}$. The assumption that $(0.20 + x) \approx 0.20$ is seen to be valid indeed.

Thus 4.2×10^{-16} M of AgI will dissolve in a 0.20 M NaI solution.

In pure water,
$K_{sp} = 8.3 \times 10^{-17} = [Ag^+][I^-] = (x)(x)$
$x = [AgI(aq)] = 9.1 \times 10^{-9}$ M (much more soluble)

19.8 $Fe(OH)_3(s) \leftrightarrows Fe^{3+}(aq) + 3OH^-(aq)$ $K_{sp} = [Fe^{3+}][OH^-]^3 = 1.6 \times 10^{-39}$

	$[Fe^{3+}]$	$[OH^-]$
I	–	0.050
C	+ x	+ 3x
E	+ x	0.050 + 3x

Substituting the above values for equilibrium concentrations into the expression for K_{sp} gives:

$$K_{sp} = 1.6 \times 10^{-39} = [Fe^{3+}][OH^-]^3 = (x)[0.050 + 3x]^3$$

We try to simplify by making the approximation that $(0.050 + 3x) \approx 0.050$:

$$1.6 \times 10^{-39} = (x)(0.050)^3 \quad \text{or } x = 1.3 \times 10^{-35}$$

Clearly the assumption that $(0.050 + 3x) \approx 0.050$ is justified.
Thus 1.3×10^{-35} M of $Fe(OH)_3$ will dissolve in a 0.050 M sodium hydroxide solution.

19.9 The expression for K_{sp} is $K_{sp} = [Ca^{2+}][SO_4^{2-}] = 2.4 \times 10^{-5}$ and the ion product for this solution would be:
$[Ca^{2+}][SO_4^{2-}] = (2.5 \times 10^{-3})(3.0 \times 10^{-2}) = 7.5 \times 10^{-5}$

Since the ion product is larger than the value of K_{sp}, a precipitate will form.

19.10 The solubility product constant is $K_{sp} = [Ag^+]^2[CrO_4^{2-}] = 1.2 \times 10^{-12}$ and the ion product for this solution would be:
$[Ag^+]^2[CrO_4^{2-}] = (4.8 \times 10^{-5})^2(3.4 \times 10^{-4}) = 7.8 \times 10^{-13}$

Since the ion product is smaller than the value of K_{sp}, no precipitate will form.

19.11 We expect $PbSO_4(s)$ since nitrates are soluble.

Because two solutions are to be mixed together, there will be a dilution of the concentrations of the various ions, and the diluted ion concentrations must be used. In general, on dilution, the following relationship is found for the concentrations of the initial solution (M_i) and the concentration of the final solution (M_f):
$M_iV_i = M_fV_f$

Thus the final or diluted concentrations are:

$$\left[Pb^{2+}\right] = \left(1.0 \times 10^{-3}\ M\right)\left(\frac{100.0\ mL}{200.0\ mL}\right) = 5.0 \times 10^{-4}\ M$$

$$\left[SO_4^{\ 2-}\right] = \left(2.0 \times 10^{-3}\ M\right)\left(\frac{100.0\ mL}{200.0\ mL}\right) = 1.0 \times 10^{-3}\ M$$

The value of the ion product for the final (diluted) solution is:
$[Pb^{2+}][SO_4^{2-}] = (5.0 \times 10^{-4})(1.0 \times 10^{-3}) = 5.0 \times 10^{-7}$

Since this is smaller than the value of K_{sp} (6.3×10^{-7}), a precipitate of $PbSO_4$ is not expected.

19.12 We expect a precipitate of $PbCl_2$ since nitrates are soluble.
We proceed as in Practice Exercise 11. $M_iV_i = M_fV_f$

$$\left[Pb^{2+}\right] = \left(0.10\ M\right)\left(\frac{50.0\ mL}{70.0\ mL}\right) = 0.071\ M$$

$$\left[Cl^-\right] = \left(0.040\ M\right)\left(\frac{20.0\ mL}{70.0\ mL}\right) = 0.011\ M$$

The value of the ion product for such a solution would be:
$[Pb^{2+}][Cl^-]^2 = (7.1 \times 10^{-2})(1.1 \times 10^{-2})^2 = 8.6 \times 10^{-6}$

Since the ion product is smaller than K_{sp}, a precipitate of $PbCl_2$ is not expected.

19.13 Consulting Table 19.2, we find that Fe^{2+} is much more soluble in acid than Hg^{2+}. We want to make the H^+ concentration large enough to prevent FeS from precipitating, but small enough that HgS *does* precipitate. First, we calculate the highest pH at which FeS will remain soluble, by using K_{spa} for FeS. (Recall that a saturated solution of $H_2S = 0.10$ M.)

$$K_{spa} = \frac{[Fe^{2+}][H_2S]}{[H^+]^2} = \frac{[0.010][0.10]}{[H^+]^2} = 6 \times 10^2$$

$[H^+] = 0.0013$ M
pH = –log $[H^+]$ = 2.9

Since Fe^{2+} is much more soluble in acid than Hg^{2+} we already know that this pH will precipitate HgS, but we can check it by using K_{spa} for HgS:

$$K_{spa} = \frac{[Hg^{2+}][H_2S]}{[H^+]^2} = \frac{[0.010][0.10]}{[H^+]^2} = 2 \times 10^{-32}$$

$[H^+] = 2.2 \times 10^{14}$ M

(This concentration is impossibly high, but it tells us that this much acid would be required to dissolve HgS at these concentrations.)

19.14 Follow the exact procedure outlined in Example 19.10.
$K_{sp} = [Ca^{2+}][CO_3^{2-}] = 4.5 \times 10^{-9}$
$K_{sp} = [Ni^{2+}][CO_3^{2-}] = 1.3 \times 10^{-7}$

$NiCO_3$ is more soluble and will precipitate when:

$$[CO_3^{2-}] = \frac{K_{sp}}{[Ni^{2+}]} = \frac{1.3 \times 10^{-7}}{0.10} = 1.3 \times 10^{-6}$$

$CaCO_3$ will precipitate when:

$$[CO_3^{2-}] = \frac{K_{sp}}{[Ca^{2+}]} = \frac{4.5 \times 10^{-9}}{0.10} = 4.5 \times 10^{-8}$$

$CaCO_3$ will precipitate and $NiCO_3$ will not precipitate if $[CO_3^{2-}]$ .4.5×10^{-8} and $[CO_3^{2-}] < 1.3 \times 10^{-6}$. Now, using the equation highlighted in example 19.10 we get:

$$[H^+]^2 = (2.4 \times 10^{-17})\left(\frac{0.030}{[CO_3^{2-}]}\right) \quad NiCO_3 \text{ will precipitate if}$$

$$[H^+]^2 = (2.4 \times 10^{-17})\left(\frac{0.030}{1.3 \times 10^{-6}}\right) = 5.5 \times 10^{-13}$$

$[H^+] = 7.4 \times 10^{-7}$ pH = 6.13

$CaCO_3$ will precipitate if :

$$[H^+]^2 = (2.4 \times 10^{-17})\left(\frac{0.030}{4.5 \times 10^{-8}}\right) = 1.6 \times 10^{-11}$$

$[H^+] = 4.0 \times 10^{-6}$ pH = 5.40

So $CaCO_3$ will precipitate and $NiCO_3$ will not if the pH is maintained between pH = 5.40 and pH = 6.13

19.15 The overall equilibrium is $AgCl(s) + 2NH_3(aq) \leftrightarrows Ag(NH_3)_2^+(aq) + Cl^-(aq)$

$$K_c = \frac{[Ag(NH_3)_2^+][Cl^-]}{[NH_3]^2}$$

In order to obtain a value for K_c for this reaction, we need to use the expressions for K_{sp} of AgCl(*s*) and the K_{form} of $Ag(NH_3)_2^+$:

$$K_{sp} = [Ag^+][Cl^-] = 1.8 \times 10^{-10}$$

$$K_{form} = \frac{[Ag(NH_3)_2^+]}{[Ag^+][NH_3]^2} = 1.6 \times 10^7$$

$$K_c = K_{sp} \times K_{form} = \frac{[Ag(NH_3)_2^+][Cl^-]}{[NH_3]^2} = 2.9 \times 10^{-3}$$

Now we may use an equilibrium table for the reaction in question:

	$[NH_3]$	$[Ag(NH_3)_2^+]$	$[Cl^-]$
I	0.10	–	–
C	–2x	+x	+x
E	0.10–2x	x	x

Substituting these values into the mass action expression gives:

$$K_c = 2.9 \times 10^{-3} = \frac{(x)(x)}{(0.10-2x)^2}$$

Take the square root of both sides to get $0.054 = \frac{(x)}{(0.10-2x)}$

Solving for x we get, $x = 4.9 \times 10^{-3}$ M. The molar solubility of AgCl in 0.10 M NH_3 is therefore 4.9×10^{-3} M.

In order to determine the solubility in pure water, we simply look at K_{sp}

$$AgCl(s) \leftrightarrows Ag^+(aq) + Cl^-(aq) \qquad K_{sp} = [Ag^+][Cl^-] = 1.8 \times 10^{-10}$$

At equilibrium; $[Ag^+] = [Cl^-] = 1.3 \times 10^{-5}$. Hence the molar solubility of AgCl in 0.10 M NH_3 is about 380 times greater than in pure water.

19.16 We will use the information gathered for the last problem. Specifically,

$$AgCl(s) + 2NH_3(aq) \leftrightarrows Ag(NH_3)_2^+(aq) + Cl^-(aq)$$

$$K_c = \frac{[Ag(NH_3)_2^+][Cl^-]}{[NH_3]^2} = 2.9 \times 10^{-3}$$

If we completely dissolve 0.20 mol of AgCl, the equilibrium $[Cl^-]$ and $[Ag(NH_3)_2^+]$ will be 0.20 M in a one liter container. The question asks, therefore, what amount of NH_3 must be initially present so that the equilibrium concentration of Cl^- is 0.20M?

	$[NH_3]$	$[Ag(NH_3)_2^+]$	$[Cl^-]$
I	Z	–	–
C	–2x	+x	+x
E	Z–2x	x	x

$$K_c = 2.9 \times 10^{-3} = \frac{(x)(x)}{(Z-2x)^2}$$

Take the square root of both sides to get;

$$0.054 = \frac{x}{Z-2x} = \frac{0.20}{Z-0.40}$$

We have substituted the known value of x. Solving for Z we get, Z = 4.1 M

Consequently, we would need to add 4.1 moles of NH_3 to a one liter container of 0.20 M AgCl in order to completely dissolve the AgCl.

Thinking It Through

T19.1 $PbCl_2 \leftrightarrows Pb^{2+} + 2Cl^-$ $K_{sp} = [Pb^{2+}][Cl^-]^2$
At equilibrium $[Pb^{2+}] = 0.016$ M
and $[Cl^-] = 0.032$ M

so $K_{sp} = (0.016)(0.032)^2 = 1.6 \times 10^{-5}$

T19.2 The precipitate that may form is $PbBr_2(s)$. To determine if a precipitate will form, a value for the reaction quotient, Q, must be calculated: $Q = [Pb^{2+}][Br^-]^2$. In performing this calculation, the dilution of the ions must be considered:
$[Pb^{2+}] = [Br^-] = 0.00500$ M. $Q = [0.00500][0.00500]^2$.
If Q > K_{sp}, a precipitate will form. K_{sp} for $PbBr_2(s)$ is 2.1×10^{-6}. Therefore, a precipitate will not form. Hence, the concentrations calculated are the diluted concentrations. Since no precipitate forms, the concentrations are not equilibrium values.

T19.3 By combining the three dissociation expressions for phosphoric acid, we come up with a single expression that relates the hydrogen ion concentration and the phosphate ion concentration:

$H_3PO_4(aq) \leftrightarrows 3H^+(aq) + PO_4^{3-}(aq)$
$K = K_{a_1} \times K_{a_2} \times K_{a_3}$

T19.4 The insoluble lead chloride will precipitate if Pb^{2+} and Cl^- ions are both present and the reaction quotient is greater than K_{sp}. To determine the maximum Pb^{2+} concentration that can be present, simply divide K_{sp} by $[Cl^-]^2$.

T19.5 $PbBr_2$ will be less soluble in a solution of 0.10 M NaBr because the values of Q and K_{sp} for lead bromide are dependent upon the $[Br^-]^2$.

T19.6 Start by determining the $[Cl^-]$ if the $[Pb^{2+}] = 0.0050$ M.

$$[Cl^-] = \sqrt{\frac{K_{sp}}{[Pb^{2+}]}} = \sqrt{\frac{1.7 \times 10^{-5}}{0.0050}} = 5.8 \times 10^{-2}$$

Then determine the concentration of chloride in a saturated solution, $K_{sp} = 4x^3$ where $x = [Pb^{2+}]$ and $[Cl^-] = 2x$, $x = \sqrt[3]{\frac{K_{sp}}{4}} = 1.6 \times 10^{-2}$ M, $[Cl^-] = 3.2 \times 10^{-2}$ M. Adding HCl to a saturated solution of $PbCl_2$ will increase the chloride ion concentration and will shift the equilibrium to the left thereby decreasing the lead ion concentration. Adding additional volume will also help to lower the lead ion concentration.

T19.7 To solve this problem, we need to determine the hydroxide ion concentration just before the more soluble compound begins to precipitate. This is easily calculated using the known concentration and the K_{sp} expression. The calculated hydroxide ion concentration is then substituted into the K_{sp} expression of the less soluble compound and the concentration of the cation in solution is determined. The K_{sp} values for both $Cd(OH)_2$ and $Mn(OH)_2$ are needed for this calculation.

T19.8 AgCl and AgBr are both insoluble compounds. However, AgBr is less soluble than AgCl. When solid AgCl is added to an aqueous solution of $MgBr_2$, some of the AgCl dissociates. The Ag^+ ion reacts with the aqueous Br^- to form insoluble AgBr. This disrupts the AgCl/Ag^+ equilibrium and additional AgCl dissociates. With sufficient stirring, and perhaps a little heat, all of the AgCl will dissolve and AgBr will precipitate.

T19.9 There are two equations to consider:

$$CaCO_3(s) \rightarrow Ca^{2+} + CO_3^{2-} \qquad K_{sp} = 4.5 \times 10^{-9}$$

$$CO_3^{2-} + H_2O \rightarrow HCO_3^- + OH^- \qquad K_b = 2.1 \times 10^{-4}$$

The net reaction is:

$$CaCO_3(s) + H_2O \rightarrow Ca^{2+} + HCO_3^- + OH^- \qquad K = 9.5 \times 10^{-13}$$

We know the $[OH^-]$ from the pH and the $[Ca^{2+}] = [HCO_3^-]$ = molar solubility.

T19.10 In case (a), the formation constant is relatively small indicating that the complex is not very stable. At the same time, the extremely small value for K_{sp} indicates that the ML_2 solid is very stable. Consequently, the solution will contain very little M^{2+}.

In case (b), the solubility of ML_2 is even smaller than in case (a). However, the large value for the formation constant indicates that any M^{2+} ions in solution will react with any ligand present to form the complex ion. As a result, more of the ML_2 solid will dissolve increasing the amount of M^{2+} in solution.

T19.11 The equilibria reactions and expressions involved in this problem are:

$$AgI(s) \leftrightarrows Ag^+(aq) + I^-(aq) \qquad K_{sp} = [Ag^+][I^-] = 8.3 \times 10^{-17}$$

$$Ag^+(aq) + 2I^-(aq) \leftrightarrows AgI_2^-(aq) \qquad K_{form} = \frac{[AgI_2^-]}{[Ag^+][I^-]^2}$$

If we combine these two equations we obtain a new reaction:

$$AgI(s) + I^-(aq)) \leftrightarrows AgI_2^-(aq) \qquad K_c = \frac{[AgI_2^-]}{[I^-]}$$

where $K_c = K_{sp} \times K_{form}$. We can rearrange this equation to obtain a means of determining the $[I^-]$ needed to dissolve the AgI(s): $[I^-] = \frac{[AgI_2^-]}{K_c}$. We can solve this equation if we know K_c and we also know $[AgI_2^-]$. Recall that there are initially 0.020 moles of AgI present in the solution. When it completely dissolves, there will be 0.020 moles of AgI_2^-. Since we know the volume of the solution, we may determine the concentration. Substituting into the expression above enables us to determine the $[I^-]$ and we may then calculate the amount of KI needed in order to obtain that concentration.

T19.12 $Mg(OH)_2(s) \leftrightharpoons Mg^{2+}(aq) + 2OH^-(aq)$
$OH^-(aq) + NH_4^+(aq) \leftrightharpoons H_2O(l) + NH_3(aq)$
Overall reaction: $Mg(OH)_2(s) + 2NH_4^+(aq) \leftrightharpoons Mg^{2+}(aq) + 2OH^-(aq) + 2NH_3(aq)$

T19.13 (a) In order, Figure C → Figure B → Figure E
(b) Figures A and D are excluded because $PbBr_2$ will precipitate before $PbCl_2$.

Review Questions

19.1 The ion product is the quantity which results from the mass action expression and it is a product of the ion concentrations. The ion product constant, also called the solubility product constant, is equal to the product of the ion concentrations for a *saturated* solution of a sparingly soluble substance.

19.2 $$K_c = \frac{[Ba^{2+}]^3[PO_4^{3-}]^2}{[Ba_3(PO_4)_2]}$$

The denominator of the mass action expression is a constant since the concentration of substance within a pure solid is constant. Consequently, we define a new constant which is the product of K_c and the concentration of the pure solid.

$$K_{sp} = [Ba^{2+}]^3[PO_4^{3-}]^2$$

19.3 The addition of a common ion to a saturated solution lowers the solubility of a sparingly soluble ionic salt. According to Le Châtelier's principle, addition of an ion to a saturated solution will shift the equilibrium so as to absorb as much of the added ion as possible. This results in the precipitation of the sparingly soluble salt. In the case of AgCl,
$K_{sp} = 1.8 \times 10^{-10}$, addition of NaCl to a saturated solution containing Ag^+ and Cl^- will result in the precipitation of AgCl(*s*). The common ion in this case is Cl^-. It is present in the solution due to the NaCl as well as the AgCl.

19.4 When sodium acetate is added to a solution of acetic acid, the equilibrium is disrupted and more acetic acid is formed. This lowers the amount of hyrodium ions in solution thus raising the pH.

19.5 When the value of the ion product is greater than K_{sp}, a precipitate will form.

19.6 K_{sp} values assume 100% dissociation of the ions, but in solution, the ions are not fully dissociated, but instead ion pairs are often formed.

19.7 The oxide ion reacts with the water to produce hydroxide ion: $O^{2-} + H_2O \rightarrow 2OH^-$

19.8 $Na_2S(s) \leftrightharpoons 2Na^+(aq) + HS^-(aq) + OH^-(aq)$

19.9 (a) $CoS(s) \leftrightharpoons Co^{2+}(aq) + HS^-(aq) + OH^-(aq)$ $\quad K_{sp} -= [CO^{2+}][HS^-][OH^-]$

(b) $CoS(s) + 2H^+(aq) \leftrightharpoons Co^{2+}(aq) + H_2S(aq)$ $\quad K_{spa} = \frac{[Co^{2+}][H_2S]}{[H^+]^2}$

19.10 As the pH of an oxalic acid solution is decreased, the hydrogen ion concentration is increased, this shifts the equilibrium to the undissociated acid, thus increasing the $H_2C_2O_4$ concentration. The opposite is true as the pH of the solution is increased.

19.11 The addition of NH_4Cl to the mixture of $Mg(OH)_2$ and water causes the $Mg(OH)_2$ to dissolve because the NH_4^+ reacts with the OH^- and shifts the equilibrium towards the dissolution of the solid.

19.12 $AgCl(s) \leftrightarrows Ag^+(aq) + Cl^-(aq)$
$Ag^+(aq) + 2NH_3(aq) \leftrightarrows Ag(NH_3)_2^+(aq)$

Silver chloride is an insoluble solid. However, any Ag^+ ions present react with added NH_3 to form the $Ag(NH_3)_2^+$ complex ion. According to Le Châtelier's Principle, as NH_3 is added to a solution containing Ag^+ ions, the complex ion forms using up the Ag^+ ions. This disrupts the equilibrium and forces AgCl to dissolve.

Upon addition of HNO_3, a strong acid, H^+ reacts with NH_3 to form NH_4^+. This disrupts the complex equilibria and causes the $Ag(NH_3)_2^+$ to dissociate and form free Ag^+ ions. Once again, the $[Ag^+]$ reaches a value sufficient in the presence of Cl^- to precipitate AgCl in accordance with Le Châtelier's Principle.

19.13 $PbCl_2(s) \leftrightarrows Pb^{2+}(aq) + 2Cl^-(aq)$
$Pb^{2+}(aq) + 3Cl^-(aq) \leftrightarrows PbCl_3^-(aq)$

$$K_{form} = \frac{[PbCl_3^-]}{[Pb^{2+}][Cl^-]^3} = \frac{\left(\text{mol } PbCl_3^- / \text{volume}\right)}{\left(\text{mol } Pb^{2+} / \text{volume}\right)\left(\text{mol } Cl^- / \text{volume}\right)^3}$$

$$K_{form} = (\text{volume})^3 \times \frac{\left(\text{mol } PbCl_3^-\right)}{\left(\text{mol } Pb^{2+}\right)\left(\text{mol } Cl^-\right)^3}$$

Notice that the above expression is the product of a ratio of mole amounts and a volume^3 term. The constant K_{form} does not change on dilution, but the volume term is changed by dilution. This means that the ratio of moles term in the above expression must change on dilution, in order to hold the product constant. If the volume is doubled, the ratio of moles would have to become smaller by a factor of 8 ($= 2^3$) in order for the entire argument to have a constant value, i.e, in order for K_{form} to remain constant on dilution.

This means that the concentrations of Pb^{2+} and Cl^- must increase as the solution containing the complex ion is diluted. Eventually the ion product for $PbCl_2$ will exceed the value of K_{sp} for $PbCl_2$, resulting in its precipitation.

Review Problems

19.14
(a) $CaF_2(s) \leftrightarrows Ca^{2+} + 2F^-$ $\quad K_{sp} = [Ca^{2+}][F^-]^2$
(b) $Ag_2CO_3(s) \leftrightarrows 2Ag^+ + CO_3^{2-}$ $\quad K_{sp} = [Ag^+]^2[CO_3^{2-}]$
(c) $PbSO_4(s) \leftrightarrows Pb^{2+} + SO_4^{2-}$ $\quad K_{sp} = [Pb^{2+}][SO_4^{2-}]$
(d) $Fe(OH)_3(s) \leftrightarrows Fe^{3+} + 3OH^-$ $\quad K_{sp} = [Fe^{3+}][OH^-]^3$
(e) $PbI_2(s) \leftrightarrows Pb^{2+} + 2I^-$ $\quad K_{sp} = [Pb^{2+}][I^-]^2$
(f) $Cu(OH)_2(s) \leftrightarrows Cu^{2+} + 2OH^-$ $\quad K_{sp} = [Cu^{2+}][OH^-]^2$

19.15
(a) $AgI(s) \leftrightarrows Ag^+ + I^-$ $\quad K_{sp} = [Ag^+][I^-]$
(b) $Ag_3PO_4(s) \leftrightarrows 3Ag^+ + PO_4^{3-}$ $\quad K_{sp} = [Ag^+]^3[PO_4]^{3-}$
(c) $PbCrO_4(s) \leftrightarrows Pb^{2+} + CrO_4^{2-}$ $\quad K_{sp} = [Pb^{2+}][CrO_4^{2-}]$
(d) $Al(OH)_3(s) \leftrightarrows Al^{3+} + 3OH^-$ $\quad K_{sp} = [Al^{3+}][OH^-]^3$
(e) $ZnCO_3(s) \leftrightarrows Zn^{2+} + CO_3^{2-}$ $\quad K_{sp} = [Zn^{2+}][CO_3^{2-}]$
(f) $Zn(OH)_2(s) \leftrightarrows Zn^{2+} + 2OH^-$ $\quad K_{sp} = [Zn^{2+}][OH^-]^2$

19.16

$$\text{\# moles BaSO}_4 = (0.00245 \text{ g BaSO}_4)\left(\frac{1 \text{ mole BaSO}_4}{233.3906 \text{ g BaSO}_4}\right)$$

$$\text{\# moles BaSO}_4 = 1.05 \times 10^{-5} \text{ moles}$$

$[Ba^{2+}] = [SO_4^{2-}] = 1.05 \times 10^{-5}$ M

$K_{sp} = [Ba^{2+}][SO_4^{2-}] = (1.05 \times 10^{-5})^2 = 1.10 \times 10^{-10}$

19.17

$$\text{\# moles AgC}_2\text{H}_3\text{O}_2 = (0.800 \text{ g AgC}_2\text{H}_3\text{O}_2)\left(\frac{1 \text{ mole AgC}_2\text{H}_3\text{O}_2}{166.9 \text{ g AgC}_2\text{H}_3\text{O}_2}\right)$$

$$= 4.79 \times 10^{-3} \text{ moles AgC}_2\text{H}_3\text{O}_2$$

$[AgC_2H_3O_2] = 4.79 \times 10^{-3}$ moles $AgC_2H_3O_2/0.100$ L $= 4.79 \times 10^{-2}$ M

One mole of Ag^+ and one mole of $C_2H_3O_2$ will be produced for every mole of $AgC_2H_3O_2$. Therefore,

$K_{sp} = [Ag^+][C_2H_3O_2^-] = (4.79 \times 10^{-2})^2 = 2.29 \times 10^{-3}$.

19.18 $BaSO_3(s) \leftrightarrows Ba^{2+} + SO_3^{2-}$ $K_{sp} = [Ba^{2+}][SO_3^{2-}]$

$K_{sp} = (0.10)(8.0 \times 10^{-6}) = 8.0 \times 10^{-7}$

In this problem, all of the Ba^{2+} comes from the $BaCl_2$.

19.19 To solve this problem we need to realize that the concentration of the solution is equal to the number of moles of solid recovered divided by the volume of the solution, i.e.,

$$[\text{CaCrO}_4] = \left(\frac{0.649 \text{ g CaCrO}_4}{156 \text{ mL}}\right)\left(\frac{1 \text{ mole CaCrO}_4}{156.1 \text{ g CaCrO}_4}\right)\left(\frac{1000 \text{ mL}}{1 \text{ L}}\right) = 0.0267 \text{ M}$$

The equilibrium for this problem is $CaCrO_4(s) \leftrightarrows Ca^{2+} + CrO_4^{2-}$

$K_{sp} = [Ca^{2+}][CrO_4^{2-}] = (0.0267)^2 = 7.13 \times 10^{-4}$

19.20 $Ag_3PO_4(s) \leftrightarrows 3Ag^+ + PO_4^{3-}$ $K_{sp} = [Ag^+]^3[PO_4^{3-}]$

$K_{sp} = [3(1.8 \times 10^{-5})]^3[1.8 \times 10^{-5}] = 2.8 \times 10^{-18}$

19.21 $Ba_3(PO_4)_2(s) \leftrightarrows 3Ba^{2+} + 2PO_4^{3-}$ $K_{sp} = [Ba^{2+}]^3[PO_4^{3-}]^2$

$K_{sp} = [3(1.4 \times 10^{-8})]^3[2(1.4 \times 10^{-8})]^2 = 5.8 \times 10^{-38}$

19.22 $PbBr_2(s) \leftrightarrows Pb^{2+} + 2Br^-$ $K_{sp} = [Pb^{2+}][Br^-]^2$

	$[Pb^{2+}]$	$[Br^-]$
I	–	–
C	+ x	+ 2x
E	+ x	+ 2x

$$K_{sp} = (x)(2x)^2 = 4x^3 = 2.1 \times 10^{-6}, \quad x = \sqrt[3]{\frac{2.1 \times 10^{-6}}{4}} = 8.1 \times 10^{-3} \text{ M}$$

19.23 $Ag_2CrO_4(s) \leftrightarrows 2Ag^+ + CrO_4^{2-}$ $K_{sp} = [Ag^+]^2[CrO_4^{2-}]$

	$[Ag^+]$	$[CrO_4^{2-}]$
I	–	–
C	+ 2x	+ x
E	+ 2x	+ x

$$K_{sp} = (2x)^2(x) = 4x^3 = 1.2 \times 10^{-12},\ x = \sqrt[3]{\frac{1.2 \times 10^{-12}}{4}} = 6.7 \times 10^{-5}\ M$$

19.24 For every mole of CO_3^{2-} produced, 2 moles of Ag^+ will be produced. Let x = $[CO_3^{2-}]$ at equilibrium and $[Ag^+]$ = 2x at equilibrium. $K_{sp} = [Ag^+]^2[CO_3^{2-}] = 8.1 \times 10^{-12} = (2x)^2(x) = 4x^3$. Solving we find $x = 1.3 \times 10^{-4}$. Thus, the molar solubility of Ag_2CO_3 is 1.3×10^{-4} moles/L.

19.25 For every mole of Pb^{2+} produced, 2 moles of I^- will be produced. Let x = $[Pb^{2+}]$ at equilibrium and $[I^-]$ = 2x at equilibrium. $K_{sp} = [Pb^{2+}][I^-]^2 = 7.9 \times 10^{-9} = (x)(2x)^2 = 4x^3$. Solving we find $x = 1.3 \times 10^{-3}$ M. Thus, the molar solubility of PbI_2 is 1.3×10^{-3} moles/L.

19.26 To solve this problem, determine the molar solubility for each compound.

LiF: let $x = [Li^+] = [F^-]$ $K_{sp} = [Li^+][F^-] = x^2 = 1.7 \times 10^{-3}$
$x = 4.1 \times 10^{-2}$ moles/L = molar solubility of LiF.

BaF_2: let $x = [Ba^{2+}]$; $[F^-]^2 = 2x$ $K_{sp} = [Ba^{2+}][F^-]^2 = (x)(2x)^2 = 1.7 \times 10^{-6}$
$4x^3 = 1.7 \times 10^{-6}$, and $x = 7.5 \times 10^{-3}$ M = molar solubility of BaF_2.

Because the molar solubility of LiF is greater than the molar solubility of BaF_2, LiF is more soluble.

19.27 To solve this problem, first determine the molar solubility for each compound.

AgCN: let $x = [Ag^+] = [CN^-]$ $K_{sp} = [Ag^+][CN^-] = x^2 = 2.2 \times 10^{-16}$
$x = 1.5 \times 10^{-8}$ moles/L = molar solubility of AgCN. The number of grams of AgCN that will dissolve in 100 mL is;

$$\#\ g = \left(\frac{1.5 \times 10^{-8}\ \text{moles}}{1000\ \text{mL}}\right)\left(\frac{133.9\ g}{1\ \text{mole}}\right)(100\ \text{mL}) = 2.0 \times 10^{-7}\ g$$

$Zn(CN)_2$: let $x = [Zn^{2+}]$, $[CN^-] = 2x$; $K_{sp} = [Zn^{2+}][CN^-]^2 = (x)(2x)^2 = 3 \times 10^{-16}$
$4x^3 = 3 \times 10^{-16}$, and $x = 4.2 \times 10^{-6}$ M = molar solubility of $Zn(CN)_2$. The number of grams of $Zn(CN)_2$ that will dissolve in 100 mL is;

$$\#\ g = \left(\frac{4.2 \times 10^{-6}\ \text{moles}}{1000\ \text{mL}}\right)\left(\frac{117.4\ g}{1\ \text{mole}}\right)(100\ \text{mL}) = 4.9 \times 10^{-5}\ g$$

More $Zn(CN)_2$ will dissolve in 100 mL of water so it has the larger solubility.

19.28 First determine the molar solubility of the MX salt.
Let $x = [M^+] = [X^-]$, $K_{sp} = [M^+][X^-] = (x)(x) = 3.2 \times 10^{-10}$
$x = 1.8 \times 10^{-5}$ M. This is the equilibrium concentration of the two ions.

For the MX_3 salt, let x = equilibrium concentration of M^{3+}, $[X^-] = 3x$.
$K_{sp} = [M^+][X^-]^3 = (x)(3x)^3 = 27x^4$. The value of x in this expression is the value determined in the first part of this problem.
So, $K_{sp} = (27)(1.8 \times 10^{-5})^4 = 2.8 \times 10^{-18}$

19.29 First determine the molar solubility of the M_2X_3 salt.

$M_2X_3\ (s) \leftrightarrows 2M^{3+}\ (aq) + 3X^{2-}(aq)$ $K_{sp} = [M^{3+}]^2[X^{2-}]^3$

	$[M^{3+}]$	$[X^{2-}]$
I	–	–
C	+2x	+3x
E	+2x	+3x

$K_{sp} = (2x)^2(3x)^3 = 2.2 \times 10^{-20} = 108x^5$ $x = 4.6 \times 10^{-5}$ M

The molar solubility of this compound is 4.6×10^{-5} moles/L

We want the molar solubility of the M_2X compound to be twice the value just calculated or 9.2×10^{-5} moles/L. We need to solve the equilibrium expression:

$M_2X\ (s) \rightleftharpoons 2M^{+}\ (aq) + X^{2-}(aq)$ $K_{sp} = [M^+]^2[X^{2-}]$

	$[M^{+}]$	$[X^{2-}]$
I	–	–
C	+2x	+x
E	+2x	+x

$K_{sp} = (2x)^2(x) = 4x^3$ where $x = 9.2 \times 10^{-5}$ M

So, $K_{sp} = 4(9.2 \times 10^{-5})^3 = 3.1 \times 10^{-12}$

19.30 $CaSO_4(s) \leftrightarrows Ca^{2+}(aq) + SO_4^{2-}(aq)$ $K_{sp} = [Ca^{2+}][SO_4^{2-}] = 2.4 \times 10^{-5}$

let $x = [Ca^{2+}] = [SO_4^{2-}]$ $K_{sp} = x^2 = 2.4 \times 10^{-5}$ and $x = 4.9 \times 10^{-3}$ M.

The molar solubility of $CaSO_4$ is 4.9×10^{-3} moles/L.

19.31 $CaCO_3(s) \leftrightarrows Ca^{2+} + CO_3^{2-}$ $K_{sp} = [Ca^{2+}][CO_3^{2-}] = 4.5 \times 10^{-9}$

let $x = [Ca^{2+}] = [CO_3^{2-}]$

$x^2 = 4.5 \times 10^{-9}$; $x = 6.7 \times 10^{-5}$ M

The solubility of $CaCO_3$ is 6.7×10^{-5} M. So, 6.7×10^{-5} moles of $CaCO_3$ dissolves in 1 L of H_2O. If 100 mL of water are available, 6.7×10^{-6} moles of $CaCO_3$ will dissolve. Converting this to #g/100 mL we multiply by the molar mass of 100 g/mole and determine that 6.7×10^{-4} g will dissolve in 100 mL of water.

19.32 (a) $CuCl(s) \leftrightarrows Cu^{+}(aq) + Cl^{-}(aq)$ $K_{sp} = [Cu^+][Cl^-] = 1.9 \times 10^{-7}$

	$[Cu^{+}]$	$[Cl^{-}]$
I	–	–
C	+x	+x
E	+x	+x

$K_{sp} = x^2 = 1.9 \times 10^{-7}$ $\therefore$ x = molar solubility = 4.4×10^{-4} M

(b) $CuCl(s) \leftrightarrows Cu^{+}(aq) + Cl^{-}(aq)$ $K_{sp} = [Cu^+][Cl^-] = 1.9 \times 10^{-7}$

	$[Cu^{+}]$	$[Cl^{-}]$
I	–	0.0200
C	+x	+x
E	+x	0.0200+x

$K_{sp} = (x)(0.0200+x) = 1.9 \times 10^{-7}$ Assume that $x \ll 0.0200$

$\therefore$ x = molar solubility = 9.5×10^{-6} M

(c) $CuCl(s) \leftrightarrows Cu^{+}(aq) + Cl^{-}(aq)$ $K_{sp} = [Cu^{+}][Cl^{-}] = 1.9 \times 10^{-7}$

	$[Cu^{+}]$	$[Cl^{-}]$
I	–	0.200
C	+x	+x
E	+x	0.200+x

$K_{sp} = (x)(0.200+x) = 1.9 \times 10^{-7}$ Assume that $x << 0.200$
$\therefore$ x = molar solubility = 9.5×10^{-7} M

(d) $CuCl(s) \leftrightarrows Cu^{+}(aq) + Cl^{-}(aq)$ $K_{sp} = [Cu^{+}][Cl^{-}] = 1.9 \times 10^{-7}$
Note that the Cl^{-} concentration equals (2)(0.150 M) since two moles of Cl^{-} are produced for every mole of $CaCl_2$.

	$[Cu^{+}]$	$[Cl^{-}]$
I	–	0.300
C	+x	+x
E	+x	0.300+x

$K_{sp} = (x)(0.300+x) = 1.9 \times 10^{-7}$ Assume that $x << 0.300$
$\therefore$ x = molar solubility = 6.3×10^{-7} M

19.33 $AuCl_3(s) \leftrightarrows Au^{3+} + 3Cl^{-}$ $K_{sp} = [Au^{3+}][Cl^{-}]^3 = 3.2 \times 10^{-25}$

(a) let $x = [Au^{3+}]$; then $[Cl^{-}] = 3x$
$K_{sp} = (x)(3x)^3 = 27x^4$

$$x = \sqrt[4]{\frac{3.2 \times 10^{-25}}{27}} = 3.3 \times 10^{-7} \text{ M}$$

The molar solubility of $AuCl_3$ is 3.3×10^{-7} M in H_2O.

(b) $[Au^{3+}] = x$; $[Cl^{-}] = 0.010 + 3x$
$K_{sp} = (x)(0.010 + 3x)^3$: Assume $3x << 0.010$
$K_{sp} = (x)(0.010)^3$
$x = 3.2 \times 10^{-19}$ M
The molar solubility of $AuCl_3$ is 3.2×10^{-19} M in 0.010 M HCl.

(c) $[Au^{3+}] = x$; $[Cl^{-}] = 0.020 + 3x$
$K_{sp} = (x)(0.020 + 3x)^3$: Assume $3x << 0.020$
$K_{sp} = (x)(0.020)^3$
$x = 4.0 \times 10^{-20}$ M
The molar solubility of $AuCl_3$ is 4.0×10^{-20} M in 0.010 M $MgCl_2$.

(d) $[Au^{3+}] = 0.010 + x$; $[Cl^{-}] = 3x$
$K_{sp} = (0.010 + x)(3x)^3$: Assume $x << 0.010$
$K_{sp} = (0.010)(3x)^3$

$$x = \sqrt[3]{\frac{3.2 \times 10^{-25}}{0.27}} = 1.1 \times 10^{-8} \text{ M}$$

The molar solubility of $AuCl_3$ is 1.1×10^{-8} M in 0.010 M $Au(NO_3)_3$.

19.34 $Mg(OH)_2 \leftrightarrows Mg^{2+} + 2OH^-$ $K_{sp} = [Mg^{2+}][OH^-]^2 = 7.1 \times 10^{-12}$

The concentration of OH^- is determined from the pH:

$pOH = 14 - 12.50 = 1.50$

$[OH^-] = 0.0316$ M

$[Mg^{2+}] = x$ $[OH^-] = 0.0316$ M

$K_{sp} = x(0.0316)^2 = 7.1 \times 10^{-12}$

$x = 7.1 \times 10^{-9}$ M

The molar solubility of $Mg(OH)_2$ is 2.2×10^{-10} M in a solution with a pH of 12.50.

19.35 $Al(OH)_3(s) \leftrightarrows Al^{3+} + 3OH^-$ $K_{sp} = [Al^{3+}][OH^-]^3 = 3 \times 10^{-34}$

The concentration of OH^- is determined from the pH:

$pOH = 14 - 9.50 = 4.50$

$[OH^-] = 3.2 \times 10^{-5}$ M

$[Al^{3+}] = x$ $[OH^-] = 3.2 \times 10^{-5}$

$K_{sp} = x(3.2 \times 10^{-5})^3 = 3 \times 10^{-34}$

$x = 9.2 \times 10^{-21}$

The molar solubility of $Al(OH)_3$ is 9.2×10^{-21} M in a solution with a pH of 9.50.

19.36 $Ag_2CrO_4\ (s) \leftrightarrows 2Ag^+\ (aq) + CrO_4^{2-}(aq)$ $K_{sp} = [Ag^+]^2[CrO_4^{2-}] = 1.2 \times 10^{-12}$

(a)

	$[Ag^+]$	$[CrO_4^{2-}]$
I	0.200	–
C	0.200+2x	+x
E	0.200+2x	+x

$K_{sp} = (0.200+2x)^2(x)$ Assume that $x << 0.200$

$1.2 \times 10^{-12} = (0.200)^2(x)$ $x = 3.0 \times 10^{-11}$

The molar solubility is 3.0×10^{-11} moles/L

(b)

	$[Ag^+]$	$[CrO_4^{2-}]$
I	–	0.200
C	2x	0.200+x
E	2x	0.200+x

$K_{sp} = (2x)^2(0.200+x)$ Assume that $x << 0.200$

$1.2 \times 10^{-12} = (2x)^2(0.200)$ $x = 1.2 \times 10^{-6}$

The molar solubility is 1.2×10^{-6} moles/L.

19.37 $MgOH_2\ (s) \leftrightarrows Mg^{2+}(aq) + 2OH^-(aq)$ $K_{sp} = [Mg^{2+}][OH^-]^2 = 7.1 \times 10^{-12}$

	$[Mg^{2+}]$	$[OH^-]$
I	–	0.20
C	+x	0.20+2x
E	+x	0.20+2x

$K_{sp} = (x)(0.20 + 2x)^2$ Assume $2x << 0.20$

$K_{sp} = (x)(0.20)^2 = 7.1 \times 10^{-12}$ $x = 1.8 \times 10^{-10}$ M

The assumption is valid and the molar solubility is 1.8×10^{-10} moles/L

19.38 $CaSO_4(s) \leftrightarrows Ca^{2+}(aq) + SO_4^{2-}(aq)$ $K_{sp} = [Ca^{2+}][SO_4^{2-}] = 2.5 \times 10^{-5}$

let $x = [Ca^{2+}]$, $[SO_4^{2-}] = 0.015$ M + x $K_{sp} = (x)(0.015 + x) = 2.4 \times 10^{-5}$

Solving the resulting quadratic equation gives $x = 1.5 \times 10^{-3}$ M.

The molar solubility of $CaSO_4$ in 0.015 M $CaCl_2$ is 1.5×10^{-3} moles/L.
(Note: If we had assumed that x << 0.015, as is usually the method we use to solve these problems, we would have determined that x = 0.0016 M. This is slightly larger than 10% of the value 0.015 so our usual assumption is not valid in this problem.)

19.39 $AgCl(s) \leftrightarrows Ag^+ + Cl^-$ $K_{sp} = [Ag^+][Cl^-]$
$[Ag^+] = K_{sp}/[Cl^-] = 1.8 \times 10^{-10}/3(0.050) = 1.2 \times 10^{-9}$ M

19.40 $HC_2H_3O_2 + H_2O \leftrightarrows H_3O^+ + C_2H_3O_2^-$

$$K_a = \frac{[H_3O^+][C_2H_3O_2^-]}{[HC_2H_3O_2]} = 1.8 \times 10^{-5}$$

First, we calculate the % ionization in water (no added sodium acetate):

	$[HC_2H_3O_2]$	$[H_3O^+]$	$[C_2H_3O_2^-]$
I	0.10	–	–
C	0.10–x	+x	+x
E	0.10 –x	x	+x

Assume x << 0.10

$$K_a = \frac{[x][x]}{[0.10]} = 1.8 \times 10^{-5} \quad x = 1.3 \times 10^{-3} = [H_3O^+]$$

% ionization = $([H^+]/[\text{total acid}]) \times 100 = (1.3 \times 10^{-3}/0.10) \times 100 = 1.3$ %

Now we calculate the % ionization using 0.050 mol/0.500L = 0.10 M for the concentration of sodium acetate:

	$[HC_2H_3O_2]$	$[H_3O^+]$	$[C_2H_3O_2^-]$
I	0.10	–	0.10
C	0.10–x	+x	+x
E	0.10 –x	x	0.10 + x

Assume x << 0.10

$$K_a = \frac{[x][0.10]}{[0.10]} = 1.8 \times 10^{-5} \quad x = 1.8 \times 10^{-5} = [H_3O^+]$$

% ionization = $([H^+]/[\text{total acid}]) \times 100 = (1.8 \times 10^{-5}/0.10) \times 100 = 0.018$ %

So the % ionization decreased by (1.3 – 0.018) = 1.3 %, using correct significant figures. (This does not mean it has no dissociation, but that the dissociation is very small compared to 1.3%.)

Using the $[H^+]$ values above, the pH initially was:
$-\log [H^+] = -\log [1.3 \times 10^{-3}] = 2.9$

After addition of the sodium acetate it was:
$-\log [H^+] = -\log [1.8 \times 10^{-5}] = 4.7$

The pH changes by (4.7 – 2.9) = +1.8 pH units

19.41 $HC_2H_3O_2 + H_2O \leftrightarrows H_3O^+ + C_2H_3O_2^-$

$$K_a = \frac{[H_3O^+][C_2H_3O_2^-]}{[HC_2H_3O_2]} = 1.8 \times 10^{-5}$$

First, we calculate the % ionization in water (no added HCl):

	$[HC_2H_3O_2]$	$[H_3O^+]$	$[C_2H_3O_2^-]$
I	0.10	–	–
C	0.10–x	+x	+x
E	0.10 –x	x	+x

Assume x << 0.10

$$K_a = \frac{[x][x]}{[0.10]} = 1.8 \times 10^{-5} \quad x = 1.3 \times 10^{-3} = [H_3O^+]$$

% ionization = $([H^+]/[\text{total acid}]) \times 100 = (1.3 \times 10^{-3}/0.10) \times 100 = 1.3$ %

Now we calculate the % ionization using 0.025 mol/0.500 L = 0.050 M for the concentration of H_3O^+:

	$[HC_2H_3O_2]$	$[H_3O^+]$	$[C_2H_3O_2^-]$
I	0.10	0.050	–
C	0.10–x	+x	+x
E	0.10 –x	0.050 + x	x

Assume x << 0.050

$$K_a = \frac{[x][0.050]}{[0.10]} = 1.8 \times 10^{-5} \quad x = 3.6 \times 10^{-5} = [C_2H_3O_2^-]$$

% ionization = $([C_2H_3O_2^-]/[\text{total acid}]) \times 100 = (3.6 \times 10^{-5}/0.10) \times 100 = 0.036$ %

So the % ionization decreased by (1.3 – 0.036) = 1.3 %, using correct significant figures. (This does not mean it has no dissociation, but that the dissociation is very small compared to 1.3%.)

Using the $[H^+]$ values above, the pH initially was:
$-\log [H^+] = -\log [1.3 \times 10^{-3}] = 2.9$

After addition of the hydrogen chloride it was:
$-\log [H^+] = -\log [0.050] = 1.3$

The pH changes by (1.3 – 2.9) = –1.6 pH units

19.42 $Fe(OH)_2(s) \leftrightarrows Fe^{2+}(aq) + 2\ OH^-(aq)$ $\qquad K_{sp} = [Fe^{2+}][OH^-]^2$

mol OH^- = 2.20 g NaOH(1 mol/40.01 g NaOH) = 0.0550 mol NaOH

$[OH^-]$ = mol OH^-/L solution = 0.0550 mol/0.250 L = 0.22 M

	$[Fe^{2+}]$	$[OH^-]$
I	–	0.22
C	+x	+2x
E	x	0.22 + 2x

We assume that x << 0.22, so that 0.22 + 2x ≈ 0.22, then we enter the equilibrium values of the above table into the K_{sp} expression:

$K_{sp} = [Fe^{2+}][OH^-]^2$
$7.9 \times 10^{-16} = x(0.22)^2$
x = molar solubility = 1.6×10^{-14} M

Next, we must determine how many moles of $Fe(OH)_2$ are formed in the reaction.
This is a limiting reactant problem.

The number of moles of OH^- is 0.0550 (see above).
The number of moles of Fe^{2+} is (.250 L)(0.10 mol/L) = 0.025 mol

From the balanced equation at the top, we need two OH^- for every one Fe^{2+}.
This would be 2(0.025 mol) = 0.050 mol OH^-. Looking at the molar quantities above, we have more than enough OH^- so, Fe^{2+} is our limiting reactant:

0.025 mol $Fe(OH)_2$ will form in 0.25 L solution. If dissolved, this would be a concentration of 0.025 mol/0.25 L = 0.10 M. But from above, the maximum molar solubility of is 1.6×10^{-14} M.

This means that remainder of $Fe(OH)_2$ in excess of this value precipitates:
$0.10 - 1.6 \times 10^{-14} \approx 0.10$ M.

This works out to 0.25 L(0.10 mol/L) = 0.025 mol $Fe(OH)_2$(89.8 g/mol)
= 2.2 g solid $Fe(OH)_2$ (essentially all of it).

The concentration of Fe^{2+} in the final solution is at its maximum,
1.6×10^{-14} M.

19.43 $Ni(OH)_2(s) \leftrightarrows Ni^{2+}(aq) + 2OH^-(aq)$
mol OH^- = 1.75 g NaOH(1 mol/40.01 g NaOH) = 0.0437 mol NaOH
First, assume that all of the ions are in solution and no precipitate has formed. At that point the concentrations are:

$$[OH^-] = \frac{0.0437 \text{ mol } OH^-}{0.250 \text{ L}} = 0.175 \text{ M } OH^-$$

$[Ni^{2+}] = 0.10$ M

Next, determine the limiting reactant:

$$\left(\frac{0.10 \text{ mol } Ni^{2+}}{1 \text{ L soln}}\right)\left(\frac{2 \text{ mol } OH^-}{1 \text{ mol } Ni^{2+}}\right) = 0.20 \text{ M } OH^-$$

Only 0.175 M OH^- is available, therefore OH^- is the limiting reagent.
Using the limiting reagent, assume that all of the OH^- has precipitated

$[OH^-] = 0.0$ M

$$[Ni^{2+}] = 0.10 \text{ M } Ni^{2+} - \left(\frac{0.175 \text{ mol } OH^-}{1 \text{ L soln}}\right)\left(\frac{1 \text{ mol } Ni^{2+}}{2 \text{ mol } OH^-}\right)$$

$$= 0.10 \text{ M } Ni^{2+} - 0.088 \text{ M } Ni^{2+} = 0.012 \text{ M } Ni^{2+}$$

From here, calculate the equilibrium concentrations:

	$[Ni^{2+}]$	$[OH^-]$
I	0.012	–
C	+x	+2x
E	0.012 + x	2x

We assume that x << 0.012, so that 0.012 + 2x ≈ 0.012, then we enter the equilibrium values of the above table into the K_{sp} expression:

$$K_{sp} = [Ni^{2+}][OH^-]^2$$

$$6 \times 10^{-16} = 0.012(x)^2$$

$$x = 2.2 \times 10^{-7}\ M$$

The amount of $Ni(OH)_2$ formed can be calculated from the amount of OH^- added to the solution since essentially all of the OH^- was precipitated:

$$\#\ g\ Ni(OH)_2 = 0.0437\ mol\ OH^- \left(\frac{1\ mol\ Ni(OH)_2}{2\ mol\ OH^-}\right)\left(\frac{92.71\ g\ Ni(OH)_2}{1\ mol\ Ni(OH)_2}\right) = 2.03\ g\ Ni(OH)_2$$

To determine the pH of the final solution, the $[OH^-]$ needs to be found:
From the ICE table, the equilibrium concentration of $OH^- = 2x$

$x = 2.2 \times 10^{-7}$ M $\qquad 2x = 4.4 \times 10^{-7}\ M = [OH^-]$

$pOH = -\log[OH^-] = -\log[4.4 \times 10^{-7}] = 6.36$

$pH = 14 - pOH = 14 - 6.36 = 7.64$

19.44 $Fe(OH)_2(s) \leftrightarrows Fe^{2+}(aq) + 2\ OH^-(aq)$ $\qquad K_{sp} = [Fe^{2+}][OH^-]^2$

$pH = 9.50$

$pOH = 14.00 - pH = 4.50$

$[OH^-] = 10^{-4.50} = 3.16 \times 10^{-5}$ M

	$[Fe^{2+}]$	$[OH^-]$
I	–	3.16×10^{-5}
C	+x	+2x
E	x	$3.16 \times 10^{-5} + 2x$

Since K_{sp} for iron(II) hydroxide is so tiny, we can safely assume that $2x << 3.16 \times 10^{-5}$, so that $(3.16 \times 10^{-5}) + 2x \approx 3.16 \times 10^{-5}$, then we enter the equilibrium values of the above table into the K_{sp} expression:

$$K_{sp} = [Fe^{2+}][OH^-]^2$$

$$7.9 \times 10^{-16} = x(3.16 \times 10^{-5})^2$$

x = molar solubility = 7.9×10^{-7} M

19.45 (a) $Ca(OH)_2(s) \leftrightarrows Ca^{2+}(aq) + 2OH^-(aq)$ $\qquad K_{sp} = [Ca^{2+}][OH^-]^2 = 6.5 \times 10^{-5}$

let $2x = [OH^-]$, $[Ca^{2+}] = x + 0.10$ $\quad K_{sp} = (0.10 + x)(2x)^2 = 6.5 \times 10^{-5}$

By successive approximations, x = 0.012

The molar solubility of $Ca(OH)_2$ in 0.10 M $CaCl_2$ is 0.012 moles/L.

(b) $Ca(OH)_2(s) \leftrightarrows Ca^{2+}(aq) + 2OH^-(aq)$ $\qquad K_{sp} = [Ca^{2+}][OH^-]^2 = 6.5 \times 10^{-5}$

let $x = [Ca^{2+}]$, $[OH^-] = 2x + 0.10$ $\quad K_{sp} = (x)(2x + 0.10)^2 = 6.5 \times 10^{-5}$

By successive approximations, x = 0.0053

The molar solubility of $Ca(OH)_2$ in 0.010 M NaOH is 0.053 moles/L.

19.46 In order for a precipitate to form, the value of the reaction quotient, Q, must be greater than the value of K_{sp}. For $PbCl_2$, $K_{sp} = 1.7 \times 10^{-5}$ (see Table 19.1).

$Q = [Pb^{2+}][Cl^-]^2 = (0.0150)(0.0120)^2 = 2.16 \times 10^{-6}$. Since $Q < K_{sp}$, no precipitate will form.

19.47 In order for a precipitate to form, the value of the reaction quotient, Q, must be greater than the value of K_{sp}. For $AgC_2H_3O_2$, $K_{sp} = 2.3 \times 10^{-3}$.

$Q = [Ag^+][C_2H_3O_2^-] = (0.015)(0.50) = 7.5 \times 10^{-3}$. Since $Q > K_{sp}$, a precipitate will form. (Note: The concentration of $C_2H_3O_2^-$ is twice the concentration of $Ca(C_2H_3O_2)_2$ since one mole of $Ca(C_2H_3O_2)_2$ produces two moles of $C_2H_3O_2^-$).

19.48 To solve this problem, determine the value for Q and apply LeChâtelier's Principle.

(a) $[Pb^{2+}] = (50.0 \text{ mL})(0.0100 \text{ moles/L})/(100.0 \text{ mL}) = 5.00 \times 10^{-3}$

$[Br^-] = (50.0 \text{ mL})(0.0100 \text{ moles/L})/(100.0 \text{ mL}) = 5.00 \times 10^{-3}$

$Q = [Pb^{2+}][Br^-]^2 = (5.00 \times 10^{-3})(5.00 \times 10^{-3})^2 = 1.25 \times 10^{-7}$

For $PbBr_2$, $K_{sp} = 2.1 \times 10^{-6}$

Since $Q < K_{sp}$, no precipitate will form.

(b) $[Pb^{2+}] = (50.0 \text{ mL})(0.0100 \text{ moles/L})/(100.0 \text{ mL}) = 5.00 \times 10^{-3}$

$[Br^-] = (50.0 \text{ mL})(0.100 \text{ moles/L})/(100.0 \text{ mL}) = 5.00 \times 10^{-2}$

$Q = [Pb^{2+}][Br^-]^2 = (5.00 \times 10^{-3})(5.00 \times 10^{-2})^2 = 1.25 \times 10^{-5}$

For $PbBr_2$, $K_{sp} = 2.1 \times 10^{-6}$

Since $Q > K_{sp}$, a precipitate will form.

19.49 In order for a precipitate to form, the value of the reaction quotient, Q, must be greater than the value of K_{sp}. For $AgC_2H_3O_2$, $K_{sp} = 2.3 \times 10^{-3}$.

$[Ag^+] = (22.0 \text{ mL})(0.100 \text{ M})/(67.0 \text{ mL}) = 3.28 \times 10^{-2}$ M

$[C_2H_3O_2^-] = (45.0 \text{ mL})(0.0260 \text{ M})/(67.0 \text{ mL}) = 1.75 \times 10^{-2}$ M

$Q = [Ag^+][C_2H_3O_2^-] = (3.28 \times 10^{-2})(1.75 \times 10^{-2}) = 5.74 \times 10^{-4}$.

Since $Q < K_{sp}$, no precipitate will form.

19.50 $AgCl(s) \leftrightarrows Ag^+ + Cl^-$ $\quad K_{sp} = [Ag^+][Cl^-] = 1.8 \times 10^{-10}$

$AgI(s) \leftrightarrows Ag^+ + I^-$ $\quad K_{sp} = [Ag^+][I^-] = 8.3 \times 10^{-17}$

When $AgNO_3$ is added to the solution, AgI will precipitate before any AgCl does due to the lower solubility of AgI. In order to answer the question, i.e., what is the $[I^-]$ when AgCl first precipitates, we need to find the minimum concentration of Ag^+ that must be added to precipitate AgCl.

Let $x = [Ag^+]$; $K_{sp} = (x)(0.050) = 1.8 \times 10^{-10}$; $x = 3.6 \times 10^{-9}$ M

When the AgCl starts to precipitate, the solution will have a $[Ag^+]$ of 3.6×10^{-9} M. Now we ask, what is the $[I^-]$ if $[Ag^+] = 3.6 \times 10^{-9}$ M?

So, $K_{sp} = [Ag^+][I^-] = (3.6 \times 10^{-9})(x) = 8.3 \times 10^{-17}$; $x = 2.3 \times 10^{-8} \text{ M} = [I^-]$

19.51 This problem is similar to 19.50, except that the K_{sp} constants are closer in value. We first determine the minimum amount of SO_4^{2-} that must be added to initiate the precipitation of $CaSO_4$. $CaSO_4$ will precipitate after $SrSO_4$ due to its larger value for K_{sp}: $K_{sp}(CaSO_4) = 2.4 \times 10^{-5}$ and $K_{sp}(SrSO_4) = 3.2 \times 10^{-7}$ (see Table 19.1)

(a) Let $x = [Ca^{2+}]$; $K_{sp} = [Ca^{2+}][SO_4^{2-}] = (0.15)(x) = 2.4 \times 10^{-5}$

$$x = [SO_4^{2-}] = 1.6 \times 10^{-4}\text{ M}$$

When the $[SO_4^{2-}] = 1.6 \times 10^{-4}$ the $CaSO_4$ will start to precipitate. Now we ask, what is the $[Sr^{2+}]$ if $[SO_4^{2-}] = 1.6 \times 10^{-4}$ M?

$SrSO_4\,(s) \leftrightarrows Sr^{2+}\,(aq) + SO_4^{2-}(aq)$ $\qquad K_{sp} = [Sr^{2+}][SO_4^{2-}]$

	$[Sr^{2+}]$	$[SO_4^{2-}]$
I	–	1.6×10^{-4}
C	+x	1.6×10^{-4}+x
E	+x	1.6×10^{-4}+x

$$K_{sp} = [Sr^{2+}][SO_4^{2-}] = (x)(1.6 \times 10^{-4} + x) = 3.2 \times 10^{-7}$$

For this problem, we must solve the quadratic equation and we determine that $x = 4.9 \times 10^{-4}$ M. Thus, the $[Sr^{2+}] = 4.9 \times 10^{-4}$ M when the $CaSO_4$ starts to precipitate.

(b) Initially the solution had a concentration of 0.15 M. The solution now has a $[Sr^{2+}] = 4.9 \times 10^{-4}$. So, the percentage of Sr^{2+} precipitated is;

$$\frac{0.15 - 4.9 \times 10^{-4}}{0.15} \times 100\% = 99.7\,\%$$

19.52 The less soluble substance is PbS. We need to determine the minimum $[H^+]$ at which CoS will precipitate.

$$K_{spa} = \frac{[Co^{2+}][H_2S]}{[H^+]^2} = \frac{(0.010)(0.1)}{[H^+]^2} = 0.5 \qquad \text{(from Table 19.2)}$$

$$[H^+] = \sqrt{\frac{(0.010)(0.1)}{0.5}} = 0.045$$

pH = –log[H^+] = 1.35. At a pH lower than 1.35, PbS will precipitate and CoS will not. At larger values of pH, both PbS and CoS will precipitate.

19.53 The less soluble substance is SnS so we will determine the maximum amount of H^+ that is permitted before MnS starts to precipitate.

$$K_{spa} = \frac{[Mn^{2+}][H_2S]}{[H^+]^2} = \frac{(0.010)(0.1)}{[H^+]^2} = 3 \times 10^7 \qquad \text{(from Table 19.2)}$$

$$[H^+] = \sqrt{\frac{(0.010)(0.1)}{3 \times 10^7}} = 6 \times 10^{-6}$$

pH = –log[H^+] = 5.2. At pH values equal to or less than 5.2, MnS will not precipitate.

19.54 $Cu(OH)_2(s) \leftrightarrows Cu^{2+}(aq) + 2\ OH^-(aq)$

$$K_{sp} = [Cu^{2+}][OH^-]^2$$
$$4.8\times10^{-20} = [0.10][OH^-]^2$$
$$[OH^-] = 6.9 \times 10^{-10}\ M$$
$$pOH = -\log[OH^-] = -\log[6.9 \times 10^{-10}] = 9.2$$
$$pH = 14.00 - pOH = 4.8$$

4.8 is the pH *below* which all the $Cu(OH)_2$ will be soluble.

$Mn(OH)_2(s) \leftrightarrows Mn^{2+}(aq) + 2\ OH^-(aq)$

$$K_{sp} = [Mn^{2+}][OH^-]^2$$
$$1.6\times10^{-13} = [0.10][OH^-]^2$$
$$[OH^-] = 1.3 \times 10^{-6}\ M$$
$$pOH = -\log[OH^-] = -\log[1.3 \times 10^{-6}] = 5.9$$
$$pH = 14.00 - pOH = 8.1$$

8.1 is the pH *below* which all the $Mn(OH)_2$ will be soluble.

Therefore, from 4.8-8.1 $Mn(OH)_2$ will be soluble, but some $Cu(OH)_2$ will precipitate out of solution.

19.55 The following reactions are possible:

$CaC_2O_4(s) \leftrightarrows Ca^{2+}(aq) + C_2O_4^{2-}(aq)$ $\quad K_{sp} = [Ca^{2+}][C_2O_4^{2-}] = 2.3 \times 10^{-9}$

$MgC_2O_4(s) \leftrightarrows Mg^{2+}(aq) + C_2O_4^{-2}(aq)$ $\quad K_{sp} = [Mg^{2+}][C_2O_4^{2-}] = 8.6 \times 10^{-5}$

$H_2C_2O_4(aq) \leftrightarrows H^+(aq) + HC_2O_4^-(aq)$ $\quad K_{a_1} = \dfrac{[H^+][HC_2O_4^-]}{[H_2C_2O_4]} = 6.5 \times 10^{-2}$

$HC_2O_4^-(aq) \leftrightarrows H^+(aq) + C_2O_4^{2-}(aq)$ $\quad K_{a_2} = \dfrac{[H^+][C_2O_4^{2-}]}{[HC_2O_4^-]} = 6.1 \times 10^{-5}$

$H_2C_2O_4(aq) \leftrightarrows 2H^+(aq) + C_2O_4^{2-}(aq)$ $\quad K_O = \dfrac{[H^+]^2[C_2O_4^{2-}]}{[H_2C_2O_4]} = 4.0 \times 10^{-6}$

Assume that the concentration of $H_2C_2O_4$ at the end of the process is 0.10 M
First, calculate the concentration of $C_2O_4^{2-}$, at which the MgC_2O_4 will precipitate

$K_{sp} = [Mg^{2+}][C_2O_4^{2-}] = 8.6 \times 10^{-5}$
$[Mg^{2+}] = 0.10\ M$
$(0.10)(x) = 8.6 \times 10^{-5}$
$x = 8.6 \times 10^{-4}\ M = [C_2O_4^{2-}]$

As long as the concentration of $C_2O_4^{2-}$ is kept below 8.6×10^{-4} M, the Mg^{2+} will remain in solution
In order for the concentration of oxalate to remain below = 8.6×10^{-4} M, the pH will be:

$H_2C_2O_4(aq) \leftrightarrows 2H^+(aq) + C_2O_4^{2-}(aq)$ $\quad K_O = \dfrac{[H^+]^2[C_2O_4^{2-}]}{[H_2C_2O_4]} = 4.0 \times 10^{-6}$

$$K_O = \frac{[H^+]^2[8.6\times10^{-4}]}{[0.10]} = 4.0 \times 10^{-6}$$

$[H^+] = 0.0216$ M
pH = 1.67

By keeping the pH > 1.67, the MgC_2O_4 will not precipitate.
To precipitate the CaC_2O_4, a similar calculation needs to be done:

$K_{sp} = [Ca^{2+}][C_2O_4^{2-}] = 2.3 \times 10^{-9}$
$[Ca^{2+}] = 0.10$ M
$(0.10)(x) = 2.3 \times 10^{-9}$
$x = 2.3 \times 10^{-8}$ M $= [C_2O_4^{2-}]$

As long as the concentration of $C_2O_4^{2-}$ is kept above 2.3×10^{-8} M, the Ca^{2+} will precipitate
In order for the concentration of oxalate to remain above = 2.3×10^{-8} M, the pH must be:

$H_2C_2O_4(aq) \leftrightarrows 2H^+(aq) + C_2O_4^{2-}(aq)$ $\qquad K_{a_2} = \frac{[H^+]^2[C_2O_4^{2-}]}{[H_2C_2O_4]} = 4.0 \times 10^{-6}$

$$K = \frac{[H^+]^2[2.3\times10^{-8}]}{[0.10]} = 4.0 \times 10^{-6}$$

$[H^+] = 4.17$ M
pH = –0.62

19.56 (a) $Cu^{2+}(aq) + 4Cl^-(aq) \leftrightarrows CuCl_4^{2-}(aq)$ $\qquad K_{form} = \frac{[CuCl_4^{2-}]}{[Cu^{2+}][Cl^-]^4}$

(b) $Ag^+(aq) + 2I^-(aq) \leftrightarrows AgI_2^-(aq)$ $\qquad K_{form} = \frac{[AgI_2^-]}{[Ag^+][I^-]^2}$

(c) $Cr^{3+}(aq) + 6NH_3(aq) \leftrightarrows Cr(NH_3)_6^{3+}(aq)$ $\qquad K_{form} = \frac{[Cr(NH_3)_6^{3+}]}{[Cr^{3+}][NH_3]^6}$

19.57 (a) $Ag(S_2O_3)_2^{3-}(aq) \leftrightarrows Ag^+(aq) + 2S_2O_3^{2-}(aq)$ $\qquad K_{inst} = \frac{[Ag^+][S_2O_3^{2-}]^2}{[Ag(S_2O_3)_2^{3-}]}$

(b) $Zn(NH_3)_4^{2+}(aq) \leftrightarrows Zn^{2+}(aq) + 4NH_3(aq)$ $\qquad K_{inst} = \frac{[Zn^{2+}][NH_3]^4}{[Zn(NH_3)_4^{2+}]}$

(c) $SnS_3^{2-}(aq) \leftrightarrows Sn^{4+}(aq) + 3S^{2-}(aq)$ $\qquad K_{inst} = \frac{[Sn^{4+}][S^{2-}]^3}{[SnS_3^{2-}]}$

19.58 (a) $Co(NH_3)_6^{3+}(aq) \leftrightarrows Co^{3+}(aq) + 6NH_3(aq)$ $K_{inst} = \dfrac{[Co^{3+}][NH_3]^6}{[Co(NH_3)_6^{3+}]}$

(b) $HgI_4^{2-}(aq) \leftrightarrows Hg^{2+}(aq) + 4I^-(aq)$ $K_{inst} = \dfrac{[Hg^{2+}][I^-]^4}{[HgI_4^{2-}]}$

(c) $Fe(CN)_6^{4-}(aq) \leftrightarrows Fe^{2+}(aq) + 6CN^-(aq)$ $K_{inst} = \dfrac{[Fe^{2+}][CN^-]^6}{[Fe(CN)_6^{4-}]}$

19.59 (a) $Hg^{2+}(aq) + 4NH_3(aq) \leftrightarrows Hg(NH_3)_4^{2+}(aq)$ $K_{form} = \dfrac{[Hg(NH_3)_4^{2+}]}{[Hg^{2+}][NH_3]^4}$

(b) $Sn^{4+}(aq) + 6F^-(aq) \leftrightarrows SnF_6^{2-}(aq)$ $K_{form} = \dfrac{[SnF_6^{2-}]}{[Sn^{4+}][F^-]^6}$

(c) $Fe^{3+}(aq) + 6CN^-(aq) \leftrightarrows Fe(CN)_6^{3-}(aq)$ $K_{form} = \dfrac{[Fe(CN)_6^{3-}]}{[Fe^{3+}][CN^-]^6}$

19.60 $K_c = K_{sp} \times K_{form} = (1.7 \times 10^{-5})(2.5 \times 10^1) = 4.3 \times 10^{-4}$

19.61 $K_c = K_{sp} \times K_{form} = (1.6 \times 10^{-14})(5.5 \times 10^{18}) = 8.8 \times 10^4$

19.62 There are two events in this net process: one is the formation of a complex ion (an equilibrium which has an appropriate value for K_{form}), and the other is the dissolving of $Fe(OH)_3$, which is governed by K_{sp} for the solid.

$Fe(OH)_3(s) \leftrightarrows Fe^{3+}(aq) + 3OH^-(aq)$ $K_{sp} = [Fe^{3+}][OH^-]^3 = 1.6 \times 10^{-39}$

$Fe^{3+}(aq) + 6CN^-(aq) \leftrightarrows Fe(CN)_6^{3-}(aq)$ $K_{form} = \dfrac{[Fe(CN)_6^{3-}]}{[Fe^{3+}][CN^-]^6} = 1.0 \times 10^{31}$

The net process is:

$Fe(OH)_3(s) + 6CN^-(aq) \leftrightarrows Fe(CN)_6^{3-}(aq) + 3OH^-(aq)$

The equilibrium constant for this process should be:

$$K_c = \frac{\left[Fe(CN)_6^{3-}\right]\left[OH^-\right]^3}{\left[CN^-\right]^6}$$

The numerical value for the above K_c is equal to the product of K_{sp} for $Fe(OH)_3(s)$ and K_{form} for $Fe(CN)_6^{3-}$, as can be seen by multiplying the mass action expressions for these two equilibria: $K_c = K_{form} \times K_{sp} = 1.6 \times 10^{-8}$

Because K_{form} is so very large, we can assume that all of the dissolved iron ion is present in solution as the complex, thus: $[Fe(CN)_6^{3-}] = 0.11$ mol/1.2 L = 0.092 M. Also the reaction stoichiometry shows that each iron ion that dissolves gives 3 OH^- ions in solution, and we have: $[OH^-] = 0.092 \times 3 = 0.28$ M. We substitute these values into the K_c expression and rearrange to get:

$$\left[CN-\right] = \sqrt[6]{\frac{\left[Fe(CN)_6^{3-}\right]\left[OH^-\right]^3}{K_c}}$$

$$= \sqrt[6]{\frac{(0.092)(0.28)^3}{1.6 \times 10^{-8}}}$$

Thus we arrive at the concentration of cyanide ion that is required in order to satisfy the mass action requirements of the equilibrium: $[CN^-] = 7.1$ mol L^{-1}. Since this concentration of CN^- must be present in 1.2 L, the number of moles of cyanide that are required is: 7.1 mol $L^{-1} \times 1.2$ L = 8.5 mol CN^-.

Additionally, a certain amount of cyanide is needed to form the complex ion. The stoichiometry requires six times as much cyanide ion as iron ion. This is 0.11 moles × 6 = 0.66 mol. This brings the total required cyanide to (8.5 + 0.66) = 9.2 mol.

9.2 mol × 49.0 g/mol = 450 g NaCN are required.

19.63 $AgBr(s) \leftrightarrows Ag^+ + Br^-$ $K_{sp} = 5.0 \times 10^{-13}$

$Ag^+ + 2S_2O_3^{2-} \leftrightarrows Ag(S_2O_3)_2^{3-}$ $K_f = 2.0 \times 10^{13}$

$AgBr(s) + 2S_2O_3^{2-} \leftrightarrows Ag(S_2O_3)_2^{3-} + Br^-$ $K_c = K_{sp} * K_f = 10$

	$[S_2O_3^{2-}]$	$[Ag(S_2O_3)_2^{3-}]$	$[Br^-]$
I	1.20	–	–
C	–2x	+x	+x
E	1.20–2x	+x	+x

Note: Since the AgBr(*s*) has a constant concentration, it may be neglected.

$$K_c = \frac{[Ag(S_2O_3)_2^{3-}][Br^-]}{[S_2O_3^{2-}]^2} = 10$$

$$K_c = \frac{x^2}{(1.20-2x)^2} = 10$$

To solve this equation, take the square root of both sides and then solve for x.

$x = 0.518\ M = [Ag(S_2O_3)_2^{3-}]$

Since 1 mole of AgBr produces 1 mole of $Ag(S_2O_3)_2^{3-}$, we can determine the number of grams of AgBr that will dissolve in 125 mL.

g AgBr = (0.125 L)(0.518 moles/L)(187.77 g/mole) = 12.2 g AgBr

19.64 The applicable equilibria are as follows:

$AgI(s) \leftrightarrows Ag^+(aq) + I^-(aq)$ $K_{sp} = [Ag^+][I^-] = 8.3 \times 10^{-17}$

$Ag^+(aq) + 2I^-(aq) \leftrightarrows AgI_2^-(aq)$ $K_{form} = \dfrac{[AgI_2^-]}{[Ag^+][I^-]^2} = 1 \times 10^{11}$

When a solution of AgI_2^- is diluted, all of the concentrations of the species in K_{form} above decrease. However, the decrease of $[I^-]$ has more effect on equilibrium because its expression is *squared*. Hence, the denominator is decreased more than the numerator in the reaction quotient, Q. The system reacts according to Le Châtelier's Principle, by moving to the left (toward reactants) to increase the value of $[I^-]$.

As the system moves to the left, more Ag^+ is created, which has an effect on the first equilibrium above. Again, Le Châtelier's Principle causes the reaction to move to the left to re-establish equilibrium, which produces AgI(*s*) precipitate.

The two equations above may be combined and K_c found as follows:

$AgI(s) + I^-(aq) \leftrightarrows AgI_2^-(aq)$ $K_c = \dfrac{[AgI_2^-]}{[I^-]} = K_{sp} \times K_{form} = 8.3 \times 10^{-6}$

To answer the second question, we make a table and fill in what we know. We begin with 1.0 M I^-. This is reduced by some amount (x) as it reacts with the silver ions, and $[AgI_2^-]$ is increased by the same amount:

	$[I^-]$	$[AgI_2^-]$
I	1.0	–
C	–x	+x
E	1.0 – x	x

Now we insert the equilibrium values into the above equation:

$$K_c = \frac{[AgI_2^-]}{[I^-]} = 8.3 \times 10^{-6}$$

$$K_c = \frac{[x]}{[1.0 - x]} = 8.3 \times 10^{-6}$$

$$x = 8.3 \times 10^{-6}$$

This value represents the change in concentration of I^- which, from the balanced equation, equals the change in concentration of AgI(*s*). The given volume is 100 mL, which allows us to find the amount of AgI reacting:

$0.100\ L(8.3 \times 10^{-6}\ mol/L) = 8.3 \times 10^{-7}$ mol AgI

8.3×10^{-7} mol AgI(234.8 g/mol) $= 1.9 \times 10^{-4}$ g AgI

19.65 The applicable equilibria are as follows:

$AgI(s) \leftrightarrows Ag^+(aq) + I^-(aq)$ $K_{sp} = [Ag^+][I^-] = 8.3 \times 10^{-17}$

$Ag^+(aq) + 2CN^-(aq) \leftrightarrows Ag(CN)_2^-(aq)$ $K_{form} = \dfrac{[Ag(CN)_2^-]}{[Ag^+][CN^-]^2} = 5.3 \times 10^{18}$

The two equations above may be combined and K_c found as follows:

$AgI(s) + 2CN^-(aq) \leftrightarrows Ag(CN)_2^-(aq) + I^-(aq)$ $\qquad K_c = \dfrac{[Ag(CN)_2^-][I^-]}{[CN^-]^2} = K_{sp} \times K_{form} = 4.4 \times 10^2$

We begin with 0.010 M CN^-. This is reduced by some amount (x) as it reacts with the silver ions, and $[AgI_2^-]$ is increased by the same amount:

	$[CN^-]$	$[Ag(CN)_2^-]$	I^-
I	0.010	–	–
C	–x	+x	+x
E	0.010 – x	x	x

Now we insert the equilibrium values into the above equation:

$$K_c = \frac{[Ag(CN)_2^-][I^-]}{[CN^-]^2} = 4.4 \times 10^2$$

$$K_c = \frac{[x][x]}{[0.010 - x]^2} = 4.4 \times 10^2$$

Take the square root of both sides and solve for x:

$$x = 9.5 \times 10^{-3}$$

This value represents the change in concentration of I^- which, from the balanced equation, equals the change in concentration of $AgI(s)$.
The molar solubility of AgI in 0.010 M KCN is 9.5×10^{-3} M.

19.66 Recall that $K_{inst} = 1/K_{form}$.

$Zn(OH)_2(s) \leftrightarrows Zn^{2+}(aq) + 2OH^-(aq)$ $\qquad K_{sp} = [Zn^{2+}][OH^-]^2 = 3.0 \times 10^{-16}$

$Zn^{2+}(aq) + 4NH_3(aq) \leftrightarrows Zn(NH_3)_4^{2+}(aq)$ $\qquad K_{form} = \dfrac{[Zn(NH_3)_4^{2+}]}{[Zn^{2+}][NH_3]^4} = ?$

Combined, this is:

$Zn(OH)_2(s) + 4NH_3(aq) \leftrightarrows Zn(NH_3)_4^{2+}(aq) + 2OH^-(aq)$

$$K_c = \frac{[Zn(NH_3)_4^{2+}][OH^-]^2}{[NH_3]^4}$$

	$[NH_3]$	$[Zn(NH_3)_4^{2+}]$	$[OH^-]$
I	1.0	–	–
C	– 4x	+x	+2x
E	1.0 – 4x	x	2x

$$K_c = \frac{[x][2x]^2}{[1.0 - 4x]^4}$$

The problem gives the molar solubility of $Zn(OH)_2$ as 5.7×10^{-3} M. This means in one liter of 1.0 M NH_3, $x = 5.7 \times 10^{-3}$ moles. Substituting this value in for x, we get $K_c = 8.1 \times 10^{-7}$.

$$K_c = K_{sp} \times K_{form}$$
$$8.1 \times 10^{-7} = 3.0 \times 10^{-16} \times K_{form}$$
$$K_{form} = 2.7 \times 10^{9}$$

$K_{inst} = 1/K_{form}$
$K_{inst} = 1/(2.7 \times 10^{9}) = 3.7 \times 10^{-10}$

19.67 $Cu(OH)_2\,(s) \leftrightarrows Cu^{2+}(aq) + 2OH^-(aq)$ $\qquad K_{sp} = [Cu^{2+}][OH^-]^2 = 4.8 \times 10^{-20}$

$Cu^{2+}(aq) + 4NH_3\,(aq) \leftrightarrows Cu(NH_3)_4{}^{2+}(aq)$ $\qquad K_{form} = \dfrac{[Cu(NH_3)_4{}^{2+}]}{[Cu^{2+}][NH_3]^4} = 1.1\times10^{13}$

Combined, this is:

$Cu(OH)_2\,(s) + 4NH_3\,(aq) \leftrightarrows Cu(NH_3)_4{}^{2+}(aq) + 2OH^-(aq)$

$$K_c = \frac{[Cu(NH_3)_4{}^{2+}][OH^-]^2}{[NH_3]^4} = 5.3 \times 10^{-7}$$

	$[NH_3]$	$[Cu(NH_3)_4{}^{2+}]$	$[OH^-]$
I	2.0	–	–
C	– 4x	+x	+2x
E	2.0 – 4x	x	2x

$$K_c = \frac{[x][2x]^2}{[2.0 - 4x]^4} = 5.3 \times 10^{-7}$$

Solve for x using successive approximations.

$x = 1.2 \times 10^{-2}$

$[Cu(NH_3)_4{}^{2+}] = 1.2 \times 10^{-2}$

Since all of the Cu^{2+} comes from the $Cu(OH)_2$, the molar solubility of $Cu(OH)_2$ is 1.2×10^{-2}.

Additional Exercises

19.68 We must first calculate the solubility in terms of # mols/L, i.e.,

$$\#\ \text{mol}/\text{L} = \left(7.05 \times 10^{-3}\ \text{g}/\text{L}\right)\left(\frac{1\ \text{mol}\ Mg(OH)_2}{58.32\ \text{g}\ Mg(OH)_2}\right) = 1.21 \times 10^{-4}\ \text{M}$$

Next, use this to establish the individual ion concentrations based on the equilibrium:

$Mg(OH)_2 \leftrightarrows Mg^{2+} + 2OH^-$

$$[Mg^{2+}] = 1.21 \times 10^{-4}\ \text{M}$$
$$[OH^-] = 2.42 \times 10^{-4}\ \text{M}$$

Finally, calculate K_{sp} using the standard expression:

$$K_{sp} = [Mg^{2+}][OH^-]^2 = (1.21 \times 10^{-4})(2.42 \times 10^{-4})^2 = 7.09 \times 10^{-12}$$

19.69 $FeS(s) + 2H^+(aq) \leftrightarrows Fe^{2+}(aq) + H_2S(aq)$ $K_{spa} = \dfrac{[Fe^{2+}][H_2S]}{[H^+]^2}$

	$[H^+]$	$[Fe^{2+}]$	$[H_2S]$
I	8	–	–
C	8–2x	+x	+x
E	8–2x	+x	+x

$$K_{spa} = \frac{[Fe^{2+}][H_2S]}{[H^+]^2} = \frac{(x)(x)}{(8-2x)^2} = 600 \quad \text{(from Table 19.2)}$$

take the square root of both sides to get; $\dfrac{x}{(8-2x)} = 24.5$

Solving gives x = 3.92 M. FeS is very soluble in 8 M acid.

19.70 In order to answer this question, we need the $[OH^-]$ at equilibrium. $Mg(OH)_2$ is a sparingly soluble compound. According to Table 19.1, $K_{sp} = 7.1 \times 10^{-12}$.

$MgOH_2(s) \leftrightarrows Mg^{2+}(aq) + 2OH^-(aq)$ $K_{sp} = [Mg^{2+}][OH^-]^2$

	$[Mg^{2+}]$	$[OH^-]$
I	–	–
C	+x	+2x
E	+x	+2x

$K_{sp} = (x)(2x)^2 = 4x^3 = 7.1 \times 10^{-12}$
$x = 1.2 \times 10^{-4}$ M, $[OH^-] = 2x = 2.4 \times 10^{-4}$ M: $pOH = -\log[OH^-] = 3.62$
The pH = 14.00 – pOH = 10.38.

19.71 The reaction that will dissolve the $Mg(OH)_2$ is:
$Mg(OH)_2 + 2H^+ \rightarrow Mg^{2+} + 2H_2)$
The concentration of H^+ in the solution, before any reaction has occurred between the acid and the $Mg(OH)_2$, is
$(0.025\ L)(0.10\ M\ HCl) = 2.5 \times 10^{-3}$ mol HCl
2.5×10^{-3} mol/1.000 L = 2.5×10^{-3} M
Since all of the H^+ will react with the solid $Mg(OH)_2$, the amount of Mg^{2+} in solution will be:

$$2.5 \times 10^{-3} \text{ mol HCl}\left(\frac{1 \text{ mol } Mg^{2+}}{2 \text{ mol } H^+}\right) = 1.3 \times 10^{-3} \text{ mol } Mg^{2+}$$

Find the equilibrium concentration of OH^- with 1.3×10^{-3} M Mg^{2+}
$K_{sp} = [Mg^{2+}][OH^-]^2 = 7.1 \times 10^{-12}$
$7.1 \times 10^{-12} = (1.3 \times 10^{-3})(x)^2$
$x = 7.4 \times 10^{-5} = [OH^-]$
$pOH = -\log[OH^-] = -\log(7.4 \times 10^{-5}) = 4.13$
pH = 14 – pOH = 14 – 4.13 = 9.87

19.72 In this problem, we have two simultaneous equilibria occurring:

$Mn(OH)_2(s) \leftrightarrows Mn^{2+}(aq) + 2OH^-(aq)$ $K_{sp} = [Mn^{2+}][OH^-]^2 = 1.6 \times 10^{-13}$

$Fe^{2+}(aq) + 2OH^-(aq) \leftrightarrows Fe(OH)_2(s)$ $K_c = \dfrac{1}{[Fe^{2+}][OH^-]^2} = 1/(7.9 \times 10^{-16}) = 1.3 \times 10^{15}$

The second equilibrium represents the opposite equation from that of K_{sp}. Therefore, its value is $1/K_{sp}$ for $Fe(OH)_2$.

Combined, and omitting spectator ions, this is:

$$Mn(OH)_2(s) + Fe^{2+}(aq) \leftrightarrows Mn^{2+}(aq) + Fe(OH)_2(s)$$

$$K_c = \frac{[Mn^{2+}]}{[Fe^{2+}]} = K_{sp\,(Mn)} \cdot K_{c\,(Fe)} = (1.6\times10^{-13})(1.3\times10^{15}) = 208$$

	$[Fe^{2+}]$	$[Mn^{2+}]$
I	0.100	–
C	–x	+x
E	0.100–x	x

$$K_c = \frac{[Mn^{2+}]}{[Fe^{2+}]}$$

$$208 = \frac{[x]}{[0.100 - x]}$$

$$20.8 - 208x = x$$

$$20.8 = 209x$$

$$x = 0.0995$$

Therefore, $[Fe^{2+}] = 0.100 - x = 0.001$ M and $[Mn^{2+}] = 0.0995$ M

Since K_{sp} for $Fe(OH)_2$ and $Mn(OH)_2$ are so small, we assume there is almost no free hydroxide ion present and therefore the pH would remain neutral, around 7.

19.73 (a) The number of moles of the two reactants are:

$0.12\text{ M } Ag^+ \times 0.050\text{ L} = 6.0 \times 10^{-3}$ moles Ag^+
$0.048\text{ M } Cl^- \times 0.050\text{ L} = 2.4 \times 10^{-3}$ moles Cl^-

The precipitation of AgCl proceeds according to the following stoichiometry: $Ag^+ + Cl^- \rightarrow AgCl(s)$. If we assume that the product is completely insoluble, then 2.4×10^{-3} moles of AgCl will be formed because Cl^- is the limiting reagent (see above.)

$$\#\text{ g AgCl} = (2.4 \times 10^{-3}\text{ mol AgCl})\left(\frac{143.3\text{ g AgCl}}{1\text{ mol AgCl}}\right) = 0.35\text{ g AgCl}$$

(b) The silver ion concentration may be determined by calculating the amount of excess silver added to the solution:

$[Ag^+] = (6.0 \times 10^{-3}\text{ moles} - 2.4 \times 10^{-2}\text{ moles})/1.00\text{ L} = 3.6 \times 10^{-2}$ M

The concentrations of nitrate and sodium ions are easily calculated since they are spectators in this reaction:

$[NO_3^-] = (0.12\text{ M})(50.0\text{ mL})/(100.0\text{ mL}) = 6.0 \times 10^{-2}$ M
$[Na^+] = (0.048\text{ M})(50.0\text{ mL})/(100.0\text{ mL}) = 2.4 \times 10^{-2}$ M

In order to determine the chloride ion concentration, we need to solve the equilibrium expression. Specifically, we need to ask what is the chloride ion concentration in a saturated solution of AgCl that has a $[Ag^+] = 3.6 \times 10^{-2}$ M.

$$AgCl(s) \leftrightarrows Ag^+ + Cl^- \qquad K_{sp} = 1.8 \times 10^{-10}$$

	$[Ag^+]$	$[Cl^-]$
I	0.036	–
C	+x	+x
E	0.036+x	+x

$$K_{sp} = [Ag^+][Cl^-] = (0.036+x)(x) = 1.8 \times 10^{-10}$$

$$x = 5.0 \times 10^{-9} \text{ M if we assume that } x \ll 0.036$$

Therefore, $[Cl^-] = 5.0 \times 10^{-9}$ M.

(c) The percentage of the silver that has precipitated is:

$$(2.4 \times 10^{-3} \text{ moles})/(6.0 \times 10^{-3} \text{ moles}) \times 100\% = 40\%$$

19.74 To solve this problem, recognize that for a solution having a density of 1.00 g mL^{-1}, 1 ppm = 1 mg L^{-1}. Therefore, the initial hard water solution has a concentration of 278 mg Ca^{2+} / 1 L solution. Converting to molar concentration:

$$\frac{\#\text{ mol } Ca^{2+}}{\text{L solution}} = \left(\frac{278 \text{ mg } Ca^{2+}}{1 \text{ L solution}}\right)\left(\frac{1 \text{ g } Ca^{2+}}{1000 \text{ mg } Ca^{2+}}\right)\left(\frac{1 \text{ mol } Ca^{2+}}{40.078 \text{ g } Ca^{2+}}\right)$$

$$= 6.94 \times 10^{-3} \text{ M } Ca^{2+}$$

The concentration of CO_3^{2-} is:

$$\frac{\#\text{ mol } CO_3^{2-}}{\text{L solution}} = \left(\frac{1.00 \text{ g } Na_2CO_3}{1 \text{ L solution}}\right)\left(\frac{1 \text{ mol } Na_2CO_3}{105.99 \text{ g } Na_2CO_3}\right)\left(\frac{1 \text{ mol } CO_3^{2-}}{1 \text{ mol } Na_2CO_3}\right)$$

$$= 9.43 \times 10^{-3} \text{ M } CO_3^{2-}$$

Comparing the concentrations of Ca^{2+} and CO_3^{2-}, we observe that Ca^{2+} is the limiting reactant. Because of the small value of Ksp, we can assume that $CaCO_3$ will precipitate using all of the available Ca^{2+} and leaving
$9.43 \times 10^{-3} - 6.94 \times 10^{-3} = 2.49 \times 10^{-3}$ M CO_3^{2-}. The question now becomes, how much Ca^{2+} will be present in a solution having a $[CO_3^{2-}] = 2.49 \times 10^{-3}$ M? Use the solubility product constant for K_{sp} to answer this question.

$$K_{sp} = 4.5 \times 10^{-9} = [Ca^{2+}][CO_3^{2-}]$$

$$[Ca^{2+}] = \frac{K_{sp}}{[CO_3^{2-}]} = \frac{4.5 \times 10^{-9}}{2.49 \times 10^{-3}} = 1.8 \times 10^{-6} \text{ M}$$

Converting back to units of ppm (mg L^{-1}) we get:

$$\# \text{ ppm } Ca^{2+} = \left(\frac{1.8 \times 10^{-6} \text{ mol } Ca^{2+}}{1 \text{ L solution}}\right)\left(\frac{40.08 \text{ g } Ca^{2+}}{1 \text{ mol } Ca^{2+}}\right)\left(\frac{1000 \text{ mg}}{1 \text{ g}}\right)$$

$$= 7.2 \times 10^{-2} \text{ ppm } Ca^{2+}$$

19.75 A saturated solution of $La_2(CO_3)_3$ satisfies the following equilibrium expression:

$$K_{sp} = 4.0 \times 10^{-34} = [La^{3+}]^2[CO_3^{2-}]^3$$

If $[La^{3+}] = 0.010$ M, then the carbonate concentration of a saturated solution is:

$$[CO_3^{2-}] = \sqrt[3]{\frac{K_{sp}}{[La^{3+}]^2}} = \sqrt[3]{\frac{4.0 \times 10^{-34}}{(0.010)^2}} = 1.6 \times 10^{-10}$$

We do the same calculation for $PbCO_3$:

$$K_{sp} = 7.4 \times 10^{-14} = [Pb^{2+}][CO_3^{2-}]$$

If $[Pb^{2+}] = 0.010$ M, then the carbonate concentration of a saturated solution is:

$$[CO_3^{2-}] = \frac{K_{sp}}{[Pb^{2+}]} = \frac{7.4 \times 10^{-14}}{0.010} = 7.4 \times 10^{-12} \text{ M}$$

Therefore, at a carbonate ion concentration between 7.4×10^{-12} M and 1.6×10^{-10} M, $PbCO_3$ will precipitate, but $La_2(CO_3)_3$ will not precipitate. The upper limit for the carbonate ion concentration is therefore 1.6×10^{-10} M.

The equilibrium we need to look at now is: $H_2CO_3(aq) \leftrightarrows 2H^+(aq) + CO_3^{2-}(aq)$

The K_a for this reaction is the product of K_{a_1} and K_{a_2} for carbonic acid. From Table 18.1 we see that $K_{a_1} = 4.5 \times 10^{-7}$ and $K_{a_2} = 4.7 \times 10^{-11}$. So the equilibrium expression and value for the reaction of interest is:

$$K_a = \frac{[H^+]^2[CO_3^{2-}]}{[H_2CO_3]} = K_{a_1} \times K_{a_2} = 2.1 \times 10^{-17}$$

This equation is rearranged and the values above and the values given in the problem are substituted in order to determine the pH range over which $PbCO_3$ will selectively precipitate:

$$[H^+] = \sqrt{\frac{K_a\,[H_2CO_3]}{[CO_3^{2-}]}} = \sqrt{\frac{(2.1 \times 10^{-17})(3.3 \times 10^{-2})}{[CO_3^{2-}]}}$$

If we substitute $[CO_3^{2-}] = 7.4 \times 10^{-12}$ M we determine $[H^+] = 3.1 \times 10^{-4}$ M and the pH = 3.51.

Substituting $[CO_3^{2-}] = 1.6 \times 10^{-10}$ M, $[H^+] = 6.6 \times 10^{-5}$ M and

pH = 4.18.

Consequently, if $[H^+] = 6.6 \times 10^{-5}$ M (pH = 4.18), $La_2(CO_3)_3$ will not precipitate but $PbCO_3$ will precipitate. At pH = 3.51 and below, neither carbonate will precipitate.

19.76 a) $Mg(OH)_2(s) \leftrightarrows Mg^{2+} + 2OH^-$

$NH_4^+ + OH^- \leftrightarrows NH_3 + H_2O$

$Mg(OH)_2(s) + 2NH_4^+(aq) \leftrightarrows Mg^{2+}(aq) + 2H_2O + 2NH_3(aq)$

b) The NH_4^+ reacts with any OH^- produced in the dissociation of $Mg(OH)_2$ thereby shifting the equilibrium to the right.

c) We want all of the $Mg(OH)_2$ to go into solution. Using the K_{sp} value for $Mg(OH)_2$, we may find the hydroxide ion concentration under these conditions:

$$Mg(OH)_2(s) \leftrightarrows Mg^{2+} + 2OH^-$$

$$K_{sp} = [Mg^{2+}][OH^-]^2$$

$$7.1 \times 10^{-12} = [0.10][OH^-]^2$$

$$[OH^-] = 8.4 \times 10^{-6}$$

Now we can use this value in the following, simultaneous equilibrium:

$$NH_4^+(aq) + OH^-(aq) \leftrightarrows H_2O + NH_3(aq)$$

$$K_c = 1/K_bNH_3 = 1/1.8 \times 10^{-5} = 5.6 \times 10^4$$

$$K_c = \frac{[NH_3]}{[OH^-][NH_4^+]} = \frac{[0.20]}{[8.4 \times 10^{-6}][NH_4^+]} = 5.6 \times 10^4$$

(We know that $[NH_3] = 0.20$ M because in the equation below 2 moles of ammonia are formed for every one mole of magnesium ion:

$$Mg(OH)_2(s) + 2NH_4^+(aq) \leftrightarrows Mg^{2+}(aq) + 2H_2O + 2NH_3(aq))$$

Solving for $[NH_4^+]$, we get 0.43 M.

So the total $[NH_3] + [NH_4^+] = 0.20 + 0.43 = 0.63$ M

One must therefore add 0.63 mol NH_4Cl to a liter of solution.

d) The resulting solution will contain 0.20 mol of NH_3. Solve the weak base equilibrium problem for NH_3. The pH = 11.28.

19.77 There are two reactions that have to be considered here: the dissociation of $CaCO_3$ in water,

$$CaCO_3(s) \rightarrow Ca^{2+}(aq) + CO_3^{2-}(aq) \qquad K_{sp} = [Ca^{2+}][CO_3^{2-}] = 4.5 \times 10^{-9}$$

and the ionization of carbonate ion in water,

$$CO_3^{2-}(aq) + H_2O \rightarrow HCO_3^-(aq) + OH^-(aq) \quad K_b = \frac{[OH^-][HCO_3^-]}{[CO_3^{2-}]} = 1.8 \times 10^{-4}$$

Assuming that all of the CO_3^{2-} reacts with the water, the net reaction is:

$$CaCO_3(s) + H_2O(l) \rightarrow Ca^{2+}(aq) + HCO_3^-(aq) + OH^-(aq)$$

$$K_c = K_{sp} \times K_b = [Ca^{2+}][HCO_3^-][OH^-] = 8.1 \times 10^{-13}$$

We can obtain $[OH^-]$ from the pH:

$$pOH = 14 - pH = 14 - 8.50 = 5.50$$

$$[OH^-] = 10^{-pOH} = 10^{-5.50} = 3.2 \times 10^{-6} \text{ M}$$

Assume that $[Ca^{2+}] = [HCO_3^-] = x$

$$K_c = 8.1 \times 10^{-13} = (x)(3.2 \times 10^{-6} \text{ M})(x) = (x)^2(3.2 \times 10^{-6} \text{ M})$$

Solve for x:

$$x = 5.0 \times 10^{-4} \text{ M} = [Ca^{2+}] = [HCO_3^-]$$

The $[Ca^{2+}]$ is equal to the molar solubility. Thus, the molar solubility of $CaCO_3$ is 5.0×10^{-4} M.

19.78 First, we must calculate the mass of $CaSO_4$ dissolved. The volume is a cylinder, with $V = h\pi r^2$:

$$h = 0.50 \text{ in } (2.54 \text{ cm}/1 \text{ inch}) = 1.27 \text{ cm}$$

r = 1/2 diameter = 0.50 cm

Therefore:

V = 1.0 cm^3, and

mass = 1.0 cm^3(0.97 g/cm^3) = 0.97 g

However, because it is a hydrate ($CaSO_4 \cdot 2H_2O$), plaster is only (136.2/172.2 = 0.79) 79% calcium sulfate, therefore the true mass of $CaSO_4$ is:

Mass = 0.97(.79) = 0.77 g $CaSO_4$

In moles, this is 0.77 g $CaSO_4$ = (1 mol/136.2g) = 0.0056 mol $CaSO_4$

Now we must calculate the volume of water necessary to dissolve 0.77 g of $CaSO_4$. We start by finding its molar solubility:

$$K_{sp} = [Ca^{2+}][SO_4^{2-}] = 2.4 \times 10^{-5}$$

$$K_{sp} = [x][x] = 2.4 \times 10^{-5}$$

$x = 4.9 \times 10^{-3}$ mol/L

So the volume of water needed is:

0.0056 mol(1 L/4.9 × 10^{-3} mol) = 1.2 L

Finally, we find the amount of time needed to produce this much water:

1.2 L (1 day/2.00 L) = 0.60 days, or about 14 hours.

19.79 $Fe(OH)_3(s) \leftrightarrows Fe^{3+} + 3OH^-$ $K_{sp} = [Fe^{3+}][OH^-]^3 = 1.6 \times 10^{-39}$

	$[Fe^{3+}]$	$[OH^-]$
I	–	1.0×10^{-7}
C	+x	+3x
E	+x	$1.0 \times 10^{-7}+3x$

Assume $3x << 1.0 \times 10^{-7}$, $K_{sp} = (x)(1.0 \times 10^{-7})^3$, solving for x we get, $x = 1.6 \times 10^{-18}$ M. Thus, 1.6×10^{-18} mol of $Fe(OH)_3$ dissolve in 1 L of water.

19.80 (a) $K_{form} = \dfrac{1}{K_{inst}} = \dfrac{1}{5.6 \times 10^{-3}} = 1.8 \times 10^2$

(b) Because this value is small, this ion is much less stable than those listed in the table.

19.81 The reaction for this problem is the formation of $Ag(NH_3)^{2+}$:

$Ag^+(aq) + 2NH_3(aq) \rightarrow Ag(NH_3)^{2+}$ $\quad K_{form} = \dfrac{[Ag(NH_3)_2^+]}{[Ag^+][NH_3]^2} = 1.6 \times 10^7$

We can rearrange this equation and substitute the values from the text to determine the $[Ag^+]$:

$$[Ag^+] = \frac{[Ag(NH_3)_2^+]}{K_{form}[NH_3]^2} = \frac{(2.8 \times 10^{-3})}{(1.6 \times 10^7)(1)^2} = 1.8 \times 10^{-10} \text{ M}$$

19.82 Let x = mols of PbI_2 that dissolve per liter;

Let y = mols of $PbBr_2$ that dissolve per liter.

Then, at equilibrium, we have

$[Pb^{2+}] = x + y$, $[I^-] = 2x$ and $[Br^-] = 2y$

We know:

$PbBr_2(s) \leftrightarrows Pb^{2+} + 2Br^-$ $\quad K_{sp} = 2.1 \times 10^{-6} = [Pb^{2+}][Br^-]^2$

$PbI_2(s) \leftrightarrows Pb^{2+} + 2I^-$ $\quad K_{sp} = 7.9 \times 10^{-9} = [Pb^{2+}][I^-]^2$

Substituting we get: $2.1 \times 10^{-6} = (x + y)(2y)^2$ and $7.9 \times 10^{-9} = (x + y)(2x)^2$

Solving for x and y we find:

$x = 4.85 \times 10^{-4}$

$y = 7.91 \times 10^{-3}$

Thus, $[Pb^{2+}] = 8.40 \times 10^{-3}$ M, $[I^-] = 9.70 \times 10^{-4}$ M and $[Br^-] = 1.58 \times 10^{-2}$ M

Note: $[Br^-] > [I^-]$ because $PbBr_2$ is more soluble than PbI_2.

19.83 Initially, both Ag^+ and $HC_2H_3O_2$ are at 1.0 M concentrations. These values will be used to determine if the $AgC_2H_3O_2$ will precipitate.

First, determine the concentration of the acetate ion from the equilibrium:

$HC_2H_3O_2 \leftrightarrows H^+ + C_2H_2O_2^-$

$$K_a = \frac{[H^+][C_2H_3O_2^-]}{[HC_2H_3O_2]} = 1.8 \times 10^{-5}$$

	$[HC_2H_3O_2]$	$[H_3O^+]$	$[C_2H_3O_2^-]$
I	1.0	–	–
C	–x	+x	+x
E	1.0 –x	x	x

Assume x << 1.0

$$K_a = \frac{[x][x]}{[1.0]} = 1.8 \times 10^{-5} \quad x = 4.2 \times 10^{-3} = [C_2H_3O_2^-]$$

Next, using the concentration of the acetate ion, determine whether or not a precipitate will form.

$AgC_2H_3O_2(s) \leftrightarrows Ag^+ + C_2H_3O_2^-$

$K_{sp} = [Ag^+][C_2H_3O_2^-]$

$Q = [Ag^+][C_2H_3O_2^-]$ before equilibrium is established

$Q = (1.0\ M)(4.2 \times 10^3\ M) = 4.2\times 10^3$

$Q > K_{sp}$ therefore a precipitate will form.

19.84 We want all of the 0.10 mol $Mg(OH)_2$ to go into solution. Using the K_{sp} value for $Mg(OH)_2$, we may find the hydroxide ion concentration under these conditions:

$$Mg(OH)_2(s) \leftrightarrows Mg^{2+} + 2OH^-$$

$$K_{sp} = [Mg^{2+}][OH^-]^2$$

$$7.1 \times 10^{-12} = [0.10][OH^-]^2$$

$$[OH^-] = 8.4 \times 10^{-6}$$

Now we can use this value in the following, simultaneous equilibrium:

$$NH_4^+(aq) + OH^-(aq) \leftrightarrows H_2O + NH_3(aq)$$

$K_c = 1/K_b(NH_3) = 1/1.8 \times 10^{-5} = 5.6 \times 10^4$

$$K_c = \frac{[NH_3]}{[OH^-][NH_4^+]} = \frac{[0.20]}{[8.4 \times 10^{-6}][NH_4^+]} = 5.6 \times 10^4$$

(We know that $[NH_3] = 0.20$ M because in the equation below 2 moles of ammonia are formed for every one mole of magnesium ion:

$$Mg(OH)_2(s) + 2NH_4^+(aq) \leftrightharpoons Mg^{2+}(aq) + 2H_2O + 2NH_3(aq))$$

Solving for $[NH_4^+]$, we get 0.43 M.

So the total $[NH_3] + [NH_4^+] = 0.20 + 0.43 = 0.63$ M

One must therefore add 0.63 mol NH_4Cl to a liter of solution.

19.85 $[Ag^+] = 0.200$ M
$[H^+] = 0.10$ M
First, the concentration of acetate ion needs to be determined at the point that the silver acetate precipitates:

$AgC_2H_3O_2(s) \leftrightharpoons Ag^+ + C_2H_3O_2^-$
$K_{sp} = [Ag^+][C_2H_3O_2^-] = 2.3 \times 10^{-3}$
Let $x = [C_2H_3O_2^-]$
$2.3 \times 10^{-3} = (0.200)(x)$
$x = 1.2 \times 10^{-2}\text{ M} = [C_2H_3O_2^-]$

When $NaC_2H_3O_2$ is added to the solution, the $C_2H_3O_2^-$ will react with the H^+ from the nitric acid to form $HC_2H_3O_2$. This will give a concentration of 0.10 M $HC_2H_3O_2$. This will then come to equilibrium
Now, find the concentration of $HC_2H_3O_2$ using the K_a for acetic acid:

$HC_2H_3O_2 \leftrightharpoons H^+ + C_2H_3O_2^-$
$[HC_2H_3O_2](0.200\text{ L}) + [C_2H_3O_2^-](0.200\text{ L}) = $ mole $NaC_2H_3O_2$ that needs to be added.
$[HC_2H_3O_2] + [H^+] = 0.10$ M which is from the nitric acid
$[HC_2H_3O_2] = 0.10\text{ M} - [H^+]$

$$K_a = \frac{[H^+][C_2H_3O_2^-]}{[HC_2H_3O_2]}$$

Let $x = [H^+]$

$$1.8 \times 10^{-5} = \frac{[x][1.2 \times 10^{Ğ2}]}{[0.10\ Ğ\ x]}$$

$x = 1.5 \times 10^{-4}\text{ M} = [H^+]$
$[HC_2H_3O_2] = 0.10\text{ M} - [H^+] = 0.10 - 1.5 \times 10^{-4}\text{ M} = 9.99 \times 10^{-4}$ M
The amount of $NaC_2H_3O_2$ to be added is:
$[HC_2H_3O_2](0.200\text{ L}) + [C_2H_3O_2^-](0.200\text{ L}) = $ mole $NaC_2H_3O_2$
$[9.99 \times 10^{-4}\text{ M}](0.200\text{ L}) + [1.2 \times 10^{-2}\text{ M}](0.200\text{ L}) = 2.6 \times 10^{-3}$ mole $NaC_2H_3O_2$
The number of grams to be added is

$$\#\text{ g } NaC_2H_3O_2 = 2.6 \times 10^{-3}\text{ mole } NaC_2H_3O_2\left(\frac{82.03\text{ g } NaC_2H_3O_2}{1\text{ mol } NaC_2H_3O_2}\right) = 0.21$$

19.86 First, let's examine the question to make clear what is happening. A solution contains 0.20 M Ag^+ ions and 0.10 M acetate ions. The ion product of these two (0.020) is less than K_{sp} for silver acetate (2.3×10^{-3}), so the silver acetate remains in solution. There are also H^+ ions (H_3O^+) in the solution as a result of the following equilibrium (OAc^- will symbolize acetate):

$$H_2O + HOAc \leftrightharpoons H_3O^+(aq) + OAc^-(aq)$$

The amount of H_3O^+ may be found by using the K_a for acetic acid:

	[HOAc]	[H_3O^+]	[OAc^-]
I	0.10	–	–
C	–x	+x	+x
E	0.10 – x	x	x

$$K_a = \frac{[H_3O^+][OAc^-]}{[HOAc]}$$

$$1.8 \times 10^{-5} = \frac{[x][x]}{[0.10 - x]}$$

$$x \approx 1.3 \times 10^{-3}$$

So $[H_3O^+] = 1.3 \times 10^{-3}$ M.

However, when F^- is added to the solution (in the form of KF), the following equilibrium takes place:

$$F^-(aq) + H_3O^+(aq) \leftrightarrows H_2O + HF$$

This depletes H_3O^+ ions from the solution, which causes the first equilibrium above to move to the right, producing more acetate ions. When the acetate ion concentration hits some minimum value (determined by K_{sp}) silver acetate will precipitate. That value may be found as follows:

$$K_{sp} = [Ag^+][OAc^-]$$

$$2.3 \times 10^{-3} = [0.20][x]$$

$$x = 0.012 \text{ mol/L}$$

So the problem becomes...How many grams of KF must be added such that the acetate concentration increases to 0.012 M? This is now a simultaneous equilibrium problem:

$H_2O + HOAc \leftrightarrows H_3O^+(aq) + OAc^-(aq)$ $\qquad K_a = 1.8 \times 10^{-5}$

$F^-(aq) + H_3O^+(aq) \leftrightarrows H_2O + HF$ $\qquad K_cÕ = \frac{1}{K_a} = \frac{1}{6.5 \times 10^{-4}} = 1.5 \times 10^3$

(Note the above equation is simply the reverse of that for the K_a of HF, so $K_c' = 1/K_a$.)

Combined, this becomes:

$$F^-(aq) + HOAc \leftrightarrows HF + OAc^-(aq)$$

$$K_c = K_a \cdot K_b = (1.8 \times 10^{-5})(1.5 \times 10^3) = 2.7 \times 10^{-2}$$

Recall that we have already found the initial concentrations of OAc^- and HOAc above. Using this information, and the fact that we want the final [OAc^-] to be 1.2×10^{-2}, we can begin to fill out the table below.

	[F^-]	[HOAc]	[HF]	[OAc^-]
I	x	0.099	–	1.3×10^{-3}
C	– 0.011	– 0.011	+ 0.011	+ 0.011
E	?	0.088	0.011	0.012

$$K_c = \frac{[HF][OAc^-]}{[F^-][HOAc]}$$

$$2.7 \times 10^{-2} = \frac{[0.011][0.012]}{[F^-][0.088]}$$

$$[F^-] = 0.056\ M$$

Placing this value into the table as the equilibrium concentration of $[F^-]$, we find the initial $[F^-]$ must be 0.056 + 0.011 = 0.067 M.

Therefore the amount of KF needed in the 200 mL solution is:

0.200 L(0.067 mol KF/L)(58.01 g KF/1 mol KF) = 0.78 g KF

19.87 Step 1: Determine the $[OH^-]$ from the NH_3 reaction with water:

$NH_4^+ + OH^- \leftrightarrows NH_3 + H_2O$

$$K_b = \frac{[NH_4^+][OH^-]}{[NH_3]}$$

$$1.8\times10^{-5} = \frac{[NH_4^+][OH^-]}{[NH_3]}$$

	$[NH_3]$	$[OH^-]$	$[NH_4^+]$
I	0.10	–	–
C	–x	+x	+x
E	0.10 –x	x	x

Assume x << 0.10

$$1.8 \times 10^{-5} = \frac{[x][x]}{[0.10]}$$

$x = 1.3 \times 10^{-3} = [OH^-]$

Step 2: Find the concentration of Mg^{2+} at the given concentration of OH^-.

The concentration of OH^- from the NH_3 is 1.3×10^{-3}, there is an additional amount of OH^- from the equilibrium of the $Mg(OH)_2$, which makes the calculation:

$K_{sp} = 7.1 \times 10^{-12} = [Mg^{2+}][OH^-]^2$

$7.1 \times 10^{-12} = (x)(1.3 \times 10^{-3} + 2x)^2$

The additional amount of OH^- can be ignored since it will be less than 1.3×10^{-3}:

Using the molar solubility of $Mg(OH)_2$ in distilled water:

$7.1 \times 10^{-12} = [Mg^{2+}][OH^-]^2$

$s = [Mg^{2+}]$ and $2s = [OH^-]$

$4s^3 = 7.1 \times 10^{-12}$

$s = 1.2 \times 10^{-4}$

The solubility of $Mg(OH)_2$ in distilled water is less than the amount of OH^- supplied by the ammonia so we a re justified in ignoring its contribution. We may now solve for x

$x = 4.2 \times 10^{-6}\ M = [Mg^{2+}]$

19.88 At its simplest, this is only a K_{sp} problem.

The concentration of Mn^{2+} is (0.400 L)(0.10 M Mn^{2+})/(0.500 L) = 0.080 M Mn^{2+}

$$K_{sp} = [Mn^+][OH^-]^2$$

$$1.6 \times 10^{-13} = [0.080][OH^-]^2$$

$$[OH^-] = 1.4 \times 10^{-6}$$

When $[OH^-] = 1.4 \times 10^{-6}$ M, $Mn(OH)_2$ precipitates.

0.100 L(2.0 mol/L) = 0.20 mol NH_3 are added to 400 mL solution which would make an initial concentration of 0.20 mol/0.500 L = 0.40 M NH_3. The following equilibrium is set up:

$H_2O + NH_3 \leftrightarrows NH_4^+(aq) + OH^-(aq)$ $\qquad K_b = 1.8 \times 10^{-5}$

The problem tells us that *all* of the Sn is precipitated as $Sn(OH)_2$:

$Sn^{2+}(aq) + 2OH^-(aq) \leftrightarrows Sn(OH)_2(s)$

The concentration of the Sn^{2+} is (0.400 L)(0.10 M Sn^{2+})/(0.500 L) = 0.080 M Sn^{2+}, just before it precipitates. The Sn^{2+} immediately uses the first 2(0.080 M) = 0.16 M OH^- which is produced from the reaction of ammonia with water above, using 0.16 M NH_3. This effectively brings our initial concentration of NH_3 to 0.24 M.

Now we determine how much NH_3 will produce $[OH^-] = 1.3 \times 10^{-6}$ M.

$H_2O + NH_3 \rightleftharpoons NH_4^+(aq) + OH^-(aq)$ $\qquad K_b = 1.8 \times 10^{-5}$

	$[NH_3]$	$[NH_4^+]$	$[OH^-]$
I	x	–	1.0×10^{-7}
C	-1.2×10^{-6}	$+1.2 \times 10^{-6}$	$+1.2 \times 10^{-6}$
E	?	1.2×10^{-6}	1.3×10^{-6}

$$K_b = \frac{[NH_4^+][OH^-]}{[NH_3]}$$

$$1.8 \times 10^{-5} = \frac{[1.2 \times 10^{-6}][1.3 \times 10^{-6}]}{[NH_3]}$$

$$[NH_3] = 8.7 \times 10^{-8}$$

Therefore the initial $[NH_3]$ should be $8.7 \times 10^{-8} + 1.2 \times 10^{-6} = 1.3 \times 10^{-6}$. So we want to reduce $[NH_3]$ by $0.24 - 1.3 \times 10^{-6} = 0.23999$ M, essentially by 0.24 M. This would require adding equimolar amounts of HCl, or:

0.500 L(0.24 mol NH_3/L)(1 mol HCl/1 mol NH_3)(36.5 g HCl/1 mol HCl)
= 4.4 g HCl.

(The difficulty here arises from the fact that such a small amount of OH^- is required to precipitate the Mg^{2+} from solution that even a minimal amount of NH_3 produces enough hydroxide ion to do so.)

19.89 Three reactions are occurring in the solution:

1) $Cu^{2+}(aq) + 4NH_3^-(aq) \rightleftharpoons Cu(NH_3)_4^{2+-}(aq)$ $\qquad K_{form} = 1.1 \times 10^{13} = \dfrac{[Cu(NH_3)_4]}{[Cu^{2+}][NH_3]^4}$

2) $NH_3(aq) + H_2O \rightleftharpoons NH_4^+(aq) + OH^-(aq)$ $K_b = 1.8 \times 10^{-5} = \frac{[NH_4^+][OH^-]}{[NH_3]}$

3) $Cu(OH)_2(s) \rightleftharpoons Cu^{2+} + 2OH^-$ $K_{sp} = 4.8 \times 10^{-20} = [Cu^{2+}][OH^-]^2$

Starting with the first reaction, using the K_{form} of $Cu(NH_3)_4^{2+}$, calculate the concentration of NH_3. The concentration of Cu^{2+} before any reaction has occurred is 0.050 M, and the concentration of NH_3 before any reaction has occurred is 0.50 M. Assume that all of the Cu^{2+} has reacted and then the reaction has come to equilibrium. Therefore, the initial concentration of Cu^{2+} is 0 and the initial concentration of NH_3 is 0.50 M – (4 × 0.05 M)

	$[Cu^{2+}]$	$[NH_3]$	$Cu(NH_3)_4^{2+}$
I	–	0.30	0.050 M
C	+ x	+ 4x	–x
E	+ 2x	0.30 + 4x	0.050 – x

$$K_{form} = 1.1 \times 10^{13} = \frac{[Cu(NH_3)_4]}{[Cu^{2+}][NH_3]^4}$$

$$K_{form} = 1.1 \times 10^{13} = \frac{[0.050 - x]}{[2x][0.30 + 4x]^4}$$ assume x << 0.050

$x = 2.8 \times 10^{-13}$
$[NH_3] = 0.30$
$[Cu^{2+}] = 5.6 \times 10^{-13}$

Next, determine the $[OH^-]$ from the reaction of NH_3 with H_2O

$NH_3(aq) + H_2O \rightleftharpoons NH_4^+(aq) + OH^-(aq)$ $K_b = 1.8 \times 10^{-5} = \frac{[NH_4^+][OH^-]}{[NH_3]}$

	$[NH_3]$	$[NH_4^+]$	OH^-
I	0.30	–	–
C	– x	+ x	+x
E	0.30 – x	x	x

$$K_b = 1.8 \times 10^{-5} = \frac{[NH_4^+][OH^-]}{[NH_3]}$$

$$K_b = 1.8 \times 10^{-5} = \frac{[x][x]}{[0.30 - x]}$$ assume x << 0.30

$x = 2.3 \times 10^{-3} = [OH^-]$

Finally, with the concentration of OH^- and the concentration of Cu^{2+}, determine if any precipitate has formed:

$Cu(OH)_2 \rightleftharpoons Cu^{2+} + 2OH^-$ $K_{sp} = 4.8 \times 10^{-20} = [Cu^{2+}][OH^-]^2$

$[OH^-] = 2.3 \times 10^{-3}$ $[Cu^{2+}] = 5.6 \times 10^{-13}$

$Q = [Cu^{2+}][OH^-]^2 = (5.6 \times 10^{-13})(2.3 \times 10^{-3})^2 = 3.0 \times 10^{-18}$

$Q > K_{sp}$ therefore a precipitate forms. Assume that all of the Cu^{2+} has reacted with the OH^- and then the solution returns to equilibrium

	$[Cu^{2+}]$	$[OH^-]$
I	–	2.3×10^{-3}
C	+x	+2x
E	+x	$2.3 \times 10^{-3} + 2x$

$$K_{sp} = 4.8 \times 10^{-20} = [Cu^{2+}][OH^-]^2$$

$$K_{sp} = 4.8 \times 10^{-20} = [x][2.3 \times 10^{-3} + 2x]^2 \qquad x << 2.3 \times 10^{-3}$$

$$x = 9.1 \times 10^{-15}\ M = [Cu^{2+}]$$

19.90 $Fe(OH)_3$ has such an exceedingly small K_{sp}, there is virtually no dissociation in pure water. Therefore, the pH would be expected to be about 7.

(If the calculations are done, this is borne out:

$$K_{sp} = [Fe^{3+}][OH^-]^3$$

$$1.6 \times 10^{-39} = [x][3x]^3$$

$$x = 8.8 \times 10^{-11}$$

$$[OH^-] = 2.6 \times 10^{-10}$$

This number is 1,000 times smaller than the hydroxide ion already present in pure water.

Practice Exercises

20.1 a) $q>0$, $w>0$ b) $q<0$, $w<0$ c) $q<0$, $w>0$ d) $q>0$, $w<0$
In case (b), $q<0$ and $w<0$ so ΔE is the most negative.

20.2 $q = -P\Delta V = -(14.0 \text{ atm})(12.0 \text{ L} - 1.0 \text{ L}) = -154 \text{ L atm}$

$\Delta E = w + q = 0$

Therefore, $q = +154$ L atm.

(The energy is converted into heat; since the heat does not leave the system the temperature increases.)

20.3 $\Delta E = q - P\Delta V$ since $q = 0$
$\Delta E = -P\Delta V$
but ΔV is negative for a compression so ΔE increases and T increases.
Energy is added to the system in the form of work.

20.4 $\Delta E° = \Delta H° - \Delta nRT = -217.1 \text{ kJ} - (-1 \text{ mol})(8.314 \text{ J mol}^{-1} \text{ K}^{-1})(298 \text{ K})$
$= -217.1 \text{ kJ} + 2.48 \text{ kJ}$
$= -214.6 \text{ kJ}$

% Difference = $(2.48/217) \times 100 = 1.14$ %

20.5 a) ΔS is negative since the products have a lower entropy, i.e. a lower freedom of movement.
b) ΔS is positive since the products have a higher entropy, i.e. a higher freedom of movement.

20.6 a) ΔS is negative since there are less gas molecules. (The product is also more complex, indicating an increase in order.)
b) ΔS is negative since there are less gas molecules. (The product is also more complex, indicating an increase in order.)

20.7 a) ΔS is negative since there is a change from a gas phase to a liquid phase. (The product is also more complex, indicating an increase in order.)
b) ΔS is negative since there are less gas molecules. (The product is also more complex, indicating an increase in order.)
c) ΔS is positive since the particles go from an ordered, crystalline state to a more disordered, aqueous state.

20.8 $\Delta S° = (\text{sum } S°[\text{products}]) - (\text{sum } S°[\text{reactants}])$

a)
$$\Delta S° = \{S°[H_2O(l)] + S°[CaCl_2(s)]\} - \{S°[CaO(s)] + 2S°[HCl(g)]\}$$
$$\Delta S° = \{1 \text{ mol} \times (69.96 \text{ J mol}^{-1} \text{ K}^{-1}) + 1 \text{ mol} \times (114 \text{ J mol}^{-1} \text{ K}^{-1})\}$$
$$- \{1 \text{ mol} \times (40 \text{ J mol}^{-1} \text{ K}^{-1}) + 2 \text{ mol} \times (186.7 \text{ J mol}^{-1} \text{ K}^{-1})\}$$
$$\Delta S° = -229 \text{ J/K}$$

b)
$$\Delta S° = \{S°[C_2H_6(g)]\} - \{S°[H_2(g)] + S°[C_2H_4(g)]\}$$
$$\Delta S° = \{1 \text{ mol} \times (229.5 \text{ J mol}^{-1} \text{ K}^{-1})\}$$
$$- \{1 \text{ mol} \times (130.6 \text{ J mol}^{-1} \text{ K}^{-1}) + 1 \text{ mol} \times (219.8 \text{ J mol}^{-1} \text{ K}^{-1})\}$$
$$\Delta S° = -120.9 \text{ J/K}$$

20.9 First, we calculate $\Delta S°$, using the data:

$\Delta S° = \{2S°[Fe_2O_3(s)]\} - \{3S°[O_2(g)] + 4S°[Fe(s)]\}$
$\Delta S° = \{2 \text{ mol} \times (90.0 \text{ J mol}^{-1} \text{ K}^{-1})\}$
$- \{3 \text{ mol} \times (205.0 \text{ J mol}^{-1} \text{ K}^{-1}) + 4 \text{ mol} \times (27 \text{ J mol}^{-1} \text{ K}^{-1})\}$
$\Delta S° = -543 \text{ J/K} = -0.543 \text{ kJ/mol}$

Next, we calculate $\Delta H°$ using the data :

$\Delta H° = (\text{sum } \Delta H_f°[\text{products}]) - (\text{sum } \Delta H_f°[\text{reactants}])$
$\Delta H° = \{2\Delta H_f°[Fe_2O_3(s)]\} - \{3\Delta H_f°[O_2(g)] + 4\Delta H_f°[Fe(s)]\}$
$\Delta H° = \{2 \text{ mol} \times (-822.2 \text{ kJ/mol})\} - \{3 \text{ mol} \times (0.0 \text{ kJ/mol}) + 4 \times (0.0 \text{ kJ/mol})\}$
$\Delta H° = -1644 \text{ kJ}$

The temperature is 25.0 + 273.15 = 298.15 K, and the calculation of $\Delta G°$ is as follows: $\Delta G° = \Delta H° - T\Delta S° = -1644 \text{ kJ} - (298.15 \text{ K})(-0.543 \text{ kJ/K}) = -1482 \text{ kJ}$

20.10 We calculate $\Delta G°_{rxn}$, using the data from Table 20.2:

a) $\Delta G°_{rxn} = \{2\Delta G_f°[NO_2(g)]\} - \{2\Delta G_f°[NO(g)] + \Delta G_f°[O_2(g)]\}$
$\Delta G°_{rxn} = \{2 \text{ mol} \times (+51.84 \text{ kJ mol}^{-1})\}$
$- \{2 \text{ mol} \times (+86.69 \text{ kJ mol}^{-1}) + 1 \text{ mol} \times (0 \text{ kJ mol}^{-1})\}$
$\Delta G°_{rxn} = -69.7 \text{ kJ/mol}$

b) $\Delta G°_{rxn} = \{\Delta G_f°[CaCl_2(s)] + 2\Delta G_f°[H_2O(g)]\}$
$- \{\Delta G_f°[Ca(OH)_2(s)] + 2\Delta G_f°[HCl(g)]\}$
$\Delta G°_{rxn} = \{1 \text{ mol} \times (-750.2 \text{ kJ mol}^{-1}) + 2 \text{ mol} \times (-228.6 \text{ kJ mol}^{-1})\}$
$- \{1 \text{ mol} \times (-896.76 \text{ kJ mol}^{-1}) + 2 \text{ mol} \times (-95.27 \text{ kJ mol}^{-1})\}$
$\Delta G°_{rxn} = -120.1 \text{ kJ/mol}$

20.11 The maximum amount of work that is available is the free energy change for the process, in this case, the standard free energy change, $\Delta G°$, since the process occurs at 25 °C.

$4Al(s) + 3O_2(g) \rightarrow 2Al_2O_3(s)$

$\Delta G° = (\text{sum } \Delta G_f°[\text{products}]) - (\text{sum } \Delta G_f°[\text{reactants}])$
$\Delta G° = 2\Delta G_f°[Al_2O_3(s)] - \{3\Delta G_f°[O_2(g)] + 4\Delta G_f°[Al(s)]\}$
$\Delta G° = 2 \text{ mol} \times (-1576.4 \text{ kJ/mol}) - \{3 \text{ mol} \times (0.0 \text{ kJ/mol}) + 4 \text{ mol} \times (0.0 \text{ kJ/mol})\}$
$\Delta G° = -3152.8$ kJ, for the reaction as written.

This calculation conforms to the reaction *as written*. This means that the above value of $\Delta G°$ applies to the equation involving *4 mol* of Al. The conversion to give energy *per mole* of aluminum is then: −3152.8 kJ/4 mol Al = −788 kJ/mol

The maximum amount of energy that may be obtained is thus 788 kJ.

20.12 For the vaporization process in particular, and for any process in general, we have:

$$\Delta G = \Delta H - T\Delta S$$

If the temperature is taken to be that at which equilibrium is obtained, that is the temperature of the boiling point (where liquid and vapor are in equilibrium with one another), then we also have the result that ΔG is equal to zero:

$$\Delta G = 0 = \Delta H - T\Delta S, \text{ or } T_{eq} = \Delta H/\Delta S$$

We know ΔH to be 60.7 kJ/mol; we need the value for ΔS in units kJ mol^{-1} K^{-1}:

$\Delta S° = (\text{sum } S°[\text{products}]) - (\text{sum } S°[\text{reactants}])$
$\Delta S° = S°[Hg(g)] - S°[Hg(l)]$
$\Delta S° = (175 \times 10^{-3} \text{ kJ mol}^{-1} \text{ K}^{-1}) - (76.1 \times 10^{-3} \text{ kJ mol}^{-1} \text{ K}^{-1})$
$\Delta S° = 98.9 \times 10^{-3} \text{ kJ mol}^{-1} \text{ K}^{-1}$

$T_{eq} = 60.7 \text{ kJ/mol} \div 98.9 \times 10^{-3} \text{ kJ/mol K} = 614 \text{ K } (341\ °C)$

20.13 $\Delta G° = (\text{sum } \Delta G_f°[\text{products}]) - (\text{sum } \Delta G_f°[\text{reactants}])$
$\Delta G° = 2\Delta G_f°[SO_3(g)] - \{2\Delta G_f°[SO_2(g)] + \Delta G_f°[O_2(g)]\}$
$\Delta G° = 2 \text{ mol} \times (-370.4 \text{ kJ/mol}) - \{2 \text{ mol} \times (-300.4 \text{ kJ/mol}) + (0.0 \text{ kJ/mol})\}$
$\Delta G° = -140.0 \text{ kJ/mol}$

Since the sign of $\Delta G°$ is negative, the reaction should be spontaneous.

20.14 $\Delta G° = \Delta H° - T\Delta S°$
$\Delta G° = \{2\Delta G_f°[HCl(g)] + \Delta G_f°[CaCO_3(s)]\}$
$\quad - \{\Delta G_f°[CaCl_2(s)] + \Delta G_f°[H_2O(g)] + \Delta G_f°[CO_2(g)]\}$
$\Delta G° = \{2 \text{ mol} \times (-95.27 \text{ kJ/mol}) + 1 \text{ mol} \times (-1128.8 \text{ kJ/mol})\}$
$\quad - \{1 \text{ mol} \times (-750.2 \text{ kJ/mol}) + 1 \text{ mol} \times (-228.6 \text{ kJ/mol}) +$
$\quad 1 \text{ mol} \times (-394.4 \text{ kJ/mol})\}$
$\Delta G° = +53.9 \text{ kJ}$

$\Delta G°$ is positive, the reaction is not spontaneous, and we do not expect to see products formed from reactants.

20.15 First, we compute the standard free energy change for the reaction, based on the:

$\Delta G° = (\text{sum } \Delta G_f°[\text{products}]) - (\text{sum } \Delta G_f°[\text{reactants}])$
$\Delta G° = \{\Delta G_f°[H_2O(g)] + \Delta G_f°[CO_2(g)] + \Delta G_f°[Na_2CO_3(s)]\} -$
$\quad \{2\Delta G_f°[NaHCO_3(s)]\}$
$\Delta G° = \{1 \text{ mol} \times (-228.6 \text{ kJ/mol}) + 1 \text{ mol} \times (-394.4 \text{ kJ/mol})$
$\quad + 1 \text{ mol} \times (-1048 \text{ kJ/mol})\} - \{2 \text{ mol} \times (-851.9 \text{ kJ/mol})\}$
$\Delta G° = +33 \text{ kJ}$

Next, we determine values for $\Delta H°$ and $\Delta S°$:
$\Delta H° = (\text{sum } \Delta H_f°[\text{products}]) - (\text{sum } \Delta H_f°[\text{reactants}])$
$\Delta H° = \{\Delta H_f°[H_2O(g)] + \Delta H_f°[CO_2(g)] + \Delta H_f°[Na_2CO_3(s)]\} - \{2\Delta H_f°[NaHCO_3(s)]\}$
$\Delta H° = \{1 \text{ mol} \times (-241.8 \text{ kJ/mol}) + 1 \text{ mol} \times (-393.5 \text{ kJ/mol})$
$\quad + 1 \text{ mol} \times (-1131 \text{ kJ/mol})\} - \{2 \text{ mol} \times (-947.7 \text{ kJ/mol})\}$
$\Delta H° = +129 \text{ kJ}$

$\Delta S° = (\text{sum } S°[\text{products}]) - (\text{sum } S°[\text{reactants}])$
$\Delta S° = \{S°[H_2O(g)] + S°[CO_2(g)] + S°[Na_2CO_3(s)]\} - \{2S°[NaHCO_3(s)]\}$
$\Delta S° = \{1 \text{ mol} \times (188.7 \text{ J mol}^{-1} \text{ K}^{-1}) + 1 \text{ mol} \times (213.6 \text{ J mol}^{-1} \text{ K}^{-1})$
$\quad + 1 \text{ mol} \times (136 \text{ J mol}^{-1} \text{ K}^{-1})\} - \{2 \text{ mol} \times (102 \text{ J mol}^{-1} \text{ K}^{-1})\}$
$\Delta S° = 334 \text{ J/K} = 0.334 \text{ kJ/K}$

Next, we assume that both $\Delta H°$ and $\Delta S°$ are independent of temperature, and use these values to determine ΔG at a temperature of 200 + 273 = 473 K:

$\Delta G°_{473} = \Delta H° - T\Delta S° = 129 \text{ kJ} - (473 \text{ K})(0.334 \text{ kJ/K}) = -29 \text{ kJ}$

At the lower of these two temperatures (25 °C), the reaction has a positive value of ΔG. At the higher of these two temperatures (200 °C), the reaction has a negative value of ΔG. Thus ΔG becomes more negative as the temperature is raised, so the reaction becomes increasingly more favorable as the temperature is increased. In other words, the position of the equilibrium will be shifted more towards products at the higher temperature.

20.16 Using the data provided we may write:

$$\Delta G = \Delta G^\circ + RT \ln\left(\frac{P_{N_2O_4}}{P_{NO_2}{}^2}\right)$$

$$= -5.40 \times 10^3 \text{ J mol}^{-1} + (8.314 \text{ J mol}^{-1} \text{ K}^{-1})(298 \text{ K}) \ln\left(\frac{0.25 \text{ atm}}{(0.60 \text{ atm})^2}\right)$$

$$= -5.40 \times 10^3 \text{ J mol}^{-1} + (-9.03 \times 10^2 \text{ J mol}^{-1})$$

$$= -6.30 \times 10^3 \text{ J mol}^{-1}$$

Since ΔG is negative, the forward reaction is spontaneous and the reaction will proceed to the right.

20.17 $\Delta G^\circ = -RT \ln K_p$
$\Delta G^\circ = -(8.314 \text{ J K}^{-1} \text{ mol}^{-1})(25 + 273 \text{ K}) \times \ln(6.9 \times 10^5) = -33 \times 10^3 \text{ J}$
$\Delta G^\circ = -33 \text{ kJ}$

20.18 $\Delta G^\circ = -RT \ln K_p$
$3.3 \times 10^3 \text{ J} = -(8.314 \text{ J K}^{-1} \text{ mol}^{-1})(298 \text{ K}) \times \ln(K_p)$
$\ln(K_p) = -3.3 \times 10^3 \text{ J}/[(8.314 \text{ J K}^{-1} \text{ mol}^{-1})(298 \text{ K})] = -1.3$
Taking the antilog of both sides of the above equation gives: $K_p = 0.26$

20.19 $\Delta G^\circ = \Delta H^\circ - T\Delta S^\circ = -92.4 \times 10^3 \text{ J} - (323 \text{ K})(-198.3 \text{ J/K}) = -2.83 \times 10^4 \text{ J}$
$\Delta G = -RT \ln K$
Thus, $-2.83 \times 10^4 \text{ J} = -(8.314 \text{ J K}^{-1} \text{ mol}^{-1})(323 \text{ K}) \times \ln(K_p)$
$\ln(K_p) = -2.38 \times 10^4 \text{ J}/[(-8.314 \text{ J K}^{-1} \text{ mol}^{-1})(323 \text{ K})] = 10.5$
Taking the antilog of both sides of the above equation gives: $K_p = 3.84 \times 10^4$

Thinking It Through

T20.1 The work is calculated by the quantity PΔV, where P is the external pressure against which the expansion takes place. We need to determine ΔV from the given information.

$P\Delta V = 1 \text{ atm} \times 5.00 \text{ L} = 5.00 \text{ L atm}$

Next, we convert from units L atm to units J by remembering that the unit Pa is equal to a Newton per meter squared, or 1 Pa = 1 N/m^2:

$$(5.00 \text{ L atm}) \times \left(\frac{101{,}325 \text{ N/m}^2}{1 \text{ atm}}\right) = 5.07 \times 10^5 \frac{\text{L N}}{\text{m}^2}$$

Finally, we convert liters to the units m^3 and multiply in order to arrive at a final answer in units joules = N m.

T20.2 This type of calculation is done using the equation from the text: $\Delta H = \Delta E + \Delta n_{gas}RT$. Here the value for Δn_{gas} is equal to $3 - 2 = +1$. We convert the mass of material that is to react to a number of moles since we are given the value of ΔH for a 2 mole reaction.

$$\Delta E = \Delta H - \Delta n_{gas}RT = -163.14\text{ kJ} - (1\text{ mol})(8.314 \times 10^{-3}\text{ kJ/mol K})(298\text{ K})$$

This answer must then be multiplied by the number of moles we calculated above. The second calculation is the same, but the temperature that is used is 498 K.

T20.3 Values of ΔS are strongly influenced by changes in state or by changes in the number of moles of high entropy substances, such as gases. Here the reaction produces three moles of gaseous products starting with only two moles of gaseous reactants. We conclude that entropy increases, meaning that ΔS has a positive value.

T20.4 The spontaneity of the reaction depends on ΔG (=ΔH – TΔS). We therefore need to know a value for ΔH and the temperature, in addition to a value for ΔS.

T20.5 The sign of ΔS should be negative if more order is found among the products. In this case, two homonuclear diatomic molecules possess more order than the two heteronuclear diatomic molecules of the reactants. The sign of ΔS should therefore be negative.

T20.6 Every substance possesses an entropy, given the symbol S°, and listed in tables such as Table 20.1. Evidently, what is called for here is the entropy change for the reaction in which one mole of HI is formed from the elements H_2 and I_2. This would be termed the ΔS° for the reaction. Its value could be calculated using equation 20.6, once values of S° for each reactant and product are known. Appendix E is also a good source for this type of information.

$$\Delta S^\circ = (\text{sum of } S^\circ \text{ of the products}) - (\text{sum of } S^\circ \text{ of the reactants})$$

$$\Delta S^\circ = S^\circ[HI(g)] - \{1/2 \times S^\circ[H_2(g)] + 1/2 \times S^\circ[I_2(s)]\}$$

T20.7 The value for ΔG° is given in the problem, and values for the free energy of formation of all other reactants and products can be obtained in Appendix E.

$$\Delta G^\circ = (\text{sum of } \Delta G_f^\circ \text{ of the products}) - (\text{sum of } \Delta G_f^\circ \text{ of the reactants})$$

$$\Delta G^\circ = \{8 \times \Delta G_f^\circ[CO_2(g)] + 10 \times \Delta G_f^\circ[H_2O(g)]\} - \{2 \times \Delta G_f^\circ[C_4H_{10}(g)] + 13 \times \Delta G_f^\circ[O_2(g)]\}$$

$$\Delta G_f^\circ[C_4H_{10}(g)] = 1/2[\ \{8 \times \Delta G_f^\circ[CO_2(g)] + 10 \times \Delta G_f^\circ[H_2O(g)]\} - \{13 \times \Delta G_f^\circ[O_2(g)]\} - \Delta G^\circ]$$

T20.8 First, it is necessary to determine values for ΔH° and ΔS° using the data of Appendix E. Once these are determined, we use the temperature T = 500 K.

$$\Delta G_T^\circ \cong \Delta H_{298\text{ K}}^\circ - T\Delta S_{298\text{ K}}^\circ$$

T20.9 The maximum work that can be obtained from the combustion of one mole under standard conditions is given by the value of ΔG° for this reaction. First, it is necessary to calculate the number of moles of butane that are actually present (using the ideal gas equation, where P = 2.00 atm, V = 10.0 L, T = 293 K). Next, the balanced equation is used to calculate the value for ΔG°, using equation 20.7 and the data of appendix E. Last, we multiply the value of ΔG° (in units kJ/mol) by the number of moles of gas that are present, in order to arrive at the number of kJ to be expected from the combustion of the sample.

T20.10 We must rearrange these equations so that they can be added together, giving a net reaction in which one mole of N_2O_5 is formed from elements only. When this has been accomplished, the value for the net ΔG° will correspond to the free energy of formation of N_2O_5. Remember to adjust the individual values for ΔG°, as was done for enthalpy when applying Hess's Law in Chapter 7.

$2HNO_3(l) \rightarrow N_2O_5(g) + H_2O(l)$	ΔG° = 37.6 kJ
$N_2(g) + 3O_2(g) + H_2(g) \rightarrow 2HNO_3(l)$	ΔG° = –159.82 kJ
$H_2O(l) \rightarrow H_2(g) + 1/2O_2(g)$	ΔG° = 237.2 kJ
$N_2(g) + 5/2O_2(g) \rightarrow N_2O_5(g)$	ΔG° = sum of the other equations

T20.11 Add the energy of each of the bonds.

T20.12 Sum the heats of formation of the gaseous elemental atoms and subtract the sum of the bond energies in the HCN molecule.

T20.13 If we multiply P in units N/m^2 (i.e., Pascals), by volume in units m^3, the result has units N × m, which is the correct set of units for a joule.

T20.14 For an ideal gas, there are no interactions among the various gas molecules. This means that no inelastic collisions can take place, and consequently, no energy transfer is possible for a system that has an unchanging temperature.

T20.15 ΔE = 0, so q = PΔV = (2.0 atm)(6.0 L – 12.0 L) = –12.0 L atm

T20.16 It is necessary to use the equation for ΔG (=ΔH – TΔS) to answer this question. Since ΔS is negative, the term –TΔS will be positive. We expect that, as the temperature rises, the value of ΔG will become increasingly less negative. Eventually the term –TΔS will become sufficiently positive to offset the exothermic nature of the reaction, causing the value for ΔG to become positive. Above this temperature, the reaction is nonspontaneous.

T20.17 We know that ΔG = –RT ln(K). We can calculate the value of the equilibrium constant because we can determine the pressure of each of the gases from the data provided if, in addition, we know the initial amount of ClNO.

T20.18 Using ΔH and ΔS we can determine ΔG and from that K. We can then set up the resulting equilibrium problem to determine the equilibrium concentration of each component of the reaction.

Review Questions

20.1 The term thermodynamics is meant to convey the two ideas of thermo (heat) and dynamics (motion), namely movement or transfer of heat.

20.2 The internal energy of a system consists of kinetic and potential energy. We cannot determine the amount of potential energy that a system has, nor can we know the precise speeds of the various particles in a system. We are, therefore, unable to determine the absolute total internal energy of a system. We can, however, determine changes in internal energy.

20.3 $\Delta E = E_{final} - E_{initial}$

20.4 If we refer to the equation in the answer to Review Question 20.3, we can see that if E_{final} is larger than $E_{initial}$, then ΔE must be positive, by definition. Thus, ΔE for an endothermic process is positive.

20.5 The first law of thermodynamics states that the internal energy may be transferred as heat or work, but it cannot be created or destroyed. The change in internal energy of a system is the sum of two terms, the amount of energy the system gains from heat transfer and the amount of energy the system gains from work transfer: $\Delta E = q + w$. ΔE is the change in internal energy, which is positive for an endothermic process, and negative for an exothermic process. q is the heat absorbed by the system during the process, and it has a positive sign if the system absorbs heat during the process. w is the work done on the system by the surroundings during the process, and it has a negative sign if the system does work on the surroundings, whereas it has a positive sign if the surroundings do work on the system during the process.

20.6 A state function is a thermodynamic quantity whose value is determined only by the state of the system currently, and is not determined by a system's prior condition or history. A change in a state function is the same regardless of the path that is used to arrive at the final state from the initial state. That is, changes in state function quantities are path–independent. In the statement of the first law, only ΔE is a state function.

20.7 ΔE is the heat of reaction at constant volume, and applies, for instance, to reactions performed in closed vessels such as bomb calorimeters. ΔH is the heat of reaction at constant pressure.

20.8 A pressure in Pascals, by definition, has the units Newtons/m^2, or N/m^2.
Thus, P in pascals times ΔV in m^3 is: $N/m^2 \times m^3 = N \bullet m$

The last result is the Newton • meter, or a force times a distance. By definition, a force of 1 N operating over a distance of 1 meter is the quantity 1 joule. Thus, the quantity $P \times \Delta V$ has the units of energy, namely joules.

20.9 $\Delta H = \Delta E + P\Delta V$

20.10 ΔH and ΔE are numerically equivalent when there is not change in the volume of the system.

20.11 Since the quantity PΔV is negative (corresponding to a decrease in volume), ΔE must be a smaller negative quantity than ΔH. ΔE is a smaller negative quantity than ΔH because, if the change is carried out at constant pressure, some energy is gained by the system as volume contracts.

20.12 A spontaneous change, in thermodynamic terms, is one for which the sign of ΔG is negative. It is a process that occurs by itself, without continued outside assistance.

20.13 Student dependent answer. Examples of spontaneous changes may include: leaves falling from a tree, open soda cans losing there carbonation, food molding in the refrigerator, dorm rooms becoming messy. Examples of non–spontaneous changes may include: dorm rooms becoming neat and tidy, class notes becoming organized, leaves being raked into a pile.

20.14 Student dependent answer. For the examples given above all of the non–spontaneous changes are endothermic, they all require the surroundings expenditure of energy. The spontaneous changes are all exothermic except, perhaps, the dorm room becoming messy. This change may be either exothermic or endothermic.

20.15 A change that is characterized by a negative value for ΔH tends to be spontaneous. The only situation where this is not true is that in which the value of the product TΔS is sufficiently negative to make the quantity $\Delta G = \Delta H - T\Delta S$ become positive.

20.16 The probability of an outcome is equal to the number of ways the outcomes can be produced divided by the total number of ways all outcomes can be produced.

20.17 Spontaneous processes tend to proceed from states of low probability to states of higher probability.

20.18 The ammonium and nitrate atoms are in a highly ordered geometry in the crystalline NH_4NO_3 sample. When NH_4NO_3 dissolves, the NH_4^+ and NO_3^- ions become randomly dispersed throughout the solvent. The increase in randomness that attends the dissolving of the solid is responsible for the process being favorable, or spontaneous, in spite of the fact that the enthalpy change is endothermic.

20.19 Entropy is a measure of the randomness of a system. An equivalent statement is that entropy is a measure of the statistical probability of a system.

20.20 (a) negative (b) negative (c) positive
(d) negative (e) negative (f) positive

20.21 The statistical probability of a system in a given state, relative to all the other states, is the same regardless of how the system happened to have been formed.

20.22 The entropy of the universe increases when a spontaneous event occurs.

20.23 Two examples might be (1) a disorganized pile of bricks and boards jumping up to produce a house, or (2) a pile of bricks falling off a cliff, and landing in the form of a perfect cube. Both of these examples are accompanied by enormous decreases in entropy (randomness), and hence are not realistic spontaneous events.

20.24 A spontaneous event occurs when ΔG is negative. Since $\Delta G = \Delta H - T\Delta S$, even if the entropy is negative for a process, the enthalpy factor, ΔH may be negative enough at a given temperature to allow the overall change in free energy to be negative.

20.25 This is the statement that the entropy of a perfect crystalline solid at 0 K is equal to zero: S = 0 at 0 K.

20.26 The entropy of a mixture must be higher than that of two separate pure materials, because the mixture is guaranteed to have a higher degree of disorder. Said another way, a mixture is more disordered than either of its two separate components. Only a pure substance can have an entropy of zero, and then only at 0 K.

20.27 Entropy increases with increasing temperature because vibrations and movements within a solid lead to greater disorder at the higher temperatures. Melting especially produces more disorder and vaporization even more.

20.28 No, glass is an amorphous solid that is really a mixture of different substances so it is not a perfect crystalline solid and does not have an entropy value of 0 at 0K.

20.29 $\Delta G = \Delta H - T\Delta S$

20.30 (a) A change is spontaneous at all temperatures only if ΔH is negative and ΔS is positive.
(b) A change is spontaneous at low temperatures but not at high temperatures only if ΔH is negative and ΔS is negative.
(c) A change is spontaneous at high temperatures but not at low temperatures only if ΔH is positive and ΔS is positive.

20.31 A change is nonspontaneous regardless of the temperature, if ΔH is positive and ΔS is negative.

20.32 By itself it is not possible to run a reaction in reverse, but if the reaction is coupled to another process which can put energy into the reaction to reverse it, it can occur.

20.33 The value of ΔG is equal to the maximum amount of work that may be obtained from any process.

20.34 A reversible process is one in which the driving force of the process is nearly completely balanced by an opposing force. The situation is a bit esoteric, since no process can be run in a completely reversible manner. Nevertheless, the closer one obtains to reversibility, the more efficient the system becomes as a source of the maximum amount of useful work that can be achieved with the process.

20.35 The slower that the energy extraction is performed, the greater is the total amount of energy that can be obtained. This is the same as saying that the most energy is available from a process that occurs reversibly.

20.36 As with other processes that are not carried out in a reversible fashion, the energy is lost to the environment and becomes unavailable for use. In addition, much of the energy is lost as heat which is used to maintain our body temperature.

20.37 A truly reversible process would take forever to occur. Thus, if we can observe an event happening, it cannot be a truly reversible one.

20.38 At equilibrium, the value of ΔG is zero.

20.39 The process of bond breaking always has an associated value for ΔH that is positive. Also, since bond breaking increases disorder, it always has an associated value for ΔS that is positive. A process such as this, with a positive value for ΔH and a positive value for ΔS, becomes spontaneous at high temperatures.

20.40 Before the heat transfer, the molecules in the hot object vibrate and move more violently than do those of the cooler object. When in contact with one another, the objects transfer heat through collisions, and eventually, some of the kinetic energy of the molecules in the hot object is transferred to the molecules of the cool object. This process of energy transfer continues until the objects have the same temperature. The heat transfer is spontaneous because the scattering of kinetic energy among the molecules of both objects is a process with a positive value for its associated ΔS.

20.41 See Figure 20.11.

20.42 See Figure 20.12.

20.43 Although a reaction may have a favorable ΔG, and therefore be a spontaneous reaction in the thermodynamic sense of the word, the rate of reaction may be too slow at normal temperatures to be observed.

20.44 ΔG depends explicitly on temperature: $\Delta G = \Delta H - T\Delta S$

20.45 $\Delta G' = \Delta H° - T\Delta S°$

20.46 As the temperature is raised, °G' will become less negative, if $\Delta H°$ is negative and $\Delta S°$ is negative. Accordingly, less product will be present at equilibrium.

20.47 $\Delta G = \Delta G° + RT \ln Q$

20.48 $\Delta G° = -RT \ln K$

20.49 This is an equilibrium constant whose numerical value is calculated from thermodynamic data, using equation 20.12.

20.50 The natural log of 1 is zero so $\Delta G° = 0$.

20.51 The amount of energy needed to break all the chemical bonds in one mole of gaseous molecules to give gaseous atoms.

20.52 It is easy to see why the conversion of a solid or liquid element to gaseous atoms is an endothermic process. In the case of elements which exist naturally as gases, most are polyatomic (the exception are the noble gases). To convert these elements to gaseous atoms will require an input of energy as bonds need to be broken.

20.53 Heat of formation is defined as the amount of energy needed to form the compound from its elements in their most stable state. $C_2(g)$ is not the naturally occurring state of carbon.

Review Problems

20.54 $\Delta E = q + w = 300\text{ J} + 700\text{ J} = +1000\text{ J}$

The overall process is endothermic, meaning that the internal energy of the system increases. Notice that both terms, q and w, contribute to the increase in internal energy of the system; the system gains heat (+q) and has work done on it (+w).

20.55 $\Delta E = q + w$
$-1455\text{ J} = 812\text{ J} + w$
$w = -2267\text{ J}$

Since w is defined to be the work done on the system by the surroundings, then in this case, a negative amount of work is done on the system by the surroundings. The system, in fact, does work on the surroundings.

20.56 work = $P \times \Delta V$
The total pressure is atmospheric pressure plus that caused by the hand pump:
$P = (30.0 + 14.7)\text{ lb/in}^2 = 44.7\text{ lb/in}^2$

Converting to atmospheres we get:
$P = 44.7\text{ lb/in}^2 \times 1\text{ atm}/14.7\text{ lb/in}^2 = 3.04\text{ atm}$

Next we convert the volume change in units in^3 to units L:
$24.0\text{ in}^3 \times (2.54\text{ cm/in})^3 \times 1\text{ L}/1000\text{ cm}^3 = 0.393\text{ L}$

Hence $P \times \Delta V = (3.04\text{ atm})(0.393\text{ L}) = 1.19\text{ L·atm}$
$1.19\text{ L·atm} \times 101.3\text{ J/L·atm} = 121\text{ J}$

20.57 This is a reaction that produces 1 mol of a single gaseous product, CO_2. Furthermore, this mole of gaseous product forms from nongaseous materials. The volume that will be occupied by this gas, once it is formed, can be found by application of Charles' Law: $22.4\text{ L} \times 298\text{ K}/273\text{ K} = 24.5\text{ L}$

The work of gas expansion on forming the products is thus:
$W = -P\Delta V = -(1.00\text{ atm})(24.5\text{ L}) = -24.5\text{ L atm}$

20.58 We use the data supplied in Appendix.

(a) $3PbO(s) + 2NH_3(g) \rightarrow 3Pb(s) + N_2(g) + 3H_2O(g)$

$$\Delta H° = \{\Delta H_f^\circ[Pb(s)] + \Delta H_f^\circ[N_2(g)] + 3\Delta H_f^\circ[H_2O(g)]\} - \{3\Delta H_f^\circ\ [PbO(s)] + 2\Delta H_f^\circ NH_3(g)]\}$$

$$\Delta H° = \{3\text{ mol} \times (0\text{ kJ/mol}) + 1\text{ mol} \times (0\text{ kJ/mol}) + 3\text{ mol} \times (-241.8\text{ kJ/mol})\} - \{3\text{ mol} \times (-217.3\text{ kJ/mol}) + 2\text{ mol} \times (-46.19\text{ kJ/mol})\}$$

$$\Delta H° = +\ 24.58\text{ kJ}$$

$$\Delta E = \Delta H° - \Delta nRT$$

$$\Delta E = 24.58\text{ kJ} - (+2\text{ mol})(8.314\text{ J/mol K})(10^{-3}\text{ kJ/J})(298\text{ K}) = 19.6\text{ kJ}$$

(b) $NaOH(s) + HCl(g) \rightarrow NaCl(s) + H_2O(l)$

$$\Delta H° = \{\Delta H_f^\circ[NaCl(s)] + \Delta H_f^\circ[H_2O(l)]\} - \{\Delta H_f^\circ[NaOH(s)] + \Delta H_f^\circ[HCl(g)]\}$$

$$\Delta H° = \{1\text{ mol} \times (-411.0\text{ kJ/mol}) + 1\text{ mol} \times (-285.9\text{ kJ/mol})\} - \{1\text{ mol} \times (-426.8\text{ kJ/mol}) + 1\text{ mol} \times (-92.30)\}$$

$$\Delta H° = -178\text{ kJ}$$

$$\Delta E = \Delta H° - \Delta nRT$$

$$\Delta E = -178\text{ kJ} - (-1)(8.314\text{ J/mol K})(10^{-3}\text{ kJ/J})(298\text{ K}) = -175\text{ kJ}$$

(c) $Al_2O_3(s) + 2Fe(s) \rightarrow Fe_2O_3(s) + 2Al(s)$

$$\Delta H° = \{\Delta H_f^\circ[Fe_2O_3(s)] + 2\Delta H_f^\circ[Al(s)]\} - \{\Delta H_f^\circ[\ Al_2O_3(s)] + 2\Delta H_f^\circ[Fe(s)]\}$$

$$\Delta H° = \{1\text{ mol} \times (-822.3\text{ kJ/mol}) + 2\text{ mol} \times (0\text{ kJ/mol})\} - \{1\text{ mol} \times (-1669.8\text{ kJ/mol}) + 2\text{ mol} \times (0\text{ kJ/mol})\}$$

$$\Delta H° = 847.6\text{ kJ}$$

$\Delta E = \Delta H°$, since the value of Δn for this reaction is zero.

(d) $2CH_4(g) \rightarrow C_2H_6(g) + H_2(g)$

$$\Delta H° = \{\Delta H_f^\circ[C_2H_6(g)] + \Delta H_f^\circ[H_2(g)]\} - \{2\Delta H_f^\circ[CH_4(g)]\}$$

$$\Delta H° = \{1\text{ mol} \times (-84.667\text{ kJ/mol}) + 1\text{ mol} \times (0.0\text{ kJ/mol})\} - \{2\text{ mol} \times (-74.848\text{ kJ/mol})\}$$

$$\Delta H° = 65.029\text{ kJ}$$

$\Delta E = \Delta H°$, since the value of Δn for this reaction is zero.

20.59 We proceed as in the answer to Review Exercise 20.58, using the data supplied in Appendix.

(a) $2C_2H_2(g) + 5O_2(g) \rightarrow 4CO_2(g) + 2H_2O(g)$

$$\Delta H° = \{2\ \Delta H_f^\circ[H_2O(g)] + 4\ \Delta H_f^\circ\ [CO_2(g)]\} - \{5\Delta H_f^\circ[O_2(g)] + 2\Delta H_f^\circ[C_2H_2(g)]\}$$

$$\Delta H° = \{2\text{ mol} \times (-242\text{ kJ/mol}) + 4\text{ mol} \times (-394\text{ kJ/mol})\} - \{5\text{ mol} \times (0.0\text{ kJ/mol}) + 2\text{ mol} \times (227\text{ kJ/mol})\}$$

$$\Delta H° = -2514\text{ kJ}$$

$$\Delta E = \Delta H° - \Delta nRT = -2514\text{ kJ} - (-1\text{ mol})(8.314\text{ J/mol K})(10^{-3}\text{ kJ/J})(298\text{ K})$$

$$\Delta E = -2512\text{ kJ}$$

(b) $C_2H_2(g) + 5N_2O(g) \rightarrow 5N_2(g) + H_2O(g) + 2CO_2(g)$

$$\Delta H° = \{5\,\Delta H_f^\circ[N_2(g)] + \Delta H_f^\circ\, H_2O(g)] + 2\Delta H_f^\circ[CO_2(g)]\}$$
$$- \{5\Delta H_f^\circ[N_2O(g)] + \Delta H_f^\circ[C_2H_2(g)]\}$$
$$\Delta H° = \{5 \text{ mol} \times (0.0 \text{ kJ/mol}) + 1 \text{ mol} \times (-242 \text{ kJ/mol}) + 2 \text{ mol} \times (-394 \text{ kJ/mol})\}$$
$$- \{5 \text{ mol} \times (81.5 \text{ kJ/mol}) + 1 \text{ mol} \times (227 \text{ kJ/mol})\}$$
$$\Delta H° = -1665 \text{ kJ}$$

$$\Delta E = \Delta H° - \Delta nRT$$
$$\Delta E = -1665 \text{ kJ} - (2 \text{ mol})(8.314 \text{ J/mol K})(10^{-3} \text{ kJ/J})(298 \text{ K})$$
$$\Delta E = -1670 \text{ kJ}$$

(c) $NH_4Cl(s) \rightarrow NH_3(g) + HCl(g)$

$$\Delta H° = \{\Delta H_f^\circ[HCl(g)] + \Delta H_f^\circ[NH_3(g)]\} - \{\Delta H_f^\circ[NH_4Cl(s)]\}$$
$$\Delta H° = \{1 \text{ mol} \times (-92.5 \text{ kJ/mol}) + 1 \text{ mol} \times (-46.0 \text{ kJ/mol})\}$$
$$- \{1 \text{ mol} \times (-314.4 \text{ kJ/mol})\}$$
$$\Delta H° = 175.9 \text{ kJ}$$

$$\Delta E = \Delta H° - \Delta nRT$$
$$\Delta E = 175.9 \text{ kJ} - (2 \text{ mol})(8.314 \text{ J/mol K})(10^{-3} \text{ kJ/J})(298 \text{ K})$$
$$\Delta E = 170.9 \text{ kJ}$$

(d) $(CH_3)_2CO(l) + 4O_2(g) \rightarrow 3CO_2(g) + 3H_2O(g)$

$$\Delta H° = \{3\Delta H_f^\circ[CO_2(g)] + 3\Delta H_f^\circ[H_2O(g)]\} - \{\Delta H_f^\circ[(CH_3)_2CO(l)] + 4\Delta H_f^\circ[O_2(g)]\}$$
$$\Delta H° = \{3 \text{ mol} \times (-394 \text{ kJ/mol}) + 3 \text{ mol} \times (-241.8 \text{ kJ/mol})\}$$
$$- \{1 \text{ mol} \times (-248.1 \text{ kJ/mol}) + 4 \times (0 \text{ kJ/mol})\}$$
$$\Delta H° = -1659 \text{ kJ}$$

$$\Delta E = \Delta H° - \Delta nRT$$
$$\Delta E = -1659 \text{ kJ} - (2 \text{ mol})(8.314 \text{ J/mol K})(10^{-3} \text{ kJ/J})(298 \text{ K})$$
$$\Delta E = -1664 \text{ kJ}$$

20.60 In general, we have the equation: $\Delta H° = (\text{sum } \Delta H_f^\circ[\text{products}]) - (\text{sum } \Delta H_f^\circ[\text{reactants}])$

(a)
$$\Delta H° = \{\Delta H_f^\circ[CaCO_3(s)]\} - \{\Delta H_f^\circ[CO_2(g)] + \Delta H_f^\circ[CaO(s)]\}$$
$$\Delta H° = \{1 \text{ mol} \times (-1207 \text{ kJ/mol})\}$$
$$- \{1 \text{ mol} \times (-394 \text{ kJ/mol}) + 1 \text{ mol} \times (-635.5 \text{ kJ/mol})\}$$
$$\Delta H° = -178 \text{ kJ} \ \therefore \text{ favored.}$$

(b)
$$\Delta H° = \{\Delta H_f^\circ[C_2H_6(g)]\} - \{\Delta H_f^\circ[C_2H_2(g)] + 2\Delta H_f^\circ[H_2(g)]\}$$
$$\Delta H° = \{1 \text{ mol} \times (-84.5 \text{ kJ/mol})\}$$
$$- \{1 \text{ mol} \times (227 \text{ kJ/mol}) + 2 \text{ mol} \times (0.0 \text{ kJ/mol})\}$$
$$\Delta H° = -311 \text{ kJ} \ \therefore \text{ favored.}$$

(c)
$$\Delta H° = \{\Delta H_f^\circ[Fe_2O_3(s)] + 3\Delta H_f^\circ[Ca(s)]\}$$
$$- \{2\Delta H_f^\circ[Fe(s)] + 3\Delta H_f^\circ[CaO(s)]\}$$
$$\Delta H° = \{1 \text{ mol} \times (-822.2 \text{ kJ/mol}) + 3 \text{ mol} \times (0.0 \text{ kJ/mol})\}$$
$$- \{2 \text{ mol} \times (0.0 \text{ kJ/mol}) + 3 \text{ mol} \times (-635.5 \text{ kJ/mol})\}$$
$$\Delta H° = +1084.3 \text{ kJ} \ \therefore \text{ not favorable from the standpoint of enthalpy alone.}$$

(d) $\Delta H° = \{\Delta H_f^\circ[H_2O(l)] + \Delta H_f^\circ[CaO(s)]\} - \{\Delta H_f^\circ[Ca(OH)_2(s)]\}$

$\Delta H° = \{1 \text{ mol} \times (-285.9 \text{ kJ/mol}) + 1 \text{ mol} \times (-635.5 \text{ kJ/mol})\}$
$- \{1 \text{ mol} \times (-986.59 \text{ kJ/mol})\}$

$\Delta H° = +65.2$ kJ ∴ not favored from the standpoint of enthalpy alone.

(e) $\Delta H° = \{2\Delta H_f^\circ[HCl(g)] + \Delta H_f^\circ[Na_2SO_4(s)]\}$
$- \{2\Delta H_f^\circ[NaCl(s)] + \Delta H_f^\circ[H_2SO_4(l)]\}$

$\Delta H° = \{2 \text{ mol} \times (-92.30 \text{ kJ/mol}) + 1 \text{ mol} \times (-1384.5 \text{ kJ/mol})\}$
$- \{2 \text{ mol} \times (-411.0 \text{ kJ/mol}) + 1 \text{ mol} \times (-811.32 \text{ kJ/mol})\}$

$\Delta H° = +64.2$ kJ ∴ not favored from the standpoint of enthalpy alone.

20.61 (a) $\Delta H° = \{2\Delta H_f^\circ[H_2O(g)] + 4\Delta H_f^\circ[CO_2(g)]\}$
$- \{5\Delta H_f^\circ[O_2(g)] + 2\Delta H_f^\circ[C_2H_2(g)]\}$

$\Delta H° = \{2 \text{ mol} \times (-242 \text{ kJ/mol}) + 4 \text{ mol} \times (-394 \text{ kJ/mol})\}$
$- \{5 \text{ mol} \times (0.0 \text{ kJ/mol}) + 2 \text{ mol} \times (227 \text{ kJ/mol})\}$

$\Delta H° = -2514$ kJ ∴ favored from the standpoint of enthalpy alone.

(b) $\Delta H° = \{5\Delta H_f^\circ[N_2(g)] + \Delta H_f^\circ[H_2O(g)] + 2\Delta H_f^\circ[CO_2(g)]\}$
$- \{5\Delta H_f^\circ[N_2O(g)] + \Delta H_f^\circ[C_2H_2(g)]\}$

$\Delta H° = \{5 \text{ mol} \times (0.0 \text{ kJ/mol}) + 1 \text{ mol} \times (-242 \text{ kJ/mol})$
$+ 2 \text{ mol} \times (-394 \text{ kJ/mol})\} - \{5 \text{ mol} \times (81.5 \text{ kJ/mol})$
$+ 1 \text{ mol} \times (227 \text{ kJ/mol})\}$

$\Delta H° = -1665$ kJ ∴ favorable from the standpoint of enthalpy.

(c) $\Delta H° = \{2\Delta H_f^\circ[Fe(s)] + \Delta H_f^\circ[Al_2O_3(s)]\} - \{2\Delta H_f^\circ[Al(s)] + \Delta H_f^\circ[Fe_2O_3(s)]\}$

$\Delta H° = \{2 \text{ mol} \times (0.0 \text{ kJ/mol}) + 1 \text{ mol} \times (-1676 \text{ kJ/mol})\}$
$- \{2 \text{ mol} \times (0.0 \text{ kJ/mol}) + 1 \text{ mol} \times (-822.2 \text{ kJ/mol})\}$

$\Delta H° = -854$ kJ ∴ favorable from the standpoint of enthalpy.

d) $\Delta H° = \{\Delta H_f^\circ[HCl(g)] + \Delta H_f^\circ[NH_3(g)]\} - \{\Delta H_f^\circ[NH_4Cl(s)]\}$

$\Delta H° = \{1 \text{ mol} \times (-92.5 \text{ kJ/mol}) + 1 \text{ mol} \times (-46.0 \text{ kJ/mol})\}$
$- \{1 \text{ mol} \times (-314.4 \text{ kJ/mol})\}$

$\Delta H° = 175.9$ kJ ∴ not favorable from the standpoint of enthalpy.

(e) $\Delta H° = \{\Delta H_f^\circ[AgCl(s)] + \Delta H_f^\circ[K(s)]\} - \{\Delta H_f^\circ[Ag(s)] + \Delta H_f^\circ[KCl(s)]\}$

$\Delta H° = \{1 \text{ mol} \times (-127.1 \text{ kJ/mol}) + 1 \text{ mol} \times (0.0 \text{ kJ/mol})\}$
$- \{1 \text{ mol} \times (0.0 \text{ kJ/mol}) + 1 \text{ mol} \times (-436.8 \text{ kJ/mol})\}$

$\Delta H° = 309.7$ kJ, ∴ not favorable from the standpoint of enthalpy.

20.62 The probability is given by the number of possibilities that lead to the desired arrangement, divided by the total number of possible arrangements.

We list each of the possible results, heads (H) or tails (T) for each of the coins, and systematically write down all of the distinct arrangements:

There is only one arrangement that gives four heads: HHHH.
There is only one arrangement that gives four tails: TTTT
Four distinct arrangements can lead to three heads and one tail:
HHHT, THHH, HTHH, HHTH
There are similarly four distinct arrangements that lead to three tails and one head:
TTTH, HTTT, THTT, TTHT

There are six distinct arrangements that can lead to two heads and two tails:

HHTT, TTHH, THHT, THTH, HTTH, HTHT

Hence, the probability of all heads (HHHH) is 1 in 16 or $1/16 = 0.0625$.
The probability of two heads and two tails is 6 in 16 or $6/16 = 0.375$.

20.63 This situation is completely analogous to that in Review Problem 20.62, since there is an equal probability of finding a given molecule in either container. This arises because the containers have equal volumes. One container is thus analogous to a "head" and the other container is analogous to a "tail." Thus, there are to be 16 possible distinct arrangements of the four molecules in the two containers. Only two of the 16 possible arrangements involve having all four molecules in one container. Hence, the probability of finding all 4 molecules in one container is 2 parts in 16, or 1/8.

There are six arrangements in which each container will have two molecules, just as there are six possible ways of obtaining two heads and two tails when tossing four coins. Thus, there is a 6/16 = 3/8 probability of finding two molecules in each container.

Since 3/8 (the probability of finding two molecules in each of two containers) is larger than 1/8 (the probability of finding all four molecules in one container), it is evident that the more likely distribution is the first. This suggests that molecules that are introduced into a container will undergo a spontaneous expansion to distribute themselves uniformly throughout the volume that is available to them.

20.64 (a) negative – since the number of moles of gaseous material decreases.
(b) negative – since the number of moles of gaseous material decreases.
(c) negative – since the number of moles of gas decreases.
(d) positive – since a gas appears where there formerly was none.

20.65 (a) ΔS is positive since randomness in a gas is higher than that in a solid.
(b) ΔS is negative. There are fewer moles of gases among the products.
(c) ΔS is negative since gaseous material (which is highly random) is replaced by a solid (which is highly ordered).
(d) ΔS is negative since the relatively random liquid reactant disappears in a process that makes only a solid.

20.66 $\Delta S° = (\text{sum } S°[\text{products}]) - (\text{sum } S°[\text{reactants}])$

(a)
$$\Delta S° = \{2S°[NH_3(g)]\} - \{3S°[H_2(g)] + S°[N_2(g)]\}$$
$$\Delta S° = \{2 \text{ mol} \times (192.5 \text{ J mol}^{-1} \text{ K}^{-1})\} - \{3 \text{ mol} \times (130.6 \text{ J mol}^{-1} \text{ K}^{-1}) + 1 \text{ mol} \times (191.5 \text{ J mol}^{-1} \text{ K}^{-1})\}$$
$\Delta S° = -198.3$ J/K $\therefore$ not spontaneous from the standpoint of entropy.

(b)
$$\Delta S° = \{S°[CH_3OH(l)]\} - \{2S°[H_2(g)] + S°[CO(g)]\}$$
$$\Delta S° = \{1 \text{ mol} \times (126.8 \text{ J mol}^{-1} \text{ K}^{-1})\} - \{2 \text{ mol} \times (130.6 \text{ J mol}^{-1} \text{ K}^{-1}) + 1 \text{ mol} \times (197.9 \text{ J mol}^{-1} \text{ K}^{-1})\}$$
$\Delta S° = -332.3$ J/K $\therefore$ not favored from the standpoint of entropy alone.

(c)
$$\Delta S° = \{6S°[H_2O(g)] + 4S°[CO_2(g)]\} - \{7S°[O_2(g)] + 2S°[C_2H_6(g)]\}$$
$$\Delta S° = \{6 \text{ mol} \times (188.7 \text{ J mol}^{-1} \text{ K}^{-1}) + 4 \text{ mol} \times (213.6 \text{ J mol}^{-1} \text{ K}^{-1})\} - \{7 \text{ mol} \times (205.0 \text{ J mol}^{-1} \text{ K}^{-1}) + 2 \text{ mol} \times (229.5 \text{ J mol}^{-1} \text{ K}^{-1})\}$$
$\Delta S° = +92.6$ J/K $\therefore$ favorable from the standpoint of entropy alone.

(d) $\Delta S° = \{2S°[H_2O(l)] + S°[CaSO_4(s)]\}$
$- \{S°[H_2SO_4(l)] + S°[Ca(OH)_2(s)]\}$
$\Delta S° = \{2 \text{ mol} \times (69.96 \text{ J mol}^{-1} \text{ K}^{-1}) + 1 \text{ mol} \times (107 \text{ J mol}^{-1} \text{ K}^{-1})\}$
$- \{1 \text{ mol} \times (157 \text{ J mol}^{-1} \text{ K}^{-1}) + 1 \text{ mol} \times (76.1 \text{ J mol}^{-1} \text{ K}^{-1})\}$
$\Delta S° = +14$ J/K ∴ favorable from the standpoint of entropy alone.

(e) $\Delta S° = \{2S°[N_2(g)] + S°[SO_2(g)]\} - \{2S°[N_2O(g)] + S°[S(s)]\}$
$\Delta S° = \{2 \text{ mol} \times (191.5 \text{ J mol}^{-1} \text{ K}^{-1}) + 1 \text{ mol} \times (248 \text{ J mol}^{-1} \text{ K}^{-1})\}$
$- \{2 \text{ mol} \times (220.0 \text{ J mol}^{-1} \text{ K}^{-1}) + 1 \text{ mol} \times (31.9 \text{ J mol}^{-1} \text{ K}^{-1})\}$
$\Delta S° = +159$ J/K ∴ favorable from the standpoint of entropy alone.

20.67 $\Delta S° = (\text{sum } S°[\text{products}]) - (\text{sum } S°[\text{reactants}])$

(a) $\Delta S° = \{S°[AgCl(s)]\} - \{1/2S°[Cl_2(g)] + S°[Ag(s)]\}$
$\Delta S° = \{1 \text{ mol} \times (96.2 \text{ J mol}^{-1} \text{ K}^{-1})\}$
$- \{1/2 \text{ mol} \times (223.0 \text{ J mol}^{-1} \text{ K}^{-1}) + 1 \text{ mol} \times (42.55 \text{ J mol}^{-1} \text{ K}^{-1})\}$
$\Delta S° = -57.9$ J/K

(b) $\Delta S° = \{S°[H_2O(g)]\} - \{1/2S°[O_2(g)] + S°[H_2(g)]\}$
$\Delta S° = \{1 \text{ mol} \times (188.7 \text{ J mol}^{-1} \text{ K}^{-1})\}$
$- \{1/2 \text{ mol} \times (205.0 \text{ J mol}^{-1} \text{ K}^{-1}) + 1 \text{ mol} \times (130.6 \text{ J mol}^{-1} \text{ K}^{-1})\}$
$\Delta S° = -44.4$ J/K

(c) $\Delta S° = \{S°[H_2O(l)]\} - \{1/2S°[O_2(g)] + S°[H_2(g)]\}$
$\Delta S° = \{1 \text{ mol} \times (69.96 \text{ J mol}^{-1} \text{ K}^{-1})\}$
$- \{1/2 \text{ mol} \times (205.0 \text{ J mol}^{-1} \text{ K}^{-1}) + 1 \text{ mol} \times (130.6 \text{ J mol}^{-1} \text{ K}^{-1})\}$
$\Delta S° = -163.1$ J/K

(d) $\Delta S° = \{S°[CO_2(g)] + S°[H_2O(g)] + S°[CaSO_4(s)]\}$
$- \{S°[CaCO_3(s)] + S°[H_2SO_4(l)]\}$
$\Delta S° = \{1 \text{ mol} \times (213.6 \text{ J mol}^{-1} \text{ K}^{-1}) + 1 \text{ mol} \times (188.7 \text{ J mol}^{-1} \text{ K}^{-1})$
$+ 1 \text{ mol} \times (107 \text{ J mol}^{-1} \text{ K}^{-1})\} - \{1 \text{ mol} \times (92.9 \text{ J mol}^{-1} \text{ K}^{-1})$
$+ 1 \text{ mol} \times (157 \text{ J mol}^{-1} \text{ K}^{-1})\}$
$\Delta S° = +259$ J/K

(e) $\Delta S° = \{S°[NH_4Cl(s)]\} - \{S°[HCl(g)] + S°[NH_3(g)]\}$
$\Delta S° = \{1 \text{ mol} \times (94.6 \text{ J mol}^{-1} \text{ K}^{-1})\}$
$- \{1 \text{ mol} \times (186.7 \text{ J mol}^{-1} \text{ K}^{-1}) + 1 \text{ mol} \times (192.5 \text{ J mol}^{-1} \text{ K}^{-1})\}$
$\Delta S° = -284.6$ J/K

20.68 The entropy change that is designated ΔS_f° is that which corresponds to the reaction in which one mole of a substance is formed from elements in their standard states. Since the value is understood to correspond to the reaction forming one mole of a single pure substance, the units may be written either J K^{-1} or $\text{J mol}^{-1} \text{ K}^{-1}$.

(a) $2C(s) + 2H_2(g) \rightarrow C_2H_4(g)$
$\Delta S° = \{S°[C_2H_4(g)]\} - \{2S°[C(s)] + 2S°[H_2(g)]\}$
$\Delta S° = \{1 \text{ mol} \times (219.8 \text{ J mol}^{-1} \text{ K}^{-1})\}$
$- \{2 \text{ mol} \times (5.69 \text{ J mol}^{-1} \text{ K}^{-1}) + 2 \text{ mol} \times (130.6 \text{ J mol}^{-1} \text{ K}^{-1})\}$
$\Delta S° = -52.8$ J/K or $-52.8 \text{ J mol}^{-1} \text{ K}^{-1}$

(b) $N_2(g) + 1/2O_2(g) \rightarrow N_2O(g)$
$\Delta S° = \{S°[N_2O(g)]\} - \{S°[N_2(g)] + 1/2S°[O_2(g)]\}$

$\Delta S° = \{1 \text{ mol} \times (220.0 \text{ J mol}^{-1} \text{ K}^{-1})\}$
$- \{1 \text{ mol} \times (191.5 \text{ J mol}^{-1} \text{ K}^{-1}) + 1/2 \text{ mol} \times (205.0 \text{ J mol}^{-1} \text{ K}^{-1})\}$
$\Delta S° = -74.0 \text{ J/K or } -74.0 \text{ J mol}^{-1} \text{ K}^{-1}$

(c) $Na(s) + 1/2Cl_2(g) \rightarrow NaCl(s)$
$\Delta S° = \{S°[NaCl(s)]\} - \{1/2S°[Cl_2(g)] + S°[Na(s)]\}$
$\Delta S° = \{1 \text{ mol} \times (72.38 \text{ J mol}^{-1} \text{ K}^{-1})\} - \{1/2 \text{ mol} \times 223.0 \text{ J mol}^{-1} \text{ K}^{-1}) + 1 \text{ mol} \times (51.0 \text{ J mol}^{-1} \text{ K}^{-1})\}$
$\Delta S° = -90.1 \text{ J/K or } -90.1 \text{ J mol}^{-1} \text{ K}^{-1}$

(d) $Ca(s) + S(s) + 3O_2(g) + 2H_2(g) \rightarrow CaSO_4{\cdot}2H_2O(s)$
$\Delta S° = \{S°[CaSO_4{\cdot}2H_2O(s)]\} - \{2S°[H_2(g)] + 3S°[O_2(g)] + S°[S(s)] + S°[Ca(s)]\}$
$\Delta S° = \{1 \text{ mol} \times (194.0 \text{ J mol}^{-1} \text{ K}^{-1})\} - \{2 \text{ mol} \times (130.6 \text{ J mol}^{-1} \text{ K}^{-1})$
$+ 3 \text{ mol} \times (205.0 \text{ J mol}^{-1} \text{ K}^{-1}) + 1 \text{ mol} \times (31.8 \text{ J mol}^{-1} \text{ K}^{-1}) + 1 \text{ mol} \times (154.8 \text{ J mol}^{-1} \text{ K}^{-1})\}$
$\Delta S° = -868.9 \text{ J/K or } -868.9 \text{ J mol}^{-1} \text{ K}^{-1}$

(e) $2H_2(g) + 2C(s) + O_2(g) \rightarrow HC_2H_3O_2(l)$
$\Delta S° = \{S°[HC_2H_3O_2(l)]\} - \{2S°[H_2(g)] + 2S°[C(s)] + S°[O_2(g)]\}$
$\Delta S° = \{1 \text{ mol} \times (160 \text{ J mol}^{-1} \text{ K}^{-1})\} - \{2 \text{ mol} \times (130.6 \text{ J mol}^{-1} \text{ K}^{-1})$
$+ 2 \text{ mol} \times (5.69 \text{ J mol}^{-1} \text{ K}^{-1}) + 1 \text{ mol} \times (205.0 \text{ J mol}^{-1} \text{ K}^{-1})\}$
$\Delta S° = -318 \text{ J/K or } -318 \text{ J mol}^{-1} \text{ K}^{-1}$

20.69 The entropy change that is designated ΔS_f° is that which corresponds to the reaction in which one mole of a substance is formed from elements in their standard states. Since the value is understood to correspond to the reaction forming one mole of a single pure substance, the units may be written either J K^{-1} or $\text{J mol}^{-1} \text{ K}^{-1}$.

(a) $2Al(s) + 3/2O_2(g) \rightarrow Al_2O_3(s)$
$\Delta S° = \{S°[Al_2O_3(s)]\} - \{2S°[Al(s)] + 3/2S°[O_2(g)]\}$
$\Delta S° = \{1 \text{ mol} \times (51.0 \text{ J mol}^{-1} \text{ K}^{-1})\}$
$- \{2 \text{ mol} \times (28.3 \text{ J mol}^{-1} \text{ K}^{-1}) + 3/2 \text{ mol} \times (205.0 \text{ J mol}^{-1} \text{ K}^{-1})\}$
$\Delta S° = -313.1 \text{ J/K or } -313.1 \text{ J mol}^{-1} \text{ K}^{-1}$

(b) $Ca(s) + C(s) + 3/2O_2(g) \rightarrow CaCO_3(s)$
$\Delta S° = \{S°[CaCO_3(s)]\} - \{S°[Ca(s)] + S°[C(s)] + 3/2S°[O_2(g)]\}$
$\Delta S° = \{1 \text{ mol} \times (92.9 \text{ J mol}^{-1} \text{ K}^{-1})\} - \{1 \text{ mol} \times (41.4 \text{ J mol}^{-1} \text{ K}^{-1})$
$+ 1 \text{ mol} \times (5.69 \text{ J mol}^{-1} \text{ K}^{-1}) + 3/2 \text{ mol} \times (205.0 \text{ J mol}^{-1} \text{ K}^{-1})\}$
$\Delta S° = -261.7 \text{ J/K or } -261.7 \text{ J mol}^{-1} \text{ K}^{-1}$

(c) $N_2(g) + 2O_2(g) \rightarrow N_2O_4(g)$
$\Delta S° = \{S°[N_2O_4(g)]\} - \{S°[N_2(g)] + 2S°[O_2(g)]\}$
$\Delta S° = \{1 \text{ mol} \times (304 \text{ J mol}^{-1} \text{ K}^{-1})\} - \{1 \text{ mol} \times (191.5 \text{ J mol}^{-1} \text{ K}^{-1})$
$+ 2 \text{ mol} \times (205.0 \text{ J mol}^{-1} \text{ K}^{-1})\}$
$\Delta S° = -298 \text{ J/K or } -298 \text{ J mol}^{-1} \text{ K}^{-1}$

(d) $1/2N_2(g) + 2H_2(g) + 1/2Cl_2(g) \rightarrow NH_4Cl(s)$
$\Delta S° = \{S°[NH_4Cl(s)]\} - \{1/2S°[N_2(g)] + 2S°[H_2(g)] + 1/2S°[Cl_2(g)]\}$
$\Delta S° = \{1 \text{ mol} \times (94.6 \text{ J mol}^{-1} \text{ K}^{-1})\} - \{1/2 \text{ mol} \times (191.5 \text{ J mol}^{-1} \text{ K}^{-1})$
$+ 2 \text{ mol} \times (130.6 \text{ J mol}^{-1} \text{ K}^{-1}) + 1/2 \text{ mol} \times (223.0 \text{ J mol}^{-1} \text{ K}^{-1})\}$
$\Delta S° = -373.9 \text{ J/K or } -373.9 \text{ J mol}^{-1} \text{ K}^{-1}$

(e) $Ca(s) + S(s) + 9/4O_2(g) + 1/2H_2(g) \rightarrow CaSO_4 \cdot 1/2H_2O(s)$

$\Delta S° = \{S°[CaSO_4 \cdot 1/2H_2O(s)]\} - \{S°[Ca(s)] + S°[S(s)] + 9/4S°[O_2(g)] + 1/2S°[H_2(g)]\}$

$\Delta S° = \{1 \text{ mol} \times (131 \text{ J mol}^{-1} \text{ K}^{-1})\} - \{1 \text{ mol} \times (41.4 \text{ J mol}^{-1} \text{ K}^{-1}) + 1 \text{ mol} \times (31.8 \text{ J mol}^{-1} \text{ K}^{-1}) + 9/4 \text{ mol} \times (205.0 \text{ J mol}^{-1} \text{ K}^{-1}) + 1/2 \text{ mol} \times (130.6 \text{ J mol}^{-1} \text{ K}^{-1})\}$

$\Delta S° = -469 \text{ J/K or } -469 \text{ J mol}^{-1} \text{ K}^{-1}$

20.70 $\Delta S° = (\text{sum } S°[\text{products}]) - (\text{sum } S°[\text{reactants}])$

$\Delta S° = \{2S°[HNO_3(l)] + S°[NO(g)]\} - \{3S°[NO_2(g)] + S°[H_2O(l)]\}$

$\Delta S° = \{2 \text{ mol} \times (155.6 \text{ J mol}^{-1} \text{ K}^{-1}) + 1 \text{ mol} \times (210.6 \text{ J mol}^{-1} \text{ K}^{-1})\} - \{3 \text{ mol} \times (240.5 \text{ J mol}^{-1} \text{ K}^{-1}) + 1 \text{ mol} \times (69.96 \text{ J mol}^{-1} \text{ K}^{-1})\}$

$\Delta S° = -269.7 \text{ J/K}$

20.71 $\Delta S° = (\text{sum } S°[\text{products}]) - (\text{sum } S°[\text{reactants}])$

$\Delta S° = \{S°[HC_2H_3O_2(l)] + S°[H_2O(l)]\} - \{S°[C_2H_5OH(l)] + S°[O_2(g)]\}$

$\Delta S° = \{1 \text{ mol} \times (160 \text{ J mol}^{-1} \text{ K}^{-1}) + 1 \text{ mol} \times (69.96 \text{ J mol}^{-1} \text{ K}^{-1})\} - \{1 \text{ mol} \times (161 \text{ J mol}^{-1} \text{ K}^{-1}) + 1 \text{ mol} \times (205.0 \text{ J mol}^{-1} \text{ K}^{-1})\}$

$\Delta S° = -136 \text{ J/K}$

20.72 The quantity ΔG_f° applies to the equation in which one mole of pure phosgene is produced from the naturally occurring forms of the elements:

$$C(s) + 1/2O_2(g) + Cl_2(g) \rightarrow COCl_2(g),\ \Delta G_f^\circ = ?$$

We can determine ΔG_f° if we can find values for ΔH_f° and ΔS_f°, because:

$\Delta G° = \Delta H° - T\Delta S°$

The value of ΔS_f° is determined using S° for phosgene in the following way:

$\Delta S_f^\circ = \{S°[COCl_2(g)]\} - \{S°[C(s)] + 1/2S°[O_2(g)] + S°[Cl_2(g)]\}$

$\Delta S_f^\circ = \{1 \text{ mol} \times (284 \text{ J mol}^{-1} \text{ K}^{-1})\} - \{1 \text{ mol} \times (5.69 \text{ J mol}^{-1} \text{ K}^{-1}) + 1/2 \text{ mol} \times (205.0 \text{ J mol}^{-1} \text{ K}^{-1}) + 1 \text{ mol} \times (223.0 \text{ J mol}^{-1} \text{ K}^{-1})\}$

$\Delta S_f^\circ = -47 \text{ J mol}^{-1} \text{ K}^{-1} \text{ or } -47 \text{ J/K}$

$\Delta G_f^\circ = \Delta H_f^\circ - T\Delta S_f^\circ = -223 \text{ kJ/mol} - (298 \text{ K})(-0.047 \text{ kJ/mol K}) = -209 \text{ kJ/mol}$

20.73 $\Delta S° = (\text{sum } S°[\text{products}]) - (\text{sum } S°[\text{reactants}])$

$2Al(s) + 3/2O_2(g) \rightarrow Al_2O_3(s)$

$\Delta S_f^\circ = \{S°[Al_2O_3(s)]\} - \{2S°[Al(s)] + 3/2S°[O_2(g)]\}$

$\Delta S_f^\circ = \{1 \text{ mol} \times (51.0 \text{ J mol}^{-1} \text{ K}^{-1})\} - \{2 \text{ mol} \times (28.3 \text{ J mol}^{-1} \text{ K}^{-1}) + 3/2 \text{ mol} \times (205.0 \text{ J mol}^{-1} \text{ K}^{-1})\}$

$\Delta S_f^\circ = -313.1 \text{ J/K or } -313.1 \text{ J mol}^{-1} \text{ K}^{-1} = -0.3131 \text{ kJ mol}^{-1} \text{ K}^{-1}$

$\Delta G_f^\circ = \Delta H_f^\circ - T\Delta S_f^\circ = -1676 \text{ kJ/mol} - (298 \text{ K})(-0.3131 \text{ kJ mol}^{-1} \text{ K}^{-1})$

$\Delta G_f^\circ = -1583 \text{ kJ/mol}$

This value agrees well with the value listed in Appendix E.

20.74 $\Delta G° = (\text{sum } \Delta G_f^\circ[\text{products}]) - (\text{sum } \Delta G_f^\circ[\text{reactants}])$

(a) $\Delta G° = \{\Delta G_f^\circ[H_2SO_4(l)]\} - \{\Delta G_f^\circ[H_2O(l)] + \Delta G_f^\circ[SO_3(g)]\}$

$\Delta G° = \{1 \text{ mol} \times (-689.9 \text{ kJ/mol})\} - \{1 \text{ mol} \times (-237.2 \text{ kJ/mol}) + 1 \text{ mol} \times (-370 \text{ kJ/mol})\}$

$\Delta G° = -82.3 \text{ kJ}$

(b) $\Delta G° = \{2\Delta G_f^\circ[NH_3(g)] + \Delta G_f^\circ[H_2O(l)] + \Delta G_f^\circ[CaCl_2(s)]\} - \{\Delta G_f^\circ[CaO(s)] + 2\Delta G_f^\circ[NH_4Cl(s)]\}$

$\Delta G° = \{2 \text{ mol} \times (-16.7 \text{ kJ/mol}) + 1 \text{ mol} \times (-237.2 \text{ kJ/mol}) + 1 \text{ mol} \times (-750.2 \text{ kJ/mol})\} - \{1 \text{ mol} \times (-604.2 \text{ kJ/mol}) + 2 \text{ mol} \times (-203.9 \text{ kJ/mol})\}$

$\Delta G° = -8.8 \text{ kJ}$

(c) $\Delta G° = \{\Delta G_f^\circ[H_2SO_4(l)] + \Delta G_f^\circ[CaCl_2(s)]\} - \{\Delta G_f^\circ[CaSO_4(s)] + \Delta G_f^\circ[HCl(g)]\}$

$\Delta G° = \{1 \text{ mol} \times (-689.9 \text{ kJ/mol}) + 1 \text{ mol} \times (-750.2 \text{ kJ/mol})\} - \{1 \text{ mol} \times (-1320.3 \text{ kJ/mol}) + 2 \text{ mol} \times (-95.27 \text{ kJ/mol})\}$

$\Delta G° = +70.7 \text{ kJ}$

(d) $\Delta G° = \{\Delta G_f^\circ[C_2H_5OH(l)]\} - \{\Delta G_f^\circ[H_2O(g)] + \Delta G_f^\circ[C_2H_4(g)]\}$

$\Delta G° = \{1 \text{ mol} \times (-174.8 \text{ kJ/mol})\} - \{1 \text{ mol} \times (-228.6 \text{ kJ/mol}) + 1 \text{ mol} \times (68.12 \text{ kJ/mol})\}$

$\Delta G° = -14.3 \text{ kJ}$

(e) $\Delta G° = \{2\Delta G_f^\circ[H_2O(l)] + \Delta G_f^\circ[SO_2(g)] + \Delta G_f^\circ[CaSO_4(s)]\} - \{2\Delta G_f^\circ[H_2SO_4(l)] + \Delta G_f^\circ[Ca(s)]\}$

$\Delta G° = \{2 \text{ mol} \times (-237.2 \text{ kJ/mol}) + 1 \text{ mol} \times (-300 \text{ kJ/mol}) + 1 \text{ mol} \times (-1320.3 \text{ kJ/mol})\} - \{2 \text{ mol} \times (-689.9 \text{ kJ/mol}) + 1 \text{ mol} \times (0.0 \text{ kJ/mol})\}$

$\Delta G° = -715 \text{ kJ}$

20.75 $\Delta G° = (\text{sum } \Delta G_f^\circ[\text{products}]) - (\text{sum } \Delta G_f^\circ[\text{reactants}])$

(a) $\Delta G° = \{\Delta G_f^\circ[H_2O(g)] + \Delta G_f^\circ[CaCl_2(s)]\} - \{2\Delta G_f^\circ[HCl(g)] + \Delta G_f^\circ[CaO(s)]\}$

$\Delta G° = \{1 \text{ mol} \times (-228.6 \text{ kJ/mol}) + 1 \text{ mol} \times (-750.2 \text{ kJ/mol})\} - \{2 \text{ mol} \times (-95.27 \text{ kJ/mol}) + 1 \text{ mol} \times (-604.2 \text{ kJ/mol})\}$

$\Delta G° = -184.1 \text{ kJ}$

(b) $\Delta G° = \{\Delta G_f^\circ[Na_2SO_4(s)] + 2\Delta G_f^\circ[HCl(g)]\} - \{\Delta G_f^\circ[H_2SO_4(l)] + 2\Delta G_f^\circ[NaCl(s)]\}$

$\Delta G° = \{1 \text{ mol} \times (-1266.83 \text{ kJ/mol}) + 2 \text{ mol} \times (-95.27 \text{ kJ/mol})\} - \{1 \text{ mol} \times (-689.9 \text{ kJ/mol}) + 2 \text{ mol} \times (-384.0 \text{ kJ/mol})\}$

$\Delta G° = +0.5 \text{ kJ}$

(c) $\Delta G° = \{\Delta G_f^\circ[NO(l)] + 2\Delta G_f^\circ[HNO_3(l)]\} - \{3\Delta G_f^\circ[NO_2(l)] + \Delta G_f^\circ[H_2O(l)]\}$

$\Delta G° = \{1 \text{ mol} \times (86.69 \text{ kJ/mol}) + 2 \text{ mol} \times (-79.91 \text{ kJ/mol})\}$
$- \{3 \text{ mol} \times (51.84 \text{ kJ/mol}) + 1 \text{ mol} \times (-237.2 \text{ kJ/mol})\}$
$\Delta G° = +8.6 \text{ kJ}$

(d) $\Delta G° = \{2\Delta G_f^\circ[Ag(s)] + \Delta G_f^\circ[CaCl_2(s)]\}$
$- \{2\Delta G_f^\circ[AgCl(s)] + \Delta G_f^\circ[Ca(s)]\}$
$\Delta G° = \{2 \text{ mol} \times (0.0 \text{ kJ/mol}) + 1 \text{ mol} \times (-750.2 \text{ kJ/mol})\}$
$- \{2 \text{ mol} \times (-109.8 \text{ kJ/mol}) + 1 \text{ mol} \times (0.0 \text{ kJ/mol})\}$
$\Delta G° = -530.6 \text{ kJ}$

(e) $\Delta G° = \{\Delta G_f^\circ[NH_4Cl(s)]\} - \{\Delta G_f^\circ[HCl(g)] + \Delta G_f^\circ[NH_3(g)]\}$
$\Delta G° = \{1 \text{ mol} \times (-203.9 \text{ kJ/mol})\}$
$- \{1 \text{ mol} \times (-95.27 \text{ kJ/mol}) + 1 \text{ mol} \times (-16.7 \text{ kJ/mol})\}$
$\Delta G° = -91.9 \text{ kJ}$

20.76 $CaSO_4{\cdot}1/2H_2O(s) + 3/2H_2O(l) \rightarrow CaSO_4{\cdot}2H_2O(s)$
$\Delta G° = (\text{sum } \Delta G_f^\circ[\text{products}]) - (\text{sum } \Delta G_f^\circ[\text{reactants}])$
$\Delta G° = \{\Delta G_f^\circ[CaSO_4{\cdot}2H_2O(s)]\} -$
$\{\Delta G_f^\circ[CaSO_4{\cdot}1/2H_2O(s)] + 3/2\Delta G_f^\circ[H_2O(l)]\}$
$\Delta G° = \{1 \text{ mol} \times (-1795.7 \text{ kJ/mol})\}$
$- \{1 \text{ mol} \times (-1435.2 \text{ kJ/mol}) + 1.5 \text{ mol} \times (-237.2 \text{ kJ/mol})\}$
$\Delta G° = -4.7 \text{ kJ}$

20.77 $COCl_2(g) + H_2O(g) \rightarrow CO_2(g) + 2HCl(g)$
$\Delta G° = (\text{sum } \Delta G_f^\circ[\text{products}]) - (\text{sum } \Delta G_f^\circ[\text{reactants}])$
$\Delta G° = \{2\Delta G_f^\circ[HCl(g)] + \Delta G_f^\circ[CO_2(g)]\} - \{\Delta G_f^\circ[H_2O(g)] + \Delta G_f^\circ[COCl_2(g)]\}$
$\Delta G° = \{2 \text{ mol} \times (-95.27 \text{ kJ/mol}) + 1 \text{ mol} \times (-394.4 \text{ kJ/mol})\}$
$- \{1 \text{ mol} \times (-228.6 \text{ kJ/mol}) + 1 \text{ mol} \times (-210 \text{ kJ/mol})\}$
$\Delta G° = -146 \text{ kJ}$

20.78 Multiply the reverse of the second equation by 2 (remembering to multiply the associated free energy change by –2), and add the result to the first equation:

$4NO(g) \rightarrow 2N_2O(g) + O_2(g),$ $\Delta G° = -139.56 \text{ kJ}$
$4NO_2(g) \rightarrow 4NO(g) + 2O_2(g),$ $\Delta G° = +139.40 \text{ kJ}$

$4NO_2(g) \rightarrow 3O_2(g) + 2N_2O(g),$ $\Delta G° = -0.16 \text{ kJ}$

This result is the reverse of the desired reaction, which must then have $\Delta G° = +0.16$ kJ

20.79 Add the reverse of the first equation to the second equation plus twice the third equation:

$CO(NH_2)_2(s) + 2NH_4Cl(s) \rightarrow COCl_2(g) + 4NH_3(g),$ $\Delta G° = +332.0 \text{ kJ}$
$COCl_2(g) + H_2O(l) \rightarrow CO_2(g) + 2HCl(g),$ $\Delta G° = -141.8 \text{ kJ}$
$2NH_3(g) + 2HCl(g) \rightarrow 2NH_4Cl(s),$ $\Delta G° = -183.9 \text{ kJ}$

$CO(NH_2)_2(s) + H_2O(l) \rightarrow 2NH_3(g) + CO_2(g),$ $\Delta G° = +6.3 \text{ kJ}$

20.80 The maximum work obtainable from a reaction is equal in magnitude to the value of ΔG for the reaction. Thus, we need only determine $\Delta G°$ for the process:

$\Delta G° = (\text{sum } \Delta G_f^\circ \text{[products]}) - (\text{sum } \Delta G_f^\circ \text{[reactants]})$

$\Delta G° = \{3\Delta G_f^\circ [H_2O(g)] + 2\Delta G_f^\circ [CO_2(g)]\} - \{3\Delta G_f^\circ [O_2(g)] + \Delta G_f^\circ [C_2H_5OH(l)]\}$

$\Delta G° = \{3 \text{ mol} \times (-228.6 \text{ kJ/mol}) + 2 \text{ mol} \times (-394.4 \text{ kJ/mol})\}$
$\quad - \{3 \text{ mol} \times (0.0 \text{ kJ/mol}) \ 1 \text{ mol} \times (-174.8 \text{ kJ/mol})\}$

$\Delta G° = -1299.8 \text{ kJ}$

20.81 We must first determine ΔG° for the reaction:

$CH_4(g) + 2O_2(g) \rightarrow CO_2(g) + 2H_2O(g)$

$\Delta G° = (\text{sum } \Delta G_f^\circ \text{[products]}) - (\text{sum } \Delta G_f^\circ \text{[reactants]})$

$\Delta G° = \{\Delta G_f^\circ [CO_2(g)] + 2\Delta G_f^\circ [H_2O(g)]\} - \{\Delta G_f^\circ [CH_4(g)] + 2\Delta G_f^\circ [O_2(g)]\}$

$\Delta G° = \{1 \text{ mol} \times (-394.4 \text{ kJ/mol}) + 2 \text{ mol} \times (-228.6 \text{ kJ/mol})\}$
$\quad - \{1 \text{ mol} \times (-50.79 \text{ kJ/mol}) + 2 \text{ mol} \times (0.0 \text{ kJ/mol})\}$

$\Delta G° = -800.8 \text{ kJ/mol}$

Next, we determine the amount of work available from the combustion of 48.0 g of CH_4:

$$\# \text{ kJ} = (48.0 \text{ g } CH_4)\left(\frac{1 \text{ mol } CH_4}{16.043 \text{ g } CH_4}\right)\left(\frac{-800.8 \text{ kJ}}{1 \text{ mol } CH_4}\right) = -2.40 \times 10^3 \text{ kJ}$$

20.82 At equilibrium, $\Delta G = 0 = \Delta H - T\Delta S$

$T_{eq} = \Delta H/\Delta S$, and assuming that ΔS is independent of temperature, we have:

$T_{eq} = (31.4 \times 10^3 \text{ J mol}^{-1}) \div (94.2 \text{ J mol}^{-1} \text{ K}^{-1}) = 333 \text{ K}$

20.83 At equilibrium, $\Delta G = 0 = \Delta H - T\Delta S$

$T_{eq} = \Delta H/\Delta S$, and assuming that ΔS and ΔH are independent of temperature, we have:

$T_{eq} = (10.0 \times 10^3 \text{ J mol}^{-1}) \div (9.50 \text{ J mol}^{-1} \text{ K}^{-1}) = 1.05 \times 10^3 \text{ K}$ or 780 °C

Apparently, ΔH and ΔS are not independent of temperature!

20.84 At equilibrium, $\Delta G = 0 = \Delta H - T\Delta S$

Thus $\Delta H = T\Delta S$, and if we assume that both ΔH and ΔS are independent of temperature, we have:

$\Delta S = \Delta H/T_{eq} = (37.7 \times 10^3 \text{ J/mol}) \div (99.3 + 273.15 \text{ K})$

$\Delta S = 101 \text{ J mol}^{-1} \text{ K}^{-1}$

20.85 We proceed as in the answer to Review Problem 20.84:

$\Delta S = \Delta H/T_{eq} = (31.9 \times 10^3 \text{ J/mol}) \div (56.2 + 273.15 \text{ K})$

$\Delta S = 96.9 \text{ J mol}^{-1} \text{ K}^{-1}$

20.86 The reaction is spontaneous if its associated value for ΔG° is negative.

$\Delta G° = (\text{sum } \Delta G_f^\circ \text{[products]}) - (\text{sum } \Delta G_f^\circ \text{[reactants]})$

$\Delta G° = \{\Delta G_f^\circ [HC_2H_3O_2(l)] + \Delta G_f^\circ [H_2O(l)] + \Delta G_f^\circ [NO(g)] + \Delta G_f^\circ [NO_2(g)]\}$
$\quad - \{\Delta G_f^\circ [C_2H_4(g)] + 2\Delta G_f^\circ [HNO_3(l)]\}$

$\Delta G° = \{1 \text{ mol} \times (-392.5 \text{ kJ/mol}) + 1 \text{ mol} \times (-237.2 \text{ kJ/mol})$
$\quad + 1 \text{ mol} \times (86.69 \text{ kJ/mol}) + 1 \text{ mol} \times (51.84 \text{ kJ/mol})\}$
$\quad - \{1 \text{ mol} \times (68.12 \text{ kJ/mol}) + 2 \text{ mol} \times (-79.9 \text{ kJ/mol})\}$

$\Delta G° = -399.5 \text{ kJ}$

Yes, the reaction is spontaneous.

20.87 We first balance each equation, and then calculate a value of $\Delta G°$. If $\Delta G°$ is a negative number, then the reaction is spontaneous.

(a) $3PbO(s) + 2NH_3(g) \rightarrow 3Pb(s) + N_2(g) + 3H_2O(g)$

$\Delta G° = \{3\Delta G_f^\circ[Pb(s)] + \Delta G_f^\circ[N_2(g)] + 3\Delta G_f^\circ[H_2O(g)]\} - \{3\Delta G_f^\circ[PbO(s)] + 2\Delta G_f^\circ[NH_3(g)]\}$

$\Delta G° = \{3 \text{ mol} \times (0.0 \text{ kJ/mol}) + 1 \text{ mol} \times (0.0 \text{ kJ/mol}) + 3 \text{ mol} \times (-228.6 \text{ kJ/mol})\} - \{3 \text{ mol} \times (-189.3 \text{ kJ/mol}) + 2 \text{ mol} \times (-16.7 \text{ kJ/mol})\}$

$\Delta G° = -84.5$ kJ $\therefore$ the reaction is spontaneous.

(b) $NaOH(s) + HCl(g) \rightarrow NaCl(s) + H_2O(l)$

$\Delta G° = \{\Delta G_f^\circ[NaCl(s)] + \Delta G_f^\circ[H_2O(l)]\} - \{\Delta G_f^\circ[NaOH(s)] + \Delta G_f^\circ[HCl(g)]\}$

$\Delta G° = \{1 \text{ mol} \times (-384.0 \text{ kJ/mol}) + 1 \text{ mol} \times (-237.2 \text{ kJ/mol})\} - \{1 \text{ mol} \times (-382 \text{ kJ/mol}) + 1 \text{ mol} \times (-95.27)\}$

$\Delta G° = -144$ kJ $\therefore$ the reaction is spontaneous.

(c) $Al_2O_3(s) + 2Fe(s) \rightarrow Fe_2O_3(s) + 2Al(s)$

$\Delta G° = \{\Delta G_f^\circ[Fe_2O_3(s)] + 2\Delta G_f^\circ[Al(s)]\} - \{\Delta G_f^\circ[Al_2O_3(s)] + 2\Delta G_f^\circ[Fe(s)]\}$

$\Delta G° = \{1 \text{ mol} \times (-741.0 \text{ kJ/mol}) + 2 \text{ mol} \times (0.0 \text{ kJ/mol})\} - \{1 \text{ mol} \times (-1576.4 \text{ kJ/mol}) + 2 \text{ mol} \times (0.0 \text{ kJ/mol})\}$

$\Delta G° = +835.4$ k $\therefore$ the reaction is not spontaneous.

(d) $2CH_4(g) \rightarrow C_2H_6(g) + H_2(g)$

$\Delta G° = \{\Delta G_f^\circ[C_2H_6(g)] + \Delta G_f^\circ[H_2(g)]\} - \{2\Delta G_f^\circ[CH_4(g)]\}$

$\Delta G° = \{1 \text{ mol} \times (-32.9 \text{ kJ/mol}) + 1 \text{ mol} \times (0.0 \text{ kJ/mol})\} - \{2 \text{ mol} \times (-50.79 \text{ kJ/mol})\}$

$\Delta G° = +68.7$ kJ $\therefore$ the reaction is not spontaneous.

20.88 $\Delta G°_T = \Delta H° - T\Delta S°$, where T = 373 K in all cases, and where the values of $\Delta H°$ and $\Delta S°$ are obtained in the usual manner, i.e.:

$\Delta H° = (\text{sum } \Delta H_f^\circ[\text{products}]) - (\text{sum } \Delta H_f^\circ[\text{reactants}])$

$\Delta S° = (\text{sum } S°[\text{products}]) - (\text{sum } S°[\text{reactants}])$

$\Delta H° = \{\Delta H_f^\circ[C_2H_6(g)]\} - \{\Delta H_f^\circ[H_2(g)] + \Delta H_f^\circ[C_2H_4(g)]\}$

$\Delta H° = \{1 \text{ mol} \times (-84.5 \text{ kJ/mol})\} - \{1 \text{ mol} \times (0.0 \text{ kJ/mol}) + 1 \text{ mol} \times (51.9 \text{ kJ/mol})\}$

$\Delta H° = -136.4$ kJ

$\Delta S° = \{S°[C_2H_6(g)]\} - \{S°[H_2(g)] + S°[C_2H_4(g)]\}$

$\Delta S° = \{1 \text{ mol} \times (229.5 \text{ J mol}^{-1} \text{ K}^{-1})\} - \{1 \text{ mol} \times (130.6 \text{ J mol}^{-1} \text{ K}^{-1}) + 1 \text{ mol} \times (219.8 \text{ J mol}^{-1} \text{ K}^{-1})\}$

$\Delta S° = -120.9 \text{ J/K} = -0.1209 \text{ kJ/K}$

$\Delta G°_{373} = \Delta H° - T\Delta S° = -136.4 \text{ kJ} - (373 \text{ K})(-0.1209 \text{ kJ/K}) = -91.3 \text{ kJ}$

20.89 $\Delta H° = \{3\Delta H_f^\circ[H_2O(g)] + 5\Delta H_f^\circ[SO_2(g)] + 2\Delta H_f^\circ[NO(g)]\} - \{5\Delta H_f^\circ[SO_3(g)] + 2\Delta H_f^\circ[NH_3(g)]\}$

$\Delta H° = \{3 \text{ mol} \times (-242 \text{ kJ/mol}) + 5 \text{ mol} \times (-297 \text{ kJ/mol}) + 2 \text{ mol} \times (90.4 \text{ kJ/mol})\} - \{5 \text{ mol} \times (-396 \text{ kJ/mol}) + 2 \text{ mol} \times (-46.0 \text{ kJ/mol})\}$

$\Delta H° = +42\ kJ$

$\Delta S° = \{3S°[H_2O(g)] + 5S°[SO_2(g)] + 2S°[NO(g)]\}$
$\quad - \{5S°[SO_3(g)] + 2S°[NH_3(g)]\}$
$\Delta S° = \{3\ mol \times (188.7\ J\ mol^{-1}\ K^{-1}) + 5\ mol \times (248.5\ J\ mol^{-1}\ K^{-1})$
$\quad + 2\ mol \times (210.6\ J\ mol^{-1}\ K^{-1})\} - \{5\ mol \times (256.2\ J\ mol^{-1}\ K^{-1})$
$\quad + 2\ mol \times (192.5\ J\ mol^{-1}\ K^{-1})\}$
$\Delta S° = +564\ J/K = 0.564\ kJ/K$

$\Delta G° = \Delta H° - T\Delta S°$
$\Delta G°_{373} = 42\ kJ - (373\ K)(0.562\ kJ/K) = -168\ kJ$

20.90 (a) $\Delta G° = \{2 \times \Delta G_f°[POCl_3(g)]\} - \{2 \times \Delta G_f°[PCl_3(g)] + 2 \times \Delta G_f°[O_2(g)]\}$
$\Delta G° = \{2\ mol \times (-1019\ kJ/mol)\}$
$\quad - \{2\ mol \times (-267.8\ kJ/mol) + 1\ mol \times (0\ kJ/mol)\}$
$\Delta G° = -1502\ kJ = -1.502 \times 10^6\ J$
$-1.502 \times 10^6\ J = -RT\ln K_p = -(8.314\ J/K\ mol)(298\ K) \times \ln K_p$
$\ln K_p = 606 \therefore \log K_p = 263$, and $K_p = 10^{263}$.

(b) $\Delta G° = \{2 \times \Delta G_f°[SO_2(g)] + 1 \times \Delta G_f°[O_2(g)]\} - \{2 \times \Delta G_f°[SO_3(g)]\}$
$\Delta G° = \{2\ mol \times (-300\ kJ/mol) + 1\ mol \times (0\ kJ/mol)\} - \{2\ mol \times (-370\ kJ/mol)\}$
$\Delta G° = 140\ kJ = 1.40 \times 10^5\ J$
$1.40 \times 10^5\ J = -RT\ln K_p = -(8.314\ J/K\ mol)(298\ K) \times \ln K_p$
$\ln K_p = -56.5$ and $K_p = 2.90 \times 10^{-25}$

20.91 (a) $\Delta G° = \{2 \times \Delta G_f°[NO(g)] + 2 \times \Delta G_f°[H_2O(g)]\}$
$\quad - \{1 \times \Delta G_f°[N_2H_4(g)] + 2 \times \Delta G_f°[O_2(g)]\}$
$\Delta G° = \{2\ mol \times (86.69\ kJ/mol) + 2\ mol \times (-228.6\ kJ/mol)\}$
$\quad - \{1\ mol \times (149.4\ kJ/mol) + 2\ mol \times (0.0\ kJ/mol)\}$
$\Delta G° = -433.2\ kJ = -4.332 \times 10^5\ J$
$-4.332 \times 10^5\ J = -RT\ln K_p = -(8.314\ J/K\ mol)(298\ K) \times \ln K_p$
$\ln K_p = 174.9$ and $K_p = 8.625 \times 10^{75}$

(b) $\Delta G° = \{2 \times \Delta G_f°[NO_2(g)] + 8 \times \Delta G_f°[H_2O(g)]\}$
$\quad - \{1 \times \Delta G_f°[N_2H_4(g)] + 6 \times \Delta G_f°[H_2O(g)]\}$
$\Delta G° = [2\ mol \times (51.84\ kJ/mol) + 8\ mol \times (-228.6\ kJ/mol)]$
$\quad - [1\ mol \times (149.4\ kJ/mol) + 6\ mol \times (-228.6\ kJ/mol)]$
$\Delta G° = -502.9\ kJ = -5.029 \times 10^5\ J$
$-5.029 \times 10^5\ J = -RT\ln K_p = -(8.314\ J/K\ mol)(298\ K) \times \ln K_p$
$\ln K_p = 203.0 \therefore K_p = 1.44 \times 10^{88}$

20.92 $\Delta G° = -RT \ln K_p$
$-9.67 \times 10^3\ J = -(8.314\ J/K\ mol)(1273\ K) \times \ln K_p$
$\ln K_p = 0.914 \therefore K_p = 2.49$

$$Q = \frac{[N_2O][O_2]}{[NO_2][NO]} = \frac{(0.015)(0.0350)}{(0.0200)(0.040)} = 0.66$$

Since the value of Q is less than the value of K, the system is not at equilibrium and must shift to the right to reach equilibrium.

20.93 $79.8 \times 10^3\ J = -(8.314\ J/K\ mol)(673\ K) \times \ln K_p$

$\ln K_p = -14.3 \quad \therefore\ K_p = 6.40 \times 10^{-7}$

We start by calculating the reaction quotient, Q. Be sure to determine the pressure of the gases using the ideal gas law.

$$Q = \frac{\left(\frac{(3.8 \times 10^{-3}\ \text{mol})(0.0821\ \frac{\text{L atm}}{\text{mol K}})(673\ \text{K})}{2.50\ \text{L}}\right)}{\left(\frac{(0.040\ \text{mol})(0.0821\ \frac{\text{L atm}}{\text{mol K}})(673\ \text{K})}{2.50\ \text{L}}\right)\left(\frac{(0.022\ \text{mol})(0.0821\ \frac{\text{L atm}}{\text{mol K}})(673\ \text{K})}{2.50\ \text{L}}\right)}$$

$$= 4.32$$

Since the value of Q is larger than the value of K_p, the system must shift to the left in order to reach equilibrium.

20.94 $\Delta G° = -RT \ln K_p$

$-50.79 \times 10^3\ J = -(8.314\ J\ K^{-1}\ mol^{-1})(298\ K) \times \ln K_p$

$\ln K_p = 20.50$

Taking the exponential of both sides of this equation gives: $K_p = 8.000 \times 10^8$

This is a favorable reaction, since the equilibrium lies far to the side favoring products and is worth studying as a method for methane production.

20.95 $T = 37 + 273 = 310\ K$

$\Delta G°_T = -RT \ln K_c$

$-33 \times 10^3\ J = -(8.314\ J\ K^{-1}\ mol^{-1})(310\ K) \times \ln K_c$

$\ln K_c = 13$

$K_c = 3.6 \times 10^5$

20.96 If $\Delta G° = 0$, $K_c = 1$. If we start with pure products, the value of Q will be infinite (there are zero reactants) and, since $Q > K_c$, the equilibrium will shift towards the reactants, i.e., the pure products will decompose to their elements.

20.97 $\Delta G°_{500} = -RT \ln K_p$

$\Delta G°_{500} = -(8.314\ J\ K^{-1}\ mol^{-1})(500\ K) \ln (6.25 \times 10^{-3})$

$\Delta G°_{500} = 2.11 \times 10^4\ J = 21.1\ kJ$

20.98 $\Delta G° = (\text{sum } \Delta G_f^\circ [\text{products}]) - (\text{sum } \Delta G_f^\circ [\text{reactants}])$

$\Delta G° = \{\Delta G_f^\circ [N_2(g)] + 2\Delta G_f^\circ [CO_2(g)]\} - \{2\Delta G_f^\circ [NO(g)] + 2\Delta G_f^\circ [CO(g)]\}$

$\Delta G° = \{1\ mol \times (0.0\ kJ/mol) + 2\ mol \times (-394.4\ kJ/mol)\}$
$\quad - \{2\ mol \times (86.69\ kJ/mol) + 2\ mol \times (-137.3\ kJ/mol)\}$

$\Delta G° = -687.6\ kJ$

$\Delta G° = -RT \ln K_p$

$-687.6 \times 10^3\ J = -(8.314\ J\ K^{-1}\ mol^{-1})(298\ K) \ln K_p$

$\ln K_p = 278$ and $\log K_p = 121 \quad \therefore K_p = 10^{121}$

20.99 First, we determine $\Delta H°$ and $\Delta S°$ by using the normal approach:

$\Delta H° = (\text{sum } \Delta H_f^\circ [\text{products}]) - (\text{sum } \Delta H_f^\circ [\text{reactants}])$

$\Delta H° = \{\Delta H_f^\circ [N_2(g)] + 2 \times \Delta H_f^\circ [CO_2(g)]\} - \{2 \times \Delta H_f^\circ [NO(g)] + 2 \times \Delta H_f^\circ [CO(g)]\}$

$\Delta H° = \{1 \text{ mol} \times (0.0 \text{ kJ/mol}) + 2 \text{ mol} \times (-394 \text{ kJ/mol})\}$
$\quad - \{2 \text{ mol} \times (90.4 \text{ kJ/mol}) + 2 \text{ mol} \times (-110 \text{ kJ/mol})\}$
$\Delta H° = -749$ kJ for the formation of one mole of N_2.

$\Delta S° = (\text{sum } S°[\text{products}]) - (\text{sum } S°[\text{reactants}])$
$\Delta S° = \{S°[N_2(g)] + 2 \times S°[CO_2(g)]\} - \{2 \times S°[NO(g)] + 2 \times S°[CO(g)]\}$
$\Delta S° = \{1 \text{ mol} \times (191.5 \text{ J K}^{-1} \text{ mol}^{-1}) + 2 \text{ mol} \times (213.6 \text{ J K}^{-1} \text{ mol}^{-1})\}$
$\quad - \{2 \text{ mol} \times (210.6 \text{ J K}^{-1} \text{ mol}^{-1}) + 2 \text{ mol} \times (197.9 \text{ J K}^{-1} \text{ mol}^{-1})\}$
$\Delta S° = -198.3$ J/K for the formation of one mole of N_2.

$\Delta G°_T = \Delta H° - T\Delta S°$
$\Delta G°_{773} = -749 \text{ kJ} - (773 \text{ K})(-0.1983 \text{ kJ/K}) = -596 \text{ kJ/mol}$

$\Delta G°_T = -RT \ln K_p$
$-596 \times 10^3 \text{ J/mol} = -(8.314 \text{ J K}^{-1} \text{ mol}^{-1})(773 \text{ K}) \ln K_p$
$\ln K_p = 92.7$

Taking the exponential of both sides of this equation gives:
$K_p = 1.82 \times 10^{40}$

20.100 This requires the breaking of three N–H single bonds:

$$NH_3 \rightarrow N + 3H$$

The enthalpy of atomization of NH_3 is thus three times the average N–H single bond energy:
$3 \times 388 \text{ kJ/mol} = 1.16 \times 10^3 \text{ kJ/mol}$

20.101 The energy released during the formation of 1 mol of acetone is equal to the sum of all of the bond energies in the molecule:
for the six C—H bonds: 412 kJ/mol × 6 mol C—H bonds
for the two C—C bonds: 348 kJ/mol × 2 mol C—C bonds
for the one C=O bond: 743 kJ/mol × 1 mol C=O bonds

Adding the above contributions we get 3.91×10^3 kJ released per mole of acetone formed.

20.102 The heat of formation for ethanol vapor describes the following change:
$2C(s) + 3H_2(g) + 1/2O_2(g) \rightarrow C_2H_5OH(g)$

This can be arrived at by adding the following thermochemical equations, using data from Table 20.3:

$3H_2(g) \rightarrow 6H(g)$	$\Delta H_1° = (6)217.89 \text{ kJ} = 1307.34 \text{ kJ}$
$2C(s) \rightarrow 2C(g)$	$\Delta H_2° = (2)716.67 \text{ kJ} = 1{,}433.34 \text{ kJ}$
$1/2O_2(g) \rightarrow O(g)$	$\Delta H_3° = (1)249.17 \text{ kJ} = 249.17 \text{ kJ}$
$6H(g) + 2C(g) + O(g) \rightarrow C_2H_5OH(g)$	$\Delta H_4° = x$
$3H_2(g) + 2C(s) + 1/2O_2(g) \rightarrow C_2H_5OH(g)$	$\Delta H_f° = (2989.85 + x) \text{ kJ}$

Since $\Delta H_f°$ is given as –235.3 kJ…
$-235.3 = 2989.85 + x$
$x = -3225.2 \text{ kJ}$

$\Delta H°_{atom}$ is the reverse reaction, so the sign will change:

$\Delta H°_{atom} = 3225.2 \text{ kJ}$

The sum of all the bond energies in the molecule should be equal to the atomization energy:

ΔH°_{atom} = 1(C-C bond) + 5(C-H bonds) + 1(O-H bond) + 1(C-O bond)

We use values from Table 20.4:

3225.2 kJ = 1(348 kJ) + 5(412 kJ) + 1(463 kJ) + 1(C-O bond)

C-O bond energy = 354 kJ/mol

20.103 $\Delta H_f^\circ[C_2H_4(g)]$ refers to the enthalpy change under standard conditions for the following reaction:

$2C_{graphite} + 2H_2(g) \rightarrow C_2H_4(g)$, $\quad \Delta H_f^\circ[C_2H_4(g)] = 52.284$ kJ/mol

We can arrive at this net reaction in an equivalent way, namely, by vaporizing all of the necessary elements to give gaseous atoms, and then allowing the gaseous atoms to form all of the appropriate bonds. The overall enthalpy of formation by this route is numerically equal to that for the above reaction, and, conveniently, the enthalpy changes for each step are available in either Table 20.3 or Table 20.4:

ΔH_f° = sum(ΔH_f°[gaseous atoms]) – sum(average bond energies in the molecule)
$\Delta H_f^\circ[C_2H_4(g)] = 52.284$ kJ/mol = $[2 \times 716.7 + 4 \times 218.0] - [4 \times 412 + \text{C=C}]$ from which we can calculate the C=C bond energy: 605 kJ/mol.

20.104 There are two C=S double bonds to be considered:
ΔH_f° = sum(ΔH_f°[gaseous atoms]) – sum(average bond energies in the molecule)
$\Delta H_f^\circ[CS_2(g)] = 115.3$ kJ/mol = $[716.67 + 2 \times 276.98] - [2 \times \text{C=S}]$
The C=S double bond energy is therefore given by the equation:
C=S = $-(115.3 - 716.67 - 2 \times 276.98) \div 2 = 577.7$ kJ/mol

20.105 ΔH_f° = sum(ΔH_f°[gaseous atoms]) – sum(average bond energies in the molecule)
$\Delta H_f^\circ[H_2S(g)] = -20.15$ kJ/mol = $[277.0 + 2 \times 218.0] - [2 \times \text{H–S}]$
H–S = $(20.15 + 277.0 + 2 \times 218.0) \div 2 = 366.6$ kJ/mol

20.106 There are six S—F bonds in the molecule:

ΔH_f° = sum(ΔH_f°[gaseous atoms]) – sum(average bond energies in the molecule)
$\Delta H_f^\circ[SF_6(g)] = -1096$ kJ/mol = $[277.0 + 6 \times 79.14] - [6 \times \text{S—F}]$
S—F = $(1096 + 277.0 + 6 \times 79.14) \div 6 = 308.0$ kJ/mol

20.107 ΔH_f° = sum(ΔH_f°[gaseous atoms]) – sum(average bond energies in the molecule)
$\Delta H_f^\circ[SF_4(g)] = [277.0 + 4 \times 79.14] - [4 \times 308.0]$
$= -638.4$ kJ/mol
The % difference is $[(718.4 - 638.4) \div 718.4] \times 100\% = 11\%$

20.108 See the method of review problems 20.100 through 20.107.
ΔH_f° = sum(ΔH_f°[gaseous atoms]) – sum(average bond energies in the molecule)

$\Delta H_f^\circ[C_2H_2(g)] = [2 \times 716.7 + 2 \times 218.0] - [2 \times 412 + 960]$
$= 85$ kJ/mol

20.109 ΔH_f° = sum(ΔH_f°[gaseous atoms]) – sum(average bond energies in the molecule)

$\Delta H_f^\circ[CCl_4(g)] = [716.67 + 4 \times 121.47] - [4 \times 338] = -149$ kJ/mol

20.110 The heat of formation of CF_4 should be more exothermic than that of CCl_4 because more energy is released on formation of a C—F bond than on formation of a C—Cl bond. Also, less energy is needed to form gaseous F atoms than to form gaseous Cl atoms.

20.111 The computed value for ΔH_f° for benzene is likely to be larger than the experimentally measured value. The reason for the difference is that the computation neglects the stabilization provided by the high degree of conjugation/resonance.

Additional Exercises

20.112 First, we calculate a value for ΔH and ΔS, using the data of Appendix E. Next, we calculate a value for ΔG_T°, using the equation ΔG = ΔH – TΔS. Last, we calculate a value for Kp, using the equation ΔG = –RT lnKp. (Recall that ln(x) = 2.303 × log(x))

(a)

$\Delta H^\circ = \{2 \times \Delta H_f^\circ[POCl_3(g)]\} - \{2 \times \Delta H_f^\circ[PCl_3(g)] + 1 \times \Delta H_f^\circ[O_2(g)]\}$

$\Delta H^\circ = \{2 \text{ mol} \times (-1109.7 \text{ kJ/mol})\}$
$- \{2 \text{ mol} \times (-287.0 \text{ kJ/mol}) + 1 \text{ mol} \times (0 \text{ kJ/mol})\}$

$\Delta H^\circ = -1645.4 \text{ kJ} = -1.6454 \times 10^6 \text{ J}$

$\Delta S^\circ = \{2 \times S^\circ[POCl_3(g)]\} - \{2 \times S^\circ[PCl_3(g)] + 1 \times S^\circ[O_2(g)]\}$

$\Delta S^\circ = \{2 \text{ mol} \times (646.5 \text{ J mol}^{-1} \text{ K}^{-1})\}$
$- \{2 \text{ mol} \times (311.8 \text{ J mol}^{-1} \text{ K}^{-1}) + 1 \text{ mol} \times (205.0 \text{ J mol}^{-1} \text{ K}^{-1})\}$

$\Delta S^\circ = 464.4 \text{ J K}^{-1}$

$\Delta G_{573}^\circ = -1.6454 \times 10^6 \text{ J} - (573 \text{ K})(464.4 \text{ J K}^{-1}) = -1.912 \times 10^6 \text{ J}$

$-1.912 \times 10^6 \text{ J} = -RT\ln K_p = -(8.314 \text{ J/K mol})(573 \text{ K}) \times \ln K_p$

$\ln K_p = 401.3$ and $\log K_p = 174.2 \ \therefore K_p = 10^{174}$

(b)

$\Delta H^\circ = \{2 \times \Delta H_f^\circ[SO_2(g)] + 1 \times \Delta H_f^\circ[O_2(g)]\} - \{2 \times \Delta H_f^\circ[SO_3(g)]\}$

$\Delta H^\circ = \{2 \text{ mol} \times (-297 \text{ kJ/mol}) + 1 \text{ mol} \times (0 \text{ kJ/mol})\} - \{2 \text{ mol} \times (-396 \text{ kJ/mol})\}$

$\Delta H^\circ = 198 \text{ kJ} = 1.98 \times 10^5 \text{ J}$

$\Delta S^\circ = \{2 \times S^\circ[SO_2(g)] + 1 \times S^\circ[O_2(g)]\} - \{2 \times S^\circ[SO_3(g)]\}$

$\Delta S^\circ = \{2 \text{ mol} \times (248.5 \text{ J mol}^{-1} \text{ K}^{-1}) + 1 \text{ mol} \times (205.0 \text{ J mol}^{-1} \text{ K}^{-1})\}$
$- \{2 \text{ mol} \times (256.2 \text{ J mol}^{-1} \text{ K}^{-1})\}$

$\Delta S^\circ = 190 \text{ J K}^{-1}$

$\Delta G_{573}^\circ = 1.98 \times 10^5 \text{ J} - (573 \text{ K})(190 \text{ J K}^{-1}) = 8.94 \times 10^4 \text{ J}$

$8.94 \times 10^4 \text{ J} = -RT\ln K_p = -(8.314 \text{ J/K mol})(573 \text{ K}) \times \ln K_p$

$\ln K_p = -18.8$ and $K_p = 6.8 \times 10^{-9}$

20.113 P(atm) × ΔV(L) = 1 L atm of work

Substituting the given information: $P \times \Delta V = 101{,}325 \text{ Pa} \times 1 \text{ dm}^3$

Now it is true that 1 dm = 0.1 m, so: $1 \text{ dm}^3 = (0.1 \text{ m})^3 = 1 \times 10^{-3} \text{ m}^3$

Also, $1 \text{ Pa} = 1 \text{ N/m}^2$

Thus we have: $P(\text{Pa}) \times \Delta V(\text{dm}^3) = (101{,}325 \text{ N/m}^2)(1 \times 10^{-3} \text{ m}^3) = 101.325 \text{ N·m}$

$101.325 \text{ N·m} \times 1 \text{ J}/1 \text{ N·m} = 101.325 \text{ J}$

20.114 Under reversible conditions, the expansion of an ideal gas produces the maximum amount of work that can be obtained from thc change. Such a situation cannot be accomplished by any real process. In an ideal gas, the gas particles are non-interacting. In a real gas, the interaction between the individual particles must be overcome so that the gas may expand. It requires an input of energy to overcome these attractive forces.

20.115 The expansion of a real gas is not reversible, and not all of the free energy change for the process is converted to work as the gas expands. Some of the energy must be used to overcome the intermolecular forces of attraction that exist between molecules. In an ideal gas, the particles are assumed to be non-interacting and this energy loss is not a concern.

20.116 No work ($P \times \Delta V$) is accomplished in a bomb calorimeter because there is no change in volume. At constant pressure, $w = -\Delta n_{gas}RT = -(18 \text{ mol} - 15 \text{ mol})(8.314 \text{ J mol}^{-1} \text{ K}^{-1})(298 \text{ K}) = -7.43 \times 10^3$ J = −7.43 kJ

20.117 We can calculate the work of the expanding gas ($P\Delta V$) if we can calculate the change in volume ΔV. Since the initial volume is given (5.00 L), we need only to calculate the final volume. For this, it is first necessary to determine the value of n, the number of moles of gas.

$$n = \frac{PV}{RT} = \frac{(4.00 \text{ atm})(5.00 \text{ L})}{\left(0.0821 \frac{\text{L atm}}{\text{mol K}}\right)(298 \text{ K})} = 0.817 \text{ mol}$$

$$V_2 = \frac{nRT}{P_2} = \frac{(0.817 \text{ mol})\left(0.0821 \frac{\text{L atm}}{\text{mol K}}\right)(298 \text{ K})}{1 \text{ atm}} = 20.0 \text{ L}$$

$$\Delta V = (20.0 \text{ L} - 5.00 \text{ L}) = 15.0 \text{ L}$$

The work of gas expansion against a constant pressure of 1 atm is then given by the quantity −$P\Delta V$: $w = -(1 \text{ atm})(15.0 \text{ L}) = -15.0$ L atm

20.118 As in exercise 20.117, we solve for the volume after the first expansion to a pressure of 2 atm:

$$V_2 = \frac{nRT}{P_2} = \frac{(0.817 \text{ mol})\left(0.0821 \frac{\text{L atm}}{\text{mol K}}\right)(298 \text{ K})}{2 \text{ atm}} = 10.0 \text{ L}$$

The work after the first stage of expansion is therefore:

$w_1 = -(2 \text{ atm})(10.0 \text{ L} - 5.00 \text{ L}) = -10.0$ L atm.

The work performed during the second stage of expansion is:

$w_2 = -(1 \text{ atm})(20.0 \text{ L} - 10.0 \text{ L}) = -10.0$ L atm.

The sum for the whole stepwise process is $w_1 + w_2 = -20.0$ L atm. Thus, we see that more work is performed in this multi–step expansion than in a single expansion.

20.119 (a) negative
(b) positive
(c) negative
(d) negative
(e) negative
(f) negative

20.120 First, review the information provided. Since the salt dissolved, the process is spontaneous, and the sign for ΔG is negative. The temperature went down indicating that this is an endothermic process and ΔH must be positive. Dissolving the solid salt increases the disorder of the system which indicates ΔS is positive. The general equation from which we must work is $\Delta G = \Delta H - T\Delta S$. Using this equation, the magnitude of $T\Delta S$ must be larger than the magnitude of ΔH in order to obtain a negative value for ΔG. However, the magnitude of ΔH is almost certainly larger than that of ΔS as can be appreciated from the fact that ΔS° values are given in J and those of $\Delta H°_f$ are given in kJ.

20.121 The value of Δn for this reaction is zero. Also, two diatomic molecules are formed in place of two other diatomic molecules. Both of these factors cause ΔS to be near zero.

20.122 There can be only one temperature for which ΔG has precisely the value zero, i.e., an equilibrium condition exists.

20.123 (a) $H_2C_2O_4(g) + 1/2O_2(g) \rightarrow H_2O(l) + 2CO_2(g)$ $\Delta H°_{comb} = -246.05$ kJ

(b) $H_2(g) + 2C(s) + 2O_2(g) \rightarrow H_2C_2O_4(s)$ $\Delta H_f^\circ = ?$ kJ mol^{-1}

(c) $\Delta H°_{comb} = \{\Delta H_f^\circ(H_2O(l)) + 2\Delta H_f^\circ(CO_2(g))\} - \{\Delta H_f^\circ(H_2C_2O_4(s)) + 1/2\Delta H_f^\circ(O_2(g))\}$

Rearranging;

$\Delta H_f^\circ(H_2C_2O_4(s)) = \{\Delta H_f^\circ(H_2O(l)) + 2\Delta H_f^\circ(CO_2(g))\}$

$-\Delta H°_{comb}$

$= -285.8 \text{ kJ} + 2(-393.5 \text{ kJ}) - (246.05 \text{ kJ}) = -826.8 \text{ kJ}$

(d) $\Delta S_f^\circ = S°(H_2C_2O_4(s)) - \{S_f°(H_2(g)) + 2S°(C(g)) + 2S°(O_2(g))\}$

$= 120.1 \text{ J/K} - \{130.6\text{J/K} + 2(5.69 \text{ J/K}) + 2(205.0 \text{ J/K})\}$

$= -431.9$J/mol K

$\Delta S°_{comb} = \{S°(H_2O(l)) + 2S°(CO_2(g))\} - \{S°(H_2C_2O_4(g)) + 1/2S°(O_2(g))\}$

$= \{69.96 \text{ J/K} + 2(213.6 \text{ J/K})\} - \{120.1 \text{ J/K} + 1/2(205.0 \text{ J/K}) = 274.6$ J/mol K

(e) $\Delta G_f^\circ = \Delta H_f^\circ - T\Delta S_f^\circ$

$= -826.8 \text{ kJ} - (298 \text{ K})(-0.4319 \text{ kJ/K}) = -698.1 \text{ kJ}$

$\Delta G_{comb}° = \Delta H_{comb}° - T\Delta S_{comb}°$

$= -246.05 \text{ kJ} - (298\text{K})(0.2746 \text{ kJ/K}) = -327.9 \text{ kJ}$

20.124 glucose + phosphate → glucose–6–phosphate + H_2O ΔG = 13.13 kJ
ATP + H_2O → ADP + phosphate ΔG = –32.22 kJ

Add these two reactions together to get:

glucose + ATP → glucose–6–phosphate + ADP ΔG = –19.09 kJ

20.125 Since all practical reactions cannot be done under perfectly reversible conditions, not all of the energy of a reaction can be released as work. Some of it shows up as heat, which must be dissipated by the cooling system.

20.126 We need to calculate the amount of energy produced when one gallon of each of these fuels is burned;

$$\text{\# mol ethanol} = (3.78 \times 10^3 \text{ mL ethanol})\left(\frac{0.7893 \text{ g ethanol}}{1 \text{ mL ethanol}}\right)\left(\frac{1 \text{ mole ethanol}}{46.07 \text{ g ethanol}}\right)$$

$$= 64.8 \text{ moles ethanol}$$

$$\text{\# kJ} = (64.8 \text{ moles ethanol})\left(\frac{-1299.8 \text{ kJ}}{1 \text{ mole ethanol}}\right) = 8.42 \times 10^4 \text{ kJ}$$

$$\text{\# mol octane} = (3.78 \times 10^3 \text{ mL octane})\left(\frac{0.7025 \text{ g octane}}{1 \text{ mL octane}}\right)\left(\frac{1 \text{ mole octane}}{114.23 \text{ g octane}}\right)$$

$$= 23.2 \text{ moles octane}$$

$$\text{\# kJ} = (23.2 \text{ moles octane})\left(\frac{-5307 \text{ kJ}}{1 \text{ mole octane}}\right) = 1.23 \times 10^5 \text{ kJ}$$

In spite of the large number of moles of ethanol in one gallon of liquid, the energy produced from the combustion of a gallon of octane is greater than the amount produced when one gallon of ethanol is burned.

20.127 The reaction is $N_2(g) \rightarrow 2N(g)$, $\Delta H° = 2\Delta H_f^\circ[N(g)] - \Delta H_f^\circ[N_2(g)]$. Since the enthalpy of formation of molecular nitrogen is defined as zero, the enthalpy change for this reaction and, consequently, the bond energy of a nitrogen molecule, is simply two times the enthalpy of formation of atomic nitrogen.

$$\text{Bond Energy} = (2 \text{ mol N atoms})\left(\frac{472.68 \text{ kJ}}{\text{mol N atoms}}\right) = 945.36 \text{ kJ}$$

A similar argument holds true for oxygen:

$$\text{Bond Energy} = (2 \text{ mol O atoms})\left(\frac{249.17 \text{ kJ}}{\text{mol N atoms}}\right) = 498.34 \text{ kJ}$$

20.128

$$\begin{array}{ccccc} C(s) & + & 2Cl_2(g) & \rightarrow & CCl_4(l) \\ \downarrow(1) & & \downarrow(2) & & \uparrow(4) \\ C(g) & + & 4Cl(g) & \underset{(3)}{\rightarrow} & CCl_4(g) \end{array}$$

Step 1: $\Delta H = 717$ kJ
Step 2: $\Delta H = 4(121.5 \text{ kJ}) = 486$ kJ
Step 3: $\Delta H_f°(CCl_4(g)) = 4(BE(C{-}Cl)) = 4(338 \text{ kJ}) = -1352$ kJ
Step 4: $\Delta H_{cond} = -29.9$ kJ

ΔH = the sum of the four steps = -179 kJ

Practice Exercises

21.1 anode: $Mg(s) \rightarrow Mg^{2+}(aq) + 2e^-$
cathode: $Fe^{2+}(aq) + 2e^- \rightarrow Fe(s)$
cell notation: $Mg(s)|Mg^{2+}(aq)||Fe^{2+}(aq)|Fe(s)$

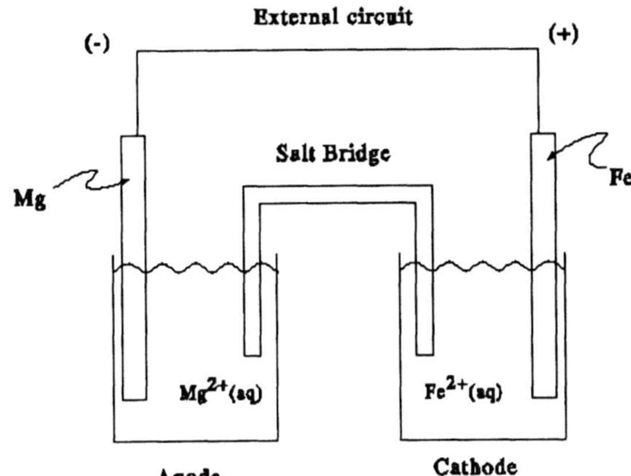

21.2 anode: $Al(s) \rightarrow Al^{3+}(aq) + 3e^-$
cathode: $Pb^{2+}(aq) + 2e^- \rightarrow Pb(s)$
overall: $3Pb^{2+}(aq) + 2Al(s) \rightarrow 2Al^{3+}(aq) + 3Pb(s)$

21.3 $E^\circ_{cell} = E^\circ_{substance\ reduced} - E^\circ_{substance\ oxidized}$

$1.93\ V = (-0.44\ V) - E^\circ_{Mg^{2+}}$
$E^\circ_{Mg2+} = -0.44 - 1.93 = -2.37\ V$

This agrees exactly with Table 21.1.

21.4 The half–reaction with the more positive value of E° (listed higher in Table 21.1) will occur as a reduction. The half–reaction having the less positive (more negative) value of E° (listed lower in Table 21.1) will be reversed and occur as an oxidation.

$Br_2(aq) + 2e^- \rightarrow 2Br^-(aq)$ reduction
$H_2SO_3(aq) + H_2O \rightarrow SO_4^{2-}(aq) + 4H^+(aq) + 2e^-$ oxidation
$Br_2(aq) + H_2SO_3(aq) + H_2O \rightarrow 2Br^-(aq) + SO_4^{2-}(aq) + 4H^+(aq)$

21.5 Either tin(II) or iron(III) will be reduced, depending on which way the reaction proceeds. Iron(III) is listed higher than tin(II) in Table 21.1 (it has a greater reduction potential), so we would expect that the reaction would not be spontaneous in the direction shown.

21.6 The half–reaction having the more positive value for E° will occur as a reduction. The other half–reaction should be reversed, so as to appear as an oxidation.

$NiO_2(s) + 2H_2O + 2e^- \rightarrow Ni(OH)_2(s) + 2OH^-(aq)$ reduction
$Fe(s) + 2OH^-(aq) \rightarrow 2e^- + Fe(OH)_2(s)$ oxidation
$NiO_2(s) + Fe(s) + 2H_2O \rightarrow Ni(OH)_2(s) + Fe(OH)_2(s)$ net reaction

$E^\circ_{cell} = E^\circ_{substance\ reduced} - E^\circ_{substance\ oxidized}$
$E^\circ_{cell} = E^\circ_{NiO_2} - E^\circ_{Fe}$
$E^\circ_{cell} = 0.49 - (-0.88) = 1.37\ V$

21.7 The half–reaction having the more positive value for E° will occur as a reduction. The other half–reaction should be reversed, so as to appear as an oxidation.

Reduction: $3 \times [MnO_4^-(aq) + 8H^+(aq) + 5e^- \rightarrow Mn^{2+}(aq) + 4H_2O]$
Oxidation: $5 \times [Cr(s) \rightarrow Cr^{3+}(aq) + 3e^-]$
Net reaction: $3MnO_4^-(aq) + 24H^+(aq) + 5Cr(s) \rightarrow 5Cr^{3+}(aq) + 3Mn^{2+}(aq) + 12H_2O$

$E°_{cell} = E°_{substance\ reduced} - E°_{substance\ oxidized}$
$E°_{cell} = E°_{MnO_4^-} - E°_{Cr}$
$E°_{cell} = 1.51\ V - (-0.74\ V) = 2.25\ V$

21.8 A reaction will occur spontaneously in the forward direction if the value of E° is positive. We therefore evaluate E° for each reaction using:

$$E°_{cell} = E°_{substance\ reduced} - E°_{substance\ oxidized}$$

a) $Br_2(aq) + 2e^- \rightarrow 2Br^-(aq)$ reduction
$Cl_2(aq) + 2H_2O \rightarrow 2HOCl(aq) + 2H^+(aq) + 2e^-$ oxidation

$E°_{cell} = E°_{Br_2} - E°_{Cl_2}$
$E°_{cell} = 1.07\ V - (1.36\ V) = -0.29\ V$, ∴ nonspontaneous

b) $2Cr^{3+}(aq) + 6e^- \rightarrow 2Cr(s)$ reduction
$3Zn(s) \rightarrow 3Zn^{2+}(aq) + 6e^-$ oxidation
$E°_{cell} = E°_{Cr^{3+}} - E°_{Zn}$
$E°_{cell} = -0.74\ V - (-0.76\ V) = +0.02\ V$, ∴ spontaneous

21.9 It was stated in Practice Exercise 3 that the reaction in Practice Exercise 1 had a standard cell potential of 1.93 V, or 1.93 J/C. Since 2 mole of e^- are involved, i.e., n = 2, we have:

$\Delta G° = -nFE°_{cell} = -(2)(96{,}500\ C)(1.93\ V) = -3.72 \times 10^5\ J = -3.72 \times 10^2\ kJ$

The cell potential in Practice Exercise 2 may be found by the difference in the standard reduction potentials of the substances involved:

$E\infty_{cell} = -0.13\ V - (-1.66\ V) = +1.53\ V$

6 moles of e^- are involved, so n = 2. Now we have:

$\Delta G° = -nFE°_{cell} = -(6)(96{,}500\ C)(1.53\ V) = -8.86 \times 10^5\ J = -8.86 \times 10^2\ kJ$

21.10 Using Equation 21.7,

$$E°_{cell} = \frac{RT}{nF} \ln K_c$$

$$-0.46\ V = \frac{(8.314\ J\ mol^{-1}K^{-1})(298\ K)}{2(96{,}500\ C\ mol^{-1})} \ln K_c$$

$$\ln K_c = -35.83$$

Taking the antilog (e^x) of both sides of the above equation gives
$K_c = 2.7 \times 10^{-16}$.

This very small value for the equilibrium constant means that the products of the reaction are not formed spontaneously. The equilibrium lies far to the left, favoring reactants, and we do not expect much product to form.

21.11 $Zn(s) \rightarrow Zn^{2+}(aq) + 2e^-$ oxidation
$Cu^{2+}(aq) + 2e^- \rightarrow 2Cu(s)$ reduction

$E°_{cell} = E°_{Cu^{2+}} - E°_{Zn}$
$E°_{cell} = +0.34\ V - (-0.76\ V) = +1.10\ V$

The Nernst equation for this cell is:

$$E_{cell} = E°_{cell} - \frac{RT}{nF} \ln \frac{[Zn^{2+}]}{[Cu^{2+}]}$$

$$E_{cell} = 1.10\ V - \frac{(8.314\ J\ mol^{-1}K^{-1})(298\ K)}{2(96{,}500\ C\ mol^{-1})} \ln \frac{[1.0]}{[0.010]}$$

$$= 1.10\ V - 0.01284(4.605)$$
$$= 1.04\ V$$

21.12 $Cu(s) \rightarrow Cu^{2+}(aq) + 2e^-$ oxidation
$Ag^+(aq) + e^- \rightarrow Ag(s)$ reduction

$E°_{cell} = E°_{Ag^+} - E°_{Cu}$
$E°_{cell} = +0.80\ V - (+0.34\ V) = +0.46\ V$

$$E_{cell} = E°_{cell} - \frac{RT}{nF} \ln \frac{[Cu^{2+}]}{[Ag^+]^2}$$

$$\ln \frac{[Cu^{2+}]}{[Ag^+]^2} = \frac{(E°_{cell} - E_{cell})}{\left(\frac{RT}{nF}\right)}$$

$$= \frac{(0.46\ V - 0.57\ V)}{0.01284}$$

$$= -8.5670$$

$$\frac{[Cu^{2+}]}{[Ag^+]^2} = e^{-8.5670} = 1.9 \times 10^{-4}$$

Since the $[Ag^+] = 1.00$ M, $[Cu^{2+}] = 1.9 \times 10^{-4}$ M

Substituting the second value into the same expression gives
$[Cu^{2+}] = 6.9 \times 10^{-13}$ M.

21.13 We are told that, in this galvanic cell, the chromium electrode is the anode, meaning that oxidation occurs at the chromium electrode.

Now in general, we have the equation:
$E°_{cell} = E°_{reduction} - E°_{oxidation}$
which becomes, in particular for this case:
$E°_{cell} = E°_{Ni^{2+}} - E°_{Cr}$

The net cell reaction is given by the sum of the reduction and the oxidation half–reactions, multiplied in each case so as to eliminate electrons from the result:

$3 \times [Ni^{2+}(aq) + 2e^- \rightarrow Ni(s)]$	reduction
$2 \times [Cr(s) \rightarrow Cr^{3+}(aq) + 3e^-]$	oxidation
$3Ni^{2+}(aq) + 2Cr(s) \rightarrow 2Cr^{3+}(aq) + 3Ni(s)$	net reaction

In this reaction, n = 6, and the Nernst equation becomes:

$$E_{cell} = E^\circ_{cell} - \frac{RT}{nF} \ln \frac{[Cr^{3+}]^2}{[Ni^{2+}]^3}$$

$$\ln \frac{[Cr^{3+}]^2}{[Ni^{2+}]^3} = \frac{(E^\circ_{cell} - E_{cell})}{\left(\frac{RT}{nF}\right)}$$

$$= \frac{(0.487 \text{ V} - 0.552 \text{ V})}{0.004279}$$

$$= -15.190$$

$$\frac{[Cr^{3+}]^2}{[Ni^{2+}]^3} = e^{-15.190} = 2.5 \times 10^{-7}$$

Substituting $[Ni^{2+}] = 1.20$ M, we solve for $[Cr^{3+}]$ and get: $[Cr^{3+}] = 6.6 \times 10^{-4}$ M.

21.14 The cathode is always where reduction occurs. We must consider which species could be candidates for reduction, then choose the species with the highest reduction potential from Table 21.1.

$Cd^{2+}(aq) + 2e^- \rightarrow 2Cd(s)$	$E° = -0.40$ V
$Sn^{2+}(aq) + 2e^- \rightarrow 2Sn(s)$	$E° = -0.14$ V
$2H_2O + 2e^- \rightarrow H_2(g) + 2OH^-(aq)$	$E° = -0.83$ V

Tin(II) has the highest reduction potential, so we would expect it to be reduced in this environment. We expect Sn(s) at the cathode.

21.15 The number of Coulombs is: $4.00 \text{ A} \times 200 \text{ s} = 800 \text{ C}$
The number of moles is:

$$\# \text{ mol } OH^- = 800 \text{ C} \times \frac{1 \text{ F}}{96{,}500 \text{ C}} \times \frac{1 \text{ mol } OH^-}{1 \text{ F}} = 8.29 \times 10^{-3} \text{ mol } OH^-$$

21.16 The number of moles of Au to be deposited is: $3.00 \text{ g Au} \div 197 \text{ g/mol} = 0.0152$ mol Au. The number of Coulombs (A × s) is:

$$\# \text{ C} = 0.0152 \text{ mol Au} \times \frac{3 \text{ F}}{1 \text{ mol Au}} \times \frac{96{,}500 \text{ C}}{1 \text{ F}} = 4.40 \times 10^3 \text{ C}$$

The number of minutes is:

$$\# \text{ min} = \frac{4.40 \times 10^3 \text{ A}\cdot\text{s}}{10.0 \text{ A}} \times \frac{1 \text{ min}}{60 \text{ s}} = 7.33 \text{ min}$$

21.17 As in Practice Exercise 16 above, the number of Coulombs is 4.40×10^3 C. This corresponds to a current of:

$$\# \text{ A} = \frac{4.40 \times 10^3 \text{ A}\cdot\text{s}}{20.0 \text{ min}} \times \frac{1 \text{ min}}{60 \text{ s}} = 3.67 \text{ A}$$

21.18 The number of Coulombs is:
$0.100 \text{ A} \times 1.25 \text{ hr}(3600 \text{ s/hr}) = 450 \text{ C}$

The number of moles of copper ions produced is:

$$\# \text{ mol } Cu^{2+} = 450 \text{ C} \times \frac{1 \text{ mol } e^-}{96{,}500 \text{ C}} \times \frac{1 \text{ mol } Cu^{2+}}{2 \text{ mol } e^-} = 0.233 \text{ mol } Cu^{2+}$$

Therefore, the increase in concentration is:
$M = \text{mol/L} = (0.233 \text{ mol } Cu^{2+})/0.125 \text{ L} = +\ 0.0187 \text{ M}$

Thinking It Through

T21.1 The overall reaction should first be separated into half–reactions. Values for the half–cell potentials are needed from Table 21.1. If the overall cell potential, calculated using Equation 21.2, is positive, the reaction will occur spontaneously.

T21.2 It is necessary to determine which half–reaction constitutes the spontaneous oxidation, and which one the reduction. In other words, standard reduction half–cell potentials are used to see which direction for this reaction gives a positive overall cell potential. This requires use of Equation 21.2. The half–cell at which reduction takes place spontaneously is assigned the positive charge.

T21.3 First, we separate the given reaction into two half–reactions:

$$2Br^- \rightarrow Br_2 + 2e^-$$
$$2e^- + I_2 \rightarrow 2I^-$$

Then Equation 21.2 is used to determine the overall cell potential, using half–cell potentials from Table 21.1. Finally, Equation 21.7 is used to calculate a value for K_c, remembering that n has the value 2 in this case.

T21.4 First, we need to convert time in minutes to time in seconds. Then 4.00 amps is multiplied by time (in seconds) to find coulombs. Using the Faraday (96,500 coul/mol), we can determine moles of electrons used. From the balanced half–reaction, we next construct the conversion factor: (1 mol Pb)/(2 mol e^-). This can be used to determine the moles of Pb that are electroplated. The atomic mass of lead is used last to convert to grams of Pb plated.

T21.5 The volume to be plated is given by the area times thickness. The mass of silver to be plated is given by volume times density. Next, we convert from mass of Ag to moles of Ag, using its atomic mass. From the half–reaction, we see that moles of electrons equal moles of silver ion. The moles of electrons required are multiplied by the Faraday to determine coulombs. Finally, the coulombs are divided by 2.00 amps (2.00 coul/s) to determine time in seconds.

T21.6 We need the equation for oxidation of water at the anode: $2H_2O \rightarrow O_2 + 4H^+ + 4e^-$
The number of moles of electrons consumed by the electrolysis can be converted to the number of moles of H^+ that are produced, because 4 moles of H^+ are produced for every 4 mol of e^- that are used. The hydrogen ion concentration is given by dividing the number of moles produced by the volume, 250 mL.

T21.7 First, calculate moles of Ag using mass (1.45 g) and the atomic mass of silver. Next, calculate coulombs from the conversion factors:

$$\frac{96{,}500 \text{ coulombs}}{1 \text{ mol } e^-} \qquad \frac{1 \text{ mol Ag}}{1 \text{ mol } e^-}$$

For the Cr^{3+} solution, we have the same number of coulombs. Thus moles of electrons consumed (and coulombs) is the same for each cell. From moles of electrons consumed, we can calculate the moles of chromium deposited, remembering that three moles of electrons are required for every mole of chromium: $Cr^{3+} + 3e^- \rightarrow Cr$

Finally, mass of Cr is determined by multiplying moles of chromium by its atomic mass.

T21.8 This question requires use of the Nernst equation. First, the overall cell reaction is written:

$$Sn + 2AgCl \rightarrow Sn^{2+} + 2Ag + 2Cl^-$$

The standard cell potential is calculated using the half–cell potentials provided. This will be the value of E°_{cell} used in Equation 21.8. The actual cell potential, $E_{cell} = 0.3114$ V, is given in the problem. The value for n is 2, and the Q term is constructed from the overall equation:

$$Q = [Sn^{2+}][Cl^-]^2$$

Since the value for $[Cl^-]$ is given, we can solve the Nernst equation for a value of $[Sn^{2+}]$.

T21.9 Once the standard cell potential is known, the value for K_c can be calculated at 25 °C using the equation:

$$E^\circ_{cell} = \frac{RT}{nF} \log K_c$$

T21.10 First, separate the overall equation into two half–reactions. Next, calculate E°_{cell}, using Equation 21.2. Next, write the expression for Q, which will be used in Equation 21.8:

$$Q = \frac{[Ag^+]^2[Ni^{2+}]}{[H^+]^4}$$

Values for $[Ag^+]$ and $[Ni^{2+}]$ are given. The $[H^+]$ is calculated from pH, since $pH = -\log[H^+]$. Finally, the above values are substituted into Equation 21.8 to determine E_{cell}.

T21.11 First, we calculate the number of coulombs involved in the electrolysis: 1.25 coul/s × 10.00 min × 60 s/min = 750 coul.

Next, the half–reaction for the reduction of water is needed:
$2H_2O + 2e^- \rightarrow H_2 + 2OH^-$

The number of coulombs is converted to moles of electrons and to moles of hydrogen, using the following conversion factors:

$$\frac{96{,}500 \text{ coulombs}}{1 \text{ mol } e^-} \qquad \frac{1 \text{ mol } H_2}{2 \text{ mol } e^-}$$

Finally, the moles of hydrogen are converted to volume of hydrogen, using the ideal gas equation, PV = nRT where

$$P_{H_2} = \left(755 \text{ torr} - P_{H_2O,\, 25^\circ C}\right)\left(\frac{1 \text{ atm}}{760 \text{ torr}}\right) \text{ and } T = 25 + 273 \text{ K.}$$

T21.12 First we calculate the number of electrons passed through the cell in 2 hours:

$$0.100\text{ A}(2.00\text{ hr})\left(\frac{3600\text{ sec}}{1\text{ hr}}\right)\left(\frac{1\text{ mol e}^-}{96{,}500\text{ C}}\right) = 7.46\times10^{-3}\text{ mol e}^-.$$

Next, determine the change in molarity for the nickel half-cell in which 2 moles of e^- are required for each mol of Ni:

$$7.46\times10^{-3}\text{ mol e}^-\left(\frac{\text{mole Ni}^{2+}}{2\text{ mol e}^-}\right) = 3.73\times10^{-3}\text{ mol Ni}^{2+}$$

$$\text{Increase in molarity of Ni}^{2+}\text{ solution} = \left(\frac{3.73\times10^{-3}\text{ mol Ni}^{2+}}{0.125\text{ L soln}}\right) = \text{M}'$$

Final molarity of Ni^{2+} solution = M' + 0.100 M

Use the same procedure to determine the change in molarity for the silver half-cell, this time, the concentration of the silver ion has decreased, not increased.

Finally, use the Nernst equation, 21.8, to determine E_{cell}.

T21.13 Set up an electrochemical cell and measure the potential of the cell containing the insoluble solid. Then using:

$$E^\circ_{cell} = \frac{0.0592}{n}\log K_c$$

we can determine the value of K_{sp}, where $K_{sp} = K_c$.

More complete answer: We need a relationship between Q and K_{sp}. This can be done using a SHE equipped electrochemical cell twice: once coupled to an AgCl–Ag metal electrode and once coupled to a Ag metal electrode.

(1) Pt, H_2(1 atm)| HCl(1.00 M), AgCl(*s*)|Ag^+

In this case we determine that the following half–cell has E° = +0.2223V, i.e.,

$$AgCl(s) + e^- \rightarrow Ag(s) + Cl^-(aq) \quad E^\circ = 0.2223\text{ V}$$

(2) Ag|$AgNO_3$(*aq*)(1.00M), HCl(1.00M)| H_2(1.00atm), Pt

In this case we determine E° = –0.7995 V, i.e.,

$$Ag(s) \rightarrow Ag^+(aq) + e^- \quad E^\circ = -0.7995\text{ V}$$

Adding these two half–cells together we get the desired solubility equilibrium and its E° value, i.e.,

$$AgCl(s) \rightarrow Ag^+(aq) + Cl^-(aq) \quad E^\circ = -0.5772\text{V}$$

Finally, this E° value is used to determine K_{sp}.

T21.14 (a) $Cu(s) + Z^{2+} \rightarrow Cu^{2+} + Z(s)$

(b) Electrons flow away from the copper electrode to the Z electrode.

(c) The K^+ ions flow toward the cell with the Z electrode and the NO_3^- ions flow toward the copper electrode.

(d) Darker, due to the increase of copper ions.

(e) The voltage will decrease as the Z^{2+} concentration decreases.

T21.15 Using the Nernst equation, 21.8, determine the value for Q

$$E_{cell} = E^\circ_{cell} - \frac{RT}{nF}\ln Q$$

$E_{cell} = 1.0764$ V

$E^\circ_{cell} = 0.7996$ V

$R = 8.314\text{ J mol}^{-1}\text{ K}^{-1}$

T = 273.15 + 25.0 = 298.2 K

N = 1

$F = 96{,}500\text{ C mol}^{-1}$

Then, determine the concentration of H^+

$$Q = \frac{[Ag^+][P_{H_2}]}{[H^+]^2}$$

$[Ag^+] = 1.000$ M
$[P_{H2}] = 1.000$ atm
$[H^+]$ = unknown

From the $[H^+]$ value, the pH of the solution can be determine by pH = $-\log[H^+]$

Review Questions

21.1 A galvanic cell is one in which a spontaneous redox reaction occurs, producing electricity. A half–cell is one of either the cathode or the anode, together with the accompanying electrolyte.

21.2 The salt bridge connects two half–cells, and allows for electrical neutrality to be maintained by a flow of appropriate ions.

21.3 These must be kept separate, because otherwise Ag^+ ions would be reduced directly by Cu metal, and no external current would be produced.

21.4 The anode is the electrode at which oxidation takes place, and the cathode is the electrode at which reduction takes place. The charges of the electrodes in a galvanic cell are opposite to those in the electrolysis cell; the cathode is positive and the anode is negative.

21.5 In both the galvanic and the electrolysis cells, the electrons move away from the anode and toward the cathode.

21.6 The anions move away from the cathode toward the anode, and the cations move away from the anode toward the cathode.

21.7 Magnesium metal is oxidized at the anode and copper ions are reduced at the cathode to give copper metal: $Mg(s) + Cu^{2+}(aq) \rightarrow Mg^{2+}(aq) + Cu(s)$. The anode half–cell is a magnesium wire dipping into a solution of Mg^{2+} ions, and the cathode half–cell is a copper wire dipping into a solution of Cu^{2+} ions. Additionally, a salt bridge must connect the two half–cell compartments.

21.8 Aluminum metal constitutes the anode: $Al(s) \rightarrow Al^{3+}(aq) + 3e^-$
The cathode is tin metal: $Sn^{2+}(aq) + 2e^- \rightarrow Sn(s)$

21.9 The charges are not balanced.

21.10 A potential is the measure of the ability to push electrons through an external circuit. Its units are the volt.

21.11 A cell potential is a standard potential only if the temperature is 25 °C, the pressure is 1 atm, and all ions have a concentration of 1 M.

21.12 The cell potential for the anode half–reaction is subtracted from the cell potential for the cathode half–cell:

$$E^\circ_{cell} = E^\circ_{substance\ reduced} - E^\circ_{substance\ oxidized}$$

$$E^\circ_{cell} = E^\circ_{reduction} - E^\circ_{oxidation}$$

21.13 An amp (A) is a Coulomb per second (C/s). A volt (V) is a joule per Coulomb (J/C). The product of amp × V × sec is: C/s × J/C × s = J (joules)

21.14 No, emf measurements require a completed circuit for current flow, and there must always be two half–cells, connected with a salt bridge.

21.15 The standard hydrogen electrode is diagramed in Figure 21.5 of the text. It consists of a platinum wire in contact with a solution having $[H^+]$ equal to 1 M, and hydrogen gas at a pressure of 1 atm is placed over the system. The half–cell potential is 0 V.

21.16 A positive reduction potential indicates that the substance is more easily reduced than the hydrogen ion. Conversely, a negative reduction potential indicates that the substance comprising the half–cell is less easily reduced than the hydrogen ion.

21.17 The difference between the reduction potentials for hydrogen and copper is a constant that is independent of the choice for the reference potential. In other words, the reduction half–cell potential for copper is to be 0.34 units higher for copper than for hydrogen, regardless of the chosen point of reference. If E° for copper is taken to be 0 V, then E° for hydrogen must be –0.34 V.

21.18 The negative terminal of the voltmeter must be connected to the anode in order to obtain correct readings of the voltage that is generated by the cell.

21.19 The metals are placed into the activity series based on their values of standard reduction potentials.

21.20 The silver will be reduced, according to the standard reduction potentials. The standard cell notation should be $Ag(s)|Ag^+(aq)||Fe^{3+}(aq)|Fe(s)$

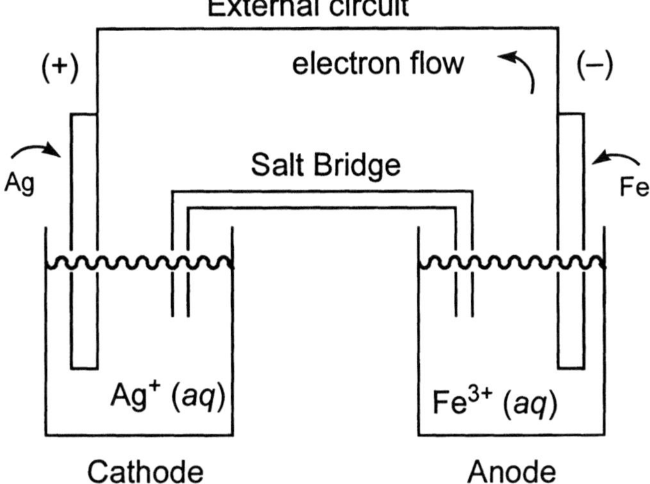

21.21

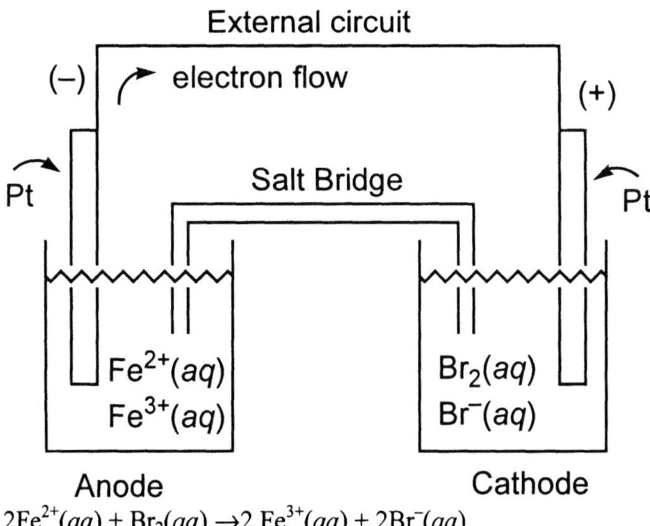

$2Fe^{2+}(aq) + Br_2(aq) \rightarrow 2\,Fe^{3+}(aq) + 2Br^-(aq)$

21.22 $\Delta G° = -nF\,E°_{cell}$

21.23 $E°_{cell} = \dfrac{0.0592}{n}\log K_c$

21.24 The Nernst equation:

$$E_{cell} = E°_{cell} - \frac{RT}{nF}\ln Q$$

$R = 8.314\ J\ mol^{-1}\ K^{-1}$

$T = 25\ °C = 298\ K$

$F = 96{,}494\ C\ mol^{-1}$

$$\frac{RT}{F} = 0.02567\ J\ C^{-1}$$

$\ln x = 2.303 \log x$

$$E_{cell} = E°_{cell} - \frac{0.02567\ J}{n\,C^{-1}}\ln Q$$

$$E_{cell} = E°_{cell} - \frac{0.0592\ J}{n\,C^{-1}}\log Q$$

If the system is at equilibrium, $Q = K_c$.

And $E_{cell} = 0$

$$0 = E°_{cell} - \frac{0.0592\ J}{n\,C^{-1}}\log K_c$$

$$E°_{cell} = \frac{0.0592\ J}{n\,C^{-1}}\log K_c$$

21.25 The maximum amount of work that can be obtained from the cell, per mole of Ag_2O reacting, is given by the absolute value of $\Delta G°$ for the reaction, where it must be remembered that the unit volt (V) is equivalent to a joule per Coulomb (J/C):

$\Delta G° = -nF\,E°_{cell} = -(2)(96{,}500\ C)(1.50\ J/C) = -2.90 \times 10^5\ J$

$$\#\ J = (0.500\ g\ Ag_2O)\left(\frac{1\ mol\ Ag_2O}{231.7\ g\ Ag_2O}\right)\left(\frac{-2.90 \times 10^5\ J}{1\ mol\ Ag_2O}\right) = -626\ J$$, the negative sign indicates that work is done on the cell.

21.26 $\Delta G = \Delta G° + RT\ln Q$

But it is also true that: $\Delta G = -nFE_{cell}$ and $\Delta G° = -nF\,E°_{cell}$, which allows the following substitution: $-nFE_{cell} = -nF\,E°_{cell} + RT\ln Q$

Dividing each side of the above equation by the quantity $-nF$ gives:

$$E_{cell} = E°_{cell} - RT/nF\ln Q$$

The various values to be used are: $R = 8.314\ J\ mol^{-1}\ K^{-1}$, $T = 298.15\ K$, $F = 96494\ C\ mol^{-1}$. Substituting these into the equation above we get

$$E_{cell} = E°_{cell} - \frac{(8.314\ J\ mol^{-1}\ K^{-1})(298.15\ K)(2.303)}{(n)(96494\ C\ mol^{-1})}\log Q$$

$$E_{cell} = E°_{cell} - \frac{0.0591}{n}\log Q$$

(Note: the factor of 2.303 is needed to convert from a natural log to a base 10 log.)

21.27 We begin by separating the reaction into its two half–reactions, in order to obtain the value of n.

$Pb(s) + SO_4^{2-}(aq) \rightarrow PbSO_4(s) + 2e^-$

$PbO_2(s) + 4H^+(aq) + SO_4^{2-}(aq) + 2e^- \rightarrow PbSO_4(s) + 2H_2O(\ell)$

Thus, n is equal to 2, and the equation that we are to use is:

$$E_{cell} = E^\circ_{cell} - \frac{0.0592}{n} \log Q$$

$$E_{cell} = 2.05\ V - \frac{0.0592}{2} \log \frac{1}{[H^+]^4[SO_4^{2-}]^2}$$

21.28 $Pb(s) + SO_4^{2-}(aq) \rightarrow PbSO_4(s) + 2e^-$ anode

$PbO_2(s) + 4H^+(aq) + SO_4^{2-}(aq) + 2e^- \rightarrow PbSO_4(s) + 2H_2O(\ell)$ cathode

Connecting six cells in series produces 12 volts.

21.29 $PbSO_4(s) + 2e^- \rightarrow Pb(s) + SO_4^{2-}(aq)$ cathode

$PbSO_4(s) + 2H_2O(\ell) \rightarrow 2e^- + PbO_2(s) + 4H^+(aq) + SO_4^{2-}(aq)$ anode

21.30 This is diagramed in Figure 21.9. The float inside the hydrometer sinks to a level that is inversely proportional to the density of the liquid that is drawn into it. This works because the concentration of the sulfuric acid (and hence the state of charge of the battery) is proportional to the concentration of sulfuric acid in the battery.

21.31 The use of calcium alloys allows the batteries to be sealed.

21.32 The anodic reaction is: $Zn(s) \rightarrow Zn^{2+}(aq) + 2e^-$

Several reactions take place at the cathode; one of the important ones is:

$2MnO_2(s) + 2NH_4^+(aq) + 2e^- \rightarrow Mn_2O_3(s) + 2NH_3(aq) + H_2O(\ell)$

21.33 $Zn(s) + 2OH^-(aq) \rightarrow ZnO(s) + H_2O(\ell) + 2e^-$ anode

$2MnO_2(s) + H_2O(\ell) + 2e^- \rightarrow Mn_2O_3(s) + 2OH^-(aq)$ cathode

21.34 $Cd(s) + 2OH^-(aq) \rightarrow Cd(OH)_2(s) + 2e^-$ anode

$2e^- + NiO_2(s) + 2H_2O(\ell) \rightarrow Ni(OH)_2(s) + 2OH^-(aq)$ cathode

The overall cell reaction on discharge of the battery is:

$Cd(s) + NiO_2(s) + 2H_2O(\ell) \rightarrow Cd(OH)_2(s) + Ni(OH)_2(s)$

On charging the battery, the reactions are reversed.

21.35 $Zn(s) + 2OH^-(aq) \rightarrow Zn(OH)_2(s) + 2e^-$ anode

$Ag_2O(s) + H_2O(\ell) + 2e^- \rightarrow 2Ag(s) + 2OH^-(aq)$ cathode

$Zn(s) + Ag_2O(s) + H_2O(\ell) \rightarrow Zn(OH)_2(s) + 2Ag(s)$ net cell reaction

21.36 Energy density, for a galvanic cell is the ratio of the energy available to the volume of the cell and the specific energy is the ratio of available energy to the weight of the cell.

21.37 Nickel–metal hydride batteries have a higher specific energy than a lead storage battery. Even though the masses are not given, these nickel–metal hydride batteries are used for portable electronic equipment, and the lighter weight of these batteries is a desirable feature.

21.38 The hydrogen is held in a metal alloy, Mg_2Ni, which has the ability to absorb and hold substantial amounts of hydrogen. The electrolyte is KOH.

21.39
$MH(s) + OH^-(aq) \rightarrow M(s) + H_2O + e^-$ anode
$NiO(OH)(s) + H_2O + e^- \rightarrow Ni(OH)_2(s) + OH^-(aq)$ cathode
$MH(s) + NiO(OH)(s) \rightarrow M(s) + Ni(OH)_2(s)$ overall reaction

The reactions are reversed upon charging.

21.40 The primary advantage of the nickel–metal battery over the nickel cadmium cell is that it can store about 50% more energy in the same volume. The nickel–metal hydride battery loses some of its capacity every day, so it cannot be stored for long periods of time. The Ni–MH battery would not be an effective wrist watch battery since it would lose more of its charge to self–discharge than it would lose to keeping time.

21.41 Lithium has the most negative reduction potential of any metal, so it is very easy to oxidize making it an excellent material for an anode, and it is a very lightweight metal. The major problem with lithium in a cell is that it reacts vigorously with water.

21.42 In a typical primary lithium cell, the electrodes are lithium as the anode and manganese(IV) oxide as the cathode.
$Li \rightarrow Li^+ + e^-$ anode
$MnO_2 + Li^+ + e^- \rightarrow LiMnO_2$ cathode
$Li + MnO_2 \rightarrow LiMnO_2$ net cell reaction

21.43 The major problem with the lithium metal batteries is that upon repeated charging and discharging of the cell, the lithium tends to form long whiskers that reach across the separator and eventually short out the cell. This would sometimes occur with catastrophic results; for example the batter organic electrolyte fluid would ignite.

21.44 The electrode materials in a typical lithium ion cell are graphite and cobalt oxide. When the cell is charged, Li^+ ions leave $LiCoO_2$ and travel through the electrolyte to the graphite. When the cell discharges, the Li^+ ions move back through the electrolyte to the cobalt oxide while electrons move through the external circuit to keep the charge in balance.

21.45
$O_2(g) + 2H_2O + 4e^- \rightarrow 4OH^-(aq)$ cathode
$H_2(g) + 2OH^-(aq) \rightarrow 2H_2O + 2e^-$ anode
$2H_2(g) + O_2(g) \rightarrow 2H_2O$ net cell reaction

21.46 Fuel cells are more efficient thermodynamically, and more of the energy of the reaction can be made available for useful work provided that the supply of reactants is maintained. The only product formed by the cell is water.

21.47 In an electrolytic cell, the cathode is negative, and the anode is positive. The opposite is true of a galvanic cell. An inert electrode is an electrode which does not chemically react in the measurement of electrochemical data.

21.48 In electrolytic conduction, charge is carried by the movement of positive and negative ions. In metallic conduction, charge is transported by the flow of electrons.

21.49 The flow of electrons in the external circuit must be accompanied by the electrolysis reaction. Otherwise the electrodes would accumulate charge, and the system would cease to function.

21.50 In direct current, electrons flow in only one direction. In alternating current, electron flow periodically reverses, flowing alternately in one direction and then in the other.

21.51 In solid NaCl, the ions are held in place and cannot move about. In molten NaCl, the crystal lattice of the solid has been destroyed; the ions are free to move, and consequently to conduct current by migrating either to the anode or to the cathode.

Anode: $2Cl^-(\ell) \rightarrow Cl_2(g) + 2e^-$

Cathode: $Na^+(\ell) + e^- \rightarrow Na(\ell)$

Net: $2Na^+(\ell) + 2Cl^-(\ell) \rightarrow 2Na(\ell) + Cl_2(g)$

21.52 oxidation: $2H_2O(\ell) \rightarrow 4H^+(aq) + 4e^- + O_2(g)$

reduction: $2H_2O(\ell) + 2e^- \rightarrow H_2(g) + 2OH^-(aq)$

21.53 It is reduction that occurs at the cathode, and near it, the pH increases due to the formation of $OH^-(aq)$. At the anode, where the oxidation of water occurs, the pH decreases due to the production of $H^+(aq)$. See the equations given in the answer to Review Question 21.52. The overall change in pH is 0 since the amount of H^+ formed and the amount of OH^- formed are equal. K_2SO_4 serves as charge carriers to balance the charge that occurs upon electrolysis of the K_2SO_4 solution.

21.54 One Faraday (F) is equivalent to one mole of electrons. Also, one Faraday is equal to 96,500 Coulombs, and a Coulomb is equivalent to an Ampere·second:

$$1\ F = 96{,}500\ C \text{ and } 1\ C = 1\ A\cdot s$$

21.55 The deposition of 0.10 mol of Cr from a Cr^{3+} solution will take longer than the deposition of 0.10 mol Cu from a Cu^{2+} solution because the Cr^{3+} requires 1.5 times as many electrons for deposition than Cu^{2+}. This is due to the difference in charges on the two ions.

21.56 The Ag^+ solution will give more metal deposited since it is in the +1 state while the Cu^{2+} solution will give half as much since the copper is in the +2 state.

21.57 Copper has a larger atomic mass than iron; therefore, the copper will deposit a greater mass of metal. Both metals are in the same +2 state.

21.58 Electroplating is a procedure by which a metal is deposited on another conducting surface. See Figure 21.21 of the text.

21.59 $Al_2O_3(s)$ is dissolved in molten cryolite, Na_3AlF_6. The liquid mixture is electrolyzed to drive the following reaction: $2Al_2O_3(\ell) \rightarrow 4Al(\ell) + 3O_2(g)$

The two half–reactions are:

anode: $2O^{2-}(\ell) \rightarrow O_2(g) + 4e^-$

cathode: $Al^{3+}(\ell) + 3e^- \rightarrow Al(\ell)$

overall cell reaction $6O^{2-} + 4Al^{3+} \rightarrow 3O_2(g) + 4Al(\ell)$

21.60 The oxygen that is formed slowly decomposes the anode.

21.61 Sea water is made basic in order to cause the precipitation of magnesium hydroxide:

$$Mg^{2+}(aq) + 2OH^-(aq) \rightarrow Mg(OH)_2(s).$$

This is purified by redissolving the hydroxide and precipitating the chloride by evaporation of water:

$$Mg(OH)_2(s) + 2HCl(aq) \rightarrow Mg^{2+}(aq) + 2Cl^-(aq) + 2H_2O(\ell)$$

The solid $MgCl_2$ that is so obtained is electrolyzed as a melt in order to recover the metal.

$$MgCl_2(\ell) \rightarrow Mg(\ell) + Cl_2(g)$$

21.62 Sodium is obtained from electrolysis of molten NaCl using the Downs cell that is diagramed in Figure 21.23 of the text. Some of the uses of sodium are to make tetraethyl lead, for sodium vapor lamps, and as a coolant in nuclear reactors.

$Na^+(\ell) + e^- \rightarrow Na(\ell)$	cathode
$Cl^-(\ell) \rightarrow 1/2Cl_2(g) + e^-$	anode
$NaCl(\ell) \rightarrow Na(\ell) + 1/2Cl_2(g)$	net reaction

21.63 This is shown in a photo and a diagram (Figure 21.24) of the text. Impure copper is the anode, which dissolves during the process. Pure copper is deposited at the cathode. Anode sludge contains precious metals, whose value makes the process cost effective.

Typical reactions occurring at the anode are:

$Cu(s) \rightarrow Cu^{2+}(aq) + 2e^-$
$Zn(s) \rightarrow Zn^{2+}(aq) + 2e^-$
$Fe(s) \rightarrow Fe^{2+}(aq) + 2e^-$

The reaction that occurs at the cathode is:

$Cu^{2+}(aq) + 2e^- \rightarrow Cu(s)$

21.64 The various methods are diagramed in Figures 21.25, 21.26 and 21.27 of the text. The physical apparatus influences the products that are obtained. The cathode reaction is the same in stirred and unstirred cells:

$$2H_2O(\ell) + 2e^- \rightarrow H_2(g) + 2OH^-(aq)$$

The unstirred anode reaction is:

$$2Cl^-(aq) \rightarrow Cl_2(g) + 2e-$$

The net reaction is:

$$2NaCl(aq) + 2H_2O(\ell) \rightarrow 2NaOH(aq) + Cl_2(g) + H_2(g)$$

In an unstirred cell, the $Cl_2(g)$ that is produced reacts with the $OH^-(aq)$ present forming $Cl^-(aq)$, $OCl^-(aq)$ and water. The anode reaction in the unstirred cell is therefore,

$$Cl^-(aq) + 2OH^-(aq) \rightarrow OCl^-(aq) + H_2O(\ell) + 2e^-$$

The net reaction in a stirred cell is:

$$NaCl(aq) + H_2O(\ell) \rightarrow NaOCl(aq) + H_2(g)$$

21.65 The diaphragm cell (Figure 21.26) offers the advantage that H_2 and Cl_2 may be captured separately, the reaction of Cl_2 and OH^- to form OCl^- is prevented, and the sodium hydroxide solution is conveniently obtained at the bottom of the apparatus. The disadvantages include the fact that the NaOH solution that is obtained is contaminated with some unreacted NaCl, and that the cell itself is complicated and difficult to maintain.

The mercury cell diagramed in Figure 21.27 offers the advantage that pure NaOH can be obtained. The disadvantage is the environmental hazard that mercury imposes.

Review Problems

21.66 (a)

anode:	$Cd(s) \rightarrow Cd^{2+}(aq) + 2e^-$
cathode:	$Au^{3+}(aq) + 3e^- \rightarrow Au(s)$
cell:	$3Cd(s) + 2Au^{3+}(aq) \rightarrow 3Cd^{2+}(aq) + 2Au(s)$

(b)

anode:	$Pb(s) + SO_4^{2-}(aq) \rightarrow PbSO_4(s) + 2e^-$
cathode:	$PbO_2(s) + SO_4^{2-}(aq) + 4H^+(aq) + 2e^- \rightarrow PbSO_4(s) + 2H_2O(\ell)$
cell:	$Pb(s) + PbO_2(s) + 2SO_4^{2-}(aq) + 4H^+(aq) \rightarrow 2PbSO_4(s) + 2H_2O(\ell)$

(c) anode: $Cr(s) \rightarrow Cr^{3+}(aq) + 3e^-$
cathode: $Cu^{2+}(aq) + 2e^- \rightarrow Cu(s)$
cell: $2Cr(s) + 3Cu^{2+}(aq) \rightarrow 2Cr^{3+}(aq) + 3Cu(s)$

21.67 (a) anode: $Zn(s) \rightarrow Zn^{2+}(aq) + 2e^-$
cathode: $Cr^{3+}(aq) + 3e^- \rightarrow Cr(s)$
cell: $3Zn(s) + 2Cr^{3+}(aq) \rightarrow 3Zn^{2+}(aq) + 2Cr(s)$

(b) anode: $Fe(s) \rightarrow Fe^{2+}(aq) + 2e^-$
cathode: $Br_2(aq) + 2e^- \rightarrow 2Br^-(aq)$
cell: $Fe(s) + Br_2(aq) \rightarrow Fe^{2+}(aq) + 2Br^-(aq)$

(c) anode: $Mg(s) \rightarrow Mg^{2+}(aq) + 2e^-$
cathode: $Sn^{2+}(aq) + 2e^- \rightarrow Sn(s)$
cell: $Mg(s) + Sn^{2+}(aq) \rightarrow Mg^{2+}(aq) + Sn(s)$

21.68 (a) $Fe(s)|Fe^{2+}(aq)||Cd^{2+}(aq)|Cd(s)$
(b) $Pt(s)|Cl^-(aq),Cl_2(g)||Br_2(aq),Br^-(aq)|Pt(s)$
(c) $Ag(s)|Ag^+(aq)||Au^{3+}(aq)|Au(s)$

21.69 (a) $Pt(s)|Fe^{2+}(aq),Fe^{3+}(aq)||NO^{3-}(aq),NO(g),H^+(aq)||Pt(s)$
(b) $Ag(s)|Ag^+(aq)||NiO_2(s)|Ni^{2+}(aq),H^+(aq)|Pt(s)$
(c) $Mg(s)|Mg^{2+}(aq)||Cd^{2+}(aq)|Cd(s)$

21.70 (a) $Sn(s)$ (b) $Br^-(aq)$ (c) $Zn(s)$ (d) $I^-(aq)$

21.71 (a) $MnO_4^-(aq)$ (b) $Au^{3+}(aq)$ (c) $PbO_2(s)$ (d) $HOCl(aq)$

21.72 (a) $E^\circ_{cell} = -0.40\ V - (-0.44)\ V = 0.04\ V$
(b) $E^\circ_{cell} = 1.07\ V - (1.36\ V) = -0.29\ V$
(c) $E^\circ_{cell} = 1.42\ V - (0.80\ V) = 0.62\ V$

21.73 (a) $E^\circ_{cell} = 0.96\ V - (0.77\ V) = 0.19\ V$
(b) $E^\circ_{cell} = 0.49\ V - (0.80\ V) = -0.31\ V$
(c) $E^\circ_{cell} = -0.40\ V - (-2.37\ V) = 1.97\ V$

21.74 The reactions are spontaneous if the overall cell potential is positive.

$E^\circ_{cell} = E^\circ_{substance\ reduced} - E^\circ_{substance\ oxidized}$

(a) $E^\circ_{cell} = 1.42\ V - (0.54\ V) = 0.88\ V$ spontaneous
(b) $E^\circ_{cell} = -0.44\ V - (0.96\ V) = -1.40\ V$ not spontaneous
(c) $E^\circ_{cell} = -0.74\ V - (-2.76\ V) = 2.02\ V$ spontaneous

21.75 A reaction is spontaneous if its net cell potential is positive:

$E^\circ_{cell} = E^\circ_{substance\ reduced} - E^\circ_{substance\ oxidized}$

(a) $E^\circ_{cell} = 1.07\ V - (1.36\ V) = -0.29\ V$ not spontaneous
(b) $E^\circ_{cell} = -0.25\ V - (-0.44\ V) = +0.19\ V$ spontaneous
(c) $E^\circ_{cell} = 1.07\ V - (0.17\ V) = +0.90\ V$ spontaneous

21.76 The given equation is separated into its two half–reactions:

$MnO_4^-(aq) + 8H^+(aq) + 5e^- \rightarrow Mn^{2+}(aq) + 4H_2O(\ell)$ reduction

$5Fe^{2+}(aq) \rightarrow 5Fe^{3+}(aq) + 5e^-$ oxidation

$$E^\circ_{cell} = E^\circ_{substance\ reduced} - E^\circ_{substance\ oxidized} = 1.51\ V - 0.77\ V = 0.74\ V$$

21.77 As in the answer to Review Problem 21.76:

$2Ag^+(aq) + 2e^- \rightarrow 2Ag(s)$ reduction

$Fe(s) \rightarrow Fe^{2+}(aq) + 2e^-$ oxidation

$$E^\circ_{cell} = E^\circ_{substance\ reduced} - E^\circ_{substance\ oxidized} = 0.80\ V - (-0.44\ V) = 1.24\ V$$

21.78 (a) $Zn(s)|Zn^{2+}(aq)||Co^{2+}(aq)|Co(s)$

$E^\circ_{cell} = -0.28\ V - (-0.76\ V) = 0.48\ V$

anode zinc

cathode cobalt

(b) $Mg(s)|Mg^{2+}(aq)||Ni^{2+}(aq)|Ni(s)$

$E^\circ_{cell} = -0.25\ V - (-2.37\ V) = 2.12\ V$

anode magnesium

cathode nickel

(c) $Sn(s)Sn^{2+}(aq)||Au^{3+}(aq)|Au(s)$

$E^\circ_{cell} = 1.42\ V - (-0.14\ V) = 1.56\ V$

anode tin

cathode gold

21.79 (a) $Cu(s)|Cu^{2+}(aq)||BrO^{3-}(aq),Br^-(aq),H^+(aq)|Pt(s)$

$E^\circ_{cell} = 1.44\ V - (0.34\ V) = 1.10\ V$

anode copper

cathode platinum

(b) $Pt(s)|Fe^{2+}(aq),Fe^{3+}(aq)||Ag^+(aq)|Ag(s)$

$E^\circ_{cell} = 0.80\ V - (0.77\ V) = 0.03\ V$

anode platinum

cathode silver

(c) $Pt(s)|NO(g)|NO_3^-(aq),H^+(aq)||MnO_4^-(aq),Mn^{2+}(aq),H^+(aq)|Pt(s)$

$E^\circ_{cell} = 1.51\ V - (0.96\ V) = 0.55\ V$

anode platinum

cathode platinum

21.80 The half–cell with the more positive E°_{cell} will appear as a reduction, and the other half–reaction is reversed, to appear as an oxidation:

$BrO_3^-(aq) + 6H^+(aq) + 6e^- \rightarrow Br^-(aq) + 3H_2O$ reduction

$3 \times (2I^-(aq) \rightarrow I_2(s) + 2e^-)$ oxidation

$BrO_3^-(aq) + 6I^-(aq) + 6H^+(aq) \rightarrow 3I_2(s) + Br^-(aq) + 3H_2O$ net reaction

$$E^\circ_{cell} = E^\circ_{subs\tan ce\ reduced} - E^\circ_{subs\tan ce\ oxided}\ \text{or}$$

$$E^\circ_{cell} = E^\circ_{reduction} - E^\circ_{oxidation} = 1.44\ V - (0.54\ V) = 0.90\ V$$

21.81 The half–reaction having the more positive reduction potential is the reduction half–reaction, and the other is reversed to become the oxidation half–reaction:

$MnO_2(s) + 4H^+(aq) + 2e^- \rightarrow Mn^{2+}(aq) + 2H_2O(\ell)$ reduction

$Pb(s) + 2Cl^-(aq) \rightarrow PbCl_2(s) + 2e^-$ oxidation

$MnO_2(s) + 4H^+(aq) + Pb(s) + 2Cl^-(aq) \rightarrow$
$Mn^{2+}(aq) + 2H_2O(\ell) + PbCl_2(aq)$ net reaction

$E^\circ_{cell} = E^\circ_{reduction} - E^\circ_{oxidation} = 1.23 - (-0.27) = 1.50\text{ V}$

21.82 The half–reaction having the more positive standard reduction potential is the one that occurs as a reduction, and the other one is written as an oxidation:

$2 \times (2HOCl(aq) + 2H^+(aq) + 2e^- \rightarrow Cl_2(g) + 2H_2O(\ell))$ reduction

$3H_2O(\ell) + S_2O_3^{2-}(aq) \rightarrow 2H_2SO_3(aq) + 2H^+(aq) + 4e^-$ oxidation

$4HOCl(aq) + 4H^+(aq) + 3H_2O(\ell) + S_2O_3^{2-}(aq) \rightarrow$
$2Cl_2(g) + 4H_2O(\ell) + 2H_2SO_3(aq) + 2H^+(aq)$

which simplifies to give the following net reaction:
$4HOCl(aq) + 2H^+(aq) + S_2O_3^{2-}(aq) \rightarrow 2Cl_2(g) + H_2O(\ell) + 2H_2SO_3(aq)$

21.83 $Br_2(aq) + 2e^- \rightarrow 2Br^-(aq)$ $E^\circ_{cell} = 1.07\text{ V}$
$I_2(s) + 2e^- \rightarrow 2I^-(aq)$ $E^\circ_{cell} = 0.54\text{ V}$

Since the first of these has the larger reduction half–cell potential, it occurs as a reduction, and the second is reversed to become an oxidation:

$Br_2(aq) + 2I^-(aq) \rightarrow I_2(s) + 2Br^-(aq)$

21.84 The two half–reactions are:

$SO_4^{2-}(aq) + 2e^- + 4H^+(aq) \rightarrow H_2SO_3(aq) + H_2O(\ell)$ reduction

$2I^-(aq) \rightarrow I_2(s) + 2e^-$ oxidation

$E^\circ_{cell} = E^\circ_{reduction} - E^\circ_{oxidation} = 0.17\text{ V} - (0.54\text{ V}) = -0.37\text{ V}$

Since the overall cell potential is negative, we conclude that the reaction is not spontaneous in the direction written.

21.85 The two half–reactions are:

$S_2O_8^{2-} + 2e^- \rightarrow 2SO_4^{2-}$ reduction
$Ni(OH)_2 + 2OH^- \rightarrow NiO_2 + 2H_2O + 2e^-$ oxidation

$E^\circ_{cell} = E^\circ_{reduction} - E^\circ_{oxidation} = 2.01\text{ V} - (0.49\text{ V}) = 1.52\text{ V}$

Since the overall cell potential is positive, we conclude that the reaction is spontaneous in the direction written.

21.86 First, separate the overall reaction into its two half–reactions:

$2Br^-(aq) \rightarrow Br_2(aq) + 2e^-$ oxidation
$I_2(s) + 2e^- \rightarrow 2I^-(aq)$ reduction

$E^\circ_{cell} = E^\circ_{reduction} - E^\circ_{oxidation} = 0.54\text{ V} - (1.07\text{ V}) = -0.53\text{ V}$

The value of n is 2: $\Delta G^\circ = -nF\,E^\circ_{cell} = -(2)(96{,}500\text{ C})(-0.53\text{ J/C})$
$= 1.0 \times 10^5\text{ J} = 1.0 \times 10^2\text{ kJ}$

21.87 Using the equation $\Delta G° = -nF E^{\circ}_{cell}$, we have $\Delta G° = -n(96{,}500 \text{ C/mol e}^-)(1.69 \text{ V})$ for which we need n. Upon writing the two half–reactions, i.e.,

$MnO_4^-(aq) + 8H^+(aq) + 5e^- \rightarrow Mn^{2+}(aq) + 4H_2O$ reduction

$HCHO_2(\ell) \rightarrow CO_2(g) + 2H^+(aq) + 2e^-$ oxidation

we see that we need to multiply the reduction half–reaction by 2 and the oxidation reaction by 5 in order to balance the eqaution:

$2MnO_4^-(aq) + 16H^+(aq) + 10e^- \rightarrow 2Mn^{2+}(aq) + 8H_2O$ reduction

$5HCHO_2(\ell) \rightarrow 5CO_2(g) + 10H^+(aq) + 10e^-$ oxidation

The net reaction has n = 10. So, $\Delta G° = -(10 \text{ mol e}^-)(96{,}500 \text{ C/mol e}^-)(1.69 \text{ V}) = -1.63 \times 10^3$ kJ.

21.88 (a) $E^{\circ}_{cell} = E^{\circ}_{reduction} - E^{\circ}_{oxidation} = 2.01 \text{ V} - (1.47 \text{ V}) = 0.54 \text{ V}$

(b) Since n = 10, $\Delta G° = -nF E^{\circ}_{cell} = -(10)(96{,}500 \text{ C})(0.54 \text{ J/C}) = -5.2 \times 10^5 \text{ J}$

$\Delta G° = -5.2 \times 10^2 \text{ kJ}$

(c)

$$E^{\circ}_{cell} = \frac{RT}{nF} \ln K_c$$

$$0.54 \text{ V} = \frac{(8.314 \text{ J mol}^{-1}\text{K}^{-1})(298 \text{ K})}{10(96{,}500 \text{ C mol}^{-1})} \ln K_c$$

$\ln K_c = 210.3$

Taking the exponential of both sides of this equation:

$K_c = 2.1 \times 10^{91}$

21.89 Ni^{2+} is reduced by two electrons and Co is oxidized by two electrons.

$E^{\circ}_{cell} = E^{\circ}_{reduction} - E^{\circ}_{oxidation} = -0.25 - (-0.28) = +0.03 \text{ V}$

$$E^{\circ}_{cell} = \frac{0.0592}{n} \log K_c$$

$+0.03 = (0.0592/2) \times \log K_c$

$\log K_c = 1$ and $K_c = 10^1 = 10$

21.90 Sn is oxidized by two electrons and Ag is reduced by two electrons:

$$E^{\circ}_{cell} = \frac{0.0592}{n} \log K_c$$

$-0.015 \text{ V} = (0.0592 \text{ V}/2) \times \log K_c$

$\log K_c = -0.51$

$K_c = \text{antilog}(-0.51) = 0.31$

21.91 First, separate the overall reaction into two half–reactions:

$2H_2O \rightarrow 4H^+ + 4e^- + O_2$ oxidation

$2 \times (Cl_2 + 2e^- \rightarrow 2Cl^-)$ reduction

$E^{\circ}_{cell} = E^{\circ}_{reduction} - E^{\circ}_{oxidation} = 1.36 - (1.23) = 0.13 \text{ V}$

$$E^{\circ}_{cell} = \frac{0.0592}{n} \log K_c$$

$+0.13 \text{ V} = (0.0592 \text{ V}/4) \log \times K_c$

$\log K_c = 8.78$

$K_c = \text{antilog}(8.78) = 6.11 \times 10^8$.

21.92 This reaction involves the oxidation of Ag by two electrons and the reduction of Ni by two electrons. The concentration of the hydrogen ion is derived from the pH of the solution: $[H^+]$ = antilog (–pH) = antilog (–5) = 1×10^{-5} M

$$E_{cell} = 2.48 \text{ V} - \frac{0.0592 \text{ V}}{2} \log \frac{[Ag^+]^2[Ni^{2+}]}{[H^+]^4}$$

$$= 2.48 \text{ V} - \frac{0.0592 \text{ V}}{2} \log \frac{[1.0 \times 10^{-2}]^2[1.0 \times 10^{-2}]}{[1.0 \times 10^{-5}]^4}$$

E_{cell} = 2.48 V – 0.41 V = 2.07 V

21.93 The following half–reactions indicate that the value of n is 30:

$$5Cr_2O_7^{2-} + 70H^+ + 30e^- \rightarrow 10Cr^{3+} + 35H_2O$$
$$3I_2 + 18H_2O \rightarrow 6IO_3^- + 36H^+ + 30e^-$$

$$E_{cell} = 0.135 \text{ V} - \frac{0.0592 \text{ V}}{30} \log \frac{[IO_3^-]^6[Cr^{3+}]^{10}}{[H^+]^{34}[Cr_2O_7^{2-}]^5}$$

$$= 0.135 \text{ V} - \frac{0.0592 \text{ V}}{30} \log \frac{[0.00010]^6[0.0010]^{10}}{[0.10]^{34}[0.010]^5}$$

$$= 0.135 \text{ V} - \frac{0.0592 \text{ V}}{30} \log 1.0 \times 10^{-10}$$

$$= 0.155 \text{ V}$$

21.94

$$E_{cell} = E^\circ_{cell} - \frac{RT}{nF} \ln \frac{[Mg^{2+}]}{[Cd^{2+}]}$$

$$E_{cell} = 1.97 - \frac{(8.314 \text{ J mol}^{-1}\text{K}^{-1})(298 \text{ K})}{2(96{,}500 \text{ C mol}^{-1})} \ln \frac{[1.00]}{[Cd^{2+}]}$$

$$1.54 \text{ V} = 1.97 \text{ V} - 0.01284 \ln \frac{1}{[Cd^{2+}]}$$

$\ln(1/[Cd^{2+}]) = 33.489$

Taking (e^x) of both sides:
$1/[Cd^{2+}] = 3.50 \times 10^{14}$
$[Cd^{2+}] = 2.86 \times 10^{-15}$ M

21.95 Since the copper half–cell is the cathode; this is the half–cell in which reduction takes place. The silver half–cell is therefore the anode, where oxidation of silver occurs. The standard cell potential is:
$E^\circ_{cell} = E^\circ_{reduction} - E^\circ_{oxidation} = 0.3419 \text{ V} - 0.2223 \text{ V} = 0.1196 \text{ V}$. The overall cell reaction is:

$Cu^{2+}(aq) + 2Ag(s) + 2Cl^-(aq) \rightarrow Cu(s) + 2AgCl(s)$, and the Nernst equation becomes:

$$E_{cell} = 0.1196\text{ V} - \frac{0.0592\text{ V}}{2}\log\frac{1}{[Cu^{2+}][Cl^-]^2}$$

If we use the values given in the exercise, we arrive at:
$0.0925\text{ V} = 0.1196\text{ V} - 0.0296\text{ V} \times \log(1/[Cl^-]^2)$, which rearranges to give: $\log(1/[Cl^-]^2) = 0.916$, $[Cl^-] = 0.349$ M

21.96 In the iron half–cell, we are initially given:
$0.0500\text{ L} \times 0.100\text{ mol/L} = 5.00 \times 10^{-3}\text{ mol } Fe^{2+}(aq)$

The precipitation of $Fe(OH)_2(s)$ consumes some of the added hydroxide ion, as well as some of the iron ion: $Fe^{2+}(aq) + 2OH^-(aq) \rightarrow Fe(OH)_2(s)$. The number of moles of OH^- that have been added to the iron half–cell is:

$$0.500\text{ mol/L} \times 0.0500\text{ L} = 2.50 \times 10^{-2}\text{ mol } OH^-$$

The stoichiometry of the precipitation reaction requires that the following number of moles of OH^- be consumed on precipitation of 5.00×10^{-3} mol of $Fe(OH)_2(s)$:
$5.00 \times 10^{-3}\text{ mol } Fe(OH)_2 \times (2\text{ mol } OH^-/\text{mol } Fe(OH)_2) = 1.00 \times 10^{-2}\text{ mol } OH^-$

The number of moles of OH^- that are unprecipitated in the iron half–cell is:
$2.50 \times 10^{-2}\text{ mol} - 1.00 \times 10^{-2}\text{ mol} = 1.50 \times 10^{-2}\text{ mol } OH^-$

Since the resulting volume is 50.0 mL + 50.0 mL, the concentration of hydroxide ion in the iron half–cell becomes, upon precipitation of the $Fe(OH)_2$:

$$[OH^-] = 1.50 \times 10^{-2}\text{ mol}/0.100\text{ L} = 0.150\text{ M } OH^-$$

We have assumed that the iron hydroxide that forms in the above precipitation reaction is completely insoluble. This is not accurate, though, because some small amount does dissolve in water according to the following equilibrium:

$$Fe(OH)_2(s) \leftrightarrows Fe^{2+}(aq) + 2OH^-(aq)$$

This means that the true $[OH^-]$ is slightly higher than 0.150 M as calculated above. Thus we must set up the usual equilibrium table, in order to analyze the extent to which $Fe(OH)_2(s)$ dissolves in 0.150 M OH^- solution:

	$[Fe^{2+}]$	$[OH^-]$
I	–	0.150
C	+x	+2x
E	+x	0.150+2x

The quantity x in the above table is the molar solubility of $Fe(OH)_2$ in the solution that is formed in the iron half–cell.

$K_{sp} = [Fe^{2+}][OH^-]^2 = (x)(0.150 + 2x)^2$

The standard cell potential is:

$$E^\circ_{cell} = E^\circ_{reduction} - E^\circ_{oxidation} = 0.3419\text{ V} - (-0.447\text{ V}) = 0.7889\text{ V}$$

The Nernst equation is:

$$E_{cell} = E^{\circ}_{cell} - \frac{RT}{nF}\ln\frac{[Fe^{2+}]}{[Cu^{2+}]}$$

$$1.175 = 0.7889 - \frac{(8.314\text{ J mol}^{-1}\text{K}^{-1})(298\text{ K})}{2(96{,}500\text{ C mol}^{-1})}\ln\frac{[Fe^{2+}]}{[1.00]}$$

$$1.175 = 0.7889 - 0.01284\ln[Fe^{2+}]$$

$\ln[Fe^{2+}] = -30.07$
$[Fe^{2+}] = 8.72 \times 10^{-14}$ M

This is the concentration of Fe^{2+} in the saturated solution, and it is the value to be used for x in the above expression for K_{sp}.

$K_{sp} = (x)(0.150 + 2x)^2 = (8.72 \times 10^{-14})[0.150 + (2)(8.72 \times 10^{-14})]^2$
$K_{sp} = 1.96 \times 10^{-15}$

21.97 The half–cell reactions and the overall cell reaction are:
$Cu^{2+}(aq) + 2e^- \rightarrow Cu(s)$ $E^{\circ}_{red} = +0.3419$ V
$H_2(g) \rightarrow 2H^+(aq) + 2e^-$ $E^{\circ}_{ox} = 0.0000$ V
$Cu^{2+}(aq) + H_2(g) \rightarrow Cu(s) + 2H^+(aq)$

(a) The standard cell potential is:
$E^{\circ}_{cell} = E^{\circ}_{reduction} - E^{\circ}_{oxidation} = 0.3419\text{ V} - 0\text{ V} = +0.3419\text{ V}$
The Nernst equation for this system is:

$$E_{cell} = E^{\circ}_{cell} - \frac{0.0592\text{ V}}{2}\log\frac{[H^+]^2}{[Cu^{2+}]}$$

which becomes, under the circumstances defined in the problem:

$$E_{cell} = E^{\circ}_{cell} - \frac{0.0592\text{ V}}{2}\log[H^+]^2$$

Rearranging the last equation gives:

$$\frac{2 \times (E_{cell} - E^{\circ}_{cell})}{0.0592\text{ V}} = -\log[H^+]^2$$

which becomes the desired relationship:

$$\frac{(E_{cell} - E^{\circ}_{cell})}{0.0592\text{ V}} = -\log[H^+] = \text{pH}$$

(b) The equation derived in the answer to part (a) of this question is conveniently rearranged to give:

$$E_{cell} = (0.0592\text{ V})(\text{pH}) + E^{\circ}_{cell} = (0.0592\text{ V})(5.15) + 0.3419\text{ V} = 0.647\text{ V}$$

(c) The equation that was derived in the answer to part (a) of this question may be used directly:

$$pH = \frac{(E_{cell} - E^{\circ}_{cell})}{0.0592\ V} = \frac{(0.645\ V - 0.3419\ V)}{0.0592\ V} = 5.12$$

21.98 1 C = 1 A•s
(a) $4.00\ A \times 600\ s = 2.40 \times 10^3\ C$
(b) $10.0\ A \times 20.0\ min \times 60\ s/min = 1.20 \times 10^4\ C$
(c) $1.50\ A \times 6.00\ hr \times 3600\ s/hr = 3.24 \times 10^4\ C$

21.99 Multiply each value times the factor 1 mol e^-/96,500 C.
(a) 0.0249 mol (b) 0.124 mol (c) 0.336 mol

21.100 (a) $Fe^{2+}(aq) + 2e^- \rightarrow Fe(s)$
$0.20\ mol\ Fe^{2+} \times 2\ mol\ e^-/mol\ Fe^{2+} = 0.40\ mol\ e^-$
(b) $Cl^-(aq) \rightarrow 1/2Cl_2(g) + e^-$
$0.70\ mol\ Cl^- \times 1\ mol\ e^-/mol\ Cl^- = 0.70\ mol\ e^-$
(c) $Cr^{3+}(aq) + 3e^- \rightarrow Cr(s)$
$1.50\ mol\ Cr^{3+} \times 3\ mol\ e^-/mol\ Cr^{3+} = 4.50\ mol\ e^-$
(d) $Mn^{2+}(aq) + 4H_2O(\ell) \rightarrow MnO_4^-(aq) + 8H^+(aq) + 5e^-$
$1.0 \times 10^{-2}\ mol\ Mn^{2+} \times 5\ mol\ e^-/mol\ Mn^{2+} = 5.0 \times 10^{-2}\ mol\ e^-$

21.101 (a) $Mg^{2+}(aq) + 2e^- \rightarrow Mg(s)$

$$\#\ mol\ e^- = (5.00\ g\ Mg)\left(\frac{1\ mol\ Mg}{24.31\ g\ Mg}\right)\left(\frac{2\ mol\ e^-}{1\ mol\ Mg}\right) = 0.411\ mol\ e^-$$

(b) $Cu^{2+}(aq) + 2e^- \rightarrow Cu(s)$

$$\#\ mol\ e^- = (41.0\ g\ Cu)\left(\frac{1\ mol\ Cu}{63.55\ g\ Cu}\right)\left(\frac{2\ mol\ e^-}{1\ mol\ Cu}\right) = 1.29\ mol\ e^-$$

21.102 $Ag^+(aq) + e^- \rightarrow Ag(s)$, and $Cr^{3+}(aq) + 3e^- \rightarrow Cr(s)$
This shows that there are three moles of electrons per mole of Cr but only one mole of electrons per mole of Ag. The number of moles of electrons involved in the silver reaction is:

$$\#\ mol\ e^- = (12.0\ g\ Ag)\left(\frac{1\ mol\ Ag}{107.9\ g\ Ag}\right)\left(\frac{1\ mol\ e^-}{1\ mol\ Ag}\right) = 0.111\ mol\ e^-$$

The amount of Cr is then:

$$\#\ mol\ Cr^{3+} = (0.111\ mol\ e^-)\left(\frac{1\ mol\ Cr^{3+}}{3\ mol\ e^-}\right) = 0.0371\ mol\ Cr^{3+}$$

21.103 $$\#\ min = (0.111\ mol\ e^-)\left(\frac{96{,}500\ coulombs}{1\ mol\ e^-}\right)\left(\frac{1\ s}{4\ amp}\right)\left(\frac{1\ Amp \bullet s}{1\ coulomb}\right)\left(\frac{1\ min}{60\ s}\right) = 44.6\ min$$

21.104 $Fe(s) + 2OH^-(aq) \rightarrow Fe(OH)_2(s) + 2e^-$
The number of Coulombs is: $12.0\ min \times 60\ s/min \times 8.00\ C/s = 5.76 \times 10^3\ C$. The number of grams of $Fe(OH)_2$ is:

$$\#\,g\,Fe(OH)_2 = \left(5.76\times10^3\ C\right)\left(\frac{1\,mol\,e^-}{96500\,C}\right)\left(\frac{1\,mol\,Fe(OH)_2}{2\,mol\,e^-}\right)\left(\frac{89.86\,g\,Fe(OH)_2}{1\,mol\,Fe(OH)_2}\right)$$

$$= 2.68\,g\,Fe(OH)_2$$

21.105 $2Cl^-(\ell) \rightarrow Cl_2(g) + 2e-$

The number of Coulombs is: 4.25 A × 35.0 min × 60 s/min = 8.92×10^3 C

The number of grams of Cl_2 that will be produced is:

$$\#\,g\,Cl_2 = \left(8.92\times10^3\ C\right)\left(\frac{1\,mol\,e^-}{96{,}500\,C}\right)\left(\frac{1\,mol\,Cl_2}{2\,mol\,e^-}\right)\left(\frac{70.91\,g\,Cl_2}{1\,mol\,Cl_2}\right) = 3.28\,g\,Cl_2$$

21.106 $Cr^{3+}(aq) + 3e^- \rightarrow Cr(s)$

The number of Coulombs that will be required is:

$$\#\,C = \left(75.0\,g\ Cr\right)\left(\frac{1\,mol\,Cr}{52.00\,g\,Cr}\right)\left(\frac{3\,mol\,e^-}{1\,mol\,Cr}\right)\left(\frac{96500\,C}{1\,mol\,e^-}\right) = 4.18\times10^5\ C$$

The time that will be required is:

$$\#\,hr = \left(4.18\times10^5\ C\right)\left(\frac{1\,s}{2.25\,C}\right)\left(\frac{1\,hr}{3600\,s}\right) = 51.5\,hr$$

21.107 The number of Coulombs that will be required is:

$$\#\,C = \left(35.0\,g\ Pb\right)\left(\frac{1\,mol\,Pb}{207.2\,g\,Pb}\right)\left(\frac{2\,mol\,e^-}{1\,mol\,Pb}\right)\left(\frac{96{,}500\,C}{1\,mol\,e^-}\right) = 3.26\times10^4\ C$$

The time that will be required is:

$$\#\,hr = \left(3.26\times10^4\ C\right)\left(\frac{1\,s}{1.50\,C}\right)\left(\frac{1\,hr}{3600\,s}\right) = 6.04\,hr$$

21.108 $Mg^{2+}(aq) + 2e^- \rightarrow Mg(\ell)$

The number of Coulombs that will be required is:

$$\#\,C = \left(60.0\,g\ Mg\right)\left(\frac{1\,mol\,Mg}{24.31\,g\,Mg}\right)\left(\frac{2\,mol\,e^-}{1\,mol\,Mg}\right)\left(\frac{96{,}500\,C}{1\,mol\,e^-}\right) = 4.76\times10^5\ C$$

The number of amperes is: 4.76×10^5 C ÷ 7200 s = 66.2 amp

21.109 $Al^{3+}(aq) + 3e^- \rightarrow Al(s)$

The number of Coulombs that are required is:

$$\#\,C = \left(409\times10^3\ g\ Al\right)\left(\frac{1\,mol\,Al}{26.98\,g\,Al}\right)\left(\frac{3\,mol\,e^-}{1\,mol\,Al}\right)\left(\frac{96{,}500\,C}{1\,mol\,e^-}\right) = 4.39\times10^9\ C$$

The number of amperes is: 4.39×10^9 C ÷ 8.64×10^4 s = 5.08×10^4 A

(Note: There are 8.64×10^4 s in 24.0 hr.)

21.110 The electrolysis of NaCl solution results in the reduction of water, together with the formation of hydroxide ion: $2H_2O(\ell) + 2e^- \rightarrow H_2(g) + 2OH^-(aq)$. The number of Coulombs is: 2.00 A × 20.0 min × 60 s/min = 2.40×10^3 C. The number of moles of OH^- is:

$$\#\,\text{mol OH}^- = \left(2.40\times10^3\ \text{C}\right)\left(\frac{1\ \text{mol e}^-}{96{,}500\ \text{C}}\right)\left(\frac{2\ \text{mol OH}^-}{2\ \text{mol e}^-}\right) = 0.0249\ \text{mol OH}^-$$

The volume of acid solution that will neutralize this much OH^- is:

$$\#\,\text{mL HCl} = \left(0.0249\ \text{mol OH}^-\right)\left(\frac{1\ \text{mol HCl}}{1\ \text{mol OH}^-}\right)\left(\frac{1000\ \text{mL HCl}}{0.620\ \text{mol HCl}}\right) = 40.1\ \text{mL HCl}$$

21.111 The electrolysis of NaCl solution results in the reduction of water, together with the formation of hydroxide ion: $2H_2O(\ell) + 2e^- \rightarrow H_2(g) + 2OH^-(aq)$. The number of seconds is: 25.0 min × 60 s/min = 1.50×10^3 s. The number of moles of OH^- is:

$$\#\,\text{mol OH}^- = \left(15.5\ \text{mL H}^+\right)\left(\frac{0.250\ \text{mol H}^+}{1000\ \text{mL H}^+}\right)\left(\frac{1\ \text{mol OH}^-}{1\ \text{mol H}^+}\right) = 3.87\times10^{-3}\ \text{mol OH}^-$$

The number of Coulombs that will form this much OH^- is:

$$\#\,\text{Coulombs} = \left(3.87\times10^{-3}\ \text{mol OH}^-\right)\left(\frac{2\ \text{mol e}^-}{2\ \text{mol OH}^-}\right)\left(\frac{96{,}500\ \text{C}}{1\ \text{mol e}^-}\right) = 3.74\times10^2\ \text{C}$$

The average current in amperes is:

$$\text{Current} = \frac{3.74\times10^2\ \text{C}}{1.50\times10^3\ \text{s}} = 0.250\ \text{A}$$

21.112 The electrolysis of NaCl solution results in the reduction of water, together with the formation of hydroxide ion: $2H_2O + 2e^- \rightarrow H_2(g) + 2OH^-(aq)$. The number of Coulombs is: 2.50 A × 15.0 min × 60 s/min = 2.25×10^3 C. The number of moles of OH^- is:

$$\#\,\text{mol OH}^- = \left(2.25\times10^3\ \text{C}\right)\left(\frac{1\ \text{mol e}^-}{96500\ \text{C}}\right)\left(\frac{2\ \text{mol OH}^-}{2\ \text{mol e}^-}\right) = 0.0233\ \text{mol OH}^-$$

The volume of acid solution that will neutralize this much OH^- is:

$$\#\,\text{mL HCl} = \left(0.0233\ \text{mol OH}^-\right)\left(\frac{1\ \text{mol HCl}}{1\ \text{mol OH}^-}\right)\left(\frac{1000\ \text{mL HCl}}{0.100\ \text{mol HCl}}\right) = 233\ \text{mL HCl}$$

21.113 $2H^+(aq) + 2e^- \rightarrow H_2(g)$
The number of Coulombs is: 15.00 min × 60 s/min × 0.750 C/s = 675 C.
The number of moles of H_2 is:

$$\#\,\text{mol H}_2 = \left(675\ \text{C}\right)\left(\frac{1\ \text{mol e}^-}{96{,}500\ \text{C}}\right)\left(\frac{1\ \text{mol H}_2}{2\ \text{mol e}^-}\right) = 3.50\times10^{-3}\ \text{mol H}_2$$

Finally, we calculate the volume of H_2 gas:

$$V = \frac{nRT}{P} = \frac{\left(0.00350\ \text{mol}\right)\left(0.0821\ \frac{\text{L atm}}{\text{mol K}}\right)\left(273\ \text{K}\right)}{\left(735\ \text{torr}\right)\left(\frac{1\ \text{atm}}{760\ \text{torr}}\right)} = 0.0811\ \text{L}$$

V = 81.1 mL

21.114 Possible cathode reactions:

$Al^{3+} + 3e^- \rightleftharpoons Al(s)$ E° = –1.66 V

$2H_2O + 2e^- \rightleftharpoons H_2(g) + 2OH^-(aq)$ E° = –0.83 V

Possible anode reactions:

$S_2O_8^{2-} + 2e^- \rightleftharpoons 2SO_4^{2-}$ E° = +2.05 V

$O_2 + 4H^+ + 4e^- \rightleftharpoons 2H_2O$ E° = +1.23 V

Cathode reaction: $2H_2O + 2e^- \rightleftharpoons H_2(g) + 2OH^-(aq)$ E° = –0.83 V

Anode reactions: $2H_2O \rightleftharpoons O_2 + 4H^+ + 4e^-$ E° = –1.23 V

Net cell reaction: $2H_2O \rightleftharpoons 2H_2(g) + O_2(g)$ E° = –2.06 V

21.115 Possible cathode reactions:

$Cd^{2+} + 2e^- \rightleftharpoons Cd(s)$ E° = –0.40 V

$2H_2O + 2e^- \rightleftharpoons H_2(g) + 2OH^-(aq)$ E° = –0.83 V

Possible anode reactions:

$O_2 + 4H^+ + 4e^- \rightleftharpoons 2H_2O$ E° = +1.23 V

$I_2(s) + 2e^- \rightleftharpoons 2I^-$ E° = +0.54 V

Cathode reaction: $Cd^{2+} + 2e^- \rightleftharpoons Cd(s)$ E° = –0.40 V

Anode reaction: $2I^- \rightleftharpoons I_2(s) + 2e^-$ E° = –0.54 V

Net reaction: $Cd^{2+} + 2I^- \rightleftharpoons I_2(s) + Cd(s)$ E° = –0.94 V

21.116 (a) Water is more readily oxidized than SO_4^{2-}, so we have:

$$2H_2O(\ell) \rightarrow 4H^+(aq) + 4e^- + O_2(g)$$

(b) Br^- is more readily oxidized than water, so we have:

$$2Br^-(aq) \rightarrow Br_2(aq) + 2e^-$$

(c) Br^-(aq) is more readily oxidized than SO_4^{2-}, so we have:

$$2Br^-(aq) \rightarrow Br_2(aq) + 2e^-$$

21.117 (a) Water is more easily reduced than K^+, so we write:

$$2H_2O(\ell) + 2e^- \rightarrow H_2(g) + 2OH^-(aq)$$

(b) Cu^{2+}(aq) is more readily reduced than water, so we write:

$$Cu^{2+}(aq) + 2e^- \rightarrow Cu(s)$$

(c) Cu^{2+}(aq) is more readily reduced than K^+(aq), so we write:

$$Cu^{2+}(aq) + 2e^- \rightarrow Cu(s)$$

21.118 The answers to the previous Review Questions guide us here:

Possible cathode reactions:

$K^+ + e^- \rightleftharpoons K(s)$ E° = –2.92 V

$Cu^{2+} + 2e^- \rightleftharpoons Cu(s)$ E° = +0.34 V

$2H_2O + 2e^- \rightleftharpoons H_2(g) + 2OH^-(aq)$ E° = –0.83 V

Cathode reaction: $Cu^{2+} + 2e^- \rightleftharpoons Cu(s)$

Possible anode reactions:

$2SO_4^{2-} \rightleftharpoons S_2O_8^{2-} + 2e^-$ E° = –2.01 V

$2Br^- \rightleftharpoons Br_2 + 2e^-$ E° = –1.07 V

$2H_2O \rightleftharpoons O_2(g) + 4H^+(aq) + 4e^-$ E° = –1.23 V

Anode reaction: $2Br^- \rightleftharpoons Br_2 + 2e^-$

Overall reaction: $Cu^{2+} + 2Br^- \rightleftharpoons Br_2 + Cu(s)$

21.119 At the cathode, where reduction occurs, we expect Cu(*s*). At the anode, where oxidation occurs, we expect $I_2(aq)$.
The net cell reaction would be $Cu^{2+}(aq) + 2I^-(aq) \rightarrow Cu(s) + I_2(aq)$

Additional Exercises

21.120 $\Delta G° = -nF E°_{cell}$, $E°_{cell} = 1.34$ V = 1.34 J/C, and n = 2
$\Delta G° = -(2)(96{,}500 \text{ C})(1.34 \text{ J/C}) = -2.59 \times 10^5$ J per mol of HgO

The maximum amount of work that can be derived from this cell, on using 1.00 g of HgO, is thus:

$$\# \text{ J} = (1.00 \text{ g HgO})\left(\frac{1 \text{ mol HgO}}{216.6 \text{ g HgO}}\right)\left(\frac{2.59 \times 10^5 \text{ J}}{1 \text{ mol HgO}}\right) = 1.20 \times 10^3 \text{ J}$$

Now, since 1 watt = 1 J s^{-1}, then 5×10^{-4} watt = 5×10^{-4} J s^{-1}, and the time required for this process is:

$$\# \text{ hr} = (1.20 \times 10^3 \text{ J})\left(\frac{1 \text{ s}}{5 \times 10^{-4} \text{ J}}\right)\left(\frac{1 \text{ min}}{60 \text{ s}}\right)\left(\frac{1 \text{ hr}}{60 \text{ min}}\right) = 6.67 \times 10^2 \text{ hr}$$

21.121 The initial numbers of moles of Ag^+ and Zn^{2+} are: 1.00 mol/L × 0.100 L = 0.100 mol. The number of Coulombs (A□s) that have been employed is: 0.10 C/s × 15.00 hr × 3600 s/hr = 5.4×10^3 C. The number of moles of electrons is: 5.4×10^3 C ÷ 96,500 C/mol = 5.6×10^{-2} mol electrons.

For Ag^+, there is 1 mol per mole of electrons, and for Zn^{2+}, there are two moles of electrons per mol of Zn. This means that the number of moles of the two ions that have been consumed or formed is given by:
5.6×10^{-2} mol e^- × 1 mol Ag^+/1 mol e^- = 5.6×10^{-2} mol Ag^+ reacted.

5.6×10^{-2} mol e^- × 1 mol Zn^{2+}/2 mol e^- = 2.8×10^{-2} mol Zn^{2+} formed

The number of moles of Ag^+ that remain is: 0.100 – 0.056 = 0.044 mol of Ag^+

The final concentration of silver ion is: $[Ag^+]$ = 0.044 mol/0.100 L = 0.44 M

The number of moles of Zn^{2+} that are present is: 0.100 + 0.028 = 0.128 mol Zn^{2+}

The final concentration of zinc ion is: $[Zn^{2+}]$ = 0.128 mol/0.100 L = 1.28 M

The standard cell potential should be: $E°_{cell} = E°_{reduction} - E°_{oxidation} = 0.80 - (-0.76) = 1.56$ V

We now apply the Nernst equation:

$$E_{cell} = E^\circ_{cell} - \frac{0.0592\ V}{2} \log \frac{1.28}{(0.44)^2}$$

$E_{cell} = 1.56\ V - 0.024\ V = 1.54\ V$

21.122 The cathode is positive in a galvanic cell, so we conclude that reduction of platinum ion takes place: $Pt^{2+}(aq) + 2e^- \rightarrow Pt(s)$ $E^\circ = ?$
The anode reaction is: $2\ Ag(s) + 2Cl^-(aq) \rightarrow 2AgCl(s) + 2e^-$ $E^\circ = -0.2223\ V$
The overall cell potential is calculated using the Nernst equation:

$$E_{cell} = E^\circ_{cell} - \frac{0.0592\ V}{2} \log \frac{1}{[Pt^{2+}][Cl^-]^2}$$

$$0.778\ V = E^\circ_{cell} - \frac{0.0592\ V}{2} \log \frac{1}{[0.0100][0.100]^2}$$

$E^\circ_{cell} = 0.778\ V + 0.118\ V = 0.896\ V$

$$E^\circ_{Pt^{2+}} - 0.2223\ V = 0.896\ V$$

$$E^\circ_{Pt^{2+}} = 0.896\ V + 0.2223\ V = 1.118\ V$$

21.123 The concentration of Br^- in solution will be 0.10 M, since the K_{sp} of AgBr is so small, the amount of dissociation of AgBr from the electrode will be negligible.
The reduction reaction will be
$AgBr(s) + e^- \leftrightarrows Ag(s) + Br^-(aq)$ $E^\circ = 0.070\ V$
The oxidation reaction will occur at the standard hydrogen electrode.
The net cell reaction is:
$2AgBr(s) + H_2 \rightarrow 2Ag(s) + 2Br^-(aq) + 2H^+(aq)$
$E^\circ_{cell} = 0.070\ V - 0.000\ V = 0.070\ V$
The potential for the constructed cell is calculated using the Nernst equation:

$$E_{cell} = E^\circ_{cell} - \frac{0.0592}{n} \log [Br^-]^2 [H^+]^2$$

$$E_{cell} = 0.070\ V - \frac{0.0592}{2} \log [0.1]^2 [1]^2$$

$E_{cell} = 0.070\ V - (-0.0592)$
$E_{cell} = 0.129\ V$

21.124 Our strategy will be thus:

Step A:
Pressure H_2 (wet) → partial pressure H_2 → mol H_2 → # e^- used
Step B:
Find total charge used = (current)(time)

Step C:
The charge per electron can be arrived at by:

Charge per e^- = total charge/total #e^- used = Step A/Step B

Step A:
Pressure H_2 (wet) → partial pressure H_2 → mol H_2 → # e^- used

The total pressure of wet hydrogen is 767 torr, but some of this is provided by water vapor. Consulting the water vapor pressure table in the appendices, we find that at 27 °C, the vapor pressure of water is 26.7 torr. Therefore pressure solely due to hydrogen gas (the partial pressure of hydrogen gas) is:

$PH_2 = 767 - 26.7 = 740$ torr
740 torr(1 atm/760 torr) = 0.974 atm

Using the ideal gas law,

PV = nRT
(0.974 atm)(0.288 L) = n(0.0821 L·atm mol^{-1}·K^{-1})(27 + 273K)
n = 0.0114 mol H_2

According to the electrolysis equation $2H^+ + 2e^- \rightleftharpoons H_2(g)$, 2 moles of electrons are required per mol of H_2 gas formed. Therefore,

electrons = 0.0114 mol H_2(2 mol e^-/1 mol H_2)(6.022 × 10^{23} electrons/mol)
= 1.35 × 10^{22} electrons

Step B:
Total charge used = (1.22 A)(1800 s) = 2200 C

Step C:
Charge per e^- = total charge/total #e^- used = 2200 C/1.35 × 10^{22} electrons
= 1.63 × 10^{-19} C per electron

(This is a fairly good estimate; the accepted value is 1.60 × 10^{-19} C.)

21.125 The oxidation reaction is:

$Fe^{2+} + 2e^- \rightarrow Fe$ $E° = -0.44$ V

The reduction reaction is:

$2H^+ + 2e^- \rightarrow H_2$ $E° = 0.00$ V

The overall cell reaction is

$2H^+ + Fe \rightarrow H_2 + Fe^{2+}$ $E^\circ_{cell} = 0.44$ V

$$E_{cell} = E^\circ_{cell} - \frac{0.0592\ V}{n}\log\frac{[Fe^{2+}]}{[H^+]^2}$$

$[Fe^{2+}] = 0.10$ M
$[H^+]$ is from the ionization of acetic acid

$$K_a = \frac{[H^+][C_2H_3O_2^-]}{[HC_2H_3O_2]} = 1.8\times 10^{-5}$$

$[H^+] = x$ $[C_2H_3O_2^-] = x$ $[HC_2H_3O_2] = 0.10 - x$

$$1.8 \times 10^{-5} = \frac{[x][x]}{[0.10 - x]}$$

Assume x is small compared to the concentration of $HC_2H_3O_2$
$x = 1.34 \times 10^{-3}$ M = $[H^+] = [C_2H_3O_2^-]$
$[HC_2H_3O_2] = 0.10$ M

n = 2

$$E_{cell} = 0.44\ V - \frac{0.0592\ V}{2}\log\frac{[0.10]}{[1.34\times10^{-3}]^2}$$

$E_{cell} = 0.44\ V - 0.14\ V$

$E_{cell} = 0.30\ V$

21.126 (a) $E^\circ_{cell} = 1.507\ V - (1.451\ V) = 0.056\ V$

(b) $\Delta G^\circ = -nFE^\circ = -(30\ mol\ e^-)\left(\frac{96,500\ C}{1\ mol\ e^-}\right)\left(\frac{0.056\ J}{1\ C}\right) = -1.62 \times 10^5\ J$

(c) $E^\circ_{cell} = \frac{0.0592}{n}\log K_c$

Since n = 30, we write:

$0.056 = 0.0592/30 \times \log K_c$

$\log K_c = 28.4$ and $K_c = 2.4 \times 10^{28}$

(d)

$$E_{cell} = E^\circ_{cell} - \frac{0.0592\ V}{2}\log Q$$

$$E_{cell} = 0.056\ V - \frac{0.0592\ V}{30}\log\frac{[Mn^{2+}]^6[ClO_3^-]^5}{[MnO_4^-]^6[Cl^-]^5[H^+]^{18}}$$

(e) $$E_{cell} = 0.056\ V - \frac{0.0592\ V}{30}\log\frac{(0.050)^6(0.110)^5}{(0.20)^6(0.0030)^5(5.62\times10^{-5})^{18}} = -0.103\ V$$

21.127 Reduction of these two metal ions should take place at different applied voltages. First, Ni should "plate out" at –0.25 V, followed by Cd at –0.40 V.

21.128 The approach is as follows:
area, thickness Cr → volume Cr → mass Cr → moles Cr → #e^- needed → current

V_{Cr} = (area)(thickness) = $(1.00\ m^2)(5.0 \times 10^{-5}\ m) = 5.0 \times 10^{-5}\ m^3$

$$\#\ e^- = (5.0\times10^{-5}\ m^3)\left(\frac{100\ cm}{1\ m}\right)^3\left(\frac{7.19\ g\ Cr}{1\ cm^3}\right)\left(\frac{1\ mol\ Cr}{52.0\ g\ Cr}\right)\left(\frac{6\ F}{1\ mol\ Cr}\right)\left(\frac{96,500\ C}{1\ F}\right)$$

$= 4.00 \times 10^6\ C$

This is done in 4.50 hr (16,200 s). So the current must be:

Current = charge/time = 4.00×10^6 C/16,200 s = 247 A

21.129 (a) First, we calculate the number of Coulombs:

$1.50\ A \times 30.0\ min \times 60\ s/min = 2.70 \times 10^3\ A□s = 2.70 \times 10^3\ C$

Then we determine the number of moles of electrons:

$$\#\ mole\ e^- = (2.70\times10^3\ C)\left(\frac{1\ mol\ e^-}{96,500\ C}\right) = 0.0280\ mol\ e^-$$

(b) $0.475\ g \div 50.9\ g/mol = 9.33 \times 10^{-3}$ mol V

(c) $(2.80 \times 10^{-2}\ \text{mol e}^-)/(9.33 \times 10^{-3}\ \text{mol V}) = 3.00\ \text{mol e}^-/\text{mol V}$
The original oxidation state was V^{3+}.

21.130 The balanced half-reactions are as follows:

$$4H_2O + Mn^{2+} \rightarrow MnO_4^- + 8H^+ + 5e^-$$

$$6e^- + 14H^+ + Cr_2O_7^{2-} \rightarrow 2Cr^{3+} + 7H_2O$$

Multiplying the top equation by 6 and the bottom by 5, and combining, we obtain:

$$24H_2O + 6Mn^{2+} \rightarrow MnO_4^- + 48H^+ + 30e^-$$

$$30e^- + 70H^+ + 5Cr_2O_7^{2-} \rightarrow 10Cr^{3+} + 35H_2O$$

Combining the two gives:

$$30e^- + 22H^+ + 5Cr_2O_7^{2-} + 6Mn^{2+} \rightarrow MnO_4^- + 10Cr^{3+} + 11H_2O + 30e^-$$

This gives n = 30.

Under standard (1.0 M) conditions,

$$Mn^{2+} \rightarrow MnO_4^- \quad E° = -1.49\ \text{V}$$
$$2Cr^{3+} \rightarrow Cr_2O_7^{2-} \quad E° = -1.33\ \text{V}$$

However in this case, the second reaction is reversed, giving:

$$Cr_2O_7^{2-} \rightarrow 2Cr^{3+} \quad E° = +1.33\ \text{V}$$

Therefore $E°_{cell} = 1.33\ \text{V} + (-1.49\ \text{V}) = -0.16\ \text{V}$.

Based on concentrations given, and the balanced chemical equation above,

$$E_{cell} = E^\circ_{cell} - \frac{RT}{nF} \ln\left(\frac{[MnO_4^-]^6 [Cr^{3+}]^{10}}{[Mn^{2+}]^5 [Cr_2O_7^{2-}]^6 [H^+]^{22}}\right)$$

$$E_{cell} = -0.16\ \text{V} - \frac{(8.314\ \text{J mol}^{-1}\text{K}^{-1})(298\ \text{K})}{30(96,500\ \text{C mol}^{-1})} \ln\left(\frac{[0.0010]^6 [0.0010]^{10}}{[0.10]^5 [0.010]^6 [10^{-6}]^{22}}\right)$$

$$E_{cell} = -0.16\ \text{V} - \frac{(8.314\ \text{J mol}^{-1}\text{K}^{-1})(298\ \text{K})}{30(96,500\ \text{C mol}^{-1})} \ln\left(\frac{[10^{-3}]^6 [10^{-3}]^{10}}{[10^{-1}]^5 [10^{-2}]^6 [10^{-6}]^{22}}\right)$$

$$E_{cell} = -0.16\ \text{V} - \left[8.558 \times 10^{-4} \ln\left(\frac{[10^{-18}][10^{-30}]}{[10^{-5}][10^{-12}][10^{-132}]}\right)\right]$$

$$E_{cell} = -0.16\text{ V} - \left[8.558 \times 10^{-4}\ \ln\left(10^{101}\right)\right]$$

$$E_{cell} = -0.16\text{ V} - \left[8.558 \times 10^{-4}(233)\right]$$

$$E_{cell} = -0.16\text{ V} - \left[0.1994\right]$$

$$E_{cell} = -0.36\text{ V}$$

(Standard temperature is assumed. H^+ concentration is obtained from the expression pH = –log [H^+].)

$$\Delta G = -nFE_{cell} = -\left(30\text{ mol e}^-\right)\left(\frac{96500\text{ C}}{1\text{ mol e}^-}\right)\left(\frac{-0.36\text{ J}}{1\text{ C}}\right) = 1.04 \times 10^6\text{ J}$$

This is 1.04×10^3 kJ.

Since ΔG is *positive*, the reaction will proceed in the reverse direction.

21.131 The half–reactions are:

$$2Cl^- \rightarrow Cl_2 + 2e^-$$

$$2H_2O + 2e^- \rightarrow H_2 + 2OH^-$$

The resulting pH (9.00) indicates a pOH of 5.00. This means that the concentration of hydroxide is: $[OH^-] = 1.00 \times 10^{-5}$ M

$1.00 \times 10^{-5}\text{ M} \times 0.500\text{ L} = 5.00 \times 10^{-6}$ moles OH^-
5.00×10^{-6} moles $OH^- \times 2\text{ mol e}^-/2\text{ mol OH}^- = 5.00 \times 10^{-6}\text{ mol e}^-$
$5.00 \times 10^{-6}\text{ mol e}^- \times 96500\text{ C/mol} = 0.482\text{ C}$
$0.482\text{ C} \div 0.500\text{ C/s} = 0.964\text{ s}$

21.132 $H_2(g) + 1/2O_2(g) \rightarrow H_2O(g)$

The theoretical maximum amount of work is given by the change in free energy:

$\Delta G = \Delta H - T\Delta S$

We can find ΔH, ΔS, and T as follows:

$\Delta H = [\Delta H^\circ_{f\ products}] - [\Delta H^\circ_{f\ reactants}] = [-241.8\text{ kJ}] - [0 + 0] = -241.8\text{ kJ}$

$\Delta S = [\Delta S^\circ_{products}] - [\Delta S^\circ_{reactants}] = [188.7\text{ J}] - [130.6\text{ J} + (1/2)205.0\text{ J}] = -44.4\text{ J}$
$= -0.0444\text{ kJ}$

$T = 110^\circ\text{C} + 273 = 373\text{ K}$

Inserting these values into the equation for ΔG, we obtain:

$\Delta G = (-241.8\text{ kJ}) - 373(-0.0444\text{ kJ}) = -224.8\text{ kJ}$

At 70% efficiency, this is $(-224.8)(0.70) = -157.4$ kJ
(The negative sign simply tells us that work is done by the system.)

1 kW = 1 kJ/s

$$\# \text{ g H}_2\text{/sec} = \left(\frac{1 \text{ kJ}}{1 \text{ sec}}\right)\left(\frac{1 \text{ mol H}_2}{157.4 \text{ kJ}}\right)\left(\frac{2.016 \text{ g H}_2}{1 \text{ mol H}_2}\right) = 0.01281 \text{ g H}_2 / \text{sec}$$

$$\# \text{ g O}_2\text{/sec} = \left(\frac{1 \text{ kJ}}{1 \text{ sec}}\right)\left(\frac{0.5 \text{ mol O}_2}{157.4 \text{ kJ}}\right)\left(\frac{32.00 \text{ g O}_2}{1 \text{ mol O}_2}\right) = 0.1065 \text{ g 0}_2 / \text{sec}$$

21.133 $1.00 \text{ A} = 1 \text{ C s}^{-1}$ $\quad 110 \text{ V} = 110 \text{ J C}^{-1}$

$$\# \text{ kJ} = 5.00 \text{ min}\left(\frac{60 \text{ s}}{1 \text{ min}}\right)\left(\frac{1 \text{ C}}{\text{s}}\right)\left(\frac{110 \text{ J}}{\text{C}}\right)\left(\frac{1 \text{ kJ}}{1000 \text{ J}}\right) = 33 \text{ kJ}$$

21.134

$$E_{cell} = E^\circ_{cell} - \frac{RT}{nF}\ln[Cl^-]$$

$$-0.0435 \text{ V} = 0.2223 \text{ V} - \frac{(8.314 \text{ J mol}^{-1}\text{K}^{-1})(298 \text{ K})}{1(96{,}500 \text{ C mol}^{-1})}\ln[Cl^-]$$

$$-0.2658 \text{ V} = \frac{(8.314 \text{ J mol}^{-1}\text{K}^{-1})(298 \text{ K})}{1(96{,}500 \text{ C mol}^{-1})}\ln[Cl^-]$$

$$-10.34 = \ln[Cl^-]$$

$$3.23\times10^{-5} = [Cl^-]$$

21.135 The half–reactions diagrammed in this problem are:

$Ag(s) \rightarrow Ag^+(aq) + e^-$	anode	$E° = -0.80$ V
$Fe^{3+}(aq) + e^- \rightarrow Fe^{2+}(aq)$	cathode	$E° = 0.77$ V

$$E^\circ_{cell} = E^\circ_{reduction} - E^\circ_{oxidation} = 0.77 \text{ V} - 0.80 \text{ V} = -0.03 \text{ V}$$

$$E_{cell} = E^\circ_{cell} - \frac{0.0592 \text{ V}}{1}\log\frac{[Fe^{2+}][Ag^+]}{[Fe^{3+}]}$$

$$E_{cell} = -0.03 \text{ V} - \frac{0.0592 \text{ V}}{1}\log\frac{[3.0\times10^{-4}][4.0\times10^{-2}]}{[1.1\times10^{-3}]}$$

$$= +0.09 \text{ V}$$

As stated, being a galvanic cell, and using conventions in cell notation, the left-side half–cell is the anode and negatively charged, and the right–side half–cell is the cathode and positively charged. The equation for the spontaneous cell reaction is $Fe^{3+}(1.1 \times 10^{-3}$ M) + Ag(s) $\rightarrow Ag^+(3.0 \times 10^{-4}$ M) + Fe^{2+}(0.040 M). This is an example of a concentration cell.

Practice Exercises

22.1 $^{226}_{88}\text{Ra} \rightarrow {}^{222}_{86}\text{Rn} + {}^{4}_{2}\text{He} + {}^{0}_{0}\gamma$

22.2 $^{90}_{38}\text{Sr} \rightarrow {}^{90}_{39}\text{Y} + {}^{0}_{-1}\text{e}$

22.3 Half life for Rn-222 = 4 days(24 h/day)(3600 s/hr) = 345,500 s
$k = \ln 2/t_{1/2} = (0.6931)/(345{,}500 \text{ s}) = 2.01 \times 10^{-6} \text{ s}^{-1}$

Activity = 6 pCi = 6×10^{-12} Ci (3.7×10^{10} dps/1Ci)
= 0.222 Bq
= 0.222 disintegrations per second

Activity = disintegrations/sec = kN
$0.222 = (2.01 \times 10^{-6} \text{ s}^{-1})N$
$N = 1.10 \times 10^{5}$ atoms Rn-222

22.4 We make use of the Inverse Square Law:

$$\frac{I_1}{I_2} = \frac{d_2^{\,2}}{d_1^{\,2}}$$

$$\frac{1.4 \text{ units}}{I_2} = \frac{(1.2 \text{ m})^2}{(10 \text{ m})^2} = 100 \text{ units (assuming 1 significant figure)}$$

Review Exercises

22.1 A radioactive substance has unstable nuclei, and emits high–energy streams of particles, and electromagnetic radiation.

22.2 We do not normally, in the course of everyday life, encounter nuclear reactions. In typical chemical reactions, the mass of the various materials that are involved is unaffected by the relatively slow velocities that these materials possess.

22.3 If the velocity of an object approaches the speed of light, then the denominator on the right side of equation 22.1 would take on a value increasingly close to zero. This would require the mass m of the object to increase, and, in the limit, become infinity.

22.4 (a) The sum of all of the energy in the universe, plus all of the mass equivalent of energy, is constant.
(b) $\Delta E = \Delta m_0 \times c^2$

22.5 A small amount of mass is converted to energy, and lost from the system, when the nucleons assemble into a stable atomic nucleus. The amount of mass that is converted into energy corresponds to the gain in stability that the nucleus achieves by loss of this amount of energy.

22.6 Naturally occurring radionuclides emit alpha particles (alpha radiation), beta particles, and gamma radiation.

22.7 (a) The alpha particle is composed of helium nuclei, and has a 2+ charge.
(b) Beta particles are electrons.
(c) Positrons are particles with a +1 charge, and have the same mass as the electron.
(d) A deuteron consists of one neutron and one proton, i.e. the nucleus of a deuterium atom.

22.8 The alpha particle is more massive than the other particles, and this, coupled with its high positive charge, makes it likely that an alpha particle will collide with something soon after it is ejected. This collision normally transforms the alpha particle into a helium atom, by gain of two electrons.

22.9 Gamma rays are emitted from the nucleus, whereas X rays arise from electronic transitions.

22.10 Electron capture of a low–level electron by the nucleus creates a low–level hole in the electron configuration that is filled by descent of an electron from an upper electronic level. The atom emits radiation corresponding to the difference in energies between the two electronic levels, and this radiation lies in the X ray region of the electromagnetic spectrum.

22.11 For each atom, the number of protons is plotted against the number of neutrons. The band is established by the location in this plot of all known stable nuclei.

22.12 Barium–140 should have the longer half-life because it has an even number of protons and neutrons.

22.13 Tin–112 has an even number of protons and neutrons. Both of these are odd for indium–112 (atomic number 49).

22.14 Lanthanum–139 is a stable nuclide because it has 82 neutrons, a magic number. Additionally, lanthanum–140 has both an odd number of protons and an odd number of neutrons which implies instability.

22.15 There must be a sufficient number of neutrons to shield the protons from one another, especially so as the number of protons becomes increasingly large.

22.16 The loss of an alpha particle is the most effective way to move the nuclide toward the band of stability, if the unstable nuclide lies to the right and above the band of stability, because the alpha particle has a relatively large mass.

22.17 The neutron-to-proton ratio of lead–164 is too low. It has too few neutrons for the number of protons in the nucleus.

22.18 The loss of a beta particle is most likely, because the net nuclear effect is the conversion of a neutron into a proton, and this reduces the neutron–to–proton ratio.

22.19 The loss of a positron is most likely, because the net nuclear effect is the conversion of a proton into a neutron, and this increases the neutron–to–proton ratio.

22.20 The net effect of electron capture is the conversion of a proton into a neutron, and this increases the neutron–to–proton ratio. Radionuclides lying below the band of stability are more likely to undergo electron capture than those lying above the band of stability.

22.21 A compound nucleus is unstable because of excess energy.

22.22 Because the compound nucleus has no memory of how it was prepared, it may decay by any path which lowers the energy. The actual pathway depends on the amount of excess energy the compound nucleus has.

22.23 ${}^{4}_{2}\text{He} + {}^{14}_{7}\text{N} \rightarrow {}^{18}_{9}\text{F}^{*} \rightarrow {}^{1}_{0}\text{n} + {}^{17}_{9}\text{F}$

22.24 The Geiger counter detects the ions that form in an enclosed gas through which radiation is passing. The radiation must have high enough energy to penetrate the tube holding the gas and the radiation must then be able to ionize the gas.

22.25 Radiation generates unstable and reactive ions within cells. These reactive ions can participate in reactions that can eventually lead to birth defects, tumors, mutations, and cancer.

22.26 (a) The common unit of radioactivity is the curie (Ci), whereas the SI unit of radioactivity is the becquerel (Bq).
(b) The unit of energy for radiation is the electron–volt, or its multiples and the SI unit is the joule.
(c) The SI unit of absorbed dose is the gray (Gy), while the rad is an older common unit.
(d) The SI unit of dose equivalent is the sievert (Sv) and the rem, or roentgen equivalent for man, is the older common unit.

22.27 This is one curie. It is also: $1.0 \text{ Ci} \times (3.7 \times 10^{10} \text{ Bq/Ci}) = 3.7 \times 10^{10} \text{ Bq}$

22.28 The sievert accounts for the type of radiation exposure and the type of tissue as well as the dose.

22.29 A short half–life is desirable in medical work in order to minimize the exposure that a patient receives from radioactive materials. This reduces radiation–caused damage to the body. If the half–life is too short, the radionuclide will not be present in the body long enough to permit useful measurements.

22.30 Alpha particles cannot reach a detector outside the body, because they cannot penetrate tissue. Furthermore, as tissues in the body capture alpha particles, they are damaged.

22.31 A sample possessing a specific element is bombarded with neutrons creating a neutron enriched sample for analysis. These samples decay by gamma emission at frequencies that are characteristic of the element in question. By analyzing the frequency of the emission, the identity of the emitting element may be determined. By analyzing the intensity of the gamma emission, the concentration may be determined.

22.32 The assumption is that the lead in a mixture being analyzed has all come from the disintegration of uranium only.

22.33 Contamination with air could introduce new sources of carbon–containing compounds which would completely invalidate any analysis.

22.34 Some of the forms of radiation that everybody experiences are cosmic rays, diagnostic X rays, radioactive pollutants, fallout from atmospheric tests of nuclear devices, and, in some cases, radiation used in medicine.

22.35 A nucleus can more easily capture a neutron than a proton because the neutron is neutral. The proton, having a positive charge, is repelled by a nucleus as it approaches.

22.36 (a) A thermal neutron is one whose kinetic energy is governed by the temperature of its surroundings, even at room temperature.
(b) Nuclear fission is the breakup of a heavy nucleus, normally caused by absorption of a neutron, to give two or more lighter nuclei, plus two or more high-energy neutrons (i.e. not simply thermal neutrons).
(c) A fissile isotope is one that is capable of undergoing fission following neutron capture.

22.37 The naturally occurring fissile isotope is uranium–235.

22.38 The fission of each uranium–235 produces more than two new neutrons, which can react further to sustain a chain reaction. In other words, more neutrons are generated by each fission event than are needed to cause another such event.

22.39 The initial isotopes of Kr and Ba that are produced by the fission of uranium–235 are neutron–rich materials. In other words, their neutron–to–proton ratios are too high, and they spontaneously emit the excess neutrons to form Kr–94 and Ba–139.

22.40 The sub–critical mass is incapable of self–capturing all of the emitted neutrons, which are necessary for sustaining a chain reaction.

22.41 A moderator in a nuclear reactor is used to slow down the fast neutrons that are produced by fission.

22.42 No critical mass can form because the atoms of the fissionable isotopes are greatly diluted by those of the non–fissile isotopes.

Review Problems

22.43 Solve the Einstein equation for Δm:
$m = \Delta E/c^2$
$1 \text{ kJ} = 1.00 \times 10^3 \text{ J} = 1.00 \times 10^3 \text{ kg m}^2 \text{ s}^{-2}$
$\Delta m = 1.\ 00 \times 10^3 \text{ kg m}^2 \text{ s}^{-2} \div (3.\ 00 \times 10^8 \text{ m/s})^2 = 1.11 \times 10^{-14} \text{ kg} = 1.11 \times 0^{-11} \text{ g}$

22.44 Equation 22.1 becomes:

$$m = \frac{1.00}{\sqrt{1 - \left(\frac{v}{c}\right)^2}}$$

On substituting the various velocities for the value of v in the above expression, and using the value $c = 3.00 \times 10^8 \text{ m s}^{-1}$, we get: (a) m = 1.01 kg (b) 3.91 kg (c) 12.3 kg

22.45 The joule is equal to one kg m^2/s^2, and this is employed directly in the Einstein equation: $\Delta m = \Delta E/c^2$, where ΔE is the enthalpy of formation of liquid water, which is available in Table 5.2.

$H_2(g) + O_2(g) \rightarrow H_2O(\ell)$, $\Delta H = -285.9 \text{ kJ/mol}$
$\Delta m = (-285.9 \times 10^3 \text{ kg m}^2/\text{s}^2) \div (3.00 \times 10^8 \text{ m/s}^2)^2 = -3.18 \times 10^{-12} \text{ kg}$
$(-3.18 \times 10^{-12} \text{ kg}) \times 1000 \text{ g/kg} \times 10^9 \text{ ng/g} = -3.18 \text{ ng}$
The negative value for the mass implies that mass is lost in the reaction.

22.46 Solve the Einstein equation for Δm:
$\Delta m = \Delta E/c^2$
$890 \text{ kJ} = 890 \times 10^3 \text{ J} = 890 \times 10^3 \text{ kg m}^2 \text{ s}^{-2}$
$\Delta m = 890 \times 10^3 \text{ kg m}^2 \text{ s}^{-2} \div (3.00 \times 10^8 \text{ m/s})^2 = 9.89 \times 10^{-12} \text{ kg}$
$= 9.89 \times 10^{-9} \text{ g} = 9.89 \text{ ng}$
Note: we have assumed three sig figs for this calculation in order to agree with the text.

22.47 The mass of the deuterium nucleus is the mass of the proton (1.00727252 u) plus that of a neutron (1.008665 u), or 2.015938 u. The difference between this calculated value and the observed value is equal to Δm:
$\Delta m = (2.015938 - 2.0135) = 2.4 \times 10^{-3} \text{ u}$
$\Delta E = \Delta mc^2 = (2.4 \times 10^{-3} \text{ u})(1.6606 \times 10^{-27} \text{ kg/u})(3.00 \times 10^8 \text{ m/s})^2$
$\Delta E = 3.6 \times 10^{-13} \text{ kg m}^2/\text{s}^2 = 3.6 \times 10^{-13} \text{ J}$

Since there are two neucleons per deuterium nucleus, we have:
$\Delta E = 3.6 \times 10^{-13} \text{ J/2 nucleons} = 1.8 \times 10^{-13} \text{ J per nucleon}$

22.48 The mass of the tritium nucleus is the mass of one proton plus that of two neutrons: 1.007276470 + 2(1.008664904) = 3.024606 u. The difference between this calculated value and the observed value is equal to Δm:
$\Delta m = 3.024606 - 3.01550 = 9.11 \times 10^{-3} \text{ u}$
$\Delta E = \Delta mc^2 = (9.11 \times 10^{-3} \text{ u})(1.6605 \times 10^{-27} \text{ kg/u})(3.00 \times 10^8 \text{ m/s})^2$
$\Delta E = 1.36 \times 10^{-12} \text{ kg m}^2/\text{s}^2 = 1.36 \times 10^{-12} \text{ J}$

Since there are three nucleons per tritium nucleus, the energy per nucleon is:
$1.36 \times 10^{-12} \text{ J/3 nucleons} = 4.53 \times 10^{-13} \text{ J per nucleon}$

22.49 (a) $^{211}_{83}Bi$ (b) $^{177}_{72}Hf$ (c) $^{216}_{84}Po$ (d) $^{19}_{9}F$

22.50 (a) $^{241}_{94}Pu$ (b) $^{146}_{57}La$ (c) $^{58}_{28}Ni$ (d) $^{68}_{31}Ga$

22.51 (a) $^{242}_{94}Pu \rightarrow ^{4}_{2}He + ^{238}_{92}U$
(b) $^{28}_{12}Mg \rightarrow ^{0}_{-1}e + ^{28}_{13}Al$
(c) $^{26}_{14}Si \rightarrow ^{0}_{1}e + ^{26}_{13}Al$
(d) $^{37}_{18}Ar + ^{0}_{-1}e \rightarrow ^{37}_{17}Cl$

22.52 (a) $^{55}_{26}Fe + ^{0}_{-1}e \rightarrow ^{55}_{25}Mn$
(b) $^{42}_{19}K \rightarrow ^{0}_{-1}e + ^{42}_{20}Ca$
(c) $^{93}_{44}Ru \rightarrow ^{0}_{1}e + ^{93}_{43}Tc$
(d) $^{251}_{98}Cf \rightarrow ^{4}_{2}He + ^{247}_{96}Cm$

22.53 (a) $^{261}_{102}No$ (b) $^{211}_{82}Pb$ (c) $^{141}_{61}Pm$ (d) $^{179}_{74}W$

22.54 (a) $^{80}_{38}Sr$ (b) $^{121}_{50}Sn$ (c) $^{50}_{25}Mn$ (d) $^{257}_{100}Fm$

22.55 $^{87}_{36}Kr \rightarrow ^{86}_{36}Kr + ^{1}_{0}n$

22.56 $^{58}_{26}Fe$

22.57 The more likely process is positron emission, because this produces a product having a higher neutron–to–proton ratio: $^{38}_{19}K \rightarrow ^{0}_{1}e + ^{38}_{18}Ar$

22.58 Electron capture is the more likely event: $^{37}_{18}Ar + ^{0}_{-1}e \rightarrow ^{37}_{17}Cl$

This produces chlorine–37, which has a magic number, i.e., 20 neutrons. Beta emission would give potassium–37, which is less stable.

22.59 Six half–life periods correspond to the fraction 1/64 of the initial material. That is, one sixty–fourth of the initial material is left after 6 half lives: $3.00 \text{ mg} \times 1/64 = 0.0469 \text{ mg}$ remaining.

22.60 After four half–life periods, one sixteenth of the original sample remains:
$(9.00 \times 10^{-9} \text{ g}) \times 1/16 = 5.62 \times 10^{-10}$ g remaining

22.61 $^{53}_{24}Cr^*$; $^{51}_{23}V + ^{2}_{1}H \rightarrow ^{53}_{24}Cr^* \rightarrow ^{1}_{1}p + ^{52}_{23}V$

22.62 $^{19}_{9}F + ^{4}_{2}He \rightarrow ^{23}_{11}Na^* \rightarrow ^{22}_{11}Na + ^{1}_{0}n$

22.63 $^{80}_{35}Br$

22.64 $^{115}_{48}Cd + ^{1}_{0}n \rightarrow ^{116}_{48}Cd + \gamma$

22.65 $^{55}_{26}Fe$; $^{55}_{25}Mn + ^{1}_{1}p \rightarrow ^{1}_{0}n + ^{55}_{26}Fe$

22.66 $^{23}_{11}Na + ^{4}_{2}He \rightarrow ^{27}_{13}Al + \gamma$

22.67 $^{70}_{30}Zn + ^{208}_{82}Pb \rightarrow ^{278}_{112}Uub \rightarrow ^{1}_{0}n + ^{277}_{112}Uub$

22.68 $^{209}_{83}Bi + ^{64}_{28}Ni \rightarrow ^{273}_{111}Uuu \rightarrow ^{1}_{0}n + ^{272}_{111}Uuu$

22.69

$$\text{Radiation} \propto \frac{1}{d^2}$$

$$\frac{I_1}{I_2} = \frac{d_2^{\,2}}{d_1^{\,2}}$$

$$d_2 = d_1\sqrt{\frac{I_1}{I_2}} = 2.0\text{m}\sqrt{\frac{2.8}{0.28}} = 6.3\text{ m}$$

22.70

$$\frac{I_1}{I_2} = \frac{d_2^{\,2}}{d_1^{\,2}}$$

$$\frac{d_2}{d_1} = \sqrt{\frac{I_1}{I_2}} = \sqrt{\frac{100}{90}} = 1.05$$

Increase the distance by 5% to decrease exposure by 10%.

22.71 This calculation makes use of the Inverse Square Law:

$$\frac{I_1}{I_2} = \frac{d_2^{\,2}}{d_1^{\,2}}$$

$$\frac{8.4\text{ rem}}{0.50\text{ rem}} = \frac{d_2^{\,2}}{(1.60\text{ m})^2}$$

$$d_2 = 6.6\text{ m}$$

22.72 This calculation makes use of the Inverse Square Law:

$$\frac{I_1}{I_2} = \frac{d_2^{\,2}}{d_1^{\,2}}$$

$$\frac{50\text{ mrem}}{I_2} = \frac{(0.50\text{ m})^2}{(4.0\text{ m})^2}$$

$$I_2 = 3.2 \times 10^3\text{ mrem or } 3.2\text{ rem}$$

22.73 Activity = kN and $t_{1/2} = \frac{\ln 2}{k}$ or $k = \frac{\ln 2}{t_{1/2}}$

$$\text{Activity} = \frac{\ln 2}{t_{1/2}}N$$

$$N = 0.20\text{ mg}\left(\frac{1\text{ g }^{241}\text{Am}}{1000\text{ mg}}\right)\left(\frac{1\text{ mol }^{241}\text{Am}}{241\text{ g }^{241}\text{Am}}\right)\left(\frac{6.022 \times 10^{23}\text{ atoms }^{241}\text{Am}}{1\text{ mol }^{241}\text{Am}}\right) = 5.0 \times 10^{17}\text{ atoms }^{241}\text{Am}$$

$$\text{Activity} = \left(\frac{\ln 2}{1.70 \times 10^5\text{ d}}\right) \times (5.0 \times 10^{17}\text{ atoms }^{241}\text{Am})$$

$$= \left(\frac{1.0\times10^{12}\text{ Am decays}}{\text{d}}\right)\left(\frac{1\text{ d}}{24\text{ h}}\right)\left(\frac{1\text{ h}}{3600\text{ s}}\right) = 2.4\times10^{7}\text{ s}^{-1}$$

$$2.4\times10^{7}\text{ s}^{-1} = 2.4\times10^{7}\text{ Bq} = 2.4\times10^{7}\text{ Bq}\left(\frac{1\text{ Ci}}{3.7\times10^{10}\text{ Bq}}\right)\left(\frac{1000\text{ mCi}}{1\text{ Ci}}\right) = 6.5\times10^{-1}\text{ mCi}$$

$$= 6.5\times10^{2}\ \mu\text{Ci}$$

22.74 Activity = kN and $t_{1/2} = \frac{\ln 2}{k}$ or $k = \frac{\ln 2}{t_{1/2}}$

$$\text{Activity} = \frac{\ln 2}{t_{1/2}}N$$

$$N = 1.00\text{ g}\left(\frac{1\text{ mol }^{90}\text{Sr}}{90\text{ g }^{90}\text{Sr}}\right)\left(\frac{6.022\times10^{23}\text{ atoms }^{90}\text{Sr}}{1\text{ mol }^{90}\text{Sr}}\right) = 6.7\times10^{21}\text{ atoms }^{90}\text{Sr}$$

$$\text{Activity} = \left(\frac{\ln 2}{1.00\times10^{4}\text{ d}}\right)\times(6.7\times10^{21}\text{ atoms }^{290}\text{Sr})$$

$$= \left(\frac{4.6\times10^{17}\text{ Sr decays}}{\text{d}}\right)\left(\frac{1\text{ d}}{24\text{ h}}\right)\left(\frac{1\text{ h}}{3600\text{ s}}\right) = 5.4\times10^{12}\text{ s}^{-1}$$

$$5.4\times10^{12}\text{ s}^{-1} = 5.4\times10^{12}\text{ Bq} = 5.4\times10^{12}\text{ Bq}\left(\frac{1\text{ Ci}}{3.7\times10^{10}\text{ Bq}}\right)\left(\frac{1000\text{ mCi}}{1\text{ Ci}}\right) = 1.5\times10^{5}\text{ mCi}$$

22.75 Activity = kN $\qquad k = \frac{\text{activity}}{N}$

$$N = \text{number of }^{131}\text{I atoms} = 1.00\text{ mg }^{131}\text{I}\left(\frac{1\text{ g }^{131}\text{I}}{1000\text{ mg }^{131}\text{I}}\right)\left(\frac{1\text{ mol }^{131}\text{I}}{131\text{ g }^{131}\text{I}}\right)\left(\frac{6.022\times10^{23}\text{ g }^{131}\text{I}}{1\text{ mol }^{131}\text{I}}\right)$$

$$= 4.60\times10^{18}\text{ atoms }^{131}\text{I}$$

$$k = \frac{4.6\times10^{12}\text{ Bq}}{4.60\times10^{18}\text{ atoms }^{131}\text{I}}\left(\frac{1\text{ s}^{-1}}{1\text{ Bq}}\right) = 1.0\times10^{-6}\text{ s}^{-1}$$

$$t_{1/2} = \frac{\ln 2}{k} = \left(\frac{\ln 2}{1.0\times10^{-6}\text{ s}^{-1}}\right) = 6.9\times10^{5}\text{ s (This is about 8 days.)}$$

22.76 Activity = kN $\qquad k = \frac{\text{activity}}{N}$

$$N = \text{number of }^{201}\text{Tl atoms} = 10.0\text{ mg }^{201}\text{Tl}\left(\frac{1\text{ g }^{201}\text{Tl}}{1000\text{ mg }^{201}\text{Tl}}\right)\left(\frac{1\text{ mol }^{201}\text{Tl}}{201\text{ g }^{201}\text{Tl}}\right)\left(\frac{6.022\times10^{23}\text{ g }^{201}\text{Tl}}{1\text{ mol }^{201}\text{Tl}}\right)$$

$$= 3.00\times10^{19}\text{ atoms }^{201}\text{Tl}$$

$$k = \frac{7.9\times10^{13}\text{ Bq}}{3.00\times10^{19}\text{ atoms }^{201}\text{Tl}}\left(\frac{1\text{ s}^{-1}}{1\text{ Bq}}\right) = 2.64\times10^{-6}\text{ s}^{-1}$$

$$t_{1/2} = \frac{\ln 2}{k} = \left(\frac{\ln 2}{2.64\times10^{-6}\text{ s}^{-1}}\right) = 2.63\times10^{5}\text{ s}$$

22.77 The chemical product is $BaCl_2$.Recall that for a first order process $k = \dfrac{0.693}{t_{1/2}}$

So $k = 0.693/30\ yr = 2.30 \times 10^{-2}/yr$. Also,

$$\ln\frac{[A]_0}{[A]_t} = kt$$

$$[A]_t = [A]_0 \exp(-kt)$$

$$\frac{[A]_0}{[A]_t} = \exp[-(2.30 \times 10^{-2}\ /yr)(150\ yr)]$$

$$\frac{[A]_0}{[A]_t} = 3.13 \times 10^{-2}$$

so 3.1% of the original sample remains.

22.78 For a first–order process, the rate constant is related to the half–life by the equation:
$k = 0.693/t_{1/2}$ Hence, $k = 0.693/28.1\ yr = 2.47 \times 10^{-2}\ yr^{-1}$

Also, for a first–order process, the concentration varies with time according to the equation:

$$\ln\frac{[A]_0}{[A]_t} = kt$$

which allows us to solve for the time, t, for the activity to decrease by the specified amount:

$$t = \frac{1}{k} \times \ln\frac{[A]_0}{[A]_t} = \frac{1}{2.47 \times 10^{-2}\ yr^{-1}} \times \ln\frac{(0.245\ Ci/g)}{(1.00 \times 10^{-6}\ Ci/g)} = 502\ yr$$

22.79 This calculation makes use of the first order rate equation, where knowing $[A]_t$, we need to calculate $[A]_0$:

$$\ln\frac{[A]_0}{[A]_t} = kt.$$

$k = 0.693/t_{1/2} = 0.693/8.07\ d = 8.59 \times 10^{-2}\ d^{-1}$

$$\ln\frac{[A]_0}{(25.6 \times 10^{-5}\ Ci/g)} = (8.59 \times 10^{-2}\ d^{-1})(28.0\ d)$$

Taking the exponential of both sides of the above equation gives:

$$\frac{[A]_0}{(25.6 \times 10^{-5}\ Ci/g)} = e^{2.41} = 11.1$$

Solving for the value of $[A]_0$ gives: $[A]_0 = 2.84 \times 10^{-3}$ Ci/g

22.80 The rate constant for the first–order process is first determined: $k = 0.693/t_{1/2} = 0.693/6.02\ hr = 0.115\ hr^{-1}$

Also, we know that: $\ln\dfrac{[A]_0}{[A]_t} = kt$

$$\ln\frac{(4.52 \times 10^{-6}\ Ci)}{[A]_t} = (0.115\ hr^{-1})(8.00\ hr) = 0.920$$

Taking the exponential of both sides of this equation gives:

$$\frac{(4.52\times10^{-6}\text{ Ci})}{[A]_t} = \text{antiln}\,(0.920) = 2.51$$

$[A]_t = 1.80 \times 10^{-6}$ Ci, the activity after 8 hours.

22.81 In order to solve this problem, it must be assumed that all of the argon–40 that is found in the rock must have come from the potassium–40, i.e., that the rock contains no other source of argon–40. If the above assumption is valid, then any argon–40 that is found in the rock represents an equivalent amount of potassium–40, since the stoichiometry is 1:1. Since equal amounts of potassium–40 and argon–40 have been found, this indicates that the amount of potassium–40 that remains is exactly half the amount that was present originally. In other words, the potassium–40 has undergone one half–life of decay by the time of the analysis. The rock is thus seen to be 1.3×10^9 years old.

22.82 In a fashion similar to that outlined in the answer to Review Problem 22.81, we conclude that, in order for the rock to be one half–life old (1.3×10^9 yr), there must be equal amounts of the two isotopes, one having been formed by decay of the other. The answer is, thus, 1.16×10^{-7} mol of potassium–40.

22.83 Using equation 22.5 we may determine how long it has been since the tree died.

$$\ln\left(\frac{^{14}C}{^{12}C}\right) = (1.2\times10^{-4})t$$

Taking the natural log we determine:

$$\ln\left(\frac{1.2\times10^{-12}}{4.8\times10^{-14}}\right) = (1.2\times10^{-4})t$$

$$t = \left(\frac{1}{1.2\times10^{-4}}\right)\ln\left(\frac{1.2\times10^{-12}}{4.8\times10^{-14}}\right) = 2.7\times10^{4}\text{ yr}$$

The tree died 2.7×10^4 years ago. This is when the volcanic eruption occurred.

22.84 Use equation 22.5

$$\ln\frac{r_0}{r_t} = kt$$

$$\ln\left(\frac{1.2\times10^{-12}}{r_t}\right) = (1.21\times10^{-4}\text{ y}^{-1})(9.0\times10^{3}\text{ y})$$

$$\left(\frac{1.2\times10^{-12}}{r_t}\right) = \exp\left[(1.21\times10^{-4}\text{ y}^{-1})(9.0\times10^{3}\text{ y})\right]$$

$$\left(\frac{1.2\times10^{-12}}{2.97}\right) = r_t = 4.0\times10^{-13}$$

22.85 $^{235}_{92}U + ^{1}_{0}n \rightarrow ^{94}_{38}Sr + ^{140}_{54}Xe + 2^{1}_{0}n.$

22.86 Both will be beta emitters. Both lie above the band of stability, and they can mover closer to it by emitting beta particles. Some of the extra neutrons produced by the fission shown in Review Problem 22.86 may bombard additional U–235 nuclei to cause further fission products to form.

Additional Exercises

22.87 (a) $^{30}_{13}Al \rightarrow {}^{0}_{-1}e + {}^{30}_{14}Si$

(b) $^{252}_{99}Es \rightarrow {}^{4}_{2}He + {}^{248}_{97}Bk$

(c) $^{93}_{42}Mo + {}^{0}_{-1}e \rightarrow {}^{93}_{41}Nb$

(d) $^{28}_{15}P \rightarrow {}^{0}_{1}e + {}^{28}_{14}Si$

22.88 The mass of the iron–56 nucleus is the mass of twenty–six protons plus the mass of thirty neutrons. The difference between this calculated value and the observed value is equal to m:
$(26 \times 1.007276470\ u) + (30 \times 1.008664904\ u) = 56.449135\ u$
$\Delta m = (56.449135\ u - 55.9349\ u) = 0.5142\ u$
$\Delta E = \Delta mc^2 = (0.5142\ u)(1.6605 \times 10^{-27}\ kg/u)(3.00 \times 10^8\ m/s)^2$
$\Delta E = 7.685 \times 10^{-11}\ kg\ m^2/s^2 = 7.685 \times 10^{-11}\ J$

Since there are fifty-six neucleons per iron–56 nucleus, we have:
$\Delta E = 7.685 \times 10^{-11}$ J/56 nucleons $= 1.372 \times 10^{-12}$ J per nucleon.

This has to be the largest binding energy possible based upon the information provided in Figure 22.1. According to this Figure, iron–56 is the most stable isotope known.

22.89 Uranium–235 has 92 protons and 143 neutrons. The mass of one nucleus is:
m: $(92 \times 1.007276470\ u) + (143 \times 1.008664904\ u) = 236.908517\ u$
$\Delta m = (236.908517\ u - 235.0439\ u) = 1.8646\ u$
$\Delta E = \Delta mc^2 = (1.8646\ u)(1.6605 \times 10^{-27}\ kg/u)(3.00 \times 10^8\ m/s)^2$
$\Delta E = 2.787 \times 10^{-10}\ kg\ m^2/s^2 = 2.787 \times 10^{-10}\ J$

Since there are 235 neucleons per uranium–235 nucleus, we have:
$\Delta E = 2.786 \times 10^{-10}$ J/235 nucleons $= 1.186 \times 10^{-12}$ J per nucleon.

22.90 (a) $^{10}_{6}C \rightarrow {}^{0}_{1}e + {}^{10}_{5}B$

(b) $^{243}_{96}Cm \rightarrow {}^{4}_{2}He + {}^{239}_{94}Pu$

(c) $^{49}_{23}V + {}^{0}_{-1}e \rightarrow {}^{49}_{22}Ti$

(d) $^{20}_{8}O \rightarrow {}^{0}_{-1}e + {}^{20}_{9}F$

22.91 None. The parent mass may be exactly the same as the daughter mass since the positron has a mass of zero.

22.92 In order to fuse, these two nuclei must overcome the repulsive force that exists between particles of like charge. This is only possible by adding huge amounts of thermal energy to a mixture of the particles or by using extreme pressure.

22.93 $^{214}_{83}Bi \rightarrow {}^{4}_{2}He + {}^{210}_{81}Tl$

$^{210}_{81}Tl \rightarrow {}^{0}_{-1}e + {}^{210}_{82}Pb$

$^{210}_{82}Pb \rightarrow {}^{0}_{-1}e + {}^{210}_{83}Bi$

$^{210}_{83}Bi \rightarrow {}^{0}_{-1}e + {}^{210}_{84}Po$

$^{210}_{84}Po \rightarrow {}^{4}_{2}He + {}^{206}_{82}Pb$

Element E is $^{206}_{82}Pb$

22.94 (a) $^{15}_{7}N$

(b) $\ln \frac{[A]_0}{[A]_t} = kt.$

$k = 0.693/t_{1/2} = 0.693/124\ s = 5.59 \times 10^{-3}\ s^{-1}$

$$\ln \frac{750\ mg}{[A]_t} = (5.59 \times 10^{-3}\ s^{-1})(300\ s)$$

$[A]_t = 140\ mg$

22.95 Positron decay

22.96 Recall that for a first order process $k = 0.693/t_{1/2}$
So $k = 0.693/1.3 \times 10^9\ yr = 5.30 \times 10^{-10}/yr$. Also,

$$\ln\frac{[A]_0}{[A]_t} = kt$$

$$t = \frac{1}{k} \ln \frac{[A]_0}{[A]_t} = \frac{1}{5.30 \times 10^{-10}\ /yr} \ln \frac{3.22 \times 10^{-5}\ mol}{2.07 \times 10^{-5}\ mol} = 8.2 \times 10^8\ yrs$$

22.97 Recall that for a first order process $k = 0.693/t_{1/2}$
So $k = 0.693/5730\ yr = 1.21 \times 10^{-4}/yr$. Also,

$$\ln\frac{[A]_0}{[A]_t} = kt$$

$$t = \frac{1}{k} \ln \frac{[A]_0}{[A]_t} = \frac{1}{1.21 \times 10^{-4}\ /yr} \ln \frac{8}{1} = 1.72 \times 10^4\ yrs$$

22.98 Because this is an equilibrium process, there is a chance that either the forward or the reverse reactions will occur. As NO reacts with NO_2 we form ONO*NO. This compound can decompose and give either NO_2 and *NO or NO and *NO_2. After sufficient time, both *NO or NO will be present in the mixture.

22.99 As time passes, equal parts of CH_3HgI and CH_3*HgI will be produced. Since half the atoms are labeled, half the products will be labeled.

22.100 This problem is similar to a dilution problem, i.e., $C_1V_1 = C_2V_2$. We will use cpm as the concentration unit. First, we need to account for the density difference.

$$\text{Methanol : concentration} = \left(\frac{580\ cpm}{g}\right)\left(\frac{0.792\ g}{mL}\right) = 459\ \frac{cpm}{mL}$$

$$\text{Coolant : concentration} = \left(\frac{29\ cpm}{g}\right)\left(\frac{0.884\ g}{mL}\right) = 26\ \frac{cpm}{mL}$$

Now we want the volume of the cooling system. Solve the following:

$$\frac{\left(\frac{459\ cpm}{mL}\right)(10.0\ mL)}{\frac{26\ cpm}{mL}} = 180\ mL$$

22.101 Start by determining an activity per mol of Cr and mol $C_2O_4^{2-}$

For Cr:

$$\left(\frac{843\text{ cpm}}{\text{g}}\right)\left(\frac{294.18\text{ g}}{1\text{ mol }K_2Cr_2O_7}\right)\left(\frac{1\text{ mol }K_2Cr_2O_7}{2\text{ mol Cr}}\right) = 1.24 \times 10^5\text{ cpm/mol Cr}$$

For oxalate:

$$\left(\frac{345\text{ cpm}}{\text{g}}\right)\left(\frac{90.04\text{ g}}{1\text{ mol}}\right) = 3.1 \times 10^4\text{ cpm/mol }C_2O_4^{2-}$$

Again this is an equilibrium problem.

$$\text{moles Cr} = \frac{165\text{ cpm}}{1.24 \times 10^5\text{ cpm/mol Cr}} = 1.33 \times 10^{-3}\text{ mole Cr}$$

$$\text{moles }C_2O_4^{2-} = \frac{83\text{ cpm}}{3.11 \times 10^4\text{ cpm/mol }C_2O_4^{2-}} = 2.67 \times 10^{-3}\text{ mole }C_2O_4^{2-}$$

It is easy to see by inspection that there are 2 moles $C_2O_4^{2-}$ for each mole of Cr(III)

22.102 Starting with the half-life of ^{131}I, 8.07 d, calculate k.

$$t_{\frac{1}{2}} = \frac{\ln 2}{k}$$

$$k = \frac{\ln 2}{t_{\frac{1}{2}}}$$

$$k = \frac{0.693}{8.07\text{ d}} = 0.0859\text{ d}^{-1}$$

Then calculate the desired activity from the mass of the thyroid and the dose:

$$A = (20\text{ g thyroid})\left(\frac{86\ \mu\text{Ci}}{\text{g}}\right) = 1720\ \mu\text{Ci}$$

$$A = 1720\ \mu\text{Ci}\left(\frac{3.7 \times 10^{10}\text{ disintegrations/sec}}{1\ \mu\text{Ci}}\right) = 6.36 \times 10^{10}\text{ disintegrations/sec}$$

Now calculate the number of ^{131}I atoms that will give this activity:

$$A = kN$$

$$6.36 \times 10^{10}\text{ disintegrations/sec} = 0.0859\text{ d}^{-1}\left(\frac{1\text{ d}}{24\text{ h}}\right)\left(\frac{1\text{ h}}{3600\text{ s}}\right)N$$

$$6.36 \times 10^{10}\text{ disintegrations/sec} = 9.94 \times 10^{-7}\,N$$

$$N = 6.40 \times 10^{16}\ ^{131}\text{I atoms}$$

$$\text{Mass of }^{131}\text{I} = 6.40 \times 10^{16}\ ^{131}\text{I atoms}\left(\frac{1\text{ mol }^{131}\text{I}}{6.022 \times 10^{23}}\right)\left(\frac{131\text{ g }^{131}\text{I}}{1\text{ mol }^{131}\text{I}}\right) = 1.39 \times 10^{-5}\text{ g }^{131}\text{I}$$

Practice Exercises

23.1 $\Delta H = \Delta H_f° = 601.7$ kJ
$\Delta S° = S°(Mg_{(s)}) + 1/2S°(O_{2(g)}) - S°(MgO_{(s)})$
= 32.5J/K + 1/2(205 J/K) – 26.9 J/K
= 108.1 J/K
= 0.108 kJ/K

$\Delta G_T° = \Delta H° - T\Delta S°$
= 601.7 kJ – T(0.108 kJ/K)

Decomposition occurs when $\Delta G<0$. Solve for $\Delta G° = 0$

$$T = \frac{601.7 \text{ kJ}}{0.108\text{kJ/K}}$$
= 5570 K

23.2 The net charge on the complex ion must first be determined. Two $S_2O_3^{2-}$ ions contribute a charge of 4–; the metal contributes a charge of 1+. The sum of these is 3–. The formula of the complex ion is therefore $[Ag(S_2O_3)_2]^{3-}$. The ammonium salt of this ion would have the formula $(NH_4)_3[Ag(S_2O_3)_2]$.

23.3 The salt must include the six hydrated water molecules. We know that Al exists as a 3+ ion and that chloride has a charge of 1–. The hydrate would have the formula $AlCl_3{\cdot}6H_2O$. The complex ion most likely has the formula $[Al(H_2O)_6]^{3+}$.

23.4 a) potassium hexacyanoferrate(III)
b) dichlorobis(ethylenediamine)chromium(III) sulfate

23.5 a) $[SnCl_6]^{2-}$
b) $(NH_4)_2[Fe(CN)_4(H_2O)_2]$

23.6 a) Since there are three ligands and $C_2O_4^{2-}$ is a bidentate ligand, the coordination number is six.
b) The coordination number is six. There are two bidentate ligands and two unidentate ligands.
c) The coordination number is six. Both $C_2O_4^{2-}$ and ethylenediamine are bidentate ligands. Since there are three bidentate ligands, the coordination number must be six.
d) EDTA is a hexadentate ligand so the coordination number is six.

Review Exercises

23.1 Na, Mg, K. Ag+ is not found since AgCl is insoluble.

23.2 The shells of sea creatures.

23.3 Many of the metal salts of these ions are very insoluble.

23.4 Carbon is used because of its abundance and low cost.

23.5 $CuO + H_2 \rightarrow Cu + H_2O$

23.6 Sodium is far too reactive. Na^+ has a high reduction potential.

23.7 Sodium and magnesium compounds melt at too high a temperature.

23.8 $\Delta G = \Delta H - T\Delta S$. Compounds decompose when $\Delta G<0$. This is more likely to occur at a high temperature.

23.9 The free energy change must be less than zero.

23.10 The kinetics of the decomposition reaction are too slow to observe.

23.11 These compounds explode because there is a large positive entropy change accompanied by release of heat upon decomposing to elements.

23.12 Metallurgy is the science and technology of metals, and is concerned with the procedures and chemical reactions that are used to separate metals from their ores and the preparation of metals for practical use.

23.13 An ore is a mineral deposit that has a desirable component in a concentration high enough to make its extraction economically profitable. The economics of the recovery operation is what distinguishes an ore form just rock.

23.14 It is not economically feasible to remove the aluminum from the aluminosilicates.

23.15 Sodium and magnesium are extracted from seawater because the concentrations are high enough, 0.6 M for Na^+ and0.06 M for Mg^{2+}, to make it economically feasible.

23.16 $Mg^{2+}(aq) + 2OH^-(aq) \rightarrow Mg(OH)_2(s)$
$Mg(OH)_2(s) + 2H^+(aq) + 2Cl^-(aq) \rightarrow Mg^{2+}(aq) + 2Cl^-(aq) + H_2O$
$MgCl_2(l) \rightarrow Mg(l) + Cl_2(g)$ (electrolysis)

23.17 Lime is CaO. It is made from $CaCO_3$ by heating:
$CaCO_3 \rightarrow CaO + CO_2$ (heating)

23.18 $CaO(s) + H_2O \rightarrow Ca(OH)_2(s)$

23.19 Gangue is the unwanted rock and sand that can be removed from an ore by washing with a stream of water.

23.20 Gold can be separated from rock and sand because gold is more dense than the rock and sand and can remain behind in the pan when the rock and sand are washed away.

23.21 The flotation process starts with the ore being crushed, mixed with water and ground into a slurry. It is then mixed with detergents and oil. The oil adheres to the sulfide ore. Air is blown through the mixture and the air bubbles bring the oil-coated ore particles to the surface where they are skimmed off.

23.22 $Cu_2S + 3O_2 \rightarrow 2Cu_2O + 2SO_2$
$2PbS + 3O_2 \rightarrow 2PbO + 2SO_2$
$SO_2 + CaCO_3 \rightarrow CaSO_3 + CO_2$
The $CaSO_3$ can be converted into sulfuric acid.

23.23 $Al_2O_3(s) + 2OH^-(aq) \rightarrow 2AlO_2^-(aq) + H_2O$
$AlO_2^-(aq) + H^+(aq) + H_2O \rightarrow Al(OH)_3(s)$
$2Al(OH)_3(s) \rightarrow Al_2O_3(s) + 3H_2O$

23.24 Metals in compounds have positive oxidation states.

23.25 Only electrolysis can provide enough energy to cause the decomposition of the compounds of sodium, magnesium and aluminum.

23.26 Carbon is a useful industrial reducing agent because is combines with the oxygen from the metal oxides to form carbon dioxide.

23.27 Coke iscoal that has been heated strongly in the absence of air. It is composed almost entirely of carbon.

23.28 $PbO + C \rightarrow 2Pb + CO_2$
$CuO + C \rightarrow 2Cu + CO_2$

23.29 $2Cu_2S + 3O_2 \rightarrow 2Cu_2O + 2SO_2$
$Cu_2S + 2Cu_2O \rightarrow 6Cu + SO_2$

23.30 A blast furnace has hot air pumped into it and at the hottest part of the furnace the temperature is 2000 °C.

23.31 The charge that is added to the blast furnace is a mixture of iron ore, limestone, and coke.

23.32 $3Fe_2O_3 + CO \rightarrow 2Fe_3O_4 + CO_2$
$Fe_3O_4 + CO \rightarrow 3FeO + CO_2$
$FeO + CO \rightarrow Fe + CO_2$
The active reducing agent in a blast furnace is carbon monoxide

23.33 Slag is $CaSiO_3$.

23.34 Refining is the purification of a metal after it has been reduced to the metallic state.

23.35 Pig iron contains impurities such as carbon, phosphorus, sulfur, manganese and other elements.. Steel contains less carbon and has other ingredients in very definite proportions.

23.36 The basic oxygen process melts the pig iron and scrap steel and then has pure oxygen blown through the molten metal. The oxygen burns off the excess carbon and oxidizes impurities to their oxides. Other metals are introduced in the proper proportions to give a product with the desired properties. The reaction vessel is then tipped to pout out its contents.

23.37 The opposite charges tend to attract electron density to region between the two nuclei, the region of a chemical bond.

23.38 The size of the charge and the ion size determine the degree of polarization. Increased polarization is caused by high positive charge and small diameter of the cation.

23.39 The ionic potential is the ratio of the charge to the radius of the cation.

23.40

$$Be^{2+}: \ \frac{2}{31} = 0.0645$$

$$Mg^{2+}: \ \frac{2}{65} = 0.0308$$

$$Al^{3+}: \ \frac{3}{50} = 0.06$$

Comparisons of the ionic potentials indicate that Be^{2+} should behave like Al^{3+}.

23.41 As melting point decreases, the covalent character of the bond increases.

23.42 This is when a molecule absorbs light and transfers charge from one atom to another one.

23.43 A white crystal is indicative of a compound that exhibits a high degree of ionic character and does not easily charge transfer.

23.44 As bonds become more covalent, charge transfer becomes easier (i.e. lower energy).

23.45 Because they are colorless, they do not absorb visible light.

23.46 CdS precipitates when these solutions are mixed. CdS is a bright yellow solid the color comes from a charge transfer absorption band.

23.47 The metal ion is a Lewis acid, accepting a pair of electrons from the ligand, which serves as a Lewis base.
(a) The Lewis acid is Cu^{2+}, and the Lewis base is H_2O.

(b) The ligand is H_2O.
(c) Water provides the donor atom, oxygen.
(d) Oxygen is the donor atom, because it is attached to the copper ion.
(e) The copper ion is the acceptor.

23.48 A ligand serves as a Lewis base because it donates the two electrons that are required for the formation of the coordinate covalent bond with the metal ion.

23.49 They are formed by the use of coordinate covalent bonds.

23.50 water, ammonia

23.51 F^-, Cl^-, Br^-, I^-

23.52

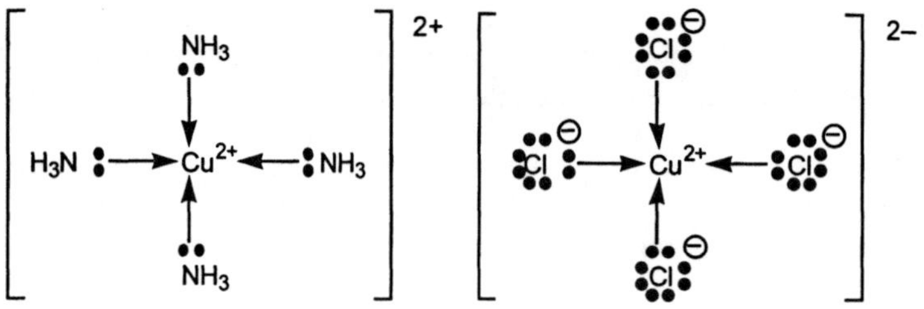

$$[Cu(NH_3)_4]^{2+} \quad [CuCl_4]^{2-}$$

23.53 A bidentate ligand must have two atoms that can form a coordinate covalent bond with a metal to give a ring including the metal.

23.54 A chelate is a compound formed between a metal ion and a polydentate ligand. The multiple sites at which a coordinate covalent bond may form enable these ligands to "grab" the metal ion very strongly.

The oxalate ion, $C_2O_4^{2-}$, is an example of a bidentate ligand that forms a chelate with a metal ion. The ligand is able to form coordinate covalent bonds using the lone pairs of electrons on each terminal oxygen.

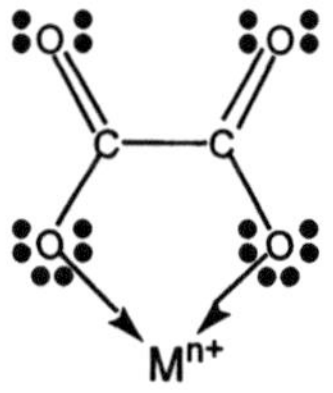

M^{n+}

23.55 EDTA has six donor atoms.

23.56 EDTA forms complexes with the sorts of metal ions that would otherwise promote spoilage.

23.57 EDTA forms complexes with the metal ions that are otherwise responsible for hardness.

23.58 The net charge is 1–, and the formula is $[Co(EDTA)]^-$.

23.59 The bonds to NH_3 are known to be stronger, because the test for copper(II) ion requires the ready displacement of water ligands by ammonia ligands to give the ion $[Cu(NH_3)_4]^{2+}$, which has a recognizable deep blue color.

23.60 $[Cr(en)_3]^{3+}$ is more stable since the ligands are bidentate.

23.61 The coordination number of a metal complex is the number of ligand atoms that are attached to the metal. The common coordination geometries of complex ions having coordination number 4 are the square plane and the tetrahedron.

23.62 Octahedral

23.63

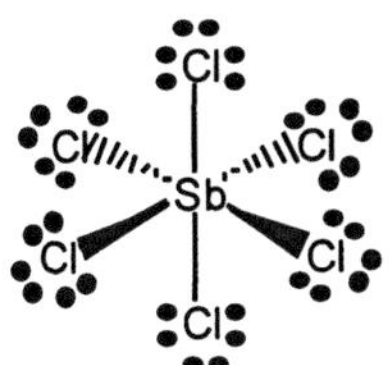

23.64

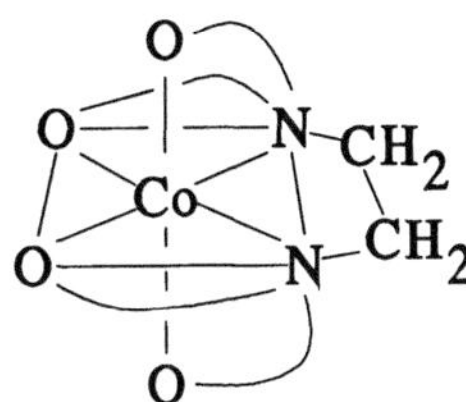

(The curved lines represent -$CH_2C{=}O$ groups.)

23.65 Isomers are compounds that have the same chemical formula, but that are also distinct and different substances, owing to differences in the arrangements of the atoms.

23.66 Stereoisomerism is the existence of isomers that differ only in the spatial orientation of their atoms.

Geometric isomerism is the existence of isomers that differ because their molecules have different geometries.

Chiral isomers are isomers that differ because one is the non-superimposable mirror image of the other.

Enantiomers are chiral isomers of one another, i.e., they are non-superimposable mirror images of one another.

23.67 In a *cis* isomer, like groups are located adjacent to one another, whereas in a *trans* isomer, like groups are located opposite one another.

23.68 Chirality requires the nonsuperimposability of mirror images.

23.69 Cisplatin is the *cis* geometrical isomer of $Pt(NH_3)_2Cl_2$. Cisplatin is an anti-cancer drug.

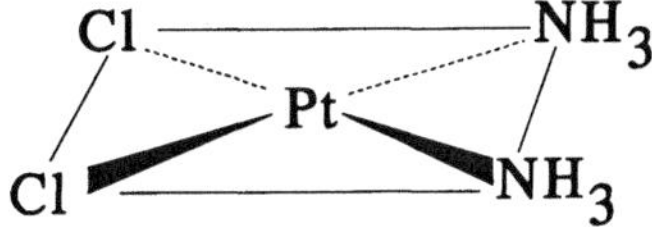

23.70 Optical isomers are chiral isomers, i.e. isomers that differ because they are non-superimposable mirror images on one another.

23.71 See Figure 23.21.

23.72 As shown in Figure 23.21, the orbitals that are oriented between the axes are: d_{xy}, d_{xz}, d_{yz}, whereas those that are oriented along the axes are: $d_{x^2-y^2}$ and d_{z^2}.

23.73 The $d_{x^2-y^2}$ and d_{z^2} orbitals are located near the ligands, whereas the d_{xy}, d_{xz}, and d_{yz}, orbitals are located between (farther from) the ligands.

23.74 See Figure 23.23.

23.75

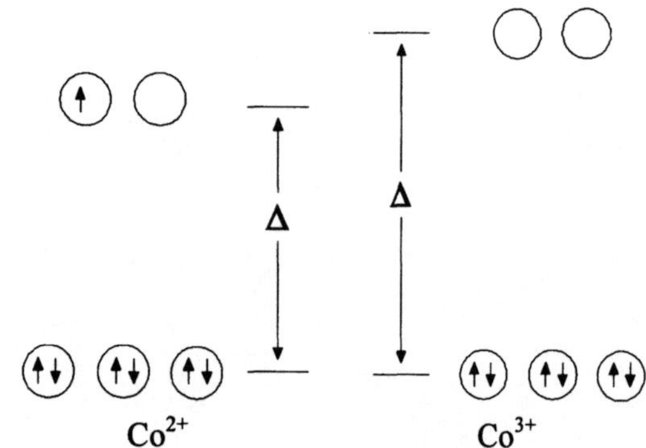

As indicated in the figure above, the oxidation of Co^{2+} to give Co^{3+} removes an electron from a higher energy level. The resulting electron configuration is more stable. Additionally, the energy separation, Δ, increases as the oxidation state increases. This also stabilizes the compound and further explains the ease of oxidation.

23.76 Various ligands may be bound to a metal, giving different complexes, all having their own characteristic value for Δ. As the value of Δ changes, so too does the frequency of light that a complex absorbs.

23.77 (a) Green-blue (b) Violet-blue

23.78 The spectrochemical series is a list of ligands, arranged in their order of increasing ability to produce a large value of Δ. This series is determined by measuring the frequencies of light that are absorbed by a series of complexes having the various ligands.

23.79 The high–spin case is the one in which the maximum number of unpaired electrons is found. The converse is true of the low–spin case; it is the one having the fewest possible number of unpaired electrons.

23.80 Both high– and low–spin configurations are possible only for the following configurations: d^4, d^5, d^6, and d^7.

23.81 There is promotion of an electron into the upper energy level, by absorption of a photon having energy equal to $E = \Delta = h\nu$:

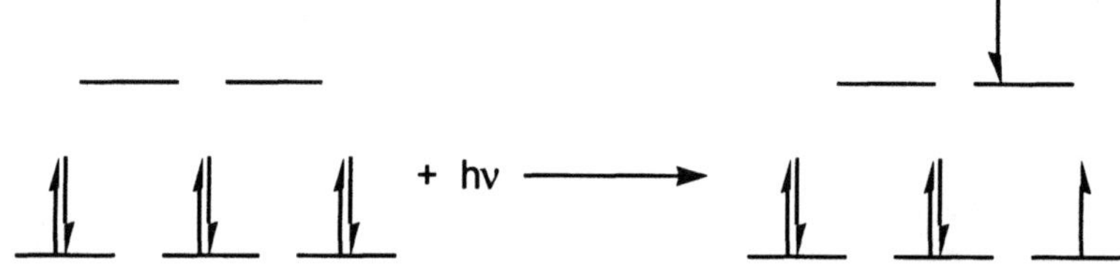

23.82 This is a complex of Co^{3+}. The complex is known to be diamagnetic, so no unpaired electrons are present. We conclude that it must be a low spin complex:

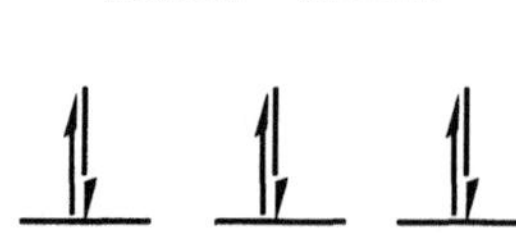

23.83 The d_{z^2}, d_{xz}, and d_{yz} orbitals will become somewhat stabilized because the given change should result in less electron-electron repulsions along the axes. Conversely, the $d_{x^2-y^2}$ and d_{xy} orbitals will become destabilized, as electron repulsions along the x and y axes are increased:

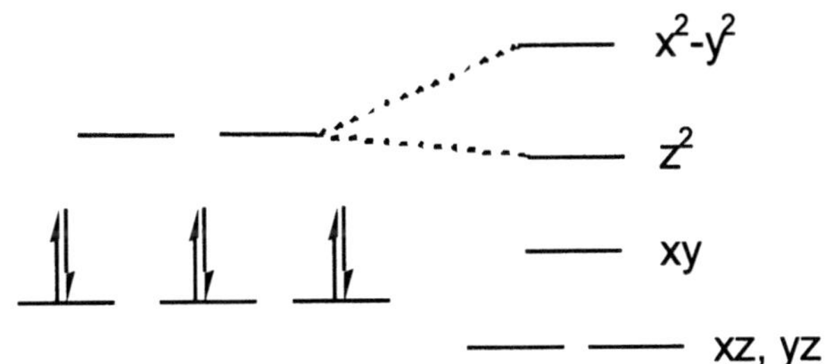

23.84 It is the porphyrin ligand that is found in chlorophyll, heme and vitamin B_{12}.

23.85 The porphyrin ring ligand by itself produces four-coordinate covalent bonds to a central metal ion having a coordination of 4.

23.86 The heme is the oxygen carrier in hemoglobin and myoglobin.

23.87 Both have square planar structures. Co^{2+}

23.88 The calcium ion is essential for the overall health and well being of the body. The most critical role calcium plays is that it is essential to bones. Calcium is necessary to maintain bone density. Insufficient levels of calcium may lead to loss in bone density and result in osteoporosis.

Review Problems

23.89 First we need ΔH° for this reaction.

$\Delta H° = 2\Delta H_f°(Hg_{(g)}) + \Delta H_f°(O_{2(g)}) - 2\Delta H_f°(HgO_{(s)})$
$= 2(61.3 \text{ kJ}) + 0 - 2(-90.8 \text{ kJ}) = 304 \text{ kJ}$

Similarly;

$\Delta S° = 2(175 \text{ J/K}) + 205 \text{ J/K} - 2(70.3 \text{ J/K}) = 414 \text{ J/K}$

Determine the temperature when ΔG = 0

$\Delta G° = \Delta H° - T\Delta S°$

$$T = \frac{\Delta H°}{\Delta S°} = \frac{304 \text{ kJ}}{0.414 \text{ kJ/K}} = 734 \text{ K}$$

23.90 $\Delta G = 0$ when $K_p = 1$

$\Delta H = -\Delta H_f°(CuO) = 155 \text{ kJ}$
$\Delta S = S°(Cu_{(s)}) + 1/2S°(O_{2(g)}) - S°(CuO_{(s)})$
$= 33.2 \text{ J/K} + 1/2(205 \text{ J/K}) - 42.6 \text{ J/K} = 93.1 \text{ J/K}$
$\Delta G = \Delta H - T\Delta S = 0$

$$T = \frac{\Delta H°}{\Delta S°} = \frac{155 \text{ kJ}}{0.0931 \text{ kJ/K}} = 1660 \text{ K}$$

23.91	(a)	Bi_2O_5	(b)	PbS	(c)	PbI_2
23.92	(a)	SnO	(b)	MgS	(c)	Li_2S
23.93	(a)	SnO	(b)	SnS		
23.94	(a)	PbS	(b)	$BeCl_2$		

23.95 (a) $CaCl_2$ (b) BeF_2

23.96 (a) $MgCl_2$ (b) $SnCl_2$

23.97 (a) HgS (b) Ag_2S

23.98 (a) SnO (b) SrS

23.99 (a) HgS (b) Ag_2S

23.100 (a) SnO (b) SrS

23.101 The net charge is –3, and the formula is $[Fe(CN)_6]^{3-}$.

23.102 This is the ion $[Ag(NH_3)_2]^+$, which can exist as the chloride salt: $[Ag(NH_3)_2]Cl$.

23.103 $[CoCl_2(en)_2]^+$

23.104 $[Cr(NH_3)_2(NO_2)_4]^-$

23.105 (a) $C_2O_4^{2-}$ oxalato (b) S^{2-} sulfido or thio
(c) Cl^- chloro (d) $(CH_3)_2NH$ dimethylamine

23.106 (a) NH_3 ammine (b) N^{3-} nitrido
(c) SO_4^{2-} sulfato (d) $C_2H_3O_2^-$ acetato

23.107 (a) hexaamminenickel(II) ion
(b) triamminetrichlorochromate(II) ion
(c) hexanitrocobaltate(III) ion
(d) diamminetetracyanomanganate(II) ion
(e) trioxalatoferrate(III) ion or trisoxalatoferrate(III) ion

23.108 (a) diiodoargentate(I) ion
(b) trisulfidostannate(IV) ion
(c) tetraaquabis(ethylenediamine)cobalt(III) sulfate
(d) pentaamminechlorochromium(III) sulfate
(e) potassium trioxalatocobaltate(III) or potassium trisoxalatocobaltate(III)

23.109 (a) $[Fe(CN)_2(H_2O)_4]^+$
(b) $Ni(C_2O_4)(NH_3)_4$
(c) $[Fe(CN)_4(H_2O)_2]^-$
(d) $K_3[Mn(SCN)_6]$
(e) $[CuCl_4]^{2-}$

23.110 (a) $[AuCl_4]^-$
(b) $[Fe(en)_2(NO_2)_2]^+$
(c) $[Co(NH_3)_4(CO_3)_2]^-$
(d) $[Fe(C_2H_3O_2)_4(en)]^{2-}$
(e) $PtCl_2(NH_3)_2$

23.111 The coordination number is six, and the oxidation number of the iron atom is +2.

23.112 $C_2O_4^{2-}$ is bidentate $\times$ 2
NO_2^- is monodentate $\times$ 2
Coordination number = 6

23.113

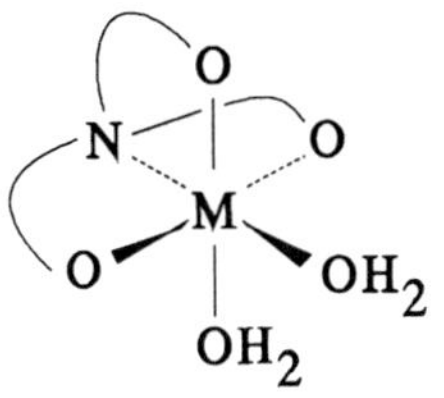

The curved lines represent $-CH_2$-CO-O- groups.

23.114 (a) The nitrogen atoms are the donor atoms.

(b) This is $2 \times 3 = 6$.

(c)

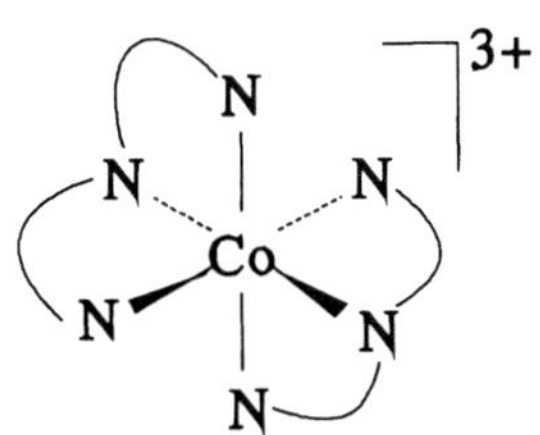

(d) $Co(dien)_2^{3+}$, due to the chelate effect

(e) H_2–N–CH_2CH_2–NH–CH_2CH_2–NH–CH_2CH_2–NH_2

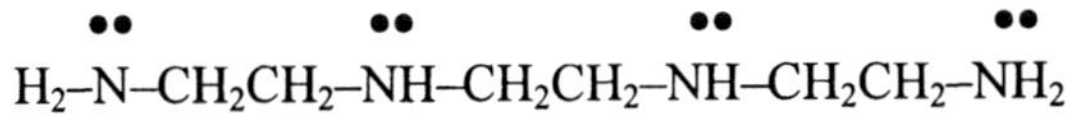

23.115 Since both are the *cis* isomer, they are identical. One can be superimposed on the other after simple rotation.

23.116 Yes, the isomers are chiral since the mirror image may not be superimposed on the original.

23.117

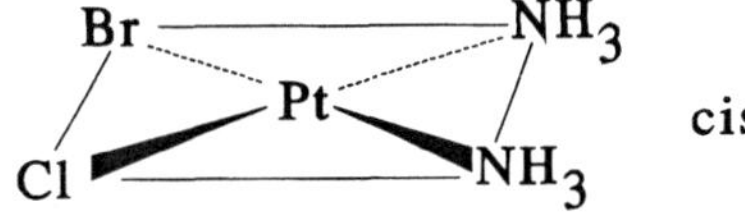

cis

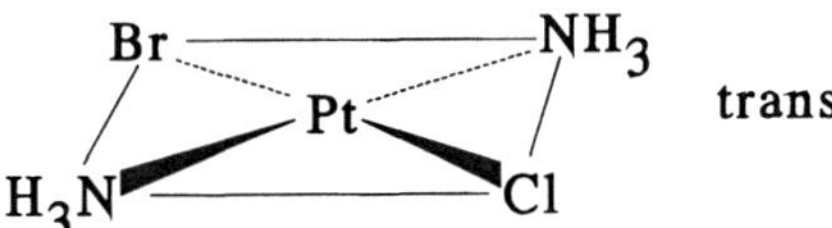

trans

23.118

23.119

23.120 The *cis* isomer is chiral:

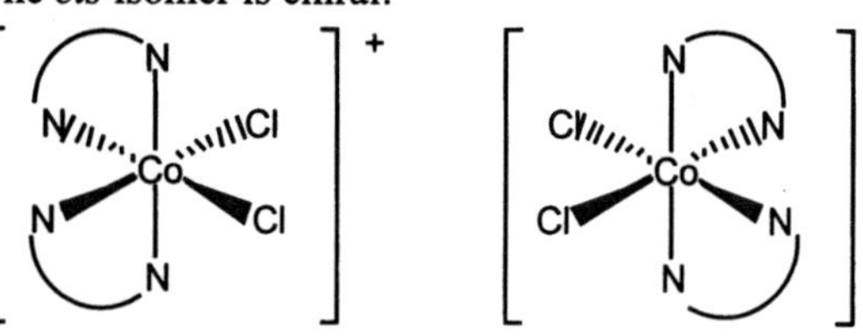

The *trans* isomer is nonchiral

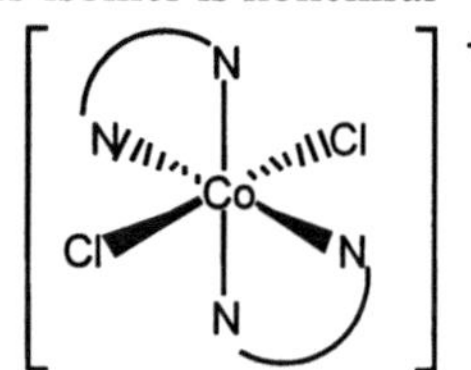

23.121 (a) $Cr(H_2O)_6^{3+}$ (b) $Cr(en)_3^{3+}$

23.122 $[Cr(CN)_6]^{3-} < [Cr(NO_2)_6]^{3-} < [Cr(en)_3]^{3+} < [Cr(NH_3)_6]^{3+} < [Cr(H_2O)_6]^{3+} < [CrF_6]^{3-} < [CrCl_6]^{3-}$

23.123 $[Cr(CN)_6]^{3-}$

23.124 (a) $[Fe(H_2O)_6]^{2+}$ (b) $[Mn(CN)_6]^{4-}$

23.125 (a) The value of Δ increases down a group. Therefore, we choose: $[RuCl(NH_3)_5]^{3+}$
(b) The value of Δ increases with oxidation state of the metal. Therefore, we choose: $[Ru(NH_3)_6]^{3+}$

23.126 Ligand A produces the larger splitting. The absorbed colors are the complements of the perceived colors:
CoA_6^{3+} – absorbed color is green
CoB_6^{3+} – absorbed color is red
The absorbed color with the highest energy (shortest wavelength) is green. We conclude that ligand A is higher in the spectrochemical series.

23.127 This is the one with the strongest field ligand, since Co^{2+} is a d^7 ion: CoA_6^{3+}

23.128 The color of CoA_6^{3+} is red. The complex with the lower oxidation state should have a smaller value for Δ. Therefore, the 2+ ion should absorb in the red, and appear blue.

23.129 This is a weak field complex of Co^{2+}, and it should be a high–spin d^7 case. It cannot be diamagnetic; even if it were low spin, we would still have one unpaired electron.

23.130 For $Fe(H_2O)_6^{3+}$, we expect a relatively small value for Δ, and we predict the high–spin case having five unpaired electrons:

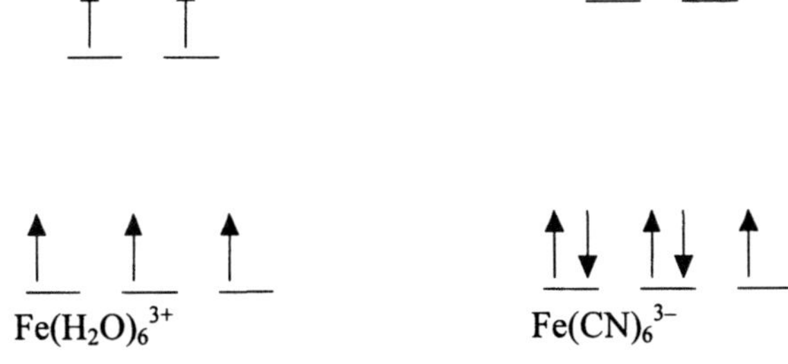

For $Fe(CN)_6^{3-}$, we expect a relatively large value for Δ, and we predict the low–spin case having one unpaired electron:

Additional Exercises

23.131 The value of the equilibrium constant when ΔG = 0 is 1. Thus, the equilibrium concentrations (pressures) are 1 atm for both species. This makes sense since we are thermally decomposing the solid so the vapor pressure of the sample should equal the atmospheric pressure.

23.132 $\Delta G = \Delta H - T\Delta S = -RT \ln K$
$\Delta H = +754.3$ kJ
$\Delta S = 28.6$ J/K + 3/2(205.0 J/K) – 78.2 J/K = 257.9 J/K

T(°C, K)	100, 373	500, 773	1000, 1273
ΔG (kJ)	658	555	426
K_p	7.09×10^{-93}	3.13×10^{-38}	3.31×10^{-18}

23.133

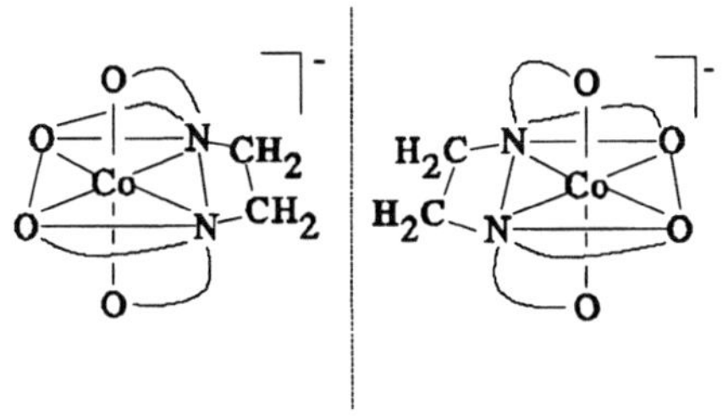

The mirror images are not superimposable.

23.134 Two different substituents on a tetrahedron cannot be arranged in isomeric ways. That is, all possibilities are the same, being superimposable on each other.

Tetrahedral form:

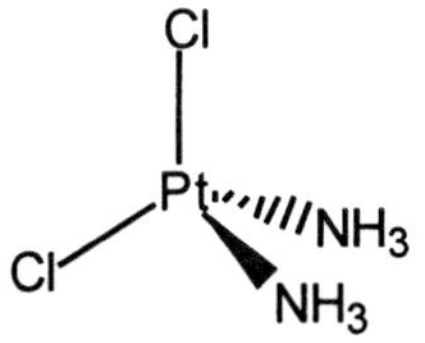

Square planar forms with the two isomers:

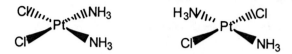

23.135 (a) The number of moles of chloride that have been precipitated is:

$$\#\text{ mol AgCl} = (0.538\text{ g AgCl})\left(\frac{1\text{ mol AgCl}}{143.32\text{ g AgCl}}\right) = 3.75 \times 10^{-3}\text{ mol AgCl}$$

The number of moles of Cr that were originally present is:

$$\#\text{ mol Cr} = (0.500\text{ g CrCl}_3\cdot 6\text{H}_2\text{O})\left(\frac{1\text{ mol CrCl}_3\cdot 6\text{H}_2\text{O}}{266.4\text{ g CrCl}_3\cdot 6\text{H}_2\text{O}}\right)\left(\frac{1\text{ mol Cr}}{1\text{ mol CrCl}_3\cdot 6\text{H}_2\text{O}}\right)$$

$$= 1.88 \times 10^{-3}\text{ mol Cr}$$

The ratio of moles of Cl^- per mole of Cr is therefore: 3.75/1.88 = 1.99.

This means that there were 2 mol of Cl^- that was free to precipitate. The other mole of chloride ion must have been bound as a ligand to the Cr. In other words, the complex ion was $[Cr(Cl)(H_2O)_5]^{2+}$.

(b) $[Cr(Cl)(H_2O)_5]Cl_2{\cdot}H_2O$

(c)

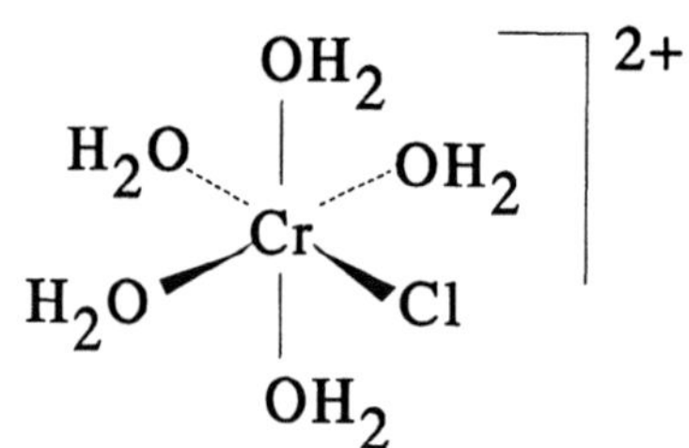

(d) There is only one isomer.

Review Questions

24.1 In most compounds, metalloids combine with nonmetals. The metalloid typically has a lower electronegativity than the nonmetal, so the metalloid exists in a positive oxidation state. To obtain the elemental form of the metalloid it must be reduced.

24.2 a) $2BCl_3(g) + 3H_2(g) \rightarrow 2B(s) + 6HCl(g)$
b) $SiO_2(s) + C(s) \rightarrow Si(s) + CO_2(g)$
c) $As_2O_3(s) + 3H_2(g) \rightarrow 2As(s) + 3H_2O(\ell)$

24.3 The atmosphere

24.4 Helium is produced as the result of electron capture of alpha particles that are formed during the radioactive decay of elements such as uranium. Radon is a radioactive element and spontaneously decomposes. Hence, very little occurs in nature.

24.5 $10Cl^- + 2MnO_4^- + 16H^+ \rightarrow 5Cl_2 + 2Mn^{2+} + 8H_2O$

24.6 a) $2KCl + I_2$
b) NR
c) NR
d) $SrF_2 + Cl_2$

24.7 Cl_2, gas, followed by air, is passed through the seawater. The Cl_2 oxidizes the Br^- to Br_2 and the air sweeps the volatile Br_2 from the solution.

24.8 Water is more easily oxidized than the fluoride ion. In aqueous solution, oxygen gas and hydrogen ions will form at the anode and sodium metal will form on the cathode.

24.9 The mixture of KF and HF has a lower melting point than KF alone. Hence, lower temperatures are used.

24.10 The free energy change for this reaction is very large and positive, making the reaction prohibitive.

24.11 $Ca_3(PO_4)_2 + 3SiO_2 + 5C \rightarrow 3CaSiO_3 + 5CO + 2P$
The C serves as the reducing agent. The presence of SiO_2 enables the formation of the low melting $CaSiO_3$. To reduce SiO_2 to Si, temperatures much higher than those used in this reaction are required.

24.12 $8I^- + SO_4^{2-} + 8H^+ \rightarrow 4I_2 + S^{2-} + 4H_2O$

24.13 Of the nonmetals, only the noble gases exist in nature as isolated atoms.

24.14 Period 2 elements form strong π bonds because the atom size is small which enables atoms to approach each other very closely. Period 3 elements (and all other elements) are too large to accommodate the close proximity required to form strong π bonds. Because the period 3 and higher elements are large and form only weak π bonds, it is much more effective to form only σ bonds than to form a σ bond and a π bond.

24.15 Because the halogens need only a single electron to complete their valence shell, these elements do not require the formation of π bonds. The formation of a single σ bond creates an extremely stable diatomic molecule.

24.16 An allotrope is a different structure or physical form of an element. An isotope is a form of an atom that differs from other atoms of the same element in the number of neutrons in the nucleus. An example is H and D, two of the isotopes of hydrogen.

24.17 The allotropes of oxygen are dioxygen, O_2, and ozone, O_3.

24.18 The two unpaired electrons are located in the π antibonding molecular orbitals.

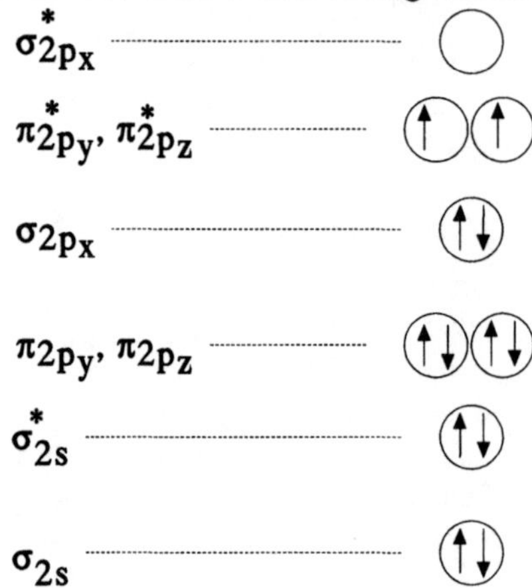

The overall bond order is $(8Be^{-} - 4ABe^{-})/2 = 2$.

24.19

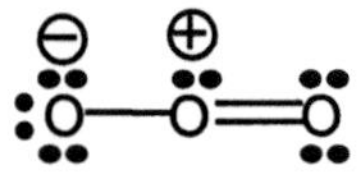

The molecule is nonlinear and uses sp^2 hybrid orbitals. There is a lone pair of electrons on the structure and, consequently, ozone is indeed a polar molecule. Additionally, the formal charges suggest that O_3 is polar.

24.20 The presence of ozone in the upper atmosphere shields the earth from harmful ultraviolet radiation.

24.21 Diamond, graphite and buckminsterfullerene

24.22 Diamond is a network covalent solid in which each carbon atom is the corner of a tetrahedron. Each carbon atom is bonded to four other carbon atoms using sp^3 hybrid orbitals.

24.23 Graphite uses sp^2 hybrid orbitals. It can be visualized as sheets of carbon atoms arranged as hexagonal rings. Each carbon atom is bonded to three other carbon atoms with bond angles of 120°. The unhybridized *p*-orbitals that remain contain a single *p*-electron. The overlap of these gives rise to weak π bonds between the planar sheets of graphite. This allows the sheets to "slide" against one another and makes graphite, among other uses, a good lubricant.

24.24 The weak π bonds that exist between the sheets of sp^2 bonded carbon are easily broken. The lubricating properties of graphite are a result of the weak attraction between parallel sheets.

24.25 The π orbitals which exist between sheets of carbon atoms form a π cloud. Electrical conduction through this cloud of electrons is very efficient. Consequently, conduction parallel to sheets of carbon atoms is very good. However, since electrons can not be transferred from one layer to another, conduction across (or perpendicular to) to the carbon sheets, is very poor.

24.26 C_{60}, or buckminsterfullerene, resembles a soccer ball. It consists of hexagonal and pentagonal arrangements of carbon atoms. This arrangement naturally assumes a spherical geometry. The molecule is named in honor of R. Buckminster Fuller, an architect. Other roughly spherical arrangements of carbon atoms are also possible. While not perfectly spherical, they resemble buckminsterfullerene and are given the general name of fullerenes.

24.27 Carbon nanotubes have the same molecular structure as graphite, but instead of forming sheets, the layer of carbon curves and forms a tube.

24.28 The clusters have the formula B_{12} and each boron atom represents the vertex of an icosahedron.

24.29 S_8

24.30 White phosphorous, P_4, is a tetrahedron. See figure 24.5.

24.31 The bond angle in P_4 is 60°. Using only p orbitals, a bond angle of 90° is expected. Because the observed bond angle is much less, the molecule is extremely reactive.

24.32 Red phosphorus consists of tetrahedra joined as in Figure 24.7. While it too is used in explosives, it is much less reactive than white phosphorus.

24.33 Black phosphorus, like graphite, exists as parallel sheets of atoms.

24.34 Silicon has a molecular structure identical to diamond, i.e., a network covalent solid arranged in a tetrahedral manner. Silicon, unlike carbon, does not form multiple bonds. Consequently, there is no graphite-like allotrope of silicon.

24.35 Direct combination of the elements and addition of protons to the conjugate base of a hydride.

24.36
(a) tetrahedral (like CH_4)
(b) pyramidal (like NH_3)
(c) bent (like H_2O)
(d) linear (like HCl)

24.37
PH_3 – pyramidal
H_2S – bent
CH_4 – tetrahedral
H_2Se – bent
Si_2H_6 – tetrahedral around each Si

24.38
$2Na + H_2 \rightarrow 2NaH$
$Ca + H_2 \rightarrow CaH_2$

24.39 The Haber process is a catalyzed reaction between hydrogen and nitrogen to make ammonia. The balanced equation is $N_2(g) + 3H_2(g) \rightarrow 2NH_3(g)$ $\Delta H° = -92$ kJ. Based on Le Chatelier's Principle, this reaction should provide more products at high pressure and low temperature. In fact, the reaction is run at high pressure and high temperature. The rate of the reaction is too slow at low temperatures and increased yields are obtained at higher temperatures.

24.40 Diborane, B_2H_6. Each boron atom is surrounded by four hydrogen atoms. The bridging hydrogen atoms are examples of two-electon three–center bonds.

24.41 Diborane uses sp^3 hybrid orbitals. The terminal hydrogens are bonded to the boron atom through the overlap of an s orbital with the sp^3 hybrid orbital. The bridging hydrogens are termed three–center bonds and consist of two electrons spread between three atoms. The bond order of these bonds is one-half.

24.42 Since the free energy of formation is positive, it is unlikely that a compound can be made by the direct combination of its elements. The equilibrium constant for the formation of a compound with a positive free energy of formation will be extremely small and the reaction is not likely to occur.

24.43 (a), (b) and (d) are the only ones with a negative free energy of formation.

24.44 In order to react with elemental nitrogen, a reaction must first overcome the large bond energy of the triple bond. Because this bond is so strong, nitrogen rarely decomposes and is a relatively inert element.

24.45 Since electronegativity increases from left to right on the periodic table, anions of nonmetals that are left of the halogens are not likely to maintain a stable noble gas configuration. These ions are likely to accept a proton in order to lower their total charge.

24.46
(a) $Ca_3P_2 + 6H_2O \rightarrow 3Ca(OH)_2 + 2PH_3$
(b) $Na_2S + 2H_2O \rightarrow H_2S + 2NaOH$
(c) $Al_4C_3 + 12H_2O \rightarrow 4Al(OH)_3 + 3CH_4$
(d) $NaF + H_2O \rightarrow NaOH + HF$

24.47 As we move from left to right in the periodic table, the base strength of the ions decreases. As we move from left to right, we need stronger Brønsted acids to protonate the ion, i.e., extremely strong bases require weak acids and weaker bases require stronger acids.

24.48 Since electronegativity decreases as we move down the periodic table, base strength also decreases as we move down.

24.49 N_3^- is a stronger base than NH_2^-. By removing additional protons from NH_2^-, the base strength increases.

24.50 Catenation is the ability of atoms of the same element to form covalent bonds with each other. The hands down champion element is carbon.

24.51 (a) hydrogen peroxide
(b) hydrazine

24.52 $3S^{2-} + S_8 \rightarrow S_5^{2-} + S_2^{2-} + S_4^{2-}$

24.53

```
[      :O:                 :O:      ] 2-
        ||                  ||
  :Ö – S – S̤ – S̤ – S̤ – S̤ – S – Ö:
        ||                  ||
       :O:                 :O:
```

24.54 Polysulfide ions are not linear because each sulfur atom consists of one or two sigma bonds and multiple unpaired electrons. These chains are best described as being zigzag shaped.

24.55

```
[      :O:              :O:      ] 4–
        ||               ||
  :O — P — O — O — P — O:
        ||               ||
       :O:              :O:
```

24.56 No, because diatomic nitrogen, N_2, is extremely stable.

24.57 P_4O_6 and P_4O_{10}. White phosphorus, P_4, is a tetrahedral molecule. P_4O_6 assumes the same geometry as P_4 but there is an oxygen atom between each phosphorus atom along the edges of the tetrahedron. P_4O_{10} is similar to P_4O_6 with an additional oxygen atom bonded to each of the phosphorus atoms at the vertices of the tetrahedron.

24.58 $P_4O_6 + 6H_2O \rightarrow 4H_3PO_3$
$P_4O_{10} + 6H_2O \rightarrow 4H_3PO_4$

24.59 The Lewis structure for NO_2 indicates a single unpaired electron on the nitrogen atom. When two NO_2 molecules dimerize, there are no unpaired electrons.

24.60 Bond formation is always an exothermic process. Bond breaking always requires the input of energy and is an endothermic process. No bonds are broken in the dimerization of NO_2 to form N_2O_4.

24.61 Since this reaction is endothermic, a bond is broken, when the reaction is heated, the reaction moves to the right.

24.62 Silicon, unlike carbon, is not likely to form multiple bonds. SiO_2 is a network covalent solid with a diamond structure. CO_2, which is isoelectronic with SiO_2, is a covalent gas that has two sets of double bonds.

24.63 Germanium does not form the simple molecular oxides GeO_2 since Ge does not form double bonds.

24.64 As_4O_6, As_4O_{10}, Sb_4O_6, Sb_4O_{10}

24.65 **Direct combination of the elements.** Using carbon as an example we can form CO_2 or CO by direct combination: $C + 1/2O_2 \rightarrow CO$, $C + O_2 \rightarrow CO_2$. The actual product is determined by experimental conditions.
Oxidation of a lower oxide by oxygen. An example of this is the combustion of carbon monoxide to produce carbon dioxide: $2CO + O_2 \rightarrow 2CO_2$.
Reaction of a nonmetal hydride with oxygen. The combustion of hydrocarbons in excess oxygen produces carbon dioxide: $CH_4 + 2O_2 \rightarrow CO_2 + 2H_2O$. (Recall that CO is produced if there is insufficient O_2 present.)

24.66 $4AsH_3 + 6O_2 \rightarrow As_4O_6 + 6H_2O$

24.67

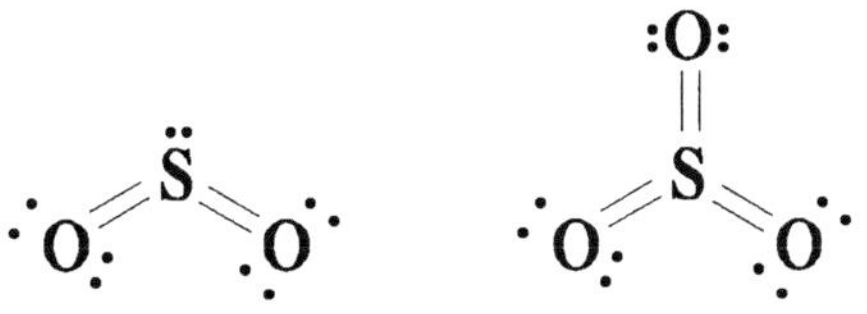

bent

trigonal planar

24.68

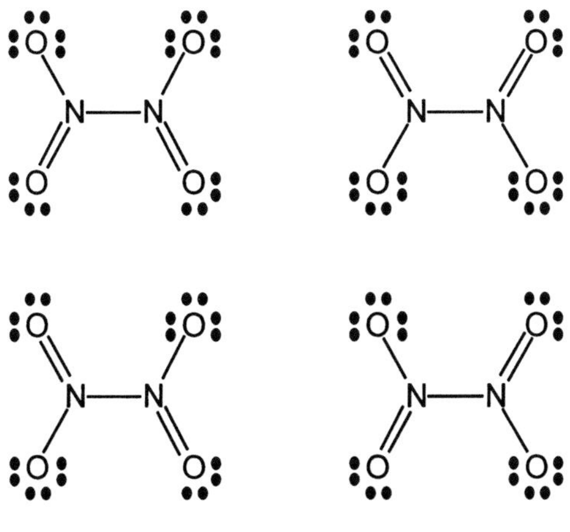

24.69 NO^+ does not have an unpaired electron as does NO. The bond order for NO^+ is 3 and the ion is isoelectronic with N_2. Consequently, it should be very stable.

24.70 a) $SO_3 + H_2O \rightarrow H_2SO_4$
b) $P_4O_6 + 6H_2O \rightarrow 4H_3PO_3$
c) $As_4O_{10} + 6H_2O \rightarrow 4H_3AsO_4$
d) $Cl_2O_7 + H_2O \rightarrow 2HClO_4$

24.71 a) $H_3PO_4 \rightarrow H^+ + H_2PO_4^-$
$H_2PO_4^- \rightarrow H^+ + HPO_4^{2-}$
$HPO_4^{2-} \rightarrow H^+ + PO_4^{3-}$
b) $H_3PO_3 \rightarrow H^+ + H_2PO_3^-$
$H_2PO_3^- \rightarrow H^+ + HPO_3^{2-}$

24.72 The Lewis pictures for H_3PO_4 and H_3PO_3 are in the text on page 1076.
NaH_2PO_4, Na_2HPO_4, Na_3PO_4
NaH_2PO_3, Na_2HPO_3

24.73

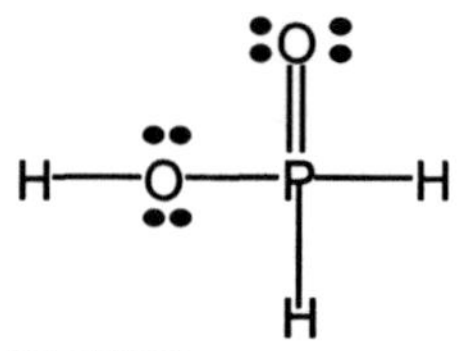

NaH_2PO_2

24.74

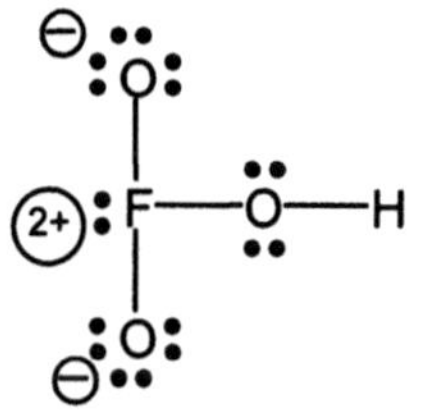

Fluorine is the most electronegative element. In the Lewis structure, the fluorine atom has a +2 formal charge. This is not likely for the most electronegative element.

24.75

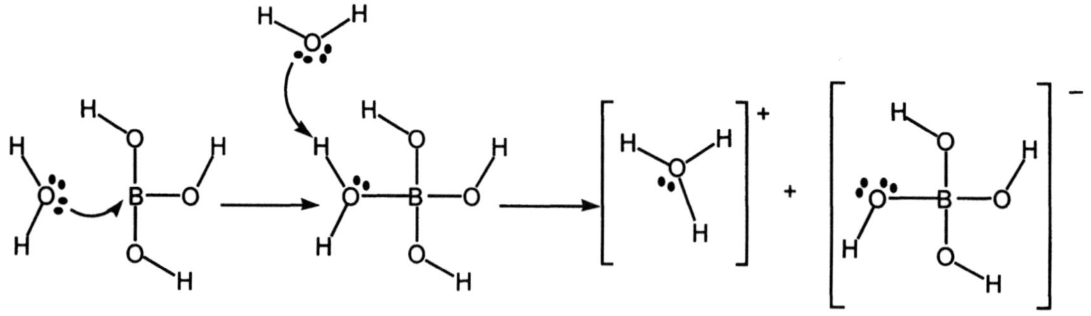

24.76 a) N_2O_3
b) N_2O_5

24.77 In a disproportionation reaction, an element in a compound is both oxidized and reduced.
$Cl_2 + 2OH^- \rightarrow OCl^- + Cl^- + H_2O$
$3OBr^- \rightarrow BrO_3^- + 2Br^-$

24.78 $3NO_2 + H_2O \rightarrow 2NO_3^- + NO + 2H^+$

24.79 $Na_2S_2O_5$.

$$\# \text{ g } Na_2S_2O_5 = 100 \text{ mL} \left(\frac{0.200 \text{ mol } NaHSO_3}{1000 \text{ mL solution}}\right)\left(\frac{1 \text{ mol } Na_2S_2O_5}{2 \text{ mol } NaHSO_3}\right)\left(\frac{190.11 \text{ g } Na_2S_2O_5}{1 \text{ mol } Na_2S_2O_5}\right) = 1.90 \text{ g } Na_2S_2O_5$$

25.80 A dehydration reaction: $2NaHSO_3 \rightarrow Na_2S_2O_5 + H_2O$

25.81

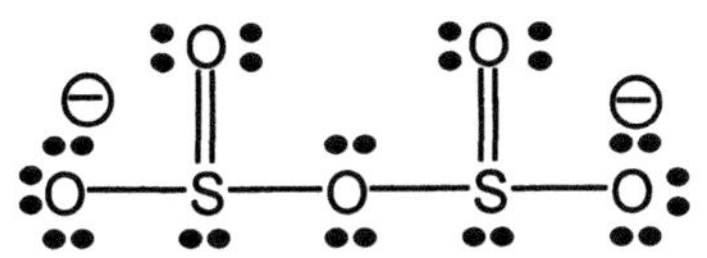

The sulfurs are trigonal pyramidal in shape

25.82 H_2SO_3 does not actually exist. When SO_2 is dissolved in water, some protons are released in the following reaction:
$SO_2 + H_2O \rightarrow HSO_3^- + H^+$

25.83 OI^- disproportionates rapidly because the O anion polarizes the I

25.84 SF_4 would use dsp^3 hybrid orbitals. The molecular structure of SF_4 is seesaw which is consistent with both VSEPR and hybrid orbital theories.

25.85 The number of fluorine atoms that bind to a central halogen atom in an interhalogen compound increases as the size of the central atom increases.

25.86 The size of the fluorine atoms surrounding the relatively small central sulfur atom in SF_6 effectively shields the central atom from attack by any molecular compound.

25.87 There are two reasons for this; first, the chlorine atoms completely shield the smaller carbon atom in CCl_4 from attack. The silicon atom is larger and the chlorine atoms do not completely shield it from attack. Second, the availability of *d*-orbitals in the silicon compound allows for a temporary bond between an attacking molecule and the $SiCl_4$.

24.88 The six N–I bonds in 2 NI_3 molecules are weaker than the one N_2 and three I_2 bonds; therefore NI_3 is unstable.
$2NI_3 \rightarrow N_2 + 3I_2$

Review Problems

24.89
a) $AlP(s) + 3H_2O(\ell) \rightarrow Al(OH)_3(s) + PH_3(g)$
b) $Mg_2C(s) + 4H_2O(\ell) \rightarrow CH_4(g) + 2Mg(OH)_2(s)$
c) $FeS(s) + 2HCl(aq) \rightarrow FeCl_2(aq) + H_2S(aq)$
d) $MgSe(s) + H_2SO_4(aq) \rightarrow H_2Se(aq) + MgSO_4(aq)$

24.90
a) $Mg_3N_2(s) + 6H_2O(\ell) \rightarrow 3Mg(OH)_2(s) + 2NH_3(aq)$
b) $2KBr(s) + H_2SO_4(conc.) \rightarrow K_2SO_4(aq) + 2HBr(aq)$
c) $Mg_2Si(s) + 4HCl(aq) \rightarrow 2MgCl_2(aq) + SiH_4(g)$
d) $CaC_2(s) + 2H_2O(\ell) \rightarrow Ca(OH)_2(s) + C_2H_2(g)$

24.91 a) $2CO + O_2 \rightarrow 2CO_2$
b) $2C_2H_6 + 7O_2 \rightarrow 4CO_2 + 6H_2O$
c) $P_4O_6 + 2O_2 \rightarrow P_4O_{10}$
d) $4NH_3 + 5O_2 \rightarrow 4NO + 6H_2O$

24.92 a) $4PH_3 + 6O_2 \rightarrow P_4O_6 + 6H_2O$
b) $2H_2S + 3O_2 \rightarrow 2SO_2 + 2H_2O$
c) $SiH_4 + 2O_2 \rightarrow SiO_2 + 2H_2O$
d) $2SbH_3 + 4O_2 \rightarrow Sb_2O_5 + 3H_2O$

24.93 $\Delta G = 0 + 2(0) - 2\text{ mol}(51.9\text{ kJ/mol}) = -103.8\text{ kJ}$

$$K_p = \frac{(P_{N_2})(P_{O_2})^2}{(P_{NO_2})^2}$$

$$\Delta G = -RT\ln K_p$$

$$K_p = \exp\left(-\frac{\Delta G}{RT}\right) = \exp\left(\frac{103.8 \times 10^3\text{ J/mol}}{(8.314\text{ J/molK})(298\text{ K})}\right) = 1.57 \times 10^{18}$$

Because K_p is so large, we expect the equilibrium to lie to the right. Since we observe that NO_2 is stable, we must conclude that the reaction kinetics are too slow to achieve equilibrium.

24.94

$$P_{NO_2} = \sqrt{\frac{(P_{N_2})(P_{O_2})^2}{K_p}} = \sqrt{\frac{(0.80)(0.20)^2}{1.57 \times 10^{18}}} = 1.43 \times 10^{-10}\text{ atm}$$

24.95 $\Delta H = 9.7\text{kJ} - 2(33.8\text{ kJ}) = -57.9\text{ kJ}$
The reaction is exothermic so heat can be considered to be a product of the reaction. If we increase the temperature, effectively adding heat (a product), Le Chatelier's Principle states that the equilibrium will shift to the left and N_2O_4 will dissociate.

24.96 $\Delta H° = \{\Delta H_f°(CO_2(g)) + 2\Delta H_f°(H_2O(g))\} - \{\Delta H_f°(CH_4(g)) + 2\Delta H_f°(O_2(g))\}$
$= \{-394\text{ kJ} + 2(-241.8\text{ kJ})\} - \{-74.9 + 2(0)\}$
$= -803\text{ kJ}$

$\Delta H° = \{\Delta H_f°(CO_2(g)) + 2\Delta H_f°(H_2O(g)) + 4\Delta H_f°(N_2(g))\} - \{\Delta H_f°(CH_4(g)) + 2\Delta H_f°(N_2O(g))\}$
$= \{-394\text{ kJ} + 2(-241.8\text{ kJ}) + 4(0)\} - \{-74.9 + 4(81.5\text{ kJ})\}$
$= -1129\text{ kJ}$

Using N_2O as an oxidizer increases the heat produced by 41%.

24.97 In each case, describe the structure without any hydrogen atoms, i.e., describe the anion structure.
(a) IO_3^- is trigonal pyramidal
(b) ClO_2^- is bent
(c) IO_6^- is octahedral
(d) ClO_4^- is tetrahedral

24.98 (a) tetrahedral
(b) bent
(c) bent
(d) tetrahedral
(e) pyramidal
(f) pyramidal

24.99 (a) pyramidal

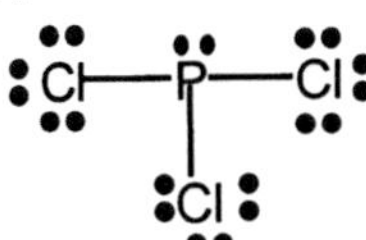

(b) t-shaped

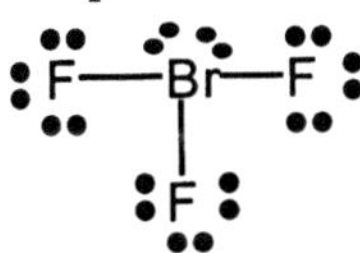

(c) octahedral

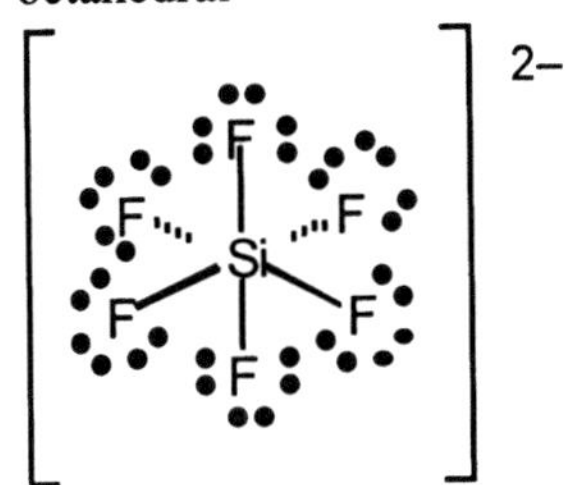

(d) planar triangular

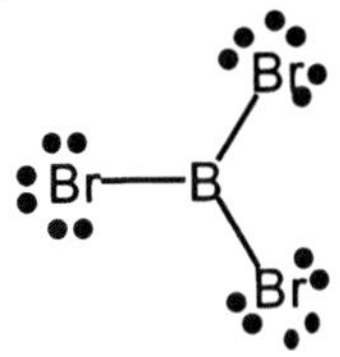

(e) square planar

$[BrF_4]^-$

24.100 (a) octahedral

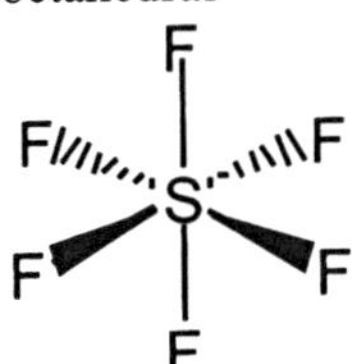

(b) see-saw

(c) bent

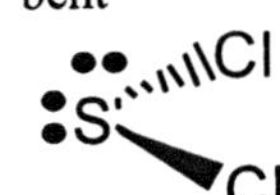

(d) linear

(e) square pyramidal

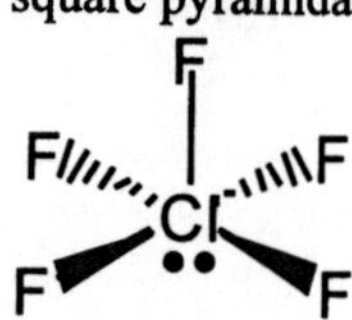

24.101 $N_2 + 3H_2 \rightarrow 2NH_3$
$4NH_3 + 5O_2 \rightarrow 4NO + 6H_2O$
$2NO + O_2 \rightarrow 2NO_2$
$3NO_2 + H_2O \rightarrow 2HNO_3 + NO$

24.102 $S + O_2 \rightarrow SO_2$
$2SO_2 + O_2 \rightarrow 2SO_3$
$SO_3 + H_2O \rightarrow H_2SO_4$

24.103

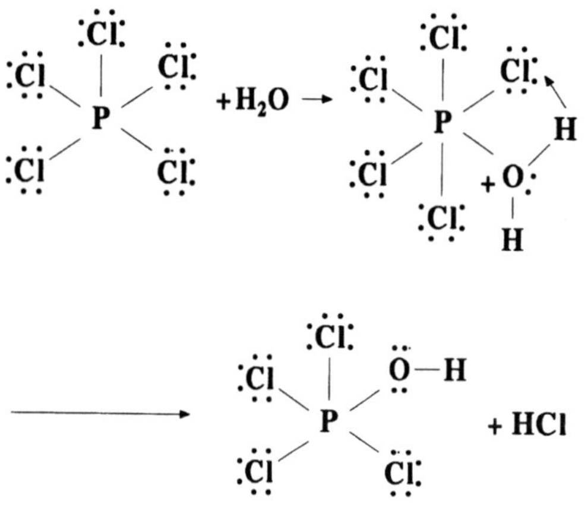

$PCl_5 + H_2O \rightarrow PCl_4(OH) + HCl$

24.104

$SF_4 + H_2O \rightarrow SF_4(OH_2)$

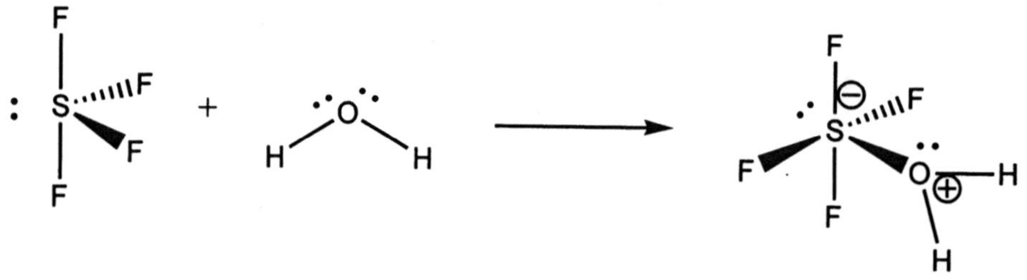

Practice Exercises

25.1 (a) 3–methylhexane
(b) 4–ethyl–2,3–dimethylheptane
(c) 5–ethyl–2,4,6–trimethyloctane

25.2 (a)

O ‖ CH_3–CH or O ‖ CH_3–COH

(depending on the strength of the oxidizing agent)

(b)

O ‖ $CH_3CH_2CCH_2CH_3$

Review Questions

25.1 The inorganic compounds of carbon are generally taken to be the oxides and cyanides of carbon, as well as carbonates and bicarbonates of the metals. Thus, the inorganic compounds in this list are (b), (d), and (e), although (c) CCl_4 might arguably be listed also.

25.2 This is due to the high strength of covalent bonds carbon forms with itself and other nonmetals at the same time, and that carbon can make four bonds.

25.3 Since ethane is H_3C–CH_3, R is CH_3, a methyl group.

25.4 (a) This is impossible, since there should be three hydrogen atoms attached to the first carbon atom.
(b) This is impossible, since there should be only two hydrogen atoms attached to the first carbon atom.
(c) This is possible.

25.5 (a) $CH_3CH_2CH_2CH_3$

(b)

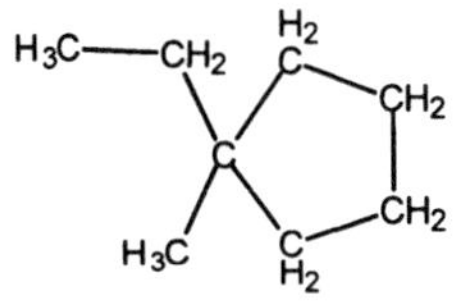

H_3C—CH_2 H_2C CH_2 C CH_2 H_3C C H_2

(c)

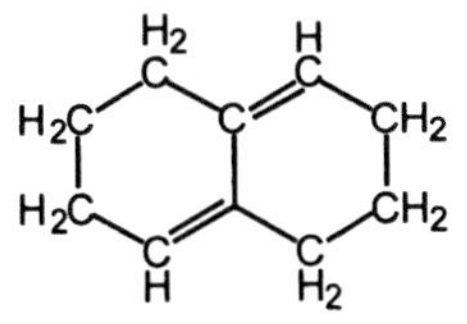

H_2 C H C H_2C C CH_2 H_2C CH_2 C H C H_2

25.6 The functional group is the characteristic portion of a molecule that sets it into one of a number of families of similar substances. It is the focus of the reactivity of the substance.

25.7 The functional groups typically impart polarity to what would be an otherwise nonpolar hydrocarbon, and they therefore give the molecule more of an opportunity for reaction than a simple hydrocarbon would have. In other words, since the functional group is the site of polarity, it is the likely site of attack by polar or ionic reactants. Amines, for instance, are Lewis and Brønsted bases, and they characteristically can be protonated.

25.8 Isomers have identical chemical formulas but different structures.

25.9 No, H_2O and D_2O have identical chemical formulas and identical structures.

25.10 Free rotation is prevented by the π–nature of the second of two bonds in the carbon–carbon double bond. Since the bond is rigid, by changing the order of the substituents, isomers can be formed.

25.11 $CH_3CH_3CH_2OH$ is more soluble in water then $CH_3CH_2CH_2CH_3$ because it can hydrogen bond with water.

25.12 Propanol, $CH_3CH_3CH_2OH$, has the higher boiling point, due to its capacity for hydrogen bonding.

25.13 (a) C, Compound C has two OH groups, and it will have the greatest capacity for hydrogen bonding.
(b) A, Compound A has no polar groups, and has only the very weak London forces.

25.14 Compound B is more soluble in water because of its polar nature and its capacity for hydrogen bonding.

25.15 Substance A has the higher boiling point. Since it is larger and has more branching, the London forces are stronger than in B.

25.16 Aldehydes and ketones do not have the capacity to form hydrogen bonds to themselves.

25.17 Acetic acid forms dimers containing two hydrogen bonds.

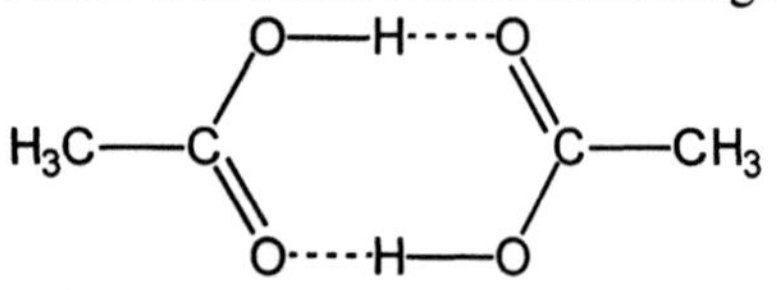

O—H····O
H_3C—C C—CH_3
O····H—O

25.18 The ester does not have the capacity to form hydrogen bonds to itself, whereas the acid readily forms a hydrogen-bonded dimer.

25.19 The amide can form hydrogen bonds and are quite polar compared to aminobutane.

25.20 The N–H bond is not as polar as the O–H bond, and the hydrogen bond of amines is therefore not as strong as that of alcohols. Also, there are more sites available for hydrogen bonding to an O–H group than to an N–H group.

25.21

(a) CH_3-C(CH_3)(CH_3)-$CH_2CH_2CH_2CH_2CH_2CH_3$

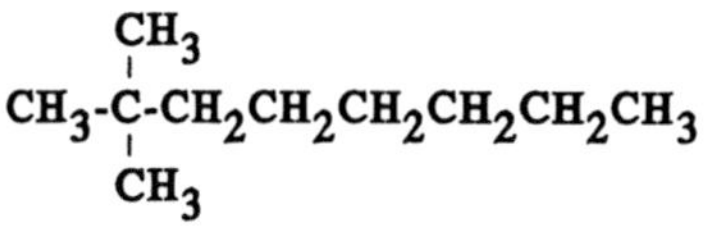

(b) CH_3, CH_3

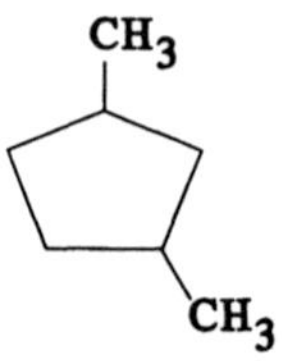

(c) CH_2CH_3, CH_2CH_3

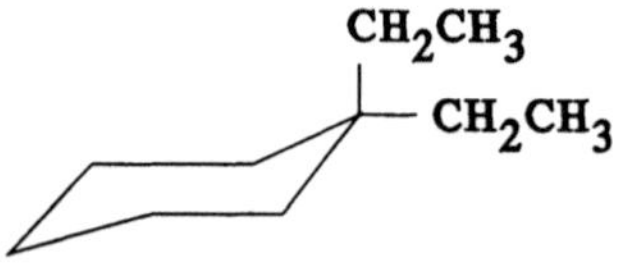

(d) H_2C, CH_2CH_3, CH_3, $(CH_3)_2HC$, CH_3

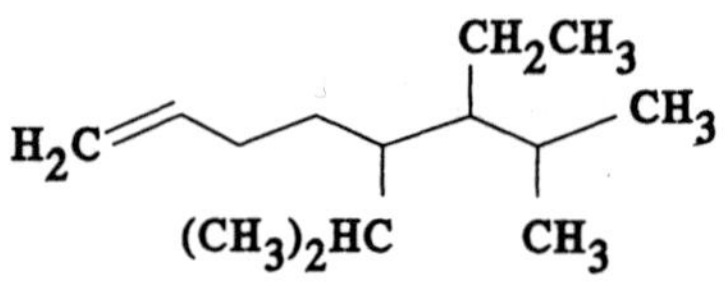

(e) H_3C, CH_3

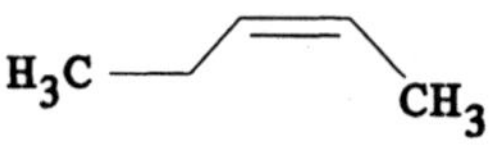

25.22 (a)

$H_3C-C(CH_3)-CH-C(=O)-H$

(b)

$H_3C-C(=O)-CH_2-CH(CH_3)-CH_2-CH_2-CH_2-CH_3$

(c)

$H_3C-CH(Cl)-C(=O)-OH$

(d)

$H_3C-CH(CH_3)-O-C(=O)-CH_3$

(e)

$H_2N-C(=O)-CH(CH_3)-CH_2-CH_3$

25.23 (a)

$H_3C-C(=O)-C(=O)-CH_3$

(b)

$O-C(=O)-CH_2-CH_2-C(=O)-O$

(c)

$H_3C-CH(NH_2)-C(=O)-H$

(d)

$CH_2-CH_2-CH_2-CH_2-CH(CH_3)-CH(-O-C(=O)-CH_2-CH_3)$ (ring)

(e)

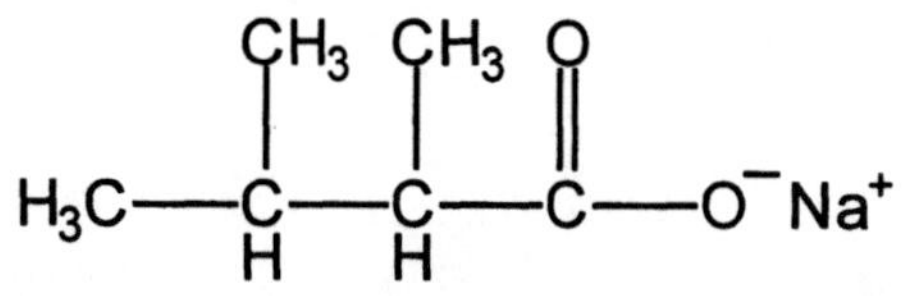

$$H_3C-\underset{H}{\overset{CH_3}{C}}-\underset{H}{\overset{CH_3}{C}}-\overset{O}{\overset{\|}{C}}-O^-\,Na^+$$

25.24 The proper IUPAC name is 2–butanol:

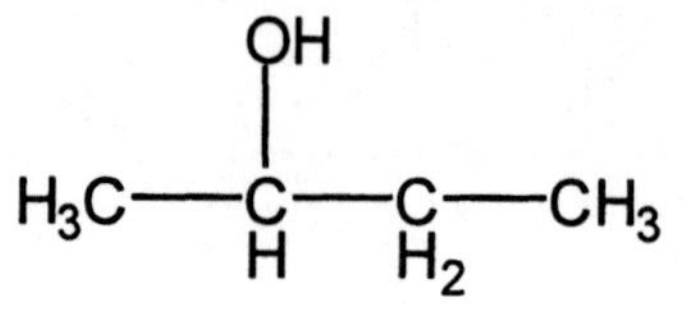

$$H_3C-\underset{H}{\overset{OH}{C}}-\underset{H_2}{C}-CH_3$$

25.25 Carbon dioxide and water.

25.26 The high electron density in the π bonds of alkenes is a desirable reaction site for a Brønsted acid.

25.27 $CH_3CH{=}CH_2 + H^+ \rightarrow CH_3CH^+{-}CH_3$

25.28

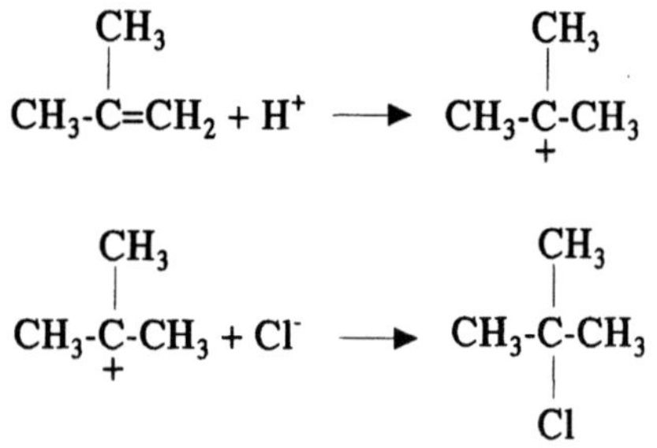

$$CH_3\text{-}\overset{CH_3}{C}{=}CH_2 + H^+ \longrightarrow CH_3\text{-}\underset{+}{\overset{CH_3}{C}}\text{-}CH_3$$

$$CH_3\text{-}\underset{+}{\overset{CH_3}{C}}\text{-}CH_3 + Cl^- \longrightarrow CH_3\text{-}\underset{Cl}{\overset{CH_3}{C}}\text{-}CH_3$$

25.29 The hydrogenation adds across the double bond the product for both isomers is butane

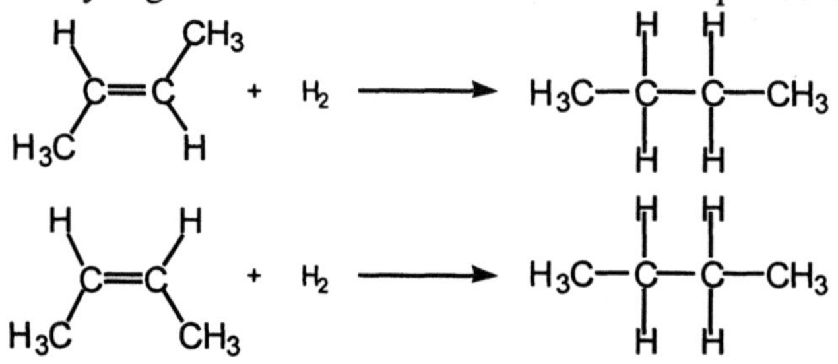

$$(H)(H_3C)C{=}C(CH_3)(H) + H_2 \longrightarrow H_3C-\underset{H}{\overset{H}{C}}-\underset{H}{\overset{H}{C}}-CH_3$$

$$(H)(H_3C)C{=}C(H)(CH_3) + H_2 \longrightarrow H_3C-\underset{H}{\overset{H}{C}}-\underset{H}{\overset{H}{C}}-CH_3$$

25.30 The oxygen atom of an alcohol group acquires another proton when the alcohol is treated with a strong acid, and this weakens all of the bonds to oxygen, including the C — O bond:

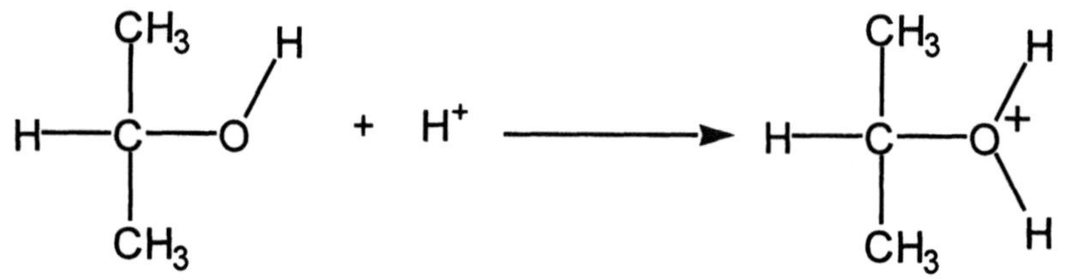

$$H-\underset{CH_3}{\overset{CH_3}{C}}-O-H + H^+ \longrightarrow H-\underset{CH_3}{\overset{CH_3}{C}}-\overset{+}{O}H_2$$

25.31

$$\begin{array}{c} CH_3 \quad H \\ | \quad / \\ H_3C - C - O \\ | \\ CH_3 \end{array}$$

2-methyl-2-propanol

25.32 Ether: Ethers are very unreactive. But, like bases, they are electron pair donors and may dissolve in a basic solution.

25.33 Amine: As derivatives of ammonia, these compounds are weakly basic and will react or dissolve in acid but as a weak base, will not dissolve in a strongly basic solution.

25.34 The lone pair on the nitrogen atom of an amide is not basic due to the electronegativity of the amide oxygen atom.

25.35 The carbonyl group in urea pulls electron density away from the nitrogen atoms making urea a neutral compound.

25.36 $CH_3CH_2CH_2NH_2 + H_2O \rightarrow CH_3CH_2CH_2NH_3^+ + OH^-$

25.37 Alkene: As a hydrocarbon, alkenes are insoluble in water. However, the high electron density around the double bonds attracts electron deficient compounds such as acids. Regions of low electron density, opposite of a double bond, attract electron rich species.

25.38 a, b, c, f

25.39 (a) $CH_3CH_2CH_2NH_3^+ Br^-$
(b) no reaction
(c) no reaction
(d) $CH_3CH_2CH_2NH_2 + H_2O$

25.40 $CH_3CH_2CO_2H + CH_3OH \overset{H^+}{\rightleftharpoons} CH_3CH_2CO_2CH_3 + H_2O$

25.41 Biochemistry is the science of living systems, emphasizing the study of cell chemistry and chemical interactions among cells.

25.42 These are materials (carbohydrates, lipids, proteins, nucleic acids, water, certain metal ions, and micronutrients, etc.), energy (supplied by lipids, carbohydrates, etc.), and information in the form of genetic codes (supplied by nucleic acids).

25.43 Carbohydrates are naturally occurring polyhydroxyaldehydes and polyhydroxyketones, or substances that give these on hydrolysis. They include the monosaccharides, the disaccharides and the polysaccharides. Examples are sucrose, starch and cellulose.

25.44 Each yields glucose on complete hydrolysis:
(a) glucose (b) glucose (c) glucose

25.45 The three forms differ from one another only in the manner in which the ring is either opened or closed, as illustrated in Figure 25.2. Furthermore, the open form (which gives the substance the characteristics of an aldehyde) is readily regenerated by the two dynamic and reversible equilibria that convert it into either the α or the β form.

25.46 The digestion of sucrose yields glucose and fructose.

25.47 The hydrolysis of lactose gives glucose and galactose.

25.48 They each give glucose:
(a) glucose (b) glucose

25.49 Amylose constitutes 20 % of starch. It is one of two kinds of glucose polymer in starch, the other being amylopectin. Amylopectin is larger than amylose, because it is composed of numerous amylose molecules linked by oxygen bridges. See page 1111 of text.

25.50 Humans lack the digestive enzyme required to hydrolyze cellulose.

25.51 Glycogen stores glucose units.

25.52 Lipids are substances that are found in living systems and that are soluble in nonpolar solvents such as ethers and benzene. The lipids include a diverse range of substances such as cholesterol, and hormones, as well as fats, oils, fatty acids, and other materials known as triacylglycerols.

25.53 Large portions of lipids are made up of nonpolar hydrocarbon-like segments.

25.54 It is soluble in nonpolar solvents, and it is a substance found in living things.

25.55 This particular type of unsaturation refers only to the presence of the C=C double bond.

25.56 No, because it has an odd number of carbon atoms.

25.57 (a) The monomers of polypeptides are alike in having the carbon – nitrogen – carbonyl repeating backbone, one linked to another by the peptide bond.
(b) The monomers of polypeptides have different side groups, since each of the various amino acids could be employed at any point in the polymer.

25.58 The peptide bond is like that found in the amide bond, a carbonyl – nitrogen bond.

25.59 Two isomeric polypeptides can differ in the amino acid sequence.

25.60 Two dissimilar polypeptides can differ in (1) the number of amino acids, (2) the identities of the amino acids, and (3) the sequence of amino acids that comprise the chain.

25.61 Although some proteins are made up of a single polypeptide, others may contain two or more associated polypeptides, an organic group, as well as a metal ion.

25.62 The final shape is determined by the exact amino acid sequence, because the side chains are of different sizes and some are hydrophillic and others are hydrophobic.

25.63 Protonation or deprotonation changes the hydrophobic-hydrophilic character of the various amino acids in the chain.

25.64 Proteins

25.65 These deactivate enzymes. Examples include nerve poisons, botulinum toxin, and heavy metals.

25.66 A particular amino acid sequence in a polypeptide is obtained by following a sort of "direction" from the gene that is responsible for a particular polypeptide synthesis. There is a one-to-one relationship between the series of side chain bases (taken in groups of three) in the DNA of a gene and the series of amino acids that are to be assembled in the polypeptide that is to be made under the "direction" of the gene. The gene directs the synthesis of hnRNA, which the cell possesses to make mRNA. It is mRNA that is the bearer of the genetic message from the nucleus of the cell to a site outside the nucleus, where the polypeptide is to be assembled. This is the process of transcription. Next, the mRNA accumulates at the ribosomes, awaiting the arrival of the correct amino acids, which are delivered by tRNA. The process of polypeptides synthesis next involves translation of the information in mRNA into a specific amino acid sequence in the growing polypeptide (Figure 25.10).

25.67 The genetic messages are carried as a sequence of side chain bases on a DNA segment.

25.68 The two strands in DNA are held together by hydrogen bonds between specific base pairs, as shown in Figures 25.7 and 25.8.

25.69 This is shown in Figures 25.7 and 25.8. Only one particular base can be found opposite another in the double helix. Both structures also have sugar–phosphate backbones.

25.70 DNA and RNA differ in the identity of the monosaccharide. In RNA it is ribose. In DNA it is deoxyribose. Also, DNA uses the bases A, T, G, and C, whereas RNA uses the bases A, U, G, and C.

25.71
(a) In DNA, A pairs with T.
(b) In RNA, A pairs with U.
(c) C pairs with G.

25.72 A segmented gene is one in which the strands of DNA are divided or split, such that the gene is a conglomerate, exons being separated by introns.

25.73 Codons are found on mRNA.

25.74 Anticodons are found on tRNA.

25.75
(a) The ribosome is the site outside the nucleus where the polypeptide is made.
(b) mRNA carries the genetic message from the DNA in the nucleus to the ribosomes.
(c) tRNA brings the necessary amino acids for polypeptide synthesis to the ribosomes.

25.76 This is hnRNA (also called ptRNA), which is used in making mRNA.

25.77 Transcription begins with DNA and ends with the synthesis of mRNA.

25.78 Translation begins with mRNA, uses tRNA, and ends with a specific polypeptide.

25.79 Viruses consist of nucleic acids and proteins. They interrupt the normal genetic activity of a host cell, and use the cell functions to multiply rapidly.

25.80 Genetic engineering is able to cause the biological synthesis of important human polypeptides, enzymes etc, by tricking bacterial plasmids into synthesizing recombinant DNA, through the introduction of human DNA.

25.81 The DNA that is present in a bacterial plasmid that has been altered (by the introduction of DNA that is foreign to the bacteria) is recombinant DNA.

Review Problems

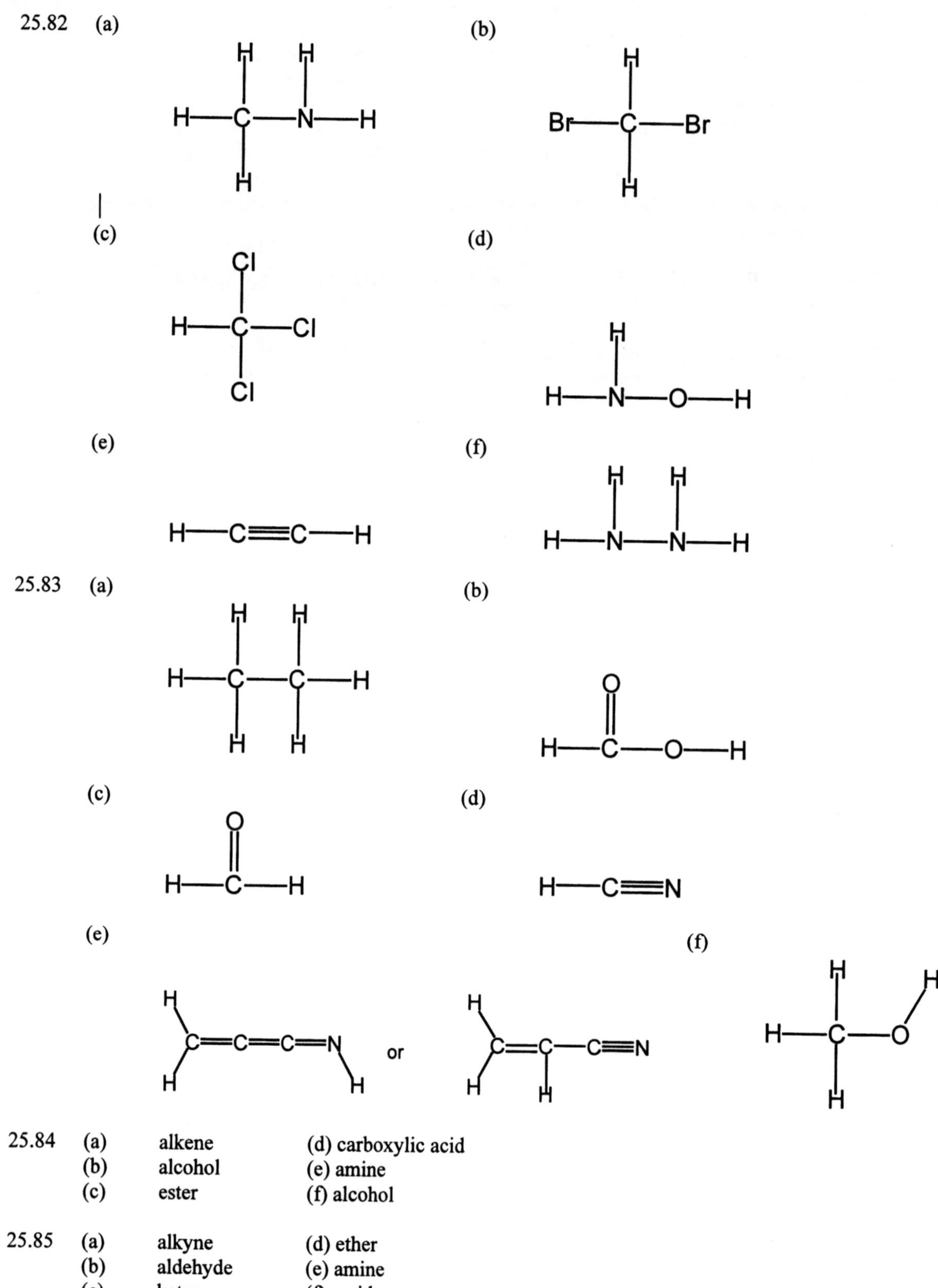

25.84 (a) alkene (d) carboxylic acid
(b) alcohol (e) amine
(c) ester (f) alcohol

25.85 (a) alkyne (d) ether
(b) aldehyde (e) amine
(c) ketone (f) amide

25.86 The saturated compounds are b, e, and f.

25.87 The saturated compounds are d and e; b, c, and f are saturated in the sense that they have no C–C multiple bonds.

25.88 (a) amine (b) amine (c) amide (d) amine, ketone

25.89
(a) ether, amine
(b) ether, amide
(c) ester, aldehyde
(d) ester, carboxylic acid

25.90
(a) These are identical, being oriented differently only.
(b) These are identical, being drawn differently only.
(c) These are unrelated, being alcohols with different numbers of carbon atoms.
(d) These are isomers, since they have the same molecular formula, but different structures.
(e) These are identical, being oriented differently only.
(f) These are identical, being drawn differently only.
(g) These are isomers, since they have the same molecular formula, but different structures.

25.91
(a) These are identical, being oriented differently only.
(b) These are isomers, since they have the same molecular formula, but different structures.
(c) These are isomers, since they have the same molecular formula, but different structures.
(d) These are identical, being oriented differently only.
(e) These are identical, being drawn differently only.
(f) These are isomers, since they have the same molecular formula, but different structures.
(g) These are unrelated.

25.92
(a) pentane
(b) 2–methylpentane
(c) 2,4–dimethylhexane

25.93
(a) 2,4–dimethylhexane
(b) 3–hexene
(c) 4–methyl–2–pentene

25.94
(a) No isomers

(b)

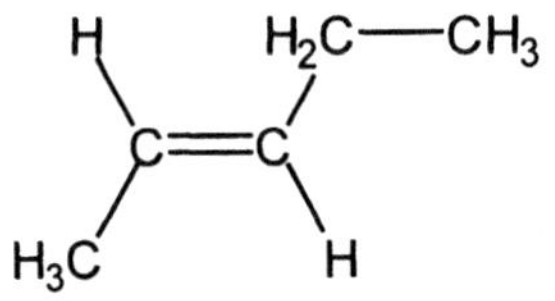

H, H_2C—CH_3, C=C, H_3C, H

trans

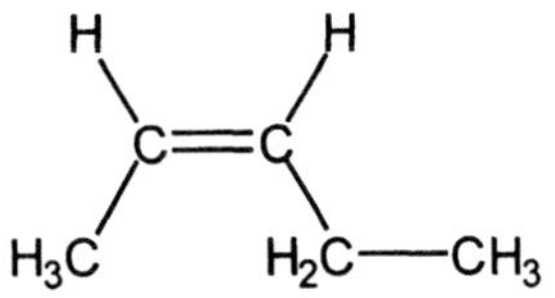

H, H, C=C, H_3C, H_2C—CH_3

cis

(c)

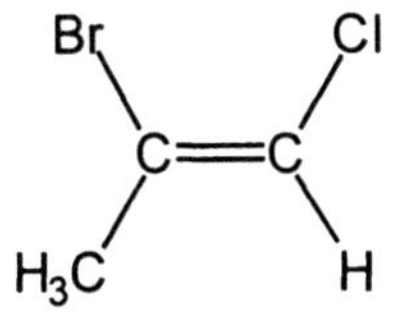

Br, Cl, C=C, H_3C, H

cis

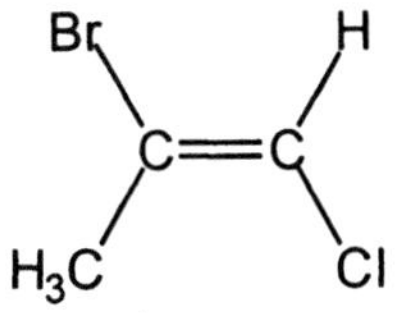

Br, H, C=C, H_3C, Cl

trans

25.95 (a) No isomers

(b)

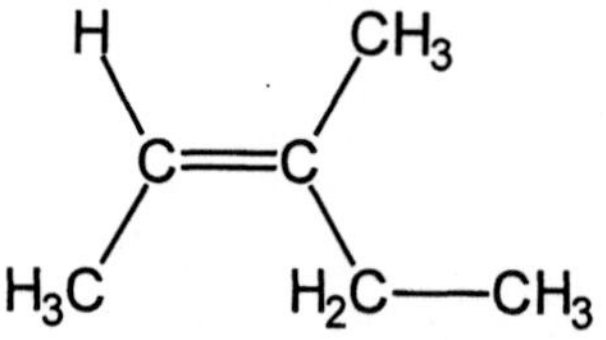

trans 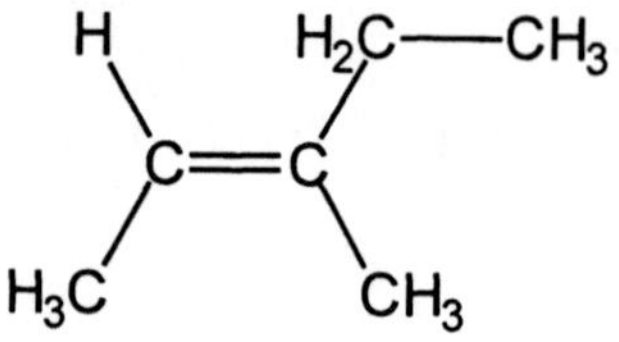 cis

(c)

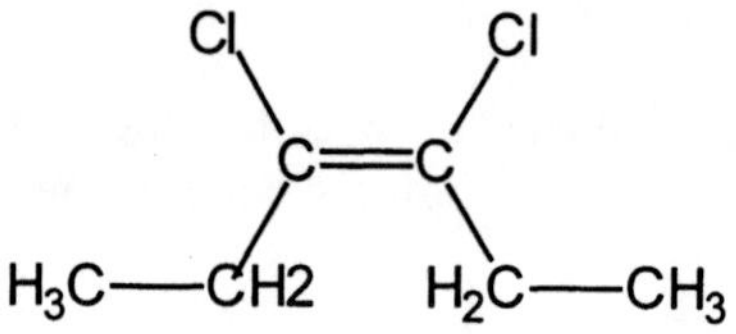

trans 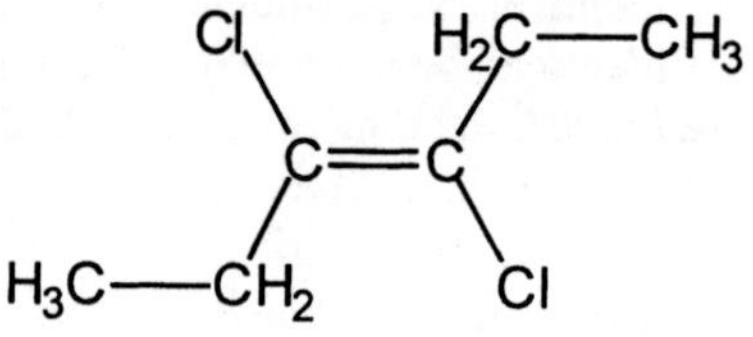 cis

25.96 (a) CH_3CH_3
(b) $ClCH_2CH_2Cl$
(c) $BrCH_2CH_2Br$
(d) CH_3CH_2Cl
(e) CH_3CH_2Br
(f) CH_3CH_2OH

25.97 (a) $CH_3CHClCH_3$
(b) CH_3CHICH_3
(c) $CH_3CHOHCH_3$

25.98 (a) $CH_3CH_2CH_2CH_3$
(b)

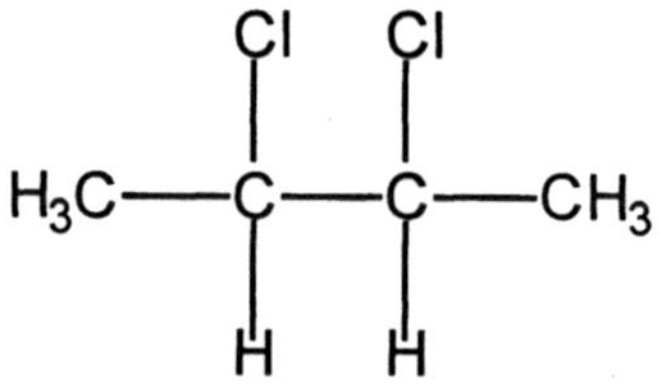

(c)

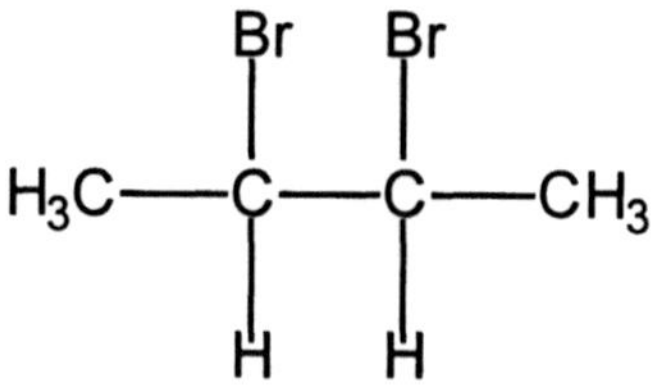

(d)

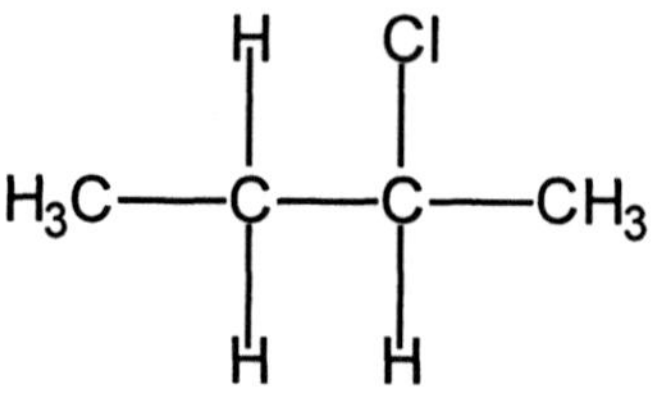

(e)

H Br
H3C—C—C—CH3
H H

(f)

H OH
H3C—C—C—CH3
H H

25.99 (a)

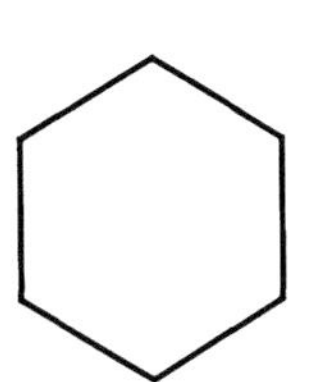

(b)

Cl
Cl

(c)

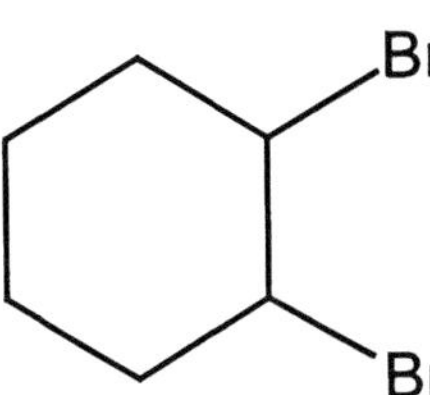

(d)

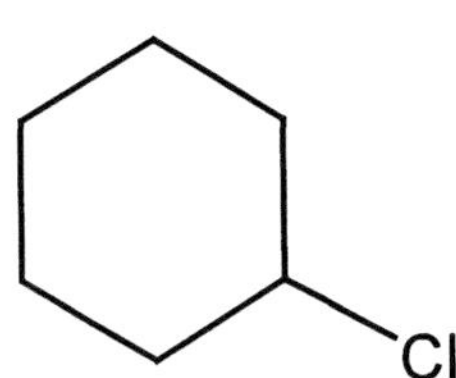

(e)

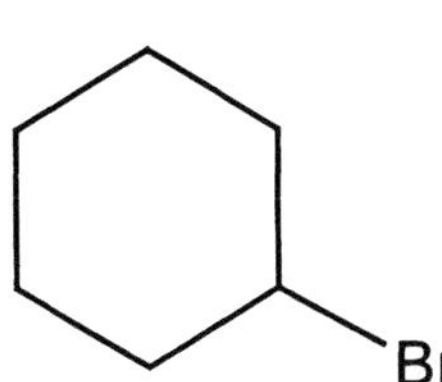

(f)

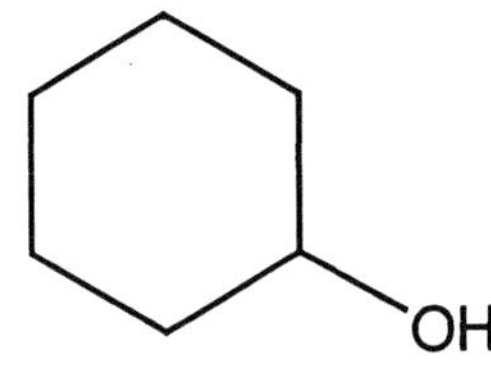

25.100 This sort of reaction would disrupt the π delocalization of the benzene ring. The subsequent loss of resonance energy would not be favorable.

25.101 Benzene does not "add" Br_2, if it did, it would add across a double bond.

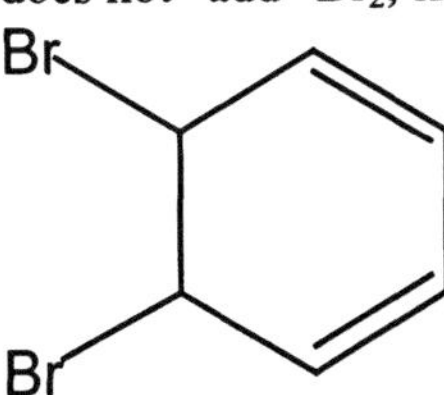

This does not occur because the loss of resonance energy would not be favorable.
Rather, in the presence of a catalyst, it gives bromobenzene:
$C_6H_6 + Br_2 \rightarrow C_6H_5Br + HBr$ ($FeBr_3$ catalyst)

25.102 CH_3OH IUPAC name = methanol; common name = methyl alcohol
CH_3CH_2OH IUPAC name = ethanol; common name = ethyl alcohol
$CH_3CH_2CH_2OH$ IUPAC name = 1–propanol; common name = propyl alcohol

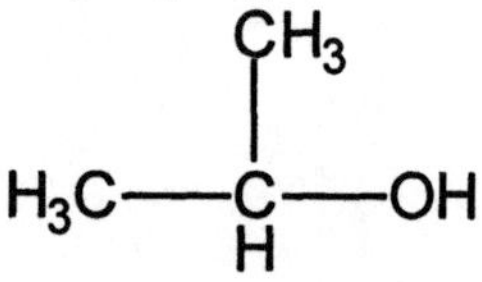

$$\begin{array}{c} CH_3 \\ | \\ H_3C{-}C{-}OH \\ H \end{array}$$

IUPAC name = 2–propanol; common name = isopropyl alcohol

25.103 $CH_3CH_2CH_2CH_2OH$ 1–butanol
$CH_3CH_2CHOHCH_3$ 2–butanol
$(CH_3)_2CHCH_2OH$ 2–methyl–1–propanol
$(CH_3)_3COH$ 2–methyl–2–propanol

25.104 $CH_3CH_2CH_2{-}O{-}CH_3$ methyl propyl ether
$CH_3CH_2{-}O{-}CH_2CH_3$ diethyl ether
$(CH_3)_2CH{-}O{-}CH_3$ methyl 2–propyl ether

25.105 Those that can be oxidized to aldehydes are:
$CH_3CH_2CH_2CH_2OH$ and $(CH_3)_2CHCH_2OH$

One can be oxidized to a ketone:
$CH_3CH_2CHOHCH_3$

25.106 (a)

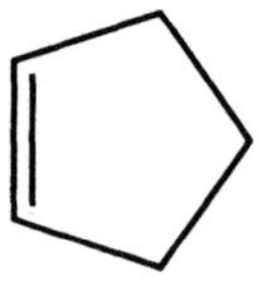

(b)

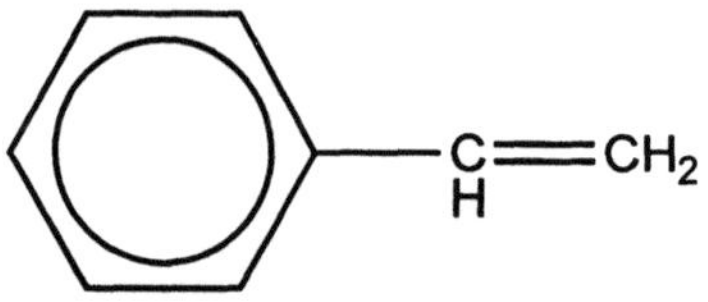

$C_6H_5{-}CH{=}CH_2$

(c)

$C_6H_5{-}CH{=}CH_2$

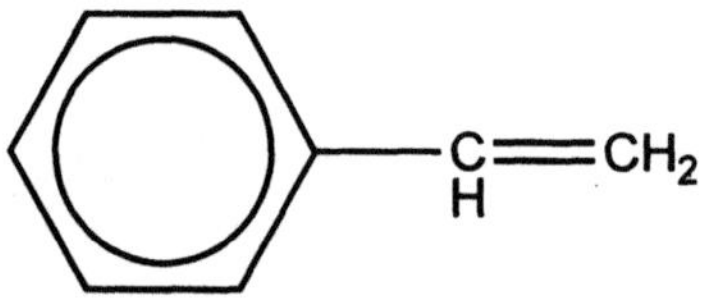

25.107

(a)

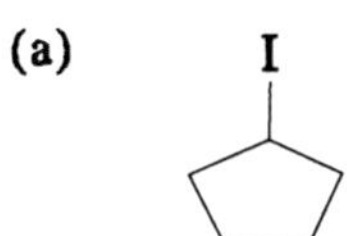

(b)

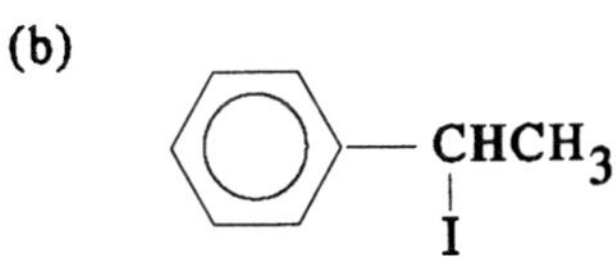

(c)

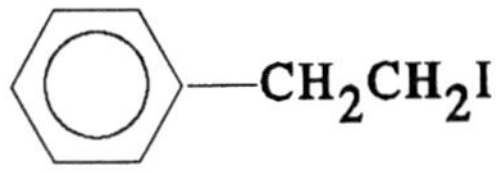

$-CH_2CH_2I$

25.108 (a)

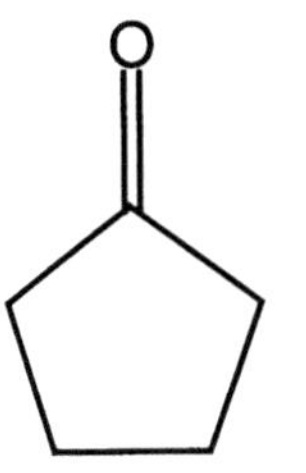

(b)

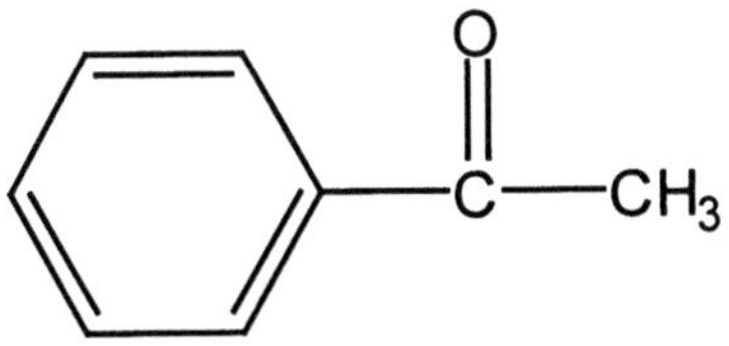

$\overset{O}{\overset{\|}{C}}-CH_3$

(c)

$\underset{H_2}{C}-\overset{O}{\overset{\|}{C}}H$

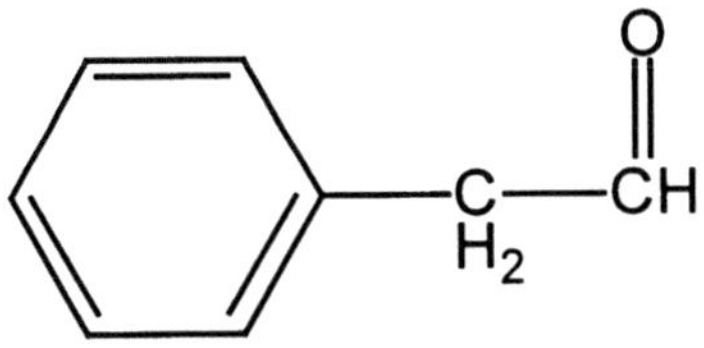

25.109 The product is propene.

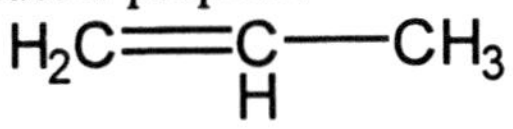

$$H_2C{=}\underset{H}{C}-CH_3$$

The mechanism is an addition of H^+ to the oxygen followed by loss of water to form a carbocation and finally the formation of the C-C double bond.

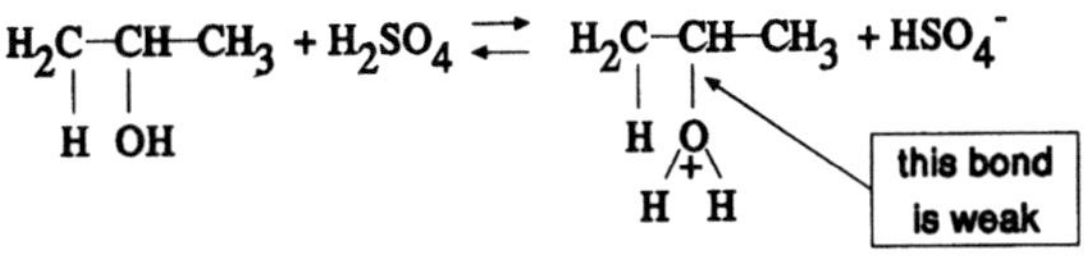

$$\underset{H}{H_2C}-\underset{OH}{CH}-CH_3 + H_2SO_4 \rightleftharpoons \underset{H}{H_2C}-\underset{\overset{+}{O}H_2}{CH}-CH_3 + HSO_4^-$$

this bond is weak

$$H_2C{-}CH(OH_2^+){-}CH_3\ (\text{with H on } H_2C) \rightleftharpoons H_2C(H){-}\overset{+}{C}H{-}CH_3 + H_2O$$

$$H_2C(H){-}\overset{+}{C}H{-}CH_3 + HSO_4^- \rightleftharpoons H_2C{=}CH{-}CH_3 + H_2SO_4$$

25.110 The elimination of water can result in a C=C double bond in two locations:

$CH_2{=}CHCH_2CH_3$	$CH_3CH{=}CHCH_3$
1–butene	2–butene

25.111 The ratio should be 3:2, since there are three hydrogen atoms on the first carbon of 2–butanol, whereas there are only two hydrogen atoms on the third carbon atom of 2–butanol.

25.112 The aldehyde is more easily oxidized. The product is:

$$H_3C{-}CH_2{-}C({=}O){-}OH$$

25.113 (a) This is C, giving:

$$H_3C{-}CH_2{-}C({=}O){-}OH$$

(b) This is B, giving:

$$H_3C{-}CH_2{-}C({=}O){-}O^-\ Na^+$$

(c) This is B, giving:

$$H_3C{-}CH_2{-}C({=}O){-}O{-}CH_3$$

(d) This is A, giving:

$$H_3C{-}CH_2{-}C({=}O){-}H$$

(e) This is A, giving: $CH_3CH{=}CH_2$

25.114 (a) $CH_3CH_2CO_2H$
(b) $CH_3CH_2CO_2H$ + CH_3OH
(c) $Na^+ + CH_3CH_2CH_2CO_2^-$ + H_2O

25.115 (a) $CH_3CH_2CONH_2$ + H_2O
(b) $CH_3CH_2CH_2CO_2H$ + NH_3
(c) $Na^+ + CH_3CH_2CO_2^-$ + CH_3OH

25.116 CH_3CO_2H + $CH_3CH_2NHCH_2CH_3$

25.117 (a) The amine neutralizes HCl:
$CH_3CH_2CH_2NH_2 + HCl \rightarrow CH_3CH_2CH_2NH_3^+ + Cl^-$

(b) The amide is hydrolyzed:

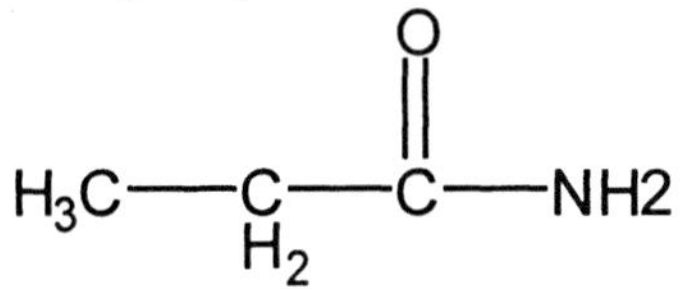

$H_3C—CH_2—C(=O)—NH2$ + H_2O → CH_3CH_2COH + NH_3

(c) The alkylammonium cation neutralizes sodium hydroxide:
$CH_3CH_2NH_3^+ + OH^- \rightarrow CH_3CH_2NH_2 + H_2O$

25.118

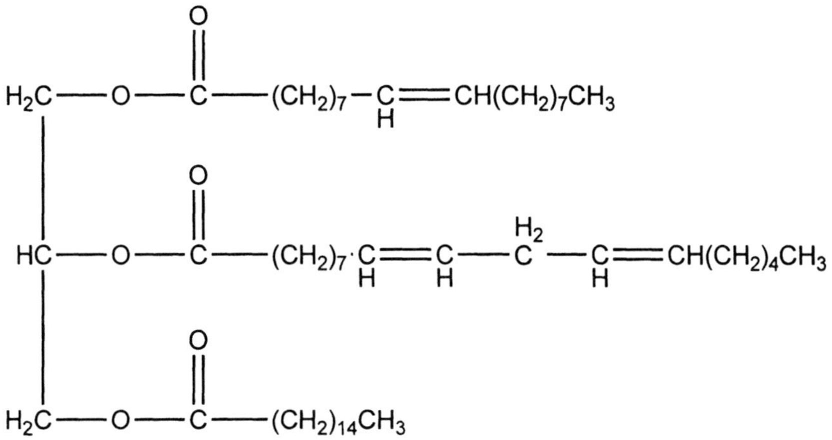

$H_2C—O—C(=O)—(CH_2)_7—CH=CH(CH_2)_7CH_3$
$HC—O—C(=O)—(CH_2)_7·CH=CH—CH_2—CH=CH(CH_2)_4CH_3$
$H_2C—O—C(=O)—(CH_2)_{14}CH_3$

2.119 These are, first, the tri-alcohol:

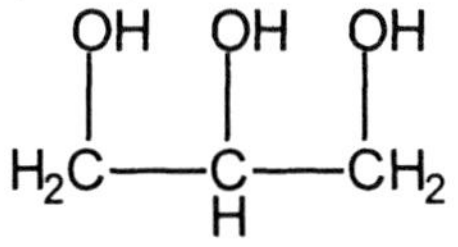

$H_2C(OH)—CH(OH)—CH_2(OH)$

and the three carboxylic acids:
linoleic acid:

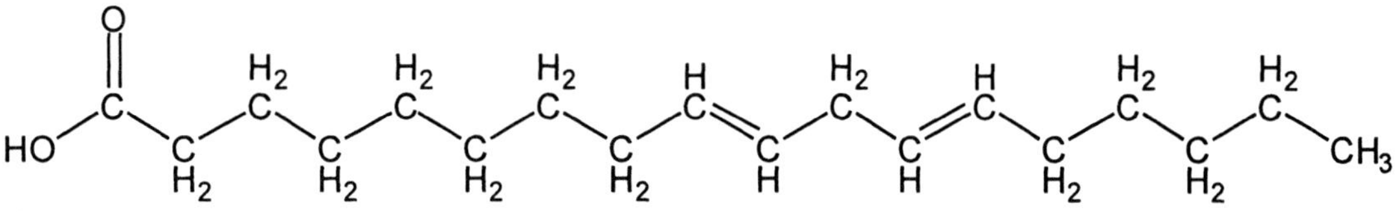

HO—C(=O)—CH2—CH2—CH2—CH2—CH2—CH2—CH2—CH=CH—CH2—CH=CH—CH2—CH2—CH2—CH2—CH3

oleic acid

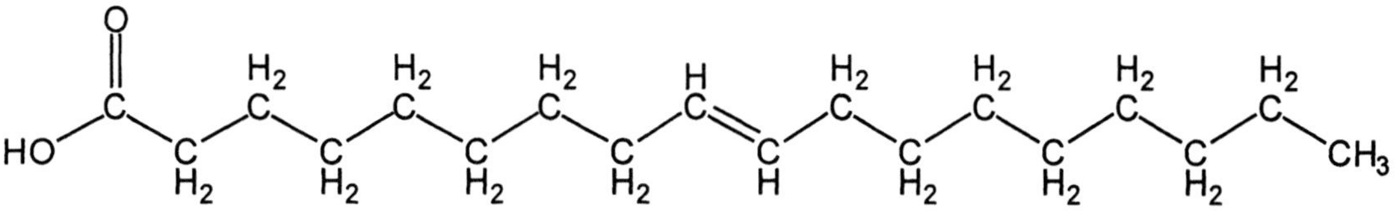

HO—C(=O)—CH2—CH2—CH2—CH2—CH2—CH2—CH2—CH=CH—CH2—CH2—CH2—CH2—CH2—CH2—CH2—CH3

myristic acid

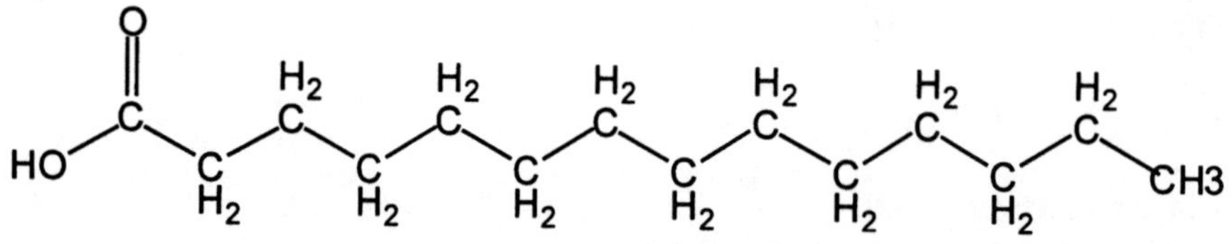

$$HO-C(=O)-CH_2-CH_2-CH_2-CH_2-CH_2-CH_2-CH_2-CH_2-CH_2-CH_2-CH_2-CH_2-CH_3$$

25.120

$$H_2C-O-C(=O)-(CH_2)_{16}CH_3$$
$$HC-O-C(=O)-(CH_2)_{12}CH_3$$
$$H_2C-O-C(=O)-(CH_2)_{16}CH_3$$

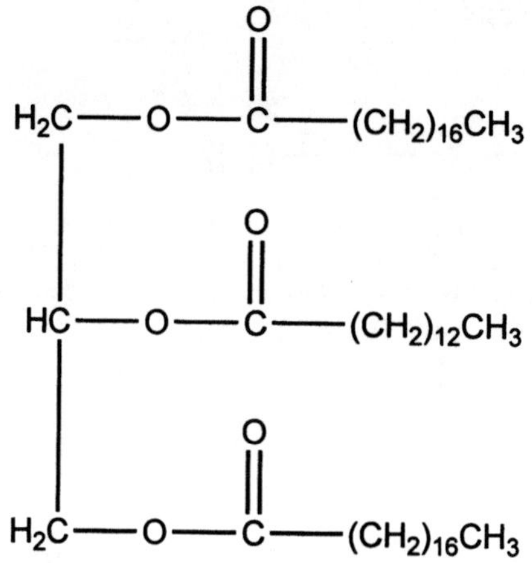

25.121 First, we have glycerol:

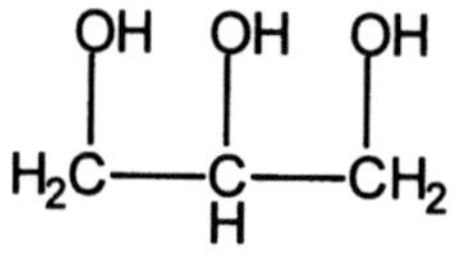

$$H_2C(OH)-CH(OH)-CH_2(OH)$$

and then three anions, as follows.

(1) linoleate anion:

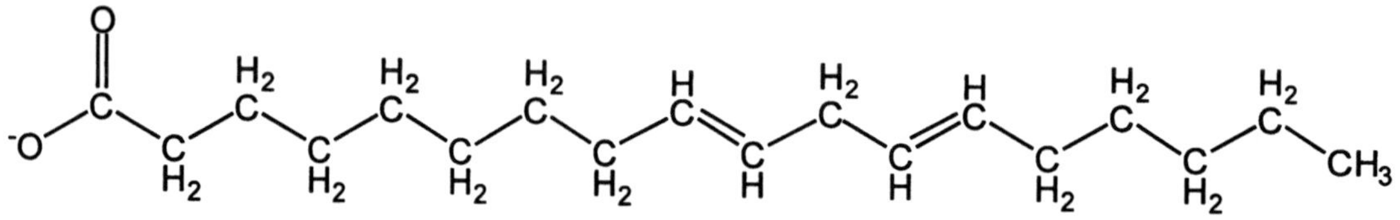

$$^{-}O-C(=O)-CH_2-CH_2-CH_2-CH_2-CH_2-CH_2-CH_2-CH=CH-CH_2-CH=CH-CH_2-CH_2-CH_2-CH_2-CH_3$$

(2) myristate anion:

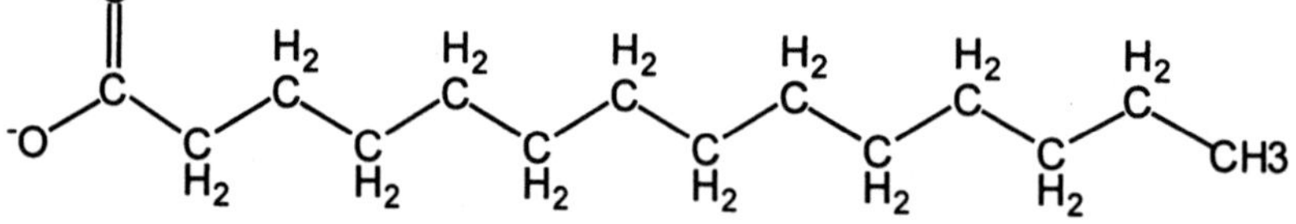

$$^{-}O-C(=O)-CH_2-CH_2-CH_2-CH_2-CH_2-CH_2-CH_2-CH_2-CH_2-CH_2-CH_2-CH_2-CH_3$$

(3) oleate anion:

$$^{-}O-C(=O)-CH_2-CH_2-CH_2-CH_2-CH_2-CH_2-CH_2-CH=CH-CH_2-CH_2-CH_2-CH_2-CH_2-CH_2-CH_2-CH_3$$

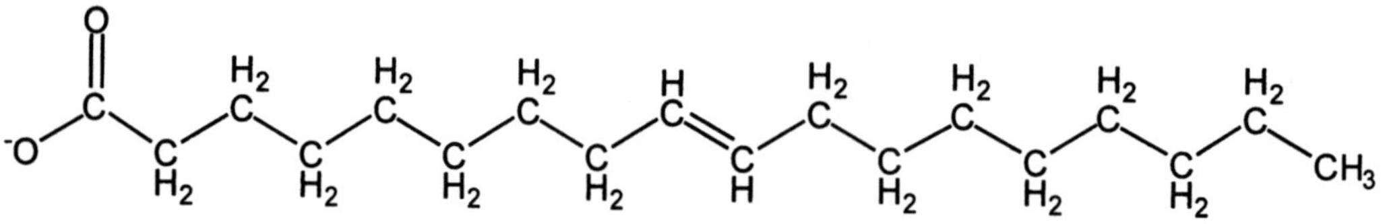

25.122 Hydrophobic sites are composed of fatty acid units. Hydrophilic sites are composed of charged units.

22.123 See Figure 25.3 and pages 1114 and 1115 of the text.

25.124

$$^{+}H_3N-CH_2-C(=O)-NH-CH_2-C(=O)-O^{-}$$

25.125

$$^{+}H_3N-CH(CH_3)-C(=O)-NH-CH(CH_3)-C(=O)-NH-CH(CH_3)-C(=O)-O^{-}$$

25.126

$$^{+}H_3N-CH_2-C(=O)-NH-CH(CH_2C_6H_5)-C(=O)-O^{-} \quad ^{+}H_3N-CH(CH_2C_6H_5)-C(=O)-NH-CH_2-C(=O)-O^{-}$$

25.127

alanine $^{+}H_3N-CH(CH_3)-C(=O)-O^{-}$

cysteine $^{+}H_3N-CH(CH_2SH)-C(=O)-O^{-}$

phenyl alanine $^{+}H_3N-CH(CH_2C_6H_5)-C(=O)-O^{-}$

Additional Exercises

25.128 (a) $(CH_3)_2CHCH_2OH$ and $(CH_3)_3COH$

(b) The one that cannot be oxidized is $(CH_3)_3COH$.

(c)

$$H_3C{-}\overset{CH_3}{\underset{+}{\overset{|}{C}}}{-}CH_3 \qquad H_3C{-}\overset{CH_3}{\overset{|}{C}H}{-}CH_2^{+}$$

(d) The first one is more stable, since it is the one that leads to the observed product. It is a tertiary carbocation.

25.129 (a) This must have been A, the primary alcohol, which is oxidized completely by an excess of dichromate to the acid. The secondary alcohol, B, would have been oxidized only to the ketone.

(b)

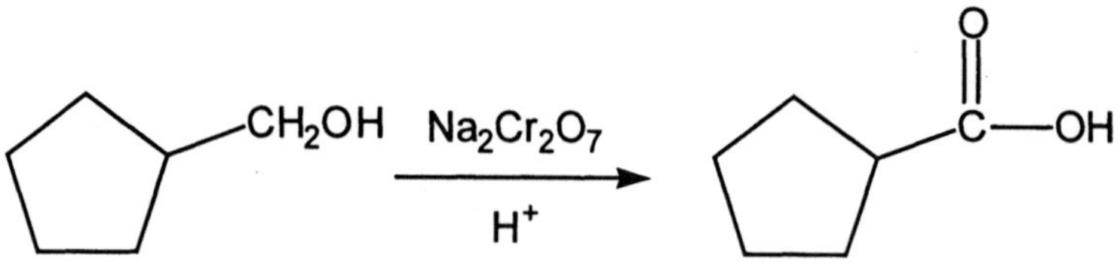

$$C_5H_9{-}CH_2OH \xrightarrow[H^+]{Na_2Cr_2O_7} C_5H_9{-}\overset{O}{\overset{\|}{C}}{-}OH$$

25.130 The dimethylamine has a higher boiling point, inspite of the lower formula mass due to the hydrogen bond between the nitrogen and the hydrogen.

25.131 (a)

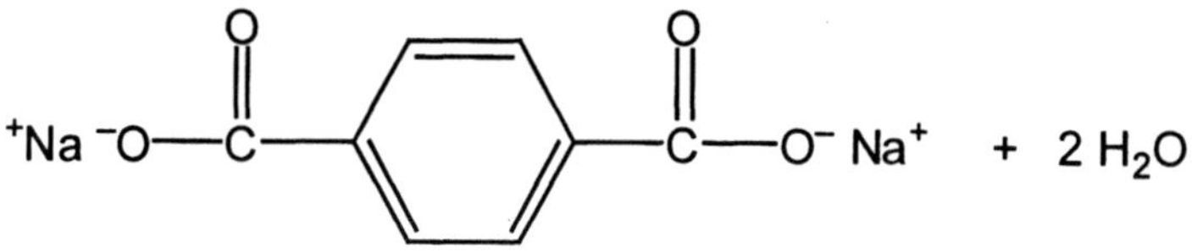

$$^{+}Na\ ^{-}O{-}\overset{O}{\overset{\|}{C}}{-}C_6H_4{-}\overset{O}{\overset{\|}{C}}{-}O^{-}\ Na^{+} \quad + \quad 2\ H_2O$$

(b)

$$H_3C{-}\overset{OH}{\overset{|}{CH}}{-}CH_2{-}CH_2{-}CH_3$$

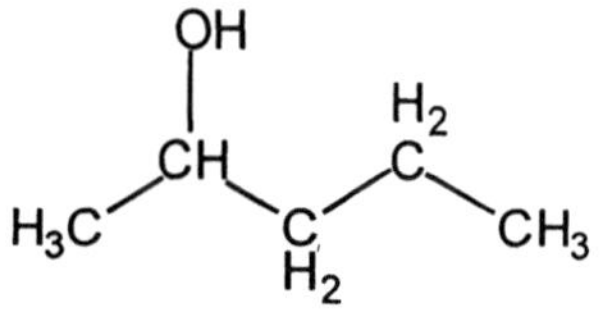

(c) $CH_3NHCH_2CH_2CH_3 + H_2O$

(d) $CH_3CH_2OCH_2CH_2CO_2H$ + CH_3OH

(e)

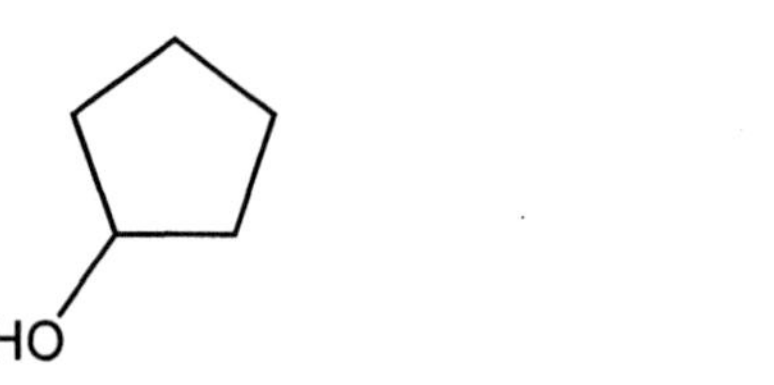

(f)

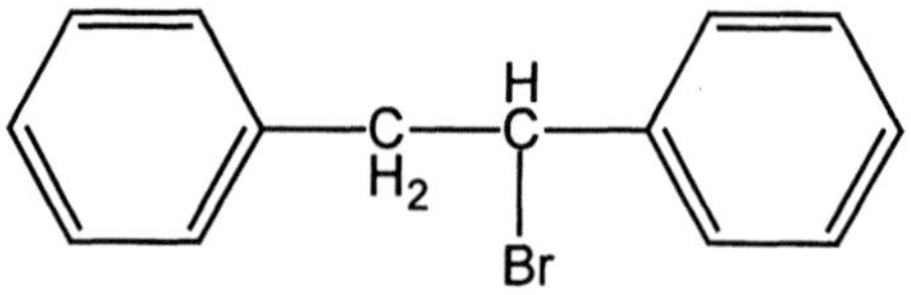

$$C_6H_5{-}CH_2{-}\overset{H}{\overset{|}{C}}(Br){-}C_6H_5$$

25.132 There are six possibilities, Gly-Ala-Phe, Gly-Phe-Ala, Ala-Gly-Phe, Ala-Phe-Gly, Phe-Gly-Ala, and Phe-Ala-Gly (GAP, GPA, AGP, APG, PGA, and PAG).

25.133 We first find the number of grams of C, H, and O that were present in the original sample:

$$\# \text{ g C} = \left(1.181 \times 10^{-3} \text{ g } CO_2\right)\left(\frac{12.01 \text{ g C}}{44.01 \text{ g } CO_2}\right) = 3.223 \times 10^{-4} \text{ g C}$$

$$\# \text{ g H} = \left(0.3653 \times 10^{-3} \text{ g } H_2O\right)\left(\frac{2.016 \text{ g H}}{18.02 \text{ g } H_2O}\right) = 4.087 \times 10^{-5} \text{ g H}$$

The mass of oxygen is given by:
5.574×10^{-4} g – 3.223×10^{-4} g – 4.087×10^{-5} g = 1.942×10^{-4} g O
The mole amounts are:
3.223×10^{-4} g C ÷ 12.01 g/mol = 2.684×10^{-5} mol C
4.087×10^{-5} g H ÷ 1.008 g/mol = 4.055×10^{-5} mol H
1.942×10^{-4} g O ÷ 16.00 g/mol = 1.214×10^{-5} mol O

The relative mole amounts are:
for C: 2.684 mol/1.214 mol = 2.211
for H: 4.055 mol/1.214 mol = 3.340
for O: 1.214 mol/1.214 mol = 1.000

These relative mole amounts correspond to the empirical formula:
$C_{20}H_{30}O_9$

25.134 The original number of moles of hydroxide are:
0.1016 M × 0.05000 L = 5.080×10^{-3} mol OH^-

The moles of hydroxide not neutralized by the organic acid are:
0.1182 M × 0.02378 L = 2.811×10^{-3} mol OH^-

Therefore, the moles of hydroxide that were neutralized by the acid were:
5.080×10^{-3} mol – 2.811×10^{-3} mol = 2.269×10^{-3} mol OH^-

This is also equal to the number of moles of the organic acid.

The formula mass is therefore:
0.2081 g/2.269×10^{-3} mol = 91.71 g/mol

This is equal to the molecular mass only if the unknown is a monoprotic acid.

Solutions To Tests of Facts and Concepts

1. (a) 24.6 cm 3 sig. fig. 0.35140 m 5 sig fig 7,424 mm 4 sig. fig.

(b) $\text{vol} = 24.6 \text{ cm} \times 0.35410\text{m}\left(\frac{100 \text{ cm}}{1 \text{ m}}\right) \times 7.424 \text{ mm}\left(\frac{1 \text{ cm}}{10 \text{ mm}}\right) = 64.6 \text{ cm}^3$

(c) $\# \text{ ft}^3 = 64.7 \text{ cm}^3\left(\frac{1 \text{ ft}}{30.48 \text{ cm}}\right)^3 = 2.12 \text{ ft}^3$

(d) $\# \text{ kg} = 64.7 \text{ cm}^3\left(\frac{7.140 \text{ g}}{1 \text{ cm}^3}\right)\left(\frac{1 \text{ kg}}{1000 \text{ g}}\right) = 0.462 \text{ kg}$

2. An atom is the smallest representative sample of an element while a molecule is the smallest representative sample of a compound. Molecules are made of atoms. A mole is a unit of measure for the amount of a substance; 6.022×10^{23} things are in a mole.

3. The atomic mass of element X is half the size of the atomic mass of element Y.

4. According to Dalton's atomic theory, a chemical reaction is simply a reordering of atoms from one combination to another. If no atoms are gained or lost and if the masses of the atoms cannot change, then the mass after the reaction must be the same as the mass before.

For the law of definite proportions, the theory states a given compound always has atoms of the same elements in the same numerical ratio. The same mass ratio would exist regardless of how many molecules were in the sample.

5. A and B do not necessarily need to be the same element. They could be two different elements with the same number of neutrons by coincidence.

6. $\# \text{ cm}^3 = 3.14 \text{ ft}^3\left(\frac{30.48 \text{ cm}}{1 \text{ ft}}\right)^3$

7. ${}^{204}_{94}\text{Pu}$ ${}^{244}_{94}\text{Pu}$

8. Protons: 28
Neutrons 32
Electrons 28

9. ${}^{23}_{11}\text{Na}^+$

10.
(a) ion particle
(b) mixture substance
(c) isotope substance
(d) atom particle
(e) compound substance
(f) molecule particle
(g) element substance
(h) nucleus particle

11.
(a) An isotope of iron Possible
(b) An atom of iron Not possible – atoms cannot be seen by eye
(c) A molecule of water Not possible – water molecules are too small to be seen by eye
(d) A mole of water Possible
(e) An ion of sodium Not possible – Na^+ is too small to be seen
(f) A formula unit of sodium chloride Not possible – one NaCl is too small to be seen.

12.

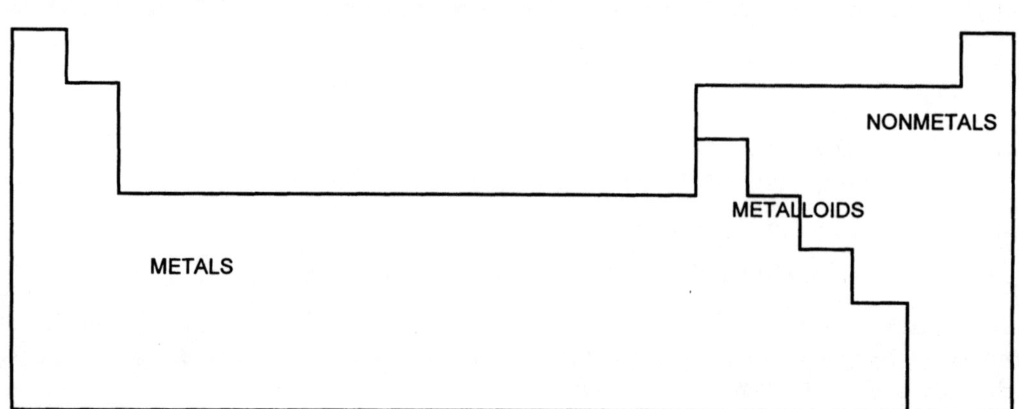

13. Zr and Hf

14.

Calcium	Alkaline earth metal
Iron	Transition metal
Helium	Noble gas
Gadolinium	Inner transition metal
Iodine	Halgen
Sodium	Alkali metal

15. Ductile is the ability to be drawn into wire.
Malleable is the ablility to be hammmered or rolled into thin sheets.

16. Mercury (m.p. –39 °C) Tungsten (m.p. 3400 °C)

17. Metalloids are semiconductors.

18. Ga, In, Sn, Tl, Ob, Bi

19.

(a)	KNO_3	(b)	$CaCO_3$
(c)	$Co_3(PO_4)_2$	(d)	$MgSO_3$
(e)	$FeBr_3$	(f)	Mg_3N_2
(g)	Al_2Se_3	(h)	$Cu(ClO_4)_2$
(i)	BrF_5	(j)	N_2O_5
(k)	$Sr(C_2H_3O_2)_2$	(l)	$(NH_4)_2Cr_2O_7$
(m)	Cu_2S		

20.
(a) sodium chlorate
(b) calcium phosphate
(c) sodium permanganate
(d) aluminum phosphide
(e) iodine trichloride
(f) phosphorous trichloride
(g) potassium chromate
(h) calcium cyanide
(i) manganese(II) chloride
(j) sodium nitrite
(k) iron(II) nitrate

21. Ionic

22. Empirical formula are written for ionc compounds since discrete molecules do not exist, the smallest set of subscripts that specify the correct ratio of the ions is used.

23. Al_2O_3, MgO, NO_2

24. $\# \text{ g of one molecule} = \left(\frac{60.2\text{ g}}{1\text{ mol}}\right)\left(\frac{1\text{ mol}}{6.022\times10^{23}\text{ molecules}}\right) = 1.00\times10^{-22}\text{ g}$

25. $\text{MM} = \left(\frac{204\text{ g}}{1.00\times10^{23}\text{ molecules}}\right)\left(\frac{6.022\times10^{23}\text{ molecules}}{1\text{ mol molecules}}\right) = 1230\text{ g mol}^{-1}$

26.
$$\begin{aligned} Fe_4[Fe(CN)_6]_3 &= 7Fe + 18C + 18N \\ &= (7\times55.85) + (18\times12.01) + (18\times14.01) \\ &= 859.3\text{ g/mole} \end{aligned}$$

27. $\# \text{ g } Cu(NO_3)_2{\cdot}3H_2O = 0.118\text{ mol}\left(\frac{241.6\text{ g } Cu(NO_3)_2\cdot3H_2O}{1\text{ mol } Cu(NO_3)_2\cdot3H_2O}\right) = 28.5\text{ g } Cu(NO_3)_2{\cdot}3H_2O$

28. $\# \text{ mol } NiI_2{\cdot}6H_2O = 15.7\text{ g } NiI_2{\cdot}6H_2O\left(\frac{1\text{ mol } NiI_2\cdot6H_2O}{420.6\text{ g } NiI_2\cdot6H_2O}\right) = 3.73\times10^{-2}\text{ mol } NiI_2{\cdot}6H_2O$

29.
$$\%\,C = \frac{0.4343\text{ g C}}{0.5866\text{ g nicotine}} \times 100\% = 74.04\%$$
$$\%\,H = \frac{0.05103\text{ g H}}{0.5866\text{ g nicotine}} \times 100\% = 8.699\%$$
$$\%\,N = \frac{0.1013\text{ g N}}{0.5866\text{ g nicotine}} \times 100\% = 17.27\%$$

30.
$$\#\text{ moles K} = (37.56\text{ g K})\left(\frac{1\text{ mole K}}{39.098\text{ g K}}\right) = 0.9607\text{ moles K}$$
$$\#\text{ moles H} = (1.940\text{ g H})\left(\frac{1\text{ mole H}}{1.00794\text{ g H}}\right) = 1.925\text{ moles H}$$
$$\#\text{ moles P} = (29.79\text{ g P})\left(\frac{1\text{ mole P}}{30.974\text{ g P}}\right) = 0.9618\text{ moles P}$$

Amount of O:
$\%\,O = 100\% - 37.56\%\,K - 1.940\%\,H - 29.79\%\,P = 30.71\%\,O$
$$\#\text{ moles O} = (30.71\text{ g O})\left(\frac{1\text{ mole O}}{15.9994\text{ g O}}\right) = 1.919\text{ moles O}$$

Now we divide each of these numbers of moles by the smallest of the three numbers, in order to obtain the simplest mole ratio among the three elements in the compound:

for K, 0.9607 moles / 0.9607 moles = 1.00
for H, 1.925 moles / 0.9607 moles = 2.00
for H, 0.9618 moles / 0.9607 moles = 1.00
for O, 1.919 moles / 0.9607 moles = 2.00

These relative mole amounts give us the empirical formula: KH_2PO_2.

31. $\# \text{ molecules } C_2H_5OH = 1.00 \text{ fl. oz.}\left(\frac{29.57 \text{ mL } C_2H_5OH}{1 \text{ fl. oz. } C_2H_5OH}\right)\left(\frac{0.798 \text{ g } C_2H_5OH}{1 \text{ mL } C_2H_5OH}\right)$

$\times\left(\frac{1 \text{ mol } C_2H_5OH}{46.07 \text{ g } C_2H_5OH}\right)\left(\frac{6.022\times10^{23} \text{ molecules } C_2H_5OH}{1 \text{ mol } C_2H_5OH}\right) = 3.08 \times 10^{23} \text{ molecules } C_2H_5OH$

32. $\# \text{ L } C_2H_6O_2 = 5.00 \times 10^{24} \text{ molecules}\left(\frac{1 \text{ mol } C_2H_6O_2}{6.022\times10^{23} \text{ molecules } C_2H_6O_2}\right)\left(\frac{62.07 \text{ g } C_2H_6O_2}{1 \text{ mol } C_2H_6O_2}\right)\left(\frac{1 \text{ mL } C_2H_6O_2}{1.11 \text{ g } C_2H_6O_2}\right)$

$\times\left(\frac{1 \text{ L } C_2H_6O_2}{1000 \text{ mL } C_2H_6O_2}\right) = 0.464 \text{ L}$

33. $\# \text{ mol } O_2 = 2.56 \text{ g } Cl_2\left(\frac{1 \text{ mol } Cl_2}{70.91 \text{ g } Cl_2}\right)\left(\frac{1 \text{ mol } Cl_2O_7}{1 \text{ mol } Cl_2}\right)\left(\frac{7 \text{ mol } O_2}{2 \text{ mol } Cl_2O_7}\right) = 0.126 \text{ mol } O_2$

$\# \text{ g } O_2 = 0.126 \text{ mol } O_2\left(\frac{16.00 \text{ g } O_2}{1 \text{ mol } O_2}\right) = 2.02 \text{ mol } O_2$

34. (a) $2Fe_2O_3 + 12HNO_3 \rightarrow 4Fe(NO_3)_3 + 6H_2O$

(b) $2C_{21}H_{30}O_2 + 55O_2 \rightarrow 42CO_2 + 30H_2O$

35. $\# \text{ mol } HNO_3 = 2.56 \text{ mol Cu}\left(\frac{8 \text{ } HNO_3}{3 \text{ mol Cu}}\right) = 6.83 \text{ mol } HNO_3$

36. $\# \text{ mol } O_2 = 56.8 \text{ g } NH_3\left(\frac{1 \text{ mol } NH_3}{17.03 \text{ g } NH_3}\right)\left(\frac{5 \text{ mol } O_2}{4 \text{ mol } NH_3}\right) = 4.17 \text{ mol } O_2$

$\# \text{ g } O_2 = 4.17 \text{ mol } O_2\left(\frac{16.00 \text{ g } O_2}{1 \text{ mol } O_2}\right) = 66.7 \text{ mol } O_2$

37. (a) $CaCO_3 \xrightarrow{\text{heat}} CaO + CO_2$

$MgCO_3 \xrightarrow{\text{heat}} MgO + CO_2$

(b) Let x = # g $CaCO_3$ and y = # g $MgCO_3$
x g $CaCO_3$ + y g $MgCO_3$ = 5.78 g sample

$\# \text{ g CaO} = x \text{ g } CaCO_3\left(\frac{1 \text{ mol } CaCO_3}{100.09 \text{ g } CaCO_3}\right)\left(\frac{1 \text{ mol CaO}}{1 \text{ mol } CaCO_3}\right)\left(\frac{52.09 \text{ g CaO}}{1 \text{ mol CaO}}\right) = 0.520x \text{ g CaO}$

$\# \text{ g MgO} = y \text{ g } MgCO_3\left(\frac{1 \text{ mol } MgCO_3}{84.31 \text{ g } MgCO_3}\right)\left(\frac{1 \text{ mol MgO}}{1 \text{ mol } MgCO_3}\right)\left(\frac{40.30 \text{ g MgO}}{1 \text{ mol MgO}}\right) = 0.478y \text{ g MgO}$

0.520x g CaO + 0.478y g MgO = 3.02 g total
x = 5.78 – y
0.520(5.78 – y) g CaO + 0.478y g MgO = 3.02 g
3.01 – 0.520y + 0.478y = 3.02
y = 0.343 g $MgCO_3$
x = 5.78 – 0.343 = 5.44 g $CaCO_3$

$$\% \ CaCO_3 = \frac{5.44 \text{ g } CaCO_3}{5.78 \text{ g sample}} \times 100\% = 94.1\% \ CaCO_3$$

$$\% \ MgCO_3 = \frac{0.343 \text{ g } MgCO_3}{5.78 \text{ g sample}} \times 100\% = 5.9\% \ MgCO_3$$

38. (a) 100% yeild adipic acid would be $\frac{12.5 \text{ g adipic acid}}{0.686} = 18.2 \text{ g}$

Amount of cyclohexene needed:

$$\# \text{ g } C_6H_{10} = 18.2 \text{ g}\left(\frac{1 \text{ mol } C_6H_{10}O_4}{146.1 \text{ g } C_6H_{10}O_4}\right)\left(\frac{3 \text{ mol } C_6H_{10}}{3 \text{ mol } C_6H_{10}O_4}\right)\left(\frac{82.14 \text{ g } C_6H_{10}}{1 \text{ mol } C_6H_{10}}\right) = 10.2 \text{ g } C_6H_{10}$$

(b) $$\# \text{ g } Na_2Cr_2O_7 \cdot 2H_2O = 18.2 \text{ g}\left(\frac{1 \text{ mol } C_6H_{10}O_4}{146.1 \text{ g } C_6H_{10}O_4}\right)\left(\frac{4 \text{ mol } Na_2Cr_2O_7 \cdot 2H_2O}{3 \text{ mol } C_6H_{10}O_4}\right)$$

$$\times\left(\frac{298.00 \text{ g } Na_2Cr_2O_7 \cdot 2H_2O}{1 \text{ mol } Na_2Cr_2O_7 \cdot 2H_2O}\right) = 49.5 \text{ g } Na_2Cr_2O_7 \cdot 2H_2O$$

39. $$\# \text{ tons ore} = 1.00 \text{ ton Fe}\left(\frac{91.5 \text{ ton Fe possible}}{100 \text{ ton Fe recovered}}\right)\left(\frac{2000 \text{ lb Fe possible}}{1 \text{ ton Fe possible}}\right)\left(\frac{454 \text{ g Fe}}{1 \text{ lb Fe}}\right)\left(\frac{1 \text{ mol Fe}}{55.8 \text{ g Fe}}\right)$$

$$\times\left(\frac{1 \text{ mol } Fe_2O_3}{1 \text{ mol Fe}}\right)\left(\frac{160 \text{ g } Fe_2O_3}{1 \text{ mol } Fe_2O_3}\right)\left(\frac{1 \text{ lb } Fe_2O_3}{454 \text{ g } Fe_2O_3}\right)\left(\frac{1 \text{ ton } Fe_2O_3}{2000 \text{ lb } Fe_2O_3}\right)\left(\frac{1 \text{ ton ore}}{0.341 \text{ ton } Fe_2O_3}\right) = 7.69 \text{ ton ore}$$

40. (a) $$\# \text{ tons seawater} = 1.0 \text{ troy ounce}\left(\frac{31.1 \text{ g Au}}{1.0 \text{ troy ounce Au}}\right)\left(\frac{1000 \text{ mg Au}}{1 \text{ g Au}}\right)\left(\frac{1 \text{ ton seawater}}{1.5 \text{ mg Au}}\right)$$

$$\times\left(\frac{65 \text{ ton seawater}}{100 \text{ ton seawater}}\right) = 1.3 \times 10^4 \text{ tons seawater}$$

(b) $$\text{breakeven point} = \frac{\$455 \text{ per troy ounce}}{1.3 \times 10^4 \text{ tons seawater}} = \$0.034 \text{ per ton seawater}$$

41. $\# \text{ g O} = 2.164 \text{ g} - (0.5259 \text{ g Fe} + 0.7345 \text{ g Cr}) = 0.9036 \text{ g O}$

$$\# \text{ mol Fe} = 0.5259 \text{ g Fe}\left(\frac{1 \text{ mole Fe}}{55.85 \text{ g Fe}}\right) = 0.009416 \text{ mol Fe}$$

$$\# \text{ mol Cr} = 0.7345 \text{ g Cr}\left(\frac{1 \text{ mole Cr}}{51.9961 \text{ g Cr}}\right) = 0.01413 \text{ mol Cr}$$

$$\# \text{ mol O} = 0.9036 \text{ g O}\left(\frac{1 \text{ mole O}}{15.9994 \text{ g O}}\right) = 0.05648 \text{ mol O}$$

Now we divide each of these numbers of moles by the smallest of the three numbers, in order to obtain the simplest mole ratio among the three elements in the compound:

for Fe, 0.009416 mol / 0.009416 mol = 1.00
for Cr, 0.01413 mol / 0.009416 mol = 1.50
for O, 0.05648 mol / 0.009416 mol = 6.00
Multiply by 2 in order to have whole numbers.
These relative mole amounts give us the empirical formula: $Fe_2Cr_3O_{12}$

To calculate the molecular mass, the molecular formula is needed.

42. The amount of water removed was

6.584 g sample – 2.889 g dry sample = 3.695 g H_2O

$$\# \text{ mol } H_2O = 3.695 \text{ g } H_2O\left(\frac{1 \text{ mol } H_2O}{18.015 \text{ g } H_2O}\right) = 0.2051 \text{ mol } H_2O$$

$$\# \text{ mol } Na_2SO_4 = 2.889 \text{ g } Na_2SO_4\left(\frac{1 \text{ mol } Na_2SO_4}{142.04 \text{ g } Na_2SO_4}\right) = 0.02034 \text{ mol } Na_2SO_4$$

Determine the mole ratio:

For Na_2SO_4 0.02034 mol / 0.02034 mol = 1

For H_2O 0.2051 mol / 0.02034 mol = 10.08

Formula $Na_2SO_4 \cdot 10H_2O$

43. First determine the amount of NH_3 that would be required to react completely with the given amount of O_2:

$$\# \text{ g } NH_3 = (58.0 \text{ g } O_2)\left(\frac{1 \text{ mol } O_2}{32.0 \text{ g } O_2}\right)\left(\frac{4 \text{ mole } NH_3}{5 \text{ moles } O_2}\right)\left(\frac{17.03 \text{ g } NH_3}{1 \text{ mole } NH_3}\right) = 24.7 \text{ g } NH_3$$

Since 45.0 g of NH_3 are supplied, O_2 is the limiting reactant.

An excess (45.0 g – 24.7 g = 20.3 g) of NH_3 is present.

The number of moles and grams of NO formed is:

$$\# \text{ mol NO} = 24.7 \text{ g } NH_3\left(\frac{1 \text{ mol } NH_3}{17.03 \text{ g } NH_3}\right)\left(\frac{4 \text{ mol NO}}{4 \text{ mol } NH_3}\right) = 1.45 \text{ mol NO}$$

$$\# \text{ g NO} = 1.45 \text{ mol NO}\left(\frac{30.01 \text{ g NO}}{1 \text{ mol NO}}\right) = 43.5 \text{ g NO}$$

1. A strong electrolyte dissociates completely in solution while a weak electrolyte does not, but has a low percentage ionization or dissociation in solution.
$HCHO_2(aq) + H_2O \leftrightarrows H_3O^+(aq) + CHO_2^-(aq)$

2. $CH_3NH_2(aq) + H_2O \leftrightarrows CH_3NH_3^+(aq) + OH^-(aq)$

3. Molecular equation: $CH_3NH_2 + HCl \leftrightarrows CH_3NH_3^+ + Cl^-$
Ionic equation: $CH_3NH_2 + H^+ + Cl^- \leftrightarrows CH_3NH_3^+ + H^+ + Cl^-$
Net ionic equation: $CH_3NH_2 + H^+ + \leftrightarrows CH_3NH_3^+ + H^+ +$

4.

(a)	$Ca_3(PO_4)_2$	insoluble	(b)	$Ni(OH)_2$	insoluble
(c)	$(NH_4)_2HPO_4$	soluble	(d)	$SnCl_2$	soluble
(e)	$Sr(NO_3)_2$	soluble	(f)	$Au(ClO_4)_3$	soluble
(g)	$Cu(C_2H_3O_2)_2$	soluble	(h)	AgBg	insoluble
(i)	KOH	soluble	(j)	Hg_2Cl_2	insoluble
(k)	$ZnSO_4$	soluble	(l)	Na_2S	soluble
(m)	$CoCO_3$	insoluble	(n)	$BaSO_3$	insoluble
(o)	MnS	insoluble			

5. (a) $CuCl_2(aq) + (NH_4)_2CO_3(aq) \rightarrow CuCO_3(s) + 2NH_4Cl(aq)$
$Cu^{2+}(aq) + 2Cl^-(aq) + 2NH_4^+(aq) + CO_3^{2-}(aq) \rightarrow CuCO_3(s) + 2NH_4^+(aq) + 2Cl^-(aq)$
$Cu^{2+}(aq) + CO_3^{2-}(aq) \rightarrow CuCO_3(s)$

(b) $2HCl(aq) + MgCO_3(s) \rightarrow MgCl_2(aq) + CO_2(g) + H_2O(l)$
$2H^+(aq) + 2Cl^-(aq) + MgCO_3(s) \rightarrow Mg^{2+}(aq) + 2Cl^-(aq) + CO_2(g) + H_2O(l)$
$2H^+(aq) + MgCO_3(s) \rightarrow Mg^{2+}(aq) + CO_2(g) + H_2O(l)$

(c) $ZnCl_2(aq) + 2AgC_2H_3O_2(aq) \rightarrow 2AgCl(s) + Zn(C_2H_3O_2)_2(aq)$
$Zn^{2+}(aq) + 2Cl^-(aq) + 2Ag^+(aq) + 2C_2H_3O_2^-(aq) \rightarrow 2AgCl(s) + Zn^{2+}(aq) + 2C_2H_3O_2(aq)$
$2Cl^-(aq) + 2Ag^+(aq) \rightarrow 2AgCl(s)$

(d) $HClO_4(aq) + NaCHO_2(aq) \rightarrow HCHO_2(aq) + NaClO_4(aq)$
$H^+(aq) + ClO_4^-(aq) + Na^+(aq) + CHO_2^-(aq) \rightarrow HCHO_2(aq) + Na^+(aq) + ClO_4^-(aq)$
$H^+(aq) + CHO_2^-(aq) \rightarrow HCHO_2(aq)$

(e) $MnO(s) + H_2SO_4(aq) \rightarrow MnSO_4(aq) + H_2O(l)$
$MnO(s) + 2H^+(aq) + SO_4^{2-}(aq) \rightarrow Mn^{2+}(aq) + SO_4^{2-}(aq) + H_2O(l)$
$MnO(s) + 2H^+(aq) \rightarrow Mn^{2+}(aq) + H_2O(l)$

(f) $FeS(s) + 2HCl(aq) \rightarrow H_2S(g) + FeCl_2(aq)$
$FeS(s) + 2H^+(aq) + 2Cl^-(aq) \rightarrow H_2S(g) + Fe^{2+}(aq) + 2Cl^-(aq)$
$FeS(s) + 2H^+(aq) \rightarrow H_2S(g) + Fe^{2+}(aq)$

6. $H_3PO_4 + 3NaOH \rightarrow Na_3PO_4 + 3H_2O$

7. Acidic oxides: P_4O_6, SeO_3, SO_2
Basic oxides: Na_2O, CaO, PbO

8.

(a)	$KHSO_3$	(b)	No acid salt
(c)	no acid salt	(d)	NaH_2PO_4, Na_2HPO_4
(e)	$NaHCO_3$		

9. (a) iodic acid (b) hypobromous acid
(c) nitrous acid (d) calcium hydrogen phosphate
(e) iron(III) hydrogen sulfate

10. (a) $HBrO_2$ (b) HIO
(c) $LiHSO_4$ (d) $HBrO_3$

11. $BaCl_2 + Na_2SO_4 \rightarrow BaSO_4 + 2NaCl$

$$\# \text{ mL } BaCl_2 = 27.0 \text{ mL } SO_4^{2-}\left(\frac{1 \text{ L } SO_4^{2-}}{1000 \text{ mL } SO_4^{2-}}\right)\left(\frac{0.600 \text{ mol } SO_4^{2-}}{1 \text{ L } SO_4^{2-}}\right)\left(\frac{1 \text{ mol } Ba^{2+}}{1 \text{ mol } SO_4^{2-}}\right)$$
$$\times\left(\frac{1 \text{ L } Ba^{2+}}{0.200 \text{ mol } Ba^{2+}}\right)\left(\frac{1000 \text{ mL } Ba^{2+}}{1 \text{ L } Ba^{2+}}\right) = 81 \text{ mL } Ba^{2+} \text{ solution}$$

12. $MgCl_2 + 2NaOH \rightarrow Mg(OH)_2 + 2NaCl$
This is a limiting reagent problem. Start with the $MgCl_2$ solution:
mL 0.420 M NaOH required for 30.0 mL 0.200 M $MgCl_2$

$$\# \text{ mL NaOH} = 30.0 \text{ mL}\left(\frac{1 \text{ L } Mg^{2+}}{1000 \text{ mL } Mg^{2+}}\right)\left(\frac{0.200 \text{ mol } Mg^{2+}}{1 \text{ L } Mg^{2+}}\right)\left(\frac{2 \text{ mol } OH^-}{1 \text{ mol } Mg^{2+}}\right)$$
$$\times\left(\frac{1 \text{ L } OH^-}{0.420 \text{ mol } OH^-}\right)\left(\frac{1000 \text{ mL } OH^-}{1 \text{ L } OH^-}\right) = 28.6 \text{ mL of } 0.420 \text{ M NaOH solution}$$

Only 25.0 mL of the 0.420 M NaOH solution are supplied; therefore the NaOH is the limiting reagent. The amount of $Mg(OH)_2$ formed is

$$\# \text{ g } Mg(OH)_2 = 25.0 \text{ mL } OH^-\left(\frac{1 \text{ L } OH^-}{1000 \text{ mL } OH^-}\right)\left(\frac{0.420 \text{ mol } OH^-}{1 \text{ L } OH^-}\right)$$
$$\times\left(\frac{1 \text{ mol } Mg(OH)_2}{2 \text{ mol } OH^Ñ}\right)\left(\frac{58.320 \text{ g } Mg(OH)_2}{1 \text{ mol } Mg(OH)_2}\right) = 0.306 \text{ g } Mg(OH)_2$$

The total volume of the solution is 30.0 mL + 25.0 mL = 55.0 mL solution
Mg^{2+}: concentration is 0.00 M
OH^-: concentration is 0.00 M
Na^+: concentration is:

$$25.0 \text{ mL } Na^+\left(\frac{1 \text{ L } Na^+}{1000 \text{ mL } Na^+}\right)\left(\frac{0.420 \text{ mol } Na^+}{1 \text{ L } Na^+}\right) = 0.0105 \text{ mol } Na^+$$
$$\left(\frac{0.0105 \text{ mol } Na^+}{55.0 \text{ mL solution}}\right)\left(\frac{1000 \text{ mL solution}}{1 \text{ L solution}}\right) = 0.191 \text{ M } Na^+$$

Cl^-: concentration is:

$$30.0 \text{ mL } MgCl_2\left(\frac{1 \text{ L } MgCl_2}{1000 \text{ mL } MgCl_2}\right)\left(\frac{0.200 \text{ mol } MgCl_2}{1 \text{ L } MgCl_2}\right)\left(\frac{2 \text{ mol } Cl^Ñ}{1 \text{ mol } MgCl_2}\right) = 0.012 \text{ mol } Cl^-$$
$$\left(\frac{0.012 \text{ mol } Cl^Ñ}{55.0 \text{ mL solution}}\right)\left(\frac{1000 \text{ mL solution}}{1 \text{ L solution}}\right) = 0.218 \text{ M } Cl^-$$

13. $M_1V_1 = M_2V_2$
$(6.00 \text{ M } HNO_3)(x \text{ mL}) = (0.150 \text{ M } HNO_3)(200 \text{ mL} + x \text{ mL})$
$6.00x \text{ mol} = 30 \text{ mol} + 0.150x \text{ mol}$

5.85x mol = 30 mol
x = 5.13
5.13 mL 6.00 M HNO_3 need to be added

14. First determine the number of moles of HSO_4^- titrated. Then, find the mass of the HSO_4, and finally the percentage by wieght of $NaHSO_4$.

$$\#\text{ mole } HSO_4^- = 24.60\text{ mL NaOH}\left(\frac{1\text{ L solution}}{1000\text{ mL solution}}\right)\left(\frac{0.105\text{ mol NaOH}}{1\text{ L solution}}\right)\left(\frac{1\text{ mol }HSO_4}{1\text{ mol NaOH}}\right)$$

$$= 2.583 \times 10^{-3}\text{ mol }HSO_4^-$$

$$\#\text{ g }NaHSO_4 = 2.583 \times 10^{-3}\text{ mol }HSO_4^-\left(\frac{1\text{ mol }NaHSO_4}{1\text{ mol }HSO_4^-}\right)\left(\frac{120.06\text{ g }NaHSO_4}{1\text{ mol }NaHSO_4}\right) = 0.3101\text{ g }NaHSO_4$$

$$\text{percentage by mass of }NaHSO_4\text{ in sample} = \frac{0.3101\text{ g }NaHSO_4}{0.500\text{ g sample}} \times 100\% = 62.0\%$$

15. To find the molarity of the KOH solution, find the moles of HCl, then the moles of KOH and finally the molarity of the KOH solution.

$$\#\text{ mole KOH} = 50.00\text{ mL HCl}\left(\frac{1\text{ L solution}}{1000\text{ mL solution}}\right)\left(\frac{0.0922\text{ mol HCl}}{1\text{ L solution}}\right)\left(\frac{1\text{ mol KOH}}{1\text{ mol HCl}}\right)$$

$$= 4.61 \times 10^{-3}\text{ mol KOH}$$

$$\text{molarity of KOH} = \left(\frac{4.610\times10^{-3}\text{ mol KOH}}{28.50\text{ mL KOH}}\right)\left(\frac{1000\text{ mL KOH}}{1\text{ L KOH}}\right) = 0.1617\text{ M KOH}$$

16. $M_1V_1 = M_2V_2$
(125 mL)(0.144 M) = (x mL)(18.0 M)
x = 1 mL
1.00 mL concentration H_2SO_4 is needed.

17. (a) $$\text{molar concentration} = \left(\frac{144\text{ g }H_3PO_4}{100\text{ mL solution}}\right)\left(\frac{1\text{ mol }H_3PO_4}{98.00\text{ g }H_3PO_4}\right)\left(\frac{1000\text{ mL solution}}{1\text{ L solution}}\right) = 14.7\text{ M }H_3PO_4$$

(b) $$\#\text{ g solution} = 50.0\text{ g }H_3PO_4\left(\frac{100\text{ mL solution}}{144\text{ g }H_3PO_4}\right)\left(\frac{1.689\text{ g solution}}{1\text{ mL solution}}\right) = 58.6\text{ g solution}$$

18. Let x = g Li_2CO_3 $\qquad x\left(\frac{1\text{ mol }Li_2CO_3}{73.89\text{ g }Li_2CO_3}\right) = \text{mol }Li_2CO_3$

Let y = g K_2CO_3 $\qquad y\left(\frac{1\text{ mol }K_2CO_3}{138.21\text{ g }K_2CO_3}\right) = \text{mol }K_2CO_3$

x + y = 4.43 g

$$2HCl + CO_3^{2-} \rightarrow H_2O + CO_2 + 2Cl^-$$

$$\#\text{ mole }CO_3^{2-} = 0.0532\text{ L HCl solution}\left(\frac{1.48\text{ mol HCl}}{1\text{ L solution}}\right)\left(\frac{1\text{ mol }CO_3^{2-}}{2\text{ mol HCl}}\right) = 0.0394\text{ mol }CO_3^{2-}$$

$$x\left(\frac{1\text{ mol }Li_2CO_3}{73.89\text{ g }Li_2CO_3}\right) + y\left(\frac{1\text{ mol }K_2CO_3}{138.21\text{ g }K_2CO_3}\right) = 0.0394\text{ mol }CO_3^{2-}$$

x = 4.43 – y

$$(4.43 - y)\left(\frac{1\text{ mol }Li_2CO_3}{73.89\text{ g }Li_2CO_3}\right) + y\left(\frac{1\text{ mol }K_2CO_3}{138.21\text{ g }K_2CO_3}\right) = 0.0394\text{ mol }CO_3^{2-}$$

Solve for y.
y = 3.26 g K_2CO_3
x = 4.43 g –3.26 = 1.17 g Li_2CO_3

19. $$\#\text{ mol } I_2 = 16.4 \text{ g NaIO}_3\left(\frac{1 \text{ mol NaIO}_3}{197.9 \text{ g NaIO}_3}\right)\left(\frac{3 \text{ mol } I_2}{1 \text{ mol NaIO}_3}\right) = 0.249 \text{ mole } I_2$$

$$\#\text{ g } I_2 = 0.249 \text{ mol } I_2\left(\frac{253.8 \text{ g } I_2}{1 \text{ mol } I_2}\right) = 63.2 \text{ g } I_2$$

20.

(a)	As	0				
(b)	H	+1	Cl	+3	O	–2
(c)	Mn	+2	Cl	–1		
(d)	V	+3	S	4+	O	–2

21. (a) $K_2Cr_2O_7 + HCl \rightarrow KCl + Cl_2 + H_2O + CrCl_3$

Reactants: K^+ Cr_2O_7 H^+ Cl^-
Products K^+ Cl^- Cl_2 H_2O Cr^{3+} Cl^-

$Cr_2O_7^{2-} \rightarrow Cr^{3+}$
$Cl^- \rightarrow Cl_2$

$Cr_2O_7^{2-} \rightarrow 2Cr^{3+}$
$2Cl^- \rightarrow Cl_2$

$Cr_2O_7^{2-} \rightarrow 2Cr^{3+} + 7H_2O$
$2Cl^- \rightarrow Cl_2$

$14H^+ + Cr_2O_7^{2-} \rightarrow 2Cr^{3+} + 7H_2O$
$2Cl^- \rightarrow Cl_2$

$6e^- + 14H^+ + Cr_2O_7^{2-} \rightarrow 2Cr^{3+} + 7H_2O$
$2Cl^- \rightarrow Cl_2 + 2e^-$

$6e^- + 14H^+ + Cr_2O_7^{2-} \rightarrow 2Cr^{3+} + 7H_2O$
$3(2Cl^- \rightarrow Cl_2 + 2e^-)$

$6e^- + 14H^+ + Cr_2O_7^{2-} \rightarrow 2Cr^{3+} + 7H_2O$
$6Cl^- \rightarrow 3Cl_2 + 6e^-$

$6e^- + 14H^+ + Cr_2O_7^{2-} + 6Cl^- \rightarrow 2Cr^{3+} + 7H_2O + 3Cl_2 + 6e^-$

$14H^+ + Cr_2O_7^{2-} + 6Cl^- \rightarrow 2Cr^{3+} + 7H_2O + 3Cl_2$

(b) $KOH + SO_2 + KMnO_4 \rightarrow K_2SO_4 + MnO_2 + H_2O$

Reactants: K^+ OH^- SO_2 K^+ MnO_4^-
Products: K^+ SO_4^{2-} MnO_2 H_2O

$SO_2 \rightarrow SO_4^{2-}$
$MnO_4^- \rightarrow MnO_2$

$2H_2O + SO_2 \rightarrow SO_4^{2-}$
$MnO_4^- \rightarrow MnO_2 + 2H_2O$

$2H_2O + SO_2 \rightarrow SO_4^{2-} + 4H^+$
$4H^+ + MnO_4^- \rightarrow MnO_2 + 2H_2O$

$2H_2O + SO_2 \rightarrow SO_4^{2-} + 4H^+ + 2e^-$
$3e^- + 4H^+ + MnO_4^- \rightarrow MnO_2 + 2H_2O$

$3(2H_2O + SO_2 \rightarrow SO_4^{2-} + 4H^+ + 2e^-)$
$2(3e^- + 4H^+ + MnO_4^- \rightarrow MnO_2 + 2H_2O)$

$6H_2O + 3SO_2 \rightarrow 3SO_4^{2-} + 12H^+ + 6e^-$
$6e^- + 8H^+ + 2MnO_4^- \rightarrow 2MnO_2 + 4H_2O$

$6H_2O + 3SO_2 + 6e^- + 8H^+ + 2MnO_4^- \rightarrow 2MnO_2 + 4H_2O + 3SO_4^{2-} + 12H^+ + 6e^-$

$2H_2O + 3SO_2 + 2MnO_4^- \rightarrow 2MnO_2 + 3SO_4^{2-} + 4H^+$

$4OH^- + 2H_2O + 3SO_2 + 2MnO_4^- \rightarrow 2MnO_2 + 3SO_4^{2-} + 4H^+ + 4OH^-$

$4OH^- + 2H_2O + 3SO_2 + 2MnO_4^- \rightarrow 2MnO_2 + 3SO_4^{2-} + 4H_2O$

$4OH^- + 3SO_2 + 2MnO_4^- \rightarrow 2MnO_2 + 3SO_4^{2-} + 2H_2O$

22. (a) $Cr_2O_7^{2-} + Br^- \rightarrow Br_2 + Cr^{3+}$

$Cr_2O_7^{2-} \rightarrow Cr^{3+}$
$Br^- \rightarrow Br_2$

$Cr_2O_7^{2-} \rightarrow 2Cr^{3+}$
$2Br^- \rightarrow Br_2$

$Cr_2O_7^{2-} \rightarrow 2Cr^{3+} + 7H_2O$
$2Br^- \rightarrow Br_2$

$14H^+ + Cr_2O_7^{2-} \rightarrow 2Cr^{3+} + 7H_2O$
$2Br^- \rightarrow Br_2$

$6e^- + 14H^+ + Cr_2O_7^{2-} \rightarrow 2Cr^{3+} + 7H_2O$
$2Br^- \rightarrow Br_2 + 2e^-$

$6e^- + 14H^+ + Cr_2O_7^{2-} \rightarrow 2Cr^{3+} + 7H_2O$
$3(2Br^- \rightarrow Br_2 + 2e^-)$

$6e^- + 14H^+ + Cr_2O_7^{2-} \rightarrow 2Cr^{3+} + 7H_2O$
$6Br^- \rightarrow 3Br_2 + 6e^-$

$6e^- + 14H^+ + Cr_2O_7^{2-} + 6Br^- \rightarrow 3Br_2 + 6e^- + 2Cr^{3+} + 7H_2O$

$14H^+ + Cr_2O_7^{2-} + 6Br^- \rightarrow 3Br_2 + 2Cr^{3+} + 7H_2O$

(b) $H_3AsO_3 + MnO_4^- \rightarrow H_2AsO_4^- + Mn^{2+}$

$H_3AsO_3 \rightarrow H_2AsO_4^-$
$MnO_4^- \rightarrow Mn^{2+}$

$H_2O + H_3AsO_3 \rightarrow H_2AsO_4^-$
$MnO_4^- \rightarrow Mn^{2+} + 4H_2O$

$H_2O + H_3AsO_3 \rightarrow H_2AsO_4^- + 3H^+$
$8H^+ + MnO_4^- \rightarrow Mn^{2+} + 4H_2O$

$H_2O + H_3AsO_3 \rightarrow H_2AsO_4^- + 3H^+ + 2e^-$
$5e^- + 8H^+ + MnO_4^- \rightarrow Mn^{2+} + 4H_2O$

$5(H_2O + H_3AsO_3 \rightarrow H_2AsO_4^- + 3H^+ + 2e^-)$
$2(5e^- + 8H^+ + MnO_4^- \rightarrow Mn^{2+} + 4H_2O)$

$5H_2O + 5\,H_3AsO_3 \rightarrow 5H_2AsO_4^- + 15H^+ + 10e^-$
$10e^- + 16H^+ + 2MnO_4^- \rightarrow 2Mn^{2+} + 8H_2O$

$5H_2O + 5\,H_3AsO_3 + 10e^- + 16H^+ + 2MnO_4^- \rightarrow 5H_2AsO_4^- + 15H^+ + 10e^- + 2Mn^{2+} + 8H_2O$

$5\,H_3AsO_3 + 1H^+ + 2MnO_4^- \rightarrow 5H_2AsO_4^- + 2Mn^{2+} + 3H_2O$

23. (a) $I^- + CrO_4^{2-} \rightarrow CrO_2^- + IO_3^-$

$I^- \rightarrow IO_3^-$
$CrO_4^{2-} \rightarrow CrO_2^-$

$3H_2O + I^- \rightarrow IO_3^-$
$CrO_4^{2-} \rightarrow CrO_2^- + 2H_2O$

$3H_2O + I^- \rightarrow IO_3^- + 6H^+$
$4H^+ + CrO_4^{2-} \rightarrow CrO_2^- + 2H_2O$

$3H_2O + I^- \rightarrow IO_3^- + 6H^+ + 6e^-$
$3e^- + 4H^+ + CrO_4^{2-} \rightarrow CrO_2^- + 2H_2O$

$3H_2O + I^- \rightarrow IO_3^- + 6H^+ + 6e^-$
$2(3e^- + 4H^+ + CrO_4^{2-} \rightarrow CrO_2^- + 2H_2O)$

$3H_2O + I^- \rightarrow IO_3^- + 6H^+ + 6e^-$
$6e^- + 8H^+ + 2CrO_4^{2-} \rightarrow 2CrO_2^- + 4H_2O$

$3H_2O + I^- + 6e^- + 8H^+ + 2CrO_4^{2-} \rightarrow 2CrO_2^- + 4H_2O + IO_3^- + 6H^+ + 6e^-$

$3H_2O + I^- + 2H^+ + 2CrO_4^{2-} \rightarrow 2CrO_2^- + 4H_2O + IO_3^-$

$2OH^- + 3H_2O + I^- + 2H^+ + 2CrO_4^{2-} \rightarrow 2CrO_2^- + 4H_2O + IO_3^- + 2OH^-$

$5H_2O + I^- + 2CrO_4^{2-} \rightarrow 2CrO_2^- + 4H_2O + IO_3^- + 2OH^-$

$H_2O + I^- + 2CrO_4^{2-} \rightarrow 2CrO_2^- + IO_3^- + 2OH^-$

(b) $SO_2 + MnO_4 \rightarrow MnO_2 + SO_4^{2-}$

$SO_2 \rightarrow SO_4^{2-}$

$MnO_4 \rightarrow MnO_2$
$2H_2O + SO_2 \rightarrow SO_4^{2-}$
$MnO_4 \rightarrow MnO_2 + 2H_2O$

$2H_2O + SO_2 \rightarrow SO_4^{2-} + 4H^+$
$4H^+ + MnO_4 \rightarrow MnO_2 + 2H_2O$

$2H_2O + SO_2 \rightarrow SO_4^{2-} + 4H^+ + 2e^-$
$4e^- + 4H^+ + MnO_4 \rightarrow MnO_2 + 2H_2O$

$2(2H_2O + SO_2 \rightarrow SO_4^{2-} + 4H^+ + 2e^-)$
$4e^- + 4H^+ + MnO_4 \rightarrow MnO_2 + 2H_2O$

$4H_2O + 2SO_2 \rightarrow 2SO_4^{2-} + 8H^+ + 4e^-$
$4e^- + 4H^+ + MnO_4 \rightarrow MnO_2 + 2H_2O$

$4H_2O + 2SO_2 + 4e^- + 4H^+ + MnO_4 \rightarrow MnO_2 + 2H_2O + 2SO_4^{2-} + 8H^+ + 4e^-$

$2H_2O + 2SO_2 + MnO_4 \rightarrow MnO_2 + 2SO_4^{2-} + 4H^+$

$4OH^- + 2H_2O + 2SO_2 + MnO_4 \rightarrow MnO_2 + 2SO_4^{2-} + 4H^+ + 4OH^-$

$4OH^- + 2H_2O + 2SO_2 + MnO_4 \rightarrow MnO_2 + 2SO_4^{2-} + 4H_2O$

$4OH^- + 2SO_2 + MnO_4 \rightarrow MnO_2 + 2SO_4^{2-} + 2H_2O$

24. Question 22.
    (a) Oxidizing agent: $Cr_2O_7^{2-}$
        Reducing agent: Br^-
    (b) Oxidizing agent: MnO_4^-
        Reducing agent: H_3AsO_3

    Question 23.
    (a) Oxidizing agent: CrO_4^{2-}
        Reducing agent: I^-
    (b) Oxidizing agent: MnO_4^-
        Reducing agent: SO_2

25. (a) $Sn(s) + 2HCl(aq) \rightarrow Sn^{2+}(aq) + H_2(g) + 2Cl^-(aq)$
    (b) $Cu(s) + 2HNO_3(\text{concd}) \rightarrow Cu^{2+}(aq) + 2NO_3^-(aq) + H_2(g)$
    (c) $Zn(s) + Cu^{2+}(aq) \rightarrow Zn^{2+}(aq) + Cu(s)$
    (d) $Ag(s) + Cu^{2+}(aq) \rightarrow$ No Reaction

26. (a) $2C_{16}H_{34} + 49O_2 \rightarrow 32CO_2 + 34H_2O$
    (b) $2C_{16}H_{34} + 33O_2 \rightarrow 32CO + 34H_2O$
    (c) $2C_{16}H_{34} + 17O_2 \rightarrow 32C + 34H_2O$

27. $C_{17}H_{35}CO_2H + 26O_2 \rightarrow 18CO_2 + 18H_2O$

28. $2CH_3SH + 6O_2 \rightarrow 2CO_2 + 4H_2O + 2SO_2$

29. (a) $2Mg + O_2 \rightarrow 2MgO$
    (b) $4Al + 3O_2 \rightarrow 2Al_2O_3$
    (c) $4P + 5O_2 \rightarrow P_4O_{10}$
    (d) $S + O_2 \rightarrow SO_2$

30. 0.238 g sample

Reactions that occurred in the titrations:

$2Cu^{2+} + 5I^- \rightarrow I_3^- + 2CuI$

$I_3^- + 2S_2O_3^{2-} \rightarrow 3I^- + S_4O_6^{2-}$

$$\# \text{ mol Cu} = 0.02862 \text{ mL } S_2O_3^{2-} \text{ solution}\left(\frac{0.01699 \text{ mol } S_2O_3^{2-}}{1 \text{ L solution}}\right)\left(\frac{1 \text{ mol } I_3^-}{2 \text{ mol } S_2O_3^{2-}}\right)\left(\frac{5 \text{ mol } I^-}{1 \text{ mol } I_3^-}\right)\left(\frac{2 \text{ mol } Cu^{2+}}{5 \text{ mol } I^-}\right)$$

$$= 4.863 \times 10^{-4} \text{ mol Cu}$$

$$\# \text{ g Cu} = 4.863 \times 10^{-4} \text{ mol Cu}\left(\frac{63.546 \text{ g Cu}}{1 \text{ mol Cu}}\right) = 0.03090 \text{ g Cu}$$

$$\text{Percentage by weight} = \frac{0.03090 \text{ g Cu}}{0.238 \text{ g sample}} \times 100\% = 12.98\%$$

31. heat capacity = (specific heat)(mass)

$= (4.184 \text{ J g}^{-1} \text{ °C}^{-1})(225 \text{ g})$

$= 941.4 \text{ J °C}^{-1}$

specific heat = $4.184 \text{ J g}^{-1} \text{ °C}^{-1}$

32. Ozone, O_3, is not the most stable form of oxygen at 25 °C and 1atm.

33. He $\# \text{J} = (5.19 \text{ J g}^{-1} \text{ °C}^{-1})(4.003 \text{ g})(1 \text{ °C}) = 20.8 \text{ J}$

N_2 $\# \text{J} = (1.04 \text{ J g}^{-1} \text{ °C}^{-1})(28.02 \text{ g})(1 \text{ °C}) = 29.1 \text{ J}$

34. $\# \text{ J released}/0.6486 \text{ g} = -(11.99 \times 10^3 \text{ J °C}^{-1})(26.746 \text{ °C} - 24.518 \text{ °C}) = 2.671 \times 10^4 \text{ J}/0.6486 \text{ g}$

$$\text{Molar heat capacity} = \left(\frac{2.671 \times 10^4 \text{ J}}{0.6486 \text{ g } C_{32}H_{64}O_2}\right)\left(\frac{480.9 \text{ g } C_{32}H_{64}O_2}{1 \text{ mol } C_{32}H_{64}O_2}\right) = 1.981 \times 10^7 \text{ J/mol } C_{32}H_{64}O_2$$

$$= 1981 \text{ kJ/mol } C_{32}H_{64}O_2$$

35. A state function does not depend on a system's history or future, and ΔH is a state function.

36. Amount of heat needed = $(4.184 \text{ J g}^{-1} \text{ °C}^{-1})(250 \text{ g } H_2O)(50.0 \text{ °C} - 25 \text{ °C}) = 26150 \text{ J}$

$$\# \text{ g } CH_4 = 26150 \text{ J}\left(\frac{1 \text{ kJ}}{1000 \text{ J}}\right)\left(\frac{1 \text{ mol } CH_4}{802.3 \text{ kJ}}\right)\left(\frac{16.04 \text{ g } CH_4}{1 \text{ mol } CH_4}\right) = 0.5228 \text{ g } CH_4$$

37.

(a)	specific heat	intensive
(b)	heat capacity	extensive
(c)	$\Delta H°_f$	intensive
(d)	ΔH°	extensive
(e)	molar heat capacity	intensive

38. Internal energy is the sum of all of the kinetic energies and potential energies of the particles within a system.
The first law of thermodynamics is that the energy of the universe is constant; it can be neither created nor destroyed but only transformed and transferred.

39. Pressure-volume work is the energy transferred as work when a system expands or contracts against the pressure exerted by the surroundings. At constant pressure, work = –PΔV.

40. The change in internal energy is equal to the "heat of reaction at constant volume" because no work is done.

41. ΔH will be larger than ΔE.

42. (a) $2CO(g) + O_2(g) \rightarrow 2CO_2(g)$ $\Delta H° = -565.96$ kJ
(b) $CO_2(g) \rightarrow CO(g) + 1/2O_2(g)$ $\Delta H° = 282.98$ kJ

43. $C_{20}H_{42}(s) + 61/2O_2(g) \rightarrow 20CO_2(g) + 21H_2O(l)$ $\Delta H°_{comb} = -1.332 \times 10^4$ kJ mol^{-1}
$\Delta H°_{comb} = -1.332 \times 10^4$ kJ mol^{-1} = [(20 mol CO_2)(–394 kJ mol^{-1}) + (21 mol H_2O)(–285.9 kJ mol^{-1})]
– [(mol $C_{20}H_{42}$)($\Delta H°_f$ $C_{20}H_{32}$) + (61/2 mol O_2)(0 kJ mnol^{-1})]
$\Delta H°_f$ $C_{20}H_{32} = 5.64 \times 10^2$ kJ mol^{-1}
$20C(s) + 16H_2(g) \rightarrow C_{20}H_{32}(s)$ $\Delta H°_f = 5.64 \times 10^2$ kJ mol^{-1}

44. (a) $H_2SO_4(l) \rightarrow 2SO_3(l) + H_2O(l)$
$\Delta H°_{reaction}$ = [(2 mol $SO_3(l)$)(–396 kJ mol^{-1}) + (1 mol $H_2O(l)$)(–285.9 kJ mol^{-1})]
– [(1 mol $H_2SO_4(l)$)(–813.8 kJ mol^{-1})]
= –264 kJ

(b) $C_2H_6(g) \rightarrow C_2H_4(g) + H_2(g)$
$\Delta H°_{reaction}$ =[(1 mol $C_2H_4(g)$)(51.9 kJ mol^{-1}) + 1 mol $H_2(g)$(0.0 kJ mol^{-1})]
– [(1 mol $C_2H_6(g)$)(–84.5 kJ mol^{-1})]
= 136.4 kJ

45. $Ca(s) + 2C(s) \rightarrow CaC_2(s)$ $\Delta H°_f$

$Ca(s) + 2H_2O(l) \rightarrow Ca(OH)_2(s) + H_2(g)$ $\Delta H° = -414.79$ kJ
$H_2(g) + 1/2O_2(g) \rightarrow H_2O(l)$ $\Delta H° = -285.9$ kJ
$Ca(OH)_2(s) \rightarrow CaO(s) + H_2O(l)$ $\Delta H° = 65.19$ kJ
$CaO(s) + 3C(s) \rightarrow CaC_2(s) + CO(g)$ $\Delta H° = 462.3$ kJ
$CO(g) \rightarrow C(s) + 1/2O_2(g)$ $\Delta H° = 110.5$ kJ

$Ca(s) + 2C(s) \rightarrow CaC_2(s)$ $\Delta H°_f = -62.7$ kJ

1. Electrons, Protons, Neutrons

Electron	$5.48579903 \times 10^{-4}$ u
Proton	1.007276470 u
Neutron	1.008664905 u

Electron	negative
Proton	positive
Neutron	neutral

2. $$\nu = \frac{c}{\lambda} = \frac{3.00 \times 10^{8}\ \text{m s}^{-1}}{500\ \text{nm}}\left(\frac{10^{9}\ \text{nm}}{1\ \text{m}}\right) = 6.00 \times 10^{14}\ \text{s}^{-1}$$

Energy of one photon of 500 nm light = $h\nu = (6.626 \times 10^{-34}\ \text{J·s})(6.00 \times 10^{14}\ \text{s}^{-1}) = 3.97 \times 10^{-19}$ J
Energy of one mole of photons of 500 nm light = $(3.97 \times 10^{-19}\ \text{J})(6.022 \times 10^{23}\ \text{photons}) = 2.39 \times 10^{5}$ J
Blue light would have more energy per photon than the green light.

3. Radio waves < microwaves < infrared light < red light < blue light < ultraviolet light < X-rays < gamma rays

4. A continuous spectrum contains a continuous unbroken distribution of light of all colors.
An atomic spectrum is the line spectrum produced when energized or excited atoms emit electromagnetic radiation.

5. Matter can be diffracted.

6. In a travelling wave the crests and troughs move in the direction the wave is moving. In a standing wave, the crests and troughs do not change position. A node is a point of zero amplitude for a wave.

7. As the number of nodes increase, the energy of the electron wave increases.

8. A wave function is a mathematical function that describes the intensity of an electron wave at a specified location in an atom.
The Greek letter: Ψ
An electron wave in an atom is an orbital.

9.

(a)	Sulfur	$n = 3$	$l = 0, 1$	$m_l = 1, 0, -1$, $m_s = 1/2$ and $-1/2$
(b)	Strontium	$n = 5$	$l = 0$	$m_l = 0$, $m_s = 1/2$ and $-1/2$
(c)	Lead	$n = 6$	$l = 0, 1$	$m_l = 1, 0$, or -1, $m_s = 1/2$ and $-1/2$
(d)	Bromine	$n = 4$	$l = 0, 1$	$m_l = 1, 0, -1$, $m_s = 1/2$ and $-1/2$
(e)	Boron	$n = 2$	$l = 0, 1$	$m_l = 1, 0$, $m_s = 1/2$ and $-1/2$

10. For $n = 4$ l is *s, p, d, f.*
The maximum number of electrons is 32.

11.

(a)	Tin	$1s^22s^22p^63s^23p^64s^23d^{10}4p^65s^24d^{10}5p^2$
(b)	Germanium	$1s^22s^22p^63s^23p^64s^23d^{10}4p^2$
(c)	Silicon	$1s^22s^22p^63s^23p^2$
(d)	Lead	$1s^22s^22p^63s^23p^64s^23d^{10}4p^65s^24d^{10}5p^66s^24f^{14}5d^{10}6p^2$
(e)	Nickel	$1s^22s^22p^63s^23p^64s^23d^8$

12.

(a)	Pb^{2+}	$1s^22s^22p^63s^23p^64s^23d^{10}4p^65s^24d^{10}5p^66s^24f^{14}5d^{10}$
(b)	Pb^{4+}	$1s^22s^22p^63s^23p^64s^23d^{10}4p^65s^24d^{10}5p^64f^{14}5d^{10}$
(c)	S^{2-}	$1s^22s^22p^63s^23p^6$
(d)	Fe^{3+}	$1s^22s^22p^63s^23p^63d^5$
(e)	Zn^{2+}	$1s^22s^22p^63s^23p^63d^{10}$

13. A paramagnetic atom or molecule has unpaired electrons
Atoms or molecules with all paired electrons are diamagnetic.

14.

(a)	Ni	$[Ar]^{18}4s^23d^8$
(b)	Cr	$[Ar]^{18}4s^23d^4$
(c)	Sr	$[Kr]^{36}4s^2$
(d)	Sb	$[Kr]^{36}4s^23d^{10}4p^3$
(e)	Po	$[Xe]^{54}6s^24f^{14}5d^{10}6p^4$

15. Ionization energy is the amount of energy required to remove an electron from an isolated gaseous atom, ion or molecule.
Electron affinity is the energy change that occurs when an electron adds to an isolated gaseous atom or ion. Metals tend to lose electrons and form cations since they have low ionization energies, and nonmetals tend to gain electrons and form anions since they have high electron affinities. The cations and anions come together to form ionic compounds.

16. For the second ionization energy, an electron is removed from a negatively charged species. It requires more energy to separate negative charges than to remove a negative charge from a neutral species.

17. Be has the greatest difference between its second and third ionization energy since an electron is being removed from a closed shell.

18.

(a)	$P^-(g) + e^- \rightarrow P^{2-}(g)$	endothermic
(b)	$Fe^{3+}(g) + e^- \rightarrow Fe^{2+}(g)$	exothermic
(c)	$Cl(g) + e^- \rightarrow Cl^-(g)$	exothermic
(d)	$S(g) + 2e^- \rightarrow S^{2-}(g)$	endothermic

19. (a)

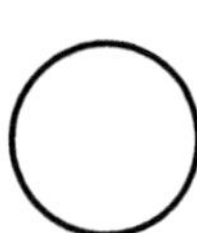

(b)

(c)

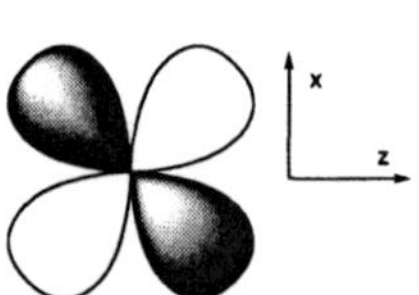

20. Electron density is the concentration of the electron's charge within a given volume.

21. Se

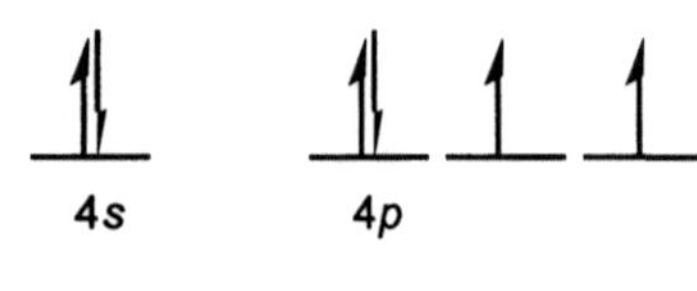

Tl

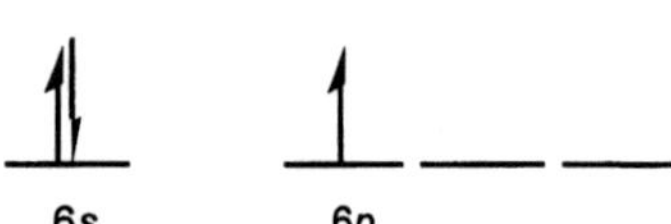

22. (a) Fe^{2+} (b) O^{-}

23. (c) Ca and F will form an ionic compound.

24.

25. (a)

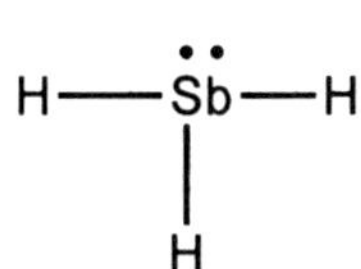

(b)

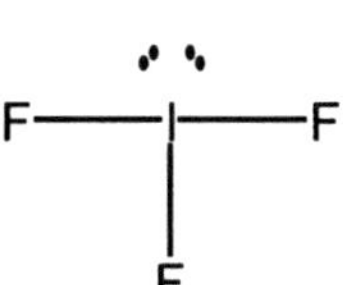

(c)

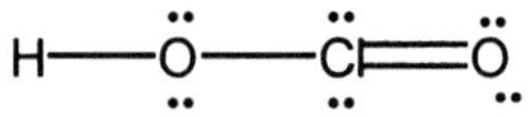

$H-\ddot{O}-\ddot{Cl}=\ddot{O}$

(d)

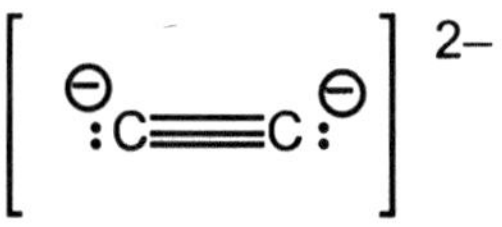

$\left[\,:C\equiv C:\,\right]^{2-}$

(e)

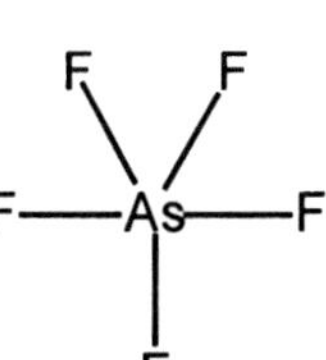

(f)

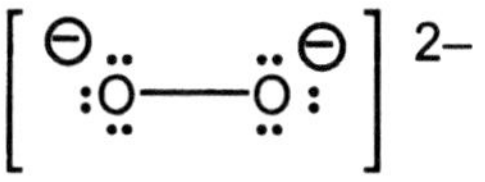

$\left[\,:\ddot{O}-\ddot{O}:\,\right]^{2-}$

(g)

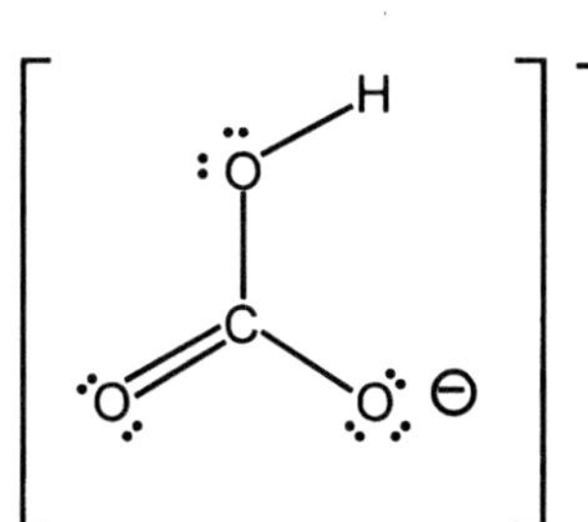

(h)

F F
F—Te—F
F F

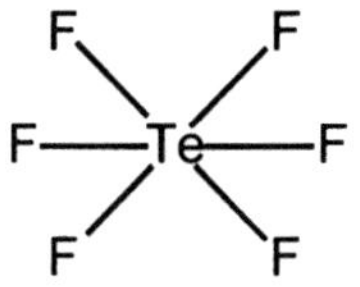

(i) 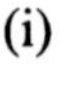

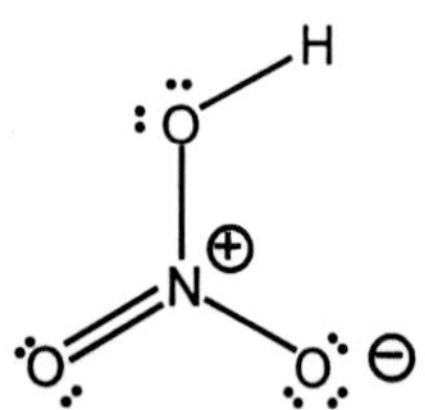

26. (a) $SbCl_3$ trigonal pyramidal
(b) IF_5 square pyramid
(c) AsH_3 trigonal pyramidal
(d) BrF_2 linear
(e) OF_2 bent

27. (a) sp^3
(b) sp^3d^2
(c) sp^3
(d) sp^3d
(e) sp^3

28. AsF_5 is nonpolar.

29.

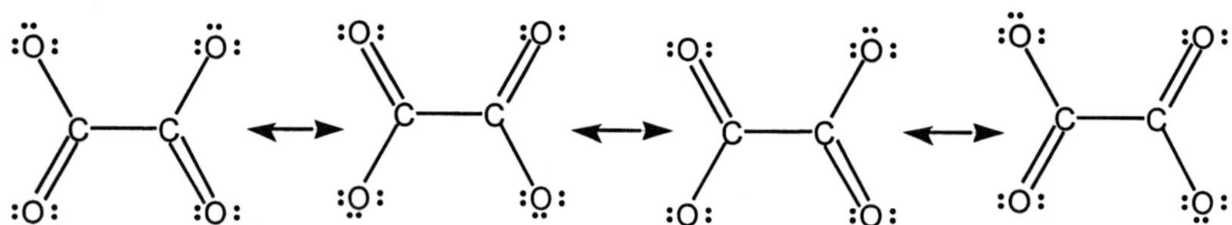

30. Overlap of orbitals is theportion of two orbitals from differentatoms that share the same space in a molecule.

31. Sigma bonds are bonds formed by the head-to-head overlap of two atomic orbitals and in which electron density becomes concentrated along and around the imagniary line joining the two nuclei.
Pi bonds are formed by the sideways overlap of a pair of *p* orbitals and that concentrate electron density into two separate regions that lie on opposite sides of a plane that contains an imagniary line joining the nuclei.
The formation of π bonds allows double and triple bonds to form.

32.

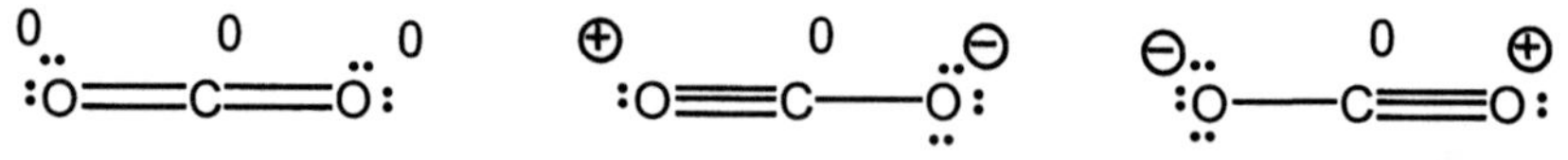

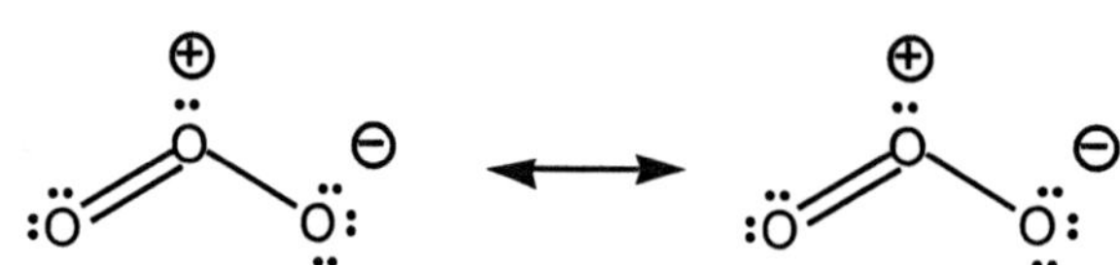

The first structure is perferred because it has the least separation of charge, and oxygen does not have a positive charge.

33. (a)

(b) Nonlinear

(c) O with the double bond is 0.
O with the single bond is –1
O in the middle is +1

(d) Ozone is polar due to the separation of charge.

34. The structure on the left is better since there is less separation of charge.
The best structure would have a double bond between the P and the O.

35. (a) Nonmetal
(b) Group VIA
(c) Any period below the second one, third or lower
(d)

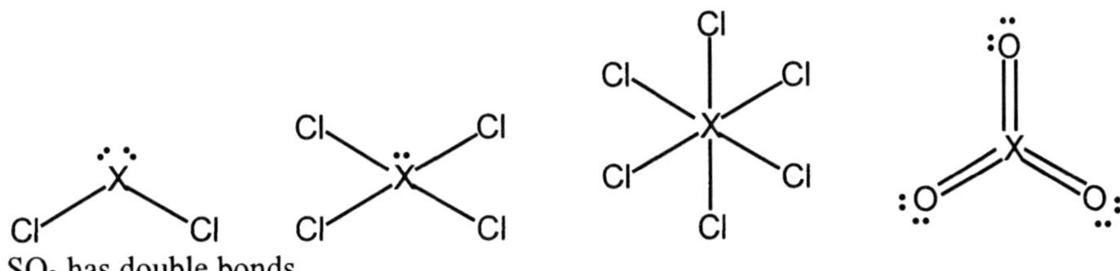

SO_3 has double bonds

(e)

XCl_2	bent	polar
XCl_4	distorted tetrahedral	polar
XCl_6	octahedral	nonpolar
XO_3	trigonal planar	nonpolar

(f)

(g) XO_2 see (f)
XO_3

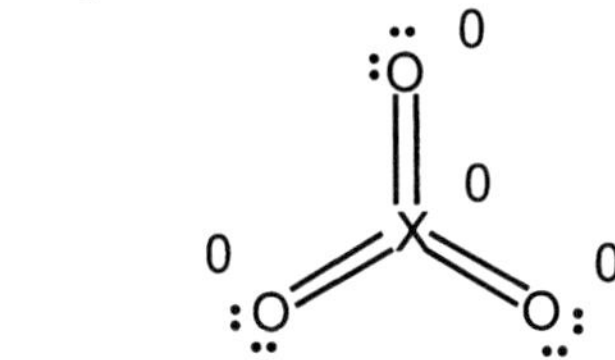

(h) XCl_4 sp^3d
XCl_6 sp^3d^2

(i) Al_2X_3

(j) The more ionic bond would be in Na_2X

(k) $5s^25p^4$

36. The very reactive metals are in group 1.
The least reactive metals are in group2 10 and 11.

37. A bonding molecular orbital introduces a buildup of electron density between nuclei and stabilizes a molecule when occupied by electrons.
An antibonding molecular orbital denies electron density to the space between nuclei and destabilizes a molecule when occupied by electrons.
Bonding molecular orbitals have electron density in the region between the nuclei while antibonding orbitals have nodes in that region.
Bonding molecular orbitals are lower in energy than antibonding molecular orbitals.

38.

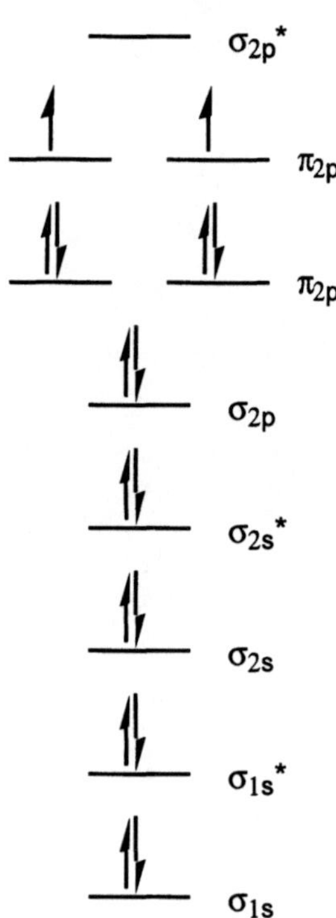

MO theory explains the two unpaired electrons in O_2 and allows for a double bond with a calculated bond order of 2.0.

39. A delocalized molecular orbital speads over more than two nuclei.
Molecular orbital thoery avoids the problem of resonance by recognizing that electron pairs can sometimes be shared among overlapping orbitals from three or more atoms.

40.

(a)	PF_3	Trigonal pyramidal
(b)	PF_4^+	Tetrahedral
(c)	PF_6^-	Octahdehral
(d)	PF_5	Trigonal bipyramid

41. PF_3 has a net dipole moment.

42. Fe^{2+} is easily oxidized to Fe^{3+} because by removing the electron, a half filled set of *d* orbitals is formed. Mn^{2+} is difficult ot oxidize to Mn^{3+} because Mn^{2+} has a half filled set of *d* orbitals, by removing that electron, the *d* orbitals will not be half filled.

43.

(a)	MgO	Mg^{2+} and O^{2-} have larger charges than Na^+ and Cl^-
(b)	BeO	Be^{2+} is smaller than Mg^{2+}, thus the ions can come closer together.
(c)	NaF	F^- is smaller than I^-, thus the ions can come closer together.
(d)	MgO	The ions are smaller and closer together than CaS.

44. Al_2O_3 has higher charges, Al^{3+} and O^{2-} than NaCl. The lattice energy for Al_2O_3 is higher than the lattice energy for NaCl, therefore it is harder to separate the ions in Al_2O_3 and this gives a higher melting point.

45. The change in atomic size is smaller among the transition elements than among the representative elements because the electtron that is added to each successive element enters the *d* orbitals. These *d* orbitals are not outer shell orbitals, but are inner shell orbitals. The amount of shielding provided by the addition of electrons to this inner 3*d* level is greater than the amount of shielding that would occur if the electrons were added to the outer shell, so the effective nuclear charge felt by the outer electrons increases more gradually.

46. (a) σ bond
(b) π bond

47. (a) 1. sp^3
2. sp^2
3. sp^2
(b) 4. 1 σ bond and 0 π bond
6. 1 σ bond and 1 π bond
(c) 4. B.O. = 1
5. B.O. = 1.5
6. B.O. = 2
(d) The carbonyl group contains carbon 3, bond 6 and the oxygen atom bound to the carbon
(e) The alcohol group conatins oxygen 1 and the hydrogen bonded to the oxygen
(f) The amine group contains nitrogen 7, the hyrogen bonded to the nitrogen and the carbons bonded to the nitrogen
(g) 1. tetrahedral
2. trigonal planar
3. trigonal planar
7. tetrahedral
(h) $C_8H_9NO_2$

1. $\frac{P_1V_1}{T_1} = \frac{P_2V_2}{T_2}$

$\frac{(748 \text{ torr})(15.5 \text{ L})}{(298 \text{ K})} = \frac{P_2(25.4 \text{ L})}{(298 \text{ K})}$

$P_2 = 456$ torr

2. $\frac{P_1V_1}{T_1} = \frac{P_2V_2}{T_2}$

$\frac{(1.00 \text{ atm})(8.95 \text{ L})}{(298 \text{ K})} = \frac{(5.56 \text{ atm})(0.895 \text{ L})}{T_2}$

$T_2 = 1660$ K

3. $\frac{P_1V_1}{T_1} = \frac{P_2V_2}{T_2}$

$\frac{(1.00 \text{ atm})(12.5 \text{ L})}{(298 \text{ K})} = \frac{P_2(12.5 \text{ L})}{(310 \text{ K})}$

$P_2 = 1.04$ atm

$P_{O_2} = (0.450) \times 1.04 \text{ atm} = 0.468 \text{ atm}\left(\frac{760 \text{ torr}}{1 \text{ atm}}\right) = 356 \text{ torr}$

$P_{N_2} = (0.550) \times 1.04 \text{ atm} = 0.572 \text{ atm}\left(\frac{760 \text{ torr}}{1 \text{ atm}}\right) = 435 \text{ torr}$

4. $w = P\Delta V$

$P = 3.0 \times 10^5 \text{ Pa} = 3.0 \times 10^5 \text{ N m}^{-2} = 3.0 \times 10^5 \text{ (kg m s}^{-2})\text{m}^{-2} = 3.0 \times 10^5 \text{ kg m}^{-1} \text{ s}^{-2}$

$\Delta V = 0.50 \text{ m}^3$

$w = (3.0 \times 10^5 \text{ kg m}^{-1} \text{ s}^{-2})(0.50 \text{ m}^3)$

$w = 1.5 \times 10^5 \text{ kg m}^2 \text{ s}^{-2} = 1.5 \times 10^5 \text{ J}$

5. $PV = nRT$

$745 \text{ torr}\left(\frac{1 \text{ atm}}{760 \text{ torr}}\right)1.92 \text{ L} = n(0.0821 \text{ L atm mol}^{-1} \text{ K}^{-1})(298 \text{ K})$

$n = 7.69 \times 10^{-2}$ mol

$MW = \frac{6.45 \text{ g}}{7.69 \times 10^{\text{Ğ}2} \text{ mol}} = 83.9 \text{ g mol}^{-1}$

The gas is Krypton with an atomic mass of 83.80 g mol^{-1}.

6.

$\frac{\text{effusion rate x}}{\text{effusion rate Xe}} = \sqrt{\frac{M_{Xe}}{M_x}}$

$M_x = M_{Xe}\left(\frac{\text{effusion rate Xe}}{\text{effusion rate x}}\right)^2$

$= 131.29 \text{ g/mol}\left(\frac{1}{2.16}\right)^2 = 28.14 \text{ g/mol}$

Formula mass = 28.1 g/mol

Element is nitrogen, N_2.

7. The size of *a* is related to the attractions between molecules, the large the value of *a*, the stronger the attractive forces. Therefore, a polar molecule would have a large *a* value than a nonpolar molecule.

8. This is a limiting reactant problem. First we need to calculate the moles of dry CO_2 that can be produced from the given quantities of $NaHCO_3$ and HCl:

$$\text{\# mol NaHCO}_3 = (20.0\text{ mL NaHCO}_3)\left(\frac{0.100\text{ mol NaHCO}_3}{1000\text{ mL solution NaHCO}_3}\right) = 0.00200\text{ mol NaHCO}_3$$

$$\text{\# mol HCl} = (30.0\text{ mL HCl})\left(\frac{0.0800\text{ mol HCl}}{1000\text{ mL solutionHCl}}\right) = 0.00240\text{ mol HCl}$$

Thus, $NaHCO_3$ is limiting and we use this to determine the moles of CO_2 that can be produced:

$$\text{\# mol CO}_2 = (0.00200\text{ mol NaHCO}_3)\left(\frac{1\text{ mol CO}_2}{1\text{ mol NaHCO}_3}\right) = 0.00200\text{ mol CO}_2$$

Next we realize that CO_2 is collected dry at a pressure of 1 atm. The volume of this dry CO_2 is calculated using the ideal gas equation:

$$V = \frac{nRT}{P} = \frac{(0.00200\text{ mol})(0.0821\ \frac{\text{L atm}}{\text{mol K}})(273.2\text{ K})}{(1\text{ atm})} = 0.0449\text{ L CO}_2 = 44.9\text{ mL CO}_2$$

9. $I^- + Cl_2 \rightarrow IO_3^- + Cl^-$

Balance the redox equation using the ion-electron method.

$I^- + 3Cl_2 + 3H_2O \rightarrow IO_3^- + 6Cl^- + 6H^+$

Calculate the number of moles of NaI:

$$\text{\# mol NaI} = 10.0\text{ mL}\left(\frac{0.10\text{ mol NaI}}{1000\text{ mL solution}}\right) = 0.00100\text{ mol NaI}$$

$$\text{\# mol Cl}_2 = 0.00100\text{ mol NaI}\left(\frac{3\text{ mol Cl}_2}{1\text{ mol I}^-}\right) = 0.00300\text{ mol Cl}_2$$

$$\text{\# mL Cl}_2 = \frac{(0.00300\text{ mol})(0.0821\ \frac{\text{L atm}}{\text{mol K}})(298.\text{ K})}{(740\text{ torr})\left(1\text{ atm}/760\text{ torr}\right)} = 0.0754\text{ L CO}_2 = 75.4\text{ mL CO}_2$$

10. First determine the number of moles of N_2, then find the number of g of KOBr.

$$\text{\# mole N}_2 = \frac{PV}{RT} = \frac{\left(738\text{ torr}\left(1\text{ atm}/760\text{ torr}\right)\right)(0.475\text{ L})}{(0.0821\ \frac{\text{L atm}}{\text{mol K}})(297.2\text{ K})} = 0.0189\text{ mol N}_2$$

$$\text{\# g KOBr} = 0.0189\text{ mol N}_2\left(\frac{3\text{ mol KOBr}}{1\text{ mol N}_2}\right)\left(\frac{135.0\text{ g KOBr}}{1\text{ mol KOBr}}\right) = 7.65\text{ g KOBr}$$

11. $5H_2O_2 + 2KMnO_4 + 3H_2SO_4 \rightarrow 5O_2 + 2MnSO_4 + K_2SO_4 + 8H_2O$

$$\text{\# mol O}_2 = \frac{PV}{RT} = \frac{\left(738\text{ torr}\left(1\text{ atm}/760\text{ torr}\right)\right)(0.375\text{ L})}{(0.0821\ \frac{\text{L atm}}{\text{mol K}})(295.2\text{ K})} = 0.0150\text{ mol O}_2$$

$$\text{\# mL KMnO}_4 = 0.0150\text{ mol O}_2\left(\frac{2\text{ mol KMnO}_4}{5\text{ mol O}_2}\right)\left(\frac{1000\text{ mL}}{0.125\text{ mol KMnO}_4}\right) = 48.0\text{ mL KMnO}_4\text{ solution}$$

12. $4HCl + MnO_2 \rightarrow Cl_2 + MnCl_2 + 2H_2O$

$$\# \text{ mol } Cl_2 = \frac{PV}{RT} = \frac{\left(742 \text{ torr}\left(1 \text{ atm}/760 \text{ torr}\right)\right)(0.525 \text{ L})}{\left(0.0821 \frac{\text{L atm}}{\text{mol K}}\right)(297.2 \text{ K})} = 0.0210 \text{ mol } Cl_2$$

$$\# \text{ mL HCl} = 0.0210 \text{ mol } Cl_2\left(\frac{4 \text{ mol HCl}}{1 \text{ mol } Cl_2}\right)\left(\frac{1000 \text{ mL}}{6.44 \text{ mol HCl}}\right) = 13.0 \text{ mL HCl solution}$$

13. partial pressure of N_2 = total pressure – pressure of H_2O = 736 torr – 18.7 torr = 717.3 torr

$$\# \text{ mol } N_2 = \frac{PV}{RT} = \frac{\left(717.3 \text{ torr}\left(1 \text{ atm}/760 \text{ torr}\right)\right)(0.248 \text{ L})}{\left(0.0821 \frac{\text{L atm}}{\text{mol K}}\right)(294.2 \text{ K})} = 9.69 \times 10^{-3} \text{ mol } N_2$$

$$\# \text{ mol } NaNO_2 = 9.69 \times 10^{-3} \text{ mol } N_2\left(\frac{1 \text{ mol } NaNO_2}{1 \text{ mol } N_2}\right) = 9.69 \times 10^{-3} \text{ mol } NaNO_2$$

$$\text{M } NaNO_2 = \frac{9.69 \times 10^{Ğ3} \text{ mol } NaNO_2}{0.425 \text{ L solution}} = 0.0228 \text{ M } NaNO_3$$

14. (a)

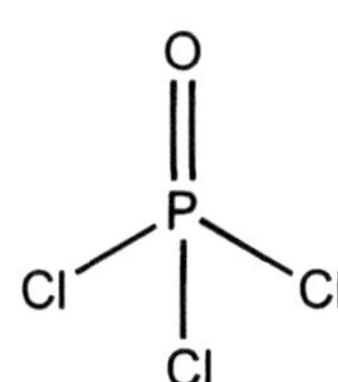

Tetrahedral

(b) The molecule is polar since the O is more electronegative than P or Cl, and pulls elecron density towards itself creating a partial negative charge.

(c) The attractive forces would be dipole-dipole attractions and London dispersion forces.

15.

(a)	$H_2O(l)$	London forces, dipole-diploe, H-bonding, covalent bonding
(b)	$CCl_4(l)$	London forces, covalent bonding
(c)	$CH_3OH(l)$	London forces, dipole-diploe, H-bonding, covalent bonding
(d)	$BrCl(l)$	London forces, dipole-diploe, covalent bonding
(e)	$NaCl(s)$	Ionic bonding
(f)	$Na_2SO_4(s)$	Ionic bonding, covalent bonding

16. Dynamic equilibrium is a condition in which two opposing processes are occurring at equal rates. For the equation:

liquid + heat $\leftrightarrows$ vapor

raising the temperature forces the reaction to proceed to the right since "heat" is part of the reactants. This is an example of Le Châtelier's principle in which a system that is a dynamic equilibrium is subjected to a disturbance that upsets the equilibrium, the system undergoes a change that counteracts the disturbance and, if possible, restores the equilibrium. By adding heat, the equilibrium is disturbed and the reaction moves to the right to restore the equilibrium.

17. Solids can have a vapor pressure. The vapor pressure of a solid would increase with increasing temperature.

18. Dimethylamine has an N–H bond which can form hydrogen bonds. These strong intermolecular attractions give dimethylamine a higher boiling point.

19. Each methanol molecule can form two hydrogen bonds since it has two lone pairs of electrons on the oxygen atom, whereas methylamine can only form one hydrogen bond with the N lone pair. Thus, methanol has the highest boiling point. Ethane cannot form hydrogen bonds, but can only has London dispersion forces. Therefore, ethane has the lowest boiling point.

20. (a) A cool breeze cools you when you are perspiring because the moving air removes the air that is surrounding your skin. The air directly around your skin has a large amount of moisture in it, but blowing the air away, more water can evaporate from you skin and this in turn cools the body.

(b) At the temperature of the cold can, the equilibrium vapor pressure of the water is lower than the partial pressure of water in the air. The air in contact with the cold can is induced to relinquish some of its water, and condensation occurs.

(c) In humid air, the rate of condensation on the skin is more nearly equal to the rate of evaporation from the skin, and the net rate of evaporation of perspiration from the skin is low. The cooling effect of the evaporation of perspiration is low, and our bodies are cooled only slowly under such conditions. In dry air, however, perspiration evaporates more rapidly, and the cooling effect is high.

(d) The source of energy for a violent thunderstorm is the energy that is released during the condensation of water vapor which then forms the clouds and rain.

(e) As the air rises, it becomes cooler, and eventually the amount of moisture in the air becomes greater than is required for equilibrium with the liquid. The air is less able to hold a given amount of vapor at the lower temperature, and condensation occurs.

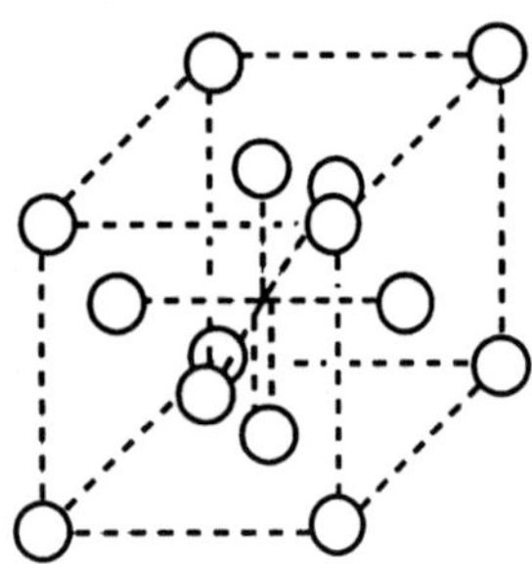

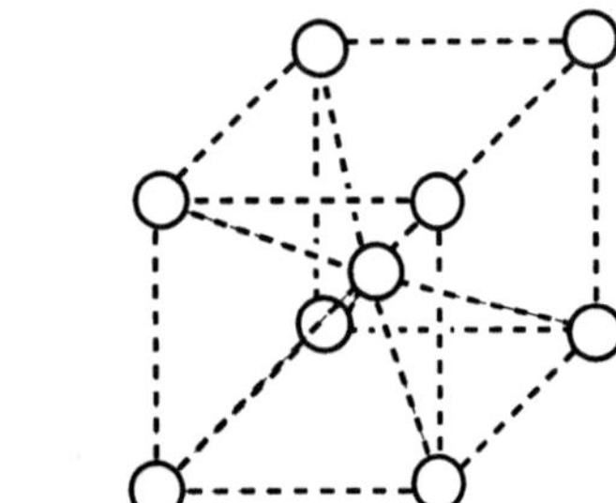

(c)

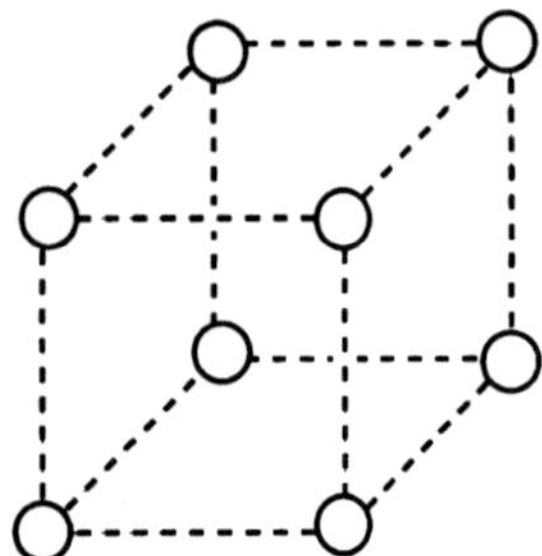

NaCl has a face centered cubic structure.

22. Face-Centered Cubic Lattice:
First determine the volume of a unit cell of Al

$$\text{\# Unit cells/cm}^3 = \left(\frac{2.70 \text{ g Al}}{1 \text{ cm}^{Ñ3}}\right)\left(\frac{1 \text{ mol Al}}{26.98 \text{ g Al}}\right)\left(\frac{6.022\times10^{23} \text{ atoms Al}}{1 \text{ mol Al}}\right)\left(\frac{1 \text{ unit cell Al}}{4 \text{ atoms Al}}\right)$$

$$= \frac{1.51\times10^{22} \text{ unit cell Al}}{\text{cm}^3}$$

$$\text{Volume of a unit cell} = \left(\left(\frac{1 \text{ cm}^3}{1.51\times10^{22} \text{ unit cell Al}}\right)\left(\frac{10^{10} \text{ pm}}{1 \text{ cm}}\right)^3\right) = 6.64 \times 10^7 \text{ pm}^3\text{/unit cell Al}$$

Length of a side of a unit cell of Al = $a^3 = 6.64 \times 10^7$ pm^3/unit cell Al
a = 405 pm
Radius of an Al atom = r

$(4r)^2 = 2a^2$
$(4r)^2 = 2(405 \text{ pm})^2$
$16r^2 = 3.28 \times 10^5 \text{ pm}^2$
$r = 143$ pm

23. The difference between the ccp, cubic closest packing, and the hcp, hexagonal closest packing, packed structures is the layering. For the ccp, the third layer is over a hole in the first layer, and the repeating layer starts with the fourth layer. The repeat pattern is A-B-C-A-B-C… For hcp, the third layer is directly over the first layer and the repeating layer starts with the third layer. The repeat pattern is A-B-A-B…

For both ccp and hcp, each atom is in contact with 12 neighboring atoms.

24. Ionic crystal

25. Covalent crystal

26.

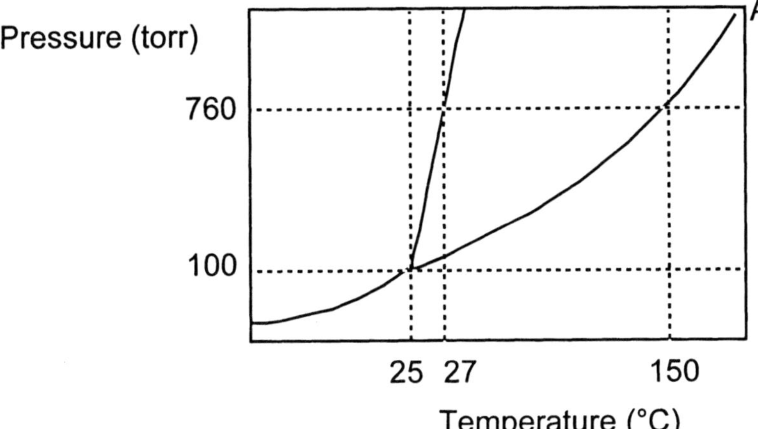

The solid is more dense than the liquid.
The critical temperature and critical pressure would be at A.
At 3- °C and 10.0 torr, the substance would be a gas.

27.

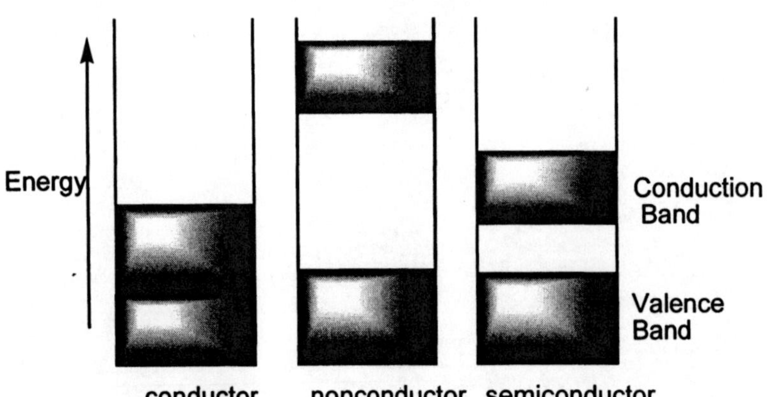

28. n-type semiconductor
Arsenic has one more electron than germanium. When the bonds are formed in the solid, there will be an electron left over that is not used in bonding. The extra electrons can enter the conduction band and move through the solid under an applied voltage as a negatively charged particle.

29. First determine the number of grams of KOH that is needed for the neutralization reaction, then find the number of grams of 4.00% solution of KOH that is needed to perform the neutralization.

$$\# \text{ g KOH} = 10.0 \text{ mL } H_2SO_4 \left(\frac{0.256 \text{ mol } H_2SO_4}{1000 \text{ mL } H_2SO_4}\right)\left(\frac{2 \text{ mol KOH}}{1 \text{ mol } H_2SO_4}\right)\left(\frac{56.11 \text{ g KOH}}{1 \text{ mol KOH}}\right) = 0.287 \text{ g KOH}$$

$$\# \text{ g 4.00\% KOH solution} = 0.287 \text{ g KOH}\left(\frac{100 \text{ g solution}}{4 \text{ g KOH}}\right) = 7.18 \text{ g solution}$$

30. Molecules that form liquid crystal phases tend to be long, thin molecules with a central rigid region and they possess strong dipoles that help them align in the liquid state.

31. Cholesteric liquid crystals reflect light that has a wavelength equal to the distance required for the twist to make one full turn.

32. The anions are small nonmetals that have large negative charges, and the cations have large positive charges.

33. (a) Sintering is heating the ceramic to a high temperature (>1000 °C) and allowing the fine particles to stick together.
(b) A xerogel is a porous solid formed when the solvent is evaporated from a wet gel.
(c) An areogel is a very porous and extremely low density solid.

34. The 4 $C_2H_5O^-$ groups leave the SiO_4^- which picks up 4 H^+ ions in an alcohol solution.

$$Si(OC_2H_5)_4 + 4\,H_2O \longrightarrow Si(OH)_4 + 4\,C_2H_5OH$$

(drawn as: C_2H_5O–Si(OC_2H_5)(OC_2H_5)–OC_2H_5 + 4 H_2O ⟶ HO–Si(OH)(OH)–OH + 4 C_2H_5OH)

Then the $SiOH_4$ groups condese and eliminate water. The network of Si–O units produce extremely small insoluble particles suspended in the alocohol and have a gel-like quality.

$$HO-Si(OH)_2-OH + HO-Si(OH)_2-OH \longrightarrow HO-Si(OH)_2-O-Si(OH)_2-OH + H_2O$$

35. Carbon nanotubes consist of tubular carbon molecules that are shaped like rolled–up sheets of graphite. They are have hexagonal rings of carbon and are capped at each end with half of a spherical fullerene molecule. Their properties include high strengths, and low densities. They can also be conductors or semiconductors depending on their structure.

36. $$\text{Molarity } Na_2CO_3 = \left(\frac{15.00 \text{ g } Na_2CO_3}{100 \text{ g solution}}\right)\left(\frac{1 \text{ mol } Na_2CO_3}{105.99 \text{ g } Na_2CO_3}\right)\left(\frac{1.160 \text{ g solution}}{1 \text{ mL solution}}\right)\left(\frac{1000 \text{ mL solution}}{1 \text{ L solution}}\right)$$
$$= 1.64 \text{ M } Na_2CO_3$$

37. $P_{O2} = (P_{total})(X_{O2})$
$P_{O2} = (760 \text{ torr})(0.211)$
$$\frac{C_1}{P_1} = \frac{C_2}{P_2}$$
$$\frac{4.30\times10^{-2} \text{ g L}^{-1}}{760 \text{ torr}} = \frac{C_2}{160 \text{ torr}}$$
$C_2 = 9.05 \times10^{-3}$ torr which is the concentration of O_2 in water in contact with air at a pressure of 585 torr

38. It is more likely to be an ionic compound and therefore soluble in water. The water molecules hydrate the ions and pull them into the solution while gasoline cannot overcome the lattice energy of the ionic compounds.

39. (a) $XY(s) \rightarrow X^+(g) + Y^-(g)$ $\Delta H = 600 \text{ kJ mol}^{-1}$
$X^+(g) + Y^-(g) \rightarrow X^+(aq) + Y^-(aq)$ $\Delta H = -610 \text{ kJ mol}^{-1}$
(b) $XY(s) \rightarrow X^+(aq) + Y^-(aq)$ $\Delta H = -10 \text{ kJ mol}^{-1}$
(c)

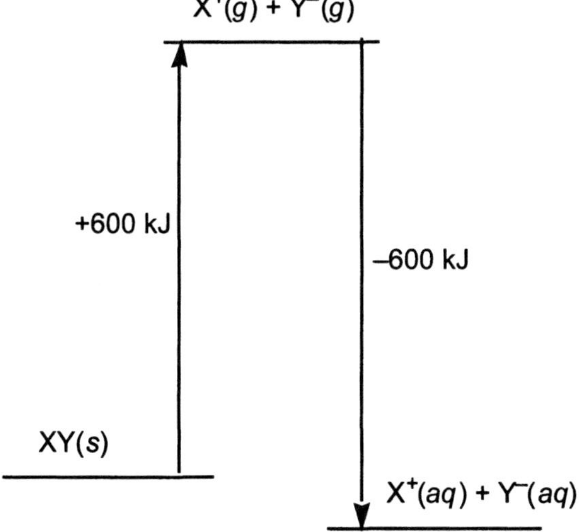

40. $$\text{\% concentration (w/w)} = \left(\frac{0.270 \text{ mol KOH}}{1 \text{ L solution}}\right)\left(\frac{56.11 \text{ g KOH}}{1 \text{ mol KOH}}\right)\left(\frac{1 \text{ L solution}}{1000 \text{ mL solution}}\right)\left(\frac{1 \text{ mL solution}}{1.01 \text{ g solution}}\right)\times 100\%$$
$$= 1.50\%$$

41. (a) $$\text{M } C_2H_5OH = \left(\frac{40 \text{ mL } C_2H_5OH}{100 \text{ mL solution}}\right)\left(\frac{0.7907 \text{ g } C_2H_5OH}{1 \text{ ml } C_2H_5OH}\right)\left(\frac{1 \text{ mol } C_2H_5OH}{46.07 \text{ g } C_2H_5OH}\right)\left(\frac{1000 \text{ mL solution}}{1 \text{ L solution}}\right)$$

$$= 6.87 \text{ M } C_2H_5OH$$

$$\text{m } C_2H_5OH = \left(\frac{40 \text{ mL } C_2H_5OH}{60 \text{ mL } H_2O}\right)\left(\frac{0.7907 \text{ g } C_2H_5OH}{1 \text{ ml } C_2H_5OH}\right)\left(\frac{1 \text{ mol } C_2H_5OH}{46.07 \text{ g } C_2H_5OH}\right)$$

$$\times \left(\frac{1 \text{ mL } H_2O}{0.9982 \text{ g } H_2O}\right)\left(\frac{1000 \text{ g } H_2O}{1 \text{ kg } H_2O}\right) = 11.5 \text{ m } C_2H_5OH$$

(b) $$\# \text{ mole } C_2H_5OH = 40 \text{ mL } C_2H_5OH\left(\frac{0.7907 \text{ g } C_2H_5OH}{1 \text{ ml } C_2H_5OH}\right)\left(\frac{1 \text{ mol } C_2H_5OH}{46.07 \text{ g } C_2H_5OH}\right) = 0.687 \text{ mol } C_2H_5OH$$

$$\# \text{ mol } H_2O = 60 \text{ mL } H_2O\left(\frac{0.9982 \text{ g } H_2O}{1 \text{ mL } H_2O}\right)\left(\frac{1 \text{ mol } H_2O}{18.02 \text{ g } H_2O}\right) = 3.32 \text{ mol } H_2O$$

$$X_{\text{ethyl alcohol}} = \frac{0.687 \text{ mol } C_2H_5OH}{0.687 \text{ mol } C_2H_5OH + 3.32 \text{ mol } H_2O} = 0.171$$

mol percent = $X_{\text{ethyl alcohol}} \times 100\% = 0.171 \times 100\% = 17.1\%$

(c) $P_{\text{ethyl alcohol}} = P°_{\text{ethyl alcohol}} \times X_{\text{ethyl alcohol}}$

$P_{\text{ethyl alcohol}} = 41.0 \text{ torr} \times 0.171 = 7.01 \text{ torr}$

$X_{\text{water}} = 1 - 0.171 = 0.829$

$P_{\text{water}} = P°_{\text{water}} \times X_{\text{water}}$

$P_{\text{water}} = 17.5 \text{ torr} \times 0.829 = 14.5 \text{ torr}$

$P_{\text{total}} = 7.01 \text{ torr} + 14.5 \text{ torr} = 21.5 \text{ torr}$

42. m of particles = 1.0 m $Al(NO_3)_3$ × 4 particles = 4.0 m particles

$\Delta T_b = K_b m$

$\Delta T_b = (0.51 \text{ °C m}^{-1})(4.0 \text{ m})$

$\Delta T_b = 2.04 \text{ °C}$

Boiling point water = 100 °C + 2.04 °C = 102.04 °C

43. (a) Since the entire amount of carbon that was present in the original sample appears among the products only as CO_2, we calculate the amount of carbon in the sample as follows:

$$\# \text{ mol C} = 1.8260 \text{ g } CO_2\left(\frac{1 \text{ mol } CO_2}{44.01 \text{ g } CO_2}\right)\left(\frac{1 \text{ mol C}}{1 \text{ mol } CO_2}\right) = 0.04149 \text{ mol C}$$

Similarly, the entire mass of hydrogen that was present in the original sample appears among the products only as H_2O. Thus the number of moles of hydrogen in the sample is:

$$\# \text{ mol H} = 0.6230 \text{ g } H_2O\left(\frac{1 \text{ mol } H_2O}{18.02 \text{ g } H_2O}\right)\left(\frac{2 \text{ mol H}}{1 \text{ mol } H_2O}\right) = 0.06915 \text{ mol H}$$

The simplest formula is obtained by dividing each of these mol amounts by the smallest:

For C: 0.04149 mol C/0.04149 mol= 1.000

For H: 0.06915 mol/0.04149 mol= 1.667

Multiply both numbers by 3

These values give us the simplest formula directly, namely C_3H_5.

(b) $\Delta T_f = K_f m$

$(178.4 \text{ °C} - 177.3 \text{ °C}) = 1.1 \text{ °C} = (37.5 \text{ °C kg–camphor}^{-1})(\text{m})$

$$\text{m} = 0.0293 \text{ m} = \frac{0.1268 \text{ g squalene}}{0.01050 \text{ kg camphor}}\left(\frac{1 \text{ mol squalene}}{x \text{ g squalene}}\right)$$

x g squalene = 412.2 g squalene
MW squalene = 412.2 g mol^{-1}
FM C_3H_5 = 41.07 g mol^{-1}
412.2 g mol^{-1} / 41.07 g mol^{-1} = 10.04
The molecular formula is 10 times the empirical formula.
Molecular formula = $C_{30}H_{50}$

Test of Facts and Concepts 15 – 19

1. rate = (0.60 L mol^{-1} s^{-1})[I^-][OCl^-]

   (a) rate = (0.60 L mol^{-1} s^{-1})[0.0100 mol L^{-1}][0.0200 mol L^{-1}]
   $= 1.2 \times 10^{-4}$ mol L^{-1} s^{-1}

   (b) rate = (0.60 L mol^{-1} s^{-1})[0.100 mol L^{-1}][0.0400 mol L^{-1}]
   $= 2.4 \times 10^{-3}$ mol L^{-1} s^{-1}

2. (a) rate = $k[A]^n[B]^m$

   First, compare the first and third experiments, in which there has been an increase in concentration by a factor of 2 for A while the concentration of B is held constant. This has caused an increase in rate of:

$$\frac{\text{rate of reaction 3}}{\text{rate of reaction 1}} = \frac{k[0.020]^n[0.010]^m}{k[0.010]^n[0.010]^m} = \frac{[0.020]^n}{[0.010]^n} = 2^n$$

$$= \frac{8.0 \times 10^{-4} \text{ mol L}^{-1}\text{ s}^{-1}}{2.0 \times 10^{-4} \text{ mol L}^{-1}\text{ s}^{-1}} = 4.00 = 2^2$$

$$n = 2$$

   rate = $k[A]^2[B]^m$

   Second, compare the second and third experiment, in which there has been an increase in concentration by a factor of 2 for B while the concentration of A is held constant. This has caused an increase in rate of:

$$\frac{\text{rate of reaction 2}}{\text{rate of reaction 3}} = \frac{k[0.020]^n[0.020]^m}{k[0.020]^n[0.010]^m} = \frac{[0.020]^m}{[0.010]^m} = 2^m$$

$$= \frac{8.0 \times 10^{-4} \text{ mol L}^{-1}\text{ s}^{-1}}{8.0 \times 10^{-4} \text{ mol L}^{-1}\text{ s}^{-1}} = 1.00 = 2^0$$

$$m = 0$$

   rate = $k[A]^2[B]^0$

   (b) The rate constant:
   8.0×10^{-4} mol L^{-1} s^{-1} = k[0.020 mol $L^{-1}]^2[0.010]^0$
   k = 2 L mol^{-1} s^{-1}

   (c) rate = 2 L mol^{-1} s^{-1}[0.017 mol $L^{-1}]^2$[0.033 mol $L^{-1}]^0$
   rate = 5.8×10^{-4} mol L^{-1} s^{-1}

3. The order of the reaction with respect to that reactant is –1

4. (a)

$$t_{1/2} = \frac{\ln 2}{k}$$

$$t_{1/2} = \frac{\ln 2}{2.41 \times 10^{Ñ6} \text{ s}^{Ñ1}} = 2.88 \times 10^5 \text{ s}$$

$$t_{1/2} = \frac{\ln 2}{2.22 \times 10^{Ñ4} \text{ s}^{Ñ1}} = 3.12 \times 10^5 \text{ s}$$

   (b) $[C_2(NO_2)_6] = 0.100$ M

$$\ln\frac{[A]_0}{[A]_t} = kt$$

$$\ln\frac{[0.100]}{[x]} = (2.41 \times 10^{-6} \text{ s}^{-1})\left(500 \text{ min}\left(\frac{60 \text{ s}}{1 \text{ min}}\right)\right)$$

   $x = 9.30 \times 10^{-2}$ mol L^{-1}

(c) $$\ln\left(\frac{k_2}{k_1}\right) = \frac{-E_a}{8.314\,\text{J mol}^{-1}\,\text{K}^{-1}}\left(\frac{1}{T_2} - \frac{1}{T_1}\right)$$

$$\ln\left(\frac{2.22\times10^{-4}\,\text{s}^{-1}}{2.41\times10^{-6}\,\text{s}^{-1}}\right) = \frac{-E_a}{8.314\,\text{J mol}^{-1}\,\text{K}^{-1}}\left(\frac{1}{373\,\text{K}} - \frac{1}{343\,\text{K}}\right)$$

$37.6\ \text{J mol}^{-1} = -E_a\,(-2.34 \times 10^{-4})$

$E_a = 1.60 \times 10^5\ \text{J mol}^{-1}$

(d) $$\ln\left(\frac{k_2}{k_1}\right) = \frac{-E_a}{8.314\,\text{J mol}^{-1}\,\text{K}^{-1}}\left(\frac{1}{T_2} - \frac{1}{T_1}\right)$$

$$\ln\left(\frac{2.22\times10^{-4}\,\text{s}^{-1}}{x\,\text{s}^{-1}}\right) = \frac{-1.60\times10^5\,\text{J mol}^{-1}}{8.314\,\text{J mol}^{-1}\,\text{K}^{-1}}\left(\frac{1}{373\,\text{K}} - \frac{1}{393\,\text{K}}\right)$$

$x\ \text{s}^{-1} = 3.07 \times 10^{-3}\ \text{s}^{-1}$

5. (a) $$t_{1/2} = \frac{1}{k \times (\text{initial concentration of reactant})}$$

$$t_{1/2} = \frac{1}{1.0\times10^{-4}\,\text{L mol}^{-1}\,\text{s}^{-1} \times (0.250\,\text{M})} = 4 \times 10^4\ \text{s}$$

(b) $$\frac{1}{[A]_t} - \frac{1}{[A]_0} = kt$$

$$\frac{1}{[A]_t} - \frac{1}{[0.250\,\text{mol L}^{-1}]} = 1.0\times10^{-4}\,\text{s}^{-1}(30\,\text{min})\left(\frac{60\,\text{s}}{1\,\text{min}}\right)$$

$[A]_{30\,\text{min}} = 0.239\ \text{mol L}^{-1}$

$[B]_{30\,\text{min}} = 0.150\ \text{M} - (0.250\ \text{M} - 0.239\ \text{M}) = 0.139\ \text{M}$

6. Reaction mechanism is an entire series of elementary processes to describe the steps involved in a reaction. Rate-determining step is the slowest step in a reaction mechanism
Elementary process is a reaction whose rate law can be written from its own chemical equation, using its coefficients as exponents for the concentration terms.

7. rate = $k[O_3]$

8.

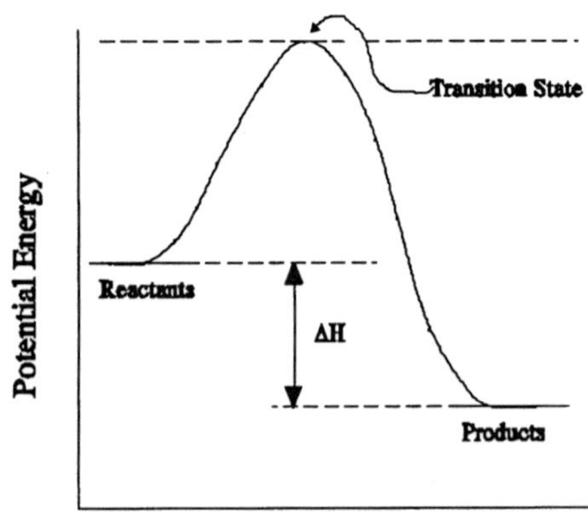

The activation energy for the forward reaction is the difference in energy between the reactants and the transition state. The activation energy for the reverse reaction is the difference in energy between the products and the transition state. The heat of reaction is ΔH.

9. A heterogeneous catalyst increases the rate of a chemical reaction by providing another reaction pathway with a lower activation energy.

10. (a) $\dfrac{[NO]^3}{[NO_2][N_2O]}$

(b) $[SO_2]$

(c) $[Ni^{2+}][CO_3^{\,2-}]$

11. The equilibrium constant is very small, therefore the reaction does not proceed appreciably to products.
[HF] = 0.010 M

	[HF]	[H_2]	[F_2]
I	0.010	–	–
C	–2x	+x	+x
E	0.010–2x	+x	+x

$$K_c = \frac{[H_2][F_2]}{[HF]^2} = \frac{[x][x]}{[0.010-2x]^2} = 1 \times 10^{-13}$$

$x << 0.010$

$$1 \times 10^{-13} = \frac{[x][x]}{[0.010]^2}$$

Take the square root of both sides and solve for x

$x = 3.2 \times 10^{-9}$

$[H_2] = [F_2] = 3.2 \times 10^{-9}$

12. $K_p = K_c(RT)^{\Delta n_g}$

$2NO_2(g) \leftrightarrows N_2O_4(g)$

$\Delta n_g = -1$

$6.5 \times 10^{-2} = K_c[(0.0821 \text{ L atm mol}^{-1} \text{ K}^{-1})(393 \text{ K})]^{-1}$

Solve for K_c

$K_c = 2.1$

13. $NO_2(g) + SO_2(g) \leftrightarrows NO(g) + SO_3(g)$

	[NO_2]	[SO_2]	[NO]	[SO_3]
I	0.0200	0.0200	–	–
C	–x	–x	+x	+x
E	0.0200–x	0.0200–x	+x	+x

$$K_c = 3.60 = \frac{[NO][SO_3]}{[NO_2][SO_2]}$$

$$K_c = 3.60 = \frac{[x][x]}{[0.0200-x][0.0200-x]}$$

Take the square root of both sides

$$1.90 = \frac{x}{0.0200 - x}$$

Solve for x.

x = 0.0131

$[NO_2] = [SO_2] = 6.90 \times 10^{-3}$ M

$[NO] = [SO_3] = 0.0131$ M

	$[NO_2]$	$[SO_2]$	[NO]	$[SO_3]$
I	6.90×10^{-3} M	6.90×10^{-3} M	0.0151	0.0151
C	+x	+x	–x	–x
E	6.90×10^{-3} +x	6.90×10^{-3} +x	0.0151–x	0.0151–x

$$K_c = 3.60 = \frac{[NO][SO_3]}{[NO_2][SO_2]}$$

$$K_c = 3.60 = \frac{[0.0151 - x][0.0151 - x]}{[0.00690 + x][0.00690 + x]}$$

Again take the square root of both sides and solve for x.

$$1.90 = \frac{0.0151 - x}{0.00690 + x}$$

$x = 6.86 \times 10^{-4}$

$[NO_2] = [SO_2] = 7.59 \times 10^{-3}$ M

$[NO] = [SO_3] = 1.44 \times 10^{-2}$ M

14. (a) Increase
    (b) Increase
    (c) Decrease
    (d) Decrease
    (e) No change

15. The concentration of OH^- is 4.7×10^{-7} g L^{-1}
    The molar concentration of OH^- is:

$$[OH^-] = \left(\frac{4.7\times10^{-7}\ g\ OH^-}{1\ L\ solution}\right)\left(\frac{1\ mol\ OH^-}{17.01\ g\ OH^-}\right) = 2.76 \times 10^{-8}\ M$$

$pOH = -\log[OH^-] = -\log(2.76 \times 10^{-8}\ M) = 7.56$

$pH = 14 - pOH = 14 - 7.56 = 6.44$

The water is acidic.

16. H_3PO_4 is a stronger acid since there are more oxygens on the P. The electrons are more electronegative and pull electron density away from the OH bonds.

17. H_2Te

18. (a) H_2SO_3
    (b) $N_2H_5^+$

19. (a) SO_3^{2-}
    (b) $N_2H_3^-$
    (c) C_5H_5N

20. CH_3NH_2 and $CH_3NH_3^+$
    NH_4^+ and NH_3

21.

$$:\ddot{O}=S(=\ddot{O}:)(-\ddot{O}:) + :\ddot{O}^{\ominus}-H \longrightarrow \left[{}^{\ominus}:\ddot{O}-S(=\ddot{O}:)(=\ddot{O}:)-\ddot{O}-H \right]^{-}$$

22. The stronger binary acid than Y would be Z. Binary acid strength increase left to right across a period.

23. (a) $C_6H_5OH(aq) + H_2O(l) \rightarrow C_6H_5O^-(aq) + H_3O^+(aq)$

(b) $$K_a = \frac{[C_6H_5O^-][H_3O^+]}{[C_6H_5OH]}$$

(c) $[C_6H_5OH] = 0.550$ M since the amount of dissociation is negligible

$[H_3O^+] = -\log(pH) = -\log(5.07) = 8.51 \times 10^{-6}$ M

$[C_6H_5O^-] = 8.51 \times 10^{-6}$ M since all of the H_3O^+ comes from the dissociation of the phenol

$$K_a = \frac{[C_6H_5O^-][H_3O^+]}{[C_6H_5OH]} = \frac{[8.51\times10^{-6}][8.51\times10^{-6}]}{[0.550]} = 1.32 \times 10^{-10}$$

$pK_a = -\log K_a = -\log(1.32 \times 10^{-10}) = 9.88$

(d) $$K_b = \frac{1\times10^{-14}}{K_a} = \frac{1\times10^{-14}}{1.32\times10^{-10}} = 7.58 \times 10^{-5}$$

$pK_b = -\log K_b = -\log(7.58 \times 10^{-5}) = 4.12$

24. (a) $pK_b = 14 - pK_a = 14 - 11.68 = 2.32$

(b) $C_7H_3SO_3^- + H_2O \leftrightarrows C_7H_3SO_3H + OH^-$

The solution will be basic

$$K_b = \frac{[C_7H_3SO_3H][OH^-]}{[C_7H_3SO_3^-]} = 4.79 \times 10^{-3}$$

	$[C_7H_3SO_3^-]$	$[C_7H_3SO_3H]$	$[OH^-]$
I	0.010	–	–
C	–x	+x	+x
E	0.010–x	+x	+x

$$4.79 \times 10^{-3} = \frac{[x][x]}{[0.010-x]}$$

x will be large compared to 0.010 M. Therefore assume that the reaction goes to completion, and then returns to equilibrium.

	$[C_7H_3SO_3^-]$	$[C_7H_3SO_3H]$	$[OH^-]$
I	–	0.010	0.010
C	+x	–x	–x
E	+x	0.010–x	0.010–x

$$4.79 \times 10^{-3} = \frac{[0.010-x][0.010-x]}{[x]}$$

$x \ll 0.010$

$$4.79 \times 10^{-3} = \frac{[0.010][0.010]}{[x]}$$

$x = 2.01 \times 10^{-4} = [OH^-]$
pOH = 3.70
pH = 10.3

25. Let Cod stand for codeine
$Cod + H_2O \leftrightarrows CodH^+ + OH^-$

$$K_b = \frac{[CodH^+][OH^-]}{[Cod]} = 1.63 \times 10^{-6}$$

	[Cod]	$[CodH^+]$	$[OH^-]$
I	0.0115	–	–
C	–x	+x	+x
E	0.0115–x	x	x

$$1.63 \times 10^{-6} = \frac{[x][x]}{[0.0115 - x]}$$

x is large compared ot 0.0115, therefore solve for x using either the quadratic equation or successive approximations
$x = 1.36 \times 10^{-4} = [OH^-]$
pOH = 5.79
pH = 14 – pH = 8.21

26. $pK_b = 3.43$
pK_a of its conjugate acid = 14 – pK_b = 14 – 3.43 = 10.57

27. (a) M ascorbic acid solution $= \left(\frac{3.12 \text{ g } H_2C_6H_6O_6}{0.125 \text{ L}}\right)\left(\frac{1 \text{ mol } H_2C_6H_6O_6}{176.1 \text{ g } H_2C_6H_6O_6}\right) = 0.142 \text{ M } H_2C_6H_6O_6$

(b) $H_2C_6H_6O_6 \leftrightarrows HC_6H_6O_6^- + H_3O^+$ $\quad K_{a1} = \frac{[HC_6H_6O_6^-][H_3O^+]}{[H_2C_6H_6O_6]} = 7.94 \times 10^{-5}$

$HC_6H_6O_6^- \leftrightarrows C_6H_6O_6^{2-} + H_3O^+$ $\quad K_{a2} = \frac{[C_6H_6O_6^{2-}][H_3O^+]}{[HC_6H_6O_6^-]} = 1.62 \times 10^{-12}$

Since all of the H_3O^+ will come from the first equilibrium reaction, it can be calculated as follows:

	$[H_2C_6H_6O_6]$	$[HC_6H_6O_6^-]$	$[H_3O^+]$
I	0.142	–	–
C	–x	+x	+x
E	0.142–x	x	x

$$7.94 \times 10^{-5} = \frac{[HC_6H_6O_6^-][H_3O^+]}{[H_2C_6H_6O_6]}$$

$$7.94 \times 10^{-5} = \frac{[x][x]}{[0.142 - x]}$$

Solve for x
$x = 3.32 \times 10^{-3} = [H_3O^+]$
$pH = -\log[H_3O^+] = -\log(3.32 \times 10^{-3}) = 2.48$

$[C_6H_6O_6^{2-}] = 1.62 \times 10^{-12}$

$[HC_6H_6O_6^-] = [H_3O^+]$ in the K_{a2} equation, these cancel

28. $$pH = pK_a + \log\frac{[A^-]_{initial}}{[HA]_{initial}}$$

$$4.50 = 4.74 + \log\frac{[A^-]_{initial}}{[HA]_{initial}}$$

$$-0.24 = \log\frac{[A^-]_{initial}}{[HA]_{initial}}$$

$$10^{-0.24} = \frac{[A^-]_{initial}}{[HA]_{initial}}$$

$$\frac{[A^-]_{initial}}{[HA]_{initial}} = 0.569$$

29. (a) $C_2H_3O_2^- + H^+ \rightarrow HC_2H_3O_2$

(b) $HC_2H_3O_2 + OH^- \rightarrow C_2H_3O_2^- + H_2O$

30. $$K_a = 1.8 \times 10^{-5} = \frac{[C_2H_3O_2^-][H_3O^+]}{[HC_2H_3O_2]}$$

$$\frac{[0.10][H_3O^+]}{[0.15]} = 1.8 \times 10^{-5}$$

$[H_3O^+] = 9.0 \times 10^{-5}$

pH = 4.05

$[HC_2H_3O_2]_{final} = (0.15 - 0.04)M = 0.11\ M$

$[C_2H_3O_2^-]_{final} = (0.10 + 0.04)M = 0.14\ M$

$$\frac{[0.14][H_3O^+]}{[0.11]} = 1.8 \times 10^{-5}$$

$[H_3O^+] = 1.4 \times 10^{-5}$

pH = 4.85

The change in pH is 0.80 pH units.

31. $pK_b = 14 - pK_a = 14 - 4.87 = 9.13$

$$K_b = \frac{[HA][OH^-]}{[A^-]} = 7.41 \times 10^{-10}$$

All of the acid has been converted into the salt at the equivalence point.

(0.115 M)(50.00 mL) = (0.100 M)(x mL)

x = 57.50 mL

total volume is 107.50 mL

$[A^-] = 0.05349\ M$

$[HA] = [OH^-] = x$

$$7.41 \times 10^{-10} = \frac{(x)(x)}{[0.05349]}$$

$x = 6.296\ M = [OH^-]$
pOH = 5.201
pH = 8.799

Thymol blue or phenolphthalein would be good indicators.

32. $K_{b1} = \frac{K_W}{K_{a_2}} = \frac{1.0\times10^{-14}}{1.6\times10^{-12}} = 6.3 \times 10^{-3}$

	$[C_6H_6O_6^{2-}]$	$[HC_6H_6O_6^-]$	$[OH^-]$
I	0.050	–	–
C	–x	+x	+x
E	0.050–x	x	x

$$\frac{[HC_6H_6O_6^-][OH^-]}{[C_6H_6O_6^{2-}]} = \frac{[x][x]}{[0.050-x]} = 6.3 \times 10^{-3}$$

x is too large to make a simplifying assumption, therefore solve for x either using the quadratic equation or by successive approximations.
$x = 0.015 = [OH^-]$
pOH = 1.82
pH = 12.18

33. Since half of the acid is neutralized the concentration of the acid is equal to the concentration of its conjugate base, the pK_a can be determined:

$$pH = pK_a + \log\frac{[A^-]}{[HA]}$$

$3.56 = pK_a + \log 1$
$pK_a = pH = 3.56$
$K_a = 10^{-pKa} = 10^{-3.56} = 2.75 \times 10^{-4}$

34. $$pH = pK_a + \log\frac{[A^-]}{[HA]}$$

$pK_a = 14 - pK_b$ $\quad\quad$ $pK_b = -\log K_b = -\log 1.8 \times 10^{-5}$
$pK_a = 14 - 4.74 = 9.26$

$$9.26 = 9.26 + \log\frac{[A^-]}{[HA]}$$

$[A^-] = [HA]$
$[NH_4^+] = 0.100$ before the addition of NaOH. In order for the two concentrations to be equal, half as much NaOH must be added:
$(0.100\ L\ NH_4^+)(0.100\ M\ NH_4^+\ \text{solution}) = 0.0100\ mol\ NH_4^+$
$(0.0100\ mol\ NH_4^+)(0.5) = 0.00500\ mol\ NaOH$

$$0.00500\ mol\ NaOH\left(\frac{40.00\ g\ NaOH}{1\ mol\ NaOH}\right) = 0.200\ g\ NaOH$$

35. $K_{sp} = [Ag^+]^2[CrO_4^{2-}]$
$[Ag^+] = 2 \times (6.7 \times 10^{-5}\ M) = 1.34 \times 10^{-4}\ M$
$[CrO_4^{2-}] = 6.7 \times 10^{-5}\ M$
$K_{sp} = (1.34 \times 10^{-4})^2(6.7 \times 10^{-5})$
$K_{sp} = 1.2 \times 10^{-12}$

36. $Mg(OH)_2(s) \leftrightarrows Mg^{2+}(aq) + OH^-(aq)$
$K_{sp} = 7.1 \times 10^{-12} = [Mg^{2+}][OH^-]^2$
$7.1 \times 10^{-12} = [x][2x]^2$
$x = 1.2 \times 10^{-4}$
$2x = 2.4 \times 10^{-4} = [OH^-]$
$pOH = -\log[OH^-] = -\log(2.4 \times 10^{-4}) = 3.62$
$pH = 14 - 3.62 = 10.38$

37. $Fe(OH)_2(s) \leftrightarrows Fe^{2+}(aq) + 2OH^-(aq)$
$k_{sp} = 7.9 \times 10^{-16} = [Fe^{2+}][OH^-]^2$
$pH = 10.00$
$pOH = 14 - 10.00 = 4.00$
$[OH^-] = 10^{-4.00} = 1 \times 10^{-4}$
$7.9 \times 10^{-16} = [Fe^{2+}][1 \times 10^{-4}]^2$
$7.9 \times 10^{-8}\ M = [Fe^{2+}]$

$$\#\ g\ L^{-1} = \left(\frac{7.9\times10^{-8}\ \text{mol Fe}^{2+}}{1\ \text{L solution}}\right)\left(\frac{1\ \text{mol Fe(OH)}_2}{1\ \text{mol Fe}^{2+}}\right)\left(\frac{89.86\ \text{g Fe(OH)}_2}{1\ \text{mol Fe(OH)}_2}\right) = 7.10 \times 10^{-6}\ g\ L^{-1}$$

38. (a) This is a limiting reagent problem.

$$\#\ \text{mL KI needed} = 30.0\ \text{mL Pb(NO}_3)_2\left(\frac{0.100\ \text{mol Pb(NO}_3)_2}{1000\ \text{mL Pb(NO}_3)_2\ \text{solution}}\right)$$
$$\times\left(\frac{2\ \text{mol KI}}{1\ \text{mol Pb(NO}_3)_2}\right)\left(\frac{1000\ \text{mL KI solution}}{0.500\ \text{mol KI}}\right) = 12.0\ \text{mL KI solution}$$

20.0 mL KI solution is supplied. KI is in excess.

$$\#\ g\ PbI_2 = 30.0\ \text{mL Pb(NO}_3)_2\left(\frac{0.100\ \text{mol Pb(NO}_3)_2}{1000\ \text{mL Pb(NO}_3)_2\ \text{solution}}\right)$$
$$\times\left(\frac{1\ \text{mol PbI}_2}{1\ \text{mol Pb(NO}_3)_2}\right)\left(\frac{461.0\ \text{g PbI}_2}{1\ \text{mol PbI}_2}\right) = 1.38\ \text{g PbI}_2$$

(b) The concentration of the spectator ions:
$[K^+] = (0.500\ M\ K^+)(20.0\ mL\ solution) \div 50.0\ mL = 0.200\ M\ K^+$
$[NO_3^-] = (0.100)(2)(30.0\ mL\ solution) \div 50.0\ mL = 0.12\ M\ NO_3^-$
$[I^-]$: Total moles of $I^- = (0.500\ M\ I^-)(0.0200\ L\ solution) = 0.0100\ mol\ I^-$

$$\text{mol I}^-\ \text{used} = 30.0\ \text{mL Pb}^{2+}\left(\frac{0.100\ \text{mol Pb}^{2+}}{1000\ \text{mL Pb}^{2+}\ \text{solution}}\right)\left(\frac{2\ \text{mol I}^-}{1\ \text{mol Pb}^{2+}}\right) = 6.00 \times 10^{-3}\ \text{mol I}^-$$

$[I^-] = (0.0100\ mol\ I^- - 6.00 \times 10^{-3}\ mol\ I^-)/(0.050\ L\ solution) = 0.0800\ M\ I^-$

$[Pb^{2+}]$: This will be the amount of Pb that is in solution after the solid PbI_2 reaches equilibrium
$PbI_2(s) \leftrightarrows Pb^{2+}(aq) + 2I^-(aq)$

	$[Pb^{2+}]$	$[I^-]$
I	–	0.0800 M
C	+x	+x
E	x	0.0800 + x

$K_{sp} = 7.9 \times 10^{-9} = [Pb^{2+}][I^-]^2 = (x)(0.0800 + x)^2$
$x << 0.0800$
$7.9 \times 10^{-9} = (x)(0.0800)^2$
$x = 1.23 \times 10^{-6}\ M = [Pb^{2+}]$

39. This is a simultaneous equilibrium problem.

$CuCO_3(s) \leftrightarrows Cu^{2+}(aq) + CO_3^{2-}(aq)$ $K_{sp} = [Cu^{2+}][CO_3^{2-}] = 2.5 \times 10^{-10}$

$Cu^{2+}(aq) + 4NH_3(aq) \leftrightarrows Cu(NH_3)_4^{2+}(aq)$ $K_{form} = \dfrac{[Cu(NH_3)_4^{2+}]}{[Cu^{2+}][NH_3]^4} = 1.1 \times 10^{13}$

$CuCO_3(s) + 4NH_3(aq) \leftrightarrows Cu(NH_3)_4^{2+}(aq) + CO_3^{2-}(aq)$

$$K_{overall} = \frac{[Cu(NH_3)_4^{2+}][CO_3^{2-}]}{[NH_3]^4} = 2.75 \times 10^3$$

All of the $Cu(NH_3)_4^{2+}$ comes from the $CuCO_3$

$$[Cu(NH_3)_4^{2+}] = 1.00\text{ g } CuCO_3\left(\frac{1\text{ mol } CuCO_3}{123.6\text{ g } CuCO_3}\right)\left(\frac{1}{1.00\text{ L solution}}\right) = 8.09 \times 10^{-3}$$

As the Cu^{2+} is used to form $Cu(NH_3)_4^{2+}$, the $[CO_3^{2-}] = [Cu(NH_3)_4^{2+}] = 8.09 \times 10^{-3}$

$$2.75 \times 10^3 = \frac{[Cu(NH_3)_4^{2+}][CO_3^{2-}]}{[NH_3]^4} = \frac{[8.09\times10^{-3}][8.09\times10^{-3}]}{[x]^4}$$

$x = 0.0124\ M\ NH_3$

$$\#\text{ mol } NH_3 = 1.00\text{ L solution}\left(\frac{0.0124\text{ mol } NH_3}{1\text{ L solution}}\right) = 0.0124\text{ mol } NH_3$$

40. Assume the solution is saturated with CO_2 and therefore the concentratin of H_2CO_3 is 0.030 M

$H_2CO_3 \leftrightarrows + 2H^+ + CO_3^{2-}$ $K = K_{a1} \times K_{a2} = (4.3 \times 10^{-7})(4.7 \times 10^{-11}) = 2.0 \times 10^{-17} = \dfrac{[H^+]^2[CO_3^{2-}]}{[H_2CO_3]}$

First, calculate the $[H^+]$ at which thePbCO$_3$ will begin to precipitate:

$PbCO_3(s) \leftrightarrows Pb^{2+}(aq) + CO_3^{2-}(aq)$ $K_{sp} = 7.4 \times 10^{-14} = [Pb^{2+}][CO_3^{2-}]$

$[Pb^{2+}] = 0.010\ M$
$[CO_3^{2-}] = (7.4 \times 10^{-14})/(0.010) = 7.4 \times 10^{-16}\ M$

$$[H^+] = \left(\frac{(2.0\times10^{-17})[H_2CO_3]}{[CO_3^{2-}]}\right)^{\frac{1}{2}} = \left(\frac{(2.0\times10^{-17})[0.030]}{(7.4\times10^{-16})}\right)^{\frac{1}{2}} = 2.8 \times 10^{-2}$$

pH = 1.55

Now, determine the $[H^+]$ at which the $BaCO_3$ will begin to precipitate:

$BaCO_3(s) \leftrightarrows Ba^{2+}(aq) + CO_3^{2-}(aq)$ $K_{sp} = 5.0 \times 10^{-9} = [Ba^{2+}][CO_3^{2-}]$

$[Ba^{2+}] = 0.010\ M$
$[CO_3^{2-}] = (5.0 \times 10^{-9})/(0.010) = 5.0 \times 10^{-7}\ M$

$$[H^+] = \left(\frac{(2.0\times10^{-17})[H_2CO_3]}{[CO_3^{2-}]}\right)^{\frac{1}{2}} = \left(\frac{(2.0\times10^{-17})[0.030]}{(5.0\times10^{-7})}\right)^{\frac{1}{2}} = 1.1 \times 10^{-6}$$

pH = 5.96

The pH range is between 1.55 and 5.96, above 5.96, $BaCO_3$ will begin to precipitate.

41. The less soluble substance is PbS. We need to determine the minimum $[H^+]$ at which NiS will precipitate.

$$K_{spa} = \frac{[Ni^{2+}][H_2S]}{[H^+]^2} = \frac{(0.100)(0.1)}{[H^+]^2} = 40 \qquad \text{(from Table 19.2)}$$

$$[H^+] = \sqrt{\frac{(0.10)(0.1)}{40}} = 0.016$$

pH = –log$[H^+]$ = 1.80. At a pH lower than 1.80, PbS will precipitate and NiS will not. At larger values of pH, both PbS and NiS will precipitate.

We also need to determine the $[H^+]$ at which PbS will state to precipitate

$$K_{spa} = \frac{[Pb^{2+}][H_2S]}{[H^+]^2} = \frac{(0.100)(0.1)}{[H^+]^2} = 3\times10^{-7} \qquad \text{(from Table 19.2)}$$

$$[H^+] = \sqrt{\frac{(0.10)(0.1)}{3\times10^{-7}}} = 182$$

pH = –log$[H^+]$ = –2.26. Any acid in water will precipitate the PbS.
The pH range is –2.26 to 1.80 to allow the PbS to precipitate without the NiS.

42. The less soluble substance is SnS. We need to determine the minimum $[H_2S]$ at which FeS will precipitate.

$$K_{spa} = \frac{[Fe^{2+}][H_2S]}{[H^+]^2} = \frac{(0.10)[H_2S]}{[1\times10^{-3}]^2} = 600 \qquad \text{(from Table 19.2)}$$

$$[H_2S] = \frac{(600)(1\times10^{-3})^2}{(0.1)} = 6\times10^{-3}$$

The concentration of Sn^{2+} can now be determined.

$$K_{spa} = \frac{[Sn^{2+}][H_2S]}{[H^+]^2} = \frac{[Sn^{2+}][6\times10^{-3}]}{[1\times10^{-3}]^2} = 1\times10^{-5} \qquad \text{(from Table 19.2)}$$

$$[Sn^{2+}] = \frac{(1\times10^{-5})(1\times10^{-3})^2}{(6\times10^{-3})} = 1.7\times10^{-9}$$

Test of Facts and Concepts 20 – 22

1. A spontaneous change will occur by itself without any external assistance.

2. State functions: E, H, S, G, and T

3. The greater the disorder or randomness, the higher is the statistical probability of the state and the higher is the entropy.

4. (a) ΔS positive
   (b) ΔS negative

5. (e) $4H(g) + 2O(g)$

6. $\Delta S° = [\text{sum of } S°_{\text{products}}] - [\text{sum of } \Delta° S_{\text{reactants}}]$
   (a) $\Delta S° = [(1 \text{ mol } H_2SO_4(l))(157 \text{ J mol}^{-1} \text{ K}^{-1})] - [(1 \text{ mol } H_2O(l))(69.96 \text{ J mol}^{-1} \text{ K}^{-1}) + (1 \text{ mol } SO_3(g))(256.2 \text{ J mol}^{-1} \text{ K}^{-1})] = -169.16 \text{ J K}^{-1}$
   (b) $\Delta S° = [(1 \text{ mol } K_2SO_4\ (s))(176 \text{ J mol}^{-1} \text{ K}^{-1}) + (2 \text{ mol } HCl(g))(186.7 \text{ J mol}^{-1} \text{ K}^{-1})] - [(2 \text{ mol } KCl(s))(82.59 \text{ J mol}^{-1} \text{ K}^{-1}) + (1 \text{ mol } H_2SO_4\ (l))(157 \text{ J mol}^{-1} \text{ K}^{-1})] = 227 \text{ J K}^{-1}$
   (c) $\Delta S° = [(1 \text{ mol } C_2H_5OH\ (l))(161 \text{ J mol}^{-1} \text{ K}^{-1})] - [(1 \text{ mol } C_2H_4(g))(219.8 \text{ J mol}^{-1} \text{ K}^{-1}) + (1 \text{ mol } H_2O(g))(188.7 \text{ J mol}^{-1} \text{ K}^{-1})] = -248 \text{ J K}^{-1}$

7. (a) $CaSO_4 \cdot 1/2H_2O(s) + 3/2H_2O(l) \rightarrow CaSO_4 \cdot 2H_2O(s)$

   $\Delta G° = (\text{sum } \Delta G_f^\circ [\text{products}]) - (\text{sum } \Delta G_f^\circ [\text{reactants}])$

   $\Delta G° = \{\Delta G_f^\circ [CaSO_4 \cdot 2H_2O(s)]\} - \{\Delta G_f^\circ [CaSO_4 \cdot 1/2H_2O(s)] + 3/2 \Delta G_f^\circ [H_2O(l)]\}$

   $\Delta G° = \{1 \text{ mol} \times (-1795.7 \text{ kJ/mol})\} - \{1 \text{ mol} \times (-1435.2 \text{ kJ/mol}) + 1.5 \text{ mol} \times (-237.2 \text{ kJ/mol})\}$

   $\Delta G° = -4.7 \text{ kJ}$

   (b) $CH_4(g) + Cl_2(g) \rightarrow CH_3Cl(g) + HCl(g)$

   $\Delta G° = (\text{sum } \Delta G_f^\circ [\text{products}]) - (\text{sum } \Delta G_f^\circ [\text{reactants}])$

   $\Delta G° = \{\Delta G_f^\circ [CH_3Cl(g)] + \{\Delta G_f^\circ [HCl(g)]\} - \{\Delta G_f^\circ [CH_4(g)] + \Delta G_f^\circ [Cl_2(g)]\}$

   $\Delta G° = \{1 \text{ mol} \times (-58.6 \text{ kJ/mol}) + 1 \text{ mol} \times (-95.27 \text{ kJ/mol})\} - \{1 \text{ mol} \times (-50.79 \text{ kJ/mol}) + 1 \text{ mol} \times (0 \text{ kJ/mol})\}$

   $\Delta G° = -\ 103.1 \text{ kJ}$

   (c) $CaSO_4(s) + CO_2(g) \rightarrow CaCO_3(s) + SO_3(g)$

   $\Delta G° = \{\Delta G_f^\circ [SO_3(g)] + \Delta G_f^\circ [CaCO_3(s)]\} - \{\Delta G_f^\circ [CaSO_4(s)] + \Delta G_f^\circ [CO_2(g)]\}$

   $\Delta G° = \{1 \text{ mol} \times (-307.4 \text{ kJ/mol}) + 1 \text{ mol} \times (-1128.8 \text{ kJ/mol})\}$
   $- \{1 \text{ mol} \times (-1320.3 \text{ kJ/mol}) + 2 \text{ mol} \times (-394.4 \text{ kJ/mol})\}$

   $\Delta G° = +672.9 \text{ kJ}$

8. (a) spontaneous
   (b) spontaneous
   (c) nonspontaneous

9. (a) $CaSO_4 \cdot 2H_2O(s) \rightarrow CaSO_4 \cdot 1/2H_2O(s) + 3/2H_2O(g)$

   $\Delta H° = \{\Delta H_f^\circ [CaSO_4 \cdot 1/2H_2O(s)] + 3/2 \times \Delta H_f^\circ [H_2O(g)]\} - \{\Delta H_f^\circ [CaSO_4 \cdot 2H_2O(s)]\}$

   $\Delta H° = \{1 \text{ mol} \times (-1573 \text{ kJ/mol}) + 3/2 \text{ mol} \times (-241.8 \text{ kJ/mol})\} - \{1 \text{ mol} \times (-2020 \text{ kJ/mol})\}$

   $\Delta H° = 84.3 \text{ kJ} = 8.43 \times 10^5 \text{ J}$

   $\Delta S° = \{S°[CaSO_4 \cdot 1/2H_2O(s)] + 3/2 \times S°[H_2O(g)]\} - \{S°[CaSO_4 \cdot 2H_2O(s)]\}$

   $\Delta S° = \{1 \text{ mol} \times (131 \text{ J mol}^{-1} \text{ K}^{-1}) + 3/2 \text{ mol} \times (188.7 \text{ J mol}^{-1} \text{ K}^{-1})\} - \{1 \text{ mol} \times (194.0 \text{ J mol}^{-1} \text{ K}^{-1})\}$

   $\Delta S° = 220.1 \text{ J K}^{-1}$

$\Delta G_{673}° = 8.43 \times 10^5\ J - (673\ K)220.1\ J\ K^{-1}$
$\Delta G_{673}° = 6.95 \times 10^5\ J = 695\ kJ$

(b) $NaOH(s) + NH_4Cl(s) \rightarrow NaCl(s) + NH_3(g) + H_2O(g)$
$\Delta H° = \{\Delta H_f^\circ[NaCl(s)] + \Delta H_f^\circ[NH_3(g)] + \Delta H_f^\circ[H_2O(g)]\} - \{\Delta H_f^\circ[NaOH(s)] + \Delta H_f^\circ[NH_4Cl(s)]\}$
$\Delta H° = \{1\ mol \times (-413\ kJ/mol) + 1\ mol \times (-46.0\ kJ/mol) + 1\ mol \times (-241.8\ kJ/mol)\}$
$- \{1\ mol \times (-426.8\ kJ/mol) + 1\ mol \times (-314.4\ kJ/mol)\}$
$\Delta H° = 40.4\ kJ = 4.04 \times 10^5\ J$

$\Delta S° = \{S°[NaCl(s)] + S°[NH_3(g)] + S°[H_2O(g)]\} - \{S°[NaOH(s)] + S°[NH_4Cl(s)]\}$
$\Delta S° = \{1\ mol \times (72.38\ J\ mol^{-1}\ K^{-1}) + 1\ mol \times (192.5\ J\ mol^{-1}\ K^{-1}) + 1\ mol \times (188.7\ J\ mol^{-1}\ K^{-1})\}$
$- \{1\ mol \times (64.18\ J\ mol^{-1}\ K^{-1}) + 1\ mol \times (94.6\ mol^{-1}\ K^{-1})\}$
$\Delta S° = 294.8\ J\ K^{-1}$

$\Delta G_{673}° = 4.04 \times 10^5\ J - (673\ K)294.8\ J\ K^{-1}$
$\Delta G_{673}° = 2.06 \times 10^5\ J = 206\ kJ$

(c) $SO_3(g) \rightarrow SO_2(g) + 1/2O_2(g)$
$\Delta H° = \{1\ mol \times \Delta H_f^\circ[SO_2(g)] + 1/2\ mol \times \Delta H_f^\circ[O_2(g)]\} - \{1\ mol \times \Delta H_f^\circ[SO_3(g)]\}$
$\Delta H° = \{1\ mol \times (-297\ kJ/mol) + 1/2\ mol \times (0\ kJ/mol)\} - \{1\ mol \times (-396\ kJ/mol)\}$
$\Delta H° = 99.0\ kJ = 9.90 \times 10^4\ J$

$\Delta S° = \{1\ mol \times S°[SO_2(g)] + 1/2\ mol \times S°[O_2(g)]\} - \{1\ mol \times S°[SO_3(g)]\}$
$\Delta S° = \{1\ mol \times (248.5\ J\ mol^{-1}\ K^{-1}) + 1/2\ mol \times (205.0\ J\ mol^{-1}\ K^{-1})\}$
$- \{1\ mol \times (256.2\ J\ mol^{-1}\ K^{-1})\}$
$\Delta S° = 94.8\ J\ K^{-1}$

$\Delta G_{673}° = 9.90 \times 10^4\ J - (673\ K)(94.5\ J\ K^{-1})$
$\Delta G_{673}° = 3.54 \times 10^4\ J = 35.4\ kJ$

10. $3NO(g) \leftrightarrows NO_2(g) + N_2O(g)$
$\Delta G° = \{1\ mol \times \Delta G°_f[NO_2(g)] + 1\ mol \times \Delta G°_f[N_2O(g)]\} - \{3\ mol \times \Delta G°_f[NO(g)]\}$
$\Delta G° = \{1\ mol \times (51.84\ kJ/mol) + 1\ mol \times (103.6\ kJ/mol)\} - \{3\ mol \times (-86.69\ kJ/mol)\}$
$\Delta G° = 416\ kJ$

$\Delta G° = RT\ln K_P$
$4.16 \times 10^5\ J = -(8.314\ J\ mol^{-1}\ K^{-1})(298\ K)(\ln K_P)$
$\ln K_P = -168$
$K_P = 1.2 \times 10^{-73}$

$K_P = K_c(RT)^{\Delta n_g}$
$1.2 \times 10^{-73} = K_c[(0.0821\ L\ atm\ mol^{-1}\ K^{-1})(298)]^{-1}$
$K_c = 2.9 \times 10^{-72}$

11. $CH_4(g) + Cl_2(g) \leftrightarrows CH_3Cl(g) + HCl(g)$

(a) $\Delta H° = \{\Delta H_f^\circ[CH_3Cl(g)] + \Delta H_f^\circ[HCl(g)]\} - \{\Delta H_f^\circ[CH_4(g)] + \Delta H_f^\circ[Cl_2(g)]\}$
$\Delta H° = \{1\ mol \times (-82.0\ kJ/mol) + 1\ mol \times (-92.3\ kJ/mol)\}$
$- \{1\ mol \times (-74.9\ kJ/mol) + 1\ mol \times (0\ kJ/mol)\}$
$\Delta H° = -99.4\ kJ = 9.94\times 10^4\ J$

$\Delta S° = \{S°[CH_3Cl(g)] + S°[HCl(g)]\} - \{S°[CH_4(g)] + S°[Cl_2(g)]\}$
$\Delta S° = \{1 \text{ mol} \times (234.2 \text{ J mol}^{-1} \text{ K}^{-1}) + 1 \text{ mol} \times (186.7 \text{ J mol}^{-1} \text{ K}^{-1})\}$
$- \{1 \text{ mol} \times (186.2 \text{ J mol}^{-1} \text{K}^{-1}) + 1 \text{ mol} \times (223.0 \text{ mol}^{-1} \text{K}^{-1})\}$
$\Delta S° = 11.7 \text{ J K}^{-1}$

$\Delta G_{473}° = 9.94 \times 10^4 \text{ J} - (473 \text{ K})11.7 \text{ J K}^{-1} = -1.05 \times 10^5 \text{ J}$
$\Delta G_{473}° = -105 \text{ kJ}$

(b) $\Delta G° = RT\ln K_P$
$-1.05 \times 10^5 \text{ J} = -(8.314 \text{ J mol}^{-1} \text{ K}^{-1})(473 \text{ K})(\ln K_P)$
$\ln K_P = 26.7$
$K_P = 3.9 \times 10^{11}$

(c) $K_P = K_c(RT)^{\Delta n_g}$
$3.9 \times 10^{11} = K_c[(0.0821 \text{ L atm mol}^{-1} \text{ K}^{-1})(473)]^{-1}$
$K_c = 1.0 \times 10^{10}$

12. atomization energy = (2 mol × N–H B.E.) + (2 mol × C–H B.E.) + (1 mol × N–C B.E.) +
(1 mol × C–C B.E.) + (1 mol × C═O B.E.) + (1 mol × C–O B.E.) + (1 mol × O–H B.E.)
= $(2 \text{ mol} \times 388 \text{ kJ mol}^{-1}) + (2 \text{ mol} \times 412 \text{ kJ mol}^{-1}) + (1 \text{ mol} \times 305 \text{ kJ mol}^{-1}) +$
$(1 \text{ mol} \times 348 \text{ kJ mol}^{-1}) + (1 \text{ mol} \times 743 \text{ kJ mol}^{-1}) + (1 \text{ mol} \times 360 \text{ kJ mol}^{-1}) +$
$(1 \text{ mol} \times 463 \text{ kJ mol}^{-1})$
= 3819 kJ

13. First calculate the atomization energy for the formation of C_2H_6, and the atomization energy for the C_2H_2 and H_2. The difference is the energy required for the reaction. Then calculate the amount of heat required for 25.0 g of C_2H_6.
atomization energy$_1$ = (6 mol C–H B.E.) + (1 mol C–C B.E.)
= $(6 \text{ mol} \times 412 \text{ kJ mol}^{-1}) + (1 \text{ mol } 348 \text{ kJ mol}^{-1})$
= 2820 kJ
atomization energy$_2$ = (2 mol C–H B.E.) + (1 mol C≡C B.E.) + (1 mol H–H B.E.)
= $(2 \text{ mol} \times 412 \text{ kJ mol}^{-1}) + (1 \text{ mol} \times 960 \text{ kJ mol}^{-1}) + (1 \text{ mol} \times 436 \text{ kJ mol}^{-1})$
= 2220 kJ
atomization energy$_1$ – atomization energy$_2$ = 2820 kJ – 2220 kJ = 600 kJ
600 kJ are absorbed.

$$\# \text{ kJ for } 25.0 \text{ g} = 25.0 \text{ g}\left(\frac{1 \text{ mol } C_2H_6}{26.04 \text{ g } C_2H_6}\right)\left(\frac{600 \text{ kJ}}{1 \text{ mol } C_2H_6}\right) = 576 \text{ kJ}$$

14. (a)

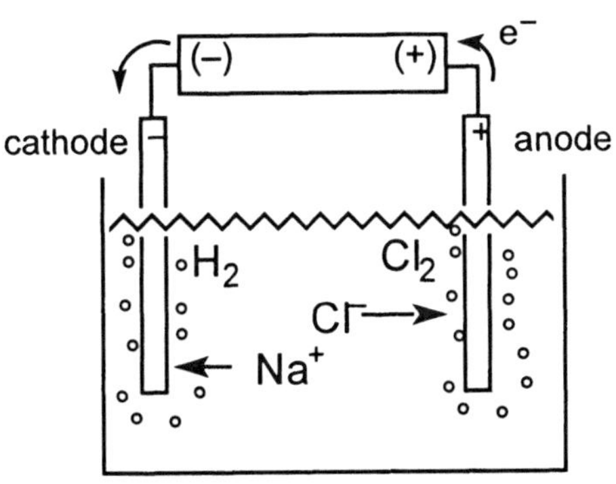

(b) Anode $2Cl^-(aq) \rightarrow Cl_2(g) + 2e^-$
Cathode $2H_2O(l) + 2e^- \rightarrow H_2(g) + 2OH^-(aq)$

(c) $2Cl^-(aq) + 2H_2O(l) \rightarrow Cl_2(g) + H_2(g) + 2OH^-(aq)$

15. (a) $\# \text{ mole } e^- = 20.0 \text{ min}\left(\frac{60\text{ s}}{1\text{ min}}\right)\left(\frac{1.00\text{ C}}{1\text{ s}}\right)\left(\frac{1\text{ mol } e^-}{9.65\times10^4\text{ C}}\right) = 1.24\times10^{-2}\text{ mol } e^-$

$\# \text{ mole } OH^- = 1.24\times10^{-2}\text{ mol } e^-\left(\frac{1\text{ mol } OH^-}{1\text{ mol } e^-}\right) = 1.24\times10^{-2}\text{ mol } OH^-$

$[OH^-] = \frac{1.24\times10^{-2}\text{ mol } OH^-}{0.250\text{ L NaCl solution}} = 0.0497\text{ M } OH^-$

$pOH = -\log[OH^-] = 1.303$

$pH = 14 - pOH$

$pH = 14 - 1.303 = 12.697$

(b) $\# \text{ mole } e^- = 10.0 \text{ min}\left(\frac{60\text{ s}}{1\text{ min}}\right)\left(\frac{5.00\text{ C}}{1\text{ s}}\right)\left(\frac{1\text{ mol } e^-}{9.65\times10^4\text{ C}}\right) = 3.11\times10^{-2}\text{ mol } e^-$

$\# \text{ mole } H_2 = 1.24\times10^{-2}\text{ mol } e^-\left(\frac{1\text{ mol } H_2}{2\text{ mol } e^-}\right) = 6.20\times10^{-3}\text{ mol } H_2$

$\# \text{ mL } H_2 = \left(\frac{(6.20\times10^{-3}\text{ mol } H_2)(0.0821\text{ L atm}^{-1}\text{ mol K}^{-1})(273\text{ K})}{1\text{ atm}}\right)\left(\frac{1000\text{ mL}}{1\text{ L}}\right) = 139\text{ mL}$

16. $\# C = 0.100\text{ g Ni}\left(\frac{1\text{ mol Ni}}{58.59\text{ g Ni}}\right)\left(\frac{2\text{ mol } e^-}{1\text{ mol Ni}}\right)\left(\frac{9.65\times10^4\text{ C}}{1\text{ mol } e^-}\right) = 3.29\times10^2\text{ C}$

$\# A = \left(\frac{3.29\times10^2\text{ C}}{20.0\text{ min}}\right)\left(\frac{1\text{ min}}{60\text{ s}}\right) = 0.274\text{ A}$

17. $Al^{3+} + 3e^- \rightarrow Al(\ell)$

The number of Coulombs that will be required is:

$\# C = (900\text{ lb Al})\left(\frac{454\text{ g Al}}{1\text{ lb Al}}\right)\left(\frac{1\text{ mol Al}}{26.98\text{ g Al}}\right)\left(\frac{3\text{ mol } e^-}{1\text{ mol Al}}\right)\left(\frac{96,500\text{ C}}{1\text{ mol } e^-}\right) = 4.38\times10^9\text{ C}$

Number of seconds is:

$\# s = 1\text{ day}\left(\frac{24\text{ hr}}{1\text{ day}}\right)\left(\frac{60\text{ min}}{1\text{ hr}}\right)\left(\frac{60\text{ s}}{1\text{ min}}\right) = 8.64\times10^4\text{ s}$

The number of amperes is: $(4.38\times10^9\text{ C}) \div (8.64\times10^4\text{ s}) = 5.07\times10^4\text{ amp}$

18. (a)

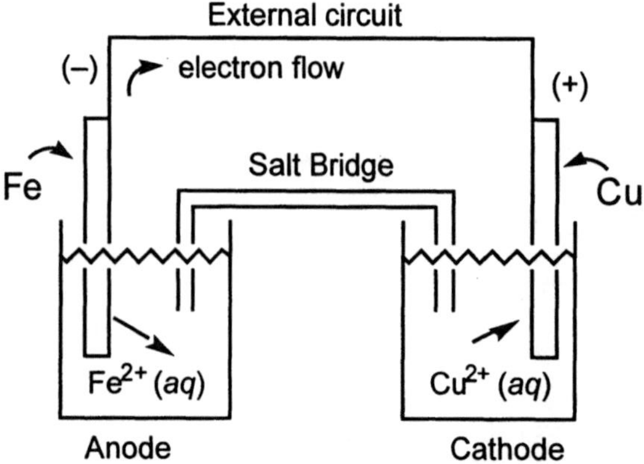

(b) $Fe(s) + Cu^{2+}(aq) \rightarrow Cu(s) + Fe^{2+}(aq)$

(c) $Fe|Fe^{2+}||Cu^{2+}|Cu$

(d) $E°_{cell} = E°_{substance\ reduced} - E°_{substance\ oxidized}$

$E°_{cell} = E°_{Cu2+} - E°_{Fe2+}$

$E°_{cell} = 0.34\ V - (-0.44\ V) = 0.78\ V$

19. The salt bridge connects two half–cells, and allows for electrical neutrality to be maintained by a flow of appropriate ions.

20. $$\#\ e^- = 50.0\ h\left(\frac{3600\ s}{1\ h}\right)\left(\frac{0.10\ C}{s}\right)\left(\frac{1\ mol\ e^-}{96{,}500\ C}\right) = 1.87 \times 10^{-1}\ mol\ e^-$$

$$\#\ mol\ Fe^{2+} = 1.87 \times 10^{-1}\ mol\ e^-\left(\frac{1\ mol\ Fe}{2\ mol\ e^-}\right) = 9.35 \times 10^{-2}\ mol\ Fe^{2+}$$

The change in concentration of Fe^{2+} will be

$$\frac{+9.35\times10^{-2}\ mol\ Fe^{+2}}{0.100\ L\ solution} = +0.935\ M\ Fe^{2+}$$

The final concentration of Fe^{2+}will be:

$1.00\ M + 0.935\ M\ Fe^{2+} = 1.94\ M\ Fe^{2+}$

The change in concentration of Cu^{2+} will be

$$\frac{-9.35\times10^{-2}\ mol\ Cu^{+2}}{0.100\ L\ solution} = -0.935\ M\ Cu^{2+}$$

The final concentration of Cu^{+} will be:

$1.00\ M - 0.935\ M\ Cu^{2+} = 0.065\ M\ Cu^{2+}$

$$E_{cell} = E^{\circ}_{cell} - \frac{RT}{nF}\ln\frac{[Fe^{2+}]}{[Cu^{2+}]}$$

$$E_{cell} = 0.78\ V - \frac{(8.314\ J\ mol^{-1}K^{-1})(298\ K)}{2(96{,}500\ C\ mol^{-1})}\ln\frac{[1.94]}{[0.065]}$$

$$= 0.78\ V - 0.01284(3.396)$$

$E_{cell} = 0.736\ V$

21. $$E°_{cell} = \frac{RT}{nF}\ln K_c$$

O_2 is reduced by four electrons and Br^- is oxidized by four electrons.

$$E^{\circ}_{cell} = E^{\circ}_{reduction} - E^{\circ}_{oxidation} = 1.23\ V - (+1.07\ V) = +0.16\ V$$

$$E^{\circ}_{cell} = \frac{0.0592}{n}\log K_c$$

$+0.16\ V = (0.0592/4) \times \log K_c$

$\log K_c = 10.8$ and $K_c = 6.47 \times 10^{10}$

22. Overall reaction:

$Cu^{2+} + Mn \rightarrow Mn^{2+} + Cu$

$E_{reduction} = +0.34\ V$

$E_{oxidation} = x$

$$E_{cell} = E°_{cell} - \frac{RT}{nF}\ln Q$$

$$1.52\ V = E°_{cell} - \frac{(8.314\ J\ mol^{-1}\ K^{-1})(298\ K)}{(2\ mol\ e^{-})(9.65 \times 10^4\ C\ mol^{-1}\ e^{-})} \ln\left(\frac{1.00}{1.00}\right)$$

$1.52\ V = E°_{cell} - (0.0128\ V)(0)$

$E^{\circ}_{cell} = E^{\circ}_{reduction} - E^{\circ}_{oxidation} = 1.52\ V$

$E°_{cell} = 1.52\ V = +0.34\ V - E°_{Mn2+}$

$E°_{Mn2+} = -1.18\ V$

23. $3Cu(s) + 2NO_3^-(aq) + 8H^+(aq) \rightarrow 2NO(g) + 4H_2O + 3Cu^{2+}(aq)$

$E^{\circ}_{cell} = E^{\circ}_{reduction} - E^{\circ}_{oxidation}$

$E_{cell} = 0.96\ V - (+0.34\ V) = 0.62\ V$

24. $E_{cell} = \frac{RT}{nF} \ln K_c$

Number of electrons transferred: 6 e^-

$$0.62\ V = \frac{(8.314\ J\ mol^{-1}\ K^{-1})(298\ K)}{(6\ mol\ e^{-})(9.65 \times 10^4\ C\ mol^{-1}\ e^{-})} \ln K_c$$

$\ln K_c = 144.9$

$k_c = 8.43 \times 10^{62}$

25. (a) $NiO_2(s) + PbO(s) + H_2O \rightarrow Ni(OH)_2(s) + PbO_2(s)$

(b) $E°_{cell} = E_{reduction} - E_{oxidation} = 0.49\ V - (+0.25\ V) = 0.24\ V$

(c) $\Delta G° = -nFE°_{cell}$

$\Delta G° = -(2\ e^-)(9.65 \times 10^4\ C\ (mol\ e^-)^{-1})(0.24\ J/C)$

$\Delta G° = -4.63 \times 10^4\ J = -46.3\ kJ$

26. (a) $2AgCl(s) + Ni(s) \rightarrow 2Ag(s) + 2Cl^-(aq) + Ni^{2+}(aq)$

(b) $E°_{cell} = E_{reduction} - E_{oxidation} = 0.222\ V - (-0.257\ V) = 0.479\ V$

(c) $E_{cell} = E°_{cell} - \frac{RT}{nF} \ln Q$

$E_{cell} = E°_{cell} - \frac{RT}{nF} \ln[Cl^-]^2[Ni^{2+}]$

(d) $$E_{cell} = 0.479\ V - \frac{(8.314\ J\ mol^{-1}\ K^{-1})(298\ K)}{(2\ mol\ e^{-})(9.65 \times 10^4\ C\ mol^{-1}\ e^{-})} \ln[0.020]^2[0.10]$$

$E_{cell} = 0.609\ V$

(e) $E_{cell} = E°_{cell} - \frac{RT}{nF} \ln[Cl^-]^2[Ni^{2+}]$

$$0.388\ V = 0.479\ V - \frac{(8.314\ J\ mol^{-1}\ K^{-1})(298\ K)}{(2\ mol\ e^{-})(9.65 \times 10^4\ C\ mol^{-1}\ e^{-})} \ln[Cl^-]^2[0.200]$$

$0.388\ V = 0.479\ V - 0.0128\ V(\ln[Cl^-]^2[0.200])$

$-0.091\ V/-0.0128\ V = \ln[Cl^-]^2[0.200]$

$7.11\ V = \ln[Cl^-]^2[0.200]$

$e^{7.11} = [Cl^-]^2[0.200]$

$(1223)/0.200 = 77.4\ M = [Cl^-]$

27. $MH(s) + OH^-(aq) \rightarrow M(s) + H_2O + e^-$ anode
$NiO(OH)(s) + H_2O + e^- \rightarrow Ni(OH)_2(s) + OH^-(aq)$ cathode
$MH(s) + NiO(OH)(s) \rightarrow M(s) + Ni(OH)_2(s)$ overall reaction

28. The electrode materials in a typical lithium ion cell are graphite and cobalt oxide. When the cell is charged, Li^+ ions leave $LiCoO_2$ and travel through the electrolyte to the graphite. When the cell discharges, the Li^+ ions move back through the electrolyte to the cobalt oxide while electrons move through the external circuit to keep the charge in balance.

29. The nucleus of a helium atom is an alpha particle. Alpha-emitting radionuclides are common among the elements near the end of the periodic table.

30. (a) ${}^{4}_{2}He$ (b) ${}^{0}_{-1}e$ (c) ${}^{0}_{1}e$

31. The rest mass of a particle of gamma radiation can be determined from the energy associated with the gamma radiation using $\Delta E = \Delta m_0c^2$. Solve for Δm_0.

32. $\Delta E = \Delta m_0c^2$
$\Delta E = (99.99900 \text{ g} - 100.00000 \text{ g})(2.99792458 \times 10^8 \text{m s}^{-1})^2$
$\Delta E = (-0.00100 \text{ g})(2.99792458 \times 10^8 \text{m s}^{-1})^2$
$$\Delta E = -8.98755 \times 10^{13} \text{ g m}^2 \text{ s}^{-2}\left(\frac{1 \text{ kg}}{1000 \text{ g}}\right) = -8.98755 \times 10^{10} \text{ kg m}^2 \text{ s}^{-2} = -8.98755 \times 10^{10} \text{ J}$$
J = (sp. heat)(mass)(ΔT)
$-8.98755 \times 10^{10} \text{ J} = (4.184 \text{ J g}^{-1} \text{ °C}^{-1})(x \text{ g})(100 \text{ °C} - 10 \text{ °C})$
$$\# \text{ L} = 2.387 \times 10^8 \text{ g}\left(\frac{1 \text{ kg}}{1000 \text{ g}}\right)\left(\frac{1 \text{ L}}{1 \text{ kg}}\right) = 2.387 \times 10^5 \text{ L}$$

33. Mass of proton = 1.007276470 u
Mass of neutron = 1.008664904 u
Mass of electrons = $5.48579903 \times 10^{-4}$ u
Total mass of particles = 2.016489954 u
$\Delta E = \Delta m_0c^2$
$\Delta E = (2.016489954 \text{ u} - 2.014102 \text{ u})(1.6605402 \times 10^{-24} \text{ g/u})(1 \text{ kg/1000 g})(2.99792458 \times 108 \text{ m s}^{-1})^2$
$\Delta E = 3.563828 \times 10^{-13}$ J

34. Electrostatic force acts over the longer distance.

35. It is relatively easy to remove the electrons from hydrogen as compared to removing the electrons from helium. In fact, alpha particles, helium nuclei, are usually from natural emitters.

36. ${}^{238}_{92}U + {}^{4}_{2}He \rightarrow {}^{239}_{94}Pu + 3\,{}^{1}_{0}n$

37. ${}^{7}_{4}Be + {}^{0}_{-1}e \xrightarrow{\text{electron capture}} {}^{7}_{3}Li + \text{X rays} + \nu$

38. ${}^{144}_{60}Nd \rightarrow {}^{4}_{2}He + {}^{140}_{58}Ce$

39. ${}^{32}_{15}P \rightarrow {}^{0}_{-1}e + {}^{32}_{16}Si$

40. Oxygen-15 is more intensely radioactive since it has a shorter half-life.

41. Ten half-lives pass in 300 years.

$$\left(\frac{1}{2}\right)^{10} = \left(\frac{1}{1024}\right) = 9.77 \times 10^{-4}$$

$(100 \text{ g } ^{137}\text{Cs})(9.77 \times 10^{-4}) = 0.0977 \text{ g } ^{137}\text{Cs}$ will remain.

42. In order to make a radionuclide, the number of protons and neutrons should both be even and there should be a magic number of protons or neutrons.

43. More protons require more neutrons to provide a compensation nuclear strong force and to dilute electrostatic proton–proton repulsions.

44. Calcium-40 is more likely to be stable since it has a magic number, 20, of neutrons and protons.

45. Because the compound nucleus has no memory of how it was prepared, it may decay by any path which lowers the energy. The actual pathway depends on the amount of excess energy the compound nucleus has.

46. $^{90}_{38}\text{Sr} \rightarrow {}^{0}_{-1}\text{e} + {}^{90}_{39}\text{Y}$ $\qquad t_{1/2} = 28.1 \text{ y}$

$$N = \#\text{ radioactive nuclei} = 36.2 \text{ mg } ^{90}\text{Sr}\left(\frac{1 \text{ g } ^{90}\text{Sr}}{1000 \text{ mg } ^{90}\text{Sr}}\right)\left(\frac{1 \text{ mol } ^{90}\text{Sr}}{90.0 \text{ g } ^{90}\text{Sr}}\right)\left(\frac{6.022\times10^{23} \text{ atoms } ^{90}\text{Sr}}{1 \text{ mol } ^{90}\text{Sr}}\right)$$

$N = 2.42 \times 10^{20}$ atoms ^{90}Sr

$$t_{1/2} = \frac{\ln 2}{k}$$

$$28.1 \text{ yr}\left(\frac{365.25 \text{ days}}{1 \text{ yr}}\right) = \frac{\ln 2}{k}$$

$k = 6.75 \times 10^{-5} \text{ d}^{-1}$

\# beta particles = $(2.42 \times 10^{20} \text{ atoms } ^{90}\text{Sr})(6.75 \times 10^{-5} \text{ d}^{-1})$

1.63×10^{-5} decays d^{-1}

47. Beta radiation possesses a greater health risk than an alpha particle because the beta particles have higher penetrating ability.

Printed in the United States
138705LV00001B/17/A

9 780471 649038